COMMONLY USED FUNCTIONS

$$\int \sin mx \sin nx \, dx = \frac{\sin (m - n)x}{2 (m - n)} - \frac{\sin (m + n)x}{2 (m + n)} \quad \text{for } m = n$$

$$\int \cos nx \, dx = \frac{\sin nx}{n}$$

$$\int \cos^2 nx \, dx = \frac{x}{2} + \frac{\sin 2nx}{4n}$$

$$\int \cos mx \cos nx \, dx = \frac{\sin (m - n)x}{2 (m - n)} + \frac{\sin (m + n)x}{2 (m + n)} \quad \text{for } m = n$$

$$\int \sin nx \cos nx \, dx = \frac{\sin^2 nx}{2n}$$

$$\int \sin mx \cos nx \, dx = - \frac{\cos (m - n)x}{2 (m - n)} - \frac{\cos (m + n)x}{2 (m + n)} \quad \text{for } m = n$$

SOME UNITS AND CONSTANTS

Quantity	Units	Equivalent
Length	1 meter (m)	3.281 feet (ft)
		39.36 inches (in)
Mass	1 kilogram (kg)	2.205 pounds (lb)
		35.27 ounces (oz)
Force	1 newton (N)	0.2248 force pounds (lbf)
Torque	1 newton-meter (N.m.)	0.738 pound-feet (lbf.ft)
Moment of inertia	1 kilogram-meter^2 (kg.m^2)	23.7 pound-feet^2 (lb.ft^2)
Power	1 watt (W)	0.7376 foot-pounds/second
		1.341×10^{-3} horsepower (hp)
Energy	1 joules (J)	1 watt-second
		0.7376 foot-pounds
		2.778×10^{-7} kilowatt-hours (kWh)
Horsepower	1 hp	746 watts
Magnetic flux	1 weber (Wb)	10^8 maxwells or lines
Magnetic flux density	1 tesla (T)	1 weber/meter^2 (Wb/m^2)
		10^4 gauss
Magnetic field intensity	1 ampere-turn/meter (At/m)	1.257×10^2 oersted
Permeability of free space	$\mu_0 = 4\pi \times 10^7$ H/m	

DATE DUE

MAY 29 2002	
MAR 1 4 2005	

Power Electronics

Circuits, Devices, and Applications

Second edition

Muhammad H. Rashid

Ph.D., Fellow IEE
Professor of Electrical Engineering
Purdue University Fort Wayne

Prentice Hall, Upper Saddle River, New Jersey 07458

Library of Congress Cataloging-in-Publication Data

Rashid, M. H.
 Power electronics: circuits, devices, and applications / Muhammad
Harunur Rashid. — 2nd ed.
  p.   cm.
 Includes bibliographical references and index.
 ISBN 0-13-678996-X
 1. Power electronics. I. Title.
TK7881.15R37   1993
621.317—dc20   92-45180
                                                        CIP

Publisher: Alan Apt
Production Editor: Mona Pompili
Cover Designer: Wanda Lubelska Design
Copy Editor: Barbara Zeiders
Prepress Buyer: Linda Behrens
Manufacturing Buyer: Dave Dickey
Supplements Editor: Alice Dworkin
Editorial Assistant: Shirley McGuire

 © 1993, 1988 by Prentice-Hall, Inc.
Simon & Schuster Company / A Viacom Company
Upper Saddle River, New Jersey 07458

IBM®PC is a registered trademark of International Business Machines Corporation.
PSpice® is a registered trademark of MicroSim Corporation.

The author and publisher of this book have used their best efforts in preparing this book. These efforts include the development, research, and testing of the theories and programs to determine their effectiveness. The author and publisher shall not be liable in any event for incidental or consequential damages in connection with, or arising out of, the furnishing, performance, or use of these programs.

Printed in the United States of America

10 9 8 7 6 5

ISBN 0-13-678996-X

PRENTICE-HALL INTERNATIONAL (UK) LIMITED, *London*
PRENTICE-HALL OF AUSTRALIA PTY. LIMITED, *Sydney*
PRENTICE-HALL CANADA, INC., *Toronto*
PRENTICE-HALL HISPANOAMERICANA, S.A., *Mexico*
PRENTICE-HALL OF INDIA PRIVATE LIMITED, *New Delhi*
PRENTICE-HALL OF JAPAN, INC., *Tokyo*
SIMON & SCHUSTER ASIA PTE. LTD., *Singapore*
EDITORA PRENTICE-HALL DO BRASIL, LTDA., *Rio de Janeiro*

To my parents, my wife Fatema
and
my children, Faeza, Farzana, and Hasan

Preface

Power Electronics is intended as a textbook for a course on "power electronics/ static power converters" for junior or senior undergraduate students in electrical and electronic engineering. It could also be used as a textbook for graduate students and could be a reference book for practicing engineers involved in the design and applications of power electronics. The prerequisites would be courses on basic electronics and basic electrical circuits. The content of *Power Electronics* is beyond the scope of a one-semester course. For an undergraduate course, Chapters 1 to 11 should be adequate to provide a strong background on power electronics. Chapters 12 to 16 could be left for other courses or included in a graduate course.

The time allocated to a course on power electronics in a typical undergraduate curriculum is normally only one semester. Power electronics has already advanced to the point where it is difficult to cover the entire subject in a one-semester course. The fundamentals of power electronics are well established and they do not change rapidly. However, the device characteristics are continuously being improved and new devices are added. *Power Electronics*, which employs the bottom-down approach, covers device characteristics conversion techniques first, and then applications. It emphasizes the fundamental principles of power conversions. This edition of power electronics is a complete revision of its first edition, and (i) features bottom-down approach, rather than top-down approach, (ii) introduces the state-of-the-art advanced Modulation Techniques, (iii) presents a new chapter on "Resonant-Pulse Inverters" and covers the state-of-the-art techniques, (iv) integrates the industry standard software, SPICE, and design examples that are verified by SPICE simulation, (v) examines converters with RL-loads, and (vi) has corrected typos, and expanded sections and/or paragraphs to add explanations. The book is divided into five parts:

1. Introduction—Chapter 1
2. Commutation techniques of SCRs and power conversion techniques—Chapters 3, 5, 6, 7, 9, 10, and 11
3. Devices—Chapters 2, 4, and 8
4. Applications—Chapters 12, 13, 14, and 15
5. Protection—Chapter 16

Topics like three-phase circuits, magnetic circuits, switching functions of converters, dc transient analysis, and Fourier analysis are reviewed in the Appendixes.

Power electronics deals with the applications of solid-state electronics for the control and conversion of electric power. Conversion techniques require the switching on and off of power semiconductor devices. Low-level electronics circuits, which normally consist of integrated circuits and discrete components, generate the required gating signals for the power devices. Integrated circuits and discrete components are being replaced by microprocessors.

An ideal power device should have no switching-on and -off limitations in terms of turn-on time, turn-off time, current, and voltage handling capabilities. Power semiconductor technology is rapidly developing fast switching power devices with increasing voltage and current limits. Power switching devices such as power BJTs, power MOSFETs, SITs, IGBTs, MCTs, SITHs, SCRs, TRIACs, GTOs, and other semiconductor devices are finding increasing applications in a wide range of products. With the availability of faster switching devices, the applications of modern microprocessors in synthesizing the control strategy for gating power devices to meet the conversion specifications are widening the scope of power electronics. The power electronics revolution has gained the momentum, since late 80s and early 90s. Within the next 30 years, power electronics will shape and condition the electricity somewhere between its generation and all its users. The potential applications of power electronics is yet to be fully explored but we've made every effort to cover as many applications as possible in this book.

Muhammad H. Rashid
Fort Wayne, Indiana

Acknowledgments

Many people have contributed to this edition and made suggestions based on their classroom experience as a professor or a student. I would like to thank the following persons for their comments and suggestions:

Mazen Abdel-Salam—King Fahd University of Petroleum and Minerals, Saudi Arabia
Ashoka K. S. Bhat—University of Victoria, Canada
Fred Brockhurst—Rose-Hulman Institution of Technology
Joseph M. Crowley—University of Illinois, Urbana–Champaign
Mehrad Ehsani—Texas A&M University
Alexander E. Emanuel—Worcester Polytechnic Institute
George Gela—The Ohio State University
Herman W. Hill—Ohio University
Wahid Hubbi—New Jersey Institute of Technology
Marrija Ilic-Spong—University of Illinois, Urbana–Champaign
Shahidul I. Khan—Concordia University, Canada
Peter Lauritzen—University of Washington
Jack Lawler—University of Tennessee
Arthur R. Miles—North Dakota State University
Medhat M. Morcos—Kansas State University
Hassan Moghbelli—Purdue University Calumet
H. Ramezani-Ferdowsi—University of Mashhad, Iran
Prasad Enjeti—Texas A & M University
Saburo Mastsusaki—TDK Corporation, Japan

It has been a great pleasure working with the editor, Alan Apt, and developmental editor, Sondra Chavez. Finally, I would thank my family for their love, patience, and understanding.

Contents

CHAPTER 6   AC VOLTAGE CONTROLLERS   190

CHAPTER 7   THYRISTOR COMMUTATION TECHNIQUES   239

CHAPTER 10   PULSE-WIDTH-MODULATED INVERTERS   356

CHAPTER 11   RESONANT PULSE CONVERTERS   414

CHAPTER 12   STATIC SWITCHES   464

CHAPTER 15   AC DRIVES   541

CHAPTER 16   PROTECTION OF DEVICES AND CIRCUITS   591

1

Introduction

1-1 APPLICATIONS OF POWER ELECTRONICS

The demand for control of electric power for electric motor drive systems and industrial controls existed for many years, and this led to early development of the Ward–Leonard system to obtain a variable dc voltage for the control of dc motor drives. Power electronics have revolutionized the concept of power control for power conversion and for control of electrical motor drives.

Power electronics combine power, electronics, and control. Control deals with the steady-state and dynamic characteristics of closed-loop systems. Power deals with the static and rotating power equipment for the generation, transmission, and distribution of electric power. Electronics deal with the solid-state devices and circuits for signal processing to meet the desired control objectives. *Power electronics* may be defined as the applications of solid-state electronics for the control and conversion of electric power. The interrelationship of power electronics with power, electronics, and control is shown in Fig. 1-1.

Power electronics is based primarily on the switching of the power semiconductor devices. With the development of power semiconductor technology, the power-handling capabilities and the switching speed of the power devices have improved tremendously. The development of microprocessors/microcomputer technology has a great impact on the control and synthesizing the control strategy for the power semiconductor devices. Modern power electronics equipment uses (1) power semiconductors that can be regarded as the muscle, and (2) microelectronics that has the power and intelligence of a brain.

Power electronics have already found an important place in modern technology and are now used in a great variety of high-power products, including heat controls, light controls, motor controls, power supplies, vehicle propulsion systems, and high-voltage direct-current (HVDC) systems. It is difficult to draw the

1

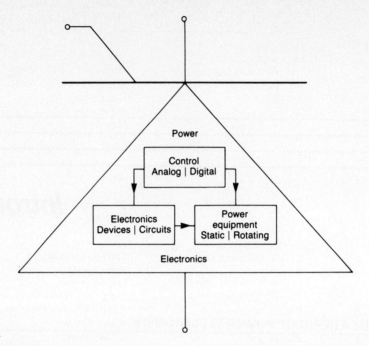

Figure 1-1 Relationship of power electronics to power, electronics, and control.

boundaries for the applications of power electronics; especially with the present trends in the development of power devices and microprocessors, the upper limit is undefined. Table 1-1 shows some applications of power electronics.

1-2 HISTORY OF POWER ELECTRONICS

The history of power electronics began with the introduction of the mercury arc rectifier in 1900. Then the metal tank rectifier, grid-controlled vacuum-tube rectifier, ignitron, phanotron, and thyratron were introduced gradually. These devices were applied for power control until the 1950s.

The first electronics revolution began in 1948 with the invention of the silicon transistor at Bell Telephone Laboratories by Bardeen, Brattain, and Schockley. Most of today's advanced electronic technologies are traceable to that invention. Modern microelectronics evolved over the years from silicon semiconductors. The next breakthrough, in 1956, was also from Bell Laboratories: the invention of the *PNPN* triggering transistor, which was defined as a thyristor or silicon-controlled rectifier (SCR).

The second electronics revolution began in 1958 with the development of the commercial thyristor by the General Electric Company. That was the beginning of a new era of power electronics. Since then, many different types of power semiconductor devices and conversion techniques have been introduced. The

TABLE 1.1 SOME APPLICATIONS OF POWER ELECTRONICS

Advertising	Magnets
Air conditioning	Mass transits
Aircraft power supplies	Mercury-arc lamp ballasts
Alarms	Mining
Appliances	Model trains
Audio amplifiers	Motor controls
Battery charger	Motor drives
Blenders	Movie projectors
Blowers	Nuclear reactor control rod
Boilers	Oil well drilling
Burglar alarms	Oven controls
Cement kiln	Paper mills
Chemical processing	Particle accelerators
Clothes dryers	People movers
Computers	Phonographs
Conveyers	Photocopies
Cranes and hoists	Photographic supplies
Dimmers	Power supplies
Displays	Printing press
Electric blankets	Pumps and compressors
Electric door openers	Radar/sonar power supplies
Electric dryers	Range surface unit
Electric fans	Refrigerators
Electric vehicles	Regulators
Electromagnets	RF amplifiers
Electromechanical electroplating	Security systems
Electronic ignition	Servo systems
Electrostatic precipitators	Sewing machines
Elevators	Solar power supplies
Fans	Solid-state contactors
Flashers	Solid-state relays
Food mixers	Space power supplies
Food warmer trays	Static circuit breakers
Forklift trucks	Static relays
Furnaces	Steel mills
Games	Synchronous machine starting
Garage door openers	Synthetic fibers
Gas turbine starting	Television circuits
Generator exciters	Temperature controls
Grinders	Timers
Hand power tools	Toys
Heat controls	Traffic signal controls
High-frequency lighting	Trains
High-voltage dc (HVDC)	TV deflections
Induction heating	Ultrasonic generators
Laser power supplies	Uninterruptible power supplies
Latching relays	Vacuum cleaners
Light dimmers	VAR compensation
Light flashers	Vending machines
Linear induction motor controls	VLF transmitters
Locomotives	Voltage regulators
Machine tools	Washing machines
Magnetic recordings	Welding

Source: Ref. 5.

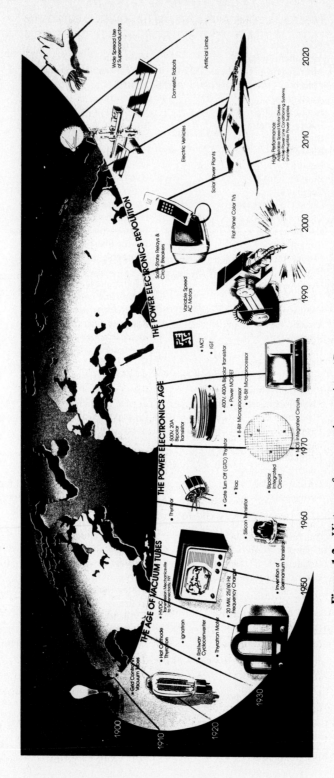

Figure 1-2 History of power electronics. (Courtesy of Tennessee Center for Research and Development.)

microelectronics revolution gave us the ability to process a huge amount of information at incredible speed. The power electronics revolution is giving us the ability to shape and control large amounts of power with ever-increasing efficiency. Due to the marriage of power electronics, the muscle, with microelectronics, the brain, many potential applications of power electronics are now emerging, and this trend will continue. Within the next 30 years, power electronics will shape and condition the electricity somewhere in the transmission line between its generation and all its users. The power electronics revolution has gained momentum since the late 1980s and early 1990s. A chronological history of power electronics is shown in Fig. 1-2.

1-3 POWER SEMICONDUCTOR DEVICES

Since the first thyristor of silicon-controlled rectifier (SCR) was developed in late 1957, there have been tremendous advances in the power semiconductor devices. Until 1970, the conventional thyristors had been exclusively used for power control in industrial applications. Since 1970, various types of power semiconductor devices were developed and became commercially available. These can be divided broadly into five types: (1) power diodes, (2) thyristors, (3) power bipolar junction transistors (BJTs), (4) power MOSFETs, and (5) insulated-gate bipolar transistors (IGBTs) and static induction transistors (SITs). The thyristors can be subdivided into eight types: (a) forced-commutated thyristor, (b) line-commutated thyristor, (c) gate-turn-off thyristor (GTO), (d) reverse-conducting thyristor (RCT), (e) static induction thyristor (SITH), (f) gate-assisted turn-off thyristor (GATT), (g) light-activated silicon-controlled rectifier (LASCR), and (h) MOS-controlled thyristors (MCTs). Static induction transistors are also commercially available.

Power diodes are of three types: general purpose, high speed (or fast recovery), and Schottky. General-purpose diodes are available up to 3000 V, 3500 A, and the rating of fast-recovery diodes can go up to 3000 V, 1000 A. The reverse recovery time varies between 0.1 and 5 μs. The fast-recovery diodes are essential for high-frequency switching of power converters. A diode has two terminals: a cathode and an anode. Schottky diodes have low on-state voltage and very small recovery time, typically nanoseconds. The leakage current increases with the voltage rating and their ratings are limited to 100 V, 300 A. A diode conducts when its anode voltage is higher than that of the cathode; and the forward voltage drop of a power diode is very low, typically 0.5 and 1.2 V. If the cathode voltage is higher than its anode voltage, a diode is said to be in a *blocking mode*. Figure 1-3 shows various configurations of general-purpose diodes, which basically fall into two types. One is called a *stud* or *stud-mounted* type and the other is called a *disk* or *press pak* or *hockey puck* type. In a stud-mounted type, either the anode or the cathode could be the stud.

A thyristor has three terminals: an anode, a cathode, and a gate. When a small current is passed through the gate terminal to cathode, the thyristor conducts, provided that the anode terminal is at a higher potential than the cathode.

Figure 1-3 Various general-purpose diode configurations. (Courtesy of Powerex, Inc.)

Once a thyristor is in a conduction mode, the gate circuit has no control and the thyristor continues to conduct. When a thyristor is in a conduction mode, the forward voltage drop is very small, typically 0.5 to 2 V. A conducting thyristor can be turned off by making the potential of the anode equal to or less than the cathode potential. The line-commutated thyristors are turned off due to the sinusoidal nature of the input voltage, and forced-commutated thyristors are turned off by an extra circuit called *commutation circuitry*. Figure 1-4 shows various configurations of phase control (or line-commutated) thyristors: stud, hockey puck, flat, and pin types.

Natural or line-commutated thyristors are available with ratings up to 6000 V, 3500 A. The *turn-off time* of high-speed reverse-blocking thyristors have been improved substantially and it is possible to have 10 to 20 μs in a 1200-V, 2000-A thyristor. The *turn-off time* is defined as the time interval between the instant when the principal current has decreased to zero after external switching of the principal voltage circuit, and the instant when the thyristor is capable of supporting a specified principal voltage without turning on [2]. RCTs and GATTs are widely used for high-speed switching, especially in traction applications. An RCT

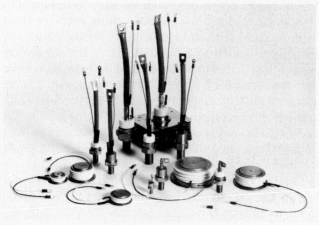

Figure 1-4 Various thyristor configurations. (Courtesy of Powerex, Inc.)

can be considered as a thyristor with an inverse-parallel diode. RCTs are available up to 2500 V, 1000 A (and 400 A in reverse conduction) with a switching time of 40 μs. GATTs are available up to 1200 V, 400 A with a switching speed of 8 μs. LASCRs, which are available up to 6000 V, 1500 A, with a switching speed of 200 to 400 μs, are suitable for high-voltage power systems, especially in HVDC. For low-power ac applications, TRIACs are widely used in all types of simple heat controls, light controls, motor controls, and ac switches. The characteristics of TRIACs are similar to two thyristors connected in inverse parallel and having only one gate terminal. The current flow through a TRIAC can be controlled in either direction.

GTOs and SITHs are self-turned-off thyristors. GTOs and SITHs are turned on by applying a short positive pulse to the gates and are turned off by the applications of short negative pulse to the gates. They do not require any commutation circuit. GTOs are very attractive for forced commutation of converters and are available up to 4000 V, 3000 A. SITHs, whose ratings can go as high as 1200 V, 300 A, are expected to be applied for medium-power converters with a frequency of several hundred kilohertz and beyond the frequency range of GTOs. Figure 1-5 shows various configurations of GTOs. An MCT can be turned "on" by a small negative voltage pulse on the MOS gate (with respect to its anode), and turned "off" by a small positive voltage pulse. It is like a GTO, except that the turn-off gain is very high. MCTs are available up to 1000 V, 100 A.

High-power bipolar transistors are commonly used in power converters at a frequency below 10 kHz and are effectively applied in the power ratings up to 1200 V, 400 A. The various configurations of bipolar power transistors are shown in Fig. 8-2. A bipolar transistor has three terminals: base, emitter, and collector. It is normally operated as a switch in the common-emitter configuration. As long as the base of an *NPN*-transistor is at a higher potential than the emitter and the base current is sufficiently large to drive the transistor in the saturation region, the transistor remains on, provided that the collector-to-emitter junction is properly biased. The forward drop of a conducting transistor is in the range 0.5 to 1.5 V. If

Figure 1-5 Gate-turn-off thyristors. (Courtesy of International Rectifiers.)

the base drive voltage is withdrawn, the transistor remains in the nonconduction (or off) mode.

Power MOSFETs are used in high-speed power converters and are available at a relatively low power rating in the range of 1000 V, 50 A at a frequency range of several tens of kilohertz. The various power MOSFETs of different sizes are shown in Fig. 8-21. IGBTs are voltage-controlled power transistors. They are inherently faster than BJTs, but still not quite as fast as MOSFETs. However, they offer far superior drive and output characteristics to those of BJTs. IGBTs are suitable for high voltage, high current, and frequencies up to 20 kHz. IGBTs are available up to 1200 V, 400 A.

A SIT is a high-power, high-frequency device. It is essentially the solid-state version of the triode vacuum tube, and is similar to a JFET. It has a low-noise, low-distortion, high-audio-frequency power capability. The turn-on and turn-off times are very short, typically 0.25 μs. The normally on-characteristic and the high on-state drop limit its applications for general power conversions. The current rating of SITs can be up to 1200 V, 300 A, and the switching speed can be as high as 100 kHz. SITs are most suitable for high-power, high-frequency applications (e.g., audio, VHF/UHF, and microwave amplifiers). The ratings of commercially available power semiconductor devices are shown in Table 1-2, where the on-voltage is the on-state voltage drop of the device at the specified current. Table 1-3 shows the v–i characteristics and the symbols of commonly used power semiconductor devices.

TABLE 1.2 RATINGS OF POWER SEMICONDUCTOR DEVICES

	Type	Voltage/current rating	Upper frequency (Hz)	Switching time (μs)	On-state resistance (Ω)
Diodes	General purpose	5000 V/5000 A	1k	100	0.16m
	High speed	3000 V/1000 A	10k	2–5	1m
	Schottky	40 V/60 A	20k	0.23	10m
Forced-turned-off thyristors	Reverse blocking	5000 V/5000 A	1k	200	0.25m
	High speed	1200 V/1500 A	10k	20	0.47m
	Reverse blocking	2500 V/400 A	5k	40	2.16m
	Reverse conducting	2500 V/1000 A	5k	40	2.1m
	GATT	1200 V/400 A	20k	8	2.24m
	Light triggered	6000 V/1500 A	400	200–400	0.53m
TRIACs		1200 V/300 A	400	200–400	3.57m
Self-turned-off thyristors	GTO	4500 V/3000 A	10k	15	2.5m
	SITH	4000 V/2200 A	20k	6.5	5.75m
Power transistors	Single	400 V/250 A	20k	9	4m
		400 V/40 A	20k	6	31m
		630 V/50 A	25k	1.7	15m
	Darlington	1200 V/400 A	10k	30	10m
SITs		1200 V/300 A	100k	0.55	1.2
Power MOSFETS	Single	500 V/8.6 A	100k	0.7	0.6
		1000 V/4.7 A	100k	0.9	2
		500 V/50 A	100k	0.6	0.4m
IGBTs	Single	1200 V/400 A	20k	2.3	60m
MCTs	Single	600 V/60 A	20k	2.2	18m

Source: Ref. 3.

TABLE 1.3 CHARACTERISTICS AND SYMBOLS OF SOME POWER DEVICES

Devices	Symbols	Characteristics
Diode		
Thyristor		
SITH		
GTO		
MCT		
TRIAC		
LASCR		
NPN BJT		
IGBT		
N-Channel MOSFET		
SIT		

The data sheets for a diode, SCR, GTO, BJT, MOSFET, IGBT, and MCT are given in Appendix G. Figure 1-6 shows the applications and frequency range of power devices. A superpower device should (1) have a zero on-state voltage, (2) withstand an infinite off-state voltage, (3) handle an infinite current, and (4) turn "on" and "off" in zero time, thereby having infinite switching speed.

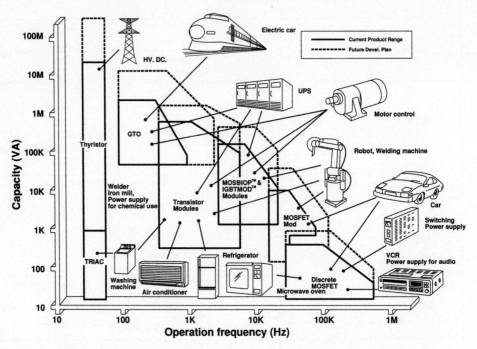

Figure 1-6 Applications of power devices. (Courtesy of Powerex, Inc.)

1-4 CONTROL CHARACTERISTICS OF POWER DEVICES

The power semiconductor devices can be operated as switches by applying control signals to the gate terminal of thyristors (and to the base of bipolar transistors). The required output is obtained by varying the conduction time of these switching devices. Figure 1-7 shows the output voltages and control characteristics of commonly used power switching devices. Once a thyristor is in a conduction mode, the gate signal of either positive or negative magnitude has no effect and this is shown in Fig. 1-7a. When a power semiconductor device is in a normal conduction mode, there is a small voltage drop across the device. In the output voltage waveforms in Fig. 1-7, these voltage drops are considered negligible, and unless specified, this assumption is made throughout the following chapters.

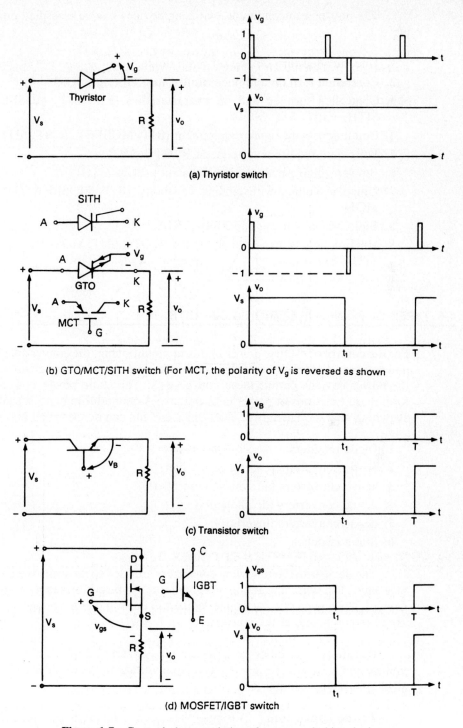

Figure 1-7 Control characteristics of power switching devices.

The power semiconductor switching devices can be classified on the basis of:

1. Uncontrolled turn on and off (e.g., diode)
2. Controlled turn on and uncontrolled turn off (e.g., SCR)
3. Controlled turn on and off characteristics (e.g., BJT, MOSFET, GTO, SITH, IGBT, SIT, MCT)
4. Continuous gate signal requirement (BJT, MOSFET, IGBT, SIT)
5. Pulse gate requirement (e.g., SCR, GTO, MCT)
6. Bipolar voltage-withstanding capability (SCR, GTO)
7. Unipolar voltage-withstanding capability (BJT, MOSFET, GTO, IGBT, MCT)
8. Bidirectional current capability (TRIAC, RCT)
9. Unidirectional current capability (SCR, GTO, BJT, MOSFET, MCT, IGBT, SITH, SIT, diode)

1-5 TYPES OF POWER ELECTRONIC CIRCUITS

For the control of electric power or power conditioning, the conversion of electric power from one form to another is necessary and the switching characteristics of the power devices permit these conversions. The static power converters perform these functions of power conversions. A converter may be considered as a switching matrix. The power electronics circuits can be classified into six types:

1. Diode rectifiers
2. ac–dc converters (controlled rectifiers)
3. ac–ac converters (ac voltage controllers)
4. dc–dc converters (dc choppers)
5. dc–ac converters (inverters)
6. Static switches

The devices in the following converters are used to illustrate the basic principles only. The switching action of a converter can be performed by more than one device. The choice of a particular device will depend on the voltage, current, and speed requirements of the converter.

Rectifiers. A diode rectifier circuit converts ac voltage into a fixed dc voltage and is shown in Fig. 1-8. The input voltage to the rectifier could be either single-phase or three-phase.

Ac–dc converters. A single-phase converter with two natural commutated thyristors is shown in Fig. 1-9. The average value of the output voltage can be controlled by varying the conduction time of thyristors or firing delay angle,

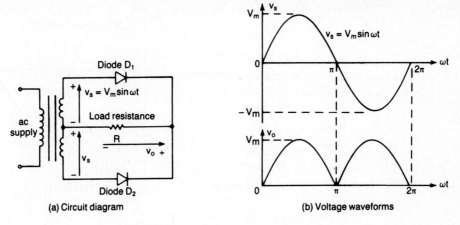

Figure 1-8 Single-phase rectifier circuit.

α. The input could be a single or three-phase source. These converters are also known as *controlled rectifiers*.

Ac–ac converters. These converters are used to obtain a variable ac output voltage from a fixed ac source and a single-phase converter with a TRIAC is shown in Fig. 1-10. The output voltage is controlled by varying the conduction time of a TRIAC or firing delay angle, α. These types of converters are also known as *ac voltage controllers*.

Dc–dc converters. A dc–dc converter is also known as a *chopper* or *switching regulator* and a transistor chopper is shown in Fig. 1-11. The average output voltage is controlled by varying the conduction time t, of transistor Q_1. If T is the chopping period, then $t_1 = \delta T$. δ is called as the *duty cycle* of the chopper.

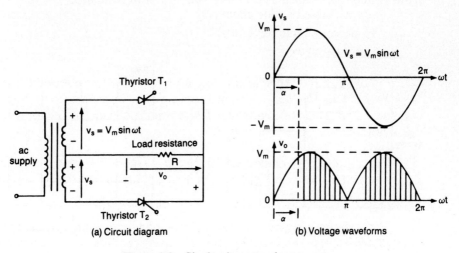

Figure 1-9 Single-phase ac–dc converter.

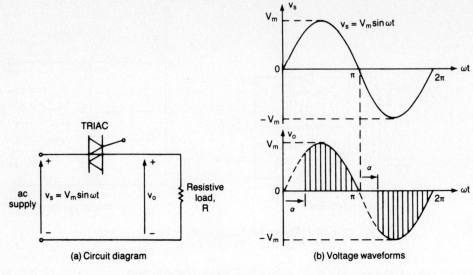

Figure 1-10 Single-phase ac–ac converter.

Dc–ac converters. A dc–ac converter is also known as an *inverter*. A single-phase transistor inverter is shown in Fig. 1-12. If transistors M_1 and M_2 conduct for one-half period and M_3 and M_4 conduct for the other half, the output voltage is of alternating form. The output voltage can be controlled by varying the conduction time of transistors.

Static switches. Since the power devices can be operated as static switches or contactors, the supply to these switches could be either ac or dc and the switches are called as *ac static switches* or *dc switches*.

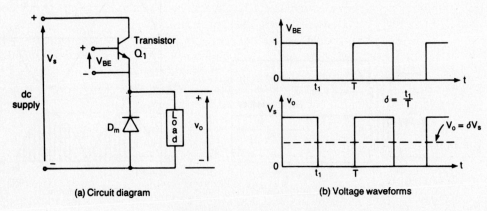

Figure 1-11 Dc–dc converter.

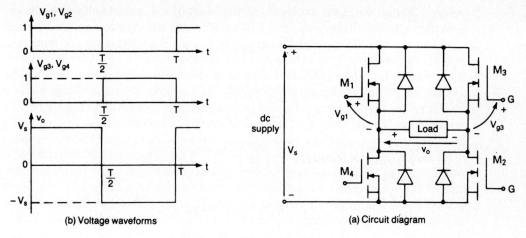

(b) Voltage waveforms

(a) Circuit diagram

Figure 1-12   Single-phase dc–ac converter.

1-6 DESIGN OF POWER ELECTRONICS EQUIPMENT

The design of a power electronics equipment can be divided into four parts:

1. Design of power circuits
2. Protection of power devices
3. Determination of the control strategy
4. Design of logic and gating circuits

In the chapters that follow, various types of power electronic circuits are described and analyzed. In the analysis, the power devices are assumed to be ideal switches unless stated otherwise; and effects of circuit stray inductance, circuit resistances, and source inductance are neglected. The practical power devices and circuits differ from these ideal conditions and the designs of the circuits are also affected. However, in the early stage of the design, the simplified analysis of a circuit is very useful to understand the operation of the circuit and to establish the characteristics and control strategy.

Before a prototype is built, the designer should investigate the effects of the circuit parameters (and devices imperfections) and should modify the design if necessary. Only after the prototype is built and tested, the designer can be confident about the validity of the design and can estimate more accurately some of the circuit parameters (e.g., stray inductance).

1-7 PERIPHERAL EFFECTS

The operations of the power converters are based mainly on the switching of power semiconductor devices; and as a result the converters introduce current and voltage harmonics into the supply system and on the output of the converters. These can cause problems of distortion of the output voltage, harmonic generation into the supply system, and interference with the communication and signaling circuits. It is normally necessary to introduce filters on the input and output of a converter system to reduce the harmonic level to an acceptable magnitude. Figure 1-13 shows the block diagram of a generalized power converter. The application of power electronics to supply the sensitive electronic loads poses a challenge on the power quality issues and raises problems and concerns to be resolved by researchers. The input and output quantities of converters could be either ac or dc. Factors such as total harmonic distortion (THD), displacement factor (HF), and input power factor (IPF) are measures of the quality of a waveform. To determine these factors, it is required to find the harmonic content of the waveforms. To evaluate the performance of a converter, the input and output voltages/currents of a converter are expressed in Fourier series. The quality of a power converter is judged by the quality of its voltage and current waveforms.

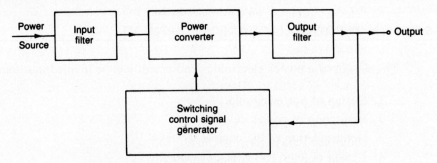

Figure 1-13   Generalized power converter system.

The control strategy for the power converters plays an important part on the harmonic generation and output waveform distortion, and can be aimed to minimize or reduce these problems. The power converters can cause radio-frequency interference due to electromagnetic radiation and the gating circuits may generate erroneous signals. This interference can be avoided by *grounded shielding*.

1-8 POWER MODULES

Power devices are available as a single unit or in a module. A power converter often requires two or four or six devices, depending on its topology. Power modules with dual (in half-bridge configuration) or quad (in full-bridge) or six (in three-phase) are available for almost all types of power devices. The modules offer the advantages of lower on-state losses, high voltage and current switching characteristics, and higher speed than that of conventional devices. Some modules even include transient protection and gate drive circuitry.

Gate drive circuits are commercially available to drive individual devices or modules. *Intelligent modules*, which are the state-of-the-art power electronics, integrate the power module and the peripheral circuit. The peripheral circuit consists of input/output isolation from and interface with the signal and high-voltage system, a drive circuit, a protection and diagnostic circuit (against excess current, short circuit, open load, overheating, excess voltage), microcomputer control, and a control power supply. The users need only to connect external (floating) power supplies. An intelligent module is also known as *smart power*. These modules are being used increasingly in power electronics [8]. Some manufacturers of devices and modules are as follows:

Advanced Power Technology, Inc.
Brown Boveri
Fuji Electric/Collmer Semiconductor, Inc.
Harris Corp.
Hitachi Ltd.
International Rectifier
Marconi Electronic Devices, Inc.
Mitsubishi Electric Corp.
Motorola, Inc.
National Semiconductors, Inc.
Nihon International Electronics Corp.
Power Integrations, Inc.
Powerex, Inc.
PowerTech, Inc.
RCA Corp.
Semikron International
Siliconix, Inc.
Tokin, Inc.
Tokyo Denki
Toshiba Corp.
Unitrode Integrated Circuits Corp.
Westcode Semiconductors Ltd.

1-10 POWER ELECTRONICS JOURNALS AND CONFERENCES

There are many professional journals and conferences in which the new developments are published. Some of them are:

IEEE Transactions on Industrial Electronics
IEEE Transactions on Industry Applications
IEEE Transactions on Power Delivery
IEEE Transactions on Power Electronics
IEE Proceedings on Electric Power
Journal of Electrical Machinery and Power Systems
Applied Power Electronics Conference (APEC)
European Power Electronics Conference (EPEC)
IEEE Industrial Electronics Conference (IECON)
IEEE Industry Applications Society Annual Meeting (IAS)
International Conference on Electrical Machines (ICEM)
International Power Electronics Conference (IPEC)
Power Conversion Intelligent Motion (PCIM)
Power Electronics Specialist Conference (PESC)

SUMMARY

As the technology for the power semiconductor devices and integrated circuits develops, the potential for the applications of power electronics becomes wider. There are already many power semiconductor devices that are commercially available; however, the development in this direction is continuing. The power converters fall generally into six categories: (1) rectifiers, (2) ac–dc converters, (3) ac–ac converters, (4) dc–dc converters, (5) dc–ac converters, and (6) static switches. The design of power electronics circuits requires designing the power and control circuits. The voltage and current harmonics that are generated by the power converters can be reduced (or minimized) with a proper choice of the control strategy.

REFERENCES

1. R. G. Hoft, "Historical review, present status and future prospects." *International Power Electronics Conference*, Tokyo, 1983, pp. 6–18.

2. General Electric, D. R. Grafham and F. B. Golden, eds., *SCR Manual*, 6th ed. Englewood Cliffs, N.J.: Prentice Hall, 1982.

3. F. Harashima, "State of the art on power electronics and electrical drives in Japan." *3rd IFAC Symposium on Control in Power Electronics and Electrical Drives*, Lausanne, Switzerland, 1983, tutorial session and survey papers, pp. 23–33.

4. B. R. Pelly, "Power semiconductor devices: a status review." *IEEE Industry Applications Society International Semiconductor Power Converter Conference*, 1982, pp. 1–19.

5. R. G. Holt, *Semiconductor Power Electronics*. New York: Van Nostrand Reinhold Company, Inc., 1986.

6. T. M. Jahns, "Designing intelligent muscle into industrial motion control." *IEEE Transactions on Industrial Electronics*, Vol. IE37, No. 5, 1990, pp. 329–341.

7. B. K. Bose, "Recent advances in power electronics." *IEEE Transactions on Power Electronics*, Vol. PE7, No. 1, 1992, pp. 2–16.

8. B. K. Bose, *Modern Power Electronics: Evolution, Technology, and Applications.* New York: IEEE Press, 1992.

REVIEW QUESTIONS

1-1. What is power electronics?

1-2. What are the various types of thyristors?

1-3. What is a commutation circuit?

1-4. What are the conditions for a thyristor to conduct?

1-5. How can a conducting thyristor be turned off?

1-6. What is a line commutation?

1-7. What is a forced commutation?

1-8. What is the difference between a thyristor and a TRIAC?

1-9. What is the gating characteristic of a GTO?

1-10. What is turn-off time of a thyristor?

1-11. What is a converter?

1-12. What is the principle of ac–dc conversion?

1-13. What is the principle of ac–ac conversion?

1-14. What is the principle of dc–dc conversion?

1-15. What is the principle of dc–ac conversion?

1-16. What are the steps involved in designing power electronics equipment?

1-17. What are the peripheral effects of power electronics equipment?

1-18. What are the differences in the gating characteristics of GTOs and thyristors?

1-19. What are the differences in the gating characteristics of thyristors and transistors?

1-20. What are the differences in the gating characteristics of BJTs and MOSFETs?

1-21. What is the gating characteristic of an IGBT?

1-22. What is the gating characteristic of an MCT?

1-23. What is the gating characteristic of a SIT?

1-24. What are the differences between BJTs and IGBTs?

1-25. What are the differences between MCTs and GTOs?

1-26. What are the differences between SITHs and GTOs?

Power semiconductor diodes

2-1 INTRODUCTION

Power semiconductor diodes play a significant role in power electronics circuits. A diode acts as a switch to perform various functions, such as switches in rectifiers, freewheeling in switching regulators, charge reversal of capacitor and energy transfer between components, voltage isolation, energy feedback from the load to the power source, and trapped energy recovery.

Power diodes can be assumed as ideal switches for most applications but practical diodes differ from the ideal characteristics and have certain limitations. The power diodes are similar to *pn*-junction signal diodes. However, the power diodes have larger power-, voltage-, and current-handling capabilities than that of ordinary signal diodes. The frequency response (or switching speed) is low compared to signal diodes.

2-2 DIODE CHARACTERISTICS

A power diode is a two-terminal *pn*-junction device and a *pn*-junction is normally formed by alloying, diffusion, and epitaxial growth. The modern control techniques in diffusion and epitaxial processes permit the desired device characteristics. Figure 2-1 shows the sectional view of a *pn*-junction and diode symbol.

When the anode potential is positive with respect to the cathode, the diode is said to be forward biased and the diode conducts. A conducting diode has a relatively small forward voltage drop across it; and the magnitude of this drop would depend on the manufacturing process and junction temperature. When the cathode potential is positive with respect to the anode, the diode is said to be reverse biased. Under reverse-biased conditions, a small reverse current (also

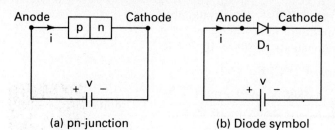

(a) pn-junction   (b) Diode symbol

Figure 2-1 *pn*-Junction and diode symbol.

known as *leakage current*) in the range of micro- or milliampere flows and this leakage current increases slowly in magnitude with the reverse voltage until the avalanche or zener voltage is reached. Figure 2-2a shows the steady-state *v–i* characteristics of a diode. For most practical purposes, a diode can be regarded as an ideal switch, whose characteristics are shown in Fig. 2-2b.

The *v–i* characteristics shown in Fig. 2-2a can be expressed by an equation known as *Schockley diode equation*, and it is given by

$$I_D = I_s(e^{V_D/nV_T} - 1) \tag{2-1}$$

where I_D = current through the diode, A

V_D = diode voltage with anode positive with respect to cathode, V

I_s = leakage (or reverse saturation) current, typically in the range 10^{-6} to 10^{-15} A

n = empirical constant known as *emission coefficient or ideality factor*, whose value varies from 1 to 2

The emission coefficient n depends on the material and the physical construction of the diode. For germanium diodes, n is considered to be 1. For silicon diodes, the predicted value of n is 2, but for most practical silicon diodes, the value of n falls in the range 1.1 to 1.8.

V_T in Eq. (2-1) is a constant called *thermal voltage* and it is given by

$$V_T = \frac{kT}{q} \tag{2-2}$$

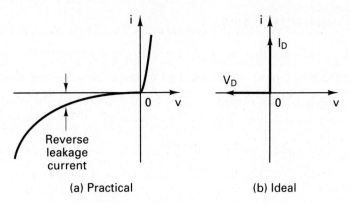

(a) Practical   (b) Ideal

Figure 2-2 *v–i* characteristics of diode.

where q = electron charge: 1.6022×10^{-19} coulomb (C)

T = absolute temperature in kelvin (K = 273 + °C)

k = Boltzmann's constant: 1.3806×10^{-23} J/K

At a junction temperature of 25°C, Eq. (2-2) gives

$$V_T = \frac{kT}{q} = \frac{1.3806 \times 10^{-23} \times (273 + 25)}{1.6022 \times 10^{-19}} \approx 25.8 \text{ mV}$$

At a specified temperature, the leakage current I_s is a constant for a given diode. The diode characteristic of Fig. 2-2a can be divided into three regions:

Forward-biased region, where $V_D > 0$

Reverse-biased region, where $V_D < 0$

Breakdown region, where $V_D < -V_{ZK}$

Forward-biased region. In the forward-biased region, $V_D > 0$. The diode current I_D is very small if the diode voltage V_D is less than a specific value V_{TD} (typically 0.7 V). The diode conducts fully if V_D is higher than this value V_{TD}, which is referred to as the *threshold voltage* or the *cut-in voltage* or the *turn-on voltage*. Thus the threshold voltage is a voltage at which the diode conducts fully.

Let us consider a small diode voltage $V_D = 0.1$ V, $n = 1$, and $V_T = 25.8$ mV. From Eq. (2-1) we can find the corresponding diode current I_D as

$$I_D = I_s(e^{V_D/nV_T} - 1) = I_s[e^{0.1/(1 \times 0.0258)} - 1] = I_s(48.23 - 1)$$

$$\approx 48.23 I_s \text{ with } 2.1\% \text{ error}$$

Therefore, for $V_D > 0.1$ V, which is usually the case, $I_D \gg I_s$, and Eq. (2-1) can be approximated within 2.1% error to

$$I_D = I_s(e^{V_D/nV_T} - 1) \approx I_s e^{V_D/nV_T} \tag{2-3}$$

Reverse-biased region. In the reverse-biased region, $V_D < 0$. If V_D is negative and $|V_D| \gg V_T$, which occurs for $V_D < -0.1$, the exponential term in Eq. (2-1) becomes negligibly small compared to unity and the diode current I_D becomes

$$I_D = I_s(e^{-|V_D|/nV_T} - 1) \approx -I_s \tag{2-4}$$

which indicates that the diode current I_D in the reverse direction is constant and equals I_s.

Breakdown region. In the breakdown region, the reverse voltage is high, usually greater than 1000 V. The magnitude of the reverse voltage exceeds a specified voltage known as the *breakdown voltage*, V_{BR}. The reverse current increases rapidly with a small change in reverse voltage beyond V_{BR}. The operation in the breakdown region will not be destructive provided that the power dissipation is within a "safe level" that is specified in the manufacturer's data sheet. However, it is often necessary to limit the reverse current in the breakdown region to limit the power dissipation within a permissible value.

Example 2-1

The forward voltage drop of a power diode is $V_D = 1.2$ V at $I_D = 300$ A. Assuming that $n = 2$ and $V_T = 25.8$ mV, find the saturation current I_s.

Solution Applying Eq. (2-1), we can find the leakage (or saturation) current I_s from

$$300 = I_s[e^{1.2/(2 \times 25.8 \times 10^{-3})} - 1]$$

which gives $I_s = 2.38371 \times 10^{-8}$ A.

2-3 REVERSE RECOVERY CHARACTERISTICS

The current in a forward-biased junction diode is due to the net effect of majority and minority carriers. Once a diode is in a forward conduction mode and then its forward current is reduced to zero (due to the natural behavior of the diode circuit or by applying a reverse voltage), the diode continues to conduct due to minority carriers which remain stored in the pn-junction and the bulk semiconductor material. The minority carriers require a certain time to recombine with opposite charges and to be neutralized. This time is called the *reverse recovery time* of the diode. Figure 2-3 shows two reverse recovery characteristics of junction diodes. The soft-recovery type is more common. The reverse recovery time is denoted as t_{rr} and is measured from the initial zero crossing of the diode current to 25% of maximum (or peak) reverse current, I_{RR}. t_{rr} consists of two components, t_a and t_b. t_a is due to charge storage in the depletion region of the junction and represents the time between the zero crossing and the peak reverse current, I_{RR}. t_b is due to charge storage in the bulk semiconductor material. The ratio t_b/t_a is known as the *softness factor*, SF. For practical purposes, one needs be concerned with the total recovery time t_{rr} and the peak value of the reverse current I_{RR}.

$$t_{rr} = t_a + t_b \tag{2-5}$$

The peak reverse current can be expressed in reverse di/dt as

$$I_{RR} = t_a \frac{di}{dt} \tag{2-6}$$

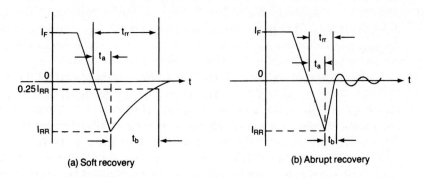

(a) Soft recovery

(b) Abrupt recovery

Figure 2-3 Reverse recovery characteristics.

Reverse recovery time t_{rr}, may be defined as the time interval between the instant the current passes through zero during the changeover from forward conduction to reverse blocking condition and the moment the reverse current has decayed to 25% of its peak reverse value i_{RR}. t_{rr} is dependent on the junction temperature, rate of fall of forward current, and the forward current prior to commutation.

Reverse recovery charge Q_{RR}, is the amount of charge carriers that flow across the diode in the reverse direction due to changeover from forward conduction to reverse blocking condition. Its value is determined from the area enclosed by the path of the reverse recovery current.

The storage charge, which is the area enclosed by the path of the recovery current, is approximately

$$Q_{RR} \cong \tfrac{1}{2} I_{RR} t_a + \tfrac{1}{2} I_{RR} t_b = \tfrac{1}{2} I_{RR} t_{rr} \tag{2-7}$$

or

$$I_{RR} \cong \frac{2 Q_{RR}}{t_{rr}} \tag{2-8}$$

Equating Eq. (2-6) to Eq. (2-8) gives

$$t_{rr} t_a = \frac{2 Q_{RR}}{di/dt} \tag{2-9}$$

If t_b is negligible as compared to t_a, which is usually the case, $t_{rr} \approx t_a$, and Eq. (2-9) becomes

$$t_{rr} \cong \sqrt{\frac{2 Q_{RR}}{di/dt}} \tag{2-10}$$

and

$$I_{RR} = \sqrt{2 Q_{RR} \frac{di}{dt}} \tag{2-11}$$

It can be noticed from Eqs. (2-10) and (2-11) that the reverse recovery time t_{rr} and the peak reverse recovery current I_{RR} depend on the storage charge Q_{RR} and the reverse (or reapplied) di/dt. The storage charge is dependent on the forward diode current I_F. The peak reverse recovery current I_{RR}, reverse charge Q_{RR}, and the softness factor are all of interest to the circuit designer, and these parameters are commonly included in the specification sheets of diodes.

If a diode is in a reverse-biased condition, a leakage current flows due to the minority carriers. Then the application of forward voltage would force the diode to carry current in the forward direction. However, it requires a certain time known as *forward recovery (or turn-on) time* before all the majority carriers over the whole junction can contribute to the current flow. If the rate of rise of the forward current is high and the forward current is concentrated to a small area of the junction, the diode may fail. Thus the forward recovery time limits the rate of the rise of the forward current and the switching speed.

Power Semiconductor Diodes   Chap. 2

Example 2-2

The reverse recovery time of a diode is $t_{rr} = 3$ μs and the rate of fall of the diode current is $di/dt = 30$ A/μs. Determine (a) the storage charge Q_{RR}, and (b) the peak reverse current I_{RR}.

Solution $t_{rr} = 3$ μs and $di/dt = 30$ A/μs.

(a) From Eq. (12-10),

$$Q_{RR} = \frac{1}{2}\frac{di}{dt}t_{rr}^2 = 0.5 \times 30 \text{ A}/\mu\text{s} \times (3 \times 10^{-6})^2 = 135 \text{ } \mu\text{C}$$

(b) From Eq. (2-11),

$$I_{RR} = \sqrt{2Q_{RR}\frac{di}{dt}} = \sqrt{2 \times 135 \times 10^{-6} \times 30 \times 10^{-6}} = 90 \text{ A}$$

2-4 POWER DIODE TYPES

Ideally, a diode should have no reverse recovery time. However, the manufacturing cost of such a diode will increase. In many applications, the effects of reverse recovery time will not be significant, and inexpensive diodes can be used. Depending on the recovery characteristics and manufacturing techniques, the power diodes can be classified into three categories. The characteristics and practical limitations of each type restrict their applications.

1. Standard or general-purpose diodes
2. Fast-recovery diodes
3. Schottky diodes

2-4.1 General-Purpose Diodes

The general-purpose rectifier diodes have relatively high reverse recovery time, typically 25 μs, and are used in low-speed applications, where recovery time is not critical (e.g., diode rectifiers and converters for a low input frequency up to 1-kHz applications and line-commutated converters). These diodes cover current ratings from less than 1 A to several thousands of amperes, with voltage ratings from 50 V to around 5 kV. These diodes are generally manufactured by diffusion. However, alloyed types of rectifiers that are used in welding power supplies are most cost-effective and rugged, and their ratings can go up to 300 A and 1000 V.

2-4.2 Fast-Recovery Diodes

The fast-recovery diodes have low recovery time, normally less than 5 μs. They are used in dc–dc and dc–ac converter circuits, where the speed of recovery is often of critical importance. These diodes cover current ratings from less than 1 A to hundreds of amperes, with voltage ratings from 50 V to around 3 kV.

For voltage ratings above 400 V, fast-recovery diodes are generally made by diffusion and the recovery time is controlled by platinum or gold diffusion. For voltage ratings below 400 V, epitaxial diodes provide faster switching speeds than that of diffused diodes. The epitaxial diodes have a narrow base width, resulting in a fast recovery time of as low as 50 ns. Fast-recovery diodes of various sizes are shown in Fig. 2-4.

2-4.3 Schottky Diodes

The charge storage problem of a *pn*-junction can be eliminated (or minimized) in a Schottky diode. It is accomplished by setting up a "barrier potential" with a contact between a metal and a semiconductor. A layer of metal is deposited on a thin epitaxial layer of *n*-type silicon. The potential barrier simulates the behavior of a *pn*-junction. The rectifying action depends on the majority carriers only, and as a result there are no excess minority carriers to recombine. The recovery effect is due solely to the self-capacitance of the semiconductor junction.

The recovered charge of a Schottky diode is much less than that of an equivalent *pn*-junction diode. Since it is due only to the junction capacitance, it is largely independent of the reverse di/dt. A Schottky diode has a relatively low forward voltage drop.

The leakage current of a Schottky diode is higher than that of a *pn*-junction diode. A Schottky diode with relatively low conduction voltage has relatively high leakage current, and vice versa. As a result, its maximum allowable voltage is generally limited to 100 V. The current ratings of Schottky diodes vary from 1 to 300 A. The Schottky diodes are ideal for high-current and low-voltage dc power supplies. However, they are also used in low-current power supplies for increased efficiency. Twenty- and 30-A dual Schottky rectifiers are shown in Fig. 2-5.

Figure 2-4 Fast-recovery diodes. (Courtesy of Powerex, Inc.)

Figure 2-5 Twenty- and 30-A dual Schottky center rectifiers. (Courtesy of International Rectifier.)

2-5 EFFECTS OF FORWARD AND REVERSE RECOVERY TIME

The importance of these parameters can be explained with Fig. 2-6a. If the switch, SW, is turned on at $t = 0$ and remains on long enough, a steady-state current of $I_0 = V_s/R$ would flow through the load and the freewheeling diode D_m

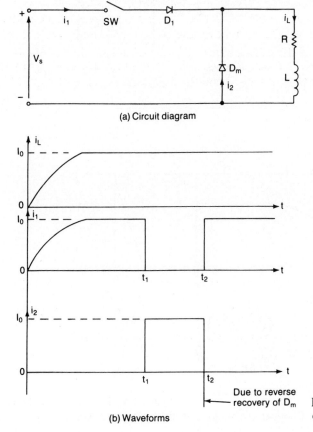

(a) Circuit diagram

(b) Waveforms

Figure 2-6 Chopper circuit without *di/dt* limiting inductor.

will be reversed biased. If the switch is turned off at $t = t_1$, diode D_m would conduct and the load current would circulate through D_m. Now, if the switch is turned on again at $t = t_2$, diode D_m would behave as a short circuit. The rate of rise of the forward current of the switch (and diode D_1) and the rate of fall of forward current of diode D_m would be very high, tending to be infinity. According to Eq. (2-11), the peak reverse current of diode D_m could be very high, and diodes D_1 and D_m may be damaged. Figure 2-6b shows the various waveforms for the diode currents. This problem is normally overcome by connecting a *di/dt* limiting inductor, L_s, as shown in Fig. 2-7a. The practical diodes require a certain turn-on time before the entire area of the junction becomes conductive and the *di/dt* must be kept low to meet the turn-on time limit. This time is sometimes known as *forward recovery time t_{rf}*.

The rate of rise of current through diode D_1, which should be the same as the rate of fall of the current through diode D_m, is

$$\frac{di}{dt} = \frac{V_s}{L_s} \tag{2-12}$$

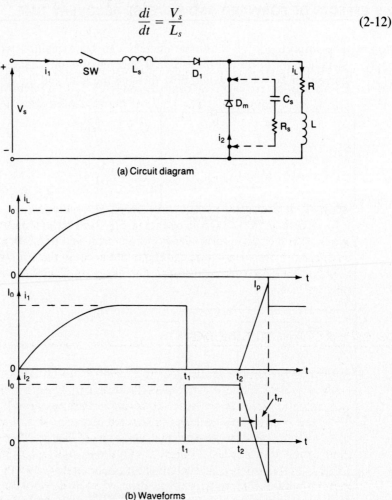

(a) Circuit diagram

(b) Waveforms

Figure 2-7 Chopper circuit with *di/dt* limiting inductor.

If t_{rr} is the reverse recovery time of D_m, the peak reverse current of D_m is

$$I_{RR} = t_{rr} \frac{di}{dt} = \frac{t_{rr} V_s}{L_s} \tag{2-13}$$

and the peak current through inductor L_s would be

$$I_p = I_0 + I_{RR} = I_0 + \frac{t_{rr} V_s}{L_s} \tag{2-14}$$

When the inductor current becomes I_p, diode D_m turns off suddenly (assuming abrupt recovery) and breaks the current flow path. Due to a highly inductive load, the load current cannot change suddenly from I_0 to I_p. The excess energy stored in L_s would induce a high reverse voltage across D_m, and this may damage diode D_m. The excess energy stored as a result of reverse recovery time is found from

$$W_R = \tfrac{1}{2} L_s [(I_0 + I_{RR})^2 - I_0^2] \tag{2-15}$$

$$W_R = \tfrac{1}{2} L_s \left[\left(I_0 + \frac{t_{rr} V_s}{L_s} \right)^2 - I_0^2 \right] \tag{2-16}$$

The waveforms for the various currents are shown in Fig. 2-7b. This excess energy can be transferred from the inductor L_s to a capacitor C_s which is connected across diode D_m. The value of the C_s can be determined from

$$\tfrac{1}{2} C_s V_c^2 = W_R$$

or

$$C_s = \frac{2W_R}{V_c^2} \tag{2-17}$$

where V_c is the allowable reverse voltage of the diode.

A resistor R_s, which is shown in Fig. 2-7a by dashed lines, is connected in series with the capacitor to damp out any transient oscillation. Equation (2-17) is an approximate one and does not take into account the effects of L_s and R_s during the transients for energy transfer. The design of C_s and R_s values is discussed in Section 15-4.

2-6 SERIES-CONNECTED DIODES

In many high-voltage applications (e.g., HVDC transmission lines), one commercially available diode cannot meet the required voltage rating, and diodes are connected in series to increase the reverse blocking capabilities.

Let us consider two series-connected diodes as shown in Fig. 2-8a. In practice, the v–i characteristics for the same type of diodes differ due to tolerances in their production process. Figure 2-11b shows two v–i characteristics for such diodes. In the forward-biased condition, both diodes conduct the same amount of current, and the forward voltage drop of each diode would be almost equal. However, in the reverse blocking condition, each diode has to carry the same leakage current, and as a result the blocking voltages will differ significantly.

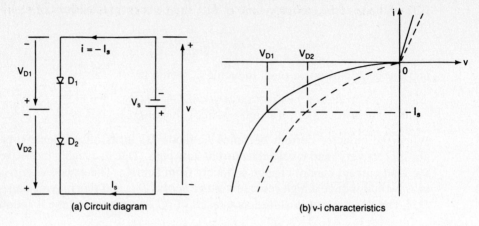

(a) Circuit diagram

(b) v-i characteristics

Figure 2-8 Two series-connected diodes with reverse bias.

A simple solution to this problem, as shown in Fig. 2-9a, is to force equal voltage sharing by connecting a resistor across each diode. Due to equal voltage sharing, the leakage current of each diode would be different, and this is shown in Fig. 2-9b. Since the total leakage current must be shared by a diode and its resistor,

$$I_s = I_{s1} + I_{R1} = I_{s2} + I_{R2} \tag{2-18}$$

But $I_{R1} = V_{D1}/R_1$ and $I_{R2} = V_{D2}/R_2 = V_{D1}/R_2$. Equation (2-18) gives the relationship between R_1 and R_2 for equal voltage sharing as

$$I_{s1} + \frac{V_{D1}}{R_1} = I_{s2} + \frac{V_{D1}}{R_2} \tag{2-19}$$

If the resistances are equal, $R = R_1 = R_2$ and the two diode voltages would be different slightly depending on the dissimilarities of the two v–i characteristics.

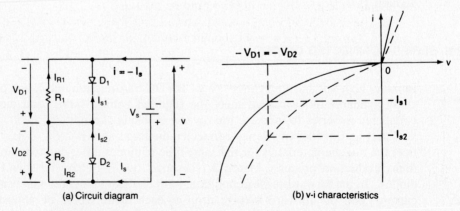

(a) Circuit diagram

(b) v-i characteristics

Figure 2-9 Series-connected diodes with steady-state voltage-sharing characteristics.

The values of V_{D1} and V_{D2} can be determined from Eqs. (2-20) and (2-21):

$$I_{s1} + \frac{V_{D1}}{R} = I_{s2} + \frac{V_{D2}}{R} \tag{2-20}$$

$$V_{D1} + V_{D2} = V_s \tag{2-21}$$

The voltage sharings under transient conditions (e.g., due to switching loads, the initial applications of the input voltage) are accomplished by connecting capacitors across each diode, which is shown in Fig. 2-10. R_s limits the rate of rise of the blocking voltage.

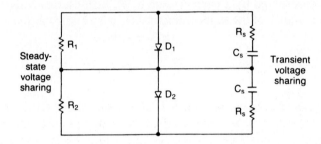

Figure 2-10 Series diodes with voltage-sharing networks under steady-state and transient conditions.

2-7 PARALLEL-CONNECTED DIODES

In high-power applications, diodes are connected in parallel to increase the current-carrying capability to meet the desired current requirements. The current sharings of diodes would be in accord with their respective forward voltage drops. Uniform current sharing can be achieved by providing equal inductances (e.g., in the leads) or by connecting current-sharing resistors (which may not be practical due to power losses); and this is depicted in Fig. 2-11. It is possible to minimize this problem by selecting diodes with equal forward voltage drops or diodes of the same type. Since the diodes are connected in parallel, the reverse blocking voltages of each diode would be the same.

The resistors of Fig. 2-11a will help current sharing under steady-state conditions. Current sharing under dynamic conditions can be accomplished by con-

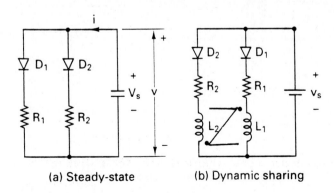

(a) Steady-state (b) Dynamic sharing

Figure 2-11 Parallel-connected diodes.

necting coupled inductors as shown in Fig. 2-11b. If the current through D_1 rises, the $L \, di/dt$ across L_1 increases, and a corresponding voltage of opposite polarity is induced across inductor L_2. The result is a low-impedance path through diode D_2 and the current is shifted to D_2. The inductors would generate voltage spikes and they may be expensive and bulky, especially at high currents.

2-8 SPICE DIODE MODEL

The SPICE model of a diode is shown in Fig. 2-12a. The diode current I_D that depends on its voltage is represented by a current source. R_s is the series resistance, and it is due to the resistance of the semiconductor. R_s, also known as *bulk*

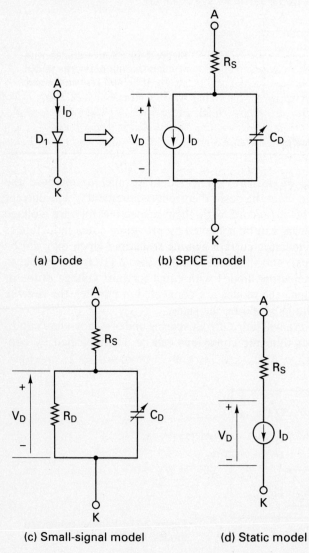

(a) Diode   (b) SPICE model

(c) Small-signal model   (d) Static model

Figure 2-12 SPICE diode model with reverse-biased diode.

resistance, is dependent on the amount of doping. The small-signal and static models that are generated by SPICE are shown in Fig. 2–12b and c, respectively. C_d is a nonlinear function of the diode voltage v_D and is equal to $C_d = dq_d/dv_D$, where q_d is the depletion-layer charge. SPICE generates the small-signal parameters from the operating point.

The SPICE model statement of a diode has the general form

```
.MODEL DNAME   D  (P1=V1 P2=V2  P3=V3 . . . . . . . .PN=VN)
```

DNAME is the model name and it can begin with any character; but its word size is normally limited to 8. D is the type symbol for diodes. P1, P2, ... and V1, V2, ... are the model parameters and their values, respectively.

Example 2-3

Two diodes are connected in series, shown in Fig. 2-9a to share a total voltage of $V_D = 5\,\text{kV}$. The reverse leakage currents of the two diodes are $I_{s1} = 30\,\text{mA}$ and $I_{s2} = 35\,\text{mA}$. (a) Find the diode voltages if the voltage-sharing resistances are equal, $R_1 = R_2 = R = 100\,\text{k}\Omega$. (b) Find the voltage-sharing resistances R_1 and R_2 if the diode voltages are equal, $V_{D1} = V_{D2} = V_D/2$. (c) Use PSpice to check your results of part (a). PSpice model parameters of the diodes are: BV=3KV and IS=30 mA for diode D_1, and IS=35 mA for diode D_2.

Solution   (a) $I_{s1} = 30\,\text{mA}$, $I_{s2} = 35\,\text{mA}$, and $R_1 = R_2 = R = 100\,\text{k}\Omega$. $V_D = V_{D1} + V_{D2}$ or $V_{D2} = V_D - V_{D1}$. From Eq. (2-19),

$$I_{s1} + \frac{V_{D1}}{R} = I_{s2} + \frac{V_{D2}}{R}$$

Substituting $V_{D2} = V_D - V_{D1}$ and solving for the diode voltage D_1, we get

$$V_{D1} = \frac{V_D}{2} + \frac{R}{2}\,(I_{s2} - I_{s1})$$

$$= \frac{5\,\text{kV}}{2} + \frac{100\,\text{k}\Omega}{2}\,(35 \times 10^{-3} - 30 \times 10^{-3}) = 2750\,\text{V}$$

(2-22)

and $V_{D2} = V_D - V_{D1} = 5\,\text{kV} - 2750 = 2250\,\text{V}$.

(b) $I_{s1} = 30\,\text{mA}$, $I_{s2} = 35\,\text{mA}$, and $V_{D1} = V_{D2} = V_D/2 = 2.5\,\text{kV}$. From Eq. (2-19),

$$I_{s1} + \frac{V_{D1}}{R_1} = I_{s2} + \frac{V_{D2}}{R_2}$$

which gives the resistance R_2 for a known value of R_1 as

$$R_2 = \frac{V_{D1}R_1}{V_{D1} - R_1(I_{s2} - I_{s1})}$$

(2-23)

Assuming that $R_1 = 100\,\text{k}\Omega$, we get

$$R_2 = \frac{2.5\,\text{kV} \times 100\,\text{k}\Omega}{2.5\,\text{kV} - 100\,\text{k}\Omega \times (35 \times 10^{-3} - 30 \times 10^{-3})} = 125\,\text{k}\Omega$$

(c) The diode circuit for PSpice simulation is shown in Fig. 2-13. The list of the circuit file is as follows:

```
Example 2-3      Diode Voltage-Sharing Circuit
VS   1   0   DC    5KV
R    1   2   0.01
R1   2   3   100K
R2   3   0   100K
D1   3   2   MOD1
D2   0   3   MOD2
.MODEL  MOD1  D  (IS=30MA  BV=3KV)    ; Diode model parameters
.MODEL  MOD2  D  (IS=35MA  BV=3KV)    ; Diode model parameters
.OP                                  ; Dc operating point analysis
.END
```

The results of PSpice simulation are

NAME	D1			D2		
ID	-3.00E-02	I_{D1}	= -30 mA	-3.50E-02	I_{D2}	= -35 mA
VD	-2.75E+03	V_{D1}	= -2750 V	-2.25E+03	V_{D2}	= -2250 V
REQ	1.00E+12	R_{D1}	= 1 GΩ	1.00E+12	R_{D2}	= 1 GΩ

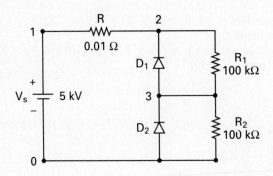

Figure 2-13 Diode circuit for PSpice simulation for Example 2-3.

SUMMARY

The characteristics of practical diodes differ from that of ideal diodes. The reverse recovery time plays a significant role, especially at high-speed switching applications. Diodes can be classified into three types: (1) general-purpose diodes, (2) fast-recovery diodes, and (3) Schottky diodes. Although a Schottky diode behaves as a *pn*-junction diode, there is no physical junction; and as a result a Schottky diode is a majority carrier device. On the other hand, a *pn*-junction diode is both a majority and a minority carrier diode.

If diodes are connected in series to increase the blocking voltage capability, voltage-sharing networks under steady-state and transient conditions are required. When diodes are connected in parallel to increase the current-carrying ability, current-sharing elements are also necessary.

REFERENCES

1. M. S. Ghausi, *Electronic Devices and Circuits.* New York: Holt, Rinehart and Winston, 1985, p. 672.
2. P. R. Gray and R. G. Meyer, *Analysis and Design of Analog Integrated Circuits.* New York: John Wiley & Sons, Inc., 1984, p. 1.
3. M. H. Rashid, *SPICE for Circuits and Electronics Using PSpice.* Englewood Cliffs, N.J.: Prentice Hall, 1990.
4. P. W. Tuinenga, *SPICE: A Guide to Circuit Simulation and Analysis Using PSPICE.* Englewood Cliffs, N.J.: Prentice Hall, 1992.
5. *PSpice Manual.* Irvine, Calif.: MicroSim Corporation, 1992.

REVIEW QUESTIONS

2-1. What are the types of power diodes?

2-2. What is a leakage current of diodes?

2-3. What is a reverse recovery time of diodes?

2-4. What is a reverse recovery current of diodes?

2-5. What is a softness factor of diodes?

2-6. What are the recovery types of diodes?

2-7. What is the cause of reverse recovery time in a *pn*-junction diode?

2-8. What is the effect of reverse recovery time?

2-9. Why is it necessary to use fast-recovery diodes for high-speed switching?

2-10. What is a forward recovery time?

2-11. What are the main differences between *pn*-junction diodes and Schottky diodes?

2-12. What are the limitations of Schottky diodes?

2-13. What is the typical reverse recovery time of general-purpose diodes?

2-14. What is the typical reverse recovery time of fast-recovery diodes?

2-15. What are the problems of series-connected diodes, and what are the possible solutions?

2-16. What are the problems of parallel-connected diodes, and what are the possible solutions?

2-17. If two diodes are connected in series with equal-voltage sharings, why do the diode leakage currents differ?

PROBLEMS

2-1. The reverse recovery time of a diode is $t_{rr} = 5$ μs, and the rate of fall of the diode current is $di/dt = 80$ A/μs. If the softness factor is SF = 0.5, determine **(a)** the storage charge Q_{RR}, and **(b)** the peak reverse current I_{RR}.

2-2. The measured values of a diode at a temperature of 25°C are

$$V_D = 1.0 \text{ V at } I_D = 50 \text{ A}$$

$$= 1.5 \text{ V at } I_D = 600 \text{ A}$$

Determine **(a)** the emission coefficient n, and **(b)** the leakage current I_s.

2-3. Two diodes are connected in series and the voltage across each diode is maintained the same by connecting a voltage-sharing resistor, such that $V_{D1} = V_{D2} = 2000$ V and $R_1 = 100$ kΩ. The *v–i* characteristics of the diodes are shown in Fig. P2-3. Determine the leakage currents of each diode and the resistance R_2 across diode D_2.

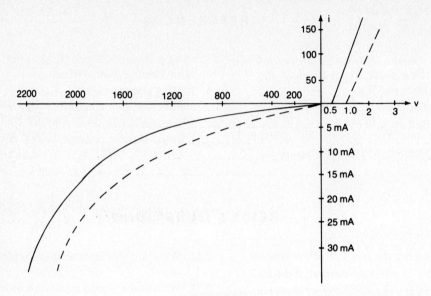

Figure P2-3

2-4. Two diodes are connected in parallel and the forward voltage drop across each diode is 1.5 V. The v–i characteristics of diodes are shown in Fig. P2-3. Determine the forward currents through each diode.

2-5. Two diodes are connected in parallel as shown in Fig. 2-11a, with current-sharing resistances. The v–i characteristics are shown in Fig. P2-3. The total current is $I_T = 200$ A. The voltage across a diode and its resistance is $v = 2.5$ V. Determine the values of resistances R_1 and R_2 if the current is shared equally by the diodes.

2-6. Two diodes are connected in series as shown in Fig. 2-9a. The resistance across the diodes is $R_1 = R_2 = 10$ kΩ. The input dc voltage is 5 kV. The leakage currents are $I_{s1} = 25$ mA and $I_{s2} = 40$ mA. Determine the voltage across the diodes.

3

Diode circuits and rectifiers

3-1 INTRODUCTION

Semiconductor diodes have found many applications in electronics and electrical engineering circuits. Diodes are also widely used in power electronics circuits for conversion of electric power. Some diode circuits that are commonly encountered in power electronics for power processing are reviewed in this chapter. The applications of diodes for conversion of power from ac to dc are introduced. Ac–dc converters are commonly known as *rectifiers*, and diode rectifiers provide a fixed dc output voltage. For the sake of simplicity the diodes are considered to be ideal. By "ideal" we mean that the reverse recovery time t_{rr} and the forward voltage drop V_D are negligible. That is, $t_{rr} = 0$ and $V_D = 0$.

3-2 DIODES WITH *RC* AND *RL* LOADS

Figure 3-1a shows a diode circuit with an *RC* load. When the switch S_1 is closed at $t = 0$, the charging current i that flows through the capacitor can be found from

$$V_s = v_R + v_c = v_R + \frac{1}{C} \int i \, dt + v_c(t = 0) \tag{3-1}$$

$$v_R = Ri \tag{3-2}$$

With initial condition $v_c(t = 0) = 0$, the solution of Eq. (3-1) (which is derived in Appendix D.1) gives the charging current i as

$$i(t) = \frac{V_s}{R} e^{-t/RC} \tag{3-3}$$

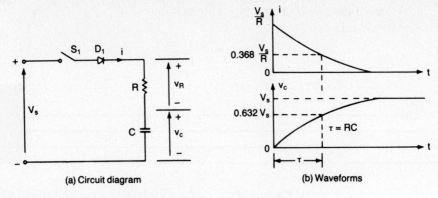

(a) Circuit diagram   (b) Waveforms

Figure 3-1   Diode circuit with RC load.

The capacitor voltage v_c is

$$v_c(t) = \frac{1}{C} \int_0^t i \, dt = V_s(1 - e^{-t/RC}) = V_s(1 - e^{-t/\tau}) \tag{3-4}$$

where $\tau = RC$ is the time constant of an RC load. The rate of change of the capacitor voltage is

$$\frac{dv_c}{dt} = \frac{V_s}{RC} e^{-t/RC} \tag{3-5}$$

and the initial rate of change of the capacitor voltage (at $t = 0$) is obtained from Eq. (3-5):

$$\left. \frac{dv_c}{dt} \right|_{t=0} = \frac{V_s}{RC} \tag{3-6}$$

A diode circuit with an RL load is shown in Fig. 3-2a. When switch S_1 is closed at $t = 0$, the current i through the inductor increases and is expressed as

$$V_s = v_L + v_R = L \frac{di}{dt} + Ri \tag{3-7}$$

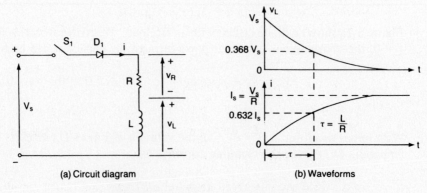

(a) Circuit diagram   (b) Waveforms

Figure 3-2   Diode circuit with RL load.

With initial condition $i(t = 0) = 0$, the solution of Eq. (3-7) (which is derived in Appendix D.2) yields

$$i(t) = \frac{V_s}{R} (1 - e^{-tR/L})$$ (3-8)

The rate of change of this current can be obtained from Eq. (3-8) as

$$\frac{di}{dt} = \frac{V_s}{L} e^{-tR/L}$$ (3-9)

and the initial rate of rise of the current (at $t = 0$) is obtained from Eq. (3-9):

$$\left.\frac{di}{dt}\right|_{t=0} = \frac{V_s}{L}$$ (3-10)

The voltage v_L across the inductor is

$$v_L(t) = L \frac{di}{dt} = V_s e^{-tR/L}$$ (3-11)

where $L/R = \tau$ is the time constant of an RL load.

The waveforms for voltage v_L and current are shown in Fig. 3-2b. If $t >> L/R$, the voltage across the inductor tends to be zero and its current reaches a steady-state value of $I_s = V_s/R$. If an attempt is then made to open switch S_1, the energy stored in the inductor ($= 0.5Li^2$) will be transformed into a high reverse voltage across the switch and diode. This energy will be dissipated in the form of sparks across the switch; and diode D_1 is likely to be damaged in this process. To overcome such a situation, a diode commonly known as a *freewheeling diode* is connected across an inductive load as shown in Fig. 3–10a.

Note. Since the current i in Figs. 3-1a and 3-2a is unidirectional and does not tend to change its polarity, the diodes have no effect on circuit operation.

Example 3-1

A diode circuit is shown in Fig. 3-3a with $R = 44 \ \Omega$ and $C = 0.1 \ \mu F$. The capacitor has an initial voltage, $V_0 = 220$ V. If switch S_1 is closed at $t = 0$, determine (a) the peak diode current, (b) the energy dissipated in the resistor R, and (c) the capacitor voltage at $t = 2 \ \mu s$.

Solution The waveforms are shown in Fig. 3-3b.

(a) Equation (3-3) can be used with $V_s = V_0$ and the peak diode current I_p is

$$I_p = \frac{V_0}{I_p} = \frac{220}{44} = 5 \text{ A}$$

(b) The energy W dissipated is

$$W = 0.5CV_0^2 = 0.5 \times 0.1 \times 10^{-6} \times 220^2 = 0.00242 \text{ J} = 2.42 \text{ mJ}$$

(c) For $RC = 44 \times 0.1 = 4.4 \ \mu s$ and $t = t_1 = 2 \ \mu s$, the capacitor voltage is

$$v_c(t = 2 \ \mu s) = V_0 e^{-t/RC} = 220 \times e^{-2/4.4} = 139.64 \text{ V}$$

Note. Since the current is unidirectional, the diode does not affect circuit operation.

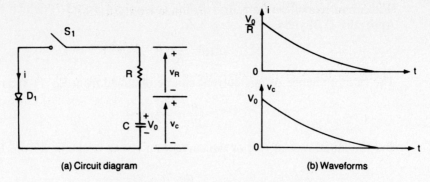

(a) Circuit diagram   (b) Waveforms

Figure 3-3   Diode circuit with RC load.

3-3 DIODES WITH *LC* AND *RLC* LOADS

A diode circuit with an LC load is shown in Fig. 3-4a. When switch S_1 is closed at $t = 0$, the charging current i of the capacitor is expressed as

$$V_s = L \frac{di}{dt} + \frac{1}{C} \int i \, dt + v_c(t = 0) \tag{3-12}$$

With initial conditions $i(t = 0) = 0$ and $v_c(t = 0) = 0$, Eq. (3-12) can be solved for the capacitor current i as (in Appendix D.3)

$$i(t) = V_s \sqrt{\frac{C}{L}} \sin \omega t \tag{3-13}$$

$$= I_p \sin \omega t \tag{3-14}$$

where $\omega = 1/\sqrt{LC}$ and the peak current I_p is

$$I_p = V_s \sqrt{\frac{C}{L}} \tag{3-15}$$

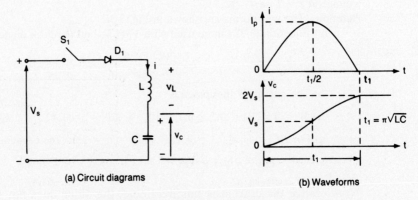

(a) Circuit diagrams   (b) Waveforms

Figure 3-4   Diode circuit with LC load.

The rate of rise of the current is obtained from Eq. (3-13) as

$$\frac{di}{dt} = \frac{V_s}{L} \cos \omega t \qquad (3\text{-}16)$$

and Eq. (3-16) gives the initial rate of rise of the current (at $t = 0$) as

$$\left.\frac{di}{dt}\right|_{t=0} = \frac{V_s}{L} \qquad (3\text{-}17)$$

The voltage v_c across the capacitor can be derived as

$$v_c(t) = \frac{1}{C} \int_0^t i \, dt = V_s(1 - \cos \omega t) \qquad (3\text{-}18)$$

At a time $t = t_1 = \pi \sqrt{LC}$, the diode current i falls to zero and the capacitor is charged to $2V_s$. The waveforms for the voltage v_L and current i are shown in Fig. 3-4b.

Example 3-2

A diode circuit with an LC load is shown in Fig. 3-5a with the capacitor having an initial voltage, $V_0 = 220$ V, capacitance, $C = 20 \, \mu$F, and inductance, $L = 80 \, \mu$H. If switch S_1 is closed at $t = 0$, determine (a) the peak current through the diode, (b) the conduction time of the diode, and (c) the steady-state capacitor voltage.

Solution (a) Using *Kirchhoff's voltage law* (KVL), we can write the equation for the current i as

$$L \frac{di}{dt} + \frac{1}{C} \int i \, dt + v_c(t = 0) = 0$$

and the current i with initial conditions of $i(t = 0) = 0$ and $v_c(t = 0) = -V_0$ is solved as

$$i(t) = V_0 \sqrt{\frac{C}{L}} \sin \omega t$$

where $\omega = 1/\sqrt{LC} = 10^6/\sqrt{20 \times 80} = 25{,}000$ rad/s. The peak current I_p is

$$I_p = V_0 \sqrt{\frac{C}{L}} = 220 \sqrt{\frac{20}{80}} = 110 \text{ A}$$

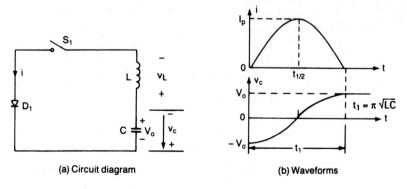

(a) Circuit diagram (b) Waveforms

Figure 3-5 Diode circuit with LC load.

(b) At $t = t_1 = \pi \sqrt{LC}$, the diode current becomes zero and the conduction time t_1 of diode is

$$t_1 = \pi \sqrt{LC} = \pi \sqrt{20 \times 80} = 125.66 \ \mu s$$

(c) The capacitor voltage can easily be shown to be

$$v_c(t) = \frac{1}{C} \int_0^t i \ dt - V_0 = -V_0 \cos \omega t$$

For $t = t_1 = 125.66 \ \mu s$, $v_c(t = t_1) = -220 \cos \pi = 220$ V.

A diode circuit with an RLC load is shown in Fig. 3-6. If switch S_1 is closed at $t = 0$, we can use the KVL to write the equation for the load current i as

$$L \frac{di}{dt} + Ri + \frac{1}{C} \int i \ dt + v_c(t = 0) = V_s \qquad (3\text{-}19)$$

with initial conditions $i(t = 0)$ and $v_c(t = 0) = V_0$. Differentiating Eq. (3-19) and dividing both sides by L gives

$$\frac{d^2i}{dt^2} + \frac{R}{L} \frac{di}{dt} + \frac{i}{LC} = 0 \qquad (3\text{-}20)$$

Under steady-state conditions, the capacitor is charged to the source voltage V_s and the steady-state current will be zero. The forced component of the current in Eq. (3-20) is also zero. The current is due to the natural component.

The characteristic equation in Laplace's domain of s is

$$s^2 + \frac{R}{L} s + \frac{1}{LC} = 0 \qquad (3\text{-}21)$$

and the roots of quadratic equation (3-21) are given by

$$s_{1,2} = -\frac{R}{2L} \pm \sqrt{\left(\frac{R}{2L}\right)^2 - \frac{1}{LC}} \qquad (3\text{-}22)$$

Let us define two important properties of a second-order circuit: the *damping factor*,

$$\alpha = \frac{R}{2L} \qquad (3\text{-}23)$$

and the *resonant frequency*,

$$\omega_0 = \frac{1}{\sqrt{LC}} \qquad (3\text{-}24)$$

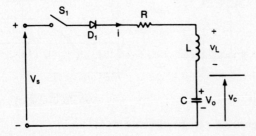

Figure 3-6 Diode circuit with RLC load.

Diode Circuits and Rectifiers   Chap. 3

Substituting these into Eq. (3-22) yields

$$s_{1,2} = -\alpha \pm \sqrt{\alpha^2 - \omega_0^2} \tag{3-25}$$

The solution for the current, which will depend on the values of α and ω_0, would follow one of the three possible cases.

Case 1. If $\alpha = \omega_0$, the roots are equal, $s_1 = s_2$, and the circuit is called *critically damped*. The solution will be of the form

$$i(t) = (A_1 + A_2 t)e^{s_1 t} \tag{3-26}$$

Case 2. If $\alpha > \omega_0$, the roots are real and the circuit is said to be *overdamped*. The solution takes the form

$$i(t) = A_1 e^{s_1 t} + A_2 e^{s_2 t} \tag{3-27}$$

Case 3. If $\alpha < \omega_0$, the roots are complex and the circuit is said to be *underdamped*. The roots are

$$s_{1,2} = -\alpha \pm j\omega_r \tag{3-28}$$

where ω_r is called the *ringing frequency* (or damped resonant frequency) and $\omega_r = \sqrt{\omega_0^2 - \alpha^2}$. The solution takes the form

$$i(t) = e^{-\alpha t}(A_1 \cos \omega_r t + A_2 \sin \omega_r t) \tag{3-29}$$

which is a *damped or decaying sinusoidal.*

Note. The constants A_1 and A_2 can be determined from the initial conditions of the circuit. The ratio of α/ω_0 is commonly known as the *damping ratio*, δ. Power electronic circuits are generally underdamped such that the circuit current becomes near sinusoidal to cause a nearly sinusoidal ac output and/or to turn off a power semiconductor device.

Example 3-3

The second-order *RLC* circuit of Fig. 3-6 has the source voltage $V_s = 220$ V, inductance $L = 2$ mH, capacitance $C = 0.05$ μF, and resistance $R = 160$ Ω. The initial value of the capacitor voltage is $V_0 = 0$. If switch s_1 is closed at $t = 0$, determine (a) an expression for the current $i(t)$, and (b) the conduction time of diode. (c) Draw a sketch of $i(t)$. (d) Use PSpice to plot the instantaneous current i for $R = 50$ Ω, 160 Ω, and 320 Ω.

Solution (a) From Eq. (3-23), $\alpha = R/2L = 160 \times 10^3/(2 \times 2) = 40{,}000$ rad/s, and from Eq. (3-24), $\omega_0 = 1/\sqrt{LC} = 10^5$ rad/s.

$$\omega_r = \sqrt{10^{10} - 16 \times 10^8} = 91{,}652 \text{ rad/s}$$

Since $\alpha < \omega_0$, it is an underdamped circuit and the solution is of the form

$$i(t) = e^{-\alpha t}(A_1 \cos \omega_r t + A_2 \sin \omega_r t)$$

At $t = 0$, $i(t = 0) = 0$ and this gives $A_1 = 0$. The solution becomes

$$i(t) = e^{-\alpha t}A_2 \sin \omega_r t$$

The derivative of $i(t)$ becomes

$$\frac{di}{dt} = \omega_r \cos \omega_r t A_2 e^{-\alpha t} - \alpha \sin \omega_r t A_2 e^{-\alpha t}$$

When the switch is closed at $t = 0$, the capacitor offers a low impedance and the inductor offers a high impedance. The initial rate of rise of the current is limited only by the inductor L. Thus at $t = 0$, the circuit di/dt is V_s/L. Therefore,

$$\left.\frac{di}{dt}\right|_{t=0} = \omega_r A_2 = \frac{V_s}{L}$$

which gives the constant as

$$A_2 = \frac{V_s}{\omega_r L} = \frac{220 \times 1000}{91,652 \times 2} = 1.2$$

The final expression for the current $i(t)$ is

$$i(t) = 1.2 \sin(91,652t)e^{-40,000t} \text{ A}$$

(b) The conduction time t_1 of the diode is obtained when $t = 0$. That is,

$$\omega_r t_1 = \pi \quad \text{or} \quad t_1 = \frac{\pi}{91,652} = 34.27 \ \mu s$$

(c) The sketch for the current waveform is shown in Fig. 3-7.
(d) The circuit for PSpice simulation [3] is shown in Fig. 3-8. The list of the circuit file is as follows:

```
Example 3-3      RLC Circuit with Diode
.PARAM  VALU = 160                     ; Define parameter VALU
.STEP  PARAM  VALU  LIST  50  160  320 ; Vary parameter    VALU
VS    1    0    PWL (0   0  INS   220V 1MS   220V) ; Piecewise linear
R     2    3    {VALU}                 ; Variable resistance
L     3    4    2MH
C     4    0    0.05UF
D1    1    2    DMOD                    ; Diode with model DMOD
.MODEL  DMOD   D(IS=2.22E-15  BV=1800V) ; Diode model parameters
.TRAN   0.1US  60US                    ; Transient analysis
.PROBE                                 ; Graphics postprocessor
.END
```

The PSpice plot of the current $I(R)$ through resistance R is shown in Fig. 3-9.

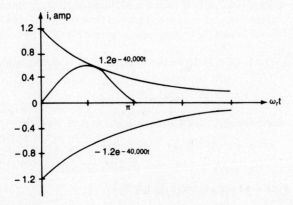

Figure 3-7 Current waveform for Example 3-3.

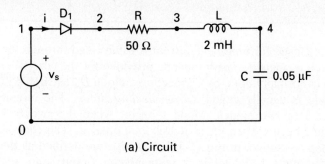

(a) Circuit

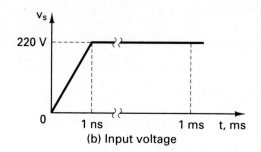

(b) Input voltage

Figure 3-8 *RLC* circuit for PSpice simulation.

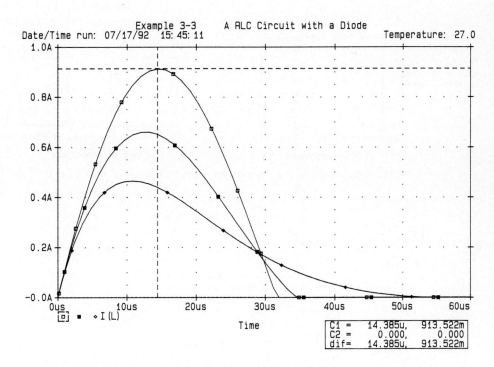

Figure 3-9 Plots for Example 3-3.

If switch S_1 in Fig. 3-2a is closed for time t_1, a current is established through the load; and then if the switch is opened, a path must be provided for the current in the inductive load. This is normally done by connecting a diode D_m as shown in Fig. 3-10a, and this diode is usually called a *freewheeling diode*. The circuit operation can be divided into two modes. Mode 1 begins when the switch is closed at $t = 0$, and mode 2 begins when the switch is then opened. The equivalent circuits for the modes are shown in Fig. 3–10b. i_1 and i_2 are defined as the instantaneous currents for mode 1 and mode 2, respectively. t_1 and t_2 are the corresponding durations of these modes.

Mode 1. During this mode, the diode current i_1, which is similar to Eq. (3-8), is

$$i_1(t) = \frac{V_s}{R} (1 - e^{-tR/L}) \tag{3-30}$$

When the switch is opened at $t = t_1$ (at the end of this mode), the current at that time becomes

$$I_1 = i_1(t = t_1) = \frac{V_s}{R} (1 - e^{t_1R/L}) \tag{3-31}$$

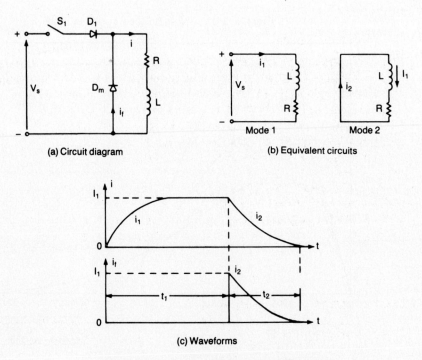

(a) Circuit diagram

(b) Equivalent circuits

(c) Waveforms

Figure 3-10 Circuit with freewheeling diode.

If the time t_1 is sufficiently long, the current reaches a steady-state value and a steady-state current of $I_s = V_s/R$ flows through the load.

Mode 2. This mode begins when the switch is opened and the load current starts to flow through the freewheeling diode D_m. Redefining the time origin at the beginning of this mode, the current through the freewheeling diode is found from

$$0 = L \frac{di_2}{dt} + Ri_2 \tag{3-32}$$

with initial condition $i_2(t = 0) = I_1$. The solution of Eq. (3-32) gives the freewheeling current $i_f = i_2$ as

$$i_2(t) = I_1 e^{-tR/L} \tag{3-33}$$

and this current decays exponentially to zero at $t = t_2$ provided that $t_2 >> L/R$. The waveforms for the currents are shown in Fig. 3-10c.

Example 3-4
_____

In Fig. 3-10a, the resistance is negligible ($R = 0$), the source voltage is $V_s = 220$ V, and the load inductance is $L = 220\ \mu\text{H}$. (a) Draw the waveform for the load current if the switch is closed for a time $t_1 = 100\ \mu\text{s}$ and is then opened. (b) Determine the energy stored in the load inductor.

Solution (a) The circuit diagram is shown in Fig. 3-11a with a zero initial current. When the switch is closed at $t = 0$, the load current rises linearly and is expressed as

$$i(t) = \frac{V_s}{L} t$$

and at $t = t_1$, $I_0 = V_s t_1/L = 220 \times 100/220 = 100$ A.

(b) When switch S_1 is opened at a time $t = t_1$, the load current starts to flow through diode D_m. Since there is no dissipative (resistive) element in the circuit, the load current will remain constant at $I_0 = 100$ A and the energy stored in the inductor is $0.5LI_0^2 = 1.1$ J. The current waveforms are shown in Fig. 3-11b.

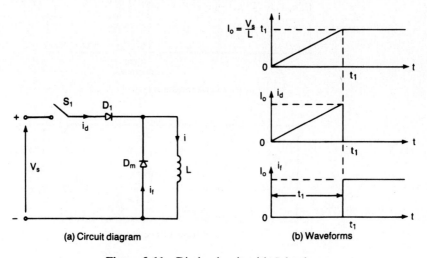

(a) Circuit diagram   (b) Waveforms

Figure 3-11 Diode circuit with L load.

In the ideal lossless circuit of Fig. 3-11a, the energy stored in the inductor is trapped there because there is no resistance in the circuit. In a practical circuit it is desirable to improve the *efficiency* by returning the stored energy into the supply source. This can be achieved by adding a second winding to the inductor and connecting a diode D_1 as shown in Fig. 3-12a. The inductor behaves as a transformer. The transformer secondary is connected such that if v_1 is positive, v_2 is negative with respect to v_1, and vice versa. The secondary winding that facilitates returning the stored energy to the source via diode D_1 is known as a *feedback winding*. Assuming a transformer with a magnetizing inductance of L_m, the equivalent circuit is as shown in Fig. 3-12b.

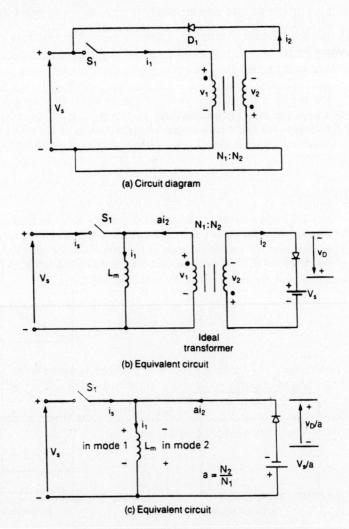

(a) Circuit diagram

(b) Equivalent circuit

(c) Equivalent circuit

Figure 3-12 Circuit with energy recovery diode.

If the diode and secondary voltage (source voltage) are referred to the primary side of the transformer, the equivalent circuit is as shown in Fig. 3-12c. i_1 and i_2 define the primary and secondary currents of the transformer, respectively. The *turns ratio* of an ideal transformer is defined as

$$a = \frac{N_2}{N_1} \qquad (3\text{-}34)$$

The circuit operation can be divided into two modes. Mode 1 begins when switch S_1 is closed at $t = 0$ and mode 2 begins when the switch is opened. The equivalent circuits for the modes are shown in Fig. 3-13a. t_1 and t_2 are the durations of mode 1 and mode 2, respectively.

Mode 1.   During this mode switch S_1 is closed at $t = 0$. Diode D_1 is reverse biased and the current through the diode (secondary current) is $ai_2 = 0$ or $i_2 = 0$. Using the KVL in Fig. 3-13a for mode 1, $V_s = (v_D - V_s)/a$, and this gives the reverse diode voltage as

$$v_D = V_s(1 + a) \qquad (3\text{-}35)$$

Assuming that there is no initial current in the circuit, the primary current is the same as the switch current i_s and is expressed as

$$V_s = L_m \frac{di_1}{dt} \qquad (3\text{-}36)$$

which gives

$$i_1(t) = i_s(t) = \frac{V_s}{L_m} t \qquad (3\text{-}37)$$

This mode is valid for $0 \le t \le t_1$ and ends when the switch is opened at $t = t_1$. At the end of this mode the primary current becomes

$$I_0 = \frac{V_s}{L_m} t_1 \qquad (3\text{-}38)$$

Mode 2.   During this mode the switch is opened, the voltage across the inductor is reversed, and the diode D_1 is forward biased. A current flows through the transformer secondary and the energy stored in the inductor is returned to the source. Using the KVL and redefining the time origin at the beginning of this mode, the primary current is expressed as

$$L_m \frac{di_1}{dt} + \frac{V_s}{a} = 0 \qquad (3\text{-}39)$$

with initial condition $i_1(t = 0) = I_0$, and we can solve the current as

$$i_1(t) = -\frac{V_s}{aL_m} t + I_0 \qquad (3\text{-}40)$$

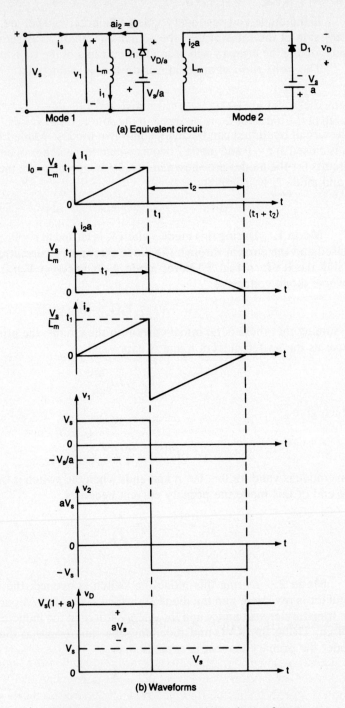

(a) Equivalent circuit

(b) Waveforms

Figure 3-13 Equivalent circuits and waveforms.

The conduction time of diode D_1 is found from the condition $i_1(t = t_2) = 0$ of Eq. (3-40) and is

$$t_2 = \frac{aL_mI_0}{V_s} = at_1 \tag{3-41}$$

Mode 2 is valid for $0 \leq t \leq t_2$. At the end of this mode at $t = t_2$, all the energy stored in the inductor L_m is returned to the source. The various waveforms for the currents and voltage are shown in Fig. 3-13b for $a = 10/6$.

Example 3-5

For the energy recovery circuit of Fig. 3-12a, the magnetizing inductance of the transformer is $L_m = 250~\mu H$, $N_1 = 10$, and $N_2 = 100$. The leakage inductances and resistances of the transformer are negligible. The source voltage is $V_s = 220$ V and there is no initial current in the circuit. If switch S_1 is closed for a time $t_1 = 50~\mu s$ and is then opened, (a) determine the reverse voltage of diode D_1, (b) calculate the peak value of primary current, (c) calculate the peak value of secondary current, (d) determine the conduction time of diode D_1, and (e) determine the energy supplied by the source.

Solution   The turns ratio is $a = N_2/N_1 = 100/10 = 10$.

(a) From Eq. (3-35) the reverse voltage of the diode,

$$v_D = V_s(1 + a) = 220 \times (1 + 10) = 2420~\text{V}$$

(b) From Eq. (3-38) the peak value of the primary current,

$$I_0 = \frac{V_s}{L_m} t_1 = 220 \times \frac{50}{250} = 44~\text{A}$$

(c) The peak value of the secondary current $I_0' = I_0/a = 44/10 = 4.4$ A.

(d) From Eq. (3-41) the conduction time of the diode

$$t_2 = \frac{aL_mI_0}{V_s} = 250 \times 44 \times \frac{10}{220} = 500~\mu s$$

(e) The source energy,

$$W = \int_0^{t_1} vi~dt = \int_0^{t_1} V_s \frac{V_s}{L_m} t~dt = \frac{1}{2} \frac{V_s^2}{L_m} t_1^2$$

Using Eq. (3-38) yields

$$W = 0.5 L_m I_0^2 = 0.5 \times 250 \times 10^{-6} \times 44^2 = 0.242~\text{J} = 242~\text{mJ}$$

3-6 SINGLE-PHASE HALF-WAVE RECTIFIERS

A *rectifier* is a circuit that converts an ac signal into a unidirectional signal. Diodes are used extensively in rectifiers. A single-phase half-wave rectifier is the simplest type, but it is not normally used in industrial applications. However, it is useful in understanding the principle of rectifier operation. The circuit diagram with a resistive load is shown in Fig. 3-14a. During the positive half-cycle of the input voltage, diode D_1 conducts and the input voltage appears across the load. During the negative half-cycle of the input voltage, the diode is in a *blocking*

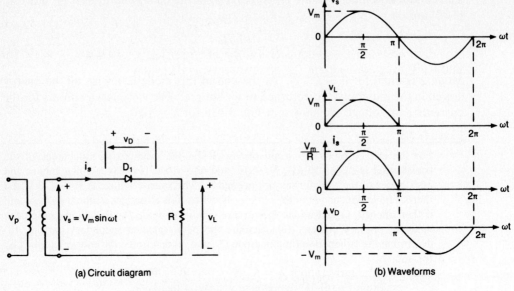

(a) Circuit diagram

(b) Waveforms

Figure 3-14   Single-phase half-wave rectifier.

condition and the output voltage is zero. The waveforms for the input voltage and output voltage are shown in Fig. 3-14b.

3-7 PERFORMANCE PARAMETERS

Although the output voltage as shown in Fig. 3-14b is dc, it is discontinuous and contains harmonics. A rectifier is a power processor that should give a dc output voltage with a minimum amount of harmonic contents. At the same time, it should maintain the input current as sinusoidal as possible and in phase with the input voltage so that the power factor is near unity. The power-processing quality of a rectifier requires the determination of harmonic contents of the input current, the output voltage, and the output current. We shall use Fourier series expansions to find the harmonic contents of voltages and currents. There are different types of rectifier circuits and the performances of a rectifier are normally evaluated in terms of the following parameters:

The *average* value of the output (load) voltage, V_{dc}

The average value of the output (load) current, I_{dc}

The output dc power,

$$P_{dc} = V_{dc}I_{dc} \tag{3-42}$$

The *rms* value of the output voltage, V_{rms}

The *rms* value of the output current, I_{rms}

The output *ac power*

$$P_{ac} = V_{rms}I_{rms} \tag{3-43}$$

The *efficiency* (or *rectification ratio*) of a rectifier, which is a figure of merit and permits us to compare the effectiveness, is defined as

$$\eta = \frac{P_{dc}}{P_{ac}} \tag{3-44}$$

The output voltage can be considered as being composed of two components: (1) the dc value, and (2) the ac component or ripple.

The *effective* (rms) value of the ac component of output voltage is

$$V_{ac} = \sqrt{V_{rms}^2 - V_{dc}^2} \tag{3-45}$$

The *form factor*, which is a measure of the shape of output voltage, is

$$FF = \frac{V_{rms}}{V_{dc}} \tag{3-46}$$

The *ripple factor*, which is a measure of the ripple content, is defined as

$$RF = \frac{V_{ac}}{V_{dc}} \tag{3-47}$$

Substituting Eq. (3-45) in Eq. (3-47), the ripple factor can be expressed as

$$RF = \sqrt{\left(\frac{V_{rms}}{V_{dc}}\right)^2 - 1} = \sqrt{FF^2 - 1} \tag{3-48}$$

The *transformer utilization factor* is defined as

$$TUF = \frac{P_{dc}}{V_s I_s} \tag{3-49}$$

where V_s and I_s are the rms voltage and rms current of the transformer secondary, respectively. Let us consider the waveforms of Fig. 3-15, where v_s is the sinusoi-

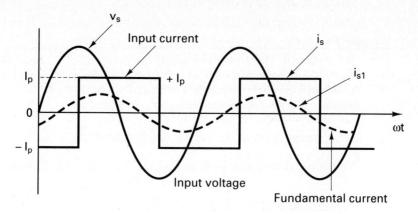

Figure 3-15 Waveforms for input voltage and current.

dal input voltage, i_s is the instantaneous input current, and i_{s1} is its fundamental component.

If ϕ is the angle between the fundamental components of the input current and voltage, ϕ is called the *displacement angle*. The *displacement factor* is defined as

$$DF = \cos \phi \qquad (3\text{-}50)$$

The *harmonic factor* of the input current is defined as

$$HF = \left(\frac{I_s^2 - I_{s1}^2}{I_{s1}^2}\right)^{1/2} = \left[\left(\frac{I_s}{I_{s1}}\right)^2 - 1\right]^{1/2} \qquad (3\text{-}51)$$

where I_{s1} is the fundamental component of the input current I_s. Both I_{s1} and I_s are expressed here in rms. The input *power factor* is defined as

$$PF = \frac{V_s I_{s1}}{V_s I_s} \cos \phi = \frac{I_{s1}}{I_s} \cos \phi \qquad (3\text{-}52)$$

Crest factor CF, which is a measure of the peak input current $I_{s(peak)}$ as compared to its rms value I_s, is often of interest in order to specify the peak current ratings of devices and components. CF of the input current is defined by

$$CF = \frac{I_{s(peak)}}{I_s} \qquad (3\text{-}53)$$

Notes
1. Harmonic factor HF is a measure of the distortion of a waveform and is also known as *total harmonic distortion* (THD).
2. If the input current i_s is purely sinusoidal, $I_{s1} = I_s$ and the power factor PF equals the displacement factor DF. The displacement angle ϕ becomes the impedance angle $\theta = \tan^{-1}(\omega L/R)$ for an *RL* load.
3. Displacement factor DF is often known as *displacement power factor* (DPF).
4. An ideal rectifier should have $\eta = 100\%$, $V_{ac} = 0$, RF = 0, TUF = 1, HF = THD = 0, and PF = DPF = 1.

Example 3-6

The rectifier in Fig. 3-14a has a purely resistive load of R. Determine (a) the efficiency, (b) the form factor, (c) the ripple factor, (d) the transformer utilization factor, (e) the peak inverse voltage (PIV) of diode D_1, and (f) the CF of the input current.
Solution The average output voltage V_{dc} is defined as

$$V_{dc} = \frac{1}{T} \int_0^T v_L(t) \, dt$$

We can notice from Fig. 3-14b that $v_L(t) = 0$ for $T/2 \leq t \leq T$. Hence we have

$$V_{dc} = \frac{1}{T} \int_0^{T/2} V_m \sin \omega t \, dt = \frac{-V_m}{\omega T}\left(\cos \frac{\omega T}{2} - 1\right)$$

But the frequency of the source is $f = 1/T$ and $\omega = 2\pi f$. Thus

$$V_{dc} = \frac{V_m}{\pi} = 0.318 V_m$$

$$I_{dc} = \frac{V_{dc}}{R} = \frac{0.318 V_m}{R}$$

(3-54)

The *root-mean-square* (rms) value of a periodic waveform is defined as

$$V_{rms} = \left[\frac{1}{T} \int_0^T v_L^2(t)\, dt \right]^{1/2}$$

For a sinusoidal voltage of $v_L(t) = V_m \sin \omega t$ for $0 \le t \le T/2$, the rms value of the output voltage is

$$V_{rms} = \left[\frac{1}{T} \int_0^{T/2} (V_m \sin \omega t)^2\, dt \right]^{1/2} = \frac{V_m}{2} = 0.5 V_m$$

$$I_{rms} = \frac{V_{rms}}{R} = \frac{0.5 V_m}{R}$$

(3-55)

From Eq. (3-42), $P_{dc} = (0.318 V_m)^2/R$, and from Eq. (3-43), $P_{ac} = (0.5 V_m)^2/R$.
  (a) From Eq. (3-44), the efficiency $\eta = (0.318 V_m)^2/(0.5 V_m)^2 = 40.5\%$.
  (b) From Eq. (3-46), the form factor FF $= 0.5 V_m/0.318 V_m = 1.57$ or 157%.
  (c) From Eq. (3-48), the ripple factor RF $= \sqrt{1.57^2 - 1} = 1.21$ or 121%.
  (d) The rms voltage of the transformer secondary is

$$V_s = \left[\frac{1}{T} \int_0^T (V_m \sin \omega t)^2\, dt \right]^{1/2} = \frac{V_m}{\sqrt{2}} = 0.707 V_m$$

(3-56)

The rms value of the transformer secondary current is the same as that of the load:

$$I_s = \frac{0.5 V_m}{R}$$

The *volt-ampere* rating (VA) of the transformer, VA $= V_s I_s = 0.707 V_m \times 0.5 V_m/R$. From Eq. (3-49) TUF $= P_{ac}/(V_s I_s) = 0.318^2/(0.707 \times 0.5) = 0.286$.
  (e) The peak reverse (or inverse) blocking voltage PIV $= V_m$.
  (f) $I_{s(peak)} = V_m/R$ and $I_s = 0.5 V_m/R$. The crest factor CF of the input current is CF $= I_{s(peak)}/I_s = 1/0.5 = 2$.

Note. $1/\text{TUF} = 1/0.286 = 3.496$ signifies that the transformer must be 3.496 times larger than that when it is being used to deliver power from a pure ac voltage. This rectifier has a high ripple factor, 121%; a low efficiency, 40.5%; and a poor TUF, 0.286. In addition, the transformer has to carry a dc current, and this results in a dc saturation problem of the transformer core.
  Let us consider the circuit of Fig. 3-14a with an *RL* load as shown in Fig. 3-16a. Due to inductive load, the conduction period of diode D_1 will extend beyond $180°$ until the current becomes zero at $\omega t = \pi + \sigma$. The waveforms for the current and voltage are shown in Fig. 3-16b. It should be noted that the average v_L

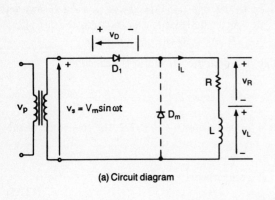

(a) Circuit diagram

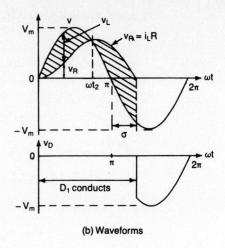

(b) Waveforms

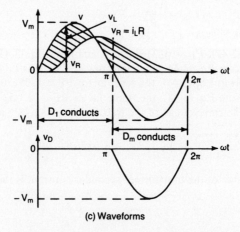

(c) Waveforms

Figure 3-16   Half-wave rectifier with *RL* load.

of the inductor is zero. The average output voltage is

$$V_{dc} = \frac{V_m}{2\pi} \int_0^{\pi+\sigma} \sin \omega t \; d(\omega t) = \frac{V_m}{2\pi} \left[-\cos \omega t\right]_0^{\pi+\sigma}$$

$$= \frac{V_m}{2\pi} \left[1 - \cos(\pi + \sigma)\right]$$

(3-57)

The average load current is $I_{dc} = V_{dc}/R$.

It can be noted from Eq. (3-57) that the average voltage (and current) can be increased by making $\sigma = 0$, which is possible by adding a freewheeling diode D_m as shown in Fig. 3-16a with dashed lines. The effect of this diode is to prevent a negative voltage appearing across the load; and as a result, the magnetic stored energy is increased. At $t = t_1 = \pi/\omega$, the current from D_1 is transferred to D_m and this process is called *commutation* of diodes and the waveforms are shown in Fig. 3-16c. Depending on the load time constant, the load current may be discontinuous. Load current i_L will be discontinuous with a resistive load and continuous with a very high inductive load. The continuity of the load current will depend on its time constant $\tau = \omega L/R$.

If the output is connected to a battery, the rectifier can be used as a battery charger. This is shown in Fig. 3-17a. For $v_s > E$, diode D_1 conducts. The angle α when the diode starts conducting can be found from the condition

$$V_m \sin \alpha = E$$

which gives

$$\alpha = \sin^{-1} \frac{E}{V_m} \tag{3-58}$$

Diode D_1 will be turned off when $v_s < E$ at

$$\beta = \pi - \alpha$$

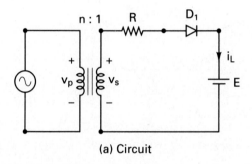

(a) Circuit

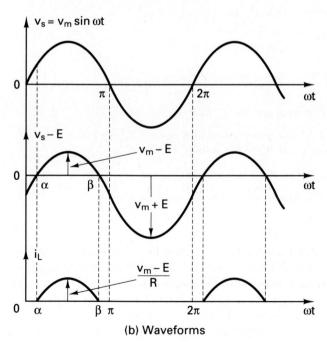

(b) Waveforms

Figure 3-17 Battery charger.

The charging current i_L, which is shown in Fig. 3-17b, can be found from

$$i_L = \frac{v_s - E}{R} = \frac{V_m \sin \omega t - E}{R} \quad \text{for } \alpha < \omega t < \beta$$

Example 3-7

The battery voltage in Fig. 3-17a is $E = 12$ V and its capacity is 100 W-h. The average charging current should be $I_{dc} = 5$ A. The primary input voltage is $V_p = 120$ V, 60 Hz and the transformer has a turns ratio of $n = 2:1$. Calculate (a) the conduction angle δ of the diode, (b) the current-limiting resistance R, (c) the power rating P_R of R, (d) the charging time h_o in hours, (e) the rectifier efficiency η, and (f) the peak inverse voltage PIV of the diode.

Solution $E = 12$ V, $V_p = 120$ V, $V_s = V_p/n = 120/2 = 60$ V, and $V_m = \sqrt{2} V_s = \sqrt{2} \times 60 = 84.85$ V.

(a) From Eq. (3-58), $\alpha = \sin^{-1}(12/84.85) = 8.13°$ or 0.1419 rad. $\beta = 180 - 8.13 = 171.87°$. The conduction angle is $\delta = \beta - \alpha = 171.87 - 8.13 = 163.74°$.

(b) The average charging current I_{dc} is

$$I_{dc} = \frac{1}{2\pi} \int_\alpha^\beta \frac{V_m \sin \omega t - E}{R} \, d(\omega t)$$

(3-59)

$$= \frac{1}{2\pi R} (2V_m \cos \alpha + 2E\alpha - \pi E), \text{ for } \beta = \pi - \alpha$$

which gives

$$R = \frac{1}{2\pi I_{dc}} (2V_m \cos \alpha + 2E\alpha - \pi E)$$

$$= \frac{1}{2\pi \times 5} (2 \times 84.85 \times \cos 8.13° + 2 \times 12 \times 0.1419 - \pi \times 12) = 4.26 \ \Omega$$

(c) The rms battery current I_{rms} is

$$I_{rms}^2 = \frac{1}{2\pi} \int_\alpha^\beta \frac{(V_m \sin \omega t - E)^2}{R^2} \, d(\omega t)$$

$$= \frac{1}{2\pi R^2} \left[\left(\frac{V_m^2}{2} + E^2 \right) (\pi - 2\alpha) + \frac{V_m^2}{2} \sin 2\alpha - 4V_m E \cos \alpha \right]$$

(3-60)

$$= 67.4$$

or $I_{rms} = \sqrt{67.4} = 8.2$ A. The power rating of R is $P_R = 8.2^2 \times 4.26 = 286.4$ W.

(d) The power delivered P_{dc} to the battery is

$$P_{dc} = EI_{dc} = 12 \times 5 = 60 \text{ W}$$

$$h_o P_{dc} = 100 \quad \text{or} \quad h_o = \frac{100}{P_{dc}} = \frac{100}{60} = 1.667 \text{ h}$$

(e) The rectifier efficiency η is

$$\eta = \frac{\text{power delivered to the battery}}{\text{total input power}} = \frac{P_{dc}}{P_{dc} + P_R} = \frac{60}{60 + 286.4} = 17.32\%$$

(f) The peak inverse voltage PIV of the diode is

$$\text{PIV} = V_m + E$$

$$= 84.85 + 12 = 96.85 \text{ V}$$

Example 3-8

The single-phase half-wave rectifier of Fig. 3-14a is connected to a source of $V_s = 120$ V, 60 Hz. Express the instantaneous output voltage $v_L(t)$ in Fourier series.

Solution The rectifier output voltage v_L may be described by a Fourier series as

$$v_L(t) = V_{dc} + \sum_{n=1,2,\ldots}^{\infty} (a_n \sin n\omega t + b_n \cos n\omega t)$$

$$V_{dc} = \frac{1}{2\pi} \int_0^{2\pi} v_L \, d(\omega t) = \frac{1}{2\pi} \int_0^{\pi} V_m \sin \omega t \, d(\omega t) = \frac{V_m}{\pi}$$

$$a_n = \frac{1}{\pi} \int_0^{2\pi} v_L \sin n\omega t \, d(\omega t) = \frac{1}{\pi} \int_0^{\pi} V_m \sin \omega t \sin n\omega t \, d(\omega t)$$

$$= \frac{V_m}{2} \quad \text{for } n = 1$$

$$= 0 \quad \text{for } n = 2, 3, 4, 5, 6, \ldots$$

$$b_n = \frac{1}{\pi} \int_0^{2\pi} v_L \cos n\omega t \, d(\omega t) = \frac{1}{\pi} \int_0^{\pi} V_m \sin \omega t \cos n\omega t \, d(\omega t)$$

$$= \frac{V_m}{\pi} \frac{1 + (-1)^n}{1 - n^2} \quad \text{for } n = 2, 4, 6, \ldots$$

$$= 0 \quad \text{for } n = 1, 3, 5, \ldots$$

Substituting a_n and b_n, the instantaneous output voltage becomes

$$v_L(t) = \frac{V_m}{\pi} + \frac{V_m}{2} \sin \omega t - \frac{2V_m}{3\pi} \cos 2\omega t - \frac{2V_m}{15\pi} \cos 4\omega t - \frac{2V_m}{35\pi} \cos 6\omega t - \ldots \quad (3\text{-}61)$$

where $V_m = \sqrt{2} \times 120 = 169.7$ V and $\omega = 2\pi \times 60 = 377$ rad/s.

3-8 SINGLE-PHASE FULL-WAVE RECTIFIERS

A full-wave rectifier circuit with a center-tapped transformer is shown in Fig. 3-18a. Each half of the transformer with its associated diode acts as a half-wave rectifier and the output of a full-wave rectifier is shown in Fig. 3-18b. Since there is no dc current flowing through the transformer, there is no dc saturation problem of transformer core. The average output voltage is

$$V_{dc} = \frac{2}{T} \int_0^{T/2} V_m \sin \omega t \, dt = \frac{2V_m}{\pi} = 0.6366 V_m \quad (3\text{-}62)$$

Instead of using a center-tapped transformer, we could use four diodes, as shown in Fig. 3-19a. During the positive half-cycle of the input voltage, the power is supplied to the load through diodes D_1 and D_2. During the negative cycle, diodes D_3 and D_4 conduct. The waveform for the output voltage is shown in Fig. 3-19b and is similar to that of Fig. 3-18b. The peak-inverse voltage of a diode is only V_m. This circuit is known as a *bridge rectifier*, and it is commonly used in industrial applications.

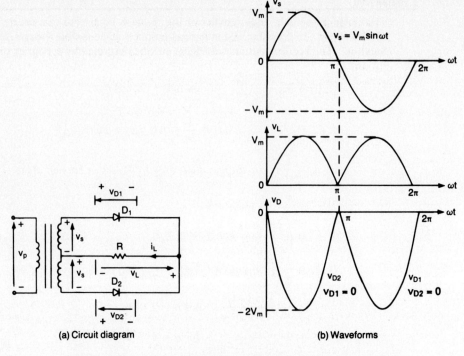

(a) Circuit diagram

(b) Waveforms

Figure 3-18 Full-wave rectifier with center-tapped transformer.

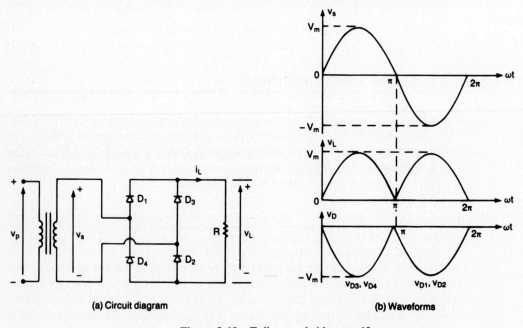

(a) Circuit diagram

(b) Waveforms

Figure 3-19 Full-wave bridge rectifier.

Example 3-9

If the rectifier in Fig. 3-18a has a purely resistive load of R, determine (a) the efficiency, (b) the form factor, (c) the ripple factor, (d) the transformer utilization factor, (e) the peak inverse voltage (PIV) of diode D_1, and (f) the CF of the input current.

Solution   From Eq. (3-62), the average output voltage is

$$V_{dc} = \frac{2V_m}{\pi} = 0.6366V_m$$

and the average load current is

$$I_{dc} = \frac{V_{dc}}{R} = \frac{0.6366V_m}{R}$$

The rms value of the output voltage is

$$V_{rms} = \left[\frac{2}{T}\int_0^{T/2} (V_m \sin \omega t)^2\, dt\right]^{1/2} = \frac{V_m}{\sqrt{2}} = 0.707V_m$$

$$I_{rms} = \frac{V_{rms}}{R} = \frac{0.707V_m}{R}$$

From Eq. (3-42) $P_{dc} = (0.6366V_m)^2/R$, and from Eq. (3-43) $P_{ac} = (0.707V_m)^2/R$.

(a) From Eq. (3-44), the efficiency $\eta = (0.6366V_m)^2/(0.707V_m)^2 = 81\%$.

(b) From Eq. (3-46), the form factor FF $= 0.707V_m/0.6366V_m = 1.11$.

(c) From Eq. (3-48), the ripple factor RF $= \sqrt{1.11^2 - 1} = 0.482$ or 48.2%.

(d) The rms voltage of the transformer secondary $V_s = V_m/\sqrt{2} = 0.707V_m$. The rms value of transformer secondary current $I_s = 0.5V_m/R$. The volt-ampere rating (VA) of the transformer, VA $= 2V_sI_s = 2 \times 0.707V_m \times 0.5V_m/R$. From Eq. (3-49),

$$\text{TUF} = \frac{0.6366^2}{2 \times 0.707 \times 0.5} = 0.5732 = 57.32\%$$

(e) The peak reverse blocking voltage, PIV $= 2V_m$.

(f) $I_{s(peak)} = V_m/R$ and $I_s = 0.707V_m/R$. The crest factor CF of the input current is CF $= I_{s(peak)}/I_s = 1/0.707 = \sqrt{2}$.

Note. The performance of a full-wave rectifier is significantly improved compared to that of a half-wave rectifier.

Example 3-10

The rectifier in Fig. 3-18a has an RL load. Use the method of Fourier series to obtain expressions for output voltage $v_L(t)$.

Solution   The rectifier output voltage may be described by a Fourier series (which is reviewed in Appendix E) as

$$v_L(t) = V_{dc} + \sum_{n=2,4,\dots}^{\infty} (a_n \cos n\omega t + b_n \sin n\omega t)$$

where

$$V_{dc} = \frac{1}{2\pi} \int_0^{2\pi} v_L(t) \, d(\omega t) = \frac{2}{2\pi} \int_0^{\pi} V_m \sin \omega t \, d(\omega t) = \frac{2V_m}{\pi}$$

$$a_n = \frac{1}{\pi} \int_0^{2\pi} v_L \cos n\omega t \, d(\omega t) = \frac{2}{\pi} \int_0^{\pi} V_m \sin \omega t \cos n\omega t \, d(\omega t)$$

$$= \frac{4V_m}{\pi} \sum_{n=2,4,\ldots}^{\infty} \frac{-1}{(n-1)(n+1)} \quad \begin{array}{l} \text{for } n = 2, 4, 6, \ldots \\ \text{for } n = 1, 3, 5, \ldots \end{array}$$

$$= 0$$

$$b_n = \frac{1}{\pi} \int_0^{2\pi} v_L \sin n\omega t \, d(\omega t) = \frac{2}{\pi} \int_0^{\pi} V_m \sin \omega t \sin n\omega t \, d(\omega t) = 0$$

Substituting the values of a_n and b_n, the expression for the output voltage is

$$v_L(t) = \frac{2V_m}{\pi} - \frac{4V_m}{3\pi} \cos 2\omega t - \frac{4V_m}{15\pi} \cos 4\omega t - \frac{4V_m}{35\pi} \cos 6\omega t - \cdots \quad \text{(3-63)}$$

Note. The output of a full-wave rectifier contains only even harmonics and the second harmonic is the most dominant one and its frequency is $2f(= 120 \text{ Hz})$. The output voltage in Eq. (3-63) can be derived by spectrum multiplication of switching function, and this is explained in Appendix C.

Example 3-11

A single-phase bridge rectifier that supplies a very high inductive load such as a dc motor is shown in Fig. 3-20a. The turns ratio of the transformer is unity. The load is such that the motor draws a ripple-free armature current of I_a as shown in Fig.

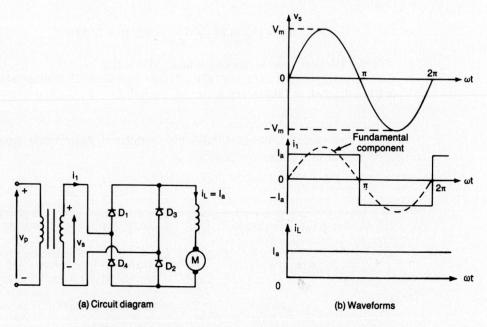

(a) Circuit diagram

(b) Waveforms

Figure 3-20 Full-wave bridge rectifier with dc motor load.

3-20b. Determine (a) the harmonic factor HF of input current, and (b) the input power factor PF of the rectifier.

Solution Normally, a dc motor is highly inductive and acts like a filter in reducing the ripple current of the load.

(a) The waveforms for the input current and input voltage of the rectifier are shown in Fig. 3-20b. The input current can be expressed in a Fourier series as

$$i_1(t) = I_{dc} + \sum_{n=1,3,\ldots}^{\infty} (a_n \cos n\omega t + b_n \sin n\omega t)$$

where

$$I_{dc} = \frac{1}{2\pi} \int_0^{2\pi} i_1(t) \, d(\omega t) = \frac{1}{2\pi} \int_0^{2\pi} I_a \, d(\omega t) = 0$$

$$a_n = \frac{1}{\pi} \int_0^{2\pi} i_1(t) \cos n\omega t \, d(\omega t) = \frac{2}{\pi} \int_0^{\pi} I_a \cos n\omega t \, d(\omega t) = 0$$

$$b_n = \frac{1}{\pi} \int_0^{2\pi} i_1(t) \sin n\omega t \, d(\omega t) = \frac{2}{\pi} \int_0^{\pi} I_a \sin n\omega t \, d(\omega t) = \frac{4I_a}{n\pi}$$

Substituting the values of a_n and b_n, the expression for the input current is

$$i_1(t) = \frac{4I_a}{\pi} \left(\frac{\sin \omega t}{1} + \frac{\sin 3\omega t}{3} + \frac{\sin 5\omega t}{5} + \cdots \right) \tag{3-64}$$

The rms value of the fundamental component of input current is

$$I_{s1} = \frac{4I_a}{\pi\sqrt{2}} = 0.90 I_a$$

The rms value of the input current is

$$I_s = \frac{4}{\pi\sqrt{2}} I_a \left[1 + \left(\frac{1}{3}\right)^2 + \left(\frac{1}{5}\right)^2 + \left(\frac{1}{7}\right)^2 + \left(\frac{1}{9}\right)^2 + \cdots \right]^{1/2} = I_a$$

From Eq. (3-51),

$$\text{HF} = \text{THD} = \left[\left(\frac{1}{0.90}\right)^2 - 1 \right]^{1/2} = 0.4843 \quad \text{or} \quad 48.43\%$$

(b) The displacement angle $\phi = 0$ and displacement factor $\text{DF} = \cos \phi = 1$. From Eq. (3-52), the power factor $\text{PF} = (I_{s1}/I_s) \cos \phi = 0.90$ (lagging).

3-9 SINGLE-PHASE FULL-WAVE RECTIFIER WITH *RL* LOAD

With a resistive load, the load current is identical in shape to the output voltage. In practice, most loads are inductive to a certain extent and the load current depends on the values of load resistance R and load inductance L. This is shown in Fig. 3-21a. A battery of voltage E is added to develop generalized equations. If $v_s = V_m \sin \omega t = \sqrt{2} \, V_s \sin \omega t$ is the input voltage, the load current i_L can be found from

$$L \frac{di_L}{dt} + Ri_L + E = \sqrt{2} \, V_s \sin \omega t$$

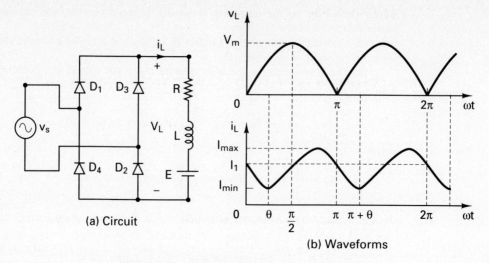

(a) Circuit

(b) Waveforms

Figure 3-21 Full-bridge rectifier with RL load.

which has a solution of the form

$$i_L = \frac{\sqrt{2}\,V_s}{Z} \sin(\omega t - \theta) + A_1 e^{-(R/L)t} - \frac{E}{R} \qquad (3\text{-}65)$$

where load impedance $Z = [R^2 + (\omega L)^2]^{1/2}$ and load impedance angle $\theta = \tan^{-1}(\omega L/R)$.

Case 1: continuous load current. The constant A_1 in Eq. (3-65) can be determined from the condition: at $\omega t = \pi$, $i_L = I_1$.

$$A_1 = \left(I_1 + \frac{E}{R} - \frac{\sqrt{2}\,V_s}{Z} \sin\theta\right) e^{(R/L)(\pi/\omega)}$$

Substitution of A_1 in Eq. (3-65) yields

$$i_L = \frac{\sqrt{2}\,V_s}{Z} \sin(\omega t - \theta) + \left(I_1 + \frac{E}{R} - \frac{\sqrt{2}\,V_s}{Z} \sin\theta\right) e^{(R/L)(\pi/\omega - t)}$$

Under a steady-state condition, $i_L(\omega t = 0) = i_L(\omega t = \pi)$. That is, $i_L(\omega t = 0) = I_1$. Applying this condition, we get the value of I_1 as

$$I_1 = \frac{\sqrt{2}\,V_s}{Z} \sin\theta \, \frac{1 + e^{-(R/L)(\pi/\omega)}}{1 - e^{-(R/L)(\pi/\omega)}} - \frac{E}{R} \qquad \text{for } I_1 \geq 0 \qquad (3\text{-}66)$$

which, after substituting in Eq. (3-66) and simplification, gives

$$i_L = \frac{\sqrt{2}\,V_s}{Z} \left[\sin(\omega t - \theta) + \frac{2}{1 - e^{-(R/L)(\pi/\omega)}} \sin\theta \, e^{-(R/L)t}\right] - \frac{E}{R}$$

$$\text{for } 0 \leq \omega t \leq \pi \text{ and } i_L \geq 0 \qquad (3\text{-}67)$$

The rms diode current can be found from Eq. (3-67) as

$$I_r = \left[\frac{1}{2\pi} \int_0^\pi i_L^2 \, d(\omega t) \right]^{1/2}$$

and the rms output current can then be determined by combining the rms current of each diode as

$$I_{rms} = (I_r^2 + I_r^2)^{1/2} = \sqrt{2} \, I_r$$

The average diode current can also be found from Eq. (3-67) as

$$I_d = \frac{1}{2\pi} \int_0^\pi i_L \, d(\omega t)$$

Case 2: discontinuous load current. The load current flows only during the period $\alpha \le \omega t \le \beta$. The diodes start to conduct at $\omega t = \alpha$ given by

$$\alpha = \sin^{-1} \frac{E}{V_m}$$

At $\omega t = \alpha$, $i_L(\omega t) = 0$ and Eq. (3-65) gives

$$A_1 = \left[\frac{E}{R} - \frac{\sqrt{2} \, V_s}{Z} \sin(\alpha - \theta) \right] e^{(R/L)(\alpha/\omega)}$$

which, after substituting in Eq. (3-65), yields the load current

$$i_L = \frac{\sqrt{2} \, V_s}{Z} \sin(\omega t - \theta) + \left[\frac{E}{R} - \frac{\sqrt{2} \, V_s}{Z} \sin(\alpha - \theta) \right] e^{(R/L)(\alpha/\omega - t)} \tag{3-68}$$

At $\omega t = \beta$, the current falls to zero, and $i_L(\omega t = \beta) = 0$. That is,

$$\frac{\sqrt{2} \, V_s}{Z} \sin(\beta - \theta) + \left[\frac{E}{R} - \frac{\sqrt{2} \, V_s}{Z} \sin(\alpha - \theta) \right] e^{(R/L)(\alpha - \beta)/\omega} = 0$$

β can be determined from this transcendental equation by an iterative (trial and error) method of solution, discussed in Section 6-5. Start with $\beta = 0$, and increase its value by a very small amount until the left-hand side of this equation becomes zero.

The rms diode current can be found from Eq. (3-68) as

$$I_r = \left[\frac{1}{2\pi} \int_\alpha^\beta i_L^2 \, d(\omega t) \right]^{1/2}$$

The average diode current can also be found from Eq. (3-68) as

$$I_d = \frac{1}{2\pi} \int_\alpha^\beta i_L \, d(\omega t)$$

Example 3-12*

The single-phase full-wave rectifier of Fig. 3-21a has $L = 6.5 \, \text{mH}$, $R = 2.5 \, \Omega$, and $E = 10 \, \text{V}$. The input voltage is $V_s = 120 \, \text{V}$ at 60 Hz. (a) Determine (1) the steady-state load current I_1 at $\omega t = 0$, (2) the average diode current I_d, (3) the rms diode current I_r, and (4) the rms output current I_{rms}. (b) Use PSpice to plot the instantaneous output current i_L. Assume diode parameters IS=2.22E−15, BV=1800V.

Solution It is not known whether the load current is continuous or discontinuous. Assume that the load current is continuous and proceed with the solution. If the assumption is not correct, the load current will be zero and then move to the case for a discontinuous current.

(a) $R = 2.5\ \Omega$, $L = 6.5$ mH, $f = 60$ Hz, $\omega = 2\pi \times 60 = 377$ rad/s, $V_s = 120$ V, $Z = [R^2 + (\omega L)^2]^{1/2} = 3.5\ \Omega$, and $\theta = \tan^{-1}(\omega L/R) = 44.43°$.

(1) The steady-state load current at $\omega t = 0$, $I_1 = 32.8$ A. Since $I_1 > 0$, the load current is continuous and the assumption is correct.

(2) The numerical integration of i_L in Eq. (3-67) yields the average diode current as $I_d = 19.61$ A.

(3) By numerical integration of i_L^2 between the limits $\omega t = 0$ and π, we get the rms diode current as $I_r = 28.5$ A.

(4) The rms output current $I_{rms} = \sqrt{2}\ I_r = \sqrt{2} \times 28.50 = 40.3$ A.

Notes

1. I_L has a minimum value of 25.2 A at $\omega t = 25.5°$ and a maximum value of 51.46 A at $\omega t = 125.25°$. I_L becomes 27.41 A at $\omega t = \theta$ and 48.2 A at $\omega t = \theta + \pi$. Therefore, the minimum value of I_L occurs approximately at $\omega t = \theta$.

2. The switching action of diodes makes the equations for currents nonlinear. A numerical method of solution for the diode currents is more efficient than the classical techniques. A computer program is used to solve for I_1, I_d, and I_r by using numerical integration. Students are encouraged to verify the results of this example and to appreciate the usefulness of numerical solution, especially in solving nonlinear equations of diode circuits.

(b) The single-phase bridge rectifier for PSpice simulation is shown in Fig. 3-22. The list of the circuit file is as follows:

```
Example 3-12    Single-Phase Bridge Rectifier with RL load
VS    1    0    SIN (0    169.7V    60HZ)
L     5    6    6.5MH
R     3    5    2.5
VX    6    4    DC    10V   ; Voltage source to measure the output current
D1    2    3    DMOD                        ; Diode model
D2    4    0    DMOD
D3    0    3    DMOD
D4    4    2    DMOD
VY    1    2        0 DC
.MODEL    DMOD    D(IS=2.22E−15   BV=1800V)    ; Diode model parameters
.TRAN    1US    32MS   16.667MS               ; Transient analysis
.PROBE                                        ; Graphics postprocessor
.END
```

The PSpice plot of instantaneous output current i_L is shown in Fig. 3-23, which gives $I_1 = 31.83$ A.

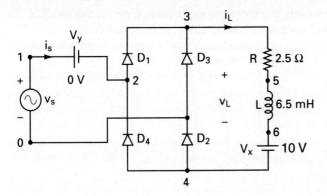

Figure 3-22 Single-phase bridge rectifier for PSpice simulation.

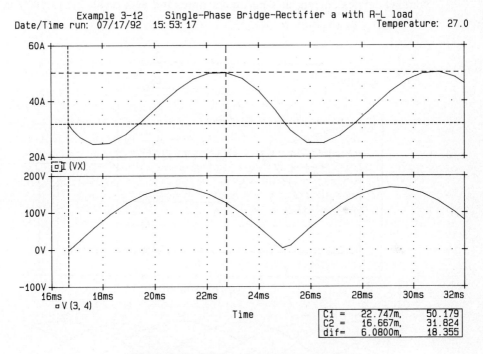

C1 =	22.747m,	50.179
C2 =	16.667m,	31.824
dif=	6.0800m,	18.355

Figure 3-23 PSpice plot for Example 3-12.

3-10 MULTIPHASE STAR RECTIFIERS

We have seen in Eq. (3-62) that the average output voltage which could be obtained from single-phase full-wave rectifiers is $0.6366V_m$ and these rectifiers are used in applications up to a power level of 15 kW. For larger power output, *three-phase* and *multiphase* rectifiers are used. The Fourier series of the output voltage given by Eq. (3-63) indicates that the output contains harmonics and the frequency of the *fundamental component* is two times the source frequency (2*f*). In practice, a filter is normally used to reduce the level of harmonics in the load; and the size of the filter decreases with the increase in frequency of the harmonics. In addition to the larger power output of multiphase rectifiers, the fundamental frequency of the harmonics is also increased and is q times the source frequency (qf). This rectifier is also known as a star rectifier.

The rectifier circuit of Fig. 3-18a can be extended to multiple phases by having multiphase windings on the transformer secondary as shown in Fig. 3-24a. This circuit may be considered as q single-phase half-wave rectifiers and can be considered as a half-wave type. The kth diode will conduct during the period when the voltage of kth phase is higher than that of other phases. The waveforms for the voltages and currents are shown in Fig. 3-24b. The conduction period of each diode is $2\pi/q$.

It can be noticed from Fig. 3-24b that the current flowing through the secondary winding is unidirectional and contains a dc component. Only one secondary winding carries current at a particular time, and as a result the primary must be

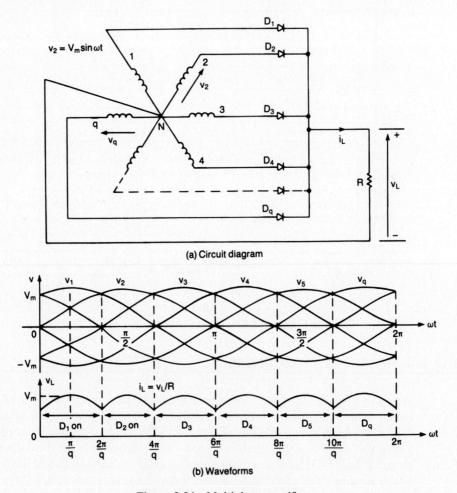

(a) Circuit diagram

(b) Waveforms

Figure 3-24 Multiphase rectifiers.

connected in delta in order to eliminate the dc component in the input side of the transformer. This minimizes the harmonic content of the primary line current.

Assuming a cosine wave from π/q to $2\pi/q$, the average output voltage for a q-phase rectifier is given by

$$V_{dc} = \frac{2}{2\pi/q} \int_0^{\pi/q} V_m \cos \omega t \, d(\omega t) = V_m \frac{q}{\pi} \sin \frac{\pi}{q} \tag{3-69}$$

$$V_{rms} = \left[\frac{2}{2\pi/q} \int_0^{\pi/q} V_m^2 \cos^2 \omega t \, d(\omega t) \right]^{1/2} \tag{3-70}$$

$$= V_m \left[\frac{q}{2\pi} \left(\frac{\pi}{q} + \frac{1}{2} \sin \frac{2\pi}{q} \right) \right]^{1/2}$$

If the load is purely resistive, the peak current through a diode is $I_m = V_m/R$ and we can find the rms value of a diode current (or transformer secondary current) as

$$I_s = \left[\frac{2}{2\pi} \int_0^{\pi/q} I_m^2 \cos^2 \omega t \, d(\omega t) \right]^{1/2} \tag{3-71}$$

$$= I_m \left[\frac{1}{2\pi} \left(\frac{\pi}{q} + \frac{1}{2} \sin \frac{2\pi}{q} \right) \right]^{1/2} = \frac{V_{rms}}{R}$$

Example 3-13

A three-phase star rectifier has a purely resistive load with R ohms. Determine (a) the efficiency, (b) the form factor, (c) the ripple factor, (d) the transformer utilization factor, (e) the peak inverse voltage PIV of each diode, and (f) the peak current through a diode if the rectifier delivers $I_{dc} = 30$ A at an output voltage of $V_{dc} = 140$ V.
Solution For a three-phase rectifier $q = 3$ in Eqs. (3-69), (3-70), and (3-71).

(a) From Eq. (3-69), $V_{dc} = 0.827V_m$ and $I_{dc} = 0.827V_m/R$. From Eq. (3-70), $V_{rms} = 0.84068V_m$ and $I_{rms} = 0.84068V_m/R$. From Eq. (3-42), $P_{dc} = (0.827V_m)^2/R$, from Eq. (3-43) $P_{ac} = (0.84068V_m)^2/R$, and from Eq. (3-44) the efficiency

$$\eta = \frac{(0.827V_m)^2}{(0.84068V_m)^2} = 96.77\%$$

(b) From Eq. (3-46), the form factor FF $= 0.84068/0.827 = 1.0165$ or 101.65%.
(c) From Eq. (3-48), the ripple factor RF $= \sqrt{1.0165^2 - 1} = 0.1824 = 18.24\%$.
(d) From Eq. (3-56), the rms voltage of the transformer secondary, $V_s = 0.707V_m$. From Eq. (3-71), the rms current of the transformer secondary,

$$I_s = 0.4854I_m = \frac{0.4854V_m}{R}$$

The volt-ampere rating (VA) of the transformer for $q = 3$ is

$$\text{VA} = 3V_sI_s = 3 \times 0.707V_m \times \frac{0.4854V_m}{R}$$

From Eq. (3-49),

$$\text{TUF} = \frac{0.827^2}{3 \times 0.707 \times 0.4854} = 0.6643$$

(e) The peak inverse voltage of each diode is equal to the peak value of the secondary line-to-line voltage. Three-phase circuits are reviewed in Appendix A. The line-to-line voltage is $\sqrt{3}$ times the phase voltage and thus $\mathrm{PIV} = \sqrt{3}\, V_m$.

(f) The average current through each diode is

$$I_d = \frac{2}{2\pi} \int_0^{\pi/q} I_m \cos \omega t \, d(\omega t) = I_m \frac{1}{\pi} \sin \frac{\pi}{q} \qquad (3\text{-}72)$$

For $q = 3$, $I_d = 0.2757 I_m$. The average current through each diode is $I_d = 30/3 = 10$ A and this gives the peak current as $I_m = 10/0.2757 = 36.27$ A.

Example 3-14

(a) Express the output voltage of a q-phase rectifier in Fig. 3-24a in Fourier series.
(b) If $q = 6$, $V_m = 170$ V, and the supply frequency is $f = 60$ Hz, determine the rms value of the dominant harmonic and its frequency.

Solution (a) The waveforms for q-pulses are shown in Fig. 3-24b and the frequency of the output is q times the fundamental component (qf). To find the constants of the Fourier series, we integrate from $-\pi/q$ to π/q and the constants are

$$b_n = 0$$

$$a_n = \frac{1}{\pi/q} \int_{-\pi/q}^{\pi/q} V_m \cos \omega t \cos n\omega t \, d(\omega t)$$

$$= \frac{qV_m}{\pi} \left\{ \frac{\sin[(n-1)\pi/q]}{n-1} + \frac{\sin[(n+1)\pi/q]}{n+1} \right\}$$

$$= \frac{qV_m}{\pi} \frac{(n+1)\sin[(n-1)\pi/q] + (n-1)\sin[(n+1)\pi/q]}{n^2 - 1}$$

After simplification and then using trigonometric relationships, we get

$$\sin(A + B) = \sin A \cos B + \cos A \sin B$$

and

$$\sin(A - B) = \sin A \cos B - \cos A \sin B$$

we get

$$a_n = \frac{2qV_m}{\pi(n^2 - 1)} \left(n \sin \frac{n\pi}{q} \cos \frac{\pi}{q} - \cos \frac{n\pi}{q} \sin \frac{\pi}{q} \right) \qquad (3\text{-}73)$$

For a rectifier with q pulses per cycle, the harmonics of the output voltage are: qth, $2q$th, $3q$th, $4q$th and Eq. (3-73) is valid for $n = 0, 1q, 2q, 3q$. The term $\sin(n\pi/q) = \sin \pi = 0$ and Eq. (3-73) becomes

$$a_n = \frac{-2qV_m}{\pi(n^2 - 1)} \left(\cos \frac{n\pi}{q} \sin \frac{\pi}{q} \right)$$

The dc component is found by letting $n = 0$ and is

$$V_{dc} = \frac{a_0}{2} = V_m \frac{q}{\pi} \sin \frac{\pi}{q} \qquad (3\text{-}74)$$

which is the same as Eq. (3-69). The Fourier series of the output voltage v_L is expressed as

$$v_L(t) = \frac{a_0}{2} + \sum_{n=q,2q,\ldots}^{\infty} a_n \cos n\omega t$$

Substituting the value of a_n, we obtain

$$v_L = V_m \frac{q}{\pi} \sin \frac{\pi}{q} \left(1 - \sum_{n=q,2q,\ldots}^{\infty} \frac{2}{n^2 - 1} \cos \frac{n\pi}{q} \cos n\omega t \right) \tag{3-75}$$

(b) For $q = 6$, the output voltage is expressed as

$$v_L(t) = 0.9549 V_m \left(1 + \frac{2}{35} \cos 6\omega t - \frac{2}{143} \cos 12\omega t + \cdots \right) \tag{3-76}$$

The sixth harmonic is the dominant one. The rms value of a sinusoidal voltage is $1/\sqrt{2}$ times its peak magnitude, and the rms of the sixth harmonic is $V_6 = 0.9549 V_m \times 2/(35 \times \sqrt{2}) = 6.56$ A and its frequency is $f_6 = 6f = 360$ Hz.

3-11 THREE-PHASE BRIDGE RECTIFIERS

A three-phase bridge rectifier is commonly used in high-power applications and it is shown in Fig. 3-25. This is a *full-wave rectifier*. It can operate with or without a transformer and gives six-pulse ripples on the output voltage. The diodes are numbered in order of conduction sequences and each one conducts for 120°. The conduction sequence for diodes is 12, 23, 34, 45, 56, and 61. The pair of diodes which are connected between that pair of supply lines having the highest amount of instantaneous line-to-line voltage will conduct. The line-to-line voltage is $\sqrt{3}$ times the phase voltage of a three-phase wye-connected source. The waveforms and conduction times of diodes are shown in Fig. 3-26.

The average output voltage is found from

$$V_{dc} = \frac{2}{2\pi/6} \int_0^{\pi/6} \sqrt{3} \, V_m \cos \omega t \, d(\omega t)$$

$$= \frac{3\sqrt{3}}{\pi} \, V_m = 1.654 V_m \tag{3-77}$$

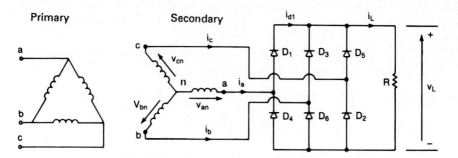

Figure 3-25 Three-phase bridge rectifier.

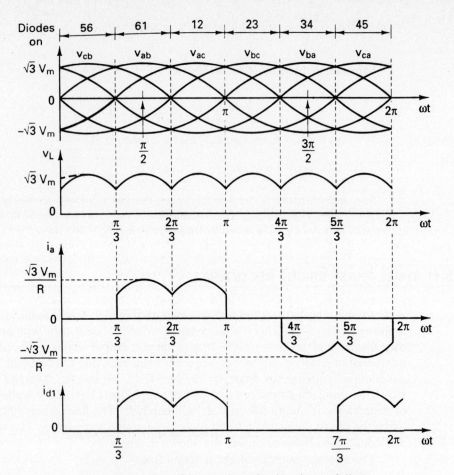

Figure 3-26 Waveforms and conduction times of diodes.

where V_m is the peak phase voltage. The rms output voltage is

$$V_{\text{rms}} = \left[\frac{2}{2\pi/6} \int_0^{\pi/6} 3V_m^2 \cos^2 \omega t \ d(\omega t) \right]^{1/2}$$

(3-78)

$$= \left(\frac{3}{2} + \frac{9\sqrt{3}}{4\pi} \right)^{1/2} V_m = 1.6554 V_m$$

If the load is purely resistive, the peak current through a diode is $I_m = \sqrt{3} \ V_m/R$ and the rms value of the diode current is

$$I_r = \left[\frac{4}{2\pi} \int_0^{\pi/6} I_m^2 \cos^2 \omega t \ d(\omega t) \right]^{1/2}$$

$$= I_m \left[\frac{1}{\pi} \left(\frac{\pi}{6} + \frac{1}{2} \sin \frac{2\pi}{6} \right) \right]^{1/2}$$

(3-79)

$$= 0.5518 I_m$$

and the rms value of the transformer secondary current,

$$I_s = \left[\frac{8}{2\pi} \int_0^{\pi/6} I_m^2 \cos^2 \omega t \, d(\omega t) \right]^{1/2}$$

$$= I_m \left[\frac{2}{\pi} \left(\frac{\pi}{6} + \frac{1}{2} \sin \frac{2\pi}{6} \right) \right]^{1/2} \tag{3-80}$$

$$= 0.7804 I_m$$

where I_m is the peak secondary line current.

Example 3-15

A three-phase bridge rectifier has a purely resistive load of R. Determine (a) the efficiency, (b) the form factor, (c) the ripple factor, (d) the transformer utilization factor, (e) the peak inverse (or reverse) voltage (PIV) of each diode, and (f) the peak current through a diode. The rectifier delivers $I_{dc} = 60$ A at an output voltage of $V_{dc} = 280.7$ V and the source frequency is 60 Hz.

Solution (a) From Eq. (3-77), $V_{dc} = 1.654 V_m$ and $I_{dc} = 1.654 V_m / R$. From Eq. (3-78), $V_{rms} = 1.6554 V_m$ and $I_{rms} = 1.6554 V_m / R$. From Eq. (3-42), $P_{dc} = (1.654 V_m)^2 / R$, from Eq. (3-43), $P_{ac} = (1.6554 V_m)^2 / R$, and from Eq. (3-44) the efficiency

$$\eta = \frac{(1.654 V_m)^2}{(1.6554 V_m)^2} = 99.83\%$$

(b) From Eq. (3-46), the form factor FF $= 1.6554/1.654 = 1.0008 = 100.08\%$.

(c) From Eq. (3-48), the ripple factor RF $= \sqrt{1.0008^2 - 1} = 0.04 = 4\%$.

(d) From Eq. (3-56), the rms voltage of the transformer secondary, $V_s = 0.707 V_m$.

From Eq. (3-80), the rms current of the transformer secondary,

$$I_s = 0.7804 I_m = 0.7804 \times \sqrt{3} \frac{V_m}{R}$$

The volt-ampere rating of the transformer,

$$VA = 3 V_s I_s = 3 \times 0.707 V_m \times 0.7804 \times \sqrt{3} \frac{V_m}{R}$$

From Eq. (3-49),

$$TUF = \frac{1.654^2}{3 \times \sqrt{3} \times 0.707 \times 0.7804} = 0.9542$$

(e) From Eq. (3-77), the peak line-to-neutral voltage is $V_m = 280.7/1.654 = 169.7$ V. The peak inverse voltage of each diode is equal to the peak value of the secondary line-to-line voltage, PIV $= \sqrt{3} V_m = \sqrt{3} \times 169.7 = 293.9$ V.

(f) The average current through each diode is

$$I_d = \frac{4}{2\pi} \int_0^{\pi/6} I_m \cos \omega t \, d(\omega t) = I_m \frac{2}{\pi} \sin \frac{\pi}{6} = 0.3183 I_m$$

The average current through each diode is $I_d = 60/3 = 20$ A, and therefore the peak current is $I_m = 20/0.3183 = 62.83$ A.

Note. This rectifier has considerably improved performances compared to that of the multiphase rectifier in Fig. 3-24 with six pulses.

Equations that are derived in Section 3-9 can be applied to determine the load current of a three-phase rectifier with an *RL* load (similar to Fig. 3-27). It can be noted from Fig. 3-26 that the output voltage becomes

$$v_{ab} = \sqrt{2}\, V_{ab} \sin \omega t \qquad \text{for} \qquad \frac{\pi}{3} \le \omega t \le \frac{2\pi}{3}$$

where V_{ab} is the line-to-line rms input voltage. The load current i_L can be found from

$$L \frac{di_L}{dt} + Ri_L + E = \sqrt{2}\, V_{ab} \sin \omega t$$

which has a solution of the form

$$i_L = \frac{\sqrt{2}\, V_{ab}}{Z} \sin(\omega t - \theta) + A_1 e^{-(R/L)t} - \frac{E}{R} \qquad (3\text{-}81)$$

where load impedance $Z = [R^2 + (\omega L)^2]^{1/2}$ and load impedance angle $\theta = \tan^{-1}(\omega L/R)$. The constant A_1 in Eq. (3-81) can be determined from the condition: at $\omega t = \pi/3$, $i_L = I_1$.

$$A_1 = \left[I_1 + \frac{E}{R} - \frac{\sqrt{2}\, V_{ab}}{Z} \sin\left(\frac{\pi}{3} - \theta\right) \right] e^{(R/L)(\pi/3\omega)}$$

Substitution of A_1 in Eq. (3-81) yields

$$i_L = \frac{\sqrt{2}\, V_{ab}}{Z} \sin(\omega t - \theta) + \left[I_1 + \frac{E}{R} - \frac{\sqrt{2}\, V_{ab}}{Z} \sin\left(\frac{\pi}{3} - \theta\right) \right] e^{(R/L)(\pi/\omega - t)} \qquad (3\text{-}82)$$

Under a steady-state condition, $i_L(\omega t = 2\pi/3) = i_L(\omega t = \pi/3)$. That is, $i_L(\omega t = 2\pi/3) = I_1$. Applying this condition, we get the value of I_1 as

$$I_1 = \frac{\sqrt{2}\, V_{ab}}{Z} \frac{\sin(2\pi/3 - \theta) - \sin(\pi/3 - \theta)e^{-(R/L)(\pi/3\omega)}}{1 - e^{-(R/L)(\pi/3\omega)}} - \frac{E}{R} \qquad \text{for } I_1 \ge 0 \qquad (3\text{-}83)$$

which, after substituting in Eq. (3-82) and simplification, gives

$$i_L = \frac{\sqrt{2}\, V_{ab}}{Z} \left[\sin(\omega t - \theta) + \frac{\sin(2\pi/3 - \theta) - \sin(\pi/3 - \theta)}{1 - e^{-(R/L)(\pi/\omega)}} e^{-(R/L)t} \right]$$

$$- \frac{E}{R} \qquad \text{for } \pi/3 \le \omega t \le 2\pi/3 \text{ and } i_L \ge 0 \qquad (3\text{-}84)$$

The rms diode current can be found from Eq. (3-84) as

$$I_r = \left[\frac{2}{2\pi} \int_{\pi/3}^{2\pi/3} i_L^2 \, d(\omega t) \right]^{1/2}$$

and the rms output current can then be determined by combining the rms current of each diode as

$$I_{rms} = (I_r^2 + I_r^2 + I_r^2)^{1/2} = \sqrt{3}\, I_r$$

The average diode current can also be found from Eq. (3-84) as

$$I_d = \frac{2}{2\pi} \int_{\pi/3}^{2\pi/3} i_L \, d(\omega t)$$

Example 3-16*

The three-phase full-wave rectifier of Fig. 3-27 has a load of $L = 1.5$ mH, $R = 2.5\ \Omega$, and $E = 10$ V. The line-to-line input voltage is $V_{ab} = 208$ V, 60 Hz. (a) Determine (1) the steady-state load current I_1 at $\omega_t = \pi/3$, (2) the average diode current I_d, (3) the rms diode current I_r, and (4) the rms output current I_{rms}. (b) Use PSpice to plot the instantaneous output current i_L. Assume diode parameters IS=2.22E−15, BV=1800V.

Solution (a) $R = 2.5\ \Omega$, $L = 1.5$ mH, $f = 60$ Hz, $\omega = 2\pi \times 60 = 377$ rad/s, $V_{ab} = 208$ V, $Z = [R^2 + (\omega L)^2]^{1/2} = 2.56\ \Omega$, and $\theta = \tan^{-1}(\omega L/R) = 12.74°$.

  (1) The steady-state load current at $\omega t = \pi/3$, $I_1 = 105.85$ A.
  (2) The numerical integration of i_L in Eq. (3-84) yields the average diode current as $I_d = 36.27$ A. Since $I_1 > 0$, the load current is continuous.
  (3) By numerical integration of i_L^2 between the limits $\omega t = \pi/3$ and $2\pi/3$, we get the rms diode current as $I_r = 62.71$ A.
  (4) The rms output current $I_{rms} = \sqrt{3}\, I_r = \sqrt{3} \times 62.71 = 108.62$ A.
  (b) The three-phase bridge rectifier for PSpice simulation is shown in Fig. 3-27. The list of the circuit file is as follows:

```
Example 3-16    Three-Phase Bridge Rectifier with RL load
VAN    8    0    SIN (0    169.7V    60HZ)
VBN    2    0    SIN (0    169.7V    60HZ    0    0    120DEG)
VCN    3    0    SIN (0    169.7V    60HZ    0    0    240DEG)
L      6    7    1.5MH
R      4    6    2.5
VX     7    5    DC    10V   ; Voltage source to measure the output current
VY     8    1    DC    0V    ; Voltage source to measure the input current
D1     1    4    DMOD                      ; Diode model
D3     2    4    DMOD
D5     3    4    DMOD
D2     5    3    DMOD
D4     5    1    DMOD
D6     5    2    DMOD
.MODEL    DMOD    D(IS=2.22E−15    BV=1800V)    ; Diode model parameters
.TRAN    10US    25MS    16.667MS    10US    ; Transient analysis
.PROBE                                   ; Graphics postprocessor
.options ITL5=0 abstol = 1.000n reltol = .01 vntol = 1.000m
.END
```

The PSpice plot of instantaneous output current i_L is shown in Fig. 3-28, which gives $I_1 = 104.89$ A.

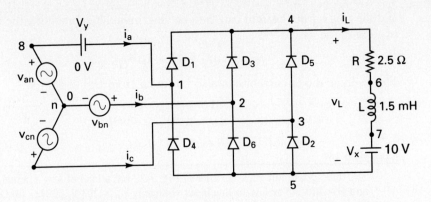

Figure 3-27 Three-phase bridge rectifier for PSpice simulation.

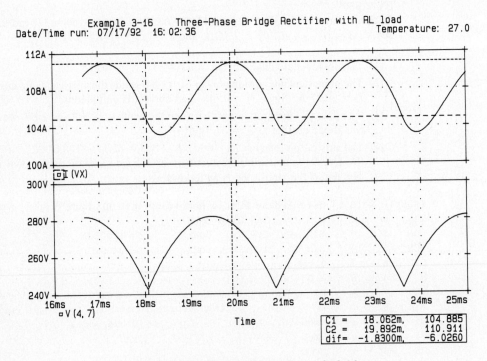

Figure 3-28 PSpice plot for Example 3-16.

3-13 RECTIFIER CIRCUIT DESIGN

The design of a rectifier involves determining the ratings of semiconductor diodes. The ratings of diodes are normally specified in terms of average current, rms current, peak current, and peak inverse voltage. There are no standard procedures for the design, but it is required to determine the shapes of the diode currents and voltages.

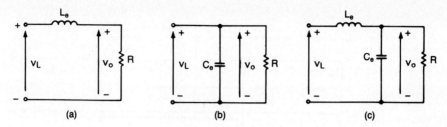

Figure 3-29 Dc filters.

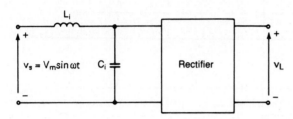

Figure 3-30 Ac filters.

We have noticed in Eqs. (3-61), (3-63), and (3-76) that the output of the rectifiers contain harmonics. Filters can be used to smooth out the dc output voltage of the rectifier and these are known as *dc filters*. The dc filters are usually of L, C, and LC type, as shown in Fig. 3-29. Due to rectification action, the input current of the rectifier contains harmonics also and an *ac filter* is used to filter out some of the harmonics from the supply system. The ac filter is normally of LC type, as shown in Fig. 3-30. Normally, the filter design requires determining the magnitudes and frequencies of the harmonics. The steps involved in designing rectifiers and filters are explained by examples.

Example 3-17

A three-phase bridge rectifier supplies a highly inductive load such that the average load current is $I_{dc} = 60$ A and the ripple content is negligible. Determine the ratings of the diodes if the line-to-neutral voltage of the Wye-connected supply is 120 V at 60 Hz.

Solution The currents through the diodes are shown in Fig. 3-31. The average current of a diode $I_d = 60/3 = 20$ A. The rms current is

$$I_r = \left[\frac{1}{2\pi} \int_{\pi/3}^{\pi} I_{dc}^2 \, d(\omega t) \right]^{1/2} = \frac{I_{dc}}{\sqrt{3}} = 34.64 \text{ A}$$

The peak inverse voltage, PIV $= \sqrt{3} \, V_m = \sqrt{3} \times \sqrt{2} \times 120 = 294$ V.

Note. The factor of $\sqrt{2}$ is used to convert rms to peak value.

Example 3-18

The current through a diode is shown in Fig. 3-32. Determine (a) the rms current, and (b) the average diode current if $t_1 = 100 \ \mu s$, $t_2 = 350 \ \mu s$, $t_3 = 500 \ \mu s$, $f = 250$ Hz, $f_s = 5$ kHz, $I_m = 450$ A, and $I_a = 150$ A.

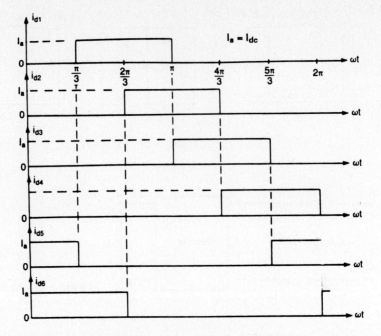

Figure 3-31 Current through diodes.

Solution (a) The rms value is defined as

$$I = \left[\frac{1}{T}\int_0^{t_1}(I_m \sin \omega_s t)^2\, dt + \frac{1}{T}\int_{t_2}^{t_3} I_a^2\, dt\right]^{1/2}$$

$$= (I_{r1}^2 + I_{r2}^2)^{1/2}$$

(3-85)

where $\omega_s = 2\pi f_s = 31{,}415.93$ rad/s, $t_1 = \pi/\omega_s = 100\ \mu s$, and $T = 1/f$.

$$I_{r1} = \left[\frac{1}{T}\int_0^{t_1}(I_m \sin \omega_s t)^2\, dt\right]^{1/2} = I_m \sqrt{\frac{ft_1}{2}}$$

$$= 50.31\ \text{A}$$

(3-86)

and

$$I_{r2} = \left(\frac{1}{T}\int_{t2}^{t3} I_a\, dt\right)^2 = I_a \sqrt{f(t_3 - t_2)}$$

$$= 29.05\ A$$

(3-87)

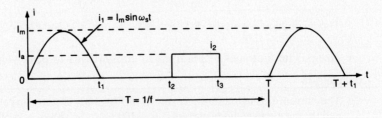

Figure 3-32 Current waveform.

Substituting Eqs. (3-86) and (3-87) in Eq. (3-85), the rms value is

$$I = \left[\frac{I_m^2 f t_1}{2} + I_a^2 f(t_3 - t_2)\right]^{1/2}$$

$$= (50.31^2 + 29.05^2)^{1/2} = 58.09 \text{ A} \tag{3-88}$$

(b) The average current is found from

$$I_d = \left[\frac{1}{T}\int_0^{t_1}(I_m \sin \omega_s t)\, dt + \frac{1}{T}\int_{t_2}^{t_3} I_a\, dt\right]$$

$$= I_{d1} + I_{d2}$$

where

$$I_{d1} = \frac{1}{T}\int_0^{t_1}(I_m \sin \omega_s t)\, dt = \frac{I_m f}{\pi f_s} \tag{3-89}$$

$$I_{d2} = \frac{1}{T}\int_{t_2}^{t_3} I_a\, dt = I_a f(t_3 - t_2) \tag{3-90}$$

Therefore, the average current becomes

$$I_{dc} = \frac{I_m f}{\pi f_s} + I_a f(t_3 - t_2) = 7.16 + 5.63 = 12.79 \text{ A}$$

Example 3-19

The single-phase bridge rectifier is supplied from a 120-V, 60-Hz source. The load resistance is $R = 500\ \Omega$. Calculate the value of a series inductor L that will limit the rms ripple current I_{ac} to less then 5% of I_{dc}.

Solution   The load impedance

$$Z = R + j(n\omega L) = \sqrt{R^2 + (n\omega L)^2}\ \underline{/\theta_n} \tag{3-91}$$

and

$$\theta_n = \tan^{-1}\frac{n\omega L}{R} \tag{3-92}$$

and the instantaneous current is

$$i_L(t) = I_{dc} - \frac{4V_m}{\pi\sqrt{R^2 + (n\omega L)^2}}\left[\frac{1}{3}\cos(2\omega t - \theta_2) + \frac{1}{15}\cos(4\omega t - \theta_4)\ldots\right] \tag{3-93}$$

where

$$I_{dc} = \frac{V_{dc}}{R} = \frac{2V_m}{\pi R}$$

Equation (3-93) gives the rms value of the ripple current as

$$I_{ac}^2 = \frac{(4V_m)^2}{2\pi^2[R^2 + (2\omega L)^2]}\left(\frac{1}{3}\right)^2 + \frac{(4V_m)^2}{2\pi^2[R^2 + (4\omega L)^2]}\left(\frac{1}{15}\right)^2 + \cdots$$

Considering only the lowest-order harmonic ($n = 2$), we have

$$I_{ac} = \frac{4V_m}{\sqrt{2}\pi\ \sqrt{R^2 + (2\omega L)^2}}\left(\frac{1}{3}\right)$$

Using the value of I_{dc} and after simplification, the ripple factor is

$$\text{RF} = \frac{I_{ac}}{I_{dc}} = \frac{0.4714}{\sqrt{1 + (2\omega L/R)^2}} = 0.05$$

For $R = 500 \, \Omega$ and $f = 60$ Hz, the inductance value is obtained as $0.4714^2 = 0.05^2$ $[1 + (4 \times 60 \times \pi L/500)^2]$ and this gives $L = 6.22$ H.

We can notice from Eq. (3-93) that an inductance in the load offers a high impedance for the harmonic currents and acts like a filter in reducing the harmonics. However, this inductance introduces a time delay of the load current with respect to the input voltage; and in the case of the single-phase half-wave rectifier, a freewheeling diode is required to provide a path for this inductive current.

Example 3-20

A single-phase bridge-rectifier is supplied from a 120-V 60-Hz source. The load resistance is $R = 500 \, \Omega$. (a) Design a C filter so that the ripple factor of the output voltage is less than 5%. (b) With the value of capacitor C in part (a), calculate the average load voltage V_{dc}.

Solution When the instantaneous voltage v_s in Fig. 3-33a is higher than the instantaneous capacitor voltage v_c, the diodes (D_1 and D_2 or D_3 and D_4) conduct; and the capacitor is then charged from the supply. If the instantaneous supply voltage v_s falls below the instantaneous capacitor voltage v_c, the diodes (D_1 and D_2 or D_3 and D_4) are reverse biased and the capacitor C_e discharges through the load resistance R. The capacitor voltage v_c varies between a minimum $V_{c(min)}$ and maximum value $V_{c(max)}$. This is shown in Fig. 3-33b.

Let us assume that t_1 is the charging time and that t_2 is the discharging time of capacitor C_e. The equivalent circuit during charging is shown in Fig. 3-33c. The capacitor charges almost instantaneously to the supply voltage v_s. The capacitor C_e will be charged to the peak supply voltage V_m, so that $v_c(t = t_1) = V_m$. Figure 3-33d shows the equivalent circuit during discharging. The capacitor discharges exponentially through R.

$$\frac{1}{C_e} \int i_L \, dt + v_c(t = 0) + Ri_L = 0$$

which, with an initial condition of $v_c(t = 0) = V_m$, gives the discharging current as

$$i_L = \frac{V_m}{R} e^{-t/RC_e}$$

The output (or capacitor) voltage v_L during the discharging period can be found from

$$v_L(t) = Ri_L = V_m e^{-t/RC_e}$$

The peak-to-peak ripple voltage $V_{r(pp)}$ can be found from

$$V_{r(pp)} = v_L(t = t_1) - v_L(t = t_2) = V_m - V_m e^{-t_2/RC_e} = V_m(1 - e^{-t_2/RC_e}) \qquad (3\text{-}94)$$

Since, $e^{-x} \approx 1 - x$, Eq. (3-94) can be simplified to

$$V_{r(pp)} = V_m\left(1 - 1 + \frac{t_2}{RC_e}\right) = \frac{V_m t_2}{RC_e} = \frac{V_m}{2fRC_e}$$

Therefore, the average load voltage V_{dc} is given by (assuming $t_2 = 1/2 \, f$)

$$V_{dc} = V_m - \frac{V_{r(pp)}}{2} = V_m - \frac{V_m}{4fRC_e} \qquad (3\text{-}95)$$

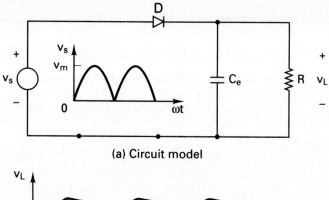

(a) Circuit model

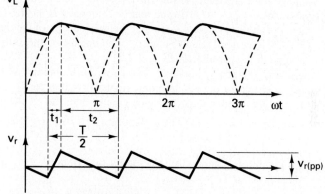

(b) Waveforms for full-wave rectifier

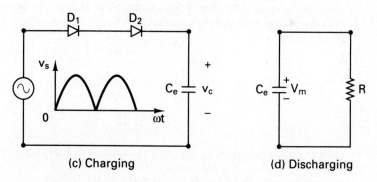

(c) Charging   (d) Discharging

Figure 3-33   Single-phase bridge rectifier with C filter.

Thus the rms output ripple voltage V_{ac} can be found approximately from

$$V_{ac} = \frac{V_{r(pp)}}{2\sqrt{2}} = \frac{V_m}{4\sqrt{2}\,fRC_e}$$

The ripple factor RF can be found from

$$RF = \frac{V_{ac}}{V_{dc}} = \frac{V_m}{4\sqrt{2}\,fRC_e}\frac{4fRC_e}{V_m(4fRC_e - 1)} = \frac{1}{\sqrt{2}\,(4fRC_e - 1)} \qquad (3\text{-}96)$$

which can be solved for C_e:

$$C_e = \frac{1}{4fR}\left(1 + \frac{1}{\sqrt{2}\,RF}\right) = \frac{1}{4 \times 60 \times 500}\left(1 + \frac{1}{\sqrt{2} \times 0.05}\right) = 126.2 \ \mu F$$

(b) From Eq. (3-95), the average load voltage V_{dc} is

$$V_{dc} = 169.7 - \frac{169.7}{4 \times 60 \times 500 \times 126.2 \times 10^{-6}} = 169.7 - 11.21 = 158.49 \ V$$

Example 3-21

An LC filter as shown in Fig. 3-29c is used to reduce the ripple content of the output voltage for a single-phase full-wave rectifier. The load resistance is $R = 40 \ \Omega$, load inductance is $L = 10$ mH, and source frequency is 60 Hz (or 377 rad/s). (a) Determine the values of L_e and C_e so that the ripple factor of the output voltage is 10%. (b) Use PSpice to calculate Fouriers components of the output voltage current v_L. Assume diode parameters IS = 2-22E $-$ 15, BV = 1800V.

Solution (a) The equivalent circuit for the harmonics is shown in Fig. 3-34. To make it easier for the nth harmonic ripple current to pass through the filter capacitor, the load impedance must be much greater than that of the capacitor. That is,

$$\sqrt{R^2 + (n\omega L)^2} >> \frac{1}{n\omega C_e}$$

This condition is generally satisfied by the relation

$$\sqrt{R^2 + (n\omega L)^2} = \frac{10}{n\omega C_e} \tag{3-97}$$

and under this condition, the effect of the load will be negligible. The rms value of the nth harmonic component appearing on the output can be found by using the voltage-divider rule and is expressed as

$$V_{on} = \left| \frac{-1/(n\omega C_e)}{(n\omega L_e) - 1/(n\omega C_e)} \right| V_n = \left| \frac{-1}{(n\omega)^2 L_e C_e - 1} \right| V_n \tag{3-98}$$

The total amount of ripple voltage due to all harmonics is

$$V_{ac} = \left(\sum_{n=2,4,6...}^{\infty} V_{on}^2 \right)^{1/2} \tag{3-99}$$

For a specified value of V_{ac} and with the value of C_e from Eq. (3-97), the value of L_e can be computed. We can simplify the computation by considering only the dominant harmonic. From Eq. (3-63) we find that the second harmonic is the dominant one and its rms value is $V_2 = 4V_m/(3\sqrt{2}\,\pi)$ and the dc value, $V_{dc} = 2V_m/\pi$.

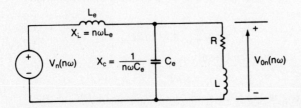

Figure 3-34 Equivalent circuit for harmonics.

Diode Circuits and Rectifiers   Chap. 3

For $n = 2$, Eqs. (3-98) and (3-99) give

$$V_{ac} = V_{o2} = \left| \frac{-1}{(2\omega)^2 L_e C_e - 1} \right| V_2$$

The value of the filter capacitor C_e is calculated from

$$\sqrt{R^2 + (2\omega L)^2} = \frac{10}{2\omega C_e}$$

or

$$C_e = \frac{10}{4\pi f \sqrt{R^2 + (4\pi f L)^2}} = 326 \; \mu F$$

From Eq. (3-47) the ripple factor is defined as

$$RF = \frac{V_{ac}}{V_{dc}} = \frac{V_{o2}}{V_{dc}} = \frac{V_2}{V_{dc}} \frac{1}{(4\pi f)^2 L_e C_e - 1} = \frac{\sqrt{2}}{3} \left| \frac{1}{[(4\pi f)^2 L_e C_e - 1]} \right| = 0.1$$

or $(4\pi f)^2 L_e C_e - 1 = 4.714$ and $L_e = 30.83$ mH.

(b) The single-phase bridge rectifier for PSpice simulation is shown in Fig. 3-35. The list of the circuit file is as follows:

```
Example 3-21    Single-Phase Bridge Rectifier with LC Filter
VS      1    0    SIN (0    169.7V    60HZ)
LE      3    8    30.83MH
CE      7    4    326UF
RX      8    7    80M                         ; Used to converge the solution
L       5    6    10MH
R       7    5    40
VX      6    4    DC    0V    ; Voltage source to measure the output current
VY      1    2    DC    0V    ; Voltage source to measure the input current
D1      2    3    DMOD                        ; Diode models
D2      4    0    DMOD
D3      0    3    DMOD
D4      4    2    DMOD
.MODEL   DMOD    D  (IS=2.22E-15  BV=1800V)  ; Diode model parameters
.TRAN    10US     50MS   33MS   50US    ; Transient analysis
.FOUR    120HZ    V(6,5)              ; Fourier analysis of output voltage
.options ITL5=0   abstol = 1.000n reltol = .01 vntol = 1.000m
.END
```

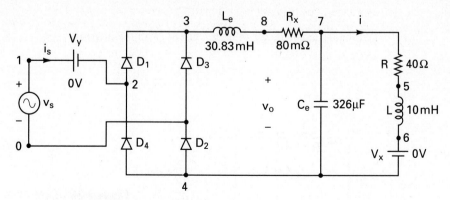

Figure 3-35   Single-phase bridge rectifier for PSpice simulation.

The results of PSpice simulation for the output voltage V(6,5) are as follows:

```
FOURIER COMPONENTS OF TRANSIENT RESPONSE V(6,5)
DC COMPONENT = 1.140973E+02
```

HARMONIC NO	FREQUENCY (HZ)	FOURIER COMPONENT	NORMALIZED COMPONENT	PHASE (DEG)	NORMALIZED PHASE (DEG)
1	1.200E+02	1.304E+01	1.000E+00	1.038E+02	0.000E+00
2	2.400E+02	6.496E-01	4.981E-02	1.236E+02	1.988E+01
3	3.600E+02	2.277E-01	1.746E-02	9.226E+01	-1.150E+01
4	4.800E+02	1.566E-01	1.201E-02	4.875E+01	-5.501E+01
5	6.000E+02	1.274E-01	9.767E-03	2.232E+01	-8.144E+01
6	7.200E+02	1.020E-01	7.822E-03	8.358E+00	-9.540E+01
7	8.400E+02	8.272E-02	6.343E-03	1.997E+00	-1.018E+02
8	9.600E+02	6.982E-02	5.354E-03	-1.061E+00	-1.048E+02
9	1.080E+03	6.015E-02	4.612E-03	-3.436E+00	-1.072E+02

```
TOTAL HARMONIC DISTORTION =    5.636070E+00 PERCENT
```

which verifies the design.

Example 3-22

An *LC* input filter as shown in Fig. 3-30 is used to reduce the input current harmonics in a single-phase full-wave rectifier of Fig. 3-20a. The load current is ripple free and its average value is I_a. If the supply frequency is $f = 60$ Hz (or 377 rad/s), determine the resonant frequency of the filter so that the total input harmonic current is reduced to 1% of the fundamental component.

Solution The equivalent circuit for the nth harmonic component is shown in Fig. 3-36. The rms value of the nth harmonic current appearing in the supply is obtained by using the current-divider rule,

$$I_{sn} = \left| \frac{1/(n\omega C_i)}{(n\omega L_i) - 1/(n\omega C_i)} \right| I_n = \left| \frac{1}{(n\omega)^2 L_i C_i - 1} \right| I_n \tag{3-100}$$

where I_n is the rms value of the nth harmonic current. The total amount of harmonic current in the supply line is

$$I_h = \left(\sum_{n=2,3,\ldots}^{\infty} I_{sn}^2 \right)^{1/2}$$

and the harmonic factor of input current (with the filter) is

$$r = \frac{I_h}{I_{s1}} = \left[\sum_{n=2,3,\ldots}^{\infty} \left(\frac{I_{sn}}{I_1} \right)^2 \right]^{1/2} \tag{3-101}$$

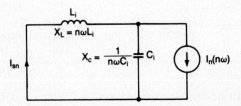

Figure 3-36 Equivalent circuit for harmonic current.

From Eq. (3-64), $I_{s1} = 4I_a/\sqrt{2}\,\pi$ and $I_n = 4I_a/(\sqrt{2}\,n\pi)$ for $n = 3, 5, 7, \ldots$ From Eqs. (3-100) and (3-101) we get

$$r^2 = \sum_{n=3,5,7,\ldots}^{\infty} \left(\frac{I_{sn}}{I_1}\right)^2 = \sum_{n=3,5,7,\ldots}^{\infty} \left|\frac{1}{n^2[(n\omega)^2 L_i C_i - 1]^2}\right| I_1 \qquad (3\text{-}102)$$

This can be solved for the value of $L_i C_i$. To simplify the calculations, if we consider only the third harmonic, $3[(3 \times 2 \times \pi \times 60)^2 L_i C_i - 1] = 1/0.01 = 100$ or $L_i C_i = 26.84 \times 10^{-6}$ and the filter frequency is $1/\sqrt{L_i C_i} = 193.02$ rad/s, or 30.72 Hz. Assuming that $C_i = 1500\ \mu\text{F}$, we obtain $L_i = 17.893$ mH.

Note. The ac filter is generally tuned to the harmonic frequency involved, but it requires a careful design to avoid the possibility of resonance with the power system. The resonant frequency of the third harmonic current is $377 \times 3 = 1131$ rad/s.

3-14 OUTPUT VOLTAGE WITH *LC* FILTER

The equivalent circuit of a full-wave rectifier with an *LC* filter is shown in Fig. 3-37a. Assume that the value of C_e is very large, so that its voltage is ripple free with an average value of $V_{o(\text{dc})}$. L_e is the total inductance, including the source or line inductance, and is generally placed at the input side to act as an ac inductance rather than a dc choke.

If V_{dc} is less than V_m, the current i_L will begin to flow at α, which is given by

$$V_{\text{dc}} = V_m \sin \alpha$$

which gives

$$\alpha = \sin^{-1}\frac{V_{\text{dc}}}{V_m} = \sin^{-1}x$$

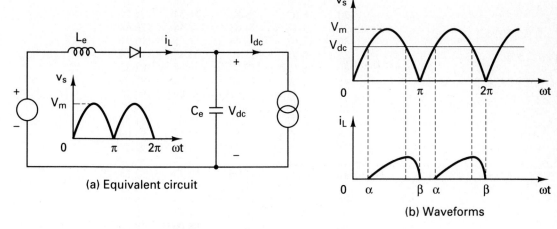

(a) Equivalent circuit

(b) Waveforms

Figure 3-37 Output voltage with *LC* filter.

where $x = V_{dc}/V_m$. The output current i_L is given by

$$L_e \frac{di_L}{dt} = V_m \sin \omega t - V_{dc}$$

which can be solved for i_L.

$$i_L = \frac{1}{L_e} \int_\alpha^{\omega t} (V_m \sin \omega t - V_{dc}) \, d(\omega t)$$

$$= \frac{V_m}{\omega L_e} (\cos \alpha - \cos \omega t) - \frac{V_{dc}}{\omega L_e} (\omega t - \alpha) \qquad \text{for } \omega t \geq \alpha \qquad (3\text{-}103)$$

The value of $\omega t = \beta$ at which the current i_L falls to zero can be found from the condition $i_L(\omega t = \beta)$.

$$\cos \beta + x\beta = \cos \alpha + x\alpha \qquad (3\text{-}104)$$

Equation (3-104) can be solved for β by iteration. Once the values of α and β are known, the average load current I_{dc} can be determined from Eq. (3-103). For $V_{dc} = 0$, the peak current that can flow through the rectifier is $I_{pk} = V_m/\omega L_e$. Normalizing I_{dc} with respect to I_{pk}, we get

$$\frac{I_{dc}}{I_{pk}} = \frac{\omega L_e I_{dc}}{V_m} = \frac{2}{2\pi} \int_\alpha^\beta \frac{\omega L_e}{V_m} i_L \, d(\omega t) \qquad (3\text{-}105)$$

TABLE 3-1 NORMALIZED LOAD CURRENT

x (%)	I_{dc}/I_{pk} (%)	I_{rms}/I_{pk} (%)	α (deg)	β (deg)
0	36.34	47.62	0	180
5	30.29	42.03	2.97	150.62
10	25.50	37.06	5.74	139.74
15	21.50	32.58	8.63	131.88
20	18.09	28.52	11.54	125.79
25	15.15	24.83	14.48	120.48
30	12.62	21.48	17.46	116.21
35	10.42	18.43	20.49	112.24
40	8.53	15.67	23.58	108.83
45	6.89	13.17	26.74	105.99
50	5.48	10.91	30.00	103.25
55	4.28	8.89	33.37	100.87
60	3.27	7.10	36.87	98.87
65	2.42	5.51	40.54	97.04
70	1.72	4.13	44.43	95.43
75	1.16	2.95	48.59	94.09
80	0.72	1.96	53.13	92.88
85	0.39	1.17	58.21	91.71
90	0.17	0.56	64.16	90.91
95	0.04	0.16	71.81	90.56
100	0	0	90.00	90.00

Normalizing I_{rms} with respect to I_{pk}, we get

$$\frac{I_{rms}}{I_{pk}} = \frac{\omega L_e I_{rms}}{V_m} = \left[\frac{2}{2\pi} \int_\alpha^\beta \left(\frac{\omega L_e}{V_m} i_L\right)^2 d(\omega t)\right]^{1/2} \tag{3-106}$$

Since α and β depend on the voltage ratio x, Eqs. (3-105) and (3-106) are dependent on x only. Table 3-1 shows the values of I_{dc}/I_{pk} and I_{rms}/I_{pk} against the voltage ratio x.

Example 3-23*

The rms input voltage to the circuit in Fig. 3-37a is 120 V, 60 Hz. (a) If the dc output voltage is $V_{dc} = 100$ V at $I_{dc} = 10$ A, determine the values of inductance L_e, α, β, and I_{rms}. (b) If $I_{dc} = 5$ A and $L_e = 6.5$ mH, use Table 3-1 to determine the values of V_{dc}, α, β, and I_{rms}.

Solution $\omega = 2\pi \times 60 = 377$ rad/s, $V_s = 120$ V, $V_m = \sqrt{2} \times 120 = 169.7$ V.

(a) Voltage ratio $x = V_{dc}/V_m = 100/169.7 = 58.93\%$; $\alpha = \sin^{-1}(x) = 36.1°$. Solving Eq. (3-104) for β gives $\beta = 99.35°$. Equation (3-105) gives the current ratio $I_{dc}/I_{pk} = 3.464\%$. Thus $I_{pk} = I_{dc}/0.03464 = 288.67$ A. The required value of inductance is

$$L_e = \frac{V_m}{(\omega I_{pk})} = \frac{169.7}{377 \times 288.67} = 1.56 \text{ mH}$$

Equation (3-106) gives the current ratio $I_{rms}/I_{pk} = 7.466\%$. Thus $I_{rms} = 0.07466 \times I_{pk} = 0.07466 \times 288.67 = 21.55$ A.

(b) $L_e = 6.5$ mH, $I_{pk} = V_m/(\omega L_e) = 169.7/(377 \times 6.5 \text{ mH}) = 69.25$ A.

$$y = \frac{I_{dc}}{I_{pk}} = \frac{5}{69.25} = 7.22\%$$

Using linear interpolation, we get

$$x = x_n + \frac{(x_{n+1} - x_n)(y - y_n)}{y_{n+1} - y_n}$$

$$= 40 + \frac{(45 - 40)(7.22 - 8.53)}{6.89 - 8.53} = 43.99\%$$

$$V_{dc} = xV_m = 0.4399 \times 169.7 = 74.66 \text{ V}$$

$$\alpha = \alpha_n + (\alpha_{n+1} - \alpha_n)(y - y_n)$$

$$= 23.58 + \frac{(26.74 - 23.58)(7.22 - 8.53)}{6.89 - 8.53} = 26.1°$$

$$\beta = \beta_n + \frac{(\beta_{n+1} - \beta_n)(y - y_n)}{y_{n+1} - y_n}$$

$$= 108.83 + \frac{(105.99 - 108.83)(7.22 - 8.53)}{6.89 - 8.53} = 106.56°$$

$$z = \frac{I_{rms}}{I_{pk}} = z_n + \frac{(z_{n+1} - z_n)(y - y_n)}{y_{n+1} - y_n}$$

$$= 15.67 + \frac{(13.17 - 15.67)(7.22 - 8.53)}{6.89 - 8.53} = 13.67\%$$

Thus $I_{rms} = 0.1367 \times I_{pk} = 0.1367 \times 69.25 = 9.47$ A.

In the derivations of the output voltages and the performance criteria of rectifiers, it was assumed that the source has no inductances and resistances. But in a practical transformer and supply, these are always present and the performances of rectifiers are slightly changed. The effect of the source inductance, which is more significant than that of resistance, can be explained with reference to Fig. 3-38.

The diode with the most positive voltage will conduct. Let us consider the point $\omega t = \pi$ where voltages v_{ac} and v_{bc} are equal as shown in Fig. 3-38. The current I_{dc} is still flowing through diode D_1. Due to the inductance L_1, the current cannot fall to zero immediately and the transfer of current cannot be on an instantaneous basis. The current i_{d1} decreases, resulting in an induced voltage across L_1 of $+v_{L1}$ and the output voltage becomes $v_L = v_{ac} + v_{L1}$. At the same time the current through D_3, i_{d3} increases from zero, inducing an equal voltage across L_2 of $-v_{L2}$ and the output voltage becomes $v_L = v_{bc} - v_{L2}$. The result is that the anode voltages of diodes D_1 and D_3 are equal; and both diodes conduct for a certain

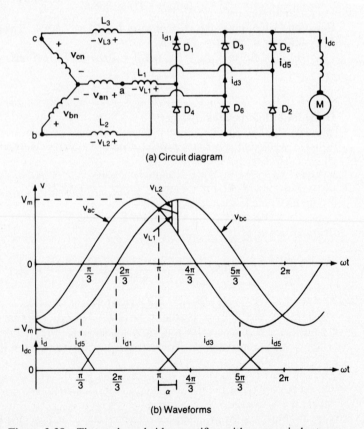

(a) Circuit diagram

(b) Waveforms

Figure 3-38 Three-phase bridge rectifier with source inductances.

Diode Circuits and Rectifiers Chap. 3

period which is called *commutation* (*or overlap*) *angle* μ. This transfer of current from one diode to another is called *commutation*. The reactance corresponding to the inductance is known as *commutating reactance*.

The effect of this overlap is to reduce the average output voltage of converters. The voltage across L_2 is

$$v_{L2} = L_2 \frac{di}{dt} \tag{3-107}$$

Assuming a linear rise of current i from 0 to I_{dc} (or a constant $di/dt = \Delta i/\Delta t$), we can write Eq. (3-107) as

$$v_{L2} \, \Delta t = L_2 \, \Delta i \tag{3-108}$$

and this is repeated six times for a three-phase bridge rectifier. Using Eq. (3-108), the average voltage reduction due to the commutating inductances is

$$V_x = \frac{1}{T} 2(v_{L1} + v_{L2} + v_{L3}) \, \Delta t = 2f(L_1 + L_2 + L_3) \, \Delta i$$
$$= 2f(L_1 + L_2 + L_3)I_{dc} \tag{3-109}$$

If all the inductances are equal and $L_c = L_1 = L_2 = L_3$, Eq. (3-109) becomes

$$V_x = 6f L_c I_{dc} \tag{3-110}$$

where f is the supply frequency in hertz.

Example 3-24

A three-phase bridge rectifier is supplied from a wye-connected 208-V 60-Hz supply. The average load current is 60 A and has negligible ripple. Calculate the percentage reduction of output voltage due to commutation if the line inductance per phase is 0.5 mH.

Solution $L_c = 0.5$ mH, $V_s = 208/\sqrt{3} = 120$ V, $f = 60$ Hz, $I_{dc} = 60$ A, and $V_m = \sqrt{2} \times 120 = 169.7$ V. From Eq. (3-77), $V_{dc} = 1.654 \times 169.7 = 280.7$ V. Equation (3-110) gives the output voltage reduction,

$$V_x = 6 \times 60 \times 0.5 \times 10^{-3} \times 60 = 10.8 \text{ V} \quad \text{or} \quad 10.8 \times \frac{100}{280.7} = 3.85\%$$

and the effective output voltage is $(280.7 - 10.8) = 266.9$ V.

Example 3-25

The diodes in the single-phase full-wave rectifier in Fig. 3-19a have a reverse recovery time of $t_{rr} = 50$ μs and the rms input voltage is $V_s = 120$ V. Determine the effect of the reverse recovery time on the average output voltage if the supply frequency is (a) $f_s = 2$ kHz, and (b) $f_s = 60$ Hz.

Solution   The reverse recovery time would affect the output voltage of the rectifier. In the full-wave rectifier of Fig. 3-19a, the diode D_1 will not be off at $\omega t = \pi$; rather, it will continue to conduct until $t = \pi/\omega + t_{rr}$. As a result of the reverse recovery time, the average output voltage would be reduced and the output voltage waveform is shown in Fig. 3-39.

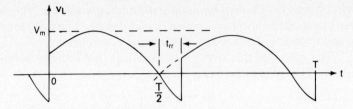

Figure 3-39   Effect of reverse recovery time on output voltage.

If the input voltage is $v = V_m \sin \omega t = \sqrt{2}\, V_s \sin \omega t$, the average output voltage reduction is

$$V_{rr} = \frac{2}{T} \int_0^{t_{rr}} V_m \sin \omega t \; dt = \frac{2V_m}{T} \left[-\frac{\cos \omega t}{\omega} \right]_0^{t_{rr}}$$

$$= \frac{V_m}{\pi}(1 - \cos \omega t_{rr})$$

$$V_m = \sqrt{2}\, V_s = \sqrt{2} \times 120 = 169.7 \text{ V}$$

(3-111)

Without any reverse recovery time, Eq. (3-62) gives the average output voltage $V_{dc} = 0.6366 V_m = 108.03$ V.

(a) For $t_{rr} = 50\ \mu$s and $f_s = 2000$ Hz, the reduction of the average output voltage is

$$V_{rr} = \frac{V_m}{\pi}(1 - \cos 2\pi f_s t_{rr})$$

$$= 0.061 V_m = 10.3 \text{ V} \quad \text{or} \quad 9.51\% \text{ of } V_{dc}$$

(b) For $t_{rr} = 50\ \mu$s and $f_s = 60$ Hz, the reduction of the output dc voltage

$$V_{rr} = \frac{V_m}{\pi}(1 - \cos 2\pi f_s t_{rr}) = 5.65 \times 10^{-5}\, V_m$$

$$= 9.6 \times 10^{-3} \text{ V} \quad \text{or} \quad 8.88 \times 10^{-3}\% \text{ of } V_{dc}$$

Note. The effect of t_{rr} is significant for high-frequency source and for the case of normal 60-Hz source, its effect can be considered negligible.

SUMMARY

In this chapter we have seen the applications of power semiconductor diodes in freewheeling action, recovering energy from inductive loads and in conversion of signal from ac to dc. There are different types of rectifiers depending on the connections of diodes and input transformer. The performance parameters of rectifiers are defined and it has been shown that the performances of rectifiers vary with their types. The rectifiers generate harmonics into the load and the supply line; and these harmonics can be reduced by filters. The performances of the rectifiers are also influenced by the source and load inductances.

REFERENCES

1. J. Schaefer, *Rectifier Circuits: Theory and Design*. New York: John Wiley & Sons, Inc., 1975.
2. R. W. Lee, *Power Converter Handbook: Theory, Design, and Application*. Peterborough, Ont.: Canadian General Electric, 1979.
3. M. H. Rashid, *SPICE for Power Electronics and Electric Power*. Englewood Cliffs, N.J.: Prentice Hall, 1993.

REVIEW QUESTIONS

3-1. What is the time constant of an *RL* circuit?

3-2. What is the time constant of an *RC* circuit?

3-3. What is the resonant frequency of an *LC* circuit?

3-4. What is the damping factor of an *RLC* circuit?

3-5. What is the difference between the resonant frequency and the ringing frequency of an *RLC* circuit?

3-6. What is a freewheeling diode, and what is its purpose?

3-7. What is the trapped energy of an inductor?

3-8. How is the trapped energy recovered by a diode?

3-9. What is the turns ratio of a transformer?

3-10. What is a rectifier? What is the difference between a rectifier and a converter?

3-11. What is the blocking condition of a diode?

3-12. What are the performance parameters of a rectifier?

3-13. What is the significance of the form factor of a rectifier?

3-14. What is the significance of the ripple factor of a rectifier?

3-15. What is the efficiency of rectification?

3-16. What is the significance of the transformer utilization factor?

3-17. What is the displacement factor?

3-18. What is the input power factor?

3-19. What is the harmonic factor?

3-20. What is the difference between a half-wave and a full-wave rectifier?

3-21. What is the dc output voltage of a single-phase half-wave rectifier?

3-22. What is the dc output voltage of a single-phase full-wave rectifier?

3-23. What is the fundamental frequency of the output voltage of a single-phase full-wave rectifier?

3-24. What are the advantages of a three-phase rectifier over a single-phase rectifier?

3-25. What are the disadvantages of a multiphase half-wave rectifier?

3-26. What are the advantages of a three-phase bridge rectifier over a six-phase star rectifier?

3-27. What are the purposes of filters in rectifier circuits?

3-28. What are the differences between ac and dc filters?

3-29. What are the effects of source inductances on the output voltage of a rectifier?

3-30. What are the effects of load inductances on the rectifier output?

3-31. What is a commutation of diodes?

3-32. What is the commutation angle of a rectifier?

PROBLEMS

3-1. The current waveforms of a capacitor are shown in Fig. P3-1. Determine the average, rms, and peak current ratings of the capacitor.

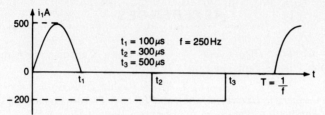

Figure P3-1

3-2. The waveforms of the current flowing through a diode are shown in Fig. P3-2. Determine the average, rms, and peak current ratings of the diode.

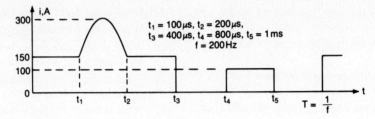

Figure P3-2

3-3. A diode circuit is shown in Fig. P3-3 with $R = 22 \, \Omega$ and $C = 10 \, \mu F$. If switch S_1 is closed at $t = 0$, determine the expression for the voltage across the capacitor and the energy lost in the circuit.

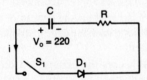

Figure P3-3

3-4. A diode circuit is shown in Fig. P3-4 with $R = 10 \, \Omega$, $L = 5$ mH, and $V_s = 220$ V. If a load current of 10 A is flowing through freewheeling diode D_m and switch S_1 is closed at $t = 0$, determine the expression for the current i through the switch.

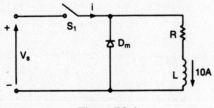

Figure P3-4

3-5. If the inductor of the circuit in Fig. 3-4 has an initial current of I_0, determine the expression for the voltage across the capacitor.

3-6. If switch S_1 of Fig. P3-6 is closed at $t = 0$, determine the expression for (**a**) the current flowing through the switch $i(t)$, and (**b**) the rate of rise of the current di/dt. (**c**) Draw sketches of $i(t)$ and di/dt. (**d**) What is the value of initial di/dt? For Fig. P3-6e, find the initial di/dt only.

3-7. The second-order circuit of Fig. 3-6 has the source voltage $V_s = 220$ V, inductance $L = 5$ mH, capacitance $C = 10 \, \mu F$, and resistance $R = 22 \, \Omega$. The initial voltage of the capacitor is $V_0 = 50$ V. If the switch is closed at $t = 0$, determine (**a**) an expression for the current, and (**b**) the conduction time of the diode. (**c**) Draw a sketch of $i(t)$.

3-8. For the energy recovery circuit of Fig. 3-12a, the magnetizing inductance of the transformer is $L_m = 150 \, \mu H$, $N_1 = 10$, and $N_2 = 200$. The leakage inductances and resistances of the transformer are negligible. The source voltage is $V_s = 200$ V and there is no initial current in the circuit. If switch S_1 is closed for a time $t_1 = 100 \, \mu s$

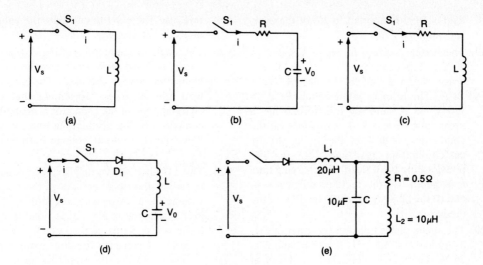

Figure P3-6

and is then opened, (**a**) determine the reverse voltage of diode D_1, (**b**) calculate the peak primary current, (**c**) calculate the peak secondary current, (**d**) determine the time for which diode D_1 conducts, and (**e**) determine the energy supplied by the source.

3-9. A single-phase bridge rectifier has a purely resistive load $R = 10\ \Omega$, the peak supply voltage $V_m = 170$ V, and the supply frequency $f = 60$ Hz. Determine the average output voltage of the rectifier if the source inductance is negligible.

3-10. Repeat Prob. 3-9 if the source inductance per phase (including transformer leakage inductance) is $L_c = 0.5$ mH.

3-11. A six-phase star rectifier has a purely resistive load of $R = 10\ \Omega$, the peak supply voltage $V_m = 170$ V, and the supply frequency $f = 60$ Hz. Determine the average output voltage of the rectifier if the source inductance is neglibile.

3-12. Repeat Prob. 3-11 if the source inductance per phase (including the transformer leakage inductance) is $L_c = 0.5$ mH.

3-13. A three-phase bridge rectifier has a purely resistive load of $R = 100\ \Omega$ and is supplied from a 280-V 60-Hz supply. The primary and secondary of the input transformer are connected in wye. Determine the av-

erage output voltage of the rectifier if the source inductances are negligible.

3-14. Repeat Prob. 3-13 if the source inductance per phase (including transformer leakage inductance) is $L_c = 0.5$ mH.

3-15. The single-phase bridge rectifier of Fig. 3-19a is required to supply an average voltage of $V_{dc} = 400$ V to a resistive load of $R = 10\ \Omega$. Determine the voltage and current ratings of diodes and transformer.

3-16. A three-phase bridge rectifier is required to supply an average voltage of $V_{dc} = 750$ V at a ripple-free current of $I_{dc} = 9000$ A. The primary and secondary of the transformer are connected in wye. Determine the voltage and current ratings of diodes and transformer.

3-17. The single-phase rectifier of Fig. 3-18a has an RL load. If the peak input voltage is $V_m = 170$ V, the supply frequency $f = 60$ Hz, and the load resistance $R = 15\ \Omega$, determine the load inductance L to limit the load current harmonic to 4% of the average value I_{dc}.

3-18. The three-phase star rectifier of Fig. 3-24a has an RL load. If the secondary peak voltage per phase is $V_m = 170$ V at 60 Hz, and the load resistance is $R = 15\ \Omega$, determine the load inductance L to limit the load current harmonics to 2% of the average value I_{dc}.

3-19. The battery voltage in Fig. 3-17a is $E = 20$ V and its capacity is 200 W-h. The average charging current should be $I_{dc} = 10$ A. The primary input voltage is $V_p = 120$ V, 60 Hz and the transformer has a turns ratio of $n = 2:1$. *Calculate* (**a**) the conduction angle δ of the diode, (**b**) the current-limiting resistance R, (**c**) the power rating P_R of R, (**d**) the charging time h in hours, (**e**) the rectifier efficiency η, and (**f**) the peak inverse voltage PIV of the diode.

3-20. The single-phase full-wave rectifier of Fig. 3-21a has $L = 4.5$ mH, $R = 5$ Ω, and $E = 20$ V. The input voltage is $V_s = 120$ V at 60 Hz. (**a**) Determine (1) the steady-state load current I_1 at $\omega t = 0$, (2) the average diode current I_d, (3) the rms diode current I_r, and (4) the rms output current I_{rms}. (**b**) Use PSpice to plot the instantaneous output current i_L. Assume diode parameters IS=2.22E−15, BV=1800V.

3-21. The three-phase full-wave rectifier of Fig. 3-25a has a load of $L = 2.5$ mH, $R = 5$ Ω, and $E = 20$ V. The line-to-line input voltage is $V_{ab} = 208$ V, 60 Hz. (**a**) Determine (1) the steady-state load current I_1 at $\omega t = \pi/3$, (2) the average diode current I_d, (3) the rms diode current I_r, and (4) the rms output current I_{rms}. (**b**) Use PSpise to plot the instantaneous output current i_L. Assume diode parameters IS=2.22E−15, BV=1800V.

3-22. A single-phase bridge rectifier is supplied from a 120-V 60-Hz source. The load resistance is $R = 200$ Ω. (**a**) Design a C-filter so that the ripple factor of the output voltage is less than 5%. (**b**) With the value of capacitor C in part (a), calculate the average load voltage V_{dc}.

3-23. Repeat Prob. 3-22 for the single-phase half-wave rectifier.

3-24. The rms input voltage to the circuit in Fig. 3-33a is 120 V, 60 Hz. (**a**) If the dc output voltage is $V_{dc} = 48$ V at $I_{dc} = 25$ A, determine the values of inductance L_e, α, β, and I_{rms}. (**b**) If $I_{dc} = 15$ A and $L_e = 6.5$ mH, use Table 3-1 to calculate the values of V_{dc}, α, β, and I_{rms}.

3-25. The single-phase rectifier of Fig. 3-18a has a resistive load of R, and a capacitor C is connected across the load. The average load current is I_{dc}. Assuming that the charging time of the capacitor is negligible compared to the discharging time, determine the rms output voltage harmonics, V_{ac}.

3-26. The LC filter shown in Fig. 3-29c is used to reduce the ripple content of the output voltage for a six-phase star rectifier. The load resistance is $R = 20$ Ω, load inductance is $L = 5$ mH, and source frequency is 60 Hz. Determine the filter parameters L_e and C_e so that the ripple factor of the output voltage is 5%.

3-27. The three-phase bridge rectifier of Fig. 3-25a has an RL load and is supplied from a wye-connected supply. (**a**) Use the method of Fourier series to obtain expressions for the output voltage $v_L(t)$ and load current $i_L(t)$. (**b**) If peak phase voltage is $V_m = 170$ V at 60 Hz and the load resistance is $R = 200$ Ω, determine the load inductance L to limit the ripple current to 2% of the average value I_{dc}.

3-28. The single-phase half-wave rectifier of Fig. 3-16a has a freewheeling diode and a ripple-free average load current of I_a. (**a**) Draw the waveforms for the currents in D_1, D_m, and the transformer primary; (**b**) express the primary current in Fourier series; and (**c**) determine the input power factor PF and harmonic factor HF of the input current at the rectifier input. Assume a transformer turns ratio of unity.

3-29. The single-phase full-wave rectifier of Fig. 3-18a has a ripple-free average load current of I_a. (**a**) Draw the waveforms for currents in D_1, D_2, and transformer primary; (**b**) express the primary current in Fourier series; and (**c**) determine the input power factor PF and harmonic factor HF of the input current at the rectifier input. Assume a transformer turns ratio of unity.

3-30. The multiphase star rectifier of Fig. 3-24a has three pulses and supplies a ripple-free average load current of I_a. The primary and secondary of the transformer are connected in wye. Assume a transformer turns ratio of unity. (a) Draw the waveforms for currents in D_1, D_2, D_3, and transformer primary; (b) express the primary current in Fourier series; and (c) determine the input power factor PF and harmonic factor HF of input current.

3-31. Repeat Prob. 3-30 if the primary of the transformer is connected in delta and secondary in wye.

3-32. The multiphase star rectifier of Fig. 3-24a has six pulses and supplies a ripple-free average load current of I_a. The primary of the transformer is connected in delta and secondary in wye. Assume a transformer turns ratio of unity. (a) Draw the waveforms for currents in D_1, D_2, D_3, and transformer primary; (b) express the primary current in Fourier series; and (c) determine the input power factor PF and harmonic factor HF of the input current.

3-33. The three-phase bridge rectifier of Fig. 3-25a supplies a ripple-free load current of I_a. The primary and secondary of the transformer are connected in wye. Assume a transformer turns ratio of unity. (a) Draw the waveforms for currents in D_1, D_3, D_5 and the secondary phase current of the transformer; (b) express the

secondary phase current in Fourier series; and (c) determine the input power factor PF and harmonic factor HF of the input current.

3-34. Repeat Prob. 3-33 if the primary of the transformer is connected in delta and secondary in wye.

3-35. Repeat Prob. 3-33 if the primary and secondary of the transformer are connected in delta.

3-36. A diode circuit is shown in Fig. P3-36, where the load current is flowing through diode D_m. If switch S_1 is closed at a time $t = 0$, determine (a) expressions for $v_c(t)$, $i_c(t)$, and $i_d(t)$; (b) time t_1 when the diode D_1 stops conducting; (c) time t_q when the voltage across the capacitor becomes zero; and (d) the time required for capacitor to recharge to the supply voltage V_s.

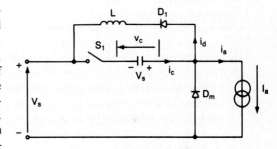

Figure P3-36

4

Thyristors

4-1 INTRODUCTION

A thyristor is one of the most important types of power semiconductor devices. Thyristors are used extensively in power electronic circuits. They are operated as bistable switches, operating from nonconducting state to conducting state. Thyristors can be assumed as ideal switches for many applications, but the practical thyristors exhibit certain characteristics and limitations.

4-2 THYRISTOR CHARACTERISTICS

A thyristor is a four-layer semiconductor device of *pnpn* structure with three *pn*-junctions. It has three terminals: anode, cathode, and gate. Figure 4-1 shows the thyristor symbol and the sectional view of three *pn*-junctions. Thyristors are manufactured by diffusion.

When the anode voltage is made positive with respect to the cathode, the junctions J_1 and J_3 are forward biased. The junction J_2 is reverse biased, and only a small leakage current flows from anode to cathode. The thyristor is then said to be in the *forward blocking* or *off-state* condition and the leakage current is known as *off-state current I_D*. If the anode-to-cathode voltage V_{AK} is increased to a sufficiently large value, the reverse-biased junction J_2 will break. This is known as *avalanche breakdown* and the corresponding voltage is called *forward break-down voltage V_{BO}*. Since the other junctions J_1 and J_3 are already forward biased, there will be free movement of carriers across all three junctions, resulting in a large forward anode current. The device will then be in a *conducting state* or *on-state*. The voltage drop would be due to the ohmic drop in the four layers and it is small, typically, 1 V. In the on-state, the anode current is limited by an external

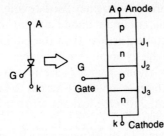

Figure 4-1 Thyristor symbol and three *pn*-junctions.

impedance or a resistance, R_L, as shown in Fig. 4-2a. The anode current must be more than a value known as *latching current I_L*, in order to maintain the required amount of carrier flow across the junction; otherwise, the device will revert to the blocking condition as the anode-to-cathode voltage is reduced. *Latching current I_L* is the minimum anode current required to maintain the thyristor in the on-state immediately after a thyristor has been turned on and the gate signal has been removed. A typical *v–i* characteristic of a thyristor is shown in Fig. 4-2b.

Once a thyristor conducts, it behaves like a conducting diode and there is no control over the device. The device will continue to conduct because there is no depletion layer on the junction J_2 due to free movements of carriers. However, if the forward anode current is reduced below a level known as the *holding current I_H*, a depletion region will develop around junction J_2 due to the reduced number of carriers and the thyristor will be in the blocking state. The holding current is in

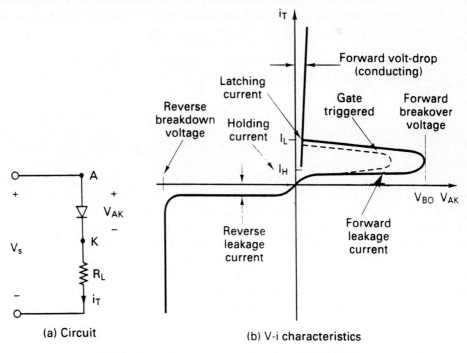

Figure 4-2 Thyristor circuit and *v–i* characteristics.

the order of milliamperes and is less than the latching current I_L. That is, $I_L > I_H$. *Holding current I_H* is the minimum anode current to maintain the thyristor in the on-state. The holding current is less than the latching current.

When the cathode voltage is positive with respect to the anode, the junction J_2 is forward biased, but junctions J_1 and J_3 are reverse biased. This is like two series-connected diodes with reverse voltage across them. The thyristor will be in the reverse blocking state and a reverse leakage current known as *reverse current*, I_R, would flow through the device.

A thyristor can be turned on by increasing the forward voltage V_{AK} beyond V_{BO}, but such a turn-on could be destructive. In practice, the forward voltage is maintained below V_{BO} and the thyristor is turned on by applying a positive voltage between its gate and cathode. This is shown in Fig. 4-2b by dashed lines. Once a thyristor is turned on by a gating signal and its anode current is greater than the holding current, the device continues to conduct due to positive feedback, even if the gating signal is removed. A thyristor is a latching device.

4-3 TWO-TRANSISTOR MODEL OF THYRISTOR

The regenerative or latching action due to a positive feedback can be demonstrated by using a two-transistor model of thyristor. A thyristor can be considered as two complementary transistors, one *pnp*-transistor, Q_1, and other *npn*-transistor, Q_2, as shown in Fig. 4-3a.

The collector current I_C of a thyristor is related, in general, to the emitter current I_E and the leakage current of the collector–base junction, I_{CBO}, as

$$I_C = \alpha I_E + I_{CBO} \qquad (4\text{-}1)$$

and the *common-base current* gain is defined as $\alpha \cong I_C/I_E$. For transistor Q_1, the emitter current is the anode current I_A, and the collector current I_{C1} can be found from Eq. (4-1):

$$I_{C1} = \alpha_1 I_A + I_{CBO1} \qquad (4\text{-}2)$$

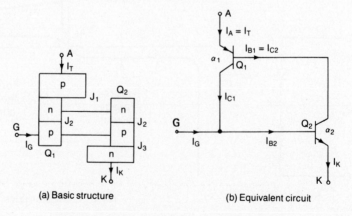

(a) Basic structure   (b) Equivalent circuit

Figure 4-3 Two-transistor model of thyristor.

where α_1 is the current gain and I_{CBO1} is the leakage current for Q_1. Similarly, for transistor Q_2, the collector current I_{C2} is

$$I_{C2} = \alpha_2 I_K + I_{CBO2} \tag{4-3}$$

where α_2 is the current gain and I_{CBO2} is the leakage current for Q_2. By combining I_{C1} and I_{C2}, we get

$$I_A = I_{C1} + I_{C2} = \alpha_1 I_A + I_{CBO1} + \alpha_2 I_K + I_{CBO2} \tag{4-4}$$

But for a gating current of I_G, $I_K = I_A + I_G$ and solving Eq. (4-4) for I_A gives

$$I_A = \frac{\alpha_2 I_G + I_{CBO1} + I_{CBO2}}{1 - (\alpha_1 + \alpha_2)} \tag{4-5}$$

The current gain α_1 varies with the emitter current $I_A = I_E$; and α_2 varies with $I_K = I_A + I_G$. A typical variation of current gain α with the emitter current I_E is shown in Fig. 4-4. If the gate current I_G is suddenly increased, say from 0 to 1 mA, this will immediately increase anode current I_A, which would further increase α_1 and α_2. α_2 will depend on I_A and I_G. The increase in the values of α_1 and α_2 would further increase I_A. Therefore, there is a regenerative or positive feedback effect. If $(\alpha_1 + \alpha_2)$ tends to be unity, the denominator of Eq. (4-5) approaches zero, resulting in a large value of anode current I_A, and the thyristor will turn on with a small gate current.

Under transient conditions, the capacitances of the pn-junctions, as shown in Fig. 4-5, will influence the characteristics of the thyristor. If a thyristor is in a blocking state, a rapidly rising voltage applied across the device would cause high current flow through the junction capacitors. The current through capacitor C_{j2} can be expressed as

$$i_{j2} = \frac{d(q_{j2})}{dt} = \frac{d}{dt}(C_{j2}V_{j2}) = V_{j2}\frac{dC_{j2}}{dt} + C_{j2}\frac{dV_{j2}}{dt} \tag{4-6}$$

where C_{j2} and V_{j2} are the capacitance and voltage of junction J_2, respectively. q_{j2} is the charge in the junction. If the rate of rise of voltage dv/dt is large, then i_{j2}

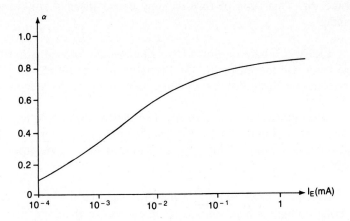

Figure 4-4 Typical variation of current gain with emitter current.

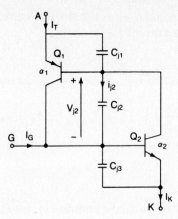

Figure 4-5 Two-transistor transient model of thyristor.

would be large and this would result in increased leakage currents I_{CBO1} and I_{CBO2}. According to Eq. (4-5), high enough values of I_{CBO1} and I_{CBO2} may cause $(\alpha_1 + \alpha_2)$ tending to unity and result in undesirable turn-on of the thyristor. However, a large current through the junction capacitors may also damage the device.

4-4 THYRISTOR TURN-ON

A thyristor is turned on by increasing the anode current. This can be accomplished in one of the following ways.

Thermals. If the temperature of a thyristor is high, there will be an increase in the number of electron–hole pairs, which would increase the leakage currents. This increase in currents would cause α_1 and α_2 to increase. Due to the regenerative action, $(\alpha_1 + \alpha_2)$ may tend to be unity and the thyristor may be turned on. This type of turn-on may cause thermal runaway and is normally avoided.

Light. If light is allowed to strike the junctions of a thyristor, the electron–hole pairs will increase; and the thyristor may be turned on. The light-activated thyristors are turned on by allowing light to strike the silicon wafers.

High voltage. If the forward anode-to-cathode voltage is greater than the forward breakdown voltage V_{BO}, sufficient leakage current will flow to initiate regenerative turn-on. This type of turn-on may be destructive and should be avoided.

dv/dt. It can be noted from Eq. (4-6) that if the rate of rise of the anode–cathode voltage is high, the charging current of the capacitive junctions may be sufficient enough to turn on the thyristor. A high value of charging current may

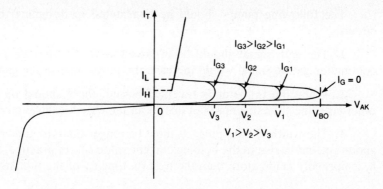

Figure 4-6 Effects of gate current on forward blocking voltage.

damage the thyristor; and the device must be protected against high dv/dt. The manufacturers specify the maximum allowable dv/dt of thyristors.

Gate current. If a thyristor is forward biased, the injection of gate current by applying positive gate voltage between the gate and cathode terminals would turn on the thyristor. As the gate current is increased, the forward blocking voltage is decreased as shown in Fig. 4-6.

Figure 4-7 shows the waveform of the anode current, following the application of gate signal. There is a time delay known as *turn-on time* t_{on} between the application of gate signal and the conduction of a thyristor. t_{on} is defined as the time interval between 10% of steady-state gate current $(0.1I_G)$ and 90% of the steady-state thyristor on-state current $(0.9I_T)$. t_{on} is the sum of *delay time* t_d and *rise time* t_r. t_d is defined as the time interval between 10% of gate current $(0.1I_G)$ and 10% of thyristor on-state current $(0.1I_T)$. t_r is the time required for the anode current to rise from 10% of on-state current $(0.1I_T)$ to 90% of on-state current $(0.9I_T)$. These times are depicted in Fig. 4-7.

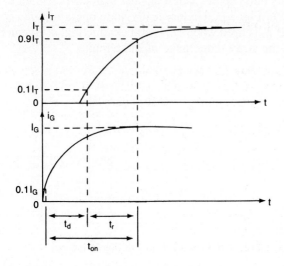

Figure 4-7 Turn-on characteristics.

The following points should be considered in designing the gate control circuit:

1. The gate signal should be removed after the thyristor is turned on. A continuous gating signal would increase the power loss in the gate junction.

2. While the thyristor is reversed biased, there should be no gate signal; otherwise, the thyristor may fail due to an increased leakage current.

3. The width of gate pulse t_G must be longer than the time required for the anode current to rise to the holding current value I_H. In practice, the pulse width t_G is normally made more than the turn-on time t_{on} of the thyristor.

Example 4-1

The capacitance of reverse-biased junction J_2 in a thyristor is $C_{J2} = 20$ pF and can be assumed to be independent of the off-state voltage. The limiting value of the charging current to turn on the thyristor is 16 mA. Determine the critical value of dv/dt.
Solution $C_{J2} = 20$ pF and $i_{J2} = 16$ mA. Since $d(C_{J2})/dt = 0$, we can find the critical value of dv/dt from Eq. (4-6):

$$\frac{dv}{dt} = \frac{i_{J2}}{C_{J2}} = \frac{16 \times 10^{-3}}{20 \times 10^{-12}} = 800 \text{ V}/\mu s$$

4-5 *di/dt* PROTECTION

A thyristor requires a minimum time to spread the current conduction uniformly throughout the junctions. If the rate of rise of anode current is very fast compared to the spreading velocity of a turn-on process, a localized "hot-spot" heating will occur due to high current density and the device may fail, as a result of excessive temperature.

The practical devices must be protected against high *di/dt*. As an example, let us consider the circuit in Fig. 4-8. Under steady-state operation, D_m conducts when thyristor T_1 is off. If T_1 is fired when D_m is still conducting, *di/dt* can be very high and limited only by the stray inductance of the circuit.

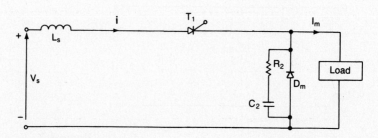

Figure 4-8   Chopper circuit with *di/dt* limiting inductors.

In practice, the di/dt is limited by adding a series inductor L_s, as shown in Fig. 4-8. The forward di/dt is

$$\frac{di}{dt} = \frac{V_s}{L_s} \tag{4-7}$$

where L_s is the series inductance, including any stray inductance.

4-6 dv/dt PROTECTION

If switch S_1 in Fig. 4-9a is closed at $t = 0$, a step voltage will be applied across thyristor T_1 and dv/dt may be high enough to turn on the device. The dv/dt can be limited by connecting capacitor C_s, as shown in Fig. 4-9a. When thyristor T_1 is turned on, the discharge current of capacitor is limited by resistor R_s as shown in Fig. 4-9b.

With an RC circuit known as a snubber circuit, the voltage across the thyristor will rise exponentially as shown in Fig. 4-9c and the circuit dv/dt can be found approximately from

$$\frac{dv}{dt} = \frac{0.632V_s}{\tau} = \frac{0.632V_s}{R_sC_s} \tag{4-8}$$

The value of snubber time constant $\tau = R_sC_s$ can be determined from Eq. (4-8) for a known value of dv/dt. The value of R_s is found from the discharge current I_{TD}.

$$R_s = \frac{V_s}{I_{TD}} \tag{4-9}$$

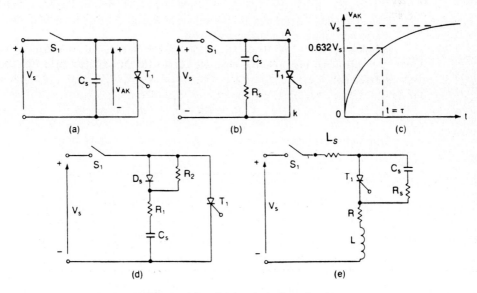

Figure 4-9 dv/dt protection circuits.

It is possible to use more than one resistor for dv/dt and discharging, as shown in Fig. 4-9d. The dv/dt is limited by R_1 and C_s. $(R_1 + R_2)$ limits the discharging current such that

$$I_{TD} = \frac{V_s}{R_1 + R_2} \tag{4-10}$$

The load can form a series circuit with the snubber network as shown in Fig. 4-9e. From Eqs. (3-23) and (3-24), the damping ratio δ of a second-order equation is

$$\delta = \frac{\alpha}{\omega_0} = \frac{R_s + R}{2} \sqrt{\frac{C_s}{L_s + L}} \tag{4-11}$$

where L_s is the stray inductance, and L and R are the load inductance and resistance, respectively.

   To limit the peak voltage overshoot applied across the thyristor, the damping ratio in the range 0.5 to 1.0 is used. If the load inductance is high, which is usually the case, R_s can be high and C_s can be small to retain the desired value of damping ratio. A high value of R_s will reduce the discharge current and a low value of C_s reduces the snubber loss. The circuits of Fig. 4-9 should be fully analyzed to determine the required value of damping ratio to limit the dv/dt to the desired value. Once the damping ratio is known, R_s and C_s can be found. The same RC network or snubber is normally used for both dv/dt protection and to suppress the transient voltage due to reverse recovery time. The transient voltage suppression is analyzed in Section 15-4.

Example 4-2

The input voltage of Fig. 4-9e is $V_s = 200$ V with load resistance of $R = 5\ \Omega$. The load and stray inductances are negligible and the thyristor is operated at a frequency of $f_s = 2$ kHz. If the required dv/dt is 100 V/μs and the discharge current is to be limited to 100 A, determine (a) the values of R_s and C_s, (b) the snubber loss, and (c) the power rating of the snubber resistor.

Solution $dv/dt = 100$ V/μs, $I_{TD} = 100$ A, $R = 5\ \Omega$, $L = L_s = 0$, and $V_s = 200$ V.

   (a) From Fig. 4-9e the charging current of the snubber capacitor can be expressed as

$$V_s = (R_s + R)i + \frac{1}{C_s} \int i\, dt + v_c(t = 0)$$

With initial condition $v_c(t = 0) = 0$, the charging current is found as

$$i(t) = \frac{V_s}{R_s + R}\, e^{-t/\tau} \tag{4-12}$$

where $\tau = (R_s + R)C_s$. The forward voltage across the thyristor is

$$v_T(t) = V_s - \frac{RV_s}{R_s + R}\, e^{-t/\tau} \tag{4-13}$$

At $t = 0$, $v_T(0) = V_s - RV_s/(R_s + R)$ and at $t = \tau$, $v_T(\tau) = V_s - 0.368RV_s/(R_s + R)$:

$$\frac{dv}{dt} = \frac{v_T(\tau) - v_T(0)}{\tau} = \frac{0.632RV_s}{C_s(R_s + R)^2} \tag{4-14}$$

From Eq. (4-9), $R_s = V_s/I_{TD} = 200/100 = 2\ \Omega$. Equation (4-14) gives

$$C_s = \frac{0.632 \times 5 \times 200 \times 10^{-6}}{(2+5)^2 \times 100} = 0.129\ \mu F$$

(b) The snubber loss

$$P_s = 0.5C_sV_s^2f_s$$

$$= 0.5 \times 0.129 \times 10^{-6} \times 200^2 \times 2000 = 5.2\ W$$

(4-15)

(c) Assuming that all the energy stored in C_s is dissipated in R_s only, the power rating of the snubber resistor is 5.2 W.

4-7 THYRISTOR TURN-OFF

A thyristor which is in the on-state can be turned off by reducing the forward current to a level below the holding current I_H. There are various techniques for turning off a thyristor, which are discussed in Chapter 7. In all the commutation techniques, the anode current is maintained below the holding current for a sufficiently long time, so that all the excess carriers in the four layers are swept out or recombined.

Due to two outer *pn*-junctions, J_1 and J_3, the turn-off characteristics would be similar to that of a diode, exhibiting reverse recovery time t_{rr} and peak reverse recovery current I_{RR}. I_{RR} can be much greater than the normal reverse blocking current, I_R. In a line-commutated converter circuit where the input voltage is alternating as shown in Fig. 4-10a, a reverse voltage appears across the thyristor immediately after the forward current goes through the zero value. This reverse voltage will accelerate the turn-off process, by sweeping out the excess carriers from *pn*-junctions J_1 and J_3. Equations (2-6) and (2-7) can be applied to calculate t_{rr} and I_{RR}.

The inner *pn*-junction J_2 will require a time known as *recombination time* t_{rc} to recombine the excess carriers. A negative reverse voltage would reduce this recombination time. t_{rc} is dependent on the magnitude of the reverse voltage. The turn-off characteristics are shown in Fig. 4-10a and b for a line-commutated circuit and forced-commutated circuit, respectively.

The turn-off time t_q is the sum of reverse recovery time t_{rr} and recombination time t_{rc}. At the end of turn-off, a depletion layer develops across junction J_2 and the thyristor recovers its ability to withstand forward voltage. In all the commutation techniques in Chapter 7, a reverse voltage is applied across the thyristor during the turn-off process.

Turn-off time t_q is the minimum value of time interval between the instant when the on-state current has decreased to zero and the instant when the thyristor is capable of withstanding forward voltage without turning on. t_q depends on the peak value of on-state current and the instantaneous on-state voltage.

Reverse recovered charge Q_{RR} is the amount of charge which has to be recovered during the turn-off process. Its value is determined from the area enclosed by the path of the reverse recovery current. The value of Q_{RR} depends on

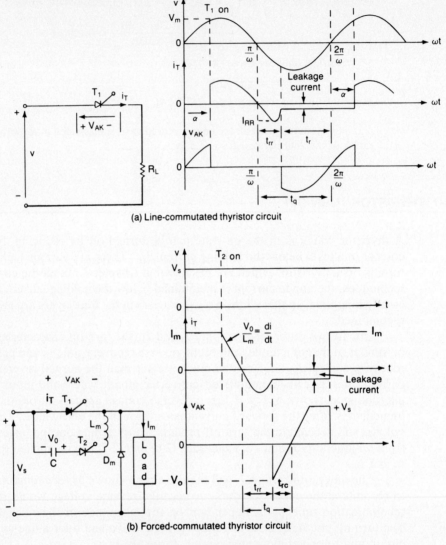

(a) Line-commutated thyristor circuit

(b) Forced-commutated thyristor circuit

Figure 4-10   Turn-off characteristics.

the rate of fall of on-state current and the peak value of on-state current before turn-off. Q_{RR} causes corresponding energy loss within the device.

4-8 THYRISTOR TYPES

Thyristors are manufactured almost exclusively by diffusion. The anode current requires a finite time to propagate to the whole area of the junction, from the point near the gate when the gate signal is initiated for turning on the thyristor. The manufacturers use various gate structures to control the di/dt, turn-on time, and turn-off time. Depending on the physical construction, and turn-on and turn-off behavior, thyristors can, broadly, be classified into nine categories:

1. Phase-control thyristors (SCRs)
2. Fast-switching thyristors (SCRs)
3. Gate-turn-off thyristors (GTOs)
4. Bidirectional triode thyristors (TRIACs)
5. Reverse-conducting thyristors (RCTs)
6. Static induction thyristors (SITHs)
7. Light-activated silicon-controlled rectifiers (LASCRs)
8. FET-controlled thyristors (FET-CTHs)
9. MOS-controlled thyristors (MCTs)

4-8.1 Phase-Control Thyristors

This type of thyristors generally operates at the line frequency and is turned off by natural commutation. The turn-off time, t_q, is of the order of 50 to 100 μs. This is most suited for low-speed switching applications and is also known as *converter thyristor*. Since a thyristor is basically a silicon-made controlled device, it is also known as *silicon-controlled rectifier* (SCR).

The on-state voltage, V_T, varies typically from about 1.15 V for 600 V to 2.5 V for 4000-V devices; and for a 5500-A 1200-V thyristor it is typically 1.25 V. The modern thyristors use an amplifying gate, where an auxiliary thyristor T_A is gated on by a gate signal and then the amplified output of T_A is applied as a gate signal to the main thyristor T_M. This is shown in Fig. 4-11. The amplifying gate permits high dynamic characteristics with typical dv/dt of 1000 V/μs and di/dt of 500 A/μs and simplifies the circuit design by reducing or minimizing di/dt limiting inductor and dv/dt protection circuits.

4-8.2 Fast-Switching Thyristors

These are used in high-speed switching applications with forced commutation (e.g., choppers in Chapter 9 and inverters in Chapter 10). They have fast turn-off time, generally in the range 5 to 50 μs, depending on the voltage range. The on-state forward drop varies approximately as an inverse function of the turn-off time t_q. This type of thyristor is also known as an *inverter thyristor*.

These thyristors have high dv/dt of typically 1000 V/μs and di/dt of 1000 A/μs. The fast turn-off and high di/dt are very important to reduce the size and

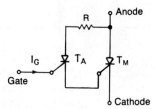

Figure 4-11 Amplifying gate thyristor.

weight of commutating and/or reactive circuit components. The on-state voltage of a 2200-A 1800-V thyristor is typically 1.7 V. Inverter thyristors with a very limited reverse blocking capability, typically 10 V, and a very fast turn-off time between 3 and 5 μs are commonly known as *asymmetrical thyristors* (ASCRs). Fast-switching thyristors of various sizes are shown in Fig. 4-12.

4-8.3 Gate-Turn-Off Thyristors

A gate-turn-off thyristor (GTO) like an SCR can be turned on by applying a positive gate signal. However, it can be turned off by a negative gate signal. A GTO is a latching device and can be built with current and voltage ratings similar to those of an SCR. A GTO is turned on by applying a short positive pulse and turned off by a short negative pulse to its gate. The GTOs have advantages over SCRs: (1) elimination of commutating components in forced commutation, resulting in reduction in cost, weight, and volume; (2) reduction in acoustic and electromagnetic noise due to the elimination of commutation chokes; (3) faster turn-off, permitting high switching frequencies; and (4) improved efficiency of converters.

In low-power applications, GTOs have the following advantages over bipolar transistors: (1) a higher blocking voltage capability; (2) a high ratio of peak controllable current to average current; (3) a high ratio of peak surge current to average current, typically 10:1; (4) a high on-state gain (anode current/gate current), typically 600; and (5) a pulsed gate signal of short duration. Under surge conditions, a GTO goes into deeper saturation due to regenerative action. On the other hand, a bipolar transistor tends to come out of saturation.

A GTO has low gain during turn-off, typically 6, and requires a relatively high negative current pulse to turn off. It has higher on-state voltage than that of SCRs. The on-state voltage of a typical 550-A 1200-V GTO is typically 3.4 V. A 160-A 200-V GTO of type 160PFT is shown in Fig. 4-13 and the junctions of this GTO are shown in Fig. 4-14.

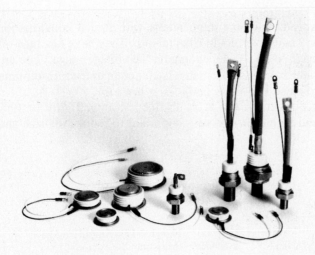

Figure 4-12 Fast-switching thyristors. (Courtesy of Powerex, Inc.)

Figure 4-13 A 160-A 200-V GTO. (Courtesy of International Rectifier.)

Controllable peak on-state current I_{TGQ} is the peak value of on-state current which can be turned off by gate control. The off-state voltage is reapplied immediately after turn-off and the reapplied dv/dt is only limited by the snubber capacitance. Once a GTO is turned off, the load current I_L, which is diverted through and charges the snubber capacitor, determines the reapplied dv/dt.

$$\frac{dv}{dt} = \frac{I_L}{C_s}$$

where C_s is the snubber capacitance.

4-8.4 Bidirectional Triode Thyristors

A TRIAC can conduct in both directions and is normally used in ac phase control (e.g., ac voltage controllers in Chapter 6). It can be considered as two SCRs

Figure 4-14 Junctions of 160-A GTO in Fig. 4-13. (Courtesy of International Rectifier.)

connected in antiparallel with a common gate connection as shown in Fig. 4-15a. The v–i characteristics are shown in Fig. 4-15c.

Since a TRIAC is a bidirectional device, its terminals cannot be designated as anode and cathode. If terminal MT_2 is positive with respect to terminal MT_1, the TRIAC can be turned on by applying a positive gate signal between gate G and terminal MT_1. If terminal MT_2 is negative with respect to terminal MT_1, it is turned on by applying negative gate signal between gate G and terminal MT_1. It is not necessary to have both polarities of gate signals and a TRIAC can be turned on with either a positive or negative gate signal. In practice, the sensitivities vary from one quadrant to another, and the TRIACs are normally operated in quadrant I^+ (positive gate voltage and gate current) or quadrant III^- (negative gate voltage and gate current).

4-8.5 Reverse-Conducting Thyristors

In many choppers and inverter circuits, an antiparallel diode is connected across an SCR in order to allow a reverse current flow due to inductive load and to improve the turn-off requirement of commutation circuit. The diode clamps the reverse blocking voltage of the SCR to 1 or 2 V under steady-state conditions. However, under transient conditions, the reverse voltage may rise to 30 V due to induced voltage in the circuit stray inductance within the device.

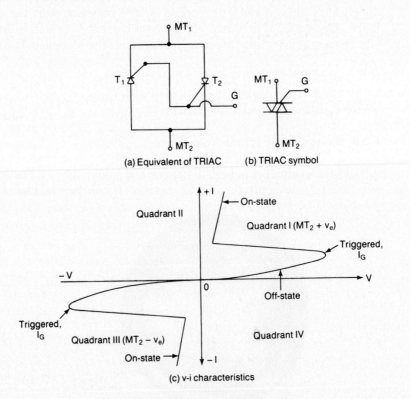

(a) Equivalent of TRIAC   (b) TRIAC symbol

(c) v-i characteristics

Figure 4-15 Characteristics of a triac.

An RCT is a compromise between the device characteristics and circuit requirement; and it may be considered as a thyristor with a built-in antiparallel diode as shown in Fig. 4-16. An RCT is also called an *asymmetrical thyristor* (ASCR). The forward blocking voltage varies from 400 to 2000 V and the current rating goes up to 500 A. The reverse blocking voltage is typically 30 to 40 V. Since the ratio of forward current through the thyristor to the reverse current of diode is fixed for a given device, their applications will be limited to specific circuit designs.

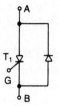

Figure 4-16 Reverse-conducting thyristor.

4-8.6 Static Induction Thyristors

The characteristics of a SITH are similar to those of a MOSFET in Chapter 8. A SITH is normally turned on by applying a positive gate voltage like normal thyristors and is turned off by application of negative voltage to its gate. A SITH is a minority carrier device. As a result, SITH has low on-state resistance or voltage drop and it can be made with higher voltage and current ratings.

A SITH has fast switching speeds and high dv/dt and di/dt capabilities. The switching time is on the order of 1 to 6 μs. The voltage rating can go up to 2500 V and the current rating is limited to 500 A. This device is extremely process sensitive, and small perturbations in the manufacturing process would produce major changes in the device characteristics.

4-8.7 Light-Activated Silicon-Controlled Rectifiers

This device is turned on by direct radiation on the silicon wafer with light. Electron–hole pairs which are created due to the radiation produce triggering current under the influence of electric field. The gate structure is designed to provide sufficient gate sensitivity for triggering from practical light sources (e.g., LED and to accomplish high di/dt and dv/dt capabilities).

The LASRCs are used in high-voltage and high-current applications [e.g., high-voltage dc (HVDC) transmission and static reactive power or volt-ampere reactive (VAR) compensation. A LASCR offers complete electrical isolation between the light-triggering source and the switching device of a power converter, which floats at a potential of as high as a few hundred kilovolts. The voltage rating of a LASCR could be as high as 4 kV at 1500 A with light-triggering power of less than 100 mW. The typical di/dt is 250 A/μs and the dv/dt could be as high as 2000 V/μs.

4-8.8 FET-Controlled Thyristors

A FET-CTH device combines a MOSFET and a thyristor in parallel as shown in Fig. 4-17. If a sufficient voltage is applied to the gate of the MOSFET, typically 3 V, a triggering current for the thyristor is generated internally. It has a high switching speed, high di/dt, and high dv/dt.

This device can be turned on like conventional thyristors, but it can not be turned off by gate control. This would find applications where optical firing is to be used for providing electrical isolation between the input or control signal and the switching device of the power converter.

4-8.9 MOS-Controlled Thyristor

A MOS-controlled thyristor (MCT) combines the features of a regenerative four-layer thyristor and a MOS-gate structure. A schematic of an MCT cell is shown in Fig. 4-18a. The equivalent circuit is shown in Fig. 4-18b and the symbol in Fig. 4-18c. The *NPNP* structure may be represented by an *NPN*-transistor Q_1 and a *PNP*-transistor Q_2. The MOS-gate structure can be represented by a *p*-channel MOSFET M_1, and an *n*-channel MOSFET M_2.

Due to an *NPNP* structure rather than the *PNPN* structure of a normal SCR, the anode serves as the reference terminal with respect to which all gate signals are applied. Let us assume that the MCT is in its forward blocking state and a negative voltage V_{GA} is applied. A *p*-channel (or an inversion layer) is formed in the *n*-doped material, causing holes to flow laterally from the *p*-emitter E_2 of Q_2 (source S_1 of *p*-channel MOSFET M_1) through the *p*-channel to the *p*-base B_1 of Q_1 (drain D_1 of *p*-channel MOSFET M_1). This hole flow is the base current for the *NPN*-transistor Q_1. The n^+-emitter E_1 of Q_1 then injects electrons that are collected in the *n*-base B_2 (and *n*-collector C_1), which causes the *p*-emitter E_2 to inject holes into the *n*-base B_2 so that the *PNP*-transistor Q_2 is turned on and latches the MCT. In short, a negative gate V_{GA} turns on the *p*-channel MOSFET M_1, thus providing the base current for the transistor Q_2.

Let us assume that the MCT is in its conduction state, and a positive voltage V_{GA} is applied. An *n*-channel is formed in the *p*-doped material, causing electrons to flow laterally from the *n*-base B_2 of Q_2 (source S_2 of *n*-channel MOSFET M_2) through the *n*-channel to the heavily doped n^+-emitter E_2 of Q_2 (drain D_2 of n^+-channel MOSFET M_2). This electron flow diverts the base current of *PNP*-transistor Q_2 so that its base–emitter junction turns off, and holes are not available for

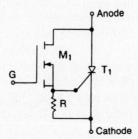

Figure 4-17 FET-controlled thyristor.

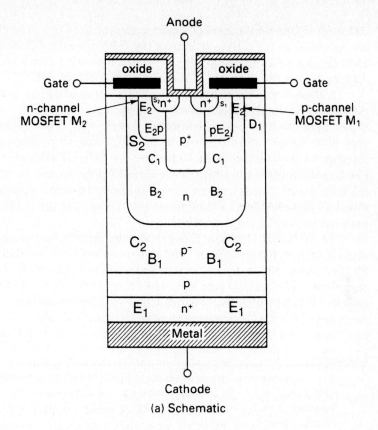

(a) Schematic

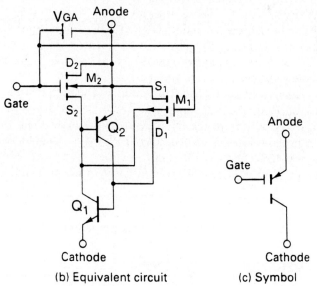

(b) Equivalent circuit   (c) Symbol

Figure 4-18 Schematic and equivalent circuit for MCTs.

collection by the p-base B_1 of Q_1 (and p-collector C_2 of Q_2). The elimination of this holes current at the p-base B_1 causes the *NPN*-transistor Q_1 to turn off, and the MCT returns to its blocking state. In short, a positive gate pulse V_{GA} diverts the current driving the base of Q_1, thereby turning off the MCT.

The MCT can be operated as a gate-controlled device if its current is less than the peak controllable current. Attempting to turn off the MCT at currents higher than its rated peak controllable current may result in destroying the device. For higher values of current, the MCT has to be commutated off like a standard SCR. The gate pulse widths are not critical for smaller device currents. For larger currents, the width of the turn-off pulse should be larger. Moreover, the gate draws a peak current during turn-off. In many applications, including inverters and choppers, a continuous gate pulse over the entire on/off period is required to avoid state ambiguity.

An MCT has (1) a low forward voltage drop during conduction; (2) a fast turn-on time, typically 0.4 μs, and a fast turn-off time, typically 1.25 μs for an MCT of 300 A, 500 V; (3) low switching losses; (4) a low reverse voltage blocking capability, and (5) a high gate input impedance, which greatly simplifies the drive circuits. It can be effectively paralleled to switch high currents with only modest deratings of the per-device current rating. It cannot easily be driven from a pulse transformer if a continuous bias is required to avoid state ambiguity.

Example 4-3

A thyristor carries a current as shown in Fig. 4-19 and the current pulse is repeated at a frequency of f_s = 50 Hz. Determine the average on-state current I_T.

Solution $I_p = I_{TM}$ = 1000 A, $T = 1/f_s$ = 1/50 = 20 ms, and $t_1 = t_2$ = 5 μs. The average on-state current is

$$I_T = \frac{1}{20,000} \, [0.5 \times 5 \times 1000 + (20,000 - 2 \times 5) \times 1000 + 0.5 \times 5 \times 1000]$$

$$= 999.5 \text{ A}$$

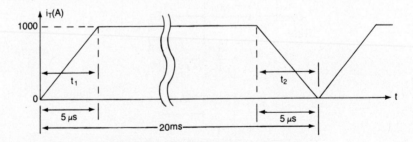

Figure 4-19   Thyristor current waveform.

4-9 SERIES OPERATION OF THYRISTORS

For high-voltage applications, two or more thyristors can be connected in series to provide the voltage rating. However, due to the production spread the characteristics of thyristors of the same type are not identical. Figure 4-20 shows the off-

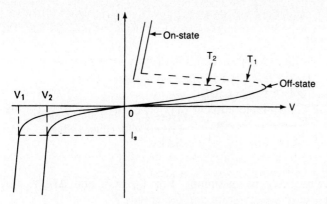

Figure 4-20 Off-state characteristics of two thyristors.

state characteristics of two thyristors. For the same off-state current, their off-state voltages differ.

In the case of diodes, only the reverse blocking voltages have to be shared, whereas for thyristors, the voltage-sharing networks are required for both reverse and off-state conditions. The voltage sharing is normally accomplished by connecting resistors across each thyristor as shown in Fig. 4-21. For equal voltage sharing, the off-state currents differ as shown in Fig. 4-22. Let there be n_s thyristors in the string. The off-state current of thyristor T_1 is I_{D1} and that of other thyristors are equal such that $I_{D2} = I_{D3} = I_{Dn}$ and $I_{D1} < I_{D2}$. Since thyristor T_1 has the least off-state current, T_1 will share higher voltage.

If I_1 is the current through the resistor R across T_1 and the currents through other resistors are equal so that $I_2 = I_3 = I_n$, the off-state current spread is

$$\Delta I_D = I_{D2} - I_{D1} = I_T - I_2 - I_T + I_1 = I_1 - I_2 \quad \text{or} \quad I_2 = I_1 - \Delta I_D$$

The voltage across T_1 is $V_{D1} = RI_1$. Using Kirchhoff's voltage law yields

$$\begin{aligned} V_s &= V_{D1} + (n_s - 1)I_2R = V_{D1} + (n_s - 1)(I_1 - \Delta I_D)R \\ &= V_{D1} + (n_s - 1)I_1R - (n_s - 1)R\,\Delta I_D \\ &= n_sV_{D1} - (n_s - 1)R\,\Delta I_D \end{aligned} \qquad (4\text{-}16)$$

Solving Eq. (4-16) for the voltage V_{D1} across T_1 gives

$$V_{D1} = \frac{V_s + (n_s - 1)R\,\Delta I_D}{n_s} \qquad (4\text{-}17)$$

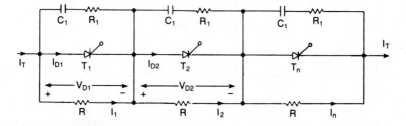

Figure 4-21 Three series-connected thyristors.

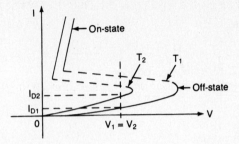

Figure 4-22 Forward leakage currents with equal voltage sharing.

V_{D1} will be maximum when ΔI_D is maximum. For $I_{D1} = 0$ and $\Delta I_D = I_{D2}$, Eq. (4-17) gives the worst-case steady-state voltage across T_1,

$$V_{DS(\max)} = \frac{V_s + (n_s - 1)RI_{D2}}{n_s} \tag{4-18}$$

During the turn-off, the differences in stored charge cause differences in the reverse voltage sharing as shown in Fig. 4-23. The thyristor with the least recovered charge (or reverse recovery time) will face the highest transient voltage. The junction capacitances which control the transient voltage distributions will not be adequate and it is normally necessary to connect a capacitor, C_1 across each thyristor as shown in Fig. 4-21. R_1 limits the discharge current. The same RC network is generally used for both transient voltage sharing and dv/dt protection.

The transient voltage across T_1 can be determined from Eq. (4-17) by applying the relationship of voltage difference,

$$\Delta V = R\,\Delta I_D = \frac{Q_2 - Q_1}{C_1} = \frac{\Delta Q}{C_1} \tag{4-19}$$

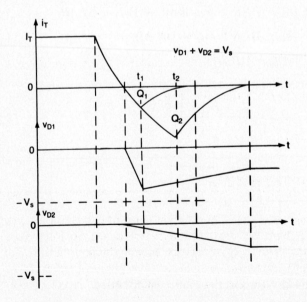

Figure 4-23 Reverse recovery time and voltage sharing.

where Q_1 is the stored charge of T_1 and Q_2 is the charge of other thyristors such that $Q_2 = Q_3 = Q_n$ and $Q_1 < Q_2$. Substituting Eq. (4-19) into Eq. (4-17) yields

$$V_{D1} = \frac{1}{n_s} \left[V_s + \frac{(n_s - 1)\, \Delta Q}{C_1} \right] \tag{4-20}$$

The worst-case transient voltage sharing which will occur when $Q_1 = 0$ and $\Delta Q = Q_2$ is

$$V_{DT(\text{max})} = \frac{1}{n_s} \left[V_s + \frac{(n_s - 1)Q_2}{C_1} \right] \tag{4-21}$$

A derating factor which is normally used to increase the reliability of the string is defined as

$$\text{DRF} = 1 - \frac{V_s}{n_s V_{DS(\text{max})}} \tag{4-22}$$

Example 4-4

Ten thyristors are used in a string to withstand a dc voltage of $V_s = 15$ kV. The maximum leakage current and recovery charge differences of thyristors are 10 mA and 150 μC, respectively. Each thyristor has a voltage-sharing resistance of $R = 56$ kΩ and capacitance of $C_1 = 0.5$ μF. Determine (a) the maximum steady-state voltage sharing $V_{DS(\text{max})}$, (b) the steady-state voltage derating factor, (c) the maximum transient voltage sharing $V_{DT(\text{max})}$, and (d) the transient voltage derating factor.

Solution $n_s = 10$, $V_s = 15$ kV, $\Delta I_D = I_{D2} = 10$ mA, and $\Delta Q = Q_2 = 150$ μC.

(a) From Eq. (4-18), the maximum steady-state voltage sharing is

$$V_{DS(\text{max})} = \frac{15{,}000 + (10 - 1) \times 56 \times 10^3 \times 10 \times 10^{-3}}{10} = 2004 \text{ V}$$

(b) From Eq. (4-22), the steady-state derating factor is

$$\text{DRF} = 1 - \frac{15{,}000}{10 \times 2004} = 25.15\%$$

(c) From Eq. (4-21), the maximum transient voltage sharing is

$$V_{DT(\text{max})} = \frac{15{,}000 + (10 - 1) \times 150 \times 10^{-6}/(0.5 \times 10^{-6})}{10} = 1770 \text{ V}$$

(d) From Eq. (4-22), the transient derating factor is

$$\text{DRF} = 1 - \frac{15{,}000}{10 \times 1770} = 15.25\%$$

4-10 PARALLEL OPERATION OF THYRISTORS

When thyristors are connected in parallel, the load current is not shared equally due to differences in their characteristics. If a thyristor carries more current than that of the others, its power dissipation increases, thereby increasing the junction temperature and decreasing the internal resistance. This, in turn, will increase its current sharing and may damage the thyristor. This thermal runaway may be avoided by having a common heat sink discussed in Chapter 16, so that all units operate at the same temperature.

A small resistance, as shown in Fig. 4-24, may be connected in series with each thyristor to force equal current sharing, but there will be considerable power loss in the series resistances. A common approach for current sharing of thyristors is to use magnetically coupled inductors as shown in Fig. 4-24b. If the current through thyristor T_1 increases, a voltage of opposite polarity will be induced in the windings of thyristor T_2 and the impedance through the path of T_2 will be reduced, thereby increasing the current flow through T_2.

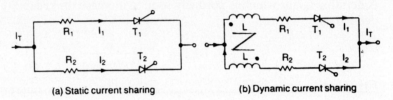

(a) Static current sharing   (b) Dynamic current sharing

Figure 4-24 Current sharing of thyristors.

4-11 THYRISTOR FIRING CIRCUITS

In thyristor converters, different potentials exist at various terminals. The power circuit is subjected to a high voltage, usually greater than 100 V, and the gate circuit is held at a low voltage, typically 12 to 30 V. An isolation circuit is required between an individual thyristor and its gate-pulse generating circuit. The isolation can be accomplished by either pulse transformers or optocouplers. An optocoupler could be a phototransistor or photo-SCR as shown in Fig. 4-25. A short pulse to the input of an infrared light-emitting diode (ILED), D_1, turns on the photo-SCR, T_1; and the power thyristor, T_L, is triggered. This type of isolation requires a separate power supply V_{cc} and increases the cost and weight of the firing circuit.

A simple isolation arrangement with pulse transformers is shown in Fig. 4-26a. When a pulse of adequate voltage is applied to the base of switching transistor Q_1, the transistor saturates and the dc voltage V_{cc} appears across the

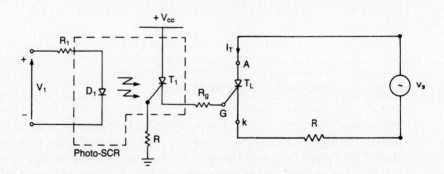

Figure 4-25 Photo-SCR coupled isolator.

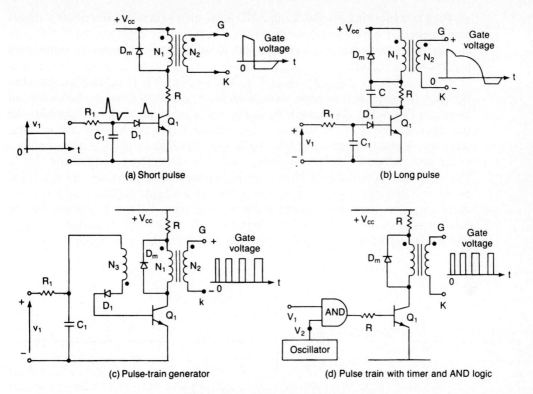

Figure 4-26 Pulse transformer isolation.

transformer primary, inducing a pulsed voltage on the transformer secondary, which is applied between the thyristor gate and cathode terminals. When the pulse is removed from the base of transistor Q_1, the transistor turns off and a voltage of opposite polarity is induced across the primary and the freewheeling diode D_m conducts. The current due to the transformer magnetic energy decays through D_m to zero. During this transient decay a corresponding reverse voltage is induced in the secondary. The pulse width can be made longer by connecting a capacitor C cross the resistor R, as shown in Fig. 4-26b. The transformer carries unidirectional current and the magnetic core will saturate, thereby limiting the pulse width. This type of pulse isolation is suitable for pulses of typically 50 μs to 100 μs.

In many power converters with inductive loads, the conduction period of a thyristor depends on the load power factor; therefore, the beginning of thyristor conduction is not well defined. In this situation, it is often necessary to trigger the thyristors continuously. However, a continuous gating increases thyristor losses. A pulse train which is preferable can be obtained with an auxiliary winding, as shown in Fig. 4-26c. When transistor Q_1 is turned on, a voltage is also induced in the auxiliary winding N_3 at the base of transistor Q_1, such that diode D_1 is reverse biased and Q_1 turns off. In the meantime, capacitor C_1 charges up through R_1 and turns on Q_1 again. This process of turn-on and turn-off continues as long as there is an input signal v_1 to the isolator. Instead of using the auxiliary winding as a blocking oscillator, an AND-logic gate with an oscillator (or a timer) could generate a pulse train as shown in Fig. 4-26d. In practice, the AND gate cannot drive transistor Q_1 directly, and a buffer stage is normally connected before the transistor.

The output of gate circuits in Fig. 4-25 or Fig. 4-26 is normally connected between the gate and cathode along with other gate-protecting components, as shown in Fig. 4-27. The resistor R_g in Fig. 4-27a increases the dv/dt capability of the thyristor, reduces the turn-off time, and increases the holding and latching currents. The capacitor C_g in Fig. 4-27b removes high-frequency noise components and increases dv/dt capability and gate delay time. The diode D_g in Fig. 4-27c protects the gate from negative voltage. However, for asymmetrical SCRs, it is desirable to have some amount of negative gate voltage to improve the dv/dt capability and also to reduce the turn-off time. All these features can be combined as shown in Fig. 4-27d, where diode D_1 allows only the positive pulses and R_1 damps out any transient oscillation and limits the gate current.

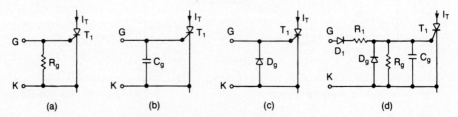

Figure 4-27 Gate protection circuits.

4-12 UNIJUNCTION TRANSISTOR

The unijunction transistor (UJT) is commonly used for generating triggering signals for SCRs. A basic UJT-triggering circuit is shown in Fig. 4-28a. A UJT has three terminals, called the emitter E, base-one B_1, and base-two B_2. Between B_1 and B_2 the unijunction has the characteristics of an ordinary resistance. This resistance is the interbase resistance R_{BB} and has values in the range 4.7 to 9.1 kΩ. The static characteristics of a UJT are shown in Fig. 4-28b.

When the dc supply voltage V_s is applied, the capacitor C is charged through resistor R since the emitter circuit of the UJT is in the open state. The time constant of the charging circuit is $\tau_1 = RC$. When the emitter voltage V_E, which is the same as the capacitor voltage v_C, reaches the *peak voltage* V_p, the UJT turns on and capacitor C will discharge through R_{B1} at a rate determined by the time constant $\tau_2 = R_{B1}C$. τ_2 is much smaller than τ_1. When the emitter voltage V_E decays to the valley point V_v, the emitter ceases to conduct, the UJT turns off, and the charging cycle is repeated. The waveforms of the emitter and triggering voltages are shown in Fig. 4-28c.

The waveform of the triggering voltage V_{B1} is identical to the discharging current of capacitor C. The triggering voltage V_{B1} should be designed to be

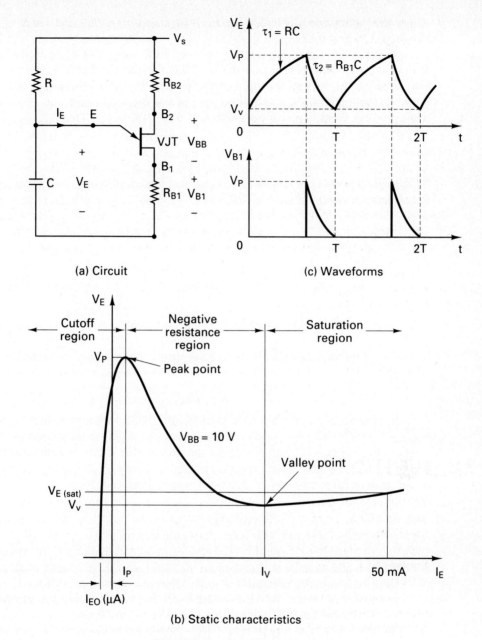

(a) Circuit

(c) Waveforms

(b) Static characteristics

Figure 4-28 UJT triggering circuit.

sufficiently large to turn on the SCR. The period of oscillation, T, is fairly independent of the dc supply voltage V_s, and is given by

$$T = \frac{1}{f} \approx RC \ln \frac{1}{1 - \eta} \tag{4-23}$$

where the parameter η is called the *intrinsic stand-off ratio*. The value of η lies between 0.51 and 0.82.

Resistor R is limited to a value between 3 kΩ and 3 MΩ. The upper limit on R is set by the requirement that the load line formed by R and V_s intersects the device characteristics to the right of the peak point but to the left of the valley point. If the load line fails to pass to the right of the peak point, the UJT will not turn on. This condition will be satisfied if $V_s - I_p R > V_p$. That is,

$$R < \frac{V_s - V_p}{I_p} \tag{4-24}$$

At the valley point $I_E = I_V$ and $V_E = V_v$ so that the condition for the lower limit on R to ensure turning off is $V_s - I_v R < V_v$. That is,

$$R > \frac{V_s - V_v}{I_v} \tag{4-25}$$

The recommended range of supply voltage V_s is from 10 to 35 V. For fixed values of η, the peak voltage V_p will vary with the voltage between the two bases, V_{BB}. V_p is given by

$$V_p = \eta V_{BB} + V_D (= 0.5 \text{ V}) \approx \eta V_s + V_D (= 0.5 \text{ V}) \tag{4-26}$$

where V_D is the one-diode forward voltage drop. The width t_g of triggering pulse is

$$t_g = R_{B1} C \tag{4-27}$$

In general, R_{B1} is limited to a value below 100 Ω, although values up to 2 or 3 kΩ are possible in some applications. A resistor R_{B2} is generally connected in series with base-two to compensate for the decrease in V_p due to temperature rise and to protect the UJT from possible thermal runaway. Resistor R_{B2} has a value of 100 Ω or greater and can be determined approximately from

$$R_{B2} = \frac{10^4}{\eta V_s} \tag{4-28}$$

Example 4-5
_____

Design the triggering circuit of Fig. 4-28a. The parameters of the UJT are $V_s = 30$ V, $\eta = 0.51$, $I_p = 10$ μA, $V_v = 3.5$ V, and $I_v = 10$ mA. The frequency of oscillation is $f = 60$ Hz, and the width of the triggering pulse is $t_g = 50$ μs.

Solution $T = 1/f = 1/60$ Hz $= 16.67$ ms. From Eq. (4-26), $V_p = 0.51 \times 30 + 0.5 = 15.8$ V. Let $C = 0.5$ μF. From Eqs. (4-24) and (4-25), the limiting values of R are

$$R < \frac{30 - 15.8}{10 \text{ } \mu A} = 1.42 \text{ M}\Omega$$

$$R > \frac{30 - 3.5}{10 \text{ mA}} = 2.65 \text{ k}\Omega$$

From Eq. (4-23), 16.67 ms $= R \times 0.5$ μF $\times \ln[1/(1 - 0.51)]$, which gives $R = 46.7$ kΩ, which falls within the limiting values. The peak gate voltage $V_{B1} = V_p = 15.8$ V.

From Eq. (4-27),

$$R_{B1} = \frac{t_g}{C} = \frac{50 \ \mu s}{0.5 \ \mu F} = 100 \ \Omega$$

From Eq. (4-28),

$$R_{B2} = \frac{10^4}{0.51 \times 30} = 654 \ \Omega$$

4-13 PROGRAMMABLE UNIJUNCTION TRANSISTOR

The programmable unijunction transistor (PUT) is a small thyristor shown in Fig. 4-29a. A PUT can be used as a relaxation oscillator as shown in Fig. 4-29b. The gate voltage V_G is maintained from the supply by the resistor divider R_1 and R_2, and determines the peak point voltage V_p. In the case of the UJT, V_p is fixed for a device by the dc supply voltage. But V_p of a PUT can be varied by varying the resistor divider R_1 and R_2. If the anode voltage V_A is less than the gate voltage V_G, the device will remain in its off-state. If V_A exceeds the gate voltage by one diode forward voltage V_D, the peak point is reached and the device turns on. The peak current I_p and the valley point current I_v both depend on the equivalent impedance on the gate $R_G = R_1 R_2/(R_1 + R_2)$ and the dc supply voltage V_s. In general, R_k is limited to a value below 100 Ω.

V_p is given by

$$V_p = \frac{R_2}{R_1 + R_2} V_s \qquad (4\text{-}29)$$

which gives the intrinsic ratio as

$$\eta = \frac{V_p}{V_s} = \frac{R_2}{R_1 + R_2} \qquad (4\text{-}30)$$

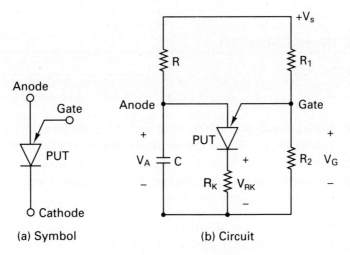

(a) Symbol   (b) Circuit

Figure 4-29 PUT triggering circuit.

R and C control the frequency along with R_1 and R_2. The period of oscillation T is given approximately by

$$T = \frac{1}{f} \approx RC \ln \frac{V_s}{V_s - V_p} = RC \ln \left(1 + \frac{R_2}{R_1}\right) \tag{4-31}$$

The gate current I_G at the valley point is given by

$$I_G = (1 - \eta) \frac{V_s}{R_G} \tag{4-32}$$

where $R_G = R_1 R_2 / (R_1 + R_2)$. R_1 and R_2 can be found from

$$R_1 = \frac{R_G}{\eta} \tag{4-33}$$

$$R_2 = \frac{R_G}{1 - \eta} \tag{4-34}$$

Example 4-6

Design the triggering circuit of Fig. 4-29b. The parameters of the PUT are $V_s = 30$ V and $I_G = 1$ mA. The frequency of oscillation is $f = 60$ Hz. The pulse width is $t_g = 50$ μs, and the peak triggering voltage is $V_{Rk} = 10$ V.

Solution $T = 1/f = 1/60$ Hz $= 16.67$ ms. The peak triggering voltage $V_{Rk} = V_p = 10$ V. Let $C = 0.5$ μF. From Eq. (4-27), $R_k = t_g/C = 50$ μs$/0.5$ μF $= 100$ Ω. From Eq. (4-30), $\eta = V_p/V_s = 10/30 = 1/3$. From Eq. (4-31), 16.67 ms $= R \times 0.5$ μF $\times \ln[30/(30 - 10)]$, which gives $R = 82.2$ kΩ. For $I_G = 1$ mA, Eq. (4-32) gives $R_G = (1 - \frac{1}{3}) \times 30/1$ mA $= 20$ kΩ. From Eq. (4-33),

$$R_1 = \frac{R_G}{\eta} = 20 \text{ k}\Omega \times \frac{3}{1} = 60 \text{ k}\Omega$$

From Eq. (4-34),

$$R_2 = \frac{R_G}{1 - \eta} = 20 \text{ k}\Omega \times \frac{3}{2} = 30 \text{ k}\Omega$$

4-14 SPICE THYRISTOR MODEL

Let us assume that the thyristor as shown in Fig. 4-30a is operated from an ac supply. This thyristor should exhibit the following characteristics.

1. It should switch to the on-state with the application of a small positive gate voltage, provided that the anode-to-cathode voltage is positive.
2. It should remain in the on-state as long as the anode current flows.
3. It should switch to the off-state when the anode current goes through zero toward the negative direction.

The switching action of the thyristor can be modeled by a voltage-controlled switch and a polynomial current source [14]. This is shown in Fig. 4-30b. The turn-on process can be explained by the following steps:

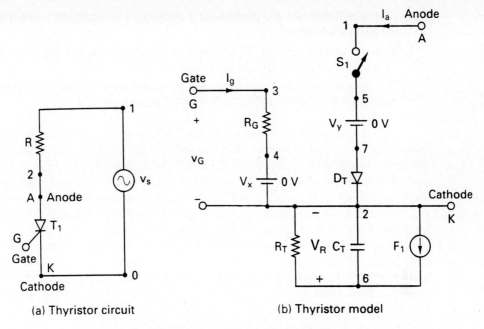

(a) Thyristor circuit **(b) Thyristor model**

Figure 4-30 SPICE thyristor model.

1. For a positive gate voltage V_g between nodes 3 and 2, the gate current is $I_g = I(VX) = V_g/R_G$.

2. The gate current I_g activates the current-controlled current source F_1 and produces a current of value $F_g = P_1I_g = P_1I(VX)$ such that $F_1 = F_g + F_a$.

3. The current source F_g produces a rapidly rising voltage V_R across resistance R_T.

4. As the voltage V_R increases above zero, the resistance R_S of the voltage-controlled switch S_1 decreases from R_{OFF} toward R_{ON}.

5. As the resistance R_S of switch S_1 decreases, the anode current $I_a = I(VY)$ increases, provided that the anode-to-cathode voltage is positive. This increasing anode current I_a produces a current $F_a = P_2I_a = P_2I(VY)$. This results in an increased value of voltage V_R.

6. This produces a regenerative condition with the switch rapidly being driven into low resistance (on-state). The switch remains on if the gate voltage V_g is removed.

7. The anode current I_a continues to flow as long as it is positive and the switch remains in the on-state.

During the turn-off, the gate current is off and $I_g = 0$. That is, $F_g = 0$, $F_1 = F_g + F_a = F_a$. The turn-off operation can be explained by the following steps:

1. As the anode current I_a goes negative, the current F_1 reverses provided that the gate voltage V_g is no longer present.

2. With a negative F_1, the capacitor C_T discharges through the current source F_1 and the resistance R_T.

3. With the fall of voltage V_R to a low level, the resistance R_S of switch S_1 increases from low (R_{ON}) to high (R_{OF}).

4. This is again a regenerative condition with the switch resistance being driven rapidly to R_{OFF} value as the voltage V_R becomes zero.

This model works well with a converter circuit in which the thyristor current falls to zero itself due to natural characteristics of the current. But for a full-wave ac–dc converter with a continuous load current discussed in Chapter 5, the current of a thyristor is diverted to another thyristor and this model may not give the true output. This problem can be remedied by adding diode D_T as shown in Fig. 4-30b. The diode prevents any reverse current flow through the thyristor resulting from the firing of another thyristor in the circuit.

This thyristor model can be used a subcircuit. Switch S_1 is controlled by the controlling voltage V_R connected between nodes 6 and 2. The switch and/or diode parameters can be adjusted to yield the desired on-state drop of the thyristor. We shall use diode parameters IS=2.2E−15, BV=1800V, TT=0, and switch parameters RON=0.0125, ROFF=10E+5, VON=0.5V, VOFF=0V. The subcircuit definition for the thyristor model *SCR* can be described as follows:

```
*     Subcircuit for ac thyristor model
.SUBCKT    SCR      1          2          3          2
*          model    anode    cathode    +control    −control
*          name                          voltage     voltage
S1     1    5    6    2    SMOD              ; Voltage-controlled switch
RG     3    4    50
VX     4    2    DC    OV
VY     5    7    DC    OV
DT     7    2    DMOD                        ; Switch diode
RT     6    2    1
CT     6    2    10UF
F1     2    6    POLY(2)    VX    VY    0    50    11
.MODEL  SMOD  VSWITCH (RON=0.0125 ROFF=10E+5 VON=0.5V VOFF=0V) ; Switch model
.MODEL  DMOD  D(IS=2.2E−15 BV=1800V TT=0)   ; Diode model parameters
.ENDS   SCR                                 ; Ends subcircuit definition
```

SUMMARY

There are nine types of thyristors. Only the GTOs, SITHs, and MCTs are gate-turn-off devices. Each type has advantages and disadvantages. The characteristics of practical thyristors differ significantly from that of ideal devices. Although there are various means of turning on thyristors, gate control is the most practical. Due to the junction capacitances and turn-on limit, thyristors must be protected from high di/dt and dv/dt failures. A snubber network is normally used to protect from high dv/dt. Due to the recovered charge, some energy is stored in the di/dt and stray inductors; and the devices must be protected from this stored energy. The switching losses of GTOs are much higher than those of normal SCRs. The snubber components of GTOs are critical to their performance.

Due to the differences in the characteristics of thyristors of the same type, the series and parallel operations of thyristors require voltage and current-sharing networks to protect them under steady-state and transient conditions. A means of isolation between the power circuit and gate circuits is necessary. A pulse transformer isolation is simple, but effective. For inductive loads, a pulse train reduces thyristor loss and is normally used for gating thyristors, instead of a continuous pulse. UJTs and PUTs are used for generating triggering pulses.

REFERENCES

1. General Electric, D. R. Grafham and F. B. Golden, eds., *SCR Manual*, 6th ed., Englewood Cliffs, N.J.: Prentice Hall, 1982.

2. D. Grant and A. Honda, *Applying International Rectifier's Gate Turn-Off Thyristors*, Application Note AN-315A. El Segundo, Calif.: International Rectifier.

3. C. K. Chu, P. B. Spisak, and D. A. Walczak, "High power asymmetrical thyristors." *IEEE Industry Applications Society Conference Record*, 1985, pp. 267–272.

4. O. Hashimoto, H. Kirihata, M. Watanabe, A. Nishiura, and S. Tagami, "Turn-on and turn-off characteristics of a 4.5-KV 3000-A gate turn-off thyristor." *IEEE Transactions on Industry Applications*, Vol. IA22, No. 3, 1986, pp. 478–482.

5. O. Hashimoto, Y. Takahashi, M. Watanabe, O. Yamada, and T. Fujihira, "2.5-kV, 2000-A monolithic gate turn-off thyristor." *IEEE Industry Applications Society Conference Record*, 1986, pp. 388–392.

6. E. Y. Ho and P. C. Sen, "Effect of gate drive on GTO thyristor characteristics." *IEEE Transactions on Industrial Electronics*, Vol. IE33, No. 3, 1986, pp. 325–331.

7. H. Fukui, H. Amano, and H. Miya, "Paralleling of gate turn-off thyristors." *IEEE Industry Applications Society Conference Record*, 1982, pp. 741–746.

8. Y. Nakamura, H. Tadano, M. Takigawa, I. Igarashi, and J. Nishizawa, "Very high speed static induction thyristor." *IEEE Transactions on Industry Applications*, Vol. IA22, No. 6, 1986, pp. 1000–1006.

9. V. A. K. Temple, "MOS controlled thyristors: a class of power devices." *IEEE Transactions on Electron Devices*, Vol. ED33, No. 10, 1986, pp. 1609–1618.

10. T. M. Jahns, R. W. De Donker, J. W. A. Wilson, V. A. K. Temple, and S. L. Watrous, "Circuit utilization characteristics of MOS-controlled thyristors." *Conference Record of the IEEE–IAS Annual Meeting*, San Diego, October 1989, pp. 1248–1254.

11. J. L. Hudgins, D. F. Blanco, S. Menhart, and W. M. Portnoy, "Comparison of the MCT and MOSFET for high frequency inverter." *Conference Record of the IEEE–IAS Annual Meeting*, San Diego, October 1989, pp. 1255–1259.

12. General Electric Company, *SCR Manual: Gate Trigger Characteristics, Ratings, and Methods*, 6th ed. Englewood Cliffs, N.J.: Prentice Hall, 1982.

13. Transistor Manual, *Unijunction Transistor Circuits*, 7th ed. Publication 450.37. Syracuse, N.Y.: General Electric Company, 1964.

14. L. J. Giacoletto, "Simple SCR and TRIAC PSPICE computer models." *IEEE Transactions on Industrial Electronics*, Vol. IE36, No. 3, 1989, pp. 451–455.

REVIEW QUESTIONS

4-1. What is the v–i characteristic of thyristors?

4-2. What is an off-state condition of thyristors?

4-3. What is an on-state condition of thyristors?

4-4. What is a latching current of thyristors?

4-5. What is a holding current of thyristors?

4-6. What is the two-transistor model of thyristors?

4-7. What are the means of turning-on thyristors?

4-8. What is turn-on time of thyristors?

4-9. What is the purpose of di/dt protection?

4-10. What is the common method of di/dt protection?

4-11. What is the purpose of dv/dt protection?

4-12. What is the common method of dv/dt protection?

4-13. What is turn-off time of thyristors?

4-14. What are the types of thyristors?

4-15. What is an SCR?

4-16. What is the difference between an SCR and a TRIAC?

4-17. What is the turn-off characteristic of thyristors?

4-18. What are the advantages and disadvantages of GTOs?

4-19. What are the advantages and disadvantages of SITHs?

4-20. What are the advantages and disadvantages of RCTs?

4-21. What are the advantages and disadvantages of LASCRs?

4-22. What is a snubber network?

4-23. What are the design considerations of snubber networks?

4-24. What is the common technique for voltage sharing of series-connected thyristors?

4-25. What are the common techniques for current sharing of parallel-connected thyristors?

4-26. What is the effect of reverse recovery time on the transient voltage sharing of parallel-connected thyristors?

4-27. What is a derating factor of series-connected thyristors?

4-28. What is a UJT?

4-29. What is the peak voltage of a UJT?

4-30. What is the valley point voltage of a UJT?

4-31. What is the intrinsic stand-off ratio of a UJT?

4-32. What is a PUT?

4-33. What are the advantages of a PUT over a UJT?

PROBLEMS

4-1. The junction capacitance of a thyristor can be assumed to be independent of off-state voltage. The limiting value of charging current to turn on the thyristor is 12 mA. If the critical value of dv/dt is 800 V/μs, determine the junction capacitance.

4-2. The junction capacitance of a thyristor is $C_{J2} = 20$ pF and can be assumed independent of off-state voltage. The limiting value of charging current to turn on the thyristor is 15 mA. If a capacitor of 0.01 μF is connected across the thyristor, determine the critical value of dv/dt.

4-3. A thyristor circuit is shown in Fig. P4-3. The junction capacitance of thyristor is $C_{J2} = 15$ pF and can be assumed to be independent of the off-state voltage. The limiting value of charging current to turn

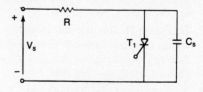

Figure P4-3

on the thyristor is 5 mA and the critical value of dv/dt is 200 V/μs. Determine the value of capacitance C_s so that the thyristor will not be turned on due to dv/dt.

4-4. The input voltage in Fig. 4-9e is $V_s = 200$ V with a load resistance of $R = 10$ Ω and a load inductance of $L = 50$ μH. If the damping ratio is 0.7 and the discharging current of the capacitor is 5 A, determine

(a) the values of R_s and C_s, and (b) the maximum dv/dt.

4-5. Repeat Prob. 4-4 if the input voltage is ac, $v_s = 179 \sin 377t$.

4-6. A thyristor carries a current as shown in Fig. P4-6. The switching frequency is $f_s = 10$ Hz. Determine the average on-state current I_T.

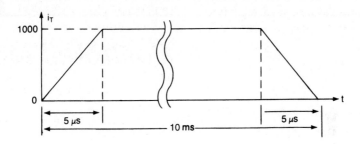

Figure P4-6

4-7. A string of thyristors is connected in series to withstand a dc voltage of $V_s = 15$ kV. The maximum leakage current and recovery charge differences of thyristors are 10 mA and 150 μC, respectively. A derating factor of 20% is applied for the steady-state and transient voltage sharings of thyristors. If the maximum steady-state voltage sharing is 1000 V, determine **(a)** the steady-state voltage-sharing resistance R for each thyristor, and **(b)** the transient voltage capacitance C_1 for each thyristor.

4-8. Two thyristors are connected in parallel to share a total load current of $I_L = 600$ A. The on-state voltage drop of one thyristor is $V_{T1} = 1.0$ at 300 A and that of other

thyristors is $V_{T2} = 1.5$ V at 300 A. Determine the values of series resistances to force current sharing with 10% difference. Total voltage $v = 2.5$ V.

4-9. Design the triggering circuit of Fig. 4-28a. The parameters of the UJT are $V_s = 20$ V, $\eta = 0.66$, $I_p = 10$ μA, $V_v = 2.5$ V, and $I_v = 10$ mA. The frequency of oscillation is $f = 1$ kHz, and the width of the gate pulse is $t_g = 40$ μs.

4-10. Design the triggering circuit of Fig. 4-29b. The parameters of the PUT are $V_s = 20$ V and $I_G = 1.5$ mA. The frequency of oscillation is $f = 1$ kHz. The pulse width is $t_g = 40$ μs, and the peak triggering pulse is $V_{Rk} = 8$ V.

5

Controlled rectifiers

5-1 INTRODUCTION

We have seen in Chapter 3 that diode rectifiers provide a fixed output voltage only. To obtain controlled output voltages, phase control thyristors are used instead of diodes. The output voltage of thyristor rectifiers is varied by controlling the delay or firing angle of thyristors. A phase-control thyristor is turned on by applying a short pulse to its gate and turned off due to *natural or line commutation*; and in case of a highly inductive load, it is turned off by firing another thyristor of the rectifier during the negative half-cycle of input voltage.

These phase-controlled rectifiers are simple and less expensive; and the efficiency of these rectifiers is, in general, above 95%. Since these rectifiers convert from ac to dc, these controlled rectifiers are also called *ac–dc converters* and are used extensively in industrial applications, especially in variable-speed drives, ranging from fractional horsepower to megawatt power level.

The phase-control converters can be classified into two types, depending on the input supply: (1) single-phase converters, and (2) three-phase converters. Each type can be subdivided into (a) semiconverter, (b) full converter, and (c) dual converter. A *semiconverter* is a one-quadrant converter and it has one polarity of output voltage and current. A *full converter* is a two-quadrant converter and the polarity of its output voltage can be either positive or negative. However, the output current of full converter has one polarity only. A *dual converter* can operate in four quadrants; and both the output voltage and current can be either positive or negative. In some applications, converters are connected in series to operate at higher voltages and to improve the input power factor.

The method of Fourier series similar to that of diode rectifiers can be applied to analyze the performances of phase-controlled converters with *RL* loads. However, to simplify the analysis, the load inductance can be assumed sufficiently high so that the load current is continuous and has negligible ripple.

130

Let us consider the circuit in Fig. 5-1a with a resistive load. During the positive half-cycle of input voltage, the thyristor anode is positive with respect to its cathode and the thyristor is said to be *forward biased*. When thyristor T_1 is fired at $\omega t = \alpha$, thyristor T_1 conducts and the input voltage appears across the load. When the input voltage starts to be negative at $\omega t = \pi$, the thyristor anode is negative with respect to its cathode and thyristor T_1 is said to be *reverse biased*; and it is turned off. The time after the input voltage starts to go positive until the thyristor is fired at $\omega t = \alpha$ is called the *delay or firing angle α*.

Figure 5-1b shows the region of converter operation, where the output voltage and current have one polarity. Figure 5-1c shows the waveforms for input voltage, output voltage, load current, and voltage across T_1. This converter is not normally used in industrial applications because its output has high ripple content and low ripple frequency. If f_s is the frequency of input supply, the lowest frequency of output ripple voltage is f_s.

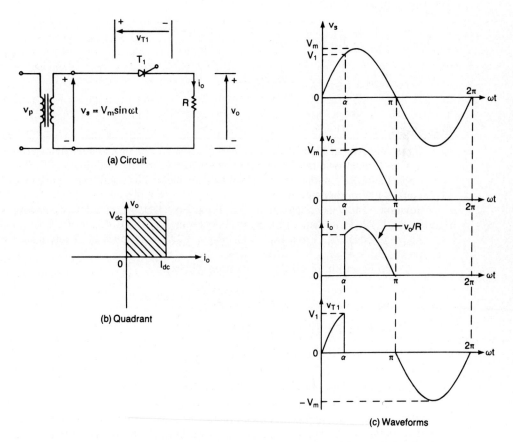

(a) Circuit

(b) Quadrant

(c) Waveforms

Figure 5-1   Single-phase thyristor converter with a resistive load.

If V_m is the peak input voltage, the average output voltage V_{dc} can be found from

$$V_{dc} = \frac{1}{2\pi} \int_\alpha^\pi V_m \sin \omega t \, d(\omega t) = \frac{V_m}{2\pi} [-\cos \omega t]_\alpha^\pi$$

$$= \frac{V_m}{2\pi} (1 + \cos \alpha) \tag{5-1}$$

and V_{dc} can be varied from V_m/π to 0 by varying α from 0 to π. The average output voltage becomes maximum when $\alpha = 0$ and the maximum output voltage V_{dm} is

$$V_{dm} = \frac{V_m}{\pi} \tag{5-2}$$

Normalizing the output voltage with respect to V_{dm}, the normalized output voltage

$$V_n = \frac{V_{dc}}{V_{dm}} = 0.5(1 + \cos \alpha) \tag{5-3}$$

The rms output voltage is given by

$$V_{rms} = \left[\frac{1}{2\pi} \int_\alpha^\pi V_m^2 \sin^2 \omega t \, d(\omega t) \right]^{1/2} = \left[\frac{V_m^2}{4\pi} \int_\alpha^\pi (1 - \cos 2\omega t) \, d(\omega t) \right]^{1/2}$$

$$= \frac{V_m}{2} \left[\frac{1}{\pi} \left(\pi - \alpha + \frac{\sin 2\alpha}{2} \right) \right]^{1/2} \tag{5-4}$$

Example 5-1

If the converter of Figure 5-1a has a purely resistive load of R and the delay angle is $\alpha = \pi/2$, determine (a) the rectification efficiency, (b) the form factor FF, (c) the ripple factor RF, (d) the transformer utilization factor TUF, and (e) the peak inverse voltage PIV of thyristor T_1.

Solution The delay angle, $\alpha = \pi/2$. From Eq. (5-1), $V_{dc} = 0.1592V_m$ and $I_{dc} = 0.1592V_m/R$. From Eq. (5-3), $V_n = 0.5$. From Eq. (5-4), $V_{rms} = 0.3536V_m$ and $I_{rms} = 0.3536V_m/R$. From Eq. (3-42), $P_{dc} = V_{dc}I_{dc} = (0.1592V_m)^2/R$ and from Eq. (3-43), $P_{ac} = V_{rms}I_{rms} = (0.3536V_m)^2/R$.

(a) From Eq. (3-44) the rectification efficiency

$$\eta = \frac{(0.1592V_m)^2}{(0.3536V_m)^2} = 20.27\%$$

(b) From Eq. (3-46), the form factor

$$FF = \frac{0.3536V_m}{0.1592V_m} = 2.221 \quad \text{or} \quad 222.1\%$$

(c) From Eq. (3-48), the ripple factor RF = $(2.221^2 - 1)^{1/2} = 1.983$ or 198.3%.

(d) The rms voltage of the transformer secondary, $V_s = V_m/\sqrt{2} = 0.707V_m$. The rms value of the transformer secondary current is the same as that of the load, $I_s = 0.3536V_m/R$. The volt-ampere rating (VA) of the transformer, VA = $V_sI_s = 0.707V_m \times 0.3536V_m/R$. From Eq. (3-49),

$$\text{TUF} = \frac{0.1592^2}{0.707 \times 0.3536} = 0.1014 \qquad \text{and} \qquad \frac{1}{\text{TUF}} = 9.86$$

(e) The peak inverse voltage PIV $= V_m$.

Note. The performance of the converter is degraded at the lower range of delay angle α.

5-3 SINGLE-PHASE SEMICONVERTERS

The circuit arrangement of a single-phase semiconverter is shown in Fig. 5-2a with a highly inductive load. The load current is assumed continuous and ripple free. During the positive half-cycle, thyristor T_1 is forward biased. When thyristor T_1 is fired at $\omega t = \alpha$, the load is connected to the input supply through T_1 and D_2 during the period $\alpha \le \omega t \le \pi$. During the period from $\pi \le \omega t \le (\pi + \alpha)$, the input voltage is negative and the freewheeling diode D_m is forward biased. D_m conducts to provide the continuity of current in the inductive load. The load current is transferred from T_1 and D_2 to D_m; and thyristor T_1 and diode D_2 are turned off. During the negative half-cycle of input voltage, thyristor T_2 is forward biased, and the firing of thyristor T_2 at $\omega t = \pi + \alpha$ will reverse bias D_m. The diode D_m is turned off and the load is connected to the supply through T_2 and D_1.

Figure 5-2b shows the region of converter operation, where both the output voltage and current have positive polarity. Figure 5-2c shows the waveforms for the input voltage, output voltage, input current, and currents through T_1, T_2, D_1, and D_2. This converter has a better power factor due to the freewheeling diode and is commonly used in applications up to 15 kW, where one-quadrant operation is acceptable.

The average output voltage can be found from

$$V_{\text{dc}} = \frac{2}{2\pi} \int_\alpha^\pi V_m \sin \omega t \, d(\omega t) = \frac{2V_m}{2\pi} [-\cos \omega t]_\alpha^\pi$$

$$= \frac{V_m}{\pi} (1 + \cos \alpha) \tag{5-5}$$

and V_{dc} can be varied from $2V_m/\pi$ to 0 by varying α from 0 to π. The maximum average output voltage is $V_{dm} = 2V_m/\pi$ and the normalized average output voltage is

$$V_n = \frac{V_{\text{dc}}}{V_{dm}} = 0.5(1 + \cos \alpha) \tag{5-6}$$

The rms output voltage is found from

$$V_{\text{rms}} = \left[\frac{2}{2\pi} \int_\alpha^\pi V_m^2 \sin^2 \omega t \, d(\omega t) \right]^{1/2} = \left[\frac{V_m^2}{2\pi} \int_\alpha^\pi (1 - \cos 2\omega t) \, d(\omega t) \right]^{1/2}$$

$$= \frac{V_m}{\sqrt{2}} \left[\frac{1}{\pi} \left(\pi - \alpha + \frac{\sin 2\alpha}{2} \right) \right]^{1/2} \tag{5-7}$$

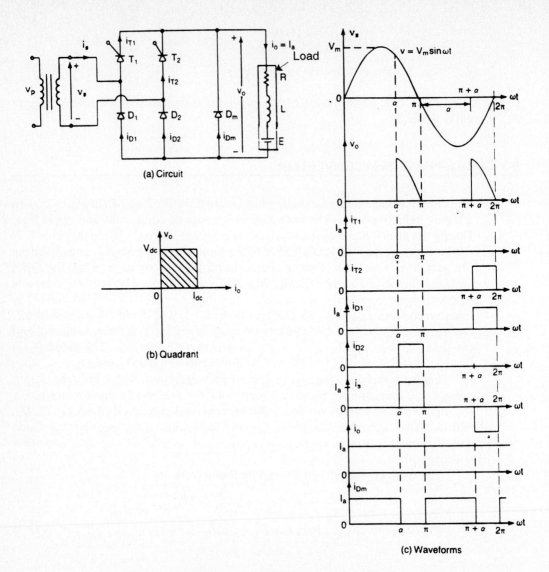

(a) Circuit

(b) Quadrant

(c) Waveforms

Figure 5-2   Single-phase semiconverter.

Example 5-2

The semiconverter in Fig. 5-2a is connected to a 120-V 60-Hz supply. The load current I_a can be assumed to be continuous and its ripple content is negligible. The turns ratio of the transformer is unity. (a) Express the input current in a Fourier series; determine the harmonic factor of input current HF, displacement factor DF, and input power factor PF. (b) If the delay angle is $\alpha = \pi/2$, calculate V_{dc}, V_n, V_{rms}, HF, DF, and PF.

Solution   (a) The waveform for input current is shown in Fig. 5-2c and the instantaneous input current can be expressed in a Fourier series as

$$i_s(t) = I_{dc} + \sum_{n=1,2,\ldots}^{\infty} (a_n \cos n\omega t + b_n \sin n\omega t) \tag{5-8}$$

where

$$I_{dc} = \frac{1}{2\pi} \int_\alpha^{2\pi} i_s(t)\ d(\omega t) = \frac{1}{2\pi} \left[\int_\alpha^\pi I_a\ d(\omega t) - \int_{\pi+\alpha}^{2\pi} I_a\ d(\omega t) \right] = 0$$

$$a_n = \frac{1}{\pi} \int_\alpha^{2\pi} i_s(t)\ \cos n\omega t\ d(\omega t)$$

$$= \frac{1}{\pi} \left[\int_\alpha^\pi I_a \cos n\omega t\ d(\omega t) - \int_{\pi+\alpha}^{2\pi} I_a \cos n\omega t\ d(\omega t) \right]$$

$$= -\frac{2I_a}{n\pi} \sin n\alpha \qquad \text{for } n = 1, 3, 5, \ldots$$

$$= 0 \qquad \text{for } n = 2, 4, 6, \ldots$$

$$b_n = \frac{1}{\pi} \int_\alpha^{2\pi} i_s(t)\ \sin n\omega t\ d(\omega t)$$

$$= \frac{1}{\pi} \left[\int_\alpha^\pi I_a \sin n\omega t\ d(\omega t) - \int_{\pi+\alpha}^{2\pi} I_a \sin n\omega t\ d(\omega t) \right]$$

$$= \frac{2I_a}{n\pi} (1 + \cos n\alpha) \qquad \text{for } n = 1, 3, 5, \ldots$$

$$= 0 \qquad \text{for } n = 2, 4, 6, \ldots$$

Since $I_{dc} = 0$, Eq. (5-8) can be written as

$$i_s(t) = \sum_{n=1,3,5,\ldots}^\infty \sqrt{2}\, I_n \sin(n\omega t + \phi_n) \tag{5-9}$$

where

$$\phi_n = \tan^{-1} \frac{a_n}{b_n} = -\frac{n\alpha}{2} \tag{5-10}$$

The rms value of the nth harmonic component of the input current is derived as

$$I_{sn} = \frac{1}{\sqrt{2}} (a_n^2 + b_n^2)^{1/2} = \frac{2\sqrt{2}\, I_a}{n\pi} \cos \frac{n\alpha}{2} \tag{5-11}$$

From Eq. (5-11), the rms value of the fundamental current is

$$I_{s1} = \frac{2\sqrt{2}\, I_a}{\pi} \cos \frac{\alpha}{2}$$

The rms input current can be calculated from Eq. (5-11) as

$$I_s = \left(\sum_{n=1,2,\ldots}^\infty I_{sn} \right)^{1/2}$$

I_s can also be determined directly from

$$I_s = \left[\frac{2}{2\pi} \int_\alpha^\pi I_a^2\ d(\omega t) \right]^{1/2} = I_a \left(1 - \frac{\alpha}{\pi} \right)^{1/2}$$

From Eq. (3-51), HF = $[(I_s/I_{s1})^2 - 1]^{1/2}$ or

$$\text{HF} = \left[\frac{\pi(\pi - \alpha)}{4(1 + \cos \alpha)} - 1 \right]^{1/2} \tag{5-12}$$

From Eqs. (3-50) and (5-10),

$$\text{DF} = \cos \phi_1 = \cos \left(-\frac{\alpha}{2} \right) \tag{5-13}$$

From Eq. (3-52),

$$\text{PF} = \frac{I_{s1}}{I_s} \cos \frac{\alpha}{2} = \frac{\sqrt{2}\,(1 + \cos \alpha)}{[\pi(\pi - \alpha)]^{1/2}} \tag{5-14}$$

(b) $\alpha = \pi/2$ and $V_m = \sqrt{2} \times 120 = 169.7$ V. From Eq. (5-5), $V_{dc} = (V_m/\pi)(1 + \cos \alpha) = 54.02$ V, from Eq. (5-6), $V_n = 0.5$ pu, and from Eq. (5-7),

$$V_{\text{rms}} = \frac{V_m}{\sqrt{2}} \left[\frac{1}{\pi} \left(\pi - \alpha + \frac{\sin 2\alpha}{2} \right) \right]^{1/2} = 84.57 \text{ V}$$

$$I_{s1} = \frac{2\sqrt{2}\,I_a}{\pi} \cos \frac{\pi}{4} = 0.6366 I_a$$

$$I_s = I_a \left(1 - \frac{\alpha}{\pi} \right)^{1/2} = 0.7071 I_a$$

$$\text{HF} = \left[\left(\frac{I_s}{I_{s1}} \right)^2 - 1 \right]^{1/2} = 0.4835 \quad \text{or} \quad 48.35\%$$

$$\phi_1 = -\frac{\pi}{4} \quad \text{and} \quad \text{DF} = \cos \left(-\frac{\pi}{2} \right) = 0.7071$$

$$\text{PF} = \frac{I_{s1}}{I_s} \cos \frac{\alpha}{2} = 0.6366 \text{ (lagging)}$$

Note. The performance parameters of the converter depend on the delay angle α.

5-3.1 Single-Phase Semiconverter with *RL* Load

In practice, a load has a finite inductance. The load current depends on the values of load resistance R, load inductance L and batttery Volgate E as shown in Fig. 5-2(a). The converter operation can be divided into two modes: mode 1 and mode 2.

Mode 1. This mode is valid for $0 \leq \omega t \leq \alpha$, during which the freewheeling diode D_m conducts. The load current i_{L1} during mode 1 is described by

$$L \frac{di_{L1}}{dt} + R i_{L1} + E = 0 \tag{5-15}$$

which, with initial condition $i_{L1}(\omega t = 0) = I_{Lo}$ in the steady state, gives

$$i_{L1} = I_{Lo} e^{-(R/L)t} - \frac{E}{R} (1 - e^{-(R/L)t}) \quad \text{for } i_{L1} \geq 0 \tag{5-16}$$

At the end of this mode at $\omega t = \alpha$, the load current becomes I_{L1}. That is,

$$I_{L1} = i_{L1}(\omega t = \alpha) = I_{Lo}e^{-(R/L)(\alpha/\omega)} - \frac{E}{R}[1 - e^{-(R/L)(\alpha/\omega)}] \qquad \text{for } I_{L1} \geq 0 \qquad (5\text{-}17)$$

Mode 2. This mode is valid for $\alpha \leq \omega t \leq \pi$ while thyristor T_1 conducts. If $v_s = \sqrt{2} \, V_s \sin \omega t$ is the input voltage, the load current i_{L2} during mode 2 can be found from

$$L\frac{di_{L2}}{dt} + Ri_{L2} + E = \sqrt{2} \, V_s \sin \omega t \qquad (5\text{-}18)$$

whose solution is of the form

$$i_{L2} = \frac{\sqrt{2} \, V_s}{Z} \sin(\omega t - \theta) + A_1 e^{-(R/L)t} - \frac{E}{R} \qquad \text{for } i_{L2} \geq 0$$

where load impedance $Z = [R^2 + (\omega L)^2]^{1/2}$ and load impedance angle $\theta = \tan^{-1}(\omega L/R)$.

Constant A_1, which can be determined from the initial condition: at $\omega t = \alpha$, $i_{L2} = I_{L1}$, is found as

$$A_1 = \left[I_{L1} + \frac{E}{R} - \frac{\sqrt{2} \, V_s}{Z} \sin(\alpha - \theta) \right] e^{(R/L)(\alpha/\omega)}$$

Substitution of A_1 yields

$$i_{L2} = \frac{\sqrt{2} \, V_s}{Z} \sin(\omega t - \theta) - \frac{E}{R} + \left[I_{L1} + \frac{E}{R} - \frac{\sqrt{2} \, V_s}{Z} \sin(\alpha - \theta) \right] e^{(R/L)(\alpha/\omega - t)}$$

$$\text{for } i_{L2} \geq 0 \qquad (5\text{-}19)$$

At the end of mode 2 in the steady-state condition: $I_{L2}(\omega t = \pi) = I_{Lo}$. Applying this condition to Eq. (5-16) and solving for I_{Lo}, we get

$$I_{Lo} = \frac{\sqrt{2} \, V_s}{Z} \frac{\sin(\pi - \theta) - \sin(\alpha - \theta)e^{(R/L)(\alpha - \pi)/\omega}}{1 - e^{-(R/L)(\pi/\omega)}} - \frac{E}{R}$$

$$\text{for } I_{Lo} \geq 0 \text{ and } \theta \leq \alpha \leq \pi \qquad (5\text{-}20)$$

The rms current of a thyristor can be found from Eq. (5-19) as

$$I_R = \left[\frac{1}{2\pi} \int_\alpha^\pi i_{L2}^2 \, d(\omega t) \right]^{1/2}$$

The average current of a thyristor can also be found from Eq. (5-19) as

$$I_A = \frac{1}{2\pi} \int_\alpha^\pi i_{L2} \, d(\omega t)$$

The rms output current can be found from Eqs. (5-16) and (5-19) as

$$I_{\text{rms}} = \left[\frac{1}{2\pi} \int_0^\alpha i_{L1}^2 \, d(\omega t) + \frac{1}{2\pi} \int_\alpha^\pi i_{L2}^2 \, d(\omega t) \right]^{1/2}$$

The average output current can be found from Eqs. (5-16) and (5-19) as

$$I_{dc} = \frac{1}{2\pi} \int_0^\alpha i_{L1} \, d(\omega t) + \frac{1}{2\pi} \int_\alpha^\pi i_{L2} \, d(\omega t)$$

Example 5-3*

The single-phase semiconverter of Fig. 5-2a has an RL load of $L = 6.5$ mH, $R = 2.5 \, \Omega$, and $E = 10$ V. The input voltage is $V_s = 120$ V (rms) at 60 Hz. Determine (a) the load current I_{Lo} at $\omega t = 0$, and the load current I_{L1} at $\omega t = \alpha = 60°$, (b) the average thyristor current I_A, (c) the rms thyristor current I_R, (d) the rms output current I_{rms}, and (e) the average output current I_{dc}.

Solution $R = 2.5 \, \Omega$, $L = 6.5$ mH, $f = 60$ Hz, $\omega = 2\pi \times 60 = 377$ rad/s, $V_s = 120$ V, $\theta = \tan^{-1}(\omega L/R) = 44.43°$, and $Z = 3.5 \, \Omega$.

(a) The steady-state load current at $\omega t = 0$, $I_{Lo} = 29.77$ A. The steady-state load current at $\omega t = \alpha$, $I_{L1} = 7.6$ A.

(b) The numerical integration of i_{L2} in Eq. (5-19) yields the average thyristor current as $I_A = 11.42$ A.

(c) By numerical integration of i_{L2}^2 between the limits $\omega t = \alpha$ to π, we get the rms thyristor current as $I_R = 20.59$ A.

(d) The rms output current $I_{rms} = 30.92$ A.

(e) The average output current $I_{dc} = 28.45$ A.

5-4 SINGLE-PHASE FULL CONVERTERS

The circuit arrangement of a single-phase full converter is shown in Fig. 5-3a with a highly inductive load so that the load current is continuous and ripple free. During the positive half-cycle, thyristors T_1 and T_2 are forward biased; and when these two thyristors are fired simultaneously at $\omega t = \alpha$, the load is connected to the input supply through T_1 and T_2. Due to the inductive load, thyristors T_1 and T_2 will continue to conduct beyond $\omega t = \pi$, even though the input voltage is already negative. During the negative half-cycle of the input voltage, thyristors T_3 and T_4 are forward biased; and firing of thyristors T_3 and T_4 will apply the supply voltage across thyristors T_1 and T_2 as reverse blocking voltage. T_1 and T_2 will be turned off due to *line* or *natural commutation* and the load current will be transferred from T_1 and T_2 to T_3 and T_4. Figure 5-3b shows the regions of converter operation and Fig. 5-3c shows the waveforms for input voltage, output voltage, and input and output currents.

During the period from α to π, the input voltage v_s and input current i_s are positive; and the power flows from the supply to the load. The converter is said to be operated in *rectification* mode. During the period from π to $\pi + \alpha$, the input voltage v_s is negative and the input current i_s is positive; and there will be reverse power flow from the load to the supply. The converter is said to be operated in *inversion mode*. This converter is extensively used in industrial applications up to 15 kW. Depending on the value of α, the average output voltage could be either positive or negative and it provides two-quadrant operation.

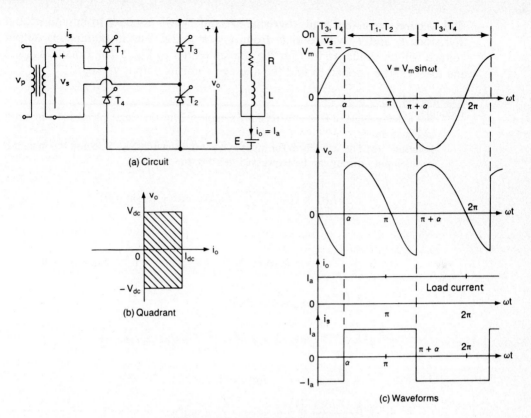

(a) Circuit

(b) Quadrant

(c) Waveforms

Figure 5-3 Single-phase full converter.

The average output voltage can be found from

$$V_{dc} = \frac{2}{2\pi} \int_{\alpha}^{\pi+\alpha} V_m \sin \omega t \, d(\omega t) = \frac{2V_m}{2\pi} \left[-\cos \omega t \right]_{\alpha}^{\pi+\alpha}$$

$$= \frac{2V_m}{\pi} \cos \alpha \tag{5-21}$$

and V_{dc} can be varied from $2V_m/\pi$ to $-2V_m/\pi$ by varying α from 0 to π. The maximum average output voltage is $V_{dm} = 2V_m/\pi$ and the normalized average output voltage is

$$V_n = \frac{V_{dc}}{V_{dm}} = \cos \alpha \tag{5-22}$$

The rms value of the output voltage is given by

$$V_{rms} = \left[\frac{2}{2\pi} \int_{\alpha}^{\pi+\alpha} V_m^2 \sin^2 \omega t \, d(\omega t) \right]^{1/2} = \left[\frac{V_m^2}{2\pi} \int_{\alpha}^{\pi+\alpha} (1 - \cos 2\omega t) \, d(\omega t) \right]^{1/2}$$

$$= \frac{V_m}{\sqrt{2}} = V_s \tag{5-23}$$

With a purely resistive load, thyristors T_1 and T_2 will conduct from α to π, and thyristors T_3 and T_4 will conduct from $\alpha + \pi$ to 2π. The instantaneous output voltage will be similar to that for the semiconverter in Fig. 5-2b. Equations (5-5) and (5-7) can be applied to find the average and rms output voltages.

Example 5-4

For a delay angle of $\alpha = \pi/3$, repeat Example 5-2 for the single-phase full converter in Fig. 5-3a.

Solution (a) The waveform for input current is shown in Fig. 5-3c and the instantaneous input current can be expressed in a Fourier series as

$$i_s(t) = I_{dc} + \sum_{n=1,2,\ldots}^{\infty} (a_n \cos n\omega t + b_n \sin n\omega t)$$

where

$$I_{dc} = \frac{1}{2\pi} \int_\alpha^{2\pi+\alpha} i_s(t)\, d(\omega t) = \frac{1}{2\pi} \left[\int_\alpha^{\pi+\alpha} I_a\, d(\omega t) - \int_{\pi+\alpha}^{2\pi+\alpha} I_a\, d(\omega t) \right] = 0$$

$$a_n = \frac{1}{\pi} \int_\alpha^{2\pi+\alpha} i_s(t) \cos n\omega t\, d(\omega t)$$

$$= \frac{1}{\pi} \left[\int_\alpha^{\pi+\alpha} I_a \cos n\omega t\, d(\omega t) - \int_{\pi+\alpha}^{2\pi+\alpha} I_a \cos n\omega t\, d(\omega t) \right]$$

$$= -\frac{4I_a}{n\pi} \sin n\alpha \qquad \text{for } n = 1,\, 3,\, 5,\, \ldots$$

$$= 0 \qquad \text{for } n = 2,\, 4,\, \ldots$$

$$b_n = \frac{1}{\pi} \int_\alpha^{2\pi+\alpha} i(t) \sin n\omega t\, d(\omega t)$$

$$= \frac{1}{\pi} \left[\int_\alpha^{\pi+\alpha} I_a \sin n\omega t\, d(\omega t) - \int_{\pi+\alpha}^{2\pi+\alpha} I_a \sin n\omega t\, d(\omega t) \right]$$

$$= \frac{4I_a}{n\pi} \cos n\alpha \qquad \text{for } n = 1,\, 3,\, 5,\, \ldots$$

$$= 0 \qquad \text{for } n = 2,\, 4,\, \ldots$$

Since $I_{dc} = 0$, the input current can be written as

$$i_s(t) = \sum_{n=1,3,5,\ldots}^{\infty} \sqrt{2}\, I_n \sin(n\omega t + \phi_n)$$

where

$$\phi_n = \tan^{-1} \frac{a_n}{b_n} = -n\alpha \tag{5-24}$$

and ϕ_n is the displacement angle of the nth harmonic current. The rms value of the nth harmonic input current is

$$I_{sn} = \frac{1}{\sqrt{2}} (a_n^2 + b_n^2)^{1/2} = \frac{4I_a}{\sqrt{2}\, n\pi} = \frac{2\sqrt{2}\, I_a}{n\pi} \tag{5-25}$$

and the rms value of the fundamental current is

$$I_{s1} = \frac{2\sqrt{2}\, I_a}{\pi}$$

The rms value of the input current can be calculated from Eq. (5-25) as

$$I_s = \left(\sum_{n=1,3,5,\ldots}^{\infty} I_{sn}^2 \right)^{1/2}$$

I_s can also be determined directly from

$$I_s = \left[\frac{2}{2\pi} \int_{\alpha}^{\pi+\alpha} I_a^2 \, d(\omega t) \right]^{1/2} = I_a$$

From Eq. (3-51) the harmonic factor is found as

$$\mathrm{HF} = \left[\left(\frac{I_s}{I_{s1}} \right)^2 - 1 \right]^{1/2} = 0.483 \quad \text{or} \quad 48.3\%$$

From Eqs. (3-50) and (5-24), the displacement factor

$$\mathrm{DF} = \cos \phi_1 = \cos(-\alpha) \tag{5-26}$$

From Eq. (3-52) the power factor is found as

$$\mathrm{PF} = \frac{I_{s1}}{I_s} \cos(-\alpha) = \frac{2\sqrt{2}}{\pi} \cos \alpha \tag{5-27}$$

(b) $\alpha = \pi/3$.

$$V_{dc} = \frac{2V_m}{\pi} \cos \alpha = 54.02 \text{ V} \qquad \text{and} \qquad V_n = 0.5 \text{ pu}$$

$$V_{rms} = \frac{V_m}{\sqrt{2}} = V_s = 120 \text{ V}$$

$$I_{s1} = \left(2\sqrt{2}\, \frac{I_a}{\pi} \right) = 0.90032 I_a \qquad \text{and} \qquad I_s = I_a$$

$$\mathrm{HF} = \left[\left(\frac{I_s}{I_{s1}} \right)^2 - 1 \right]^{1/2} = 0.4834 \quad \text{or} \quad 48.34\%$$

$$\phi_1 = -\alpha \qquad \text{and} \qquad \mathrm{DF} = \cos -\alpha = \cos \frac{-\pi}{3} = 0.5$$

$$\mathrm{PF} = \frac{I_{s1}}{I_s} \cos(-\alpha) = 0.45 \text{ (lagging)}$$

Note. The fundamental component of input current is always 90.03% of I_a and the harmonic factor remains constant at 48.34%.

5-4.1 Single-Phase Full Converter with *RL* Load

The operation of the converter in Fig. 5-3a can be divided into two identical modes: mode 1 when T_1 and T_2 conduct, and mode 2 when T_3 and T_4 conduct. The

---

output currents during these modes are similar and we need to consider only one mode to find the output current i_L.

Mode 1 is valid for $\alpha \leq \omega t \leq (\alpha + \pi)$. If $v_s = \sqrt{2}\, V_s \sin \omega t$ is the input voltage, Eq. (5-18) can be solved with the initial condition: at $\omega t = \alpha$, $i_L = I_{Lo}$. Equation (4-19) gives i_L as

$$i_L = \frac{\sqrt{2}\, V_s}{Z} \sin(\omega t - \theta) - \frac{E}{R}$$

$$+ \left[I_{Lo} + \frac{E}{R} - \frac{\sqrt{2}\, V_s}{Z} \sin(\alpha - \theta) \right] e^{(R/L)(\alpha/\omega - t)}$$

(5-28)

At the end of mode 1 in the steady-state condition $i_L(\omega t = \pi + \alpha) = I_{L1} = I_{Lo}$. Applying this condition to Eq. (5-28) and solving for I_{Lo}, we get

$$I_{Lo} = I_{L1} = \frac{\sqrt{2}\, V_s}{Z} \frac{-\sin(\alpha - \theta) - \sin(\alpha - \theta)e^{-(R/L)(\pi)/\omega}}{1 - e^{-(R/L)(\pi/\omega)}} - \frac{E}{R}$$

$$\text{for } I_{Lo} \geq 0 \qquad (5\text{-}29)$$

The critical value of α at which I_o becomes zero can be solved for known values of θ, R, L, E, and V_s by an iterative method. The rms current of a thyristor can be found from Eq. (5-28) as

$$I_R = \left[\frac{1}{2\pi} \int_\alpha^{\pi+\alpha} i_L^2 \, d(\omega t) \right]^{1/2}$$

The rms output current can then be determined from

$$I_{\text{rms}} = (I_R^2 + I_R^2)^{1/2} = \sqrt{2}\, I_R$$

The average current of a thyristor can also be found from Eq. (5-28) as

$$I_A = \frac{1}{2\pi} \int_\alpha^{\pi+\alpha} i_L \, d(\omega t)$$

The average output current can be determined from

$$I_{\text{dc}} = I_A + I_A = 2I_A$$

Example 5-5*

The single-phase full converter of Fig. 5-3a has a RL load having $L = 6.5$ mH, $R = 0.5\ \Omega$, and $E = 10$ V. The input voltage is $V_s = 120$ V at (rms) 60 Hz. Determine (a) the load current I_{Lo} at $\omega t = \alpha = 60°$, (b) the average thyristor current I_A, (c) the rms thyristor current I_R, (d) the rms output current I_{rms}, and (e) the average output current I_{dc}.

Solution $\alpha = 60°$, $R = 0.5\ \Omega$, $L = 6.5$ mH, $f = 60$ Hz, $\omega = 2\pi \times 60 = 377$ rad/s, $V_s = 120$ V, and $\theta = \tan^{-1}(\omega L/R) = 78.47°$.

(a) The steady-state load current at $\omega t = \alpha$, $I_{Lo} = 49.34$ A.

(b) The numerical integration of i_L in Eq. (5-28) yields the average thyristor current as $I_A = 44.05$ A.

(c) By numerical integration of i_L^2 between the limits $\omega t = \alpha$ to $\pi + \alpha$, we get the rms thyristor current as $I_R = 63.71$ A.

(d) The rms output current $I_{\text{rms}} = \sqrt{2}\, I_R = \sqrt{2} \times 63.71 = 90.1$ A.

(e) The average output current $I_{\text{dc}} = 2I_A = 2 \times 44.04 = 88.1$ A.

We have seen in Section 5-4 that single-phase full converters with inductive loads allow only two-quadrant operation. If two of these full converters are connected back to back as shown in Fig. 5-4a, both the output voltage and the load current flow can be reversed. The system will provide four-quadrant operation and is called a *dual converter*. Dual converters are normally used in high-power variable-speed drives. If α_1 and α_2 are the delay angles of converters 1 and 2, respectively, the corresponding average output voltages are V_{dc1} and V_{dc2}. The delay angles are controlled such that one converter operates as a rectifier and the other converter operates as an inverter; but both converters produce the same average output voltage. Figure 5-4b shows the output waveforms for two converters, where the two average output voltages are the same. Figure 5-4c shows the $v-i$ characteristics of a dual converter.

From Eq. (5-21) the average output voltages are

$$V_{dc1} = \frac{2V_m}{\pi} \cos \alpha_1 \tag{5-30}$$

and

$$V_{dc2} = \frac{2V_m}{\pi} \cos \alpha_2 \tag{5-31}$$

Since one converter is rectifying and the other one is inverting,

$$V_{dc1} = -V_{dc2} \quad \text{or} \quad \cos \alpha_2 = -\cos \alpha_1 = \cos(\pi - \alpha_1)$$

Therefore,

$$\alpha_2 = \pi - \alpha_1 \tag{5-32}$$

Since the instantaneous output voltages of the two converters are out of phase, there will be an instantaneous voltage difference and this will result in circulating current between the two converters. This circulating current will not flow through the load and is normally limited by a *circulating current reactor* L_r as shown in Fig. 5-4a.

If v_{o1} and v_{o2} are the instantaneous output voltages of converters 1 and 2, respectively, the circulating current can be found by integrating the instantaneous voltage difference starting from $\omega t = 2\pi - \alpha_1$. Since the two average output voltages during the interval $\omega t = \pi + \alpha_1$ to $2\pi - \alpha_1$ are equal and oppositive their contributions to the instantaneous circulating current i_r is zero.

$$i_r = \frac{1}{\omega L_r} \int_{2\pi - \alpha_1}^{\omega t} v_r \, d(\omega t) = \frac{1}{\omega L_r} \int_{2\pi - \alpha_1}^{\omega t} (v_{o1} + v_{o2}) \, d(\omega t)$$

$$= \frac{V_m}{\omega L_r} \left[\int_{2\pi - \alpha_1}^{\omega t} - \sin \omega t \, d(\omega t) - \int_{2\pi - \alpha_1}^{\omega t} \sin \omega t \, d(\omega t) \right] \tag{5-33}$$

$$= \frac{2V_m}{\omega L_r} (\cos \omega t - \cos \alpha_1)$$

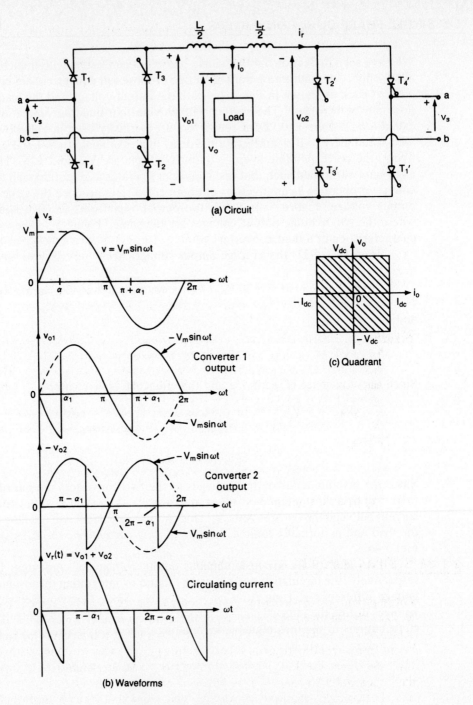

Figure 5-4 Single-phase dual converter.

The instantaneous circulating current depends on the delay angle. For $\alpha_1, = 0$, its magnitude becomes minimum when $\omega t = n\pi$, $n = 0, 2, 4, \ldots$, and maximum when $\omega t = n\pi$, $n = 1, 3, 5, \ldots$. If the peak load current is I_p, one of the converters that controls the power flow may carry a peak current of $(I_p + 4V_m/\omega L_r)$.

The dual converters can be operated with or without a circulating current. In case of operation without circulating current, only one converter operates at a time and carries the load current; and the other converter is completely blocked by inhibiting gate pulses. However, the operation with circulating current has the following advantages:

1. The circulating current maintains continuous conduction of both converters over the whole control range, independent of the load.

2. Since one converter always operates as a rectifier and the other converter operates as an inverter, the power flow in either direction at any time is possible.

3. Since both converters are in continuous conduction, the time response for changing from one quadrant operation to another is faster.

Example 5-6

The single-phase dual converter in Fig. 5-4a is operated from a 120-V 60-Hz supply and the load resistance is $R = 10\ \Omega$. The circulating inductance is $L_r = 40$ mH; delay angles are $\alpha_1 = 60°$ and $\alpha_2 = 120°$. Calculate the peak circulating current and the peak current of converter 1.

Solution $\omega = 2\pi \times 60 = 377$ rad/s, $\alpha_1 = 60°$, $V_m = \sqrt{2} \times 120 = 169.7$ V, $f = 60$ Hz, and $L_r = 40$ mH. For $\omega t = 2\pi$ and $\alpha_1 = \pi/3$, Eq. (5-33) gives the peak circulating current

$$I_r(\text{max}) = \frac{2V_m}{\omega L_r}(1 - \cos \alpha_1) = \frac{169.7}{377 \times 0.04} = 11.25 \text{ A}$$

The peak load current, $I_p = 169.71/10 = 16.97$ A. The peak current of converter 1 is $(16.97 + 11.25) = 28.22$ A.

5-6 SINGLE-PHASE SERIES CONVERTERS

For high-voltage applications, two or more converters can be connected in series to share the voltage and also to improve the power factor. Figure 5-5a shows two semiconverters that are connected in series. Each secondary has the same number of turns, and the turns ratio between the primary and secondary is $N_p/N_s = 2$. If α_1 and α_2 are the delay angles of converter 1 and converter 2, respectively, the maximum output voltage V_{dm} is obtained when $\alpha_1 = \alpha_2 = 0$.

In two-converter systems, one converter is operated to obtain output voltage from 0 to $V_{dm}/2$ and the other converter is bypassed through its freewheeling diode. To obtain output voltage from $V_{dm}/2$ to V_{dm}, one converter is fully turned on (at delay angle, $\alpha_1 = 0$) and the delay angle of other converter, α_2, is varied. Figure 5-5b shows the output voltage, input currents to converters, and the input

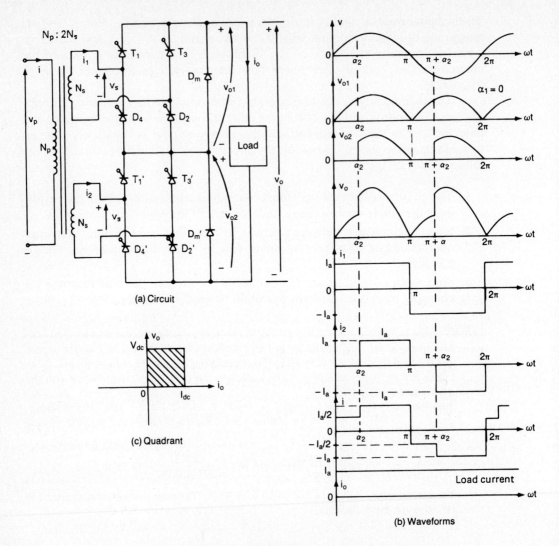

Figure 5-5 Single-phase series semiconverters.

current from the supply when both the converters are operating with a highly inductive load.

From Eq. (5-5), the average output voltages of two semiconverters are

$$V_{dc1} = \frac{V_m}{\pi} (1 + \cos \alpha_1)$$

$$V_{dc2} = \frac{V_m}{\pi} (1 + \cos \alpha_2)$$

The resultant output voltage of converters is

$$V_{dc} = V_{dc1} + V_{dc2} = \frac{V_m}{\pi} (2 + \cos \alpha_1 + \cos \alpha_2) \qquad (5\text{-}34)$$

The maximum average output voltage for $\alpha_1 = \alpha_2 = 0$ is $V_{dm} = 4V_m/\pi$. If converter 1 is operating: $0 \le \alpha_1 \le \pi$ and $\alpha_2 = \pi$, then

$$V_{dc} = V_{dc1} + V_{dc2} = \frac{V_m}{\pi}(1 + \cos \alpha_1) \tag{5-35}$$

and the normalized average output voltage is

$$V_n = \frac{V_{dc}}{V_{dm}} = 0.25(1 + \cos \alpha_1) \tag{5-36}$$

If both converters are operating: $\alpha_1 = 0$ and $0 \le \alpha_2 \le \pi$, then

$$V_{dc} = V_{dc1} + V_{dc2} = \frac{V_m}{\pi}(3 + \cos \alpha_2) \tag{5-37}$$

and the normalized average output voltage is

$$V_n = \frac{V_{dc}}{V_{dm}} = 0.25(3 + \cos \alpha_2) \tag{5-38}$$

Figure 5-6a shows two full converters that are connected in series and the turns ratio between the primary and secondary is $N_p/N_s = 2$. Due to the fact that there are no freewheeling diodes, one of the converters cannot be bypassed and both converters must operate at the same time.

In rectification mode, one converter is fully advanced ($\alpha_1 = 0$) and the delay angle of the other converter, α_2, is varied from 0 to π to control the dc output voltage. Figure 5-6b shows the input voltage, output voltages, input currents to the converters, and input supply current. Comparing Fig. 5-6b with Fig. 5-2b, we can notice that the input current from the supply is similar to that of semiconverter. As a result the power factor of this converter is improved, but the power factor is less than that of series semiconverters.

In the inversion mode, one converter is fully retarded, $\alpha_2 = \pi$, and the delay angle of the other converter, α_1, is varied from 0 to π to control the average output voltage. Figure 5-6d shows the v–i characteristics of series full converters.

From Eq. (5-21), the average output voltages of two full converters are

$$V_{dc1} = \frac{2V_m}{\pi} \cos \alpha_1$$

$$V_{dc2} = \frac{2V_m}{\pi} \cos \alpha_2$$

The resultant average output voltage is

$$V_{dc} = V_{dc1} + V_{dc2} = \frac{2V_m}{\pi}(\cos \alpha_1 + \cos \alpha_2) \tag{5-39}$$

The maximum average output voltage for $\alpha_1 = \alpha_2 = 0$ is $V_{dm} = 4V_m/\pi$. In the rectification mode, $\alpha_1 = 0$ and $0 \le \alpha_2 \le \pi$; then

$$V_{dc} = V_{dc1} + V_{dc2} = \frac{2V_m}{\pi}(1 + \cos \alpha_2) \tag{5-40}$$

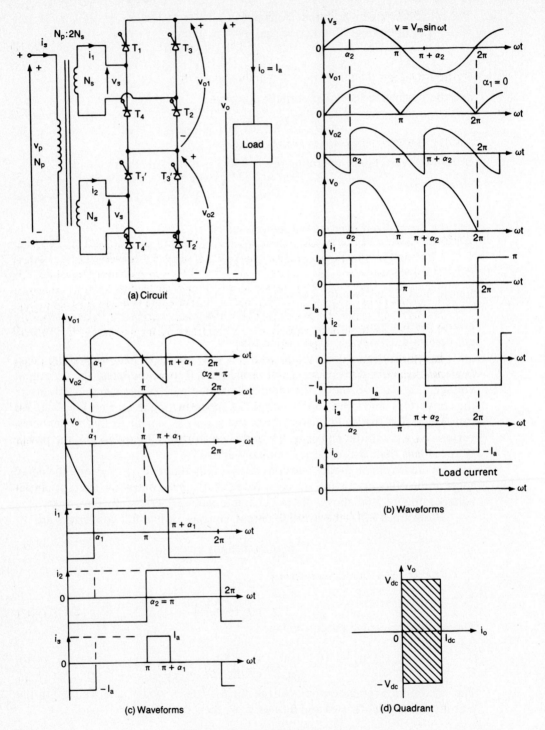

Figure 5-6 Single-phase full converters.

and the normalized dc output voltage is

$$V_n = \frac{V_{dc}}{V_{dm}} = 0.5(1 + \cos \alpha_2) \tag{5-41}$$

In the inversion mode, $0 \le \alpha_1 \le \pi$ and $\alpha_2 = \pi$; then

$$V_{dc} = V_{dc1} + V_{dc2} = \frac{2V_m}{\pi} (\cos \alpha_1 - 1) \tag{5-42}$$

and the normalized average output voltage is

$$V_n = \frac{V_{dc}}{V_{dm}} = 0.5(\cos \alpha_1 - 1) \tag{5-43}$$

Example 5-7

The load current (with an average value of I_a) of series full converters in Fig. 5-6a is continuous and the ripple content is negligible. The turns ratio of the transformer is $N_p/N_s = 2$. The converters operate in rectification mode such that $\alpha_1 = 0$ and α_2 varies from 0 to π. (a) Express the input supply current in Fourier series, determine the harmonic factor of input current HF, displacement factor DF, and input power factor PF. (b) If the delay angle is $\alpha_2 = \pi/2$ and the peak input voltage is $V_m = 162$ V, calculate V_{dc}, V_n, V_{rms}, HF, DF, and PF.

Solution (a) The waveform for input current is shown in Fig. 5-6b and the instantaneous input supply current can be expressed in a Fourier series as

$$i_s(t) = \sum_{n=1,2,\ldots}^{\infty} \sqrt{2} I_n \sin (n\omega t + \phi_n) \tag{5-44}$$

where $\phi_n = -n\alpha_2/2$. Equation (5-11) gives the rms value of the nth harmonic input current

$$I_{sn} = \frac{4I_a}{\sqrt{2} \, n\pi} \cos \frac{n\alpha_2}{2} = \frac{2\sqrt{2} \, I_a}{n\pi} \cos \frac{n\alpha_2}{2} \tag{5-45}$$

The rms value of fundamental current is

$$I_{s1} = \frac{2\sqrt{2} \, I_a}{\pi} \cos \frac{\alpha_2}{2} \tag{5-46}$$

The rms input current is found as

$$I_s = I_a \left(1 - \frac{\alpha_2}{\pi}\right)^{1/2} \tag{5-47}$$

From Eq. (3-51),

$$\text{HF} = \left[\frac{\pi(\pi - \alpha_2)}{4(1 + \cos \alpha_2)} - 1\right]^{1/2} \tag{5-48}$$

From Eqs. (3-50),

$$\text{DF} = \cos \phi_1 = \cos \left(-\frac{\alpha}{2}\right) \tag{5-49}$$

From Eq. (3-52),

$$\text{PF} = \frac{I_{s1}}{I_s} \cos \frac{\alpha_2}{2} = \frac{\sqrt{2} \, (1 + \cos \alpha_2)}{[\pi(\pi - \alpha_2)]^{1/2}} \tag{5-50}$$

(b) $\alpha_1 = 0$ and $\alpha_2 = \pi/2$. From Eq. (5-41),

$$V_{dc} = \left(2 \times \frac{162}{\pi}\right) \left(1 + \cos \frac{\pi}{2}\right) = 103.13 \text{ V}$$

From Eq. (5-42), $V_n = 0.5$ pu and

$$V_{rms}^2 = \frac{2}{2\pi} \int_{\alpha_2}^{\pi} (2V_m)^2 \sin^2 \omega t \, d(\omega t)$$

$$V_{rms} = \sqrt{2} \, V_m \left[\frac{1}{\pi}\left(\pi - \alpha_2 + \frac{\sin 2\alpha_2}{2}\right)\right]^{1/2} = V_m = 162 \text{ V}$$

$$I_{s1} = I_a \frac{2\sqrt{2}}{\pi} \cos \frac{\pi}{4} = 0.6366 I_a \qquad \text{and} \qquad I_s = 0.7071 I_a$$

$$\text{HF} = \left[\left(\frac{I_s}{I_{s1}}\right)^2 - 1\right]^{1/2} = 0.4835 \quad \text{or} \quad 48.35\%$$

$$\phi_1 = -\frac{\pi}{4} \qquad \text{and} \qquad \text{DF} = \cos\left(-\frac{\pi}{4}\right) = 0.7071$$

$$\text{PF} = \frac{I_{s1}}{I_s} \cos(-\phi_1) = 0.6366 \text{ (lagging)}$$

Note. The performance of series full converters is similar to that of single-phase semiconverters.

5-7 THREE-PHASE HALF-WAVE CONVERTERS

Three-phase converters provide higher average output voltage, and in addition the frequency of the ripples on the output voltage is higher compared to that of single-phase converters. As a result, the filtering requirements for smoothing out the load current and load voltage are simpler. For these reasons, three-phase converters are used extensively in high-power variable-speed drives. Three single-phase half-wave converters in Fig. 5-1a can be connected to form a three-phase half-wave converter, as shown in Fig. 5-7a.

When thyristor T_1 is fired at $\omega t = \pi/6 + \alpha$, the phase voltage v_{an} appears across the load until thyristor T_2 is fired at $\omega t = 5\pi/6 + \alpha$. When thyristor T_2 is fired, thyristor T_1 is reverse biased, because the line-to-line voltage, $v_{ab} (= v_{an} - v_{bn})$, is negative and T_1 is turned off. The phase voltage v_{bn} appears across the load until thyristor T_3 is fired at $\omega t = 3\pi/2 + \alpha$. When thyristor T_3 is fired, T_2 is turned off and v_{cn} appears across the load until T_1 is fired again at the beginning of next cycle. Figure 5-7b shows the v–i characteristics of the load and this is a two-quadrant converter. Figure 5-7c shows the input voltages, output voltage, and the current through thyristor T_1 for a highly inductive load. For a resistive load and

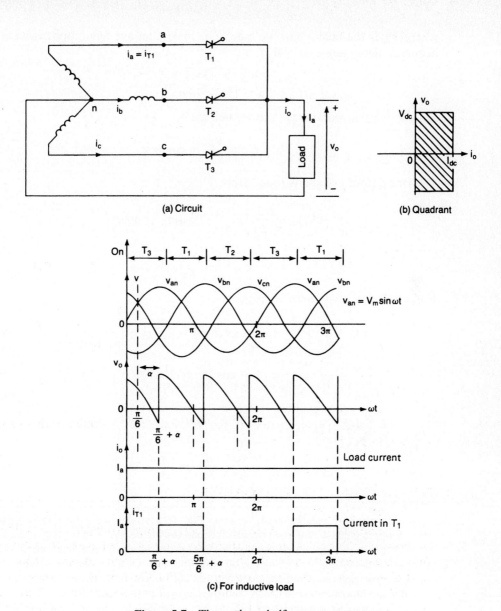

(a) Circuit

(b) Quadrant

(c) For inductive load

Figure 5-7 Three-phase half-wave converter.

$\alpha > \pi/6$, the load current would be discontinuous and each thyristor is self-commutated when the polarity of its phase voltage is reversed. The frequency of output ripple voltage is $3f_s$. This converter is not normally used in practical systems, because the supply currents contain dc components.

If the phase voltage is $v_{an} = V_m \sin \omega t$, the average output voltage for a continuous load current is

$$V_{dc} = \frac{3}{2\pi} \int_{\pi/6+\alpha}^{5\pi/6+\alpha} V_m \sin \omega t \, d(\omega t) = \frac{3\sqrt{3} \, V_m}{2\pi} \cos \alpha \qquad (5\text{-}51)$$

where V_m is the peak phase voltage. The maximum average output voltage that occurs at delay angle, $\alpha = 0$ is

$$V_{dm} = \frac{3\sqrt{3}\ V_m}{2\pi}$$

and the normalized average output voltage is

$$V_n = \frac{V_{dc}}{V_{dm}} = \cos\alpha \qquad (5\text{-}52)$$

The rms output voltage is found from

$$V_{rms} = \left[\frac{3}{2\pi}\left[\int_{\pi/6+\alpha}^{5\pi/6+\alpha} V_m^2\ \sin^2 \omega t\ d(\omega t)\right]\right]^{1/2}$$

$$= \sqrt{3}\ V_m \left(\frac{1}{6} + \frac{\sqrt{3}}{8\pi}\cos 2\alpha\right)^{1/2} \qquad (5\text{-}53)$$

For a resistive load and $\alpha \geq \pi/6$:

$$V_{dc} = \frac{3}{2\pi}\int_{\pi/6+\alpha}^{\pi} V_m \sin \omega t\ d(\omega t) = \frac{3V_m}{2\pi}\left[1 + \cos\left(\frac{\pi}{6} + \alpha\right)\right] \qquad (5\text{-}51a)$$

$$V_n = \frac{V_{dc}}{V_{dm}} = \frac{1}{\sqrt{3}}\left[1 + \cos\left(\frac{\pi}{6} + \alpha\right)\right] \qquad (5\text{-}52a)$$

$$V_{rms} = \left[\frac{3}{2\pi}\int_{\pi/6+\alpha}^{\pi} V_m^2\ \sin^2 \omega t\ d(\omega t)\right]^{1/2}$$

$$= \sqrt{3}\ V_m \left[\frac{5}{24} - \frac{\alpha}{4\pi} + \frac{1}{8\pi}\sin\left(\frac{\pi}{3} + 2\alpha\right)\right]^{1/2} \qquad (5\text{-}53a)$$

Example 5-8*

A three-phase half-wave converter in Fig. 5-7a is operated from a three-phase Y-connected 208-V 60-Hz supply and the load resistance is $R = 10\ \Omega$. If it is required to obtain an average output voltage of 50% of the maximum possible output voltage, calculate (a) the delay angle α, (b) the rms and average output currents, (c) the average and rms thyristor currents, (d) the rectification efficiency, (e) the transformer utilization factor TUF, and (f) the input power factor PF.

Solution The phase voltage is $V_s = 208/\sqrt{3} = 120.1$ V, $V_m = \sqrt{2}\ V_s = 169.83$ V, $V_n = 0.5$, and $R = 10\ \Omega$. The maximum output voltage is

$$V_{dm} = \frac{3\sqrt{3}\ V_m}{2\pi} = 3\sqrt{3} \times \frac{169.83}{2\pi} = 140.45\ \text{V}$$

The average output voltage, $V_{dc} = 0.5 \times 140.45 = 70.23$ V.

(a) For a resistive load, the load current is continuous if $\alpha \leq \pi/6$ and Eq. (5-52) gives $V_n \geq \cos(\pi/6) = 86.6\%$. With a resistive load and 50% output, the load current is discontinuous. From Eq. (5-52a), $0.5 = (1/\sqrt{3})\ [1 + \cos(\pi/6 + \alpha)]$, which gives the delay angle as $\alpha = 67.7°$.

(b) The average output current, $I_{dc} = V_{dc}/R = 70.23/10 = 7.02$ A. From Eq. (5-53a), $V_{rms} = 94.74$ V and the rms load current, $I_{rms} = 94.74/10 = 9.47$ A.

(c) The average current of a thyristor, $I_A = I_{dc}/3 = 7.02/3 = 2.34$ A and the rms current of a thyristor, $I_R = I_{rms}/\sqrt{3} = 9.47/\sqrt{3} = 5.47$ A.

(d) From Eq. (3-44) the rectification efficiency is $= 70.23 \times 7.02/(94.74 \times 9.47) = 54.95\%$.

(e) The rms input line current is the same as the thyristor rms current, and the input volt-ampere rating, $VI = 3V_sI_s = 3 \times 120.1 \times 5.47 = 1970.84$ W. From Eq. (3-49), TUF $= 70.23 \times 7.02/1970.84 = 0.25$ or 25%.

(f) The output power, $P_o = I^2_{rms}R = 9.47^2 \times 10 = 896.81$ W. The input power factor, PF $= 896.81/1970.84 = 0.455$ (lagging).

Note. Due to the delay angle, α, the fundamental component of input line current is also delayed with respect to the input phase voltage.

5-8 THREE-PHASE SEMICONVERTERS

Three-phase semiconverters are used in industrial applications up to the 120-kW level, where one-quadrant operation is required. The power factor of this converter decreases as the delay angle increases, but it is better than that of three-phase half-wave converters. Figure 5-8a shows a three-phase semiconverter with a highly inductive load and the load current has a negligible ripple content.

Figure 5-8b shows the waveforms for input voltages, output voltage, input current, and the current through thyristors and diodes. The frequency of output voltage is $3f_s$. The delay angle, α, can be varied from 0 to π. During the period $\pi/6 \le \omega t < 7\pi/6$, thyristor T_1 is forward biased. If T_1 is fired at $\omega t = (\pi/6 + \alpha)$, T_1 and D_1 conduct and the line-to-line voltage v_{ac} appears across the load. At $\omega t = 7\pi/6$, v_{ac} starts to be negative and the freewheeling diode D_m conducts. The load current continues to flow through D_m; and T_1 and D_1 are turned off.

If there were no freewheeling diode, T_1 would continue to conduct until thyristor T_2 fired at $\omega t = 5\pi/6 + \alpha$ and the freewheeling action would be accomplished through T_1 and D_2. If $\alpha \le \pi/3$, each thyristor conducts for $2\pi/3$ and the freewheeling diode D_m does not conduct. The waveforms for a three-phase semiconverter with $\alpha \le \pi/3$ are shown in Fig. 5-9.

If we define the three line-neutral voltages as follows:

$$v_{an} = V_m \sin \omega t$$

$$v_{bn} = V_m \sin \left(\omega t - \frac{2\pi}{3} \right)$$

$$v_{cn} = V_m \sin \left(\omega t + \frac{2\pi}{3} \right),$$

the corresponding line-to-line voltages are

$$v_{ac} = v_{an} - v_{cn} = \sqrt{3} \, V_m \sin \left(\omega t - \frac{\pi}{6} \right)$$

$$v_{ba} = v_{bn} - v_{an} = \sqrt{3} \, V_m \sin \left(\omega t - \frac{5\pi}{6} \right)$$

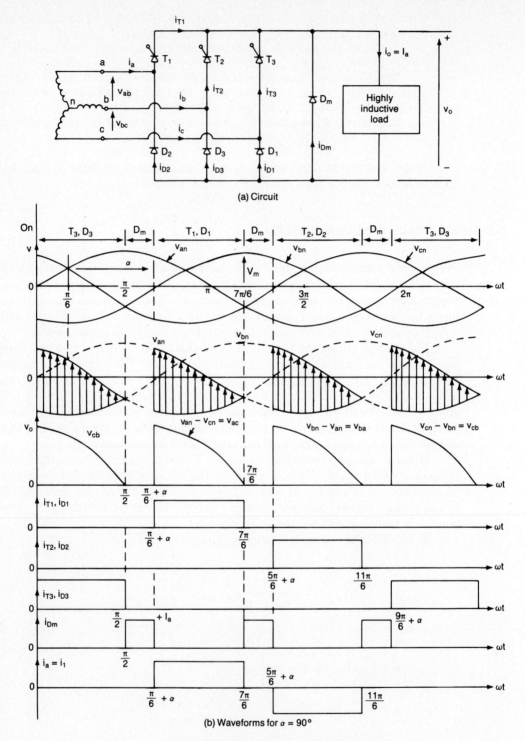

Figure 5-8 Three-phase semiconverter.

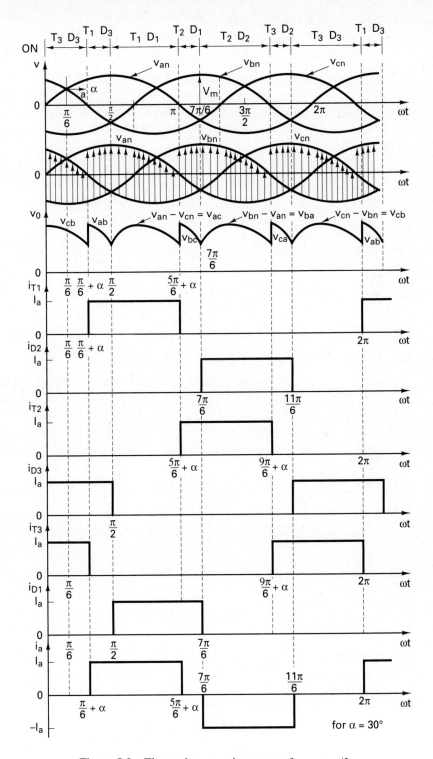

Figure 5-9   Three-phase semiconverter for $\alpha \le \pi/3$.

$$v_{cb} = v_{cn} - v_{bn} = \sqrt{3}\ V_m \sin\left(\omega t + \frac{\pi}{2}\right)$$

$$v_{ab} = v_{an} - v_{bn} = \sqrt{3}\ V_m \sin\left(\omega t + \frac{\pi}{6}\right)$$

where V_m is the peak phase voltage of a wye-connected source.

For $\alpha \geq \pi/3$, and discontinuous output voltage: the average output voltage is found from

$$V_{dc} = \frac{3}{2\pi} \int_{\pi/6+\alpha}^{7\pi/6} v_{ac}\ d(\omega t) = \frac{3}{2\pi} \int_{\pi/6+\alpha}^{7\pi/6} \sqrt{3}\ V_m \sin\left(\omega t - \frac{\pi}{6}\right) d(\omega t)$$

$$= \frac{3\sqrt{3}\ V_m}{2\pi}(1 + \cos \alpha) \tag{5-54}$$

The maximum average output voltage that occurs at a delay angle of $\alpha = 0$ is $V_{dm} = 3\sqrt{3}\ V_m/\pi$ and the normalized average output voltage is

$$V_n = \frac{V_{dc}}{V_{dm}} = 0.5(1 + \cos \alpha) \tag{5-55}$$

The rms output voltage is found from

$$V_{rms} = \left[\frac{3}{2\pi} \int_{\pi/6+\alpha}^{7\pi/6} 3V_m^2 \sin^2\left(\omega t - \frac{\pi}{6}\right) d(\omega t)\right]^{1/2}$$

$$= \sqrt{3}\ V_m \left[\frac{3}{4\pi}\left(\pi - \alpha + \frac{1}{2}\sin 2\alpha\right)\right]^{1/2} \tag{5-56}$$

For $\alpha \leq \pi/3$, and continuous output voltage:

$$V_{dc} = \frac{3}{2\pi}\left[\int_{\pi/6+\alpha}^{\pi/2} v_{ab}\ d(\omega t) + \int_{\pi/2}^{5\pi/6+\alpha} v_{ac}\ d(\omega t)\right] = \frac{3\sqrt{3}\ V_m}{2\pi}(1 + \cos \alpha) \tag{5-54a}$$

$$V_n = \frac{V_{dc}}{V_{dm}} = 0.5(1 + \cos \alpha) \tag{5-55a}$$

$$V_{rms} = \left[\frac{3}{2\pi} \int_{\pi/6+\alpha}^{\pi/2} v_{ab}^2\ d(\omega t) + \int_{\pi/2}^{5\pi/6+\alpha} v_{ac}^2\ d(\omega t)\right]^{1/2} \tag{5-56a}$$

$$= \sqrt{3}\ V_m \left[\frac{3}{4\pi}\left(\frac{2\pi}{3} + \sqrt{3}\cos^2 \alpha\right)\right]^{1/2}$$

Example 5-9

Repeat Example 5-8 for the three-phase semiconverter in Fig. 5-8a.
Solution The phase voltage is $V_s = 208/\sqrt{3} = 120.1$ V, $V_m = \sqrt{2}\ V_s = 169.83$, $V_n = 0.5$, and $R = 10\ \Omega$. The maximum output voltage is

$$V_{dm} = \frac{3\sqrt{3}\ V_m}{\pi} = 3\sqrt{3} \times \frac{169.83}{\pi} = 280.9\ V$$

The average output voltage $V_{dc} = 0.5 \times 280.9 = 140.45$ V.

(a) For $\alpha \geq \pi/3$ and Eq. (5-55) gives $V_n \leq (1 + \cos \pi/3)/2 = 75\%$. With a resistive load and 50% output, the output voltage is discontinuous. From Eq. (5-55), $0.5 = 0.5(1 + \cos \alpha)$, which gives the delay angle $\alpha = 90°$.

(b) The average output current $I_{dc} = V_{dc}/R = 140.45/10 = 14.05$ A. From Eq. (5-56),

$$V_{rms} = \sqrt{3} \times 169.83 \left[\frac{3}{4\pi} \left(\pi - \frac{\pi}{2} + 0.5 \sin 2 \times 90° \right) \right]^{1/2} = 180.13 \text{ V}$$

and the rms load current $I_{rms} = 180.13/10 = 18.01$ A.

(c) The average current of a thyristor $I_A = I_{dc}/3 = 14.05/3 = 4.68$ A and the rms current of a thyristor $I_R = I_{rms}/\sqrt{3} = 18.01/\sqrt{3} = 10.4$ A.

(d) From Eq. (3-44) the rectification efficiency is

$$\eta = \frac{140.45 \times 14.05}{180.13 \times 18.01} = 0.608 \text{ or } 60.8\%$$

(e) Since a thyristor conducts for $2\pi/3$, the rms input line current is $I_s = I_{rms}\sqrt{\frac{2}{3}} = 14.71$ A. The input voltampere rating, VI $= 3V_s I_s = 3 \times 120.1 \times 14.71 = 5300$. From Eq. (3-49), TUF $= 140.45 \times 14.05/5300 = 0.372$.

(f) The output power $P_o = I_{rms}^2 R = 18.01^2 \times 10 = 3243.6$ W. The input power factor is PF $= 3243.6/5300 = 0.612$ (lagging).

Note. The power factor is better than that of three-phase half-wave converters.

5-8.1 Three-Phase Semiconverter with *RL* Load

The output voltage of the three-phase semiconverter in Fig. 5-8a would be continuous or discontinuous depending on the value of delay angle α. In either case the output waveform can be divided into two intervals.

Case 1: continuous output voltage. For $\alpha \leq \pi/3$, the waveform of the output voltage is shown in Fig. 5-9.

Interval 1 for $\pi/6 + \alpha \leq \omega t \leq \pi/2$: Thyristor T_1 and diode D_3 conduct. The output voltage becomes

$$v_o = v_{ab} = \sqrt{2} \, V_{ab} \sin\left(\omega t + \frac{\pi}{6} \right) \qquad \text{for } \frac{\pi}{6} + \alpha \leq \omega t \leq \frac{\pi}{2}$$

where V_{ab} is the line-to-line (rms) input voltage. The load current i_{L1} during interval 1 can be found from

$$L \frac{di_{L1}}{dt} + R i_{L1} + E = \sqrt{2} \, V_{ab} \sin\left(\omega t + \frac{\pi}{6} \right) \qquad \text{for } \frac{\pi}{6} + \alpha \leq \omega t \leq \frac{\pi}{2}$$

with the boundary conditions $i_{L1}(\omega t = \pi/6 + \alpha) = I_{Lo}$ and $i_{L1}(\omega t = \pi/2) = I_{L1}$.

Interval 2 for $\pi/2 \leq \omega t \leq 5\pi/6 + \alpha$: Thyristor T_1 and diode D_1 conduct. The output voltage becomes

$$v_o = v_{ac} = \sqrt{2} \, V_{ac} \sin\left(\omega t - \frac{\pi}{6} \right) \qquad \text{for } \frac{\pi}{2} \leq \omega t \leq \frac{5\pi}{6} + \alpha$$

The load current i_{L2} during interval 2 can be found from

$$L\frac{di_{L2}}{dt} + Ri_{L2} + E = \sqrt{2}\, V_{ac} \sin\left(\omega t - \frac{\pi}{6}\right) \qquad \text{for } \frac{\pi}{2} \leq \omega t \leq \frac{5\pi}{6} + \alpha$$

with the boundary conditions $i_{L2}(\omega t = \pi/2) = I_{L1}$ and $i_{L2}(\omega t = 5\pi/6 + \alpha) = I_{Lo}$.

Case 2: discontinuous output voltage. For $\alpha \geq \pi/3$, the waveform of the output voltage is shown in Fig. 5-8b.

Interval 1 for $\pi/2 \leq \omega t \leq \pi/6 + \alpha$: Diode D_m conducts. The output voltage is zero, $v_o = 0$ for $\pi/2 \leq \omega t \leq \pi/6 + \alpha$. The load current i_{L1} during interval 1 can be found from

$$L\frac{di_{L1}}{dt} + Ri_{L1} + E = 0 \qquad \text{for } \frac{\pi}{2} \leq \omega t \leq \frac{\pi}{6} + \alpha$$

with the boundary conditions $i_{L1}(\omega t = \pi/2) = I_{Lo}$ and $i_{L1}(\omega t = \pi/6 + \alpha) = I_{L1}$.

Interval 2 for $\pi/6 + \alpha \leq \omega t \leq 7\pi/6$: Thyristor T_1 and diode D_1 conduct. The output voltage becomes

$$v_o = v_{ac} = \sqrt{2}\, V_{ac} \sin\left(\omega t - \frac{\pi}{6}\right) \qquad \text{for } \frac{\pi}{6} + \alpha \leq \omega t \leq \frac{7\pi}{6}$$

where V_{ac} is the line-to-line (rms) input voltage. The load current i_{L2} during interval 2 can be found from

$$L\frac{di_{L2}}{dt} + Ri_{L2} + E = \sqrt{2}\, V_{ac} \sin\left(\omega t - \frac{\pi}{6}\right) \qquad \text{for } \frac{\pi}{6} + \alpha \leq \omega t \leq \frac{7\pi}{6}$$

with the boundary conditions $i_{L2}(\omega t = \pi/6 + \alpha) = I_{L1}$ and $i_{L2}(\omega t = 7\pi/6) = I_{Lo}$.

5-9 THREE-PHASE FULL CONVERTERS

Three-phase converters are extensively used in industrial applications up to the 120-kW level, where two-quadrant operation is required. Figure 5-10a shows a full-converter circuit with a highly inductive load. This circuit is known as a three-phase bridge. The thyristors are fired at an interval of $\pi/3$. The frequency of output ripple voltage is $6f_s$ and the filtering requirement is less than that of three-phase semi- and half-wave converters. At $\omega t = \pi/6 + \alpha$, thyristor T_6 is already conducting and thyristor T_1 is turned on. During interval $(\pi/6 + \alpha) \leq \omega t \leq (\pi/2 + \alpha)$, thyristors T_1 and T_6 conduct and the line-to-line voltage, $v_{ab}(= v_{an} - v_{bn})$ appears across the load. At $\omega t = \pi/2 + \alpha$, thyristor T_2 is fired and thyristor T_6 is reversed biased immediately. T_6 is turned off due to natural commutation. During interval $(\pi/2 + \alpha) \leq \omega t \leq (5\pi/6 + \alpha)$, thyristors T_1 and T_2 conduct and the line-to-line voltage, v_{ac} appears across the load. If the thyristors are numbered as shown in Fig. 5-10a, the firing sequence is 12, 23, 34, 45, 56, and 61. Figure 5-10b shows the waveforms for input voltage, output voltage, input current, and currents through thyristors.

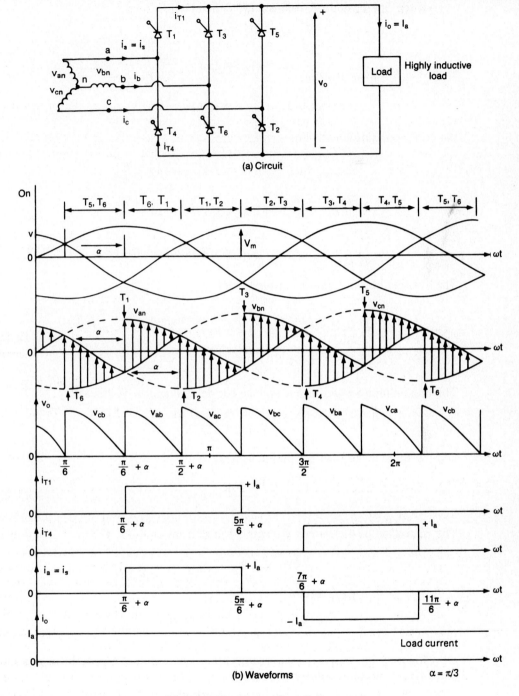

Figure 5-10 Three-phase full converter.

If the line-to-neutral voltages are defined as

$$v_{an} = V_m \sin \omega t$$

$$v_{bn} = V_m \sin \left(\omega t - \frac{2\pi}{3} \right)$$

$$v_{cn} = V_m \sin \left(\omega t + \frac{2\pi}{3} \right)$$

the corresponding line-to-line voltages are

$$v_{ab} = v_{an} - v_{bn} = \sqrt{3} \, V_m \sin \left(\omega t + \frac{\pi}{6} \right)$$

$$v_{bc} = v_{bn} - v_{cn} = \sqrt{3} \, V_m \sin \left(\omega t - \frac{\pi}{2} \right)$$

$$v_{ca} = v_{cn} - v_{an} = \sqrt{3} \, V_m \sin \left(\omega t + \frac{\pi}{2} \right)$$

The average output voltage is found from

$$V_{dc} = \frac{3}{\pi} \int_{\pi/6+\alpha}^{\pi/2+\alpha} v_{ab} \, d(\omega t) = \frac{3}{\pi} \int_{\pi/6+\alpha}^{\pi/2+\alpha} \sqrt{3} \, V_m \sin \left(\omega t + \frac{\pi}{6} \right) d(\omega t)$$

$$= \frac{3\sqrt{3} \, V_m}{\pi} \cos \alpha$$

(5-57)

The maximum average output voltage for delay angle, $\alpha = 0$ is

$$V_{dm} = \frac{3\sqrt{3} \, V_m}{\pi}$$

and the normalized average output voltage is

$$V_n = \frac{V_{dc}}{V_{dm}} = \cos \alpha \tag{5-58}$$

The rms value of the output voltage is found from

$$V_{rms} = \left[\frac{3}{\pi} \int_{\pi/6+\alpha}^{\pi/2+\alpha} 3 V_m^2 \sin^2 \left(\omega t + \frac{\pi}{6} \right) d(\omega t) \right]^{1/2}$$

(5-59)

$$= \sqrt{3} \, V_m \left(\frac{1}{2} + \frac{3\sqrt{3}}{4\pi} \cos 2\alpha \right)^{1/2}$$

Figure 5-10b shows the waveforms for $\alpha = \pi/3$. For $\alpha > \pi/3$, the instantaneous output voltage v_o will have a negative part. Since the current through thyristors cannot be negative, the load current will always be positive. Thus, with a resistive load, the instantaneous load voltage cannot be negative, and the full converter will behave as a semiconverter.

A three-phase bridge gives a six-pulse output voltage. For high-power applications such as high-voltage dc transmission and dc motor drives, a 12-pulse output is generally required to reduce the output ripples and to increase the ripple frequencies. Two six-pulse bridges can be combined either in series or in parallel to produce an effective 12-pulse output. Two configurations are shown in Fig. 5-11. A 30° phase shift between secondary windings can be accomplished by connecting one secondary in wye (Y) and the other in delta (Δ).

Example 5-10

Repeat Example 5-8 for the three-phase full converter in Fig. 5-10a.
Solution The phase voltage $V_s = 208/\sqrt{3} = 120.1$ V, $V_m = \sqrt{2} V_s = 169.83$, $V_n = 0.5$, and $R = 10$ Ω. The maximum output voltage $V_{dm} = 3\sqrt{3} V_m/\pi = 3\sqrt{3} \times 169.83/\pi = 280.9$ V. The average output voltage $V_{dc} = 0.5 \times 280.9 = 140.45$ V.
   (a) From Eq. (5-58), $0.5 = \cos \alpha$, and the delay angle $\alpha = 60°$.
   (b) The average output current $I_{dc} = V_{dc}/R = 140.45/10 = 14.05$ A. From Eq. (5-59),

$$V_{rms} = \sqrt{3} \times 169.83 \left[\frac{1}{2} + \frac{3\sqrt{3}}{4\pi} \cos(2 \times 60°) \right]^{1/2} = 159.29 \text{ V}$$

and the rms current $I_{rms} = 159.29/10 = 15.93$ A.
   (c) The average current of a thyristor $I_A = I_{dc}/3 = 14.05/3 = 4.68$ A, and the rms current of a thyristor $I_R = I_{rms}\sqrt{2/6} = 15.93\sqrt{2/6} = 9.2$ A.
   (d) From Eq. (3-44) the rectification efficiency is

$$\eta = \frac{140.45 \times 14.05}{159.29 \times 15.93} = 0.778 \quad \text{or} \quad 77.8\%$$

   (e) The rms input line current $I_s = I_{rms}\sqrt{4/6} = 13$ A and the input volt-ampere rating, $VI = 3V_s I_s = 3 \times 120.1 \times 13 = 4683.9$ W. From Eq. (3-49), TUF = $140.45 \times 14.05/4683.9 = 0.421$.
   (f) The output power $P_o = I_{rms}^2 R = 15.93^2 \times 10 = 2537.6$ W. The power factor PF = $2537.6/4683.9 = 0.542$ (lagging).

Note. The power factor is less than that of three-phase semiconverters, but higher than that of three-phase half-wave converters.

Example 5-11

The load current of a three-phase full converter in Fig. 5-10a is continuous with a negligible ripple content. (a) Express the input current in Fourier series, and determine the harmonic factor HF of input current, the displacement factor DF, and the input power factor PF. (b) If the delay angle $\alpha = \pi/3$, calculate V_n, HF, DF, and PF.
Solution (a) The waveform for input current is shown in Fig. 5-10b and the instantaneous input current of a phase can be expressed in a Fourier series as

$$i_s(t) = I_{dc} + \sum_{n=1,2,\ldots}^{\infty} (a_n \cos n\omega t + b_n \sin n\omega t)$$

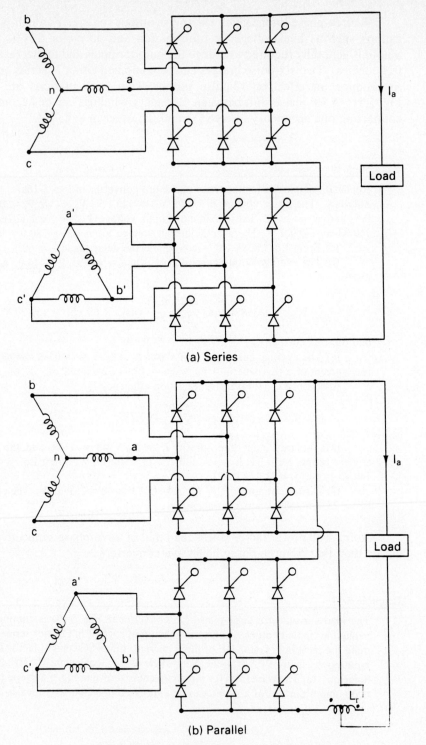

(a) Series

(b) Parallel

Figure 5-11 Configurations for 12-pulse output.

where

$$a_o = I_{dc} = \frac{1}{2\pi} \int_0^{2\pi} i_s(t)\, d(\omega t) = 0$$

$$a_n = \frac{1}{\pi} \int_0^{2\pi} i_s(t) \cos n\omega t\, d(\omega t)$$

$$= \frac{1}{\pi} \left[\int_{\pi/6+\alpha}^{5\pi/6+\alpha} I_a \cos n\omega t\, d(\omega t) - \int_{7\pi/6+\alpha}^{11\pi/6+\alpha} I_a \cos n\omega t\, d(\omega t) \right]$$

$$= -\frac{4I_a}{n\pi} \sin \frac{n\pi}{3} \sin n\alpha \qquad \text{for } n = 1, 3, 5, \ldots$$

$$= 0 \qquad \text{for } n = 2, 4, 6, \ldots$$

$$b_n = \frac{1}{\pi} \int_0^{2\pi} i_s(t) \sin n\omega t\, d(\omega t)$$

$$= \frac{1}{\pi} \left[\int_{\pi/6+\alpha}^{5\pi/6+\alpha} I_a \sin n\omega t\, d(\omega t) - \int_{7\pi/6+\alpha}^{11\pi/6+\alpha} I_a \sin n\omega t\, d(\omega t) \right]$$

$$= \frac{4I_a}{n\pi} \sin \frac{n\pi}{3} \cos n\alpha \qquad \text{for } n = 1, 3, 5, \ldots$$

$$= 0 \qquad \text{for } n = 2, 4, 6, \ldots$$

Since $I_{dc} = 0$, the input current can be written as

$$i_s(t) = \sum_{n=1,3,5,\ldots}^{\infty} \sqrt{2}\, I_{sn} \sin(n\omega t + \phi_n)$$

where

$$\phi_n = \tan^{-1} \frac{a_n}{b_n} = -n\alpha \qquad (5\text{-}60)$$

The rms value of the nth harmonic input current is given by

$$I_{sn} = \frac{1}{\sqrt{2}} (a_n^2 + b_n^2)^{1/2} = \frac{2\sqrt{2}\, I_a}{n\pi} \sin \frac{n\pi}{3} \qquad (5\text{-}61)$$

The rms value of the fundamental current is

$$I_{s1} = \frac{\sqrt{6}}{\pi} I_a = 0.7797 I_a$$

The rms input current

$$I_s = \left[\frac{2}{2\pi} \int_{\pi/6+\alpha}^{5\pi/6+\alpha} I_a^2\, d(\omega t) \right]^{1/2} = I_a \sqrt{\frac{2}{3}} = 0.8165 I_a$$

$$\text{HF} = \left[\left(\frac{I_s}{I_{s1}} \right)^2 - 1 \right]^{1/2} = \left[\left(\frac{\pi}{3} \right)^2 - 1 \right]^{1/2} = 0.3108 \quad \text{or} \quad 31.08\%$$

$$\text{DF} = \cos \phi_1 = \cos(-\alpha)$$

$$\text{PF} = \frac{I_{s1}}{I_s} \cos(-\alpha) = \frac{3}{\pi} \cos \alpha = 0.9549\, \text{DF}$$

(b) For $\alpha = \pi/3$, $V_n = \cos(\pi/3) = 0.5$ pu, HF = 31.08%, DF = $\cos 60° = 0.5$, and PF = 0.478 (lagging).

Note. If we compare the power factor with that of Example 5-8, where the load is purely resistive, we can notice that the input power factor depends on the power factor of the load.

5-9.1 Three-Phase Full Converter with *RL* Load

From Fig. 5-10b the output voltage is

$$v_o = v_{ab} = \sqrt{2}\, V_{ab} \sin\left(\omega t + \frac{\pi}{6}\right) \qquad \text{for } \frac{\pi}{6} + \alpha \le \omega t \le \frac{\pi}{2} + \alpha$$

$$= \sqrt{2}\, V_{ab} \sin \omega t' \qquad \text{for } \frac{\pi}{3} + \alpha \le \omega t' \le \frac{2\pi}{3} + \alpha$$

where $\omega t' = \omega t + \pi/6$, and V_{ab} is the line-to-line (rms) input voltage. Choosing v_{ab} as the time reference voltage, the load current i_L can be found from

$$L \frac{di_L}{dt} + R i_L + E = \sqrt{2}\, V_{ab} \sin \omega t' \qquad \text{for } \frac{\pi}{3} + \alpha \le \omega t' \le \frac{2\pi}{3} + \alpha$$

whose solution from Eq. (3-81) is

$$i_L = \frac{\sqrt{2}\, V_{ab}}{Z} \sin(\omega t' - \theta) - \frac{E}{R}$$

$$+ \left[I_{L1} + \frac{E}{R} - \frac{\sqrt{2}\, V_{ab}}{Z} \sin\left(\frac{\pi}{3} + \alpha - \theta\right) \right] e^{(R/L)[(\pi/3+\alpha)/\omega - t']} \tag{5-62}$$

where $Z = [R^2 + (\omega L)^2]^{1/2}$ and $\theta = \tan^{-1}(\omega L/R)$. Under a steady-state condition, $i_L(\omega t' = 2\pi/3 + \alpha) = i_L(\omega t' = \pi/3 + \alpha) = I_{L1}$. Applying this condition to Eq. (5-62), we get the value of I_{L1} as

$$I_{L1} = \frac{\sqrt{2}\, V_{ab}}{Z} \frac{\sin(2\pi/3 + \alpha - \theta) - \sin(\pi/3 + \alpha - \theta)e^{-(R/L)(\pi/3\omega)}}{1 - e^{-(R/L)(\pi/3\omega)}}$$

$$- \frac{E}{R} \qquad \text{for } I_{L1} \ge 0 \tag{5-63}$$

Example 5-12*

The three-phase full converter of Fig. 5-10a has a load of $L = 1.5$ mH, $R = 2.5\ \Omega$, and $E = 10$ V. The line-to-line input voltage is $V_{ab} = 208$ V (rms), 60 Hz. The delay angle is $\alpha = \pi/3$. Determine (a) the steady-state load current I_{L1} at $\omega t' = \pi/3 + \alpha$ (or $\omega t = \pi/6 + \alpha$), (b) the average thyristor current I_A, (c) the rms thyristor current I_R, (d) the rms output current I_{rms}, and (e) the average output current I_{dc}.
Solution $\alpha = \pi/3$, $R = 2.5\ \Omega$, $L = 1.5$ mH, $f = 60$ Hz, $\omega = 2\pi \times 60 = 377$ rad/s, $V_{ab} = 208$ V, $Z = [R^2 + (\omega L)^2]^{1/2} = 2.56\ \Omega$, and $\theta = \tan^{-1}(\omega L/R) = 12.74°$.
    (a) The steady-state load current at $\omega t' = \pi/3 + \alpha$, $I_{L1} = 20.49$ A.
    (b) The numerical integration of i_L in Eq. (3-62), between the limits $\omega t' = \pi/3 + \alpha$ to $2\pi/3 + \alpha$, gives the average thyristor current, $I_A = 17.42$ A.

(c) By numerical integration of i_L^2, between the limits $\omega t' = \pi/3 + \alpha$ to $2\pi/3 + \alpha$, gives the rms thyristor current, $I_R = 31.32$ A.

(d) The rms output current $I_{rms} = \sqrt{3}\, I_R = \sqrt{3} \times 31.32 = 54.25$ A.

(e) The average output current $I_{dc} = 3I_A = 3 \times 17.42 = 52.26$ A.

5-10 THREE-PHASE DUAL CONVERTERS

In many variable-speed drives, the four-quadrant operation is generally required and three-phase dual converters are extensively used in applications up to the 2000-kW level. Figure 5-12a shows three-phase dual converters where two three-phase converters are connected back to back. We have seen in Section 5-5 that due to the instantaneous voltage differences between the output voltages of converters, a circulating current flows through the converters. The circulating current is normally limited by circulating reactor, L_r as shown in Fig. 5-12a. The two converters are controlled in such a way that if α_1 is the delay angle of converter 1, the delay angle of converter 2 is $\alpha_2 = \pi - \alpha_1$. Figure 5-10b shows the waveforms for input voltages, output voltages, and the voltage across inductor L_r. The operation of each converter is identical to that of a three-phase full converter. During the interval $(\pi/6 + \alpha_1) \le \omega t \le (\pi/2 + \alpha_1)$, the line-to-line voltage v_{ab} appears across the output of converter 1, and v_{bc} appears across converter 2.

If the line-to-neutral voltages are defined as

$$v_{an} = V_m \sin \omega t$$

$$v_{bn} = V_m \sin \left(\omega t - \frac{2\pi}{3} \right)$$

$$v_{cn} = V_m \sin \left(\omega t + \frac{2\pi}{3} \right)$$

the corresponding line-to-line voltages are

$$v_{ab} = v_{an} - v_{bn} = \sqrt{3}\, V_m \sin \left(\omega t + \frac{\pi}{6} \right)$$

$$v_{bc} = v_{bn} - v_{cn} = \sqrt{3}\, V_m \sin \left(\omega t - \frac{\pi}{2} \right)$$

$$v_{ca} = v_{cn} - v_{an} = \sqrt{3}\, V_m \sin \left(\omega t + \frac{5\pi}{6} \right)$$

If v_{o1} and v_{o2} are the output voltages of converters 1 and 2, respectively, the instantaneous voltage across the inductor during interval $(\pi/6 + \alpha_1) \le \omega t \le (\pi/2 + \alpha_1)$ is

$$v_r = v_{o1} + v_{o2} = v_{ab} - v_{bc}$$

$$= \sqrt{3}\, V_m \left[\sin \left(\omega t + \frac{\pi}{6} \right) - \sin \left(\omega t - \frac{\pi}{2} \right) \right] \tag{5-64}$$

$$= 3V_m \cos \left(\omega t - \frac{\pi}{6} \right)$$

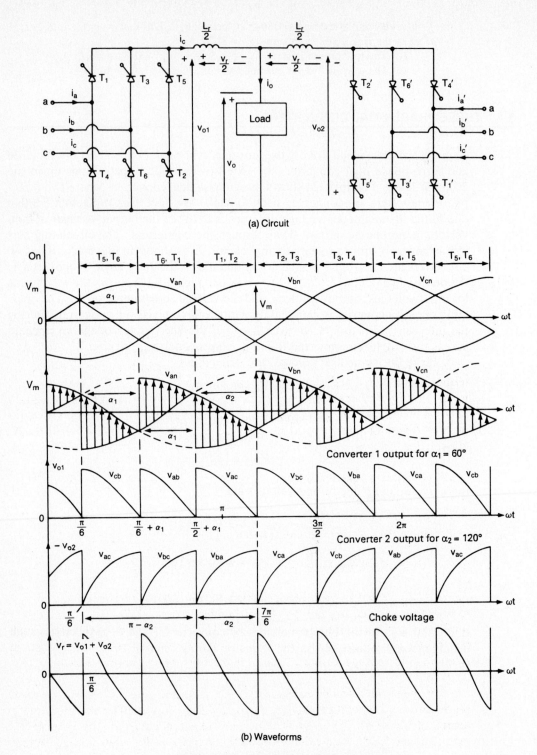

Figure 5-12 Three-phase dual converter.

The circulating current can be found from

$$i_r(t) = \frac{1}{\omega L_r} \int_{\pi/6 + \alpha_1}^{\omega t} v_r \, d(\omega t) = \frac{1}{\omega L_r} \int_{\pi/6 + \alpha_1}^{\omega t} 3V_m \cos\left(\omega t - \frac{\pi}{6}\right) d(\omega t)$$

$$= \frac{3V_m}{\omega L_r} \left[\sin\left(\omega t - \frac{\pi}{6}\right) - \sin \alpha_1 \right]$$

(5-65)

The circulating current depends on delay angle α_1 and on inductance L_r. This current becomes maximum when $\omega t = 2\pi/3$ and $\alpha_1 = 0$. Even without any external load, the converters would be continuously running due to the circulating current as a result of ripple voltage across the inductor. This allows smooth reversal of load current during the change over from one quadrant operation to another and provides fast dynamic responses, especially for electrical motor drives.

5-11 POWER FACTOR IMPROVEMENTS

The power factor of phase-controlled converters depends on delay angle α, and is in general low, especially at the low output voltage range. These converters generate harmonics into the supply. Forced commutations can improve the input power factor and reduce the harmonics levels. These forced-commutation techniques are becoming attractive to ac–dc conversion. With the advancement of power semiconductor devices (e.g., gate-turn-off thyristors), the forced commutation can be implemented in practical systems. In this section the basic techniques of forced commutation for ac–dc converters are discussed, and these can be classified as follows:

1. Extinction angle control
2. Symmetrical angle control
3. Pulse-width modulation
4. Sinusoidal pulse-width modulation

5-11.1 Extinction Angle Control

Figure 5-13a shows a single-phase semiconverter, where thyristors T_1 and T_2 are replaced by switches S_1 and S_2. The switching actions of S_1 and S_2 can be performed by gate-turn-off thyristors (GTOs). The characteristics of GTOs are such that a GTO can be turned on by applying a short positive pulse to its gate as in the case of normal thyristors and can be turned off by applying a short negative pulse to its gate.

In an extinction angle control, switch S_1 is turned on at $\omega t = 0$ and is turned off by forced commutation at $\omega t = \pi - \beta$. Switch S_2 is turned on at $\omega t = \pi$ and is turned off at $\omega t = (2\pi - \beta)$. The output voltage is controlled by varying the extinction angle, β. Figure 5-13b shows the waveforms for input voltage, output voltage, input current, and the current through thyristor switches. The fundamen-

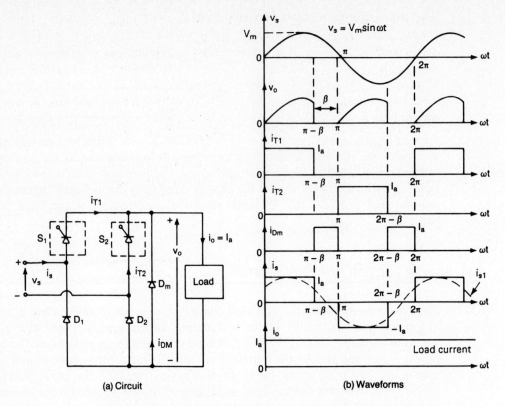

(a) Circuit   (b) Waveforms

Figure 5-13   Single-phase forced-commutated semiconverter.

tal component of input current leads the input voltage, and the displacement factor (and power factor) is leading. In some applications, this feature may be desirable to simulate a capacitive load and to compensate for line voltage drops.

The average output voltage is found from

$$V_{dc} = \frac{2}{2\pi} \int_0^{\pi - \beta} V_m \sin \omega t \, d(\omega t) = \frac{V_m}{\pi} (1 + \cos \beta) \qquad (5\text{-}66)$$

and V_{dc} can be varied from $2V_m/\pi$ to 0 by varying β from 0 to π. The rms output voltage is given by

$$V_{rms} = \left[\frac{2}{2\pi} \int_0^{\pi - \beta} V_m^2 \sin^2 \omega t \, d(\omega t) \right]^{1/2}$$

$$= \frac{V_m}{\sqrt{2}} \left[\frac{1}{\pi} \left(\pi - \beta + \frac{\sin 2\beta}{2} \right) \right]^{1/2} \qquad (5\text{-}67)$$

Figure 5-14a shows a single-phase full converter, where thyristors T_1, T_2, T_3, and T_4 are replaced by forced-commutated switches S_1, S_2, S_3, and S_4. Each switch conducts for 180°. Switches S_1 and S_2 are both on from $\omega t = 0$ to $\omega t = \pi - \beta$ and supply power to the load during the positive half-cycle of the input voltage.

Similarly, switches S_3 and S_4 are both on from $\omega t = \pi$ to $\omega t = 2\pi - \beta$ and supply power to the load during the negative half-cycle of the input voltage. For an inductive load, the freewheeling path for the load current must be provided by switches S_1S_4 or S_3S_2. The firing swquence would be 12,14, 34, and 32. Figure 5-14b shows the waveforms for input voltage, output voltage, input current, and the current through switches. Each switch conducts for 180° and this converter is operated as a semiconverter. The freewheeling action is accomplished through two switches of the same arm. The average and rms output voltage are expressed by Eq. (5-66) and Eq. (5-67), respectively.

The performance of semi- and full converters with extinction angle control are similar to those with phase-angle control, except the power factor is leading. With phase-angle control, the power factor is lagging.

5-11.2 Symmetrical Angle Control

The symmetrical angle control allows one-quadrant operation and Fig. 5-13a shows a single-phase semiconverter with forced-commutated switches S_1 and S_2.

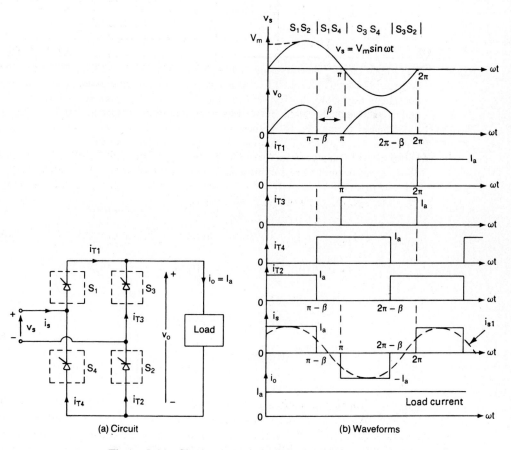

(a) Circuit   (b) Waveforms

Figure 5-14   Single-phase forced-commutated full converter.

Switch S_1 is turned on at $\omega t = (\pi - \beta)/2$ and is turned off at $\omega t = (\pi + \beta)/2$. Switch S_2 is turned on at $\omega t = (3\pi - \beta)/2$ and off at $\omega t = (3\pi + \beta)/2$. The output voltage is controlled by varying conduction angle β. The gate signals are generated by comparing half-sine waves with a dc signal as shown in Fig. 5-15b. Figure 5-15a shows the waveforms for input voltage, output voltage, input current, and the current through switches. The fundamental component of input current is in phase with the input voltage and the displacement factor is unity. Therefore, the power factor is improved.

The average output voltage is found from

$$V_{dc} = \frac{2}{2\pi} \int_{(\pi-\beta)/2}^{(\pi+\beta)/2} V_m \sin \omega t \, d(\omega t) = \frac{2V_m}{\pi} \sin \frac{\beta}{2} \tag{5-68}$$

and V_{dc} can be varied from $2V_m/\pi$ to 0 by varying β from π to 0. The rms output voltage is given by

$$V_{rms} = \left[\frac{2}{2\pi} \int_{(\pi-\beta)/2}^{(\pi+\beta)/2} V_m^2 \sin^2 \omega t \, d(\omega t) \right]^{1/2}$$

$$= \frac{V_m}{\sqrt{2}} \left[\frac{1}{\pi} (\beta + \sin \beta) \right]^{1/2} \tag{5-69}$$

Example 5-13
_____

The single-phase full converter in Fig. 5-14a is operated with symmetrical angle control. The load current with an average value of I_a, is continuous, where the ripple content is negligible. (a) Express the input current of converter in Fourier series, and determine the harmonic factor HF of input current, displacement factor DF, and input power factor PF. (b) If the conduction angle is $\beta = \pi/3$ and the peak input voltage is $V_m = 169.83$ V, calculate V_{dc}, V_{rms}, HF, DF, and PF.

Solution (a) The waveform for input current is shown in Fig. 5-15a and the instantaneous input current can be expressed in Fourier series as

$$i_s(t) = I_{dc} + \sum_{n=1,2,\ldots}^{\infty} (a_n \cos n\omega t + b_n \sin n\omega t)$$

where

$$a_o = I_{dc} = \frac{1}{2\pi} \left[\int_{(\pi-\beta)/2}^{(\pi+\beta)/2} I_a \, d(\omega t) - \int_{(3\pi-\beta)/2}^{(3\pi+\beta)/2} I_a \, d(\omega t) \right] = 0$$

$$a_n = \frac{1}{\pi} \int_0^{2\pi} i_s(t) \cos n\omega t \, d(\omega t) = 0$$

$$b_n = \frac{1}{\pi} \int_0^{2\pi} i_s(t) \sin n\omega t \, d(\omega t) = \frac{4I_a}{n\pi} \sin \frac{n\beta}{2} \qquad \text{for } n = 1, 3, \ldots$$

$$= 0 \qquad \text{for } n = 2, 4, \ldots$$

Since $I_{dc} = 0$, the input current can be written as

$$i_s(t) = \sum_{n=1,3,5,\ldots}^{\infty} \sqrt{2} I_n \sin(n\omega t + \phi_n) \tag{5-70}$$

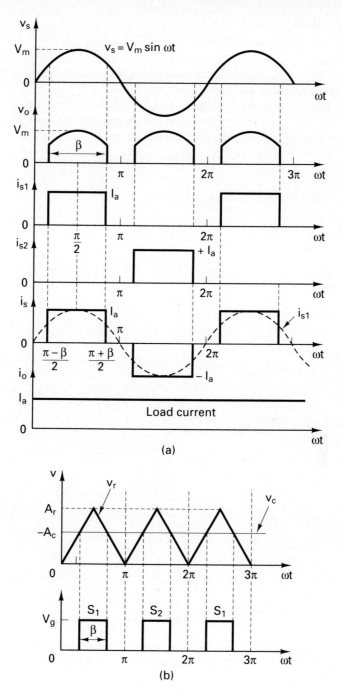

(a)

(b)

Figure 5-15 Symmetrical angle control.

where

$$\phi_n = \tan^{-1} \frac{a_n}{b_n} = 0 \tag{5-71}$$

The rms value of the nth harmonic input current is given as

$$I_{sn} = \frac{1}{\sqrt{2}} (a_n^2 + b_n^2)^{1/2} = \frac{2\sqrt{2}\, I_a}{n\pi} \sin \frac{n\beta}{2} \tag{5-72}$$

The rms value of the fundamental current is

$$I_{s1} = \frac{2\sqrt{2}\, I_a}{\pi} \sin \frac{\beta}{2} \tag{5-73}$$

The rms input current is found as

$$I_s = I_a \sqrt{\frac{\beta}{\pi}} \tag{5-74}$$

$$\text{HF} = \left[\left(\frac{I_s}{I_{s1}} \right)^2 - 1 \right]^{1/2} = \left[\frac{\pi\beta}{4(1 - \cos \beta)} - 1 \right]^{1/2} \tag{5-75}$$

$$\text{DF} = \cos \phi_1 = 1 \tag{5-76}$$

$$\text{PF} = \left(\frac{I_{s1}}{I_s} \right) \text{DF} = \frac{2\sqrt{2}}{\sqrt{\beta\pi}} \sin \frac{\beta}{2} \tag{5-77}$$

(b) $\beta = \pi/3$ and DF = 1.0. From Eq. (5-68),

$$V_{\text{dc}} = \left(2 \times \frac{169.83}{\pi} \right) \sin \frac{\pi}{6} = 54.06 \text{ V}$$

From Eq. (5-69),

$$V_{\text{rms}} = \frac{169.83}{\sqrt{2}} \left(\frac{\beta + \sin \beta}{\pi} \right)^{1/2} = 93.72 \text{ V}$$

$$I_{s1} = I_a \left(\frac{2\sqrt{2}}{\pi} \right) \sin \frac{\pi}{6} = 0.4502 I_a$$

$$I_s = I_a \sqrt{\frac{\beta}{\pi}} = 0.5774 I_a$$

$$\text{HF} = \left[\left(\frac{I_s}{I_{s1}} \right)^2 - 1 \right]^{1/2} = 0.803 \quad \text{or} \quad 80.3\%$$

$$\text{PF} = \frac{I_{s1}}{I_s} = 0.7797 \text{ (lagging)}$$

Note. The power factor is improved significantly, even higher than that of the single-phase series full converter in Fig. 5-6a. However, the harmonic factor is increased.

5-11.3 Pulse-Width-Modulation Control

If the output voltage of single-phase semi- or full converters is controlled by varying the delay angle, extinction angle, or symmetrical angle, there is only one

---

pulse per half-cycle in the input current of the converter, and as a result the lowest-order harmonic is the third. It is difficult to filter out the lower-order harmonic current. In pulse-width-modulation (PWM) control, the converter switches are turned on and off several times during a half-cycle and the output voltage is controlled by varying the width of pulses. The gate signals are generated by comparing a triangular wave with a dc signal as shown in Fig. 5-16b. Figure 5-16a shows the input voltage, output voltage, and input current. The lower-order harmonics can be eliminated or reduced by selecting the number of pulses per half-cycle. However increasing the number of pulses would also increase the magnitude of higher-order harmonics, which could easily be filtered out.

The output voltage and the performance parameters of the converter can be determined in two steps: (1) by considering only one pair of pulses such that if one pulse starts at $\omega t = \alpha_1$ and ends at $\omega t = \alpha_1 + \delta_1$, the other pulse starts at $\omega t = \pi + \alpha_1$ and ends at $\omega t = (\pi + \alpha_1 + \delta_1)$, and (2) by combining the effects of all pairs. If mth pulse starts at $\omega t = \alpha_m$ and its width is δ_m, the average output voltage due to p number of pulses is found from

$$
\begin{aligned}
V_{\text{dc}} &= \sum_{m=1}^{p} \left[\frac{2}{2\pi} \int_{\alpha_m}^{\alpha_m + \delta_m} V_m \sin \omega t \, d(\omega t) \right] \\
&= \frac{V_m}{\pi} \sum_{m=1}^{p} [\cos \alpha_m - \cos(\alpha_m + \delta_m)]
\end{aligned}
\tag{5-78}
$$

If the load current with an average value of I_a is continuous and has negligible ripple, the instantaneous input current can be expressed in a Fourier series as

$$
i_s(t) = I_{\text{dc}} + \sum_{n=1,3,\dots}^{\infty} (a_n \cos n\omega t + b_n \sin n\omega t)
\tag{5-79}
$$

Due to symmetry of the input current waveform, there will be no even harmonics and I_{dc} should be zero and the coefficients of Eq. (5-79) are

$$
\begin{aligned}
a_n &= \frac{1}{\pi} \int_0^{2\pi} i_s(t) \cos n\omega t \, d(\omega t) \\
&= \sum_{m=1}^{p} \left[\frac{1}{\pi} \int_{\alpha_m}^{\alpha_m + \delta_m} I_a \cos n\omega t \, d(\omega) - \frac{1}{\pi} \int_{\pi + \alpha_m}^{\pi + \alpha_m + \delta_m} I_a \cos n\omega t \, d(\omega t) \right] = 0 \\
b_n &= \frac{1}{\pi} \int_0^{2\pi} i_s(t) \sin n\omega t \, d(\omega t) \\
&= \sum_{m=1}^{p} \left[\frac{1}{\pi} \int_{\alpha_m}^{\alpha_m + \delta_m} I_a \sin n\omega t \, d(\omega t) - \frac{1}{\pi} \int_{\pi + \alpha_m}^{\pi + \alpha_m + \delta_m} I_a \sin n\omega t \, d(\omega t) \right] \\
&= \frac{2I_a}{n\pi} \sum_{m=1}^{p} [\cos n\alpha_m - \cos n(\alpha_m + \delta_m)] \qquad \text{for } n = 1, 3, 5, \dots
\end{aligned}
\tag{5-80}
$$

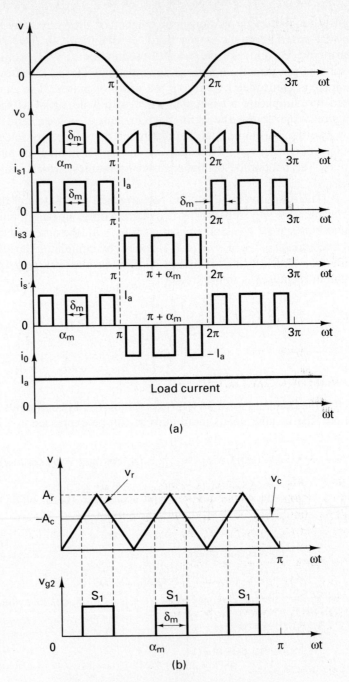

Figure 5-16 Pulse-width-modulation control.

Equation (5-79) can be rewritten as

$$i_s(t) = \sum_{n=1,3,\ldots}^{\infty} \sqrt{2}\, I_n \sin(n\omega t + \phi_n) \qquad (5\text{-}81)$$

where $\phi_n = \tan^{-1}(a_n/b_n) = 0$ and $I_n = (a_n^2 + b_n^2)^{1/2}/\sqrt{2} = b_n/\sqrt{2}$.

5-11.4 Sinusoidal Pulse-Width Modulation

The widths of pulses can be varied to control the output voltage. If there are p pulses per half-cycle with the equal width, the maximum width of a pulse is π/p. However, the pulse widths of pulses could be different. It is possible to choose the widths of pulses in such a way that certain harmonics could be eliminated. There are different methods of varying the widths of pulses and the most common one is the sinusoidal pulse-width modulation (SPWM). In sinusoidal PWM control as shown in Fig. 5-17, the pulse widths are generated by comparing a triangular reference voltage v_r of amplitude A_r and frequency f_r with a carrier half-sinusoidal voltage v_c of variable amplitude A_c and frequency $2f_s$. The sinusoidal voltage v_c is in phase with the input phase voltage v_s and has twice the supply frequency f_s. The widths of the pulses (and the output voltage) are varied by changing the amplitude A_c or the modulation index M from 0 to 1. The *modulation*

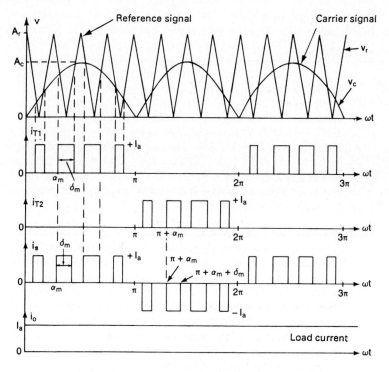

Figure 5-17 Sinusoidal pulse-width control.

index is defined as

$$M = \frac{A_c}{A_r} \qquad (5\text{-}82)$$

In a sinusoidal PWM control, the displacement factor is unity and the power factor is improved. The lower-order harmonics are eliminated or reduced. For example, with four pulses per half-cycle the lowest-order harmonic is the fifth; and with six pulses per half-cycle, the lowest-order harmonic is the seventh. Computer programs can be used to evaluate the performances of uniform PWM and sinusoidal PWM control, respectively.

5-12 DESIGN OF CONVERTER CIRCUITS

The design of converter circuits requires determining the ratings of thyristors and diodes. The thyristors and diodes are specified by the average current, rms current, peak current, and peak inverse voltage. In the case of controlled rectifiers, the current ratings of devices depend on the delay (or control) angle. The ratings of power devices must be designed under the worst-case condition and this occurs when the converter delivers the maximum average output voltage V_{dm}.

The output of converters contains harmonics that depend on the control (or delay) angle and the worst-case condition generally prevails under the minimum output voltage. Input and output filters must be designed under the minimum output voltage condition. The steps involved in designing the converters and filters are similar to those of rectifier circuit design in Section 3-13.

Example 5-14

A three-phase full converter is operated from a three-phase 230-V 60-Hz supply. The load is highly inductive and the average load current is $I_a = 150$ A with negligible ripple content. If the delay angle is $\alpha = \pi/3$, determine the ratings of thyristors.
Solution The waveforms for thyristor currents are shown in Fig. 5-10b. $V_s = 230/\sqrt{3} = 132.79$ V, $V_m = 187.79$ V, and $\alpha = \pi/3$. From Eq. (5-57), $V_{dc} = 3(\sqrt{3}/\pi) \times 187.79 \times \cos(\pi/3) = 155.3$ V. The output power $P_{dc} = 155.3 \times 150 = 23{,}295$ W. The average current through a thyristor $I_A = 150/3 = 50$ A. The rms current through a thyristor $I_R = 150\sqrt{2/6} = 86.6$ A. The peak current through a thyristor $I_{PT} = 150$ A. The peak inverse voltage is the peak amplitude of line-to-line voltage PIV $= \sqrt{3}\,V_m = \sqrt{3} \times 187.79 = 325.27$ V.

Example 5-15*

A single-phase full converter as shown in Fig. 5-18 uses delay-angle control and is supplied from a 120-V 60-Hz supply. (a) Use the method of Fourier series to obtain expressions for output voltage $v_o(t)$ and load current $i_o(t)$ as a function of decay angle α. (b) If $\alpha = \pi/3$, $E = 10$ V, $L = 20$ mH, and $R = 10\ \Omega$, determine the rms value of lowest-order harmonic current in the load. (c) If in part (b), a filter capacitor is connected across the load, determine the capacitor value to reduce the lowest-order harmonic current to 10% of the value without the capacitor. (d) Use PSpice to plot the output voltage and the load current and to compute the total harmonic distortion (THD) of the load current and the input power factor (PF) with the output filter capacitor in part (c).

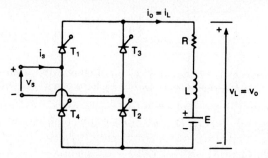

Figure 5-18 Single-phase full converter with RL load.

Solution (a) The waveform for output voltage is shown in Fig. 5-3c. The frequency of output voltage is twice that of the main supply. The instantaneous output voltage can be expressed in a Fourier series as

$$v_o(t) = V_{dc} + \sum_{n=2,4,\ldots}^{\infty} (a_n \cos n\omega t + b_n \sin n\omega t) \tag{5-83}$$

where

$$V_{dc} = \frac{1}{2\pi} \int_\alpha^{2\pi+\alpha} V_m \sin \omega t \, d(\omega t) = \frac{2V_m}{\pi} \cos \alpha$$

$$a_n = \frac{2}{\pi} \int_\alpha^{\pi+\alpha} V_m \sin \omega t \cos n\omega t \, d(\omega t) = \frac{2V_m}{\pi} \left[\frac{\cos(n+1)\alpha}{n+1} - \frac{\cos(n-1)\alpha}{n-1} \right]$$

$$b_n = \frac{2}{\pi} \int_\alpha^{\pi+\alpha} V_m \sin \omega t \sin n\omega t \, d(\omega t) = \frac{2V_m}{\pi} \left[\frac{\sin(n+1)\alpha}{n+1} - \frac{\sin(n-1)\alpha}{n-1} \right]$$

The load impedance

$$Z = R + j(n\omega L) = [R^2 + (n\omega L)^2]^{1/2} \,\underline{/\theta_n}$$

and $\theta_n = \tan^{-1}(n\omega L/R)$. Dividing $v_o(t)$ of Eq. (5-83) by load impedance Z and simplifying the sine and cosine terms give the instantaneous load current as

$$i_o(t) = I_{dc} + \sum_{n=2,4,\ldots}^{\infty} \sqrt{2} I_n \sin(n\omega t + \phi_n - \theta_n) \tag{5-84}$$

where $I_{dc} = (V_{dc} - E)/R$, $\phi_n = \tan^{-1}(a_n/b_n)$, and

$$I_n = \frac{1}{\sqrt{2}} \frac{(a_n^2 + b_n^2)^{1/2}}{\sqrt{R^2 + (n\omega L)^2}}$$

(b) If $\alpha = \pi/3$, $E = 10$ V, $L = 20$ mH, $R = 10$ Ω, $\omega = 2\pi \times 60 = 377$ rad/s, $V_m = \sqrt{2} \times 120 = 169.71$ V, and $V_{dc} = 54.02$ V.

$$I_{dc} = \frac{54.02 - 10}{10} = 4.40 \text{ A}$$

$$a_2 = -0.833, \; b_2 = -0.866, \; \phi_2 = -223.9°, \; \theta_2 = 56.45°$$

$$a_4 = -0.433, \; b_4 = -0.173, \; \phi_4 = 111.79°, \; \theta_4 = 71.65°$$

$$a_6 = -0.029, \; b_6 = 0.297, \; \phi_6 = -5.5°, \; \theta_6 = 77.53°$$

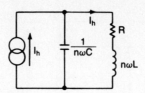

Figure 5-19 Equivalent circuit for harmonics.

$$i_L(t) = 4.4 + \frac{2V_m}{\pi[R^2 + (n\omega L)^2]^{1/2}} [1.2 \sin(2\omega t + 223.9° - 56.45°)$$

$$+ 0.47 \sin(4\omega t + 111.79° - 71.65°) + 0.3 \sin(6\omega t - 5.5° - 77.53°) + \cdot \cdot \cdot]$$

$$= 4.4 + \frac{2 \times 169.71}{\pi[10^2 + (7.54n)^2]^{1/2}} [1.2 \sin(2\omega t + 167.45°)$$

$$+ 0.47 \sin(4\omega t + 40.14°) + 0.3 \sin(6\omega t - 80.03°) + \cdot \cdot \cdot] \tag{5-85}$$

The second harmonic is the lowest one and its rms value is

$$I_2 = \frac{2 \times 169.71}{\pi[10^2 + (7.54 \times 2)^2]^{1/2}} \left(\frac{1.2}{\sqrt{2}}\right) = 5.07 \text{ A}$$

(c) Figure 5-19 shows the equivalent circuit for the harmonics. Using the current-divider rule, the harmonic current through the load is given by

$$\frac{I_h}{I_n} = \frac{1/(n\omega C)}{\{R^2 + [n\omega L - 1/(n\omega C)]^2\}^{1/2}}$$

For $n = 2$ and $\omega = 377$,

$$\frac{I_h}{I_n} = \frac{1/(2 \times 377C)}{\{10^2 + [2 \times 7.54 - 1/(2 \times 377C)]^2\}^{1/2}} = 0.1$$

and this gives $C = -670 \ \mu F$ or 793 μF. Thus $C = 793 \ \mu F$.

(d) The peak supply voltage $V_m = 169.7$ V. For $\alpha_1 = 60°$, time delay $t_1 = (60/360) \times (1000/60 \text{ Hz}) \times 1000 = 2777.78 \ \mu s$ and time delay $t_2 = (240/360) \times (1000/60 \text{ Hz}) \times 1000 = 11,111.1 \ \mu s$. The single-phase full-converter circuit for PSpice simulation is shown in Fig. 5-20a. The gate voltages V_{g1}, V_{g2}, V_{g3}, and V_{g4}

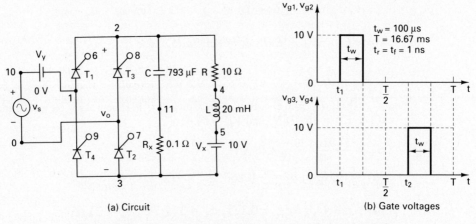

(a) Circuit

(b) Gate voltages

Figure 5-20 Single-phase full converter for PSpice simulation.

for thyristors are shown in Fig. 5-20b. The subcircuit definition for the thyristor model SCR is described in Section 4-14.

The list of the circuit file is as follows:

```
Example 5-15    Single-Phase Full Converter
VS    10   0    SIN (0    169.7V    60HZ)
Vg1   6    2    PULSE (0V   10V    2777.8US    1NS   1NS   100US   16666.7US)
Vg2   7    0    PULSE (0V   10V    2777.8US    1NS   1NS   100US   16666.7US)
Vg3   8    2    PULSE (0V   10V    11111.1US   1NS   1NS   100US   16666.7US)
Vg4   9    1    PULSE (0V   10V    11111.1US   1NS   1NS   100US   16666.7US)
R     2    4    10
L     4    5    20MH
C     2    11   793UF
RX    11   3    0.1          ; Added to help convergence
VX    5    3    DC    10V    ; Load battery voltage
VY    10   1    DC    0V     ; Voltage source to measure supply current
*     Subcircuit calls for thyristor model
XT1   1    2    6    2    SCR        ; Thyristor T1
XT3   0    2    8    2    SCR        ; Thyristor T3
XT2   3    0    7    0    SCR        ; Thyristor T2
XT4   3    1    9    1    SCR        ; Thyristor T4
*     Subcircuit SCR which is missing must be inserted
.TRAN    10US   35MS   16.67MS        ; Transient analysis
.PROBE                                ; Graphics postprocessor
.options abstol = 1.00u reltol = 1.0 m  vntol = 0.1  ITL5=10000
.FOUR    120HZ    I(VX)               ; Fourier analysis
.END
```

The PSpice plots of the output voltage V(2,3) and the load current I(VX) are shown in Fig. 5-21.

The Fourier components of the load current are:

FOURIER COMPONENTS OF TRANSIENT RESPONSE I(VX)

DC COMPONENT = 1.147163E+01

HARMONIC NO	FREQUENCY (HZ)	FOURIER COMPONENT	NORMALIZED COMPONENT	PHASE (DEG)	NORMALIZED PHASE (DEG)
1	1.200E+02	2.136E+00	1.000E+00	−1.132E+02	0.000E+00
2	2.400E+02	4.917E−01	2.302E−01	1.738E+02	2.871E+02
3	3.600E+02	1.823E−01	8.533E−02	1.199E+02	2.332E+02
4	4.800E+02	9.933E−02	4.650E−02	7.794E+01	1.912E+02
5	6.000E+02	7.140E−02	3.342E−02	2.501E+01	1.382E+02
6	7.200E+02	4.339E−02	2.031E−02	−3.260E+01	8.063E+01
7	8.400E+02	2.642E−02	1.237E−02	−7.200E+01	4.123E+01
8	9.600E+02	2.248E−02	1.052E−02	−1.126E+02	6.192E+01
9	1.080E+03	2.012E−02	9.420E−03	−1.594E+02	−4.617E+01

TOTAL HARMONIC DISTORTION = 2.535750E+01 PERCENT

To find the input power factor, we need to find the Fourier components of the input current, which are the same as the current through source VY.

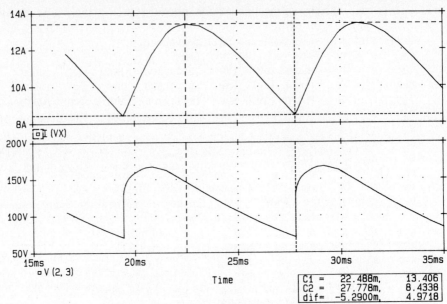

Figure 5-21 SPICE plots for Example 5-15.

FOURIER COMPONENTS OF TRANSIENT RESPONSE I(VY)
DC COMPONENT = 1.013355E-02

HARMONIC NO	FREQUENCY (HZ)	FOURIER COMPONENT	NORMALIZED COMPONENT	PHASE (DEG)	NORMALIZED PHASE (DEG)
1	6.000E+01	2.202E+01	1.000E+00	5.801E+01	0.000E+00
2	1.200E+02	2.073E-02	9.415E-04	4.033E+01	-1.768E+01
3	1.800E+02	1.958E+01	8.890E-01	-3.935E+00	-6.194E+01
4	2.400E+02	2.167E-02	9.841E-04	-1.159E+01	-6.960E+01
5	3.000E+02	1.613E+01	7.323E-01	-5.968E+01	-1.177E+02
6	3.600E+02	2.218E-02	1.007E-03	-6.575E+01	-1.238E+02
7	4.200E+02	1.375E+01	6.243E-01	-1.077E+02	-1.657E+02
8	4.800E+02	2.178E-02	9.891E-04	-1.202E+02	-1.783E+02
9	5.400E+02	1.317E+01	5.983E-01	-1.542E+02	-2.122E+02

TOTAL HARMONIC DISTORTION = 1.440281E+02 PERCENT

Total harmonic distortion of input current, THD = 144% = 1.44

Displacement angle, $\phi_1 = 58.01°$

Displacement factor, DF = $\cos \phi_1 = \cos(-58.01) = 0.53$ (lagging)

$$\text{PF} = \frac{I_{s1}}{I_s} \cos \phi_1 = \frac{1}{[1 + (\%\text{THD}/100)^2]^{1/2}} \cos \phi_1 \qquad (5\text{-}86)$$

$$= \frac{1}{(1 + 1.44^2)^{1/2}} \times 0.53 = 0.302 \text{ (lagging)}$$

Notes

1. The analyses above are valid only if the delay angle α is greater than α_0, which is given by

$$\alpha_0 = \sin^{-1} \frac{E}{V_m} = \sin^{-1} \frac{10}{169.71} = 3.38°$$

2. Due to the filter capacitor C, a high peak charging current flows from the source, and the THD of the input current has a high value of 144%.

3. Without the capacitor C, the load current becomes discontinuous, the peak second harmonic load current is $i_{2(peak)} = 5.845$ A, I_{dc} is 6.257 A, the THD of the load current is 14.75%, and the THD of the input current is 15.66%.

4. An LC filter is normally used to limit the peak supply current, and this is illustrated in Example 5-16.

Example 5-16*

The equivalent circuit of the single-phase full converter with an LC filter is shown in Fig. 5-22a. The input voltage is 120 V (rms), 60 Hz. The delay angle is $\alpha = \pi/3$. The dc output voltage is $V_{dc} = 100$ V at $I_{dc} = 10$ A. Determine the values of inductance L_e, β, and the rms inductor current I_{rms}.

Solution $\omega = 2\pi \times 60 = 377$ rad/s, $V_{dc} = 100$ V, $V_s = 120$ V, $V_m = \sqrt{2} \times 120 = 169.7$ V, $V_{dc} = 100$ V. Voltage ratio $x = V_{dc}/V_m = 100/169.7 = 58.93\%$. Let us assume that the value of C_e is very large, so that its voltage is ripple free with an average value of V_{dc}. L_e is the total inductance, including the source or line inductance. If V_{dc} is less than V_m and the thyristor T is fired at $\omega t = \alpha > \alpha_0$, the load current i_L will begin to flow. For $v_s = V_{dc} = V_m \sin \alpha_0$, α_0 is given by

$$\alpha_0 = \sin^{-1} \frac{V_{dc}}{V_m} = \sin^{-1} x = 36.1°$$

Using Eq. (3-103), the load current i_L is given by

$$i_L = \frac{V_m}{\omega L_e} (\cos \alpha - \cos \omega t) - \frac{V_{dc}}{\omega L_e} (\omega t - \alpha) \qquad \text{for } \omega t \geq \alpha \qquad (5\text{-}87)$$

The value of $\omega t = \beta$ at which the current i_L falls to zero can be found from the condition $i_L(\omega t = \beta) = 0$. That is,

$$\cos \alpha - \cos \beta - x(\beta - \alpha) = 0 \qquad (5\text{-}88)$$

Solving Eq. (5-88) for β by iteration gives $\beta = 97°$. Using Eq. (3-105), we get the normalized average current ratio, $I_{dc}/I_{pk} = 1.865\%$. Thus $I_{pk} = I_{dc}/0.01865 = 536.2$ A. The required value of inductance is

$$L_e = \frac{V_m}{\omega I_{pk}} = \frac{169.7}{377 \times 536.2} = 0.84 \text{ mH}$$

Using Eq. (3-106), we get the normalized rms current ratio $I_{rms}/I_{pk} = 4.835\%$. Thus $I_{rms} = 0.04835 \times I_{pk} = 0.04835 \times 536.2 = 25.92$ A.

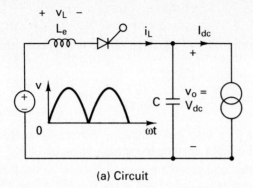

(a) Circuit

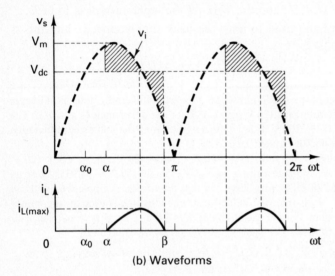

(b) Waveforms

Figure 5-22 Equivalent circuit with *LC* filter.

5-13 EFFECTS OF LOAD AND SOURCE INDUCTANCES

We can notice from Eq. (5-85) that the load current harmonics depend on load inductances. In Example 5-10 the input power factor is calculated for a purely resistive load and in Example 5-11 for a highly inductive load. We can also notice that the input power factor depends on the load power factor.

In the derivations of output voltages and the performance criteria of converters, we have assumed that the source has no inductances and resistances. Normally, the values of line resistances are small and can be neglected. The amount of voltage drop due to source inductances is equal to that of rectifiers and does not change due to the phase control. Equation (3-107) can be applied to calculate the voltage drop due to the line commutating reactance L_c. If all the line inductances are equal, Eq. (3-110) gives the voltage drop as $V_{6x} = 6fL_cI_{dc}$ for a three-phase full converter.

The voltage drop is not dependent on delay angle α_1 under normal operation. However, the commutation (or overlap) angle μ will vary with the delay angle. As the delay angle is increased, the overlap angle becomes smaller. This is illustrated in Fig. 5-23. The volt-time integral as shown by crosshatched areas is equal to $I_{dc}L_c$ and is independent of voltages. As the commutating phase voltage increases, the time required to commutate gets smaller, but the "volt-seconds" remain the same.

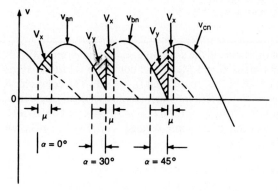

Figure 5-23 Relationship between delay angle and overlap angle.

If V_x is the average voltage drop per commutation due to overlap and V_y is the average voltage reduction due to phase-angle control, the average output voltage for a delay angle of α is

$$V_{dc}(\alpha) = V_{dc}(\alpha = 0) - V_y = V_{dm} - V_y \qquad (5\text{-}89)$$

and

$$V_y = V_{dm} - V_{dc}(\alpha) \qquad (5\text{-}90)$$

where V_{dm} = maximum possible average output voltage. The average output voltage with overlap angle μ and two commutations is

$$V_{dc}(\alpha + \mu) = V_{dc}(\alpha = 0) - 2V_x - V_y = V_{dm} - 2V_x - V_y \qquad (5\text{-}91)$$

Substituting V_y from Eq. (5-90) into Eq. (5-91) we can write the voltage drop due to overlap as

$$2V_x = 2f_s I_{dc} L_c = V_{dc}(\alpha) - V_{dc}(\alpha + \mu) \qquad (5\text{-}92)$$

The overlap angle μ can be determined from Eq. (5-92) for known values of load current I_{dc}, commutating inductance L_c, and delay angle α. It should be noted that Eq. (5-92) is applicable only to a single-phase full converter.

Example 5-17*

A three-phase full converter is supplied from a three-phase 230-V 60-Hz supply. The load current is continuous and has negligible ripple. If the average load current I_{dc} = 150 A and the commutating inductance L_c = 0.1 mH, determine the overlap angle when (a) $\alpha = 10°$, (b) $\alpha = 30°$, and (c) $\alpha = 60°$.
Solution $V_m = \sqrt{2} \times 230/\sqrt{3} = 187.79$ V and $V_{dm} = 3\sqrt{3}\, V_m/\pi = 310.61$ V. From Eq. (5-57), $V_{dc}(\alpha) = 310.6 \cos \alpha$ and

$$V_{dc}(\alpha + \mu) = 310.61 \cos(\alpha + \mu)$$

For a three-phase converter, Eq. (5-92) can be modified to

$$6V_x = 6f_s I_{dc} L_c = V_{dc}(\alpha) - V_{dc}(\alpha + \mu)$$

$$6 \times 60 \times 150 \times 0.1 \times 10^{-3} = 310.61[\cos \alpha - \cos(\alpha + \mu)]$$

(5-93)

(a) For $\alpha = 10°$, $\mu = 4.66°$.
(b) For $\alpha = 30°$, $\mu = 1.94°$.
(c) For $\alpha = 60°$, $\mu = 1.14°$.

Example 5-18

The holding current of thyristors in the single-phase full converter of Fig. 5-3a is $I_H = 500$ mA and the delay time is $t_d = 1.5$ μs. The converter is supplied from a 120-V 60-Hz supply and has a load of $L = 10$ mH and $R = 10$ Ω. The converter is operated with a delay angle of $\alpha = 30°$. Determine the minimum value of gate pulse width, t_G.

Solution $I_H = 500$ mA $= 0.5$ A, $t_d = 1.5$ μs, $\alpha = 30° = \pi/6$, $L = 10$ mH, and $R = 10$ Ω. The instantaneous value of the input voltage is $v_s(t) = V_m \sin \omega t$, where $V_m = \sqrt{2} \times 120 = 169.7$ V.

At $\omega t = \alpha$,

$$V_1 = v_s(\omega t = \alpha) = 169.7 \times \sin \frac{\pi}{6} = 84.85 \text{ V}$$

The rate of rise of anode current di/dt at the instant of triggering is approximately

$$\frac{di}{dt} = \frac{V_1}{L} = \frac{84.85}{10 \times 10^{-3}} = 8485 \text{ A/s}$$

If di/dt is assumed constant for a short time after the gate triggering, the time t_1 required for the anode current to rise to the level of holding current is calculated from $t_1 \times (di/dt) = I_H$ or $t_1 \times 8485 = 0.5$ and this gives $t_1 = 0.5/8485 = 58.93$ μs. Therefore, the minimum width of the gate pulse is

$$t_G = t_1 + t_d = 58.93 + 1.5 = 60.43 \text{ } \mu\text{s}$$

5-14 GATING CIRCUITS

The generation of gating signals for thyristors of ac–dc converters requires (1) detecting zero crossing of the input voltage, (2) appropriate phase shifting of signals, (3) pulse shaping to generate pulses of short duration, and (4) pulse isolation through pulse transformers or optocouplers. The block diagram for gating circuit of a single-phase full converter is shown in Fig. 5-24.

SUMMARY

In this chapter we have seen that average output voltage (and output power) of ac–dc converters can be controlled by varying the conduction time of power devices. Depending on the types of supply, the converters could be single-phase or three-phase. For each type of supply, they can be half-wave, semi-, or full converters. The semi- and full converters are used extensively in practical appli-

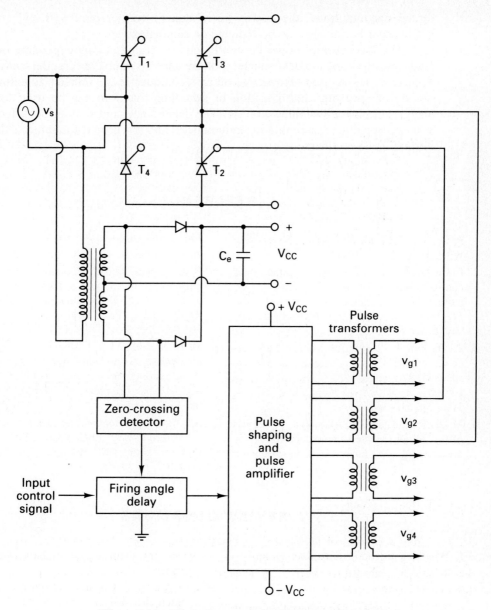

Figure 5-24 Block diagram for a thyristor gating circuit.

cations. Although semiconverters provide better input power factor than that of full converters, these converters are only suitable for one-quadrant operation. Full converters and dual converters allow two-quadrant and four-quadrant operations, respectively. Three-phase converters are normally used in high-power applications and the frequency of output ripples is higher.

The input power factor, which is dependent on the load, can be improved and the voltage rating can be increased by series connection of converters. With forced commutations, the power factor can be further improved and certain lower-order harmonics can be reduced or eliminated.

The load current could be continuous or discontinuous depending on the load-time constant and delay angle. For the analysis of converters, the method of Fourier series is used. However, other techniques (e.g., transfer function approach or spectrum multiplication of switching function) can be used for the analysis of power switching circuits. The delay-angle control does not affect the voltage drop due to commutating inductances, and this drop is the same as that of normal diode rectifiers.

REFERENCES

1. P. C. Sen, *Thyristor DC Drives*. New York: Wiley-Interscience, 1981.

2. P. C. Sen and S. R. Doradla, "Evaluation of control schemes for thyristor controlled dc motors." *IEEE Transactions on Industrial Electronics and Control Instrumentation,* Vol. IEC125, No. 3, 1978, pp. 247–255.

3. P. D. Ziogas, "Optimum voltage and harmonic control PWM techniques for 3-phase static UPS systems." *IEEE Transactions on Industry Applications,* Vol. IA16, No. 4, 1980, pp. 542–546.

4. M. H. Rashid, and M. Aboudina, "Analysis of forced-commutated techniques for ac–dc converters." *1st European Conference on Power Electronics and Applications,* Brussels, October 16–18, 1985, pp. 2.263–2.266.

5. P. D. Ziogas, and P. Photiadis, "An exact output current analysis of ideal static PWM inverters." *IEEE Transactions on Industry Applications,* Vol. IA119, No. 2, 1983, pp. 281–295.

6. M. H. Rashid and A. I. Maswood, "Analysis of 3-phase ac–dc converters under unbalanced supply conditions." *IEEE Industry Applications Conference Record,* 1985, pp. 1190–1194.

7. A. D. Wilcox, *Engineering Design for Electrical Engineers*. Englewood Cliffs, N.J.: Prentice Hall, 1990; Chapter 10, "Power module," by M. H. Rashid.

REVIEW QUESTIONS

5-1. What is a natural or line commutation?

5-2. What is a controlled rectifier?

5-3. What is a converter?

5-4. What is a delay-angle control of converters?

5-5. What is a semiconverter? Draw two semiconverter circuits.

5-6. What is a full converter? Draw two full-converter circuits.

5-7. What is a dual converter? Draw two dual-converter circuits.

5-8. What is the principle of phase control?

5-9. What are the effects of removing the free-wheeling diode in single-phase semiconverters?

5-10. Why is the power factor of semiconverters better than that of full converters?

5-11. What is the cause of circulating current in dual converters?

5-12. Why is a circulating current inductor required in dual converters?

5-13. What are the advantages and disadvantages of series converters?

5-14. How is the delay angle of one converter related to the delay angle of the other converter in a dual-converter system?

5-15. What is the inversion mode of converters?

5-16. What is the rectification mode of converters?

5-17. What is the frequency of the lowest-order harmonic in three-phase semiconverters?

5-18. What is the frequency of the lowest-order harmonic in three-phase full converters?

5-19. What is the frequency of the lowest-order harmonic in a single-phase semiconverter?

5-20. How are gate-turn-off thyristors turned on and off?

5-21. How is a phase-control thyristor turned on and off?

5-22. What is a forced commutation? What are the advantages of forced commutation for ac–dc converters?

5-23. What is extinction-angle control of converters?

5-24. What is symmetrical-angle control of converters?

5-25. What is pulse-width-modulation control of converters?

5-26. What is sinusoidal pulse-width-modulation control of a converter?

5-27. What is the modulation index?

5-28. How is the output voltage of a phase-control converter varied?

5-29. How is the output voltage of a sinusoidal PWM control converter varied?

5-30. Does the commutation angle depend on the delay angle of converters?

5-31. Does the voltage drop due to commutating inductances depend on the delay angle of converters?

5-32. Does the input power factor of converters depend on the load power factor?

5-33. Do the output ripple voltages of converters depend on the delay angle?

PROBLEMS

5-1. A single-phase half-wave converter in Fig. 5-1a is operated from a 120-V 60-Hz supply. If the load resistive load is $R = 10 \ \Omega$ and the delay angle is $\alpha = \pi/3$, determine **(a)** the efficiency, **(b)** the form factor, **(c)** the ripple factor, **(d)** the transformer utilization factor, and **(e)** the peak inverse voltage (PIV) of thyristor T_1.

5-2. A single-phase half-wave converter in Fig. 5-1a is operated from a 120-V 60-Hz supply and the load resistive load is $R = 10 \ \Omega$. If the average output voltage is 25% of the maximum possible average output voltage, calculate **(a)** the delay angle, **(b)** the rms and average output currents, **(c)** the average and rms thyristor currents, and **(d)** the input power factor.

5-3. A single-phase half-converter in Fig. 5-1a is supplied from a 120-V 60-Hz supply and a freewheeling diode is connected across the load. The load consists of series-connected resistance, $R = 10 \ \Omega$, inductance $L = 5$ mH, and battery voltage, $E =$

20 V. **(a)** Express the instantaneous output voltage in a Fourier series, and **(b)** determine the rms value of the lowest-order output harmonic current.

5-4. A single-phase semiconverter in Fig. 5-2a is operated from a 120-V 60-Hz supply. The load current with an average value of I_a is continuous with negligible ripple content. The turns ratio of the transformer is unity. If the delay angle is $\alpha = \pi/3$, calculate **(a)** the harmonic factor of input current, **(b)** the displacement factor, and **(c)** the input power factor.

5-5. Repeat Prob. 5-2 for the single-phase semiconverter in Fig. 5-2a.

5-6. The single-phase semiconverter in Fig. 5-2a is operated from a 120-V 60-Hz supply. The load consists of series-connected resistance $R = 10 \ \Omega$, inductance $L = 5$ mH, and battery voltage $E = 20$ V. **(a)** Express the output voltage in a Fourier series, and **(b)** determine the rms value of the lowest-order output harmonic current.

5-7. Repeat Prob. 5-4 for the single-phase full converter in Fig. 5-3a.

5-8. Repeat Prob. 5-2 for the single-phase full converter in Fig. 5-3a.

5-9. Repeat Prob. 5-6 for the single-phase full converter in Fig. 5-3a.

5-10. The dual converter in Fig. 5-4a is operated from a 120-V 60-Hz supply and delivers ripple-free average current of $I_{dc} = 20$ A. The circulating inductance is $L_r = 5$ mH, and the delay angles are $\alpha_1 = 30°$ and $\alpha_2 = 150°$. Calculate the peak circulating current and the peak current of converter 1.

5-11. A single-phase series semiconverter in Fig. 5-5a is operated from a 120-V 60-Hz supply and the load resistance is $R = 10\ \Omega$. If the average output voltage is 75% of the maximum possible average output voltage, calculate **(a)** the delay angles of converters, **(b)** the rms and average output currents, **(c)** the average and rms thyristor currents, and **(d)** the input power factor.

5-12. A single-phase series semiconverter in Fig. 5-5a is operated from a 120-V 60-Hz supply. The load current with an average value of I_a is continuous and the ripple content is negligible. The turns ratio of the transformer is $N_p/N_s = 2$. If delay angles are $\alpha_1 = 0$ and $\alpha_2 = \pi/3$, calculate **(a)** the harmonic factor of input current, **(b)** the displacement factor, and **(c)** the input power factor.

5-13. Repeat Prob. 5-11 for the single-phase series full converter in Fig. 5-6a.

5-14. Repeat Prob. 5-12 for the single-phase series full converter in Fig. 5-6a.

5-15. The three-phase half-wave converter in Fig. 5-7a is operated from a three-phase wye-connected 220-V 60-Hz supply and a freewheeling diode is connected across the load. The load current with an average value of I_a is continuous and the ripple content is negligible. If the delay angle $\alpha = \pi/3$, calculate **(a)** the harmonic factor of input current, **(b)** the displacement factor, and **(c)** the input power factor.

5-16. The three-phase half-wave converter in Fig. 5-7a is operated from a three-phase

wye-connected 220-V 60-Hz supply and the load resistance is $R = 10\ \Omega$. If the average output voltage is 25% of the maximum possible average output voltage, calculate **(a)** the delay angle, **(b)** the rms and average output currents, **(c)** the average and rms thyristor currents, **(d)** the rectification efficiency, **(e)** the transformer utilization factor, and **(f)** the input power factor.

5-17. The three-phase half-wave converter in Fig. 5-7a is operated from a three-phase wye-connected 220-V 60-Hz supply and a freewheeling diode is connected across the load. The load consists of series-connected resistance $R = 10\ \Omega$, inductance $L = 5$ mH, and battery voltage $E = 20$ V. **(a)** Express the instantaneous output voltage in a Fourier series, and **(b)** determine the rms value of the lowest-order harmonic on the output current.

5-18. The three-phase semiconverter in Fig. 5-8a is operated from a three-phase wye-connected 220-V 60-Hz supply. The load current with an average value of I_a is continuous with negligible ripple content. The turns ratio of the transformer is unity. If the delay angle is $\alpha = 2\pi/3$, calculate **(a)** the harmonic factor of input current, **(b)** the displacement factor, and **(c)** the input power factor.

5-19. Repeat Prob. 5-16 for the three-phase semiconverter in Fig. 5-8a.

5-20. Repeat Prob. 5-19 if the average output voltage is 90% of the maximum possible output voltage.

5-21. Repeat Prob. 5-17 for the three-phase semiconverter in Fig. 5-8a.

5-22. Repeat Prob. 5-18 for the three-phase full converter in Fig. 5-10a.

5-23. Repeat Prob. 5-16 for the three-phase full converter in Fig. 5-10a.

5-24. Repeat Prob. 5-17 for the three-phase full converter in Fig. 5-10a.

5-25. The three-phase dual converter in Fig. 5-12a is operated from a three-phase wye-connected 220-V 60-Hz supply and the load resistance, $R = 10\ \Omega$. The circulating inductance $L_r = 5$ mH, and the delay angles are $\alpha_1 = 60°$ and $\alpha_2 = 120°$. Calculate the peak circulating current and the peak current of converters.

5-26. The single-phase semiconverter of Fig. 5-2a has an RL load of $L = 1.5$ mH, $R = 1.5$ Ω, and $E = 0$ V. The input voltage is $V_s = 120$ V (rms) at 60 Hz. **(a)** Determine (1) the load current I_o at $\omega t = 0$, and the load current I_1 at $\omega t = \alpha = 30°$, (2) the average thyristor current I_A, (3) the rms thyristor current I_R, (4) the rms output current I_{rms}, and (5) the average output current I_{dc}. **(b)** Use SPICE to check your results.

5-27. The single-phase full converter of Fig. 5-3a has an RL load having $L = 4.5$ mH, $R = 1.5$ Ω, and $E = 10$ V. The input voltage is $V_s = 120$ V at (rms) 60 Hz. **(a)** Determine (1) the load current I_o at $\omega t = \alpha = 30°$, (2) the average thyristor current I_A, (3) the rms thyristor current I_R, (4) the rms output current I_{rms}, and (5) the average output current I_{dc}. **(b)** Use SPICE to check your results.

5-28. The three-phase full converter of Fig. 5-10a has a load of $L = 1.5$ mH, $R = 1.5$ Ω, and $E = 0$ V. The line-to-line input voltage is $V_{ab} = 208$ V (rms), 60 Hz. The delay angle is $\alpha = \pi/6$. **(a)** Determine (1) the steady-state load current I_1 at $\omega t' = \pi/3 + \alpha$ (or $\omega t = \pi/6 + \alpha$), (2) the average thyristor current I_A, (3) the rms thyristor current I_R, (4) the rms output current I_{rms}, and (5) the average output current I_{dc}. **(b)** Use SPICE to check your results.

5-29. The single-phase semiconverter in Fig. 5-13a is operated from a 120-V 60-Hz supply and uses an extinction angle control. The load current with an average value of I_a is continuous and has negligible ripple content. If the extinction angle is $\beta = \pi/3$, calculate **(a)** the outputs V_{dc} and V_{rms}, **(b)** the harmonic factor of input current, **(c)** the displacement factor, and **(d)** the input power factor.

5-30. Repeat Prob. 5-29 for the single-phase full converter in Fig. 5-14a.

5-31. Repeat Prob. 5-18 if symmetrical-angle control is used.

5-32. Repeat Prob. 5-18 if extinction-angle control is used.

5-33. The single-phase semiconverter in Fig. 5-13a is operated with a sinusoidal PWM control and is supplied from a 120-V 60-Hz supply. The load current with an average value of I_a is continuous with negligible ripple content. There are five pulses per half-cycle and the pulses are $\alpha_1 = 7.93°$, $\delta_1 = 5.82°$; $\alpha_2 = 30°$, $\delta_2 = 16.25°$; $\alpha_3 = 52.07°$, $\delta_3 = 127.93°$; $\alpha_4 = 133.75°$, $\delta_4 = 16.25°$; and $\alpha_5 = 166.25°$, $\delta_5 = 5.82°$. Calculate **(a)** the V_{dc} and V_{rms}, **(b)** the harmonic factor of input current, **(c)** the displacement factor, and **(d)** the input power factor.

5-34. Repeat Prob. 5-33 for five pulses per half-cycle with equal pulse width, $M = 0.8$.

5-35. A three-phase semiconverter is operated from a three-phase wye-connected 220-V 60-Hz supply. The load current is continuous and has negligible ripple. The average load current is $I_{dc} = 150$ A and commutating inductance per phase is $L_c = 0.5$ mH. Determine the overlap angle if **(a)** $\alpha = \pi/6$, and **(b)** $\alpha = \pi/3$.

5-36. The input voltage to the equivalent circuit in Fig. 5-22a is 120 V (rms), 60 Hz. The delay angle is $\alpha = \pi/6$. The dc output voltage is $V_{dc} = 75$ V at $I_{dc} = 10$ A. Determine the values of inductance L_e, β, and the rms inductor current I_{rms}.

5-37. The holding current of thyristors in the three-phase full converter in Fig. 5-10a is $I_H = 200$ mA and the delay time is 2.5 μs. The converter is supplied from a three-phase wye-connected 208-V 60-Hz supply and has a load of $L = 8$ mH and $R = 2$ Ω; it is operated with a delay angle of $\alpha = 60°$. Determine the minimum width of gate pulse width, t_G.

5-38. Repeat Prob. 5-37 if $L = 0$.

6

Ac voltage controllers

6-1 INTRODUCTION

If a thyristor switch is connected between ac supply and load, the power flow can be controlled by varying the rms value of ac voltage applied to the load; and this type of power circuit is known as an *ac voltage controller*. The most common applications of ac voltage controllers are: industrial heating, on-load transformer tap changing, light controls, speed control of polyphase induction motors, and ac magnet controls. For power transfer, two types of control are normally used:

1. On–off control
2. Phase-angle control

In on–off control, thyristor switches connect the load to the ac source for a few cycles of input voltage and then disconnect it for another few cycles. In phase control, thyristor switches connect the load to the ac source for a portion of each cycle of input voltage.

The ac voltage controllers can be classified into two types: (1) single-phase controllers and (2) three-phase controllers. Each type can be subdivided into (a) unidirectional or half-wave control and (b) bidirectional or full-wave control. There are various configurations of three-phase controllers depending on the connections of thyristor switches.

Since the input voltage is ac, thyristors are line commutated; and phase-control thyristors, which are relatively inexpensive and slower than fast-switching thyristors, are normally used. For applications up to 400 Hz, if TRIACs are available to meet the voltage and current ratings of a particular application, TRIACs are more commonly used. The thyristor commutation techniques are discussed in Chapter 7.

190

Due to line or natural commutation, there is no need of extra commutation circuitry and the circuits for ac voltage controllers are very simple. Due to the nature of output waveforms, the analysis for the derivations of explicit expressions for the performance parameters of circuits is not simple, especially for phase-angle-controlled converters with RL loads. For the sake of simplicity, resistive loads are considered in this chapter to compare the performances of various configurations. However, the practical loads are of RL type and should be considered in the design and analysis of ac voltage controllers.

6-2 PRINCIPLE OF ON–OFF CONTROL

The principle of on–off control can be explained with a single-phase full-wave controller as shown in Fig. 6-1a. The thyristor switch connects the ac supply to load for a time t_n; the switch is turned off by a gate pulse inhibiting for time t_0. The on-time, t_n, usually consists of an integral number of cycles. The thyristors are turned on at the zero-voltage crossings of ac input voltage. The gate pulses for thyristors T_1 and T_2 and the waveforms for input and output voltages are shown in Fig. 6-1b.

This type of control is applied in applications which have a high mechanical inertia and high thermal time constant (e.g., industrial heating and speed control

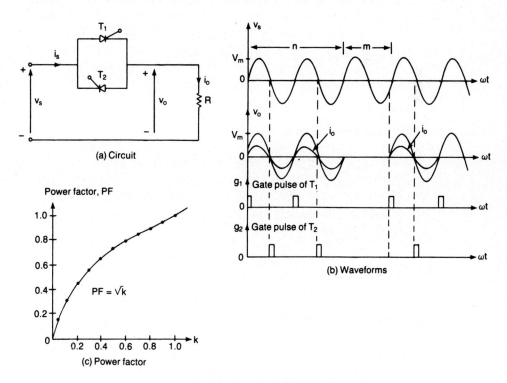

(a) Circuit

(b) Waveforms

(c) Power factor

Figure 6-1 On–off control.

of motors). Due to zero-voltage and zero-current switching of thyristors, the harmonics generated by switching actions are reduced.

For a sinusoidal input voltage, $v_s = V_m \sin \omega t = \sqrt{2} \, V_s \sin \omega t$. If the input voltage is connected to load for n cycles and is disconnected for m cycles, the rms output (or load) voltage can be found from

$$V_o = \left[\frac{n}{2\pi(n + m)} \int_0^{2\pi} 2V_s^2 \sin^2 \omega t \, d(\omega t) \right]^{1/2}$$

(6-1)

$$= V_s \sqrt{\frac{n}{m + n}} = V_s \sqrt{k}$$

where $k = n/(m + n)$ and k is called the *duty cycle*. V_s is the rms phase voltage. The circuit configurations for on–off control are similar to those of phase control and the performance analysis is also similar. For these reasons, the phase-control techniques are only discussed and analyzed in this chapter.

Example 6-1

An ac voltage controller in Fig. 6-1a has a resistive load of $R = 10 \ \Omega$ and the rms input voltage is $V_s = 120$ V, 60 Hz. The thyristors switch is on for $n = 25$ cycles and is off for $m = 75$ cycles. Determine (a) the rms output voltage V_o, (b) the input power factor PF, and (c) the average and rms current of thyristors.
Solution $R = 10 \ \Omega$, $V_s = 120$ V, $V_m = \sqrt{2} \times 120 = 169.7$ V, and $k = n/(n + m) = 25/100 = 0.25$.

(a) From Eq. (6-1), the rms value of output voltage is

$$V_o = V_s \sqrt{k} = V_s \sqrt{\frac{n}{m + n}} = 120 \sqrt{\frac{25}{100}} = 60 \text{ V}$$

and the rms load current is $I_o = V_o/R = 60/10 = 6.0$ A.

(b) The load power is $P_o = I_o^2 R = 6^2 \times 10 = 360$ W. Since the input current is the same as the load current, the input volt-amperes is

$$\text{VA} = V_s I_s = V_s I_o = 120 \times 6 = 720 \text{ W}$$

The input power factor is

$$\text{PF} = \frac{P_o}{\text{VA}} = \sqrt{\frac{n}{m + n}} = \sqrt{k}$$

(6-2)

$$= \sqrt{0.25} = \frac{360}{720} = 0.5 \text{ (lagging)}$$

(c) The peak thyristor current is $I_m = V_m/R = 169.7/10 = 16.97$ A. The average current of thyristors is

$$I_A = \frac{n}{2\pi(m + n)} \int_0^{\pi} I_m \sin \omega t \, d(\omega t) = \frac{I_m n}{\pi(m + n)} = \frac{k I_m}{\pi}$$

(6-3)

$$= \frac{16.97}{\pi} \times 0.25 = 1.33 \text{ A}$$

The rms current of thyristors is

$$I_R = \left[\frac{n}{2\pi(m + n)} \int_0^{\pi} I_m^2 \sin^2 \omega t \, d(\omega t) \right]^{1/2} = \frac{I_m}{2} \sqrt{\frac{n}{m + n}} = \frac{I_m \sqrt{k}}{2}$$

(6-4)

$$= \frac{16.97}{2} \times \sqrt{0.25} = 4.24 \text{ A}$$

Notes

1. The power factor and output voltage vary with the square root of the duty cycle. The power factor is poor at the low value of the duty cycle, k, and is shown in Fig. 6-1c.

2. If T is the period of the input voltage, $(m + n)T$ is the period of on–off control. $(m + n)T$ should be less than the mechanical or thermal time constant of the load, and is usually less than 1 s, but not in hours or days. The sum of m and n is generally around 100.

3. If Eq. (6-2) is used to determine the power factor with m and n in days, it will give erroneous results. For example, if $m = 3$ days and $n = 3$ days, Eq. (6-2) gives PF = $[3/(3 + 3)]^{1/2} = 0.707$, which is not physically possible. Because if the controller is on for 3 days and off for 3 days, the power factor will be dependent on the load impedance angle θ.

6-3 PRINCIPLE OF PHASE CONTROL

The principle of phase control can be explained with reference to Fig. 6-2a. The power flow to the load is controlled by delaying the firing angle of thyristor T_1. Figure 6-2b illustrates the gate pulses of thyristor T_1 and the waveforms for the input and output voltages. Due to the presence of diode D_1, the control range is limited and the effective rms output voltage can only be varied between 70.7 and 100%. The output voltage and input current are asymmetrical and contain a dc component. If there is an input transformer, it may cause a saturation problem. This circuit is a single-phase half-wave controller and is suitable only for low-power resistive loads, such as heating and lighting. Since the power flow is controlled during the positive half-cycle of input voltage, this type of controller is also known as a *unidirectional controller*.

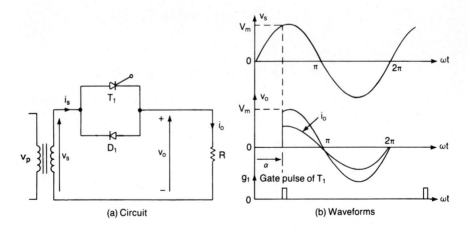

(a) Circuit

(b) Waveforms

Figure 6-2 Single-phase angle control.

If $v_s = V_m \sin \omega t = \sqrt{2}\, V_s \sin \omega t$ is the input voltage and the delay angle of thyristor T_1 is $\omega t = \alpha$, the rms output voltage is found from

$$V_o = \left\{ \frac{1}{2\pi} \left[\int_\alpha^\pi 2V_s^2 \sin^2 \omega t\, d(\omega t) + \int_\pi^{2\pi} 2V_s^2 \sin^2 \omega t\, d(\omega t) \right] \right\}^{1/2}$$

$$= \left\{ \frac{2V_s^2}{4\pi} \left[\int_\alpha^\pi (1 - \cos 2\omega t)\, d(\omega t) + \int_\pi^{2\pi} (1 - \cos 2\omega t)\, d(\omega t) \right] \right\}^{1/2} \qquad (6\text{-}5)$$

$$= V_s \left[\frac{1}{2\pi} \left(2\pi - \alpha + \frac{\sin 2\alpha}{2} \right) \right]^{1/2}$$

The average value of output voltage is

$$V_{dc} = \frac{1}{2\pi} \left[\int_\alpha^\pi \sqrt{2}\, V_s \sin \omega t\, d(\omega t) + \int_\pi^{2\pi} \sqrt{2}\, V_s \sin \omega t\, d(\omega t) \right]$$

$$= \frac{\sqrt{2}\, V_s}{2\pi} (\cos \alpha - 1) \qquad (6\text{-}6)$$

If α is varied from 0 to π, V_o varies from V_s to $V_s/\sqrt{2}$ and V_{dc} varies from 0 to $-\sqrt{2}\, V_s/\pi$.

Example 6-2

A single-phase ac voltage controller in Fig. 6-2a has a resistive load of $R = 10\ \Omega$ and the input voltage is $V_s = 120$ V, 60 Hz. The delay angle of thyristor T_1 is $\alpha = \pi/2$. Determine (a) the rms value of output voltage V_o, (b) the input power factor PF, and (c) the average input current.

Solution $R = 10\ \Omega$, $V_s = 120$ V, $\alpha = \pi/2$, and $V_m = \sqrt{2} \times 120 = 169.7$ V.

(a) From Eq. (6-5), the rms value of the output voltage

$$V_o = 120 \sqrt{\frac{3}{4}} = 103.92\ \text{V}$$

(b) The rms load current

$$I_o = \frac{V_o}{R} = \frac{103.92}{10} = 10.392\ \text{A}$$

The load power

$$P_o = I_o^2 R = 10.392^2 \times 10 = 1079.94\ \text{W}$$

Since the input current is the same as the load current, the input volt-ampere rating is

$$\text{VA} = V_s I_s = V_s I_o = 120 \times 10.392 = 1247.04\ \text{VA}$$

The input power factor

$$\text{PF} = \frac{P_o}{\text{VA}} = \frac{V_o}{V_s} = \left[\frac{1}{2\pi} \left(2\pi - \alpha + \frac{\sin 2\alpha}{2} \right) \right]^{1/2} \qquad (6\text{-}7)$$

$$= \sqrt{\frac{3}{4}} = \frac{1079.94}{1247.04} = 0.866\ \text{(lagging)}$$

(c) From Eq. (6-6), the average output voltage

$$V_{dc} = -120 \times \frac{\sqrt{2}}{2\pi} = -27 \text{ V}$$

and the average input current

$$I_D = \frac{V_{dc}}{R} = -\frac{27}{10} = -2.7 \text{ A}$$

Note. The negative sign of I_D signifies that the input current during the positive half-cycle is less than that during the negative half-cycle. If there is an input transformer, the transformer core may be saturated. The unidirectional control is not normally used in practice.

6-4 SINGLE-PHASE BIDIRECTIONAL CONTROLLERS WITH RESISTIVE LOADS

The problem of dc input current can be prevented by using bidirectional (or full-wave) control, and a single-phase full-wave controller with a resistive load is shown in Fig. 6-3a. During the positive half-cycle of input voltage, the power flow is controlled by varying the delay angle of thyristor T_1; and thyristor T_2 controls the power flow during the negative half-cycle of input voltage. The firing pulses of T_1 and T_2 are kept 180° apart. The waveforms for the input voltage, output voltage, and gating signals for T_1 and T_2 are shown in Fig. 6-3b.

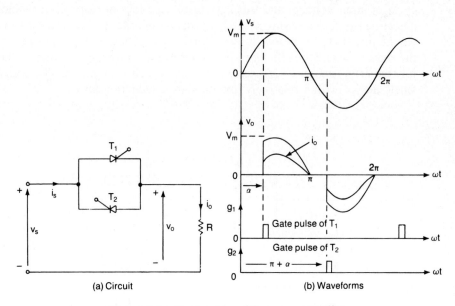

(a) Circuit (b) Waveforms

Figure 6-3 Single-phase full-wave controller.

If $v_s = \sqrt{2}\, V_s \sin \omega t$ is the input voltage, and the delay angles of thyristors T_1 and T_2 are equal $(\alpha_1 = \alpha_2 = \alpha)$, the rms output voltage can be found from

$$V_o = \left\{ \frac{2}{2\pi} \int_\alpha^\pi 2V_s^2 \sin^2 \omega t \; d(\omega t) \right\}^{1/2}$$

$$= \left\{ \frac{4V_s^2}{4\pi} \int_\alpha^\pi (1 - \cos 2\omega t) \; d(\omega t) \right\}^{1/2} \tag{6-8}$$

$$= V_s \left[\frac{1}{\pi} \left(\pi - \alpha + \frac{\sin 2\alpha}{2} \right) \right]^{1/2}$$

By varying α from 0 to π, V_o can be varied from V_s to 0.

In Fig. 6-3a, the gating circuits for thyristors T_1 and T_2 must be isolated. It is possible to have a common cathode for T_1 and T_2 by adding two diodes as shown in Fig. 6-4. Thyristor T_1 and diode D_1 conduct together during the positive half-cycle; and thyristor T_2 and diode D_2 conduct during the negative half-cycle. Since this circuit can have a common terminal for gating signals of T_1 and T_2, only one isolation circuit is required, but at the expense of two power diodes. Due to two power devices conducting at the same time, the conduction losses of devices would increase and efficiency would be reduced.

A single-phase full-wave controller can also be implemented with one thyristor and four diodes as shown in Fig. 6-5a. The four diodes act as a bridge rectifier. The voltage across thyristor T_1, and its current, are always unidirectional. With a resistive load, the thyristor current would fall to zero due to natural commutation in every half-cycle, as shown in Fig. 6-5b. However, if there is a large inductance in the circuit, thyristor T_1 may not be turned off in every half-cycle of input voltage, and this may result in a loss of control. It would require detecting the zero crossing of the load current in order to guarantee turn-off of the conducting thyristor before firing the next one. Three power devices conduct at the same time and the efficiency is also reduced. The bridge rectifier and thyristor (or transistor) act as a *bidirectional switch*, which is commercially available as a single device with a relatively low on-state conduction loss.

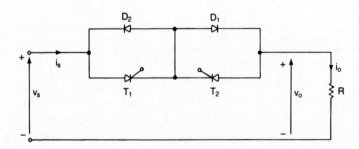

Figure 6-4 Single-phase full-wave controller with common cathode.

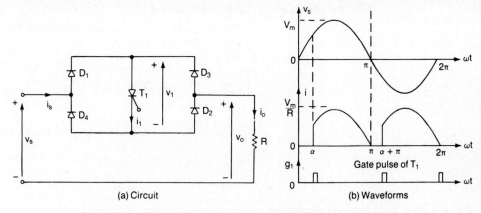

(a) Circuit   (b) Waveforms

Figure 6-5   Single-phase full-wave controller with one thyristor.

Example 6-3

A single-phase full-wave ac voltage controller in Fig. 6-3a has a resistive load of $R = 10\ \Omega$ and the input voltage is $V_s = 120$ V (rms), 60 Hz. The delay angles of thyristors T_1 and T_2 are equal: $\alpha_1 = \alpha_2 = \alpha = \pi/2$. Determine (a) the rms output voltage V_o, (b) the input power factor PF, (c) the average current of thyristors I_A, and (d) the rms current of thyristors I_R.

Solution $R = 10\ \Omega$, $V_s = 120$ V, $\alpha = \pi/2$, and $V_m = \sqrt{2} \times 120 = 169.7$ V.

(a) From Eq. (6-8), the rms output voltage

$$V_o = \frac{120}{\sqrt{2}} = 84.85\ \text{V}$$

(b) The rms value of load current, $I_o = V_o/R = 84.85/10 = 8.485$ A and the load power, $P_o = I_o^2 R = 8.485^2 \times 10 = 719.95$ W. Since the input current is the same as the load current, the input volt-ampere rating

$$\text{VA} = V_s I_s = V_s I_o = 120 \times 8.485 = 1018.2\ \text{W}$$

The input power factor

$$\text{PF} = \frac{P_o}{\text{VA}} = \frac{V_o}{V_s} = \left[\frac{1}{\pi}\left(\pi - \alpha + \frac{\sin 2\alpha}{2}\right)\right]^{1/2}$$

$$= \frac{1}{\sqrt{2}} = \frac{719.95}{1018.2} = 0.707\ \text{(lagging)}$$

(6-9)

(c) The average thyristor current

$$I_A = \frac{1}{2\pi R}\int_\alpha^\pi \sqrt{2}\ V_s \sin \omega t\ d(\omega t)$$

$$= \frac{\sqrt{2}\ V_s}{2\pi R}(\cos \alpha + 1)$$

(6-10)

$$= \sqrt{2} \times \frac{120}{2\pi \times 10} = 2.7\ \text{A}$$

(d) The rms value of the thyristor current

$$I_R = \left[\frac{1}{2\pi R^2} \int_\alpha^\pi 2V_s^2 \sin^2 \omega t \; d(\omega t)\right]^{1/2}$$

$$= \left[\frac{2V_s^2}{4\pi R^2} \int_\alpha^\pi (1 - \cos 2\omega t) \; d(\omega t)\right]^{1/2}$$

$$= \frac{V_s}{\sqrt{2} \; R} \left[\frac{1}{\pi}\left(\pi - \alpha + \frac{\sin 2\alpha}{2}\right)\right]^{1/2} \tag{6-11}$$

$$= \frac{120}{2 \times 10} = 6A$$

6-5 SINGLE-PHASE CONTROLLERS WITH INDUCTIVE LOADS

Section 6-4 deals with the single-phase controllers with resistive loads. In practice, most loads are inductive to a certain extent. A full-wave controller with an *RL* load is shown in Fig. 6-6a. Let us assume that thyristor T_1 is fired during the positive half-cycle and carries the load current. Due to inductance in the circuit, the current of thyristor T_1 would not fall to zero at $\omega t = \pi$, when the input voltage starts to be negative. Thyristor T_1 will continue to conduct until its current i_1 falls to zero at $\omega t = \beta$. The conduction angle of thyristor T_1 is $\delta = \beta - \alpha$ and depends on the delay angle α and the power factor angle of load θ. The waveforms for the thyristor current, gating pulses, and input voltage are shown in Fig. 6-6b.

If $v_s = \sqrt{2} \; V_s \sin \omega t$ is the instantaneous input voltage and the delay angle of thyristor T_1 is α, thyristor current i_1 can be found from

$$L \frac{di_1}{dt} + Ri_1 = \sqrt{2} \; V_s \sin \omega t \tag{6-12}$$

The solution of Eq. (6-12) is of the form

$$i_1 = \frac{\sqrt{2} \; V_s}{Z} \sin(\omega t - \theta) + A_1 e^{-(R/L)t} \tag{6-13}$$

where load impedance $Z = [R^2 + (\omega L)^2]^{1/2}$ and load angle $\theta = \tan^{-1}(\omega L/R)$.

The constant A_1 can be determined from the initial condition: at $\omega t = \alpha$, $i_1 = 0$. From Eq. (6-13) A_1 is found as

$$A_1 = -\frac{\sqrt{2} \; V_s}{Z} \sin(\alpha - \theta)e^{(R/L)(\alpha/\omega)} \tag{6-14}$$

Substitution of A_1 from Eq. (6-14) in Eq. (6-13) yields

$$i_1 = \frac{\sqrt{2} \; V_s}{Z} [\sin(\omega t - \theta) - \sin(\alpha - \theta)e^{(R/L)(\alpha/\omega - t)}] \tag{6-15}$$

The angle β, when current i_1 falls to zero and thyristor T_1 is turned off, can be found from the condition $i_1(\omega t = \beta) = 0$ in Eq. (6-15) and is given by the relation

$$\sin(\beta - \theta) = \sin(\alpha - \theta)e^{(R/L)(\alpha - \beta)/\omega} \tag{6-16}$$

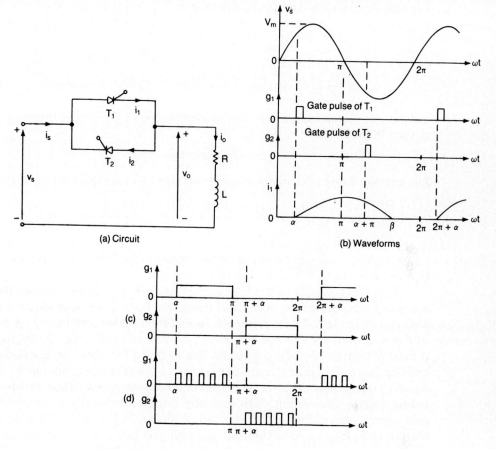

(a) Circuit

(b) Waveforms

(c)

(d)

Figure 6-6 Single-phase full-wave controller with *RL* load.

The angle β, which is also known as an *extinction angle,* can be determined from this transcendental equation and it requires an iterative method of solution. Once β is known, the conduction angle δ of thyristor T_1 can be found from

$$\delta = \beta - \alpha \qquad (6\text{-}17)$$

The rms output voltage

$$V_o = \left[\frac{2}{2\pi} \int_\alpha^\beta 2V_s^2 \sin^2 \omega t \; d(\omega t) \right]^{1/2}$$

$$= \left[\frac{4V_s^2}{4\pi} \int_\alpha^\beta (1 - \cos 2\omega t) \; d(\omega t) \right]^{1/2} \qquad (6\text{-}18)$$

$$= V_s \left[\frac{1}{\pi} \left(\beta - \alpha + \frac{\sin 2\alpha}{2} - \frac{\sin 2\beta}{2} \right) \right]^{1/2}$$

The rms thyristor current can be found from Eq. (6-15) as

$$I_R = \left[\frac{1}{2\pi} \int_\alpha^\beta i_1^2 \, d(\omega t)\right]^{1/2}$$

$$= \frac{V_s}{Z} \left[\frac{1}{\pi} \int_\alpha^\beta \{\sin(\omega t - \theta) - \sin(\alpha - \theta)e^{(R/L)(\alpha/\omega - t)}\}^2 \, d(\omega t)\right]^{1/2} \quad (6-19)$$

and the rms output current can then be determined by combining the rms current of each thyristor as

$$I_o = (I_R^2 + I_R^2)^{1/2} = \sqrt{2} \, I_R \quad (6-20)$$

The average value of thyristor current can also be found from Eq. (6-15) as

$$I_A = \frac{1}{2\pi} \int_\alpha^\beta i_1 \, d(\omega t)$$

$$= \frac{\sqrt{2} \, V_s}{2\pi Z} \int_\alpha^\beta [\sin(\omega t - \theta) - \sin(\alpha - \theta)e^{(R/L)(\alpha/\omega - t)}] \, d(\omega t) \quad (6-21)$$

The gating signals of thyristors could be short pulses for a controller with resistive loads. However, such short pulses are not suitable for inductive loads. This can be explained with reference to Fig. 6-6b. When thyristor T_2 is fired at $\omega t = \pi + \alpha$, thyristor T_1 is still conducting due to load inductance. By the time the current of thyristor T_1 falls to zero and T_1 is turned off at $\omega t = \beta = \alpha + \delta$, the gate pulse of thyristor T_2 has already ceased and consequently, T_2 will not be turned on. As a result, only thyristor T_1 will operate, causing asymmetrical waveforms of output voltage and current. This difficulty can be resolved by using continuous gate signals with a duration of $(\pi - \alpha)$ as shown in Fig. 6-6c. As soon as the current of T_1 falls to zero, thyristor T_2 (with gate pulses as shown in Fig. 6-6c) would be turned on. However, a continuous gate pulse increases the switching loss of thyristors and requires a larger isolating transformer for the gating circuit. In practice, a train of pulses with short durations as shown in Fig. 6-6d are normally used to overcome these problems.

Equation (6-15) indicates that the load voltage (and current) will be sinusoidal if the delay angle, α, is less than the load angle, θ. If a is greater than θ, the load current would be discontinuous and nonsinusoidal.

Notes

1. If $\alpha = \theta$, from Eq. (6-16),

$$\sin(\beta - \theta) = \sin(\beta - \alpha) = 0 \quad (6-22)$$

and

$$\beta - \alpha = \delta = \pi \quad (6-23)$$

2. Since the conduction angle, δ cannot exceed π and the load current must pass through zero, the delay angle α may not be less than θ and the control range of delay angle is

$$\theta \leq \alpha \leq \pi \quad (6-24)$$

3. If $\alpha \leq \theta$ and the gate pulses of thyristors are of long duration, the load current would not change with α, but both thyristors would conduct for π. Thyristor T_1 would turn on at $\omega t = \theta$ and thyristor T_2 would turn on at $\omega t = \pi + \theta$.

Example 6-4*

The single-phase full-wave controller in Fig. 6-6a supplies an *RL* load. The input rms voltage is $V_s = 120$ V, 60 Hz. The load is such that $L = 6.5$ mH and $R = 2.5\ \Omega$. The delay angles of thyristors are equal: $\alpha_1 = \alpha_2 = \pi/2$. Determine (a) the conduction angle of thyristor T_1, δ; (b) the rms output voltage V_o; (c) the rms thyristor current I_R; (d) the rms output current I_o; (e) the average current of a thyristor I_A; and (f) the input power factor PF.

Solution $R = 2.5\ \Omega$, $L = 6.5$ mH, $f = 60$ Hz, $\omega = 2\pi \times 60 = 377$ rad/s, $V_s = 120$ V, $\alpha = 90°$, and $\theta = \tan^{-1}(\omega L/R) = 44.43°$.

(a) The extinction angle can be determined from the solution of Eq. (6-16) and an iterative solution yields $\beta = 220.35°$. The conduction angle is $\delta = \beta - \alpha = 220.43 - 90 = 130.43°$.

(b) From Eq. (6-18), the rms output voltage is $V_o = 68.09$ V.

(c) Numerical integration of Eq. (6-19) between the limits $\omega t = \alpha$ to β gives the rms thyristor current as $I_R = 15.07$ A.

(d) From Eq. (6.20), $I_o = \sqrt{2} \times 15.07 = 21.3$ A.

(e) Numerical integration of Eq. (6-21) yields the average thyristor current as $I_A = 8.23$ A.

(f) The output power $P_o = 21.3^2 \times 2.5 = 1134.2$ W, and the input volt-ampere rating VA $= 120 \times 21.3 = 2556$ W; therefore,

$$\text{PF} = \frac{P_o}{\text{VA}} = \frac{1134.200}{2556} = 0.444 \text{ (lagging)}$$

Note. The switching action of thyristors makes the equations for currents nonlinear. A numerical method of solution for the thyristor conduction angle and currents is more efficient than classical techniques. A computer program is used to solve this example. Students are encouraged to verify the results of this example and to appreciate the usefulness of a numerical solution, especially in solving nonlinear equations of thyristor circuits.

6-6 THREE-PHASE HALF-WAVE CONTROLLERS

The circuit diagram of a three-phase half-wave (or unidirectional) controller is shown in Fig. 6-7 with a wye-connected resistive load. The current flow to the load is controlled by thyristors T_1, T_3, and T_5; and the diodes provide the return current path. The firing sequence of thyristors is T_1, T_3, T_5. For the current to flow through the power controller, at least one thyristor must conduct. If all the devices were diodes, three diodes would conduct at the same time and the conduction angle of each diode would be 180°. We may recall that a thyristor will conduct if its anode voltage is higher than that of its cathode and it is fired. Once a thyristor starts conducting, it would be turned off only when its current falls to zero.

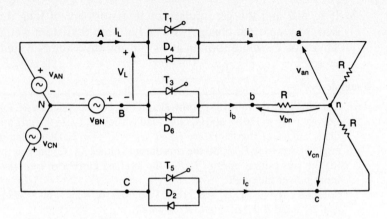

Figure 6-7 Three-phase unidirectional controller.

If V_s is the rms value of input phase voltage and we define the instantaneous input phase voltages as

$$v_{AN} = \sqrt{2}\, V_s \sin \omega t$$

$$v_{BN} = \sqrt{2}\, V_s \sin \left(\omega t - \frac{2\pi}{3} \right)$$

$$v_{CN} = \sqrt{2}\, V_s \sin \left(\omega t - \frac{4\pi}{3} \right)$$

then the input line voltages are

$$v_{AB} = \sqrt{6}\, V_s \sin \left(\omega t + \frac{\pi}{6} \right)$$

$$v_{BC} = \sqrt{6}\, V_s \sin \left(\omega t - \frac{\pi}{2} \right)$$

$$v_{CA} = \sqrt{6}\, V_s \sin \left(\omega t - \frac{7\pi}{6} \right)$$

The waveforms for input voltages, conduction angles of devices, and output voltages are shown in Fig. 6-8 for $\alpha = 60°$ and $\alpha = 150°$. It should be noted that the conduction intervals shown in Fig. 6-8 by dashed lines are not scaled, but have equal widths of 30°. For $0 \le \alpha < 60°$, either two or three devices can conduct at the same time and the possible combinations are (1) two thyristors and one diode, (2) one thyristor and one diode, and (3) one thyristor and two diodes. If three devices conduct, a normal three-phase operation occurs as shown in Fig. 6-9a and the output voltage of a phase is the same as the input phase voltage, for example,

$$v_{an} = v_{AN} = \sqrt{2}\, V_s \sin \omega t \qquad (6\text{-}25)$$

On the other hand, if two devices conduct at the same time, the current flows only through two lines and the third line can be considered open circuited. The line-to-line voltage would appear across two terminals of the load as shown in Fig. 6-9b

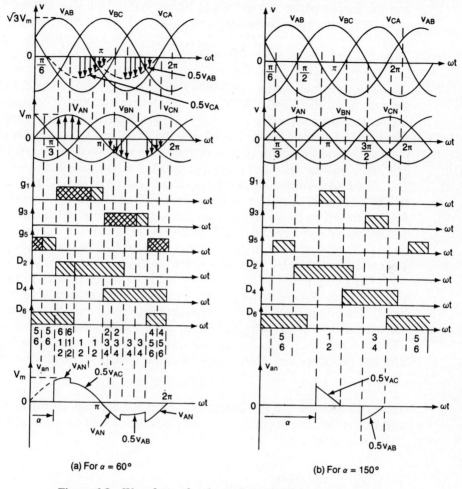

(a) For $\alpha = 60°$ (b) For $\alpha = 150°$

Figure 6-8 Waveforms for three-phase unidirectional controller.

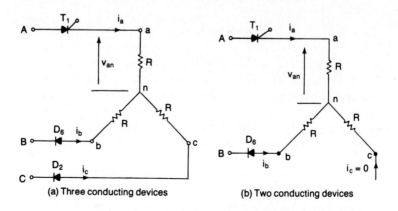

(a) Three conducting devices   (b) Two conducting devices

Figure 6-9 Wye-connected resistive load.

and the output phase voltage would be one-half of the line voltage (e.g., with terminal c being open circuited),

$$v_{an} = \frac{v_{AB}}{2} = \frac{\sqrt{3}\sqrt{2}\ V_s}{2} \sin\left(\omega t + \frac{\pi}{6}\right) \qquad (6\text{-}26)$$

The waveform for an output phase voltage (e.g., v_{an}) can be drawn directly from the input phase and line voltages by noting that v_{an} would correspond to v_{AN} if three devices conduct, to $v_{AB}/2$ (or $v_{AC}/2$) if two devices conduct, and to zero if terminal a is open circuited. For $60° \le \alpha < 120°$, at any time only one thyristor is conducting and the return path is shared by one or two diodes. For $120° \le \alpha < 210°$, only one thyristor and one diode conduct at the same time.

The extinction angle β of a thyristor can be delayed beyond $180°$ (e.g., β of T_1 is $210°$ for $\alpha = 30°$ as shown in Fig. 6-8b). For $\alpha = 60°$, the extinction angle β is delayed to $180°$ as shown in Fig. 6-8a. This is due to the fact that an output phase voltage may depend on the input line-to-line voltage. When v_{AB} becomes zero at $\omega t = 150°$, the current of thyristor T_1 can continue to flow until v_{CA} becomes zero at $\omega t = 210°$ and a delay angle of $\alpha = 210°$ gives zero output voltage (and power).

The gating pulses of thyristors should be continuous, and for example, the pulse of T_1 should end at $\omega t = 210°$. In practice, the gate pulses consist of two parts. The first pulse of T_1 starts anywhere between 0 and $150°$ and ends at $\omega t = 150°$, the second pulse, which can start at $\omega t = 150°$, always ends at $\omega t = 210°$. This allows the current to flow through thyristor T_1 during the period $150° \le \omega t \le 210°$ and increases the control range of output voltage. The range of delay is

$$0 \le \alpha \le 210° \qquad (6\text{-}27)$$

The expression for the rms output phase voltage depends on the range of delay angle. The rms output voltage for a wye-connected load can be found as follows. For $0 \le \alpha < 90°$:

$$V_o = \left[\frac{1}{2\pi}\int_0^{2\pi} v_{an}^2\ d(\omega t)\right]^{1/2}$$

$$= \sqrt{6}\ V_s \left\{\frac{1}{2\pi}\left[\int_\alpha^{2\pi/3} \frac{\sin^2 \omega t}{3}\ d(\omega t) + \int_{\pi/2}^{\pi/2+\alpha} \frac{\sin^2 \omega t}{4}\ d(\omega t)\right.\right.$$

$$\left.\left. + \int_{2\pi/3+\alpha}^{4\pi/3} \frac{\sin^2 \omega t}{3}\ d(\omega t) + \int_{3\pi/2}^{3\pi/2+\alpha} \frac{\sin^2 \omega t}{4}\ d(\omega t) + \int_{4\pi/3+\alpha}^{2\pi} \frac{\sin^2 \omega t}{3}\ d(\omega t)\right]\right\}^{1/2} \qquad (6\text{-}28)$$

$$= \sqrt{3}\ V_s \left[\frac{1}{\pi}\left(\frac{\pi}{3} - \frac{\alpha}{4} + \frac{\sin 2\alpha}{8}\right)\right]^{1/2}$$

For $90° \le \alpha < 120°$:

$$V_o = \sqrt{6}\ V_s \left\{\frac{1}{2\pi}\left[\int_\alpha^{2\pi/3} \frac{\sin^2 \omega t}{3}\ d(\omega t) + \int_{\pi/2}^{\pi} \frac{\sin^2 \omega t}{4}\ d(\omega t)\right.\right.$$

$$\left.\left. + \int_{2\pi/3+\alpha}^{4\pi/3} \frac{\sin^2 \omega t}{3}\ d(\omega t) + \int_{3\pi/2}^{2\pi} \frac{\sin^2 \omega t}{4}\ d(\omega t) + \int_{4\pi/3+\alpha}^{2\pi} \frac{\sin^2 \omega t}{3}\ d(\omega t)\right]\right\}^{1/2} \qquad (6\text{-}29)$$

$$= \sqrt{3}\ V_s \left[\frac{1}{\pi}\left(\frac{11\ \pi}{24} - \frac{\alpha}{2}\right)\right]^{1/2}$$

For $120° \leq \alpha < 210°$:

$$V_o = \sqrt{6} \, V_s \left\{ \frac{1}{2\pi} \left[\int_{\pi/2-2\pi/3+\alpha}^{\pi} \frac{\sin^2 \omega t}{4} \, d(\omega t) + \int_{3\pi/2-2\pi/3+\alpha}^{2\pi} \frac{\sin^2 \omega t}{4} \, d(\omega t) \right] \right\}^{1/2}$$

(6-30)

$$= \sqrt{3} \, V_s \left[\frac{1}{\pi} \left(\frac{7\pi}{24} - \frac{\alpha}{4} + \frac{\sin 2\alpha}{16} - \frac{\sqrt{3} \cos 2\alpha}{16} \right) \right]^{1/2}$$

In the case of a delta-connected load, the output phase voltage would be the same as the line-to-line voltage. However, the line current of the load would depend on the number of devices conducting at the same time. If three devices conduct, the line and phase currents would follow the normal relationship of a three-phase system, as shown in Fig. 6-10a. If the current in phase a is $i_{ab} = I_m \sin \omega t$, the line current will be $i_a = i_{ab} - i_{ca} = \sqrt{3} \, I_m \sin(\omega t - \pi/6)$. If two devices conduct at the same time, one terminal of load can be considered open-circuited as shown in Fig. 6-10b and $i_{ca} = i_{bc} = -i_{ab}/2$. The line current of load would be $i_a = i_{ab} - i_{ca} = (3I_m/2) \sin \omega t = 1.5 I_m \sin \omega t$.

The power devices can be connected together as shown in Fig. 6-11. This arrangement, which permits assembly as a compact unit, is possible only if the neutral of load is accessible.

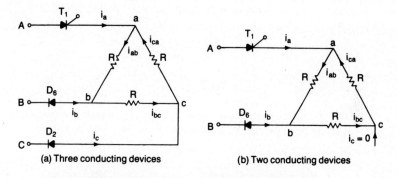

(a) Three conducting devices   (b) Two conducting devices

Figure 6-10   Delta-connected resistive load.

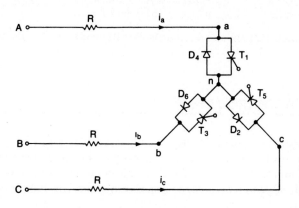

Figure 6-11   Alternative arrangement of three-phase unidirectional controller.

Example 6-5*

The three-phase unidirectional controller in Fig. 6-7 supplies a wye-connected resistive load of $R = 10\ \Omega$ and the line-to-line input voltage is 208 V (rms), 60 Hz. The delay is $\alpha = \pi/3$. Determine (a) the rms output phase voltage V_o, (b) the input power factor PF, and (c) expressions for the instantaneous output voltage of phase a.

Solution $V_L = 208$ V, $V_s = V_L/\sqrt{3} = 208/\sqrt{3} = 120$ V, $\alpha = \pi/3$, and $R = 10\ \Omega$.

(a) From (6-28) the rms output phase voltage is $V_o = 110.86$ V.

(b) The rms phase current of the load $I_a = 110.86/10 = 11.086$ A and the output power

$$P_o = 3I_a^2 R = 3 \times 11.086^2 \times 10 = 3686.98\ \text{W}$$

Since the load is connected in wye, the phase current is equal to the line current, $I_L = I_a = 11.086$ A. The input volt-ampere rating

$$\text{VA} = 3V_s I_L = 3 \times 120 \times 11.086 = 3990.96\ \text{VA}$$

The power factor

$$\text{PF} = \frac{P_o}{\text{VA}} = \frac{3686.98}{3990.96} = 0.924\ \text{(lagging)}$$

(c) If the input phase voltage is taken as the reference and is $v_{AN} = 120\sqrt{2}$ $\sin \omega t = 169.7 \sin \omega t$, the instantaneous input line voltages are

$$v_{AB} = 208\sqrt{2} \sin\left(\omega t + \frac{\pi}{6}\right) = 294.2 \sin\left(\omega t + \frac{\pi}{6}\right)$$

$$v_{BC} = 294.2 \sin\left(\omega t - \frac{\pi}{2}\right)$$

$$v_{CA} = 294.2 \sin\left(\omega t - \frac{7\pi}{6}\right)$$

The instantaneous output phase voltage, v_{an}, which depends on the number of conducting devices, can be determined from Fig. 6-8a as follows:

For $0 \le \omega t < \pi/3$:	$v_{an} = 0$
For $\pi/3 \le \omega t < 4\pi/6$:	$v_{an} = v_{AN} = 169.7 \sin \omega t$
For $4\pi/6 \le \omega t < \pi$:	$v_{an} = v_{AC}/2 = -v_{CA}/2 = 147.1 \sin(\omega t - 7\pi/6 - \pi)$
For $\pi \le \omega t < 4\pi/2$:	$v_{an} = v_{AN} = 169.7 \sin \omega t$
For $4\pi/2 \le \omega t < 5\pi/3$:	$v_{an} = v_{AB}/2 = 147.1 \sin(\omega t + \pi/6)$
For $5\pi/3 \le \omega t < 2\pi$:	$v_{an} = v_{AN} = 169.7 \sin \omega t$

Note. The power factor of this power controller depends on the delay angle α.

6-7 THREE-PHASE FULL-WAVE CONTROLLERS

The unidirectional controllers, which contain dc input current and higher harmonic content due to the asymmetrical nature of the output voltage waveform, are not normally used in ac motor drives; a three-phase bidirectional control is commonly used. The circuit diagram of a three-phase full-wave (or bidirectional) controller is shown in Fig. 6-12 with a wye-connected resistive load. The operation of this controller is similar to that of a half-wave controller, except that the return current path is provided by thyristors T_2, T_4, and T_6 instead of diodes. The firing sequence of thyristors is T_1, T_2, T_3, T_4, T_5, T_6.

If we define the instantaneous input phase voltages as

$$v_{AN} = \sqrt{2}\,V_s \sin \omega t$$

$$v_{BN} = \sqrt{2}\,V_s \sin \left(\omega t - \frac{2\pi}{3}\right)$$

$$v_{CN} = \sqrt{2}\,V_s \sin \left(\omega t - \frac{4\pi}{3}\right)$$

the instantaneous input line voltages are

$$v_{AB} = \sqrt{6}\,V_s \sin \left(\omega t + \frac{\pi}{6}\right)$$

$$v_{BC} = \sqrt{6}\,V_s \sin \left(\omega t - \frac{\pi}{2}\right)$$

$$v_{CA} = \sqrt{6}\,V_s \sin \left(\omega t - \frac{7\pi}{6}\right)$$

The waveforms for the input voltages, conduction angles of thyristors, and output phase voltages are shown in Fig. 6-13 for $\alpha = 60°$ and $\alpha = 120°$. For $0 \le \alpha < 60°$, immediately before the firing of T_1, two thyristors conduct. Once T_1 is fired, three thyristors conduct. A thyristor turns off when its current attempts to reverse. The conditions alternate between two and three conducting thyristors.

For $60° \le \alpha < 90°$, only two thyristors conduct at any time. For $90° \le \alpha < 150°$, although two thyristors conduct at any time, there are periods when no thyristors are on. For $\alpha \ge 150°$, there is no period for two conducting thyristors and the output voltage becomes zero at $\alpha = 150°$. The range of delay angle is

$$0 \le \alpha \le 150° \tag{6-31}$$

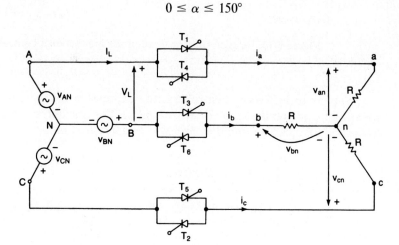

Figure 6-12   Three-phase bidirectional controller.

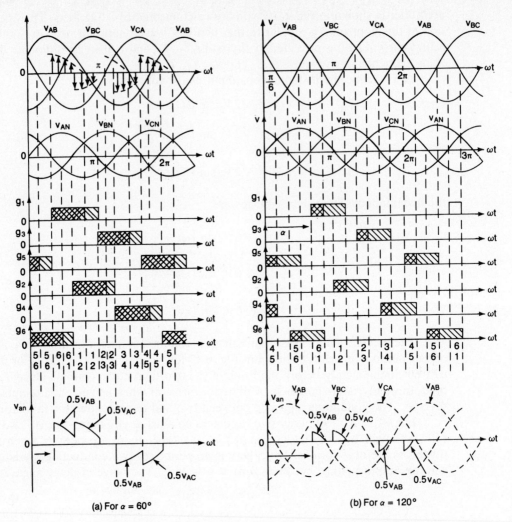

Figure 6-13 Waveforms for three-phase bidirectional controller.

Similar to half-wave controllers, the expression for the rms output phase voltage depends on the range of delay angles. The rms output voltage for a wye-connected load can be found as follows. For $0 \le \alpha < 60°$:

$$V_o = \left[\frac{1}{2\pi} \int_0^{2\pi} v_{an}^2 \, d(\omega t) \right]^{1/2}$$

$$= \sqrt{6} V_s \left\{ \frac{2}{2\pi} \left[\int_\alpha^{\pi/3} \frac{\sin^2 \omega t}{3} \, d(\omega t) + \int_{\pi/4}^{\pi/2+\alpha} \frac{\sin^2 \omega t}{4} \, d(\omega t) \right. \right.$$

$$\left. \left. + \int_{\pi/3+\alpha}^{2\pi/3} \frac{\sin^2 \omega t}{3} \, d(\omega t) + \int_{\pi/2}^{\pi/2+\alpha} \frac{\sin^2 \omega t}{4} \, d(\omega t) \right. \right. \tag{6-32}$$

$$+ \int_{2\pi/3+\alpha}^{\pi} \frac{\sin^2 \omega t}{3} \, d(\omega t) \bigg] \bigg\}^{1/2}$$

$$= \sqrt{6} V_s \left[\frac{1}{\pi} \left(\frac{\pi}{6} - \frac{\alpha}{4} + \frac{\sin 2\alpha}{8} \right) \right]^{1/2}$$

For $60° \leq \alpha < 90°$:

$$V_o = \sqrt{6} V_s \left[\frac{2}{2\pi} \left\{ \int_{\pi/2-\pi/3+\alpha}^{5\pi/6-\pi/3+\alpha} \frac{\sin^2 \omega t}{4} \, d(\omega t) + \int_{\pi/2-\pi/3+\alpha}^{5\pi/6-\pi/3+\alpha} \frac{\sin^2 \omega t}{4} \, d(\omega t) \right\} \right]^{1/2}$$

$$= \sqrt{6} V_s \left[\frac{1}{\pi} \left(\frac{\pi}{12} + \frac{3 \sin 2\alpha}{16} + \frac{\sqrt{3} \cos 2\alpha}{16} \right) \right]^{1/2}$$

(6-33)

For $90° \leq \alpha \leq 150°$:

$$V_0 = \sqrt{6} V_s \left\{ \frac{2}{2\pi} \left[\int_{\pi/2-\pi/3+\alpha}^{\pi} \frac{\sin^2 \omega t}{4} \, d(\omega t) + \int_{\pi/2-\pi/3+\alpha}^{\pi} \frac{\sin^2 \omega t}{4} \, d(\omega t) \right] \right\}^{1/2}$$

$$= \sqrt{6} V_s \left[\frac{1}{\pi} \left(\frac{5\pi}{24} - \frac{\alpha}{4} + \frac{\sin 2\alpha}{16} + \frac{\sqrt{3} \cos 2\alpha}{16} \right) \right]^{1/2}$$

(6-34)

The power devices of a three-phase bidirectional controller can be connected together as shown in Fig. 6-14. This arrangement is also known as *tie control* and allows assembly of all thyristors as one unit.

Example 6-6*

Repeat Example 6-5 for the three-phase bidirectional controller in Fig. 6-12.
Solution $V_L = 208$ V, $V_s = V_L/\sqrt{3} = 208/\sqrt{3} = 120$ V, $\alpha = \pi/3$, and $R = 10 \, \Omega$.
(a) From (6-32) the rms output phase voltage is $V_o = 100.9$ V.
(b) The rms phase current of the load is $I_a = 100.9/10 = 10.09$ A and the output power is

$$P_o = 3I_a^2 R = 3 \times 10.09^2 \times 10 = 3054.24 \text{ W}$$

Since the load is connected in wye, the phase current is equal to the line current, $I_L = I_a = 10.09$ A. The input volt-ampere

$$\text{VA} = 3 V_s I_L = 3 \times 120 \times 10.09 = 3632.4 \text{ VA}$$

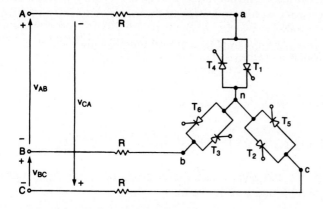

Figure 6-14   Arrangement for three-phase bidirectional tie control.

The power factor

$$PF = \frac{P_o}{VA} = \frac{3054.24}{3632.4} = 0.84 \text{ (lagging)}$$

(c) If the input phase voltage is taken as the reference and is $v_{AN} = 120\sqrt{2} \sin \omega t = 169.7 \sin \omega t$, the instantaneous input line voltages are

$$v_{AB} = 208\sqrt{2} \sin \left(\omega t + \frac{\pi}{6} \right) = 294.2 \sin \left(\omega t + \frac{\pi}{6} \right)$$

$$v_{BC} = 294.2 \sin \left(\omega t - \frac{\pi}{2} \right)$$

$$v_{CA} = 294.2 \sin \left(\omega t - \frac{7\pi}{6} \right)$$

The instantaneous output phase voltage, v_{an}, which depends on the number of conducting devices, can be determined from Fig. 6-13a as follows:

For $0 \leq \omega t < \pi/3$: $v_{an} = 0$

For $\pi/3 \leq \omega t < 2\pi/3$: $v_{an} = v_{AB}/2 = 147.1 \sin(\omega t + \pi/6)$

For $2\pi/3 \leq \omega t < \pi$: $v_{an} = v_{AC}/2 = -v_{CA}/2 = 147.1 \sin(\omega t - 7\pi/6 - \pi)$

For $\pi \leq \omega t < 4\pi/3$: $v_{an} = 0$

For $4\pi/3 \leq \omega t < 5\pi/3$: $v_{an} = v_{AB}/2 = 147.1 \sin(\omega t + \pi/6)$

For $5\pi/3 \leq \omega t < 2\pi$: $v_{an} = v_{AC}/2 = 147.1 \sin(\omega t - 7\pi/6 - \pi)$

Note. The power factor, which depends on the delay angle α, is in general poor compared to that of the half-wave controller.

6-8 THREE-PHASE BIDIRECTIONAL DELTA-CONNECTED CONTROLLERS

If the terminals of a three-phase system are accessible, the control elements (or power devices) and load may be connected in delta as shown in Fig. 6-15. Since the phase current in a normal three-phase system is only $1/\sqrt{3}$ of the line current, the current ratings of thyristors would be less than that if thyristors (or control elements) were placed in the line.

Let us assume that the instantaneous line-to-line voltages are

$$v_{AB} = v_{ab} = \sqrt{2} V_s \sin \omega t$$

$$v_{BC} = v_{bc} = \sqrt{2} V_s \sin \left(\omega t - \frac{2\pi}{3} \right)$$

$$v_{CA} = v_{ca} = \sqrt{2} V_s \sin \left(\omega t - \frac{4\pi}{3} \right)$$

The input line voltages, phase and line currents, and thyristor gating signals are shown in Fig. 6-16 for $\alpha = 120°$ and a resistive load.

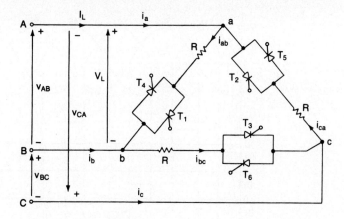

Figure 6-15 Delta-connected three-phase controller.

For resistive loads, the rms output phase voltage can be determined from

$$V_o = \left[\frac{1}{2\pi} \int_\alpha^{2\pi} v_{ab}^2 \, d(\omega t)\right]^{1/2} = \left[\frac{2}{2\pi} \int_\alpha^\pi 2 \, V_s^2 \sin \omega t \, d(\omega t)\right]^{1/2}$$

$$= V_s \left[\frac{1}{\pi} \left(\pi - \alpha + \frac{\sin 2\alpha}{2}\right)\right]^{1/2}$$

(6-35)

The maximum output voltage would be obtained when $\alpha = 0$, and the control range of delay angle is

$$0 \leq \alpha \leq \pi$$

(6-36)

The line currents, which can be determined from the phase currents, are

$$i_a = i_{ab} - i_{ca}$$
$$i_b = i_{bc} - i_{ab}$$
$$i_c = i_{ca} - i_{bc}$$

(6-37)

We can notice from Fig. 6-16 that the line currents depend on the delay angle and may be discontinuous. The rms value of line and phase currents for the load circuits can be determined by numerical solution or Fourier analysis. If I_n is the rms value of the nth harmonic component of a phase current, the rms value of phase current can be found from

$$I_{ab} = (I_1^2 + I_3^2 + I_5^2 + I_7^2 + I_9^2 + I_{11}^2 + \cdots + I_n^2)^{1/2}$$

(6-38)

Due to the delta connection, the triplen harmonic components (i.e., those of order $n = 3m$, where m is an odd integer) of the phase currents would flow around the delta and would not appear in the line. This is due to the fact that the zero-sequence harmonics are in phase in all three phases of load. The rms line current becomes

$$I_a = \sqrt{3} \, (I_1^2 + I_5^2 + I_7^2 + I_{11}^2 + \cdots + I_n^2)^{1/2}$$

(6-39)

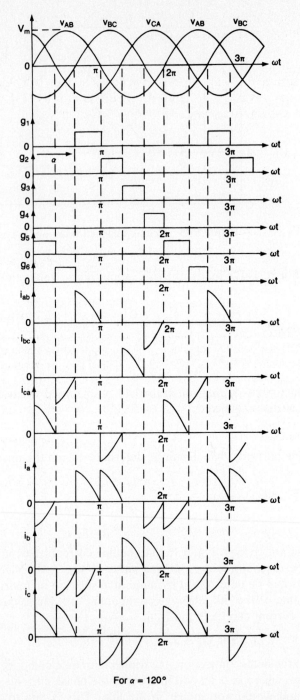

For $\alpha = 120°$

Figure 6-16 Waveforms for delta-connected controller.

As a result, the rms value of line current would not follow the normal relationship of a three-phase system such that

$$I_a < \sqrt{3}\, I_{ab} \tag{6-40}$$

An alternative form of delta-connected controllers which requires only three thyristors and simplifies the control circuitry is shown in Fig. 6-17. This arrangement is also known as a *polygon-connected controller*.

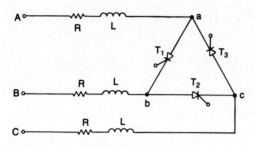

Figure 6-17 Three-phase three-thyristor controller.

Example 6-7

The three-phase bidirectional delta-connected controller in Fig. 6-15 has a resistive load of $R = 10\ \Omega$. The line-to-line voltage is $V_s = 208$ V (rms), 60 Hz, and the delay angle is $\alpha = 2\pi/3$. Determine (a) the rms output phase voltage V_o; (b) the expressions for instantaneous currents i_a, i_{ab}, and i_{ca}; (c) the rms output phase current I_{ab} and rms line current I_a; (d) the input power factor PF; and (e) the rms current of a thyristor I_R.

Solution $V_L = V_s = 208$ V, $\alpha = 2\pi/3$, $R = 10\ \Omega$, and peak value of phase current, $I_m = \sqrt{2} \times 208/10 = 29.4$ A.

(a) From Eq. (6-35), $V_o = 92$ V.

(b) Assuming i_{ab} as the reference phasor and $i_{ab} = I_m \sin \omega t$, the instantaneous currents are:

For $0 \le \omega t < \pi/3$:
$$I_{ab} = 0$$
$$i_{ca} = I_m \sin(\omega t - 4\pi/3)$$
$$i_a = i_{ab} - i_{ca} = -I_m \sin(\omega t - 4\pi/3)$$

For $\pi/3 < \omega t < 2\pi/3$: $i_{ab} = i_{ca} = i_a = 0$

For $2\pi/3 < \omega t < \pi$:
$$i_{ab} = I_m \sin \omega t$$
$$i_{ca} = 0$$
$$i_a = i_{ab} - i_{ca} = I_m \sin \omega t$$

For $\pi < \omega t < 4\pi/3$:
$$i_{ab} = 0$$
$$i_{ca} = I_m \sin(\omega t - 4\pi/3)$$
$$i_a = i_{ab} - i_{ca} = -I_m \sin(\omega t - 4\pi/3)$$

For $4\pi/3 < \omega t < 5\pi/3$: $i_{ab} = i_{ca} = i_a = 0$

For $5\pi/3 < \omega t < 2\pi$:
$$i_{ab} = I_m \sin \omega t$$
$$i_{ca} = 0$$
$$i_a = i_{ab} - i_{ca} = I_m \sin \omega t$$

(c) The rms values of i_{ab} and i_a are determined by numerical integration using a computer program. Students are encouraged to verify the results.

$$I_{ab} = 9.32 \text{ A} \qquad I_L = I_a = 13.18 \text{ A} \qquad \frac{I_a}{I_{ab}} = \frac{13.18}{9.32} = 1.1414 \neq \sqrt{3}$$

(d) The output power

$$P_o = 3I_{ab}^2 R = 3 \times 9.32^2 \times 10 = 2605.9$$

The volt-amperes

$$\text{VA} = 3V_s I_{ab} = 3 \times 208 \times 9.32 = 5815.7$$

The power factor

$$\text{PF} = \frac{P_o}{\text{VA}} = \frac{2605.9}{5815.7} = 0.448 \text{ (lagging)}$$

(e) The thyristor current can be determined from the phase current:

$$I_R = \frac{I_{ab}}{\sqrt{2}} = \frac{9.32}{\sqrt{2}} = 6.59 \text{ A}$$

Notes.

1. $V_o = I_{ab}R = 9.32 \times 10 = 93.2$ V, whereas Eq. (6-35) gives 92 V. The difference is due to rounding off in the numerical solution.
2. For the ac voltage controller of Fig. 6-17, the line current I_a is not related to the phase current I_{ab} by a factor of $\sqrt{3}$. This is due to the discontinuity of the load current in the presence of the ac voltage controller.

6-9 SINGLE-PHASE TRANSFORMER TAP CHANGERS

Thyristors can be used as static switches for on-load tap changing of transformers. The static tap changers have the advantage of very fast switching action. The changeover can be controlled to cope with load conditions and is smooth. The circuit diagram of a single-phase transformer tap changer is shown in Fig. 6-18. Although a transformer may have multiple secondary windings, only two secondary windings are shown, for the sake of simplicity.

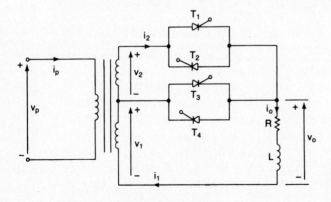

Figure 6-18 Single-phase transformer tap changer.

The turns ratio of the input transformer are such that if the primary instantaneous voltage is

$$v_p = \sqrt{2}\, V_s \sin \omega t = \sqrt{2}\, V_p \sin \omega t$$

the secondary instantaneous voltages are

$$v_1 = \sqrt{2}\, V_1 \sin \omega t$$

and

$$v_2 = \sqrt{2}\, V_2 \sin \omega t$$

A tap changer is most commonly used for resistive heating loads. When only thyristors T_3 and T_4 are alternately fired with a delay angle of $\alpha = 0$, the load voltage is held at a reduced level of $V_o = V_1$. If full output voltage is required, only thyristors T_1 and T_2 are alternately fired with a delay angle of $\alpha = 0$ and the full voltage is $v_o = V_1 + V_2$.

The gating pulses of thyristors can be controlled to vary the load voltage. The rms value of load voltage, V_o, can be varied within three possible ranges:

$$0 < V_o < V_1$$

$$0 < V_o < (V_1 + V_2)$$

and

$$V_1 < V_o < (V_1 + V_2)$$

Control range 1: $0 \le V_o \le V_1$.   To vary the load voltage within this range, thyristors T_1 and T_2 are turned off. Thyristors T_3 and T_4 can operate as a single-phase voltage controller. The instantaneous load voltage v_0 and load current i_0 are shown in Fig. 6-19c for a resistive load. The rms load voltage which can be determined from Eq. (6-8) load is

$$V_o = V_1 \left[\frac{1}{\pi} \left(\pi - \alpha + \frac{\sin 2\alpha}{2} \right) \right]^{1/2} \tag{6-41}$$

and the range of delay angle is $0 \le \alpha \le \pi$.

Control range 2: $0 \le V_o \le (V_1 + V_2)$.   Thyristors T_3 and T_4 are turned off. Thyristors T_1 and T_2 operate as a single-phase voltage controller. Figure 6-19d shows the load voltage v_0 and load current i_0 for a resistive load. The rms load voltage can be found from

$$V_o = (V_1 + V_2) \left[\frac{1}{\pi} \left(\pi - \alpha + \frac{\sin 2\alpha}{2} \right) \right]^{1/2} \tag{6-42}$$

and the range of delay angle is $0 \le \alpha \le \pi$.

Control range 3: $V_1 < V_o < (V_1 + V_2)$.   Thyristor T_3 is turned on at $\omega t = 0$ and the secondary voltage v_1 appears across the load. If thyristor T_1 is turned on at $\omega t = \alpha$, thyristor T_3 is reverse biased due to secondary voltage v_2, and T_3 is

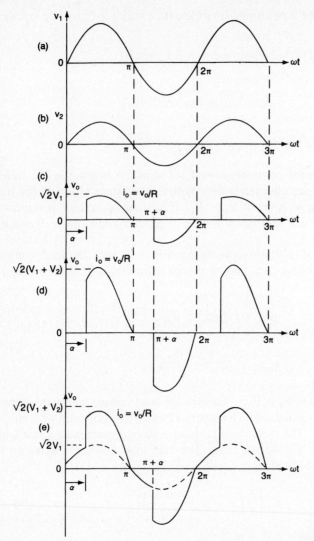

Figure 6-19 Waveforms for transformer tap changer.

turned off. The voltage appearing across the load is $(v_1 + v_2)$. At $\omega t = \pi$, T_1 is self-commutated and T_4 is turned on. The secondary voltage v_1 appears across the load until T_2 is fired at $\omega t = \pi + \alpha$. When T_2 is turned on at $\omega t = \pi + \alpha$, T_4 is turned off due to reverse voltage v_2 and the load voltage is $(v_1 + v_2)$. At $\omega t = 2\pi$, T_2 is self-commutated, T_3 is turned on again and the cycle is repeated. The instantaneous load voltage v_0 and load current i_0 are shown in Fig. 6-19e for a resistive load.

A tap changer with this type of control is also known as a *synchronous tap changer*. It uses two-step control. A part of secondary voltage v_2 is superimposed on a sinusoidal voltage v_1. As a result, the harmonic contents are less that which would be obtained by a normal phase delay as discussed above for control range 2. The rms load voltage can be found from

$$V_o = \left[\frac{1}{2\pi} \int_0^{2\pi} v_0^2 \, d(\omega t) \right]^{1/2}$$

$$= \left\{ \frac{2}{2\pi} \left[\int_0^{\alpha} 2V_1^2 \sin^2 \omega t \, d(\omega t) + \int_{\alpha}^{\pi} 2(V_1 + V_2)^2 \sin^2 \omega t \, d(\omega t) \right] \right\}^{1/2} \quad (6\text{-}43)$$

$$= \left[\frac{V_1^2}{\pi} \left(\alpha - \frac{\sin 2\alpha}{2} \right) + \frac{(V_1 + V_2)^2}{\pi} \left(\pi - \alpha + \frac{\sin 2\alpha}{2} \right) \right]^{1/2}$$

With RL loads, the gating circuit of a synchronous tap changer requires a careful design. Let us assume that thyristors T_1 and T_2 are turned off, while thyristors T_3 and T_4 are turned on during the alternate half-cycle at the zero crossing of the load current. The load current would then be

$$i_o = \frac{\sqrt{2}\,V_1}{Z} \sin(\omega t - \theta)$$

where $Z = [R^2 + (\omega L)^2]^{1/2}$ and $\theta = \tan^{-1}(\omega L/R)$.

The instantaneous load current i_0 is shown in Fig. 6-20a. If T_1 is then turned on at $\omega t = \alpha$, where $\alpha < \theta$, the second winding of transformer would be short circuited because thyristor T_3 is still conducting and carrying current due to the inductive load. Therefore, the control circuit should be designed so that T_1 is not turned on until T_3 turns off and $i_0 \geq 0$. Similarly, T_2 should not be turned on until T_4 turns off and $i_0 \leq 0$. The waveforms of load voltage v_0 and load current i_0 are shown in Fig. 6-20b for $\alpha > \theta$.

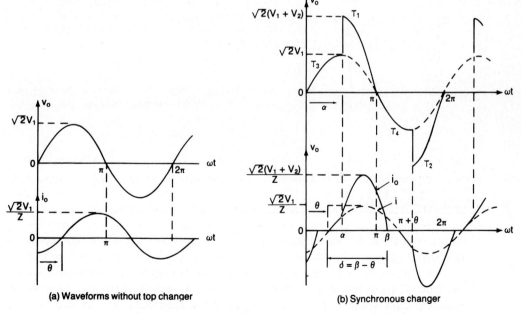

(a) Waveforms without top changer

(b) Synchronous changer

Figure 6-20   Voltage and current waveforms for RL load.

Example 6-8*

The circuit in Fig. 6-18 is controlled as a synchronous tap changer. The primary voltage is 240 V (rms), 60 Hz. The secondary voltages are $V_1 = 120$ V and $V_2 = 120$ V. If the load resistance is $R = 10\ \Omega$ and the rms load voltage is 180 V, determine (a) the delay angle of thyristors T_1 and T_2, (b) the rms current of thyristors T_1 and T_2, (c) the rms current of thyristors T_3 and T_4, and (d) the input power factor PF.

Solution $V_o = 180$ V, $V_p = 240$ V, $V_1 = 120$ V, $V_2 = 120$ V, and $R = 10\ \Omega$.

(a) The required value of delay angle α for $V_o = 180$ V can be found from Eq. (6-43) in two ways: (1) plot V_o against α and find the required value of α, or (2) use an iterative method of solution. A computer program is used to solve Eq. (6-43) for α by iteration and this gives $\alpha = 98°$.

(b) The rms current of thyristors T_1 and T_2 can be found from Eq. (6-42):

$$I_{R1} = \left[\frac{1}{2\pi R^2} \int_\alpha^\pi 2(V_1 + V_2)^2 \sin^2 \omega t\ d(\omega t) \right]^{1/2}$$

$$= \frac{V_1 + V_2}{\sqrt{2}R} \left[\frac{1}{\pi} \left(\pi - \alpha + \frac{\sin 2\alpha}{2} \right) \right]^{1/2} \qquad (6\text{-}44)$$

$$= 10.9\ \text{A}$$

(c) The rms current of thyristors T_3 and T_4 is found from

$$I_{R3} = \left[\frac{1}{2\pi R^2} \int_0^\alpha 2V_1^2 \sin^2 \omega t\ d(\omega t) \right]^{1/2}$$

$$= \frac{V_1}{\sqrt{2}R} \left[\frac{1}{\pi} \left(\alpha - \frac{\sin 2\alpha}{2} \right) \right]^{1/2} \qquad (6\text{-}45)$$

$$= 6.5\ \text{A}$$

(d) The rms current of a second (top) secondary winding is $I_2 = \sqrt{2}\ I_{R1} = 15.4$ A. The rms current of the first (lower) secondary winding, which is the total rms current of thyristors T_1, T_2, T_3, and T_4, is

$$I_1 = [(\sqrt{2}\ I_{R1})^2 + (\sqrt{2}\ I_{R3})^2]^{1/2} = 17.94\ \text{A}$$

The volt-ampere rating of primary or secondary, VA $= V_1 I_1 + V_2 I_2 = 120 \times 17.94 + 12 \times 15.4 = 4000.8$. The load power, $P_o = V_o^2/R = 3240$ W, and the power factor

$$\text{PF} = \frac{P_o}{\text{VA}} = \frac{3240}{4000.8} = 0.8098\ \text{(lagging)}$$

6-10 CYCLOCONVERTERS

The ac voltage controllers provide a variable output voltage, but the frequency of the output voltage is fixed and in addition the harmonic content is high, especially at a low output voltage range. A variable output voltage at variable frequency can be obtained from two-stage conversions: fixed ac to variable dc (e.g., controlled rectifiers) and variable dc to variable ac at variable frequency (e.g., inverters, which are discussed in Chapter 10). However, cycloconverters can eliminate the need of one or more intermediate converters. A cycloconverter is a direct-frequency changer that converts ac power at one frequency to ac power at another frequency by ac–ac conversion, without an intermediate conversion link.

The majority of cycloconverters are naturally commutated and the maximum output frequency is limited to a value that is only a fraction of the source frequency. As a result the major applications of cycloconverters are low speed, ac motor drives in the range up to 15,000 kW with frequencies from 0 to 20 Hz. Ac drives are discussed in Chapter 15.

With the development of power conversion techniques and modern control methods, inverter-fed ac motor drives are taking over cycloconverter-fed drives. However, recent advancements in fast-switching power devices and microprocessors permit synthesizing and implementing advanced conversion strategies for forced-commutated direct-frequency changers (FCDFCs) to optimize the efficiency and reduce the harmonic contents [1,2]. The switching functions of FCDFCs can be programmed to combine the switching functions of ac–dc and dc–ac converters. Due to the nature of complex derivations involved in FCDFCs, forced-commutated cycloconverters will not be discussed further.

6-10.1 Single-Phase Cycloconverters

The principle of operation of single-phase/single-phase cycloconverters can be explained with the help of Fig. 6-21a. The two single-phase controlled converters are operated as bridge rectifiers. However, their delay angles are such that the output voltage of one converter is equal and opposite to that of the other converter. If converter P is operating alone, the average output voltage is positive and if converter N is operating, the output voltage is negative. Figure 6-21b shows the waveforms for the output voltage and gating signals of positive and negative converters, with the positive converter on for time $T_0/2$ and the negative converter operating for time $T_0/2$. The frequency of the output voltage is $f_o = 1/T_0$.

If α_p is the delay angle of positive converter, the delay angle of negative converter is $\alpha_n = \pi - \alpha_p$. The average output voltage of the positive converter is equal and opposite to that of the negative converter.

$$V_{dc2} = -V_{dc1} \qquad (6-46)$$

Similar to dual converters in Sections 5-5 and 5-10, the instantaneous values of two output voltages may not be equal. It is possible for large harmonic currents to circulate within the converters.

The circulating current can be eliminated by suppressing the gate pulses to the converter not delivering load current. A single-phase cycloconverter with center-tap transformer as shown in Fig. 6-22 has an intergroup reactor, which maintains a continuous current flow and also limits the circulating current.

Example 6-9*

The input voltage to the cycloconverter in Fig. 6-21a is 120 V (rms), 60 Hz. The load resistance is 5 Ω and the load inductance is $L = 40$ mH. The frequency of the output voltage is 20 Hz. If the converters are operated as semiconverters such that $0 \le \alpha \le$

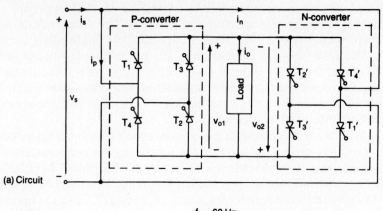

(a) Circuit

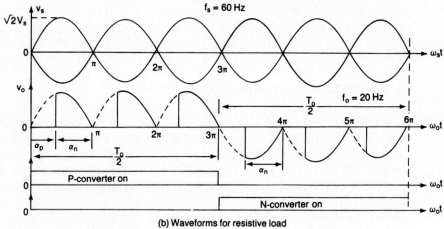

(b) Waveforms for resistive load

Figure 6-21   Single-phase/single-phase cycloconverter.

π and the delay angle is $\alpha_p = 2\pi/3$, determine (a) the rms value of output voltage V_o, (b) the rms current of each thyristor I_R, and (c) the input power factor PF.

Solution $V_s = 120$ V, $f_s = 60$ Hz, $f_o = 20$ Hz, $R = 5$ Ω, $L = 40$ mH, $\alpha_p = 2\pi/3$, $\omega_0 = 2\pi \times 20 = 125.66$ rad/s, and $X_L = \omega_0 L = 5.027$ Ω.

(a) For $0 \le \alpha \le \pi$, Eq. (6-8) gives the rms output voltage

$$V_o = V_s \left[\frac{1}{\pi} \left(\pi - \alpha + \frac{\sin 2\alpha}{2} \right) \right]^{1/2}$$

(6-47)

$$= 53 \text{ V}$$

(b) $Z = [R^2 + (\omega_0 L)^2]^{1/2} = 7.09$ Ω and $\theta = \tan^{-1}(\omega_0 L/R) = 45.2°$. The rms load current, $I_o = V_o/Z = 53/7.09 = 7.48$ A. The rms current through each converter, $I_p = I_N = I_o/\sqrt{2} = 5.29$ A and the rms current through each thyristor, $I_R = I_p/\sqrt{2} = 3.74$ A.

(c) The rms input current, $I_s = I_0 = 7.48$ A, the volt-ampere rating VA $= V_s I_s = 897.6$ VA, and the output power, $P_o = V_o I_o \cos \theta = 53 \times 7.48 \times \cos 45.2° = 279.35$ W. From Eq. (6-8), the input power factor,

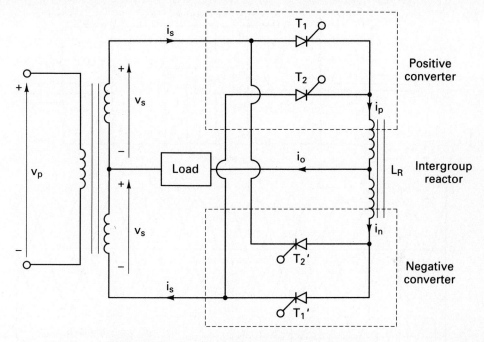

Figure 6-22 Cycloconverter with intergroup reactor.

$$ \text{PF} = \frac{P_o}{V_s I_s} = \frac{V_o \cos \theta}{V_s} = \cos \theta \left[\frac{1}{\pi} \left(\pi - \alpha + \frac{\sin 2\alpha}{2} \right) \right]^{1/2} \tag{6-48} $$

$$ = \frac{279.35}{897.6} = 0.311 \text{ (lagging)} $$

Note. Equation (6-48) does not include the harmonic content on the output voltage and gives the approximate value of power factor. The actual value will be less than that given by Eq. (6-48). Equations (6-47) and (6-48) are also valid for resistive loads.

6-10.2 Three-Phase Cycloconverters

The circuit diagram of a three-phase/single-phase cycloconverter is shown in Fig. 6-23a. The two ac–dc converters are three-phase controlled rectifiers. The synthesis of output waveform for an output frequency of 12 Hz is shown in Fig. 6-23b. The positive converter operates for half the period of output frequency and the negative converter operates for the other half-period. The analysis of this cycloconverter is similar to that of single-phase/single-phase cycloconverters.

The control of ac motors requires a three-phase voltage at variable frequency. The cycloconverter in Fig. 6-23a can be extended to provide three-phase output by having six three-phase converters as shown in Fig. 6-24a. Each phase consists of six thyristors as shown in Fig. 6-24b, and a total of 18 thyristors are

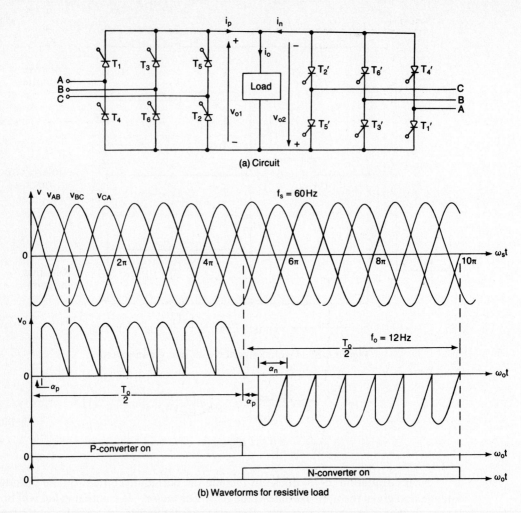

(a) Circuit

(b) Waveforms for resistive load

Figure 6-23 Three-phase/single-phase cycloconverter.

required. If six full-wave three-phase converters are used, 36 thyristors would be required.

6-10.3 Reduction of Output Harmonics

We can notice from Figs. 6-21b and 6-23b that the output voltage is not purely sinusoidal, and as a result the output voltage contains harmonics. Equation (6-48) shows that the input power factor depends on the delay angle of thyristors and is poor, especially at the low output voltage range.

The output voltage of cycloconverters is basically made up of segments of input voltage(s) and the average value of a segment depends on the delay angle for that segment. If the delay angles of segments were varied in such a way that the average values of segments correspond as closely as possible to the variations of

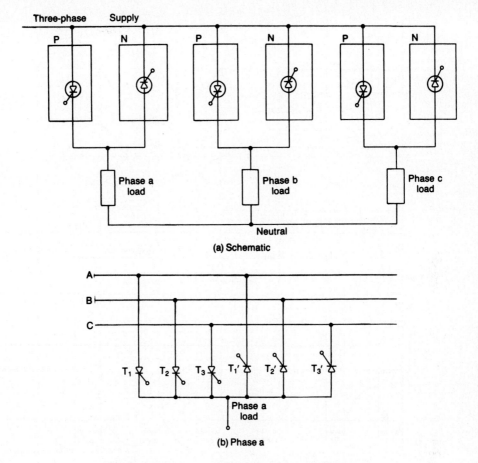

Figure 6-24 Three-phase/three-phase cycloconverter.

desired sinusoidal output voltage, the harmonics on the output voltage can be minimized. Equation (5-21) indicates that the average output voltage of a segment is a cosine function of delay angle. The delay angles for segments can be generated by comparing a cosine signal at source frequency ($v_c = \sqrt{2}\, V_s \cos \omega_s t$) with an ideal sinusoidal reference voltage at the output frequency ($v_r = \sqrt{2}\, V_r \sin \omega_0 t$). Figure 6-25 shows the generation of gating signals for the thyristors of the cycloconverter in Fig. 6-23a.

The maximum average voltage of a segment (which occurs for $\alpha_p = 0$) should be equal to the peak value of output voltage; for example, from Eq. (5-21),

$$V_p = \frac{2\sqrt{2}\, V_s}{\pi} = \sqrt{2}\, V_o \tag{6-49}$$

which gives the rms value of output voltage as

$$V_o = \frac{2V_s}{\pi} = \frac{2V_p}{\pi} \tag{6-50}$$

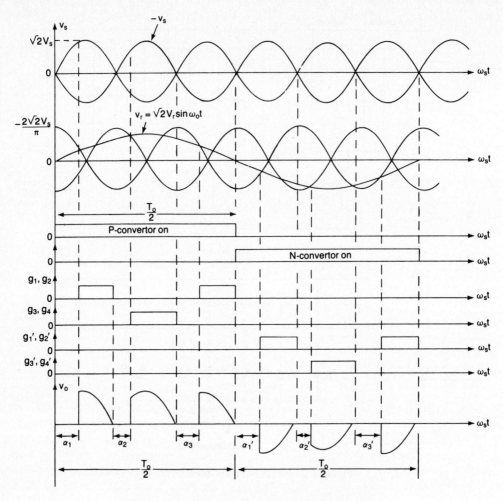

Figure 6-25 Generation of thyristor gating signals.

Example 6-10

Repeat Example 6-9 if the delay angles of the cycloconverter are generated by comparing a cosine signal at the source frequency with a sinusoidal signal at the output frequency as shown in Fig. 6-25.

Solution $V_s = 120$ V, $f_s = 60$ Hz, $f_o = 20$ Hz, $R = 5\ \Omega$, $L = 40$ mH, $\alpha_p = 2\pi/3$, $\omega_0 = 2\pi \times 20 = 125.66$ rad/s, and $X_L = \omega_0 L = 5.027\ \Omega$.

(a) From Eq. (6-50), the rms value of the output voltage

$$V_o = \frac{2V_s}{\pi} = 0.6366V_s = 0.6366 \times 120 = 76.39\ \text{V}$$

(b) $Z = [R^2 + (\omega_0 L)^2]^{1/2} = 7.09\ \Omega$ and $\theta = \tan^{-1}(\omega_0 L/R) = 45.2°$. The rms load current $I_o = V_o/Z = 76.39/7.09 = 10.77$ A. The rms current through each converter, $I_p = I_N = I_L/\sqrt{2} = 7.62$ A, and the rms current through each thyristor, $I_R = I_p/\sqrt{2} = 5.39$ A.

(c) The rms input current $I_s = I_o = 10.77$ A, the volt-ampere rating VA = $V_s I_s = 1292.4$ VA, and the output power

$$P_o = V_o I_o \cos\theta = 0.6366 V_s I_o \cos\theta = 579.73 \text{ W.}$$

The input power factor

$$\text{PF} = 0.6366 \cos\theta$$

$$= \frac{579.73}{1292.4} = 0.449 \text{ (lagging)}$$

(6-51)

Note. Equation (6-51) shows that the input power factor is independent of delay angle, α and depends only on the load angle θ. But for normal phase angle control, the input power factor is dependent on both delay angle, α, and load angle, θ. If we compare Eq. (6-48) with Eq. (6-51), there is a critical value of delay angle α_c, which is given by

$$\left[\frac{1}{\pi} \left(\pi - \alpha_c + \frac{\sin 2\alpha_c}{2} \right) \right]^{1/2} = 0.6366$$

(6-52)

For $\alpha < \alpha_c$, the normal delay-angle control would exhibit a better power factor and the solution of Eq. (6-52) yields $\alpha_c = 98.59°$.

6-11 AC VOLTAGE CONTROLLERS WITH PWM CONTROL

It has been shown in Section 5-11 that the input power factor of controlled rectifiers can be improved by the pulse-width-modulation (PWM) type of control. The naturally commutated thyristor controllers introduce lower-order harmonics in both the load and supply side and have low input power factor. The performance of ac voltage controllers can be improved by PWM control. The circuit configuration of a single-phase ac voltage controller for PWM control is shown in Fig. 6-26a. The gating signals of the switches are shown in Fig. 6-26b. Switches S_1 and S_2 are turned on and off several times during the positive and negative half-

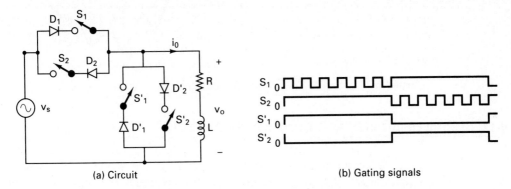

(a) Circuit (b) Gating signals

Figure 6-26 Ac voltage controller for PWM control.

cycles of the input voltage, respectively. S_1' and S_2' provide the freewheeling paths for the load current, while S_1 and S_2, respectively, are in the off state. The diodes prevent reverse voltages appearing across the switches.

The output voltage is shown in Fig. 6-27a. For a resistive load, the load current will resemble the output voltage. With an *RL* load, the load current will rise in the positive or negative direction when switch S_1 or S_2 is turned on, respectively. Similarly, the load current will fall when either S_1' or S_2' is turned on. The load current is shown in Fig. 6-27b with an *RL* load.

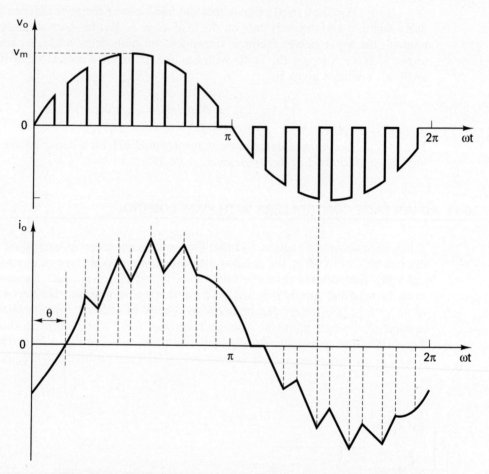

Figure 6-27 Output voltage and load current of ac voltage controller.

6-12 DESIGN OF AC VOLTAGE-CONTROLLER CIRCUITS

The ratings of power devices must be designed for the worst-case condition, which occurs when the converter delivers the maximum rms value of output voltage V_o. The input and output filters must also be designed for worst-case conditions. The output of a power controller contains harmonics, and the delay angle for the worst-case condition of a particular circuit arrangement should be determined. The steps involved in designing the power circuits and filters are similar to those of rectifier circuit design in Section 3-11.

Example 6-11

A single-phase full-wave ac voltage controller in Fig. 6-3a controls power flow from a 230-V 60-Hz ac source into a resistive load. The maximum desired output power is 10 kW. Calculate (a) the maximum rms current rating of thyristors I_{RM}, (b) the maximum average current rating of thyristors I_{AM}, (c) the peak current of thyristors I_p, and (d) the peak value of thyristor voltage V_p.

Solution $P_o = 10,000$ W, $V_s = 230$ V, and $V_m = \sqrt{2} \times 230 = 325.3$ V. The maximum power will be delivered when the delay angle is $\alpha = 0$. From Eq. (6-8), the rms value of output voltage $V_o = V_s = 230$ V, $P_o = V_o^2/R = 230^2/R = 10,000$, and load resistance is $R = 5.29$ Ω.

(a) The maximum rms value of load current $I_{oM} = V_o/R = 230/5.29 = 43.48$ A, and the maximum rms value of thyristor current $I_{RM} = I_{oM}/\sqrt{2} = 30.75$ A.

(b) From Eq. (6-10), the maximum average current of thyristors,

$$I_{AM} = \frac{\sqrt{2} \times 230}{\pi \times 5.29} = 19.57 \text{ A}$$

(c) The peak thyristor current $I_p = V_m/R = 325.3/5.29 = 61.5$ A.
(d) The peak thyristor voltage $V_p = V_m = 325.3$ V.

Example 6-12*

A single-phase full-wave controller in Fig. 6-6a controls power to an RL load and the source voltage is 120 V (rms), 60 Hz. (a) Use the method of Fourier series to obtain expressions for output voltage $v_o(t)$ and load current $i_o(t)$ as a function of delay angle α. (b) Determine the delay angle for the maximum amount of lowest-order harmonic current in the load. (c) If $R = 5$ Ω, $L = 10$ mH, and $\alpha = \pi/2$, determine the rms value of third harmonic current. (d) If a capacitor is connected across the load (Fig. 6-28), calculate the value of capacitance to reduce the third harmonic current to 10% of the value without the capacitor.

Solution (a) The waveform for the input voltage is shown in Fig. 6-6b. The instantaneous output voltage as shown in Fig. 6-28b can be expressed in Fourier series as

$$v_o(t) = V_{dc} + \sum_{n=1,2,\dots}^{\infty} a_n \cos n\omega t + \sum_{n=1,2,\dots}^{\infty} b_n \sin n\omega t \qquad (6\text{-}53)$$

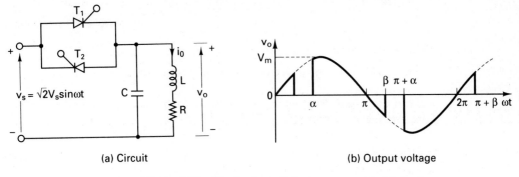

(a) Circuit   (b) Output voltage

Figure 6-28 Single-phase full converter with RL load.

where

$$V_{dc} = \frac{1}{2\pi} \int_0^{2\pi} V_m \sin \omega t \; d(\omega t) = 0$$

$$a_n = \frac{1}{\pi} \left[\int_\alpha^\beta \sqrt{2} \; V_s \sin \omega t \cos n\omega t \; d(\omega t) + \int_{\pi+\alpha}^{\pi+\beta} \sqrt{2} \; V_s \sin \omega t \cos n\omega t \; d(\omega t) \right]$$

$$= \frac{\sqrt{2} \; V_s}{2\pi} \left[\frac{\begin{array}{c} \cos(1-n)\alpha - \cos(1-n)\beta \\ + \cos(1-n)(\pi+\alpha) - \cos(1+n)(\pi+\beta) \end{array}}{1-n} \right.$$

$$\left. + \frac{\cos(1+n)\alpha - \cos(1+n)\beta + \cos(1+n)(\pi+\alpha) - \cos(1+n)(\pi+\beta)}{1+n} \right]$$

$$\text{for } n = 3, 5, \ldots \qquad (6\text{-}54)$$

$$= 0 \qquad \text{for } n = 2, 4, \ldots$$

$$b_n = \frac{1}{\pi} \left[\int_\alpha^\beta \sqrt{2} \; V_s \sin \omega t \sin n\omega t \; d(\omega t) + \int_{\pi+\alpha}^{\pi+\beta} \sqrt{2} \; V_s \sin \omega t \sin n\omega t \; d(\omega t) \right]$$

$$= \frac{\sqrt{2} \; V_s}{2\pi} \left[\frac{\begin{array}{c} \sin(1-n)\beta - \sin(1-n)\alpha \\ + \sin(1-n)(\pi+\beta) - \sin(1-n)(\pi+\alpha) \end{array}}{1-n} \right.$$

$$\left. - \frac{\sin(1+n)\beta - \sin(1+n)\alpha + \sin(1+n)(\pi+\beta) - \sin(1+n)(\pi+\alpha)}{1+n} \right]$$

$$\text{for } n = 3, 5, \ldots \qquad (6\text{-}55)$$

$$= 0 \qquad \text{for } n = 2, 4, \ldots$$

$$a_1 = \frac{1}{\pi} \left[\int_\alpha^\beta \sqrt{2} \; V_s \sin \omega t \cos \omega t \; d(\omega t) + \int_{\pi+\alpha}^{\pi+\beta} \sqrt{2} \; V_s \sin \omega t \cos \omega t \; d(\omega t) \right]$$

$$= \frac{\sqrt{2} \; V_s}{2\pi} [\sin^2 \beta - \sin^2 \alpha + \sin^2(\pi+\beta) - \sin^2(\pi+\alpha)] \qquad \text{for } n = 1 \quad (6\text{-}56)$$

$$b_1 = \frac{1}{\pi} \left[\int_\alpha^\beta \sqrt{2} \; V_s \sin^2 \omega t \; d(\omega t) + \int_{\pi+\alpha}^{\pi+\beta} \sqrt{2} \; V_s \sin^2 \omega t \; d(\omega t) \right]$$

$$= \frac{\sqrt{2} \; V_s}{2\pi} \left[2(\beta - \alpha) - \frac{\sin 2\beta - \sin 2\alpha + \sin 2(\pi+\beta) - \sin 2(\pi+\alpha)}{2} \right]$$

$$\text{for } n = 1 \qquad (6\text{-}57)$$

The load impedance

$$Z = R + j(n\omega L) = [R^2 + (n\omega L)^2]^{1/2} \; \underline{/\theta_n}$$

and $\theta_n = \tan^{-1}(n\omega L/R)$. Dividing $v_o(t)$ in Eq. (6-53) by load impedance Z and simplifying the sine and cosine terms give the load current as

$$i_o(t) = \sum_{n=1,3,5,\ldots}^\infty \sqrt{2} \; I_n \sin(n\omega t - \theta_n + \phi_n) \qquad (6\text{-}58)$$

where $\phi_n = \tan^{-1}(a_n/b_n)$ and

$$I_n = \frac{1}{\sqrt{2}} \frac{(a_n^2 + b_n^2)^{1/2}}{[R^2 + (n\omega L)^2]^{1/2}}$$

(b) The third harmonic is the lowest-order harmonic. The calculation of third harmonic for various values of delay angle shows that it becomes maximum for $\alpha = \pi/2$.

(c) For $\alpha = \pi/2$, $L = 6.5$ mH, $R = 2.5\ \Omega$, $\omega = 2\pi \times 60 = 377$ rad/s and $V_s = 120$ V. From Example 6-4, we get the extinction angle as $\beta = 220.43°$. For known values of α, β, R, L, and V_s, a_n and b_n of the Fourier series in Eq. (6-53) and the load current i_o of Eq. (6-58) can be calculated. The load current is given by

$$i_o(t) = 28.93 \sin(\omega t - 44.2° - 18°) + 7.96 \sin(3\omega t - 71.2° + 68.7°)$$

$$+ 2.68 \sin(5\omega t - 78.5° - 68.6°) + 0.42 \sin(7\omega t - 81.7° + 122.7°)$$

$$+ 0.59 \sin(9\omega t - 83.5° - 126.3°) + \cdots$$

The rms value of third harmonic current is

$$I_3 = \frac{7.96}{\sqrt{2}} = 5.63 \text{ A}$$

(d) Figure 6-29 shows the equivalent circuit for harmonic current. Using the current-divider rule, the harmonic current through load is given by

$$\frac{I_h}{I_n} = \frac{X_c}{[R^2 + (n\omega L - X_c)^2]^{1/2}}$$

where $X_c = 1/(n\omega C)$. For $n = 3$ and $\omega = 377$,

$$\frac{I_h}{I_n} = \frac{X_c}{[2.5^2 + (3 \times .377 \times 6.5 - X_c)^2]^{1/2}} = 0.1$$

which yields $X_c = -.858$ or 0.7097. Since X_c cannot be negative, $X_c = 0.7097 = 1/(3 \times 377C)$ or $C = 1245.94\ \mu$F.

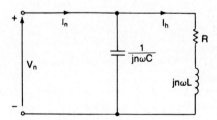

Figure 6-29 Equivalent circuit for harmonic current.

Example 6-13

The single-phase ac voltage controller in Fig. 6-6a has a load of $R = 2.5\ \Omega$ and $L = 6.5$ mH. The supply voltage is 120 V (rms), 60 Hz. The delay angle is $\alpha = \pi/2$. Use PSpice to plot the output voltage and the load current and to compute the total harmonic distortion (THD) of the output voltage and output current and the input power factor (PF).

Solution The load current of ac voltage controllers is ac type, and the current of a thyristor is always reduced to zero. There is no need for the diode D_T in Fig. 4-30b, and the thyristor model can be simplified to Fig. 6-30. This model can be used as a subcircuit.

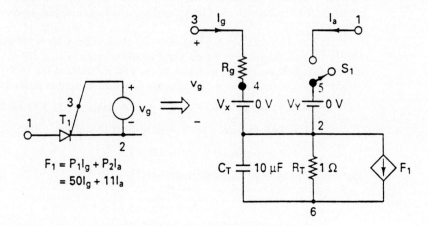

Figure 6-30 Ac thyristor SPICE model.

The subcircuit definition for the thyristor model SCR can be described as follows:

```
*      Subcircuit for ac thyristor model
.SUBCKT    SCR      1        2        3       2
*          model    anode    cathode  +control  -control
*          name                       voltage   voltage
S1    1    5    6    2    SMOD        ; Switch
RG    3    4    50
VX    4    2    DC    0V
VY    5    2    DC    0V
RT    2    6    1
CT    6    2    10UF
F1    2    6    POLY(2)    VX    VY    0    50    11
.MODEL    SMOD    VSWITCH    (RON=0.01 ROFF=10E+5 VON=0.1V VOFF=0V)
.ENDS    SCR                         ; Ends subcircuit definition
```

The peak supply voltage $V_m = 169.7$ V. For $\alpha_1 = \alpha_2 = 90°$, time delay $t_1 = (90/360) \times (1000/60 \text{ Hz}) \times 1000 = 4166.7 \ \mu s$. A series snubber with $C_s = 0.1 \ \mu F$ and $R_s = 750 \ \Omega$ is connected across the thyristor to cope with the transient voltage due to the inductive load. The single-phase ac voltage controller for PSpice simulation is shown in Fig. 6-31a. The gate voltages V_{g1} and V_{g2} for thyristors are shown in Fig. 6-31b.

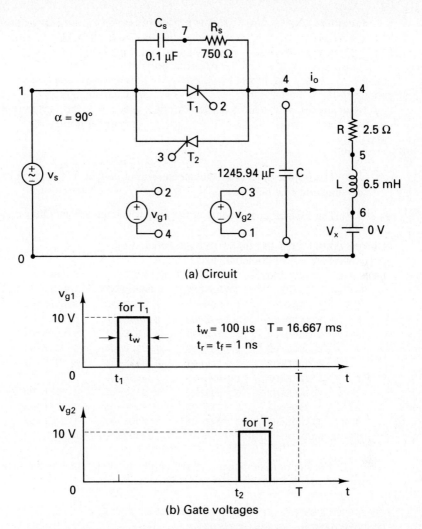

(a) Circuit

(b) Gate voltages

Figure 6-31 Single-phase ac voltage controller for PSpice simulation.

The list of the circuit file is as follows:

```
Example 6-13      Single-Phase AC Voltage Controller
VS     1    0     SIN (0    169.7V    60HZ)
Vg1    2    4     PULSE (0V   10V   4166.7US   1NS   1NS   100US   16666.7US)
Vg2    3    1     PULSE (0V   10V   12500.0US  1NS   1NS   100US   16666.7US)
R      4    5     2.5
L      5    6     6.5MH
VX     6    0     DC    0V      ; Voltage source to measure the load current
* C    4    0     1245.94UF   ; Output filter capacitance  ; Load filter
CS     1    7     0.1UF
RS     7    4     750
```

```
*    Subcircuit call for thyristor model
XT1   1   4   2   4   SCR        ; Thyristor T1
XT2   4   1   3   1   SCR        ; Thyristor T2
*    Subcircuit SCR which is missing must be inserted
.TRAN   10US   33.33MS           ; Transient analysis
.PROBE                           ; Graphics postprocessor
.options abstol = 1.00n  reltol = 1.0m  vntol = 1.0m  ITL5=10000
.FOUR   60HZ   V(4)              ; Fourier analysis
.END
```

The PSpice plots of instantaneous output voltage V(4) and load current I(VX) are shown in Fig. 6-32.

The Fourier components of the output voltage are as follows:

```
FOURIER COMPONENTS OF TRANSIENT RESPONSE V(4)
DC COMPONENT =    1.784608E−03
```

HARMONIC NO	FREQUENCY (HZ)	FOURIER COMPONENT	NORMALIZED COMPONENT	PHASE (DEG)	NORMALIZED PHASE (DEG)
1	6.000E+01	1.006E+02	1.000E+00	−1.828E+01	0.000E+00
2	1.200E+02	2.764E−03	2.748E−05	6.196E+01	8.024E+01
3	1.800E+02	6.174E+01	6.139E−01	6.960E+01	8.787E+01
4	2.400E+02	1.038E−03	1.033E−05	6.731E+01	8.559E+01
5	3.000E+02	3.311E+01	3.293E−01	−6.771E+01	−4.943E+01
6	3.600E+02	1.969E−03	1.958E−05	1.261E+02	1.444E+02
7	4.200E+02	6.954E+00	6.915E−02	1.185E+02	1.367E+02
8	4.800E+02	3.451E−03	3.431E−05	1.017E+02	1.199E+02
9	5.400E+02	1.384E+01	1.376E−01	−1.251E+02	−1.068E+02

```
TOTAL HARMONIC DISTORTION =    7.134427E+01 PERCENT
```

The Fourier components of the output current, which is the same as the input current, are as follows:

```
FOURIER COMPONENTS OF TRANSIENT RESPONSE I(VX)
DC COMPONENT =   −2.557837E−03
```

HARMONIC NO	FREQUENCY (HZ)	FOURIER COMPONENT	NORMALIZED COMPONENT	PHASE (DEG)	NORMALIZED PHASE (DEG)
1	6.000E+01	2.869E+01	1.000E+00	−6.253E+01	0.000E+00
2	1.200E+02	4.416E−03	1.539E−04	−1.257E+02	−6.319E+01
3	1.800E+02	7.844E+00	2.735E−01	−2.918E+00	5.961E+01
4	2.400E+02	3.641E−03	1.269E−04	−1.620E+02	−9.948E+01
5	3.000E+02	2.682E+00	9.350E−02	−1.462E+02	−8.370E+01
6	3.600E+02	2.198E−03	7.662E−05	1.653E+02	2.278E+02
7	4.200E+02	4.310E−01	1.503E−02	4.124E+01	1.038E+02
8	4.800E+02	1.019E−03	3.551E−05	1.480E+02	2.105E+02
9	5.400E+02	6.055E−01	2.111E−02	1.533E+02	2.158E+02

```
TOTAL HARMONIC DISTORTION =    2.901609E+01 PERCENT
```

Total harmonic distortion of input current THD = 29.01% = 0.2901

Displacement angle $\phi_1 = -62.53°$

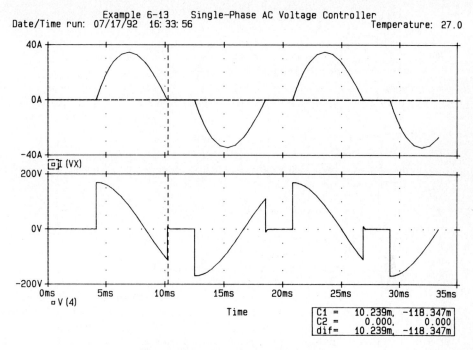

C1 =	10.239m,	−118.347m
C2 =	0.000,	0.000
dif=	10.239m,	−118.347m

Figure 6-32 Plots for Example 6-13

Displacement factor DF = $\cos \phi_1$ = cos(−62.53) = 0.461 (lagging)

From Eq. (5-86), the input power factor

$$PF = \frac{1}{(1 + THD^2)^{1/2}} \cos \phi_1 = \frac{1}{(1 + 0.2901^2)^{1/2}} \times 0.461 = 0.443 \text{ (lagging)}$$

6-13 EFFECTS OF SOURCE AND LOAD INDUCTANCES

In the derivations of output voltages, we have assumed that the source has no inductance. The effect of any source inductance would be to delay the turn-off of thyristors. Thyristors would not turn off at the zero crossing of input voltage as shown in Fig. 6-33b, and gate pulses of short duration may not be suitable. The harmonic contents on the output voltage would also increase.

We have seen in Section 6-5 that the load inductance plays a significant part on the performance of power controllers. Although the output voltage is a pulsed waveform, the load inductance tries to maintain a continuous current flow as shown in Figs. 6-6b and 6-33b. We can also notice from Eqs. (6-48) and (6-52) that the input power factor of power converter depends on the load power factor. Due to the switching characteristics of thyristors, any inductance in the circuit makes the analysis more complex.

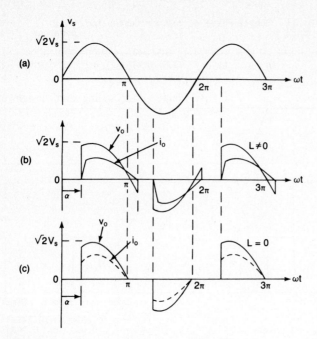

Figure 6-33 Effects of load inductance on load current and voltage.

SUMMARY

The ac voltage controller can use on–off control or phase-angle control. The on–off control is more suitable for systems having a high time constant. Due to the dc component on the output of unidirectional controllers, bidirectional controllers are normally used in industrial applications. Due to the switching characteristics of thyristors, an inductive load makes the solutions of equations describing the performance of controllers more complex and an iterative method of solution is more convenient. The input power factor of controllers, which varies with delay angle, is generally poor, especially at the low output range. The ac voltage controllers can be used as transformer static tap changers.

The voltage controllers provide an output voltage at a fixed frequency. Two phase-controlled rectifiers connected as dual converters can be operated as direct-frequency changers known as *cycloconverters*. With the development of fast-switching power devices, the forced commutation of cycloconverters is possible; however, it requires synthesizing the switching functions for power devices [1,2].

REFERENCES

1. P. D. Ziogas, S. I. Khan, and M. H. Rashid, "Some improved forced commutated cycloconverter structures." *IEEE Transactions on Industry Applications*, Vol. IA121, No. 5, 1985, pp. 1242–1253.

2. M. Venturi, "A new sine wave in sine wave out conversion technique eliminates reactive elements." *Proceedings Powercon 7*, 1980, pp. E3-1–E3-13.

3. L. Gyugi and B. R. Pelly, *Static Power Frequency Changes: Theory, Performance, and Applications.* New York: Wiley-Interscience, 1976.

4. B. R. Pelly, *Thyristor-Phase Controlled Converters and Cycloconverters.* New York: Wiley-Interscience, 1971.

5. "IEEE standard definition and requirements for thyristor ac power controllers," *IEEE Standard*, No. 428-1981, 1981.

6. S. A. Hamed and B. J. Chalmers, "New method of analysis and performance prediction for thyristor voltage-controlled RL loads." *IEEE Proceedings*, Vol. 134, Pt. B, No. 6, 1987, pp. 339–347.

7. S. A. Hamed, "Modeling and design of transistor-controlled AC voltage regulators." *International Journal of Electronics*, Vol. 69, No. 3, 1990, pp. 421–434.

REVIEW QUESTIONS

6-1. What are the advantages and disadvantages of on–off control?

6-2. What are the advantages and disadvantages of phase-angle control?

6-3. What are the effects of load inductance on the performance of ac voltage controllers?

6-4. What is the extinction angle?

6-5. What are the advantages and disadvantages of unidirectional controllers?

6-6. What are the advantages and disadvantages of bidirectional controllers?

6-7. What is a tie control arrangement?

6-8. What are the steps involved in determining the output voltage waveforms of three-phase unidirectional controllers?

6-9. What are the steps involved in determining the output voltage waveforms of three-phase bidirectional controllers?

6-10. What are the advantages and disadvantages of delta-connected controllers?

6-11. What is the control range of the delay angle for single-phase unidirectional controllers?

6-12. What is the control range of the delay angle for single-phase bidirectional controllers?

6-13. What is the control range of the delay angle for three-phase unidirectional controllers?

6-14. What is the control range of the delay angle for three-phase bidirectional controllers?

6-15. What are the advantages and disadvantages of transformer tap changers?

6-16. What are the methods for output voltage control of transformer tap changers?

6-17. What is a synchronous tap changer?

6-18. What is a cycloconverter?

6-19. What are the advantages and disadvantages of cycloconverters?

6-20. What are the advantages and disadvantages of ac voltage controllers?

6-21. What is the principle of operation of cycloconverters?

6-22. What are the effects of load inductance on the performance of cycloconverters?

6-23. What are the three possible arrangements for a single-phase full-wave ac voltage controller?

6-24. What are the advantages of sinusoidal harmonic reduction techniques for cycloconverters?

6-25. What are the gate signal requirements of thyristors for voltage controllers with *RL* loads?

6-26. What are the effects of source and load inductances?

6-27. What are the conditions for the worst-case design of power devices for ac voltage controllers?

6-28. What are the conditions for the worst-case design of load filters for ac voltage controllers?

PROBLEMS

6-1. The ac voltage controller in Fig. 6-1a is used for heating a resistive load of $R = 5\ \Omega$ and the input voltage is $V_s = 120$ V (rms), 60 Hz. The thyristor switch is on for $n = 125$ cycles and is off for $m = 75$ cycles. Determine **(a)** the rms output voltage V_o, **(b)** the input power factor PF, and **(c)** the average and rms thyristor currents.

6-2. The ac voltage controller in Fig. 6-1a uses on–off control for heating a resistive load of $R = 4\ \Omega$ and the input voltage is $V_s = 208$ V (rms), 60 Hz. If the desired output power is $P_o = 3$ kW, determine **(a)** the duty cycle k, and **(b)** the input power factor PF.

6-3. The single-phase half-wave ac voltage controller in Fig. 6-2a has a resistive load of $R = 5\ \Omega$ and the input voltage is $V_s = 120$ V (rms), 60 Hz. The delay angle of thyristor T_1 is $\alpha = \pi/3$. Determine **(a)** the rms output voltage V_o, **(b)** the input power factor PF, and **(c)** the average input current.

6-4. The single-phase half-wave ac voltage controller in Fig. 6-2a has a resistive load of $R = 5\ \Omega$ and the input voltage is $V_s = 208$ V (rms), 60 Hz. If the desired output power is $P_o = 2$ kW, calculate **(a)** the delay angle α, and **(b)** the input power factor PF.

6-5. The single-phase full-wave ac voltage controller in Fig. 6-3a has a resistive load of $R = 5\ \Omega$ and the input voltage is $V_s = 120$ V (rms), 60 Hz. The delay angles of thyristors T_1 and T_2 are equal: $\alpha_1 = \alpha_2 = \alpha = 2\pi/3$. Determine **(a)** the rms output voltage V_o, **(b)** the input power factor PF, **(c)** the average current of thyristors I_A, and **(d)** the rms current of thyristors I_R.

6-6. The single-phase full-wave ac voltage controller in Fig. 6-3a has a resistive load of $R = 1.5\ \Omega$ and the input voltage is $V_s = 120$ V (rms), 60 Hz. If the desired output power is $P_o = 7.5$ kW, determine **(a)** the delay angles of thyristors T_1 and T_2, **(b)**

the rms output voltage V_o, **(c)** the input power factor PF, **(d)** the average current of thyristors I_A, and **(e)** the rms current of thyristors I_R.

6-7. The load of an ac voltage controller is resistive, with $R = 1.5\ \Omega$. The input voltage is $V_s = 120$ V (rms), 60 Hz. Plot the power factor against the delay angle for single-phase half-wave and full-wave controllers.

6-8. The single-phase full-wave controller in Fig. 6-6a supplies an RL load. The input voltage is $V_s = 120$ V (rms) at 60 Hz. The load is such that $L = 5$ mH and $R = 5\ \Omega$. The delay angles of thyristor T_1 and thyristor T_2 are equal, where $\alpha = \pi/3$. Determine **(a)** the conduction angle of thyristor T_1, δ; **(b)** the rms output voltage V_o; **(c)** the rms thyristor current I_R; **(d)** the rms output current I_o; **(e)** the average current of a thyristor I_A; and **(f)** the input power factor PF.

6-9. The single-phase full-wave controller in Fig. 6-6a supplies an RL load. The input voltage is $V_s = 120$ V at 60 Hz. Plot the power factor, PF, against the delay angle, α, for **(a)** $L = 5$ mH and $R = 5\ \Omega$, and **(b)** $R = 5\ \Omega$ and $L = 0$.

6-10. The three-phase unidirectional controller in Fig. 6-7 supplies a wye-connected resistive load with $R = 5\ \Omega$ and the line-to-line input voltage is 208 V (rms), 60 Hz. The delay angle is $\alpha = \pi/6$. Determine **(a)** the rms output phase voltage V_o, **(b)** the input power, and **(c)** the expressions for the instantaneous output voltage of phase a.

6-11. The three-phase unidirectional controller in Fig. 6-7 supplies a wye-connected resistive load with $R = 2.5\ \Omega$ and the line-to-line input voltage is 208 V (rms), 60 Hz. If the desired output power is $P_o = 12$ kW, calculate **(a)** the delay angle α, **(b)** the rms output phase voltage V_o, and **(c)** the input power factor PF.

6-12. The three-phase unidirectional controller in Fig. 6-7 supplies a wye-connected resistive load with $R = 5\ \Omega$ and the line-to-line input voltage is 208 V (rms), 60 Hz. The delay angle is $\alpha = 2\pi/3$. Determine **(a)** the rms output phase voltage V_o, **(b)** the input power factor PF, and **(c)** the expressions for the instantaneous output voltage of phase a.

6-13. Repeat Prob. 6-10 for the three-phase bidirectional controller in Fig. 6-12.

6-14. Repeat Prob. 6-11 for the three-phase bidirectional controller in Fig. 6-12.

6-15. Repeat Prob. 6-12 for the three-phase bidirectional controller in Fig. 6-12.

6-16. The three-phase bidirectional controller in Fig. 6-12 supplies a wye-connected load of $R = 5\ \Omega$ and $L = 10$ mH. The line-to-line input voltage is 208 V, 60 Hz. The delay angle is $\alpha = \pi/2$. Plot the line current for the first cycle after the controller is switched on.

6-17. A three-phase ac voltage controller supplies a wye-connected resistive load of $R = 5\ \Omega$ and the line-to-line input voltage is $V_s = 208$ V at 60 Hz. Plot the power factor PF against the delay angle α for **(a)** the half-wave controller in Fig. 6-7, and **(b)** the full-wave controller in Fig. 6-12.

6-18. A three-phase bidirectional delta-connected controller in Fig. 6-15 has a resistive load of $R = 5\ \Omega$. If the line-to-line voltage is $V_s = 208$ V, 60 Hz and the delay angle $\alpha = \pi/3$, determine **(a)** the rms output phase voltage V_o, **(b)** the expressions for instantaneous currents i_a, i_{ab}, and i_{ca}; **(c)** the rms output phase current I_{ab} and rms output line current I_a; **(d)** the input power factor PF; and **(e)** the rms current of thyristors I_R.

6-19. The circuit in Fig. 6-18 is controlled as a synchronous tap changer. The primary voltage is 208 V, 60 Hz. The secondary voltages are $V_1 = 120$ V and $V_2 = 88$ V. If the load resistance is $R = 5\ \Omega$ and the rms load voltage is 180 V, determine **(a)** the delay angles of thyristors T_1 and T_2, **(b)** the rms current of thyristors T_1 and T_2, **(c)**

the rms current of thyristors T_3 and T_4, and **(d)** the input power factor PF.

6-20. The input voltage to the single-phase/single-phase cycloconverter in Fig. 6-21a is 120 V, 60 Hz. The load resistance is 2.5 Ω and load inductance is $L = 40$ mH. The frequency of output voltage is 20 Hz. If the delay angle of thyristors is $\alpha_p = 2\pi/4$, determine **(a)** the rms output voltage, **(b)** the rms current of each thyristor, and **(c)** the input power factor PF.

6-21. Repeat Prob. 6-20 if $L = 0$.

6-22. For Prob. 6-20, plot the power factor against the delay angle α.

6-23. Repeat Prob. 6-20 for the three-phase/single-phase cycloconverter in Fig. 6-23a, $L = 0$.

6-24. Repeat Prob. 6-20 if the delay angles are generated by comparing a cosine signal at source frequency with a sinusoidal reference signal at output frequency as shown in Fig. 6-25.

6-25. For Prob. 6-24, plot the input power factor against the delay angle.

6-26. The single-phase full-wave ac voltage controller in Fig. 6-5a controls the power from a 208-V 60-Hz ac source into a resistive load. The maximum desired output power is 10 kW. Calculate **(a)** the maximum rms current rating of thyristor, **(b)** the maximum average current rating of thyristor, and **(c)** the peak thyristor voltage.

6-27. The three-phase full-wave ac voltage controller in Fig. 6-12 is used to control the power from a 2300-V 60-Hz ac source into a delta-connected resistive load. The maximum desired output power is 100 kW. Calculate **(a)** the maximum rms current rating of thyristors I_{RM}, **(b)** the maximum average current rating of thyristors I_{AM}, and **(c)** the peak value of thyristor voltage V_p.

6-28. The single-phase full-wave controller in Fig. 6-6a controls power to an RL load and the source voltage is 208 V, 60 Hz. The load is $R = 5\ \Omega$ and $L = 6.5$ mH. **(a)** Determine the rms value of third harmonic current. **(b)** If a capacitor is connected

---

across the load, calculate the value of capacitance to reduce the third harmonic current in the load to 5% of the load current, $\alpha = \pi/3$. (c) Use PSpice to plot the output voltage and the load current, and to compute the total harmonic distortion (THD) of the output voltage and output current, and the input power factor (PF) with and without the output filter capacitor in part (b).

7

Thyristor commutation techniques

7-1 INTRODUCTION

A thyristor is normally switched on by applying a pulse of gate signal. When a thyristor is in a conduction mode, its voltage drop is small, ranging from 0.25 to 2 V, and it is neglected in this chapter. Once the thyristor is turned on and the output requirements are satisfied, it is usually necessary to turn it off. The turn-off means that the forward conduction of the thyristor has ceased and the reapplication of a positive voltage to the anode will not cause current flow without applying the gate signal. *Commutation* is the process of turning off a thyristor, and it normally causes transfer of current flow to other parts of the circuit. A commutation circuit normally uses additional components to accomplish the turn-off. With the development of thyristors, many commutation circuits have been developed and the objective of all the circuits is to reduce the turn-off process of the thyristors.

With the availability of high-speed power semiconductor devices such as power transistors, GTOs, and IGBTs, the thyristor circuits are relatively less used in power converters. However, thyristors play a major role in high-voltage and high-current applications, generally above 500 A and 1 kV. The commutation techniques use *LC* resonance (or an underdamped *RLC* circuit) to force the current and/or the voltage of a thyristor to zero, thereby turning off a power device.

Power electronics uses semiconductor devices as switches for turning power "on" and "off" to the load. Situations similar to commutation circuits often occur in many power electronics circuits. The study of commutation techniques reveals the transient voltage and current waveforms of *LC* circuits under various conditions. It helps in understanding the dc transient phenomena under switching conditions.

There are many techniques to commutate a thyristor. However, these can be broadly classified into two types:

1. Natural commutation
2. Forced commutation

7-2 NATURAL COMMUTATION

If the source (or input) voltage is ac, the thyristor current goes through a natural zero, and a reverse voltage appears across the thyristor. The device is then automatically turned off due to the natural behavior of the source voltage. This is known as *natural commutation* or *line commutation*. In practice, the thyristor is triggered synchronously with the zero crossing of the positive input voltage in every cycle in order to provide a continuous control of power. This type of commutation is applied in ac voltage controllers, phase-controlled rectifiers, and cycloconverters. Figure 7-1a shows the circuit arrangement for natural commutation and Fig. 7-1b shows the voltage and current waveforms with a delay angle, $\alpha = 0$. The *delay angle* α is defined as the angle between the zero crossing of the input voltage and the instant the thyristor is fired.

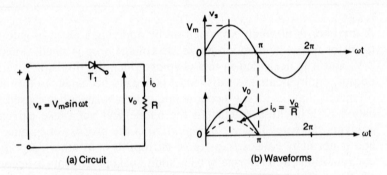

(a) Circuit   (b) Waveforms

Figure 7-1   Thyristor with natural commutation.

7-3 FORCED COMMUTATION

In some thyristor circuits, the input voltage is dc and the forward current of the thyristor is forced to zero by an additional circuitry called *commutation circuit* to turn off the thyristor. This technique is called *forced commutation* and normally applied in dc–dc converters (choppers) and dc-ac converters (inverters). The forced commutation of a thyristor can be achieved by seven ways and can be classified as:

1. Self-commutation
2. Impulse commutation

3. Resonant pulse commutation
4. Complementary commutation
5. External pulse commutation
6. Load-side commutation
7. Line-side commutation

This classification of forced commutations is based on the arrangement of the commutation circuit components and the manner in which the current of a thyristor is forced to zero. The commutation circuit normally consists of a capacitor, an inductor, and one or more thyristor(s) and/or diode(s).

7-3.1 Self-Commutation

In this type of commutation, a thyristor is turned off due to the natural characteristics of the circuit. Let us consider the circuit in Fig. 7-2a with the assumption that the capacitor is initially uncharged. When thyristor T_1 is switched on, the capacitor charging current i is given by

$$V_s = v_L + v_c = L \frac{di}{dt} + \frac{1}{C} \int i \, dt + v_c(t = 0) \tag{7-1}$$

With initial conditions $v_c(t = 0) = 0$ and $i(t = 0) = 0$, the solution of Eq. (7-1) (which is derived in Appendix D, Section D.3) gives the charging current i as

$$i(t) = V_s \sqrt{\frac{C}{L}} \sin \omega_m t \tag{7-2}$$

and the capacitor voltage as

$$v_c(t) = V_s(1 - \cos \omega_m t) \tag{7-3}$$

where $\omega_m = 1/\sqrt{LC}$. After time $t = t_0 = \pi\sqrt{LC}$, the charging current becomes zero and thyristor T_1 is switched off itself. Once thyristor T_1 is fired, there is a delay of t_0 seconds before T_1 is turned off and t_0 may be called the *commutation time* of the circuit. This method of turning off a thyristor is called *self-commutation* and thyristor T_1 is said to be self-commutated. When the circuit current falls to zero, the capacitor is charged to $2V_s$. The waveforms are shown in Fig. 7-2b.

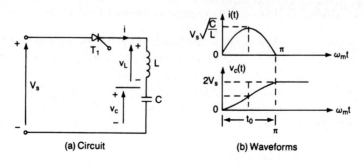

(a) Circuit   (b) Waveforms

Figure 7-2   Self-commutation circuit.

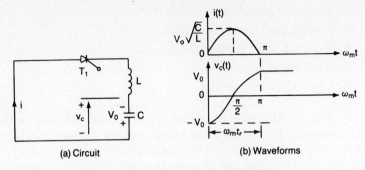

(a) Circuit   (b) Waveforms

Figure 7-3   Self-commutation circuit.

Figure 7-3a shows a typical circuit where the capacitor has an initial voltage of $-V_0$. When thyristor T_1 is fired, the current that will flow through the circuit is given by

$$L \frac{di}{dt} + \frac{1}{C} \int i \, dt + v_c(t = 0) = 0 \qquad (7\text{-}4)$$

With initial voltage $v_c(t = 0) = -V_0$ and $i(t = 0) = 0$, Eq. (7-4) gives the capacitor current as

$$i(t) = V_0 \sqrt{\frac{C}{L}} \sin \omega_m t \qquad (7\text{-}5)$$

and the capacitor voltage as

$$v_c(t) = -V_0 \cos \omega_m t \qquad (7\text{-}6)$$

After time $t = t_r = t_0 = \pi \sqrt{LC}$, the current becomes zero and the capacitor voltage is reversed to V_0. t_r is called the *reversing time*. The waveforms are shown in Fig. 7-3b.

Example 7-1

A thyristor circuit is shown in Fig. 7-4. If thyristor T_1 is switched on at $t = 0$, determine the conduction time of thyristor T_1 and the capacitor voltage after T_1 is turned off. The circuit parameters are $L = 10 \, \mu H$, $C = 50 \, \mu F$, and $V_s = 200$ V. The inductor carries an initial current of $I_m = 250$ A.

Solution   The capacitor current is expressed as

$$L \frac{di}{dt} + \frac{1}{C} \int i \, dt + v_c(t = 0) = V_s$$

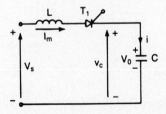

Figure 7-4   Self-commutated thyristor circuit.

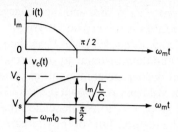

Figure 7-5 Current and voltage waveforms.

with initial current $i(t = 0) = I_m$ and $v_c(t = 0) = V_0 = V_s$. The voltage and current of the capacitor (from Appendix D, Section D.3) are

$$i(t) = I_m \cos \omega_m t$$

and

$$v_c(t) = I_m \sqrt{\frac{L}{C}} \sin \omega_m t + V_s$$

At $t = t_0 = 0.5 \times \pi \sqrt{LC}$, the commutation period ends and the capacitor voltage becomes

$$v_c(t = t_0) = V_c = V_s + I_m \sqrt{\frac{L}{C}}$$
$$= V_s + \Delta V \tag{7-7}$$

where ΔV is the overvoltage of the capacitor and depends on the initial current of the inductor, I_m, which is, in most cases, the load current. Figure 7-4 shows a typical equivalent circuit during the commutation process. For $C = 50\ \mu F$, $L = 10\ \mu H$, $V_s = 200$ V, and $I_m = 250$ A, $\Delta V = 111.8$ V, $V_c = 200 + 111.8 = 311.8$ V, and $t_0 = 35.12\ \mu s$. The current and voltage waveforms are shown in Fig. 7-5.

7-3.2 Impulse Commutation

An impulse-commutated circuit is shown in Fig. 7-6. It is assumed that the capacitor is initially charged to a voltage of $-V_0$ with the polarity shown.

Let us assume that thyristor T_1 is initially conducting and carrying a load current of I_m. When the auxiliary thyristor T_2 is fired, thyristor T_1 is reversed biased by the capacitor voltage, and T_1 is turned off. The current through thyristor T_1 will cease to flow and the capacitor would carry the load current. The

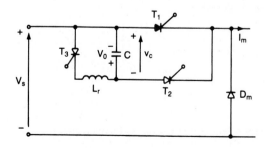

Figure 7-6 Impulse-commutated circuit.

capacitor will discharge from $-V_0$ to zero and then charge to the dc input voltage V_s when the capacitor current falls to zero and thyristor T_2 turns off. The charge reversal of the capacitor from $V_0 (= V_s)$ to $- V_0$ is then done by firing thyristor T_3. Thyristor T_3 is self-commutated similar to the circuit in Fig. 7-3.

The equivalent circuit during the commutation period is shown in Fig. 7-7a. The thyristor and capacitor voltages are shown in Fig. 7-7b. The time required for the capacitor to discharge from $-V_0$ to zero is called the *circuit turn-off time* t_{off} and must be greater than the turn-off time of the thyristor, t_q. t_{off} is also called the *available turn-off time*. The discharging time will depend on the load current and assuming a constant load current of I_m, t_{off} is given by

$$V_0 = \frac{1}{C} \int_0^{t_{off}} I_m \, dt = \frac{I_m t_{off}}{C}$$

or

$$t_{off} = \frac{V_0 C}{I_m} \qquad (7\text{-}8)$$

Since a reverse voltage of V_0 is applied across thyristor T_1 immediately after firing of thyristor T_2, this is known as *voltage commutation*. Due to the use of auxiliary thyristor T_2, this type of commutation is also called *auxiliary commutation*. Thyristor T_1 is sometimes known as the *main thyristor* because it carries the load current.

It can be noticed from Eq. (7-8) that the circuit turn-off time t_{off} is inversely proportional to the load current; and at a very light load (or low load current) the turn-off time will be large. On the other hand, at a high load current the turn-off time will be small. In an ideal commutation circuit, the turn-off time should be independent of the load current in order to guarantee the commutation of thyristor T_1. The discharging of the capacitor can be accelerated by connecting a diode D_1 and an inductor L_1 across the main thyristor as shown in Fig. 7-8; and this is illustrated in Example 7-3.

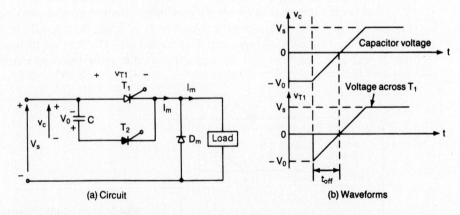

(a) Circuit   (b) Waveforms

Figure 7-7 Equivalent circuit and waveforms.

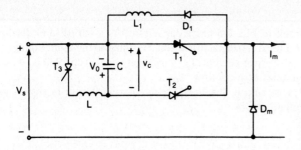

Figure 7-8 Impulse-commutated circuit with accelerated recharging.

Example 7-2

An impulse-commutated thyristor circuit is shown in Fig. 7-9. Determine the available turn-off time of the circuit if $V_s = 200$ V, $R = 10$ Ω, $C = 5$ μF, and $V_0 = V_s$.
Solution The equivalent circuit during the commutation period is shown in Fig. 7-10. The voltage across the commutation capacitor is given by

$$v_c = \frac{1}{C} \int i \, dt + v_c(t = 0)$$

$$V_s = v_c + Ri$$

The solution of these equations with initial voltage $v_c(t = 0) = -V_0 = -V_s$ gives the capacitor voltage as

$$v_c(t) = V_s(1 - 2e^{-t/RC})$$

The turn-off time t_{off}, which can be found if the condition $v_c(t = t_{off}) = 0$ is satisfied, is solved as

$$t_{off} = RC \ln(2)$$

For $R = 10$ Ω and $C = 5$ μF, $t_{off} = 34.7$ μs.

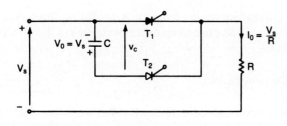

Figure 7-9 Impulse-commutated circuit with resistive load.

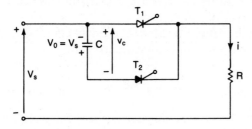

Figure 7-10 Equivalent circuit for Example 7-2.

Example 7-3

The commutation circuit in Fig. 7-8 has a capacitance $C = 20$ μF and discharging inductor $L_1 = 25$ μH. The initial capacitor voltage is equal to the input voltage. That is, $V_0 = V_s = 200$ V. If the load current, I_m, varies between 50 and 200 A, determine the variations of the circuit turn-off time, t_{off}.

Solution The equivalent circuit during the commutation period is shown in Fig. 7-11. The defining equations are

$$i_c = i + I_m$$

$$v_c = \frac{1}{C} \int i_c \, dt + v_c(t = 0)$$

$$= -L_1 \frac{di}{dt} = -L_1 \frac{di_c}{dt}$$

The initial conditions $i_c(t = 0) = I_m$ and $v_c(t = 0) = -V_0 = -V_s$. The solutions of these equations yield the capacitor current (from Appendix D, Section D.3) as

$$i_c(t) = V_0 \sqrt{\frac{C}{L_1}} \sin \omega_1 t + I_m \cos \omega_1 t$$

The voltage across the capacitor is expressed as

$$v_c(t) = I_m \sqrt{\frac{L_1}{C}} \sin \omega_1 t - V_0 \cos \omega_1 t \tag{7-9}$$

where $\omega_1 = 1/\sqrt{L_1 C}$. The available turn-off time or circuit turn-off time is obtained from the condition $v_c(t = t_{\text{off}}) = 0$ and is solved as

$$t_{\text{off}} = \sqrt{CL_1} \tan^{-1} \left(\frac{V_0}{I_m} \sqrt{\frac{C}{L_1}} \right) \tag{7-10}$$

For $C = 20$ μF, $L_1 = 25$ μH, $V_0 = 200$ V, and $I_m = 50$ A, $t_{\text{off}} = 29.0$ μs. For $C = 20$ μF, $L_1 = 25$ μH, $V_0 = 200$ V, and $I_m = 100$ A, $t_{\text{off}} = 23.7$ μs. For $C = 20$ μF, $L_1 = 25$ μH, $V_0 = 200$ V, and $I_m = 200$ A, $t_{\text{off}} = 16.3$ μs.

Note. As the load current increases from 50 A to 200 A, the turn-off time decreases from 29 μs to 16.3 μs. The use of an extra diode makes the turn-off time less dependent on the load.

7-3.3 Resonant Pulse Commutation

The resonant pulse commutation can be explained with Fig. 7-12a. Figure 7-12b shows the waveforms for the capacitor current and voltage. The capacitor is initially charged with the polarity as shown and thyristor T_1 is in the conduction mode carrying a load current of I_m.

When commutation thyristor T_2 is fired, a resonant circuit is formed by L, C, T_1 and T_2. The resonant current can be derived as

$$i(t) = V_0 \sqrt{\frac{C}{L}} \sin \omega_m t$$

$$= I_p \sin \omega_m t \tag{7-11}$$

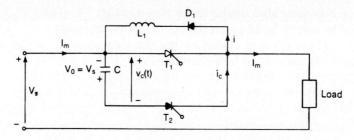

Figure 7-11 Equivalent circuit for Example 7-3.

and the capacitor voltage is

$$v_c(t) = -V_0 \cos \omega_m t \qquad (7\text{-}12)$$

where I_p is the peak permissible value of resonant current.

Due to the resonant current, the forward current of thyristor T_1 is reduced to zero at $t = t_1$, when the resonant current equals the load current I_m. The time t_1 must satisfy the condition $i(t = t_1) = I_m$ in Eq. (7-11) and is found as

$$t_1 = \sqrt{LC} \sin^{-1}\left(\frac{I_m}{V_0}\sqrt{\frac{L}{C}}\right) \qquad (7\text{-}13)$$

The corresponding value of the capacitor voltage is

$$v_c(t = t_1) = -V_1 = -V_0 \cos \omega_m t_1 \qquad (7\text{-}14)$$

The current through thyristor T_1 will cease to flow and the capacitor will recharge at a rate determined by the load current I_m. The capacitor will discharge from $-V_1$ to zero and its voltage will then rise to the dc source voltage V_s, in which case diode D_m starts conducting and a situation similar to the circuit in Fig. 7-4 exists with a time duration of t_0. This is shown in Fig. 7-12b. The energy stored in inductor L due to the peak load current I_m is transferred to the capacitor, causing it to be overcharged, and the capacitor voltage V_0 can be calculated from Eq. (7-7). The capacitor voltage is reversed from $V_c(= V_0)$ to $-V_0$ by firing T_3. T_3

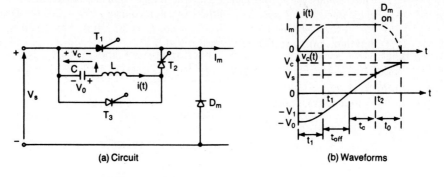

(a) Circuit   **(b) Waveforms**

Figure 7-12 Resonant pulse commutation.

is self-commutated similar to the circuit in Fig. 7-3. This circuit may not be stable due to energy build up on the commutation capacitor.

The equivalent circuit for the charging period is similar to Fig. 7-7a. From Eq. (7-8), the circuit turn-off time is

$$t_{\text{off}} = \frac{CV_1}{I_m} \qquad (7\text{-}15)$$

Let us define a parameter x which is the ratio of the peak resonant current I_p to the peak load current I_m. Then

$$x = \frac{I_p}{I_m} = \frac{V_0}{I_m} \sqrt{\frac{C}{L}} \qquad (7\text{-}16)$$

To reduce the forward current of T_1 to zero, the value of x must be greater than 1.0. In practice, the value of L and C are chosen such that $x = 1.5$. The value of t_1 in Eq. (7-13) is normally small and $V_1 \approx V_0$. The value of t_{off} obtained from Eq. (7-15) should be approximately equal to that obtained from Eq. (7-8). At time t_2, the capacitor current falls to load current I_m. During time t_c, capacitor C discharges and recharges to supply V_s. During time t_0, the energy stored in inductor L is returned to capacitor C, causing the capacitor to be overcharged with respect to supply voltage V_s.

Due to the fact that a resonant pulse of current is used to reduce the forward current of thyristor T_1 to zero, this type of commutation is also known as *current commutation*. It can be noticed from Eq. (7-15) that the circuit turn-off time t_{off} is also dependent on the load current. The discharging of the capacitor voltage can be accelerated by connecting diode D_2 as shown in Fig. 7-13a. However, once the current of thyristor T_1 is reduced to zero, the reverse voltage appearing across T_1 is the forward voltage drop of diode D_2, which is small. This will make the thyristor's recovery process slow and it would be necessary to provide longer reverse bias time than that if the diode D_2 were not present. The capacitor current $i_c(t)$ and capacitor voltage $v_c(t)$ are shown in Fig. 7-13b.

Example 7-4*

The resonant pulse commutation circuit in Fig. 7-12a has capacitance $C = 30~\mu\text{F}$ and inductance $L = 4~\mu\text{H}$. The initial capacitor voltage is $V_0 = 200$ V. Determine the circuit turn-off time t_{off} if the load current I_m is (a) 250 A, and (b) 50 A.

Solution (a) $I_m = 250$ A. From Eq. (7-13),

$$t_1 = \sqrt{4 \times 30}~\sin^{-1}\left(\frac{250}{200}\sqrt{\frac{4}{30}}\right) = 5.192~\mu\text{s}$$

$$\omega_m = \frac{1}{\sqrt{LC}} = 91{,}287.1~\text{rad/s} \qquad \text{and} \qquad \omega_m t_1 = 0.474~\text{rad}$$

From Eq. (7-14), $V_1 = 200\cos(0.474~\text{rad}) = 177.95$ V, and from Eq. (7-15),

$$t_{\text{off}} = 30 \times \frac{177.95}{250} = 21.35~\mu\text{s}$$

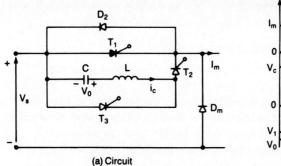
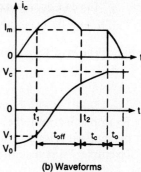

(a) Circuit   (b) Waveforms

Figure 7-13 Resonant pulse commutation with accelerating diode.

(b) $I_m = 50$ A.

$$t_1 = \sqrt{4 \times 30} \ \sin^{-1}\left(\frac{50}{200}\sqrt{\frac{4}{30}}\right) = 1.0014 \ \mu s$$

$$\omega_m = \frac{1}{\sqrt{LC}} = 91{,}287.1 \ \text{rad/s} \qquad \text{and} \qquad \omega_m t_1 = 0.0914 \ \text{rad}$$

$$-V_1 = -200 \cos(0.0914 \ \text{rad}) = -199.16 \ \text{V}$$

$$t_{\text{off}} = 30 \times \frac{199.16}{50} = 119.5 \ \mu s$$

Example 7-5*

Repeat Example 7-4 if an antiparallel diode D_2 is connected across thyristor T_1 as shown in Fig. 7-13a.

Solution (a) $I_m = 250$ A. When thyristor T_2 is fired, a resonant pulse of current flows through the capacitor and the forward current of thyristor T_1 is reduced to zero at time $t = t_1 = 5.192 \ \mu s$. The capacitor current $i_c(t)$ at this time is equal to the load current $I_m = 250$ A. After the current of T_1 is reduced to zero, the resonant oscillation continues through diode D_2 until the resonant current falls back to the load current level at time t_2. This is shown in Fig. 7-13b.

$$t_2 = \pi\sqrt{LC} - t_1 = \pi\sqrt{4 \times 30} - 5.192 = 29.22 \ \mu s$$

$$\omega_m = 91{,}287.1 \ \text{rad/s} \qquad \text{and} \qquad \omega_m t_2 = 2.667 \ \text{rad}$$

From Eq. (7-14) the capacitor voltage at $t = t_2$ is

$$v_c(t = t_2) = V_2 = -200 \cos(2.667 \ \text{rad}) = 177.9 \ \text{V}$$

The reverse-bias time of thyristor T_1 is

$$t_{\text{off}} = t_2 - t_1 = 29.22 - 5.192 = 24.03 \ \mu s$$

(b) $I_m = 50$ A.

$$t_1 = 1.0014 \ \mu s$$

$$t_2 = \pi\sqrt{LC} - t_1 = \pi\sqrt{4 \times 30} - 1.0014 = 33.41 \ \mu s$$

$$\omega_m = 91{,}287.1 \ \text{rad/s} \qquad \text{and} \qquad \omega_m t_2 = 3.05 \ \text{rad}$$

The capacitor voltage at $t = t_2$ is

$$v_c(t = t_2) = V_2 = -200 \cos(3.05 \text{ rad}) = 199.1 \text{ V}$$

The reverse-bias time of thyristor T_1 is

$$t_{\text{off}} = t_2 - t_1 = 33.41 - 1.0014 = 32.41 \ \mu s$$

Note. It can be noticed by comparing the reverse-bias times with those of Example 7-4 that the addition of a diode makes t_q less dependent on the load current variations. However, for a higher load current (e.g., $I_m = 250$ A), t_{off} in Example 7-4 is less than that of Example 7-5.

7-3.4 Complementary Commutation

A complementary commutation is used to transfer current between two loads and such an arrangement is shown in Fig. 7-14. The firing of one thyristor commutates the other one.

When thyristor T_1 is fired, the load with R_1 is connected to the supply voltage, V_s, and at the same time the capacitor C is charged to V_s through the other load with R_2. The polarity of capacitor C is as shown in Fig. 7-14. When thyristor T_2 is fired, the capacitor is then placed across thyristor T_1 and the load with R_2 is connected to the supply voltage, V_s. T_1 is reverse biased and is turned off by impulse commutation. Once thyristor T_1 is switched off, the capacitor voltage is reversed to $-V_s$ through R_1, T_2, and the supply. If thyristor T_1 is fired again, thyristor T_2 is turned off and the cycle is repeated. Normally, the two thyristors conduct with equal time intervals. The waveforms for voltages and currents are shown in Fig. 7-15 for $R_1 = R_2 = R$. Since each thyristor is switched off due to impulse commutation, this type of commutation is sometimes known as *complementary impulse commutation*.

Example 7-6

The circuit in Fig. 7-14 has load resistances of $R_1 = R_2 = R = 5 \ \Omega$, capacitance, $C = 10 \ \mu F$, and supply voltage, $V_s = 100$ V. Determine the circuit turn-off time, t_{off}.

Solution Assuming that the capacitor is charged to supply voltage V_s in the previous commutation of a complementary thyristor, the equivalent circuit during the commutation period is similar to Fig. 7-10. The current through the capacitor is given by

$$V_s = \frac{1}{C} \int i \ dt + v_c(t = 0) + Ri$$

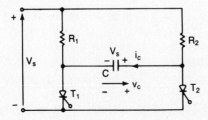

Figure 7-14 Complementary commutation circuit.

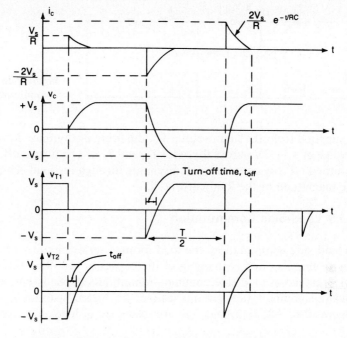

Figure 7-15 Waveforms for the circuit of Fig. 7-14.

With $v_c(t = 0) = -V_0 = -V_s$, the solution of this equation gives the capacitor current i as

$$i(t) = \frac{2V_s}{R} e^{-t/RC}$$

The capacitor voltage is obtained as

$$v_c(t) = V_s(1 - 2e^{-t/RC})$$

The turn-off time t_{off} can be found if the condition $v_c(t = t_q) = 0$ is satisfied and is solved as

$$t_{\text{off}} = RC \ln(2)$$

For $R = 5$ Ω and $C = 10$ μF, $t_{\text{off}} = 34.7$ μs.

7-3.5 External Pulse Commutation

A pulse of current is obtained from an external voltage to turn off a conducting thyristor. Figure 7-16 shows a thyristor circuit using an external pulse commutation and two supply sources. V_s is the voltage of the main supply and V is the voltage of the auxiliary source.

If thyristor T_3 is fired, the capacitor will charge from the auxiliary source. Assuming that the capacitor is initially uncharged, a resonant current pulse of peak $V\sqrt{C/L}$, which is similar to the circuit in Fig. 7-2, will flow through T_3 and the capacitor is charged to 2V. If thyristor T_1 is conducting and a load current

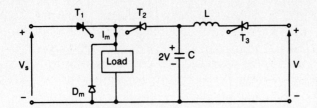

Figure 7-16 External pulse commutation.

is supplied from the main source V_s, the firing of thyristor T_2 will apply a reverse voltage of $V_s - 2V$ across thyristor T_1; and T_1 will be turned off. Once thyristor T_1 is turned off, the capacitor will discharge through the load at a rate determined by the magnitude of the load current, I_m.

7-3.6 Load-Side Commutation

In load-side commutation, the load forms a series circuit with the capacitor; and the discharging and recharging of the capacitor are done through the load. The performance of load-side commutation circuits depends on the load and in addition the commutation circuits cannot be tested without connecting the load. Figures 7-6, 7-8, 7-12, and 7-13 are examples of load-side commutation.

7-3.7 Line-Side Commutation

In this type of commutation, the discharging and recharging of the capacitor are not accomplished through the load and the commutation circuit can be tested without connecting the load. Figure 7-17a shows such a circuit.

When thyristor T_2 is fired, capacitor C is charged to $2V_s$ and T_2 is self-commutated similar to the circuit in Fig. 7-2. Thyristor T_3 is fired to reverse the

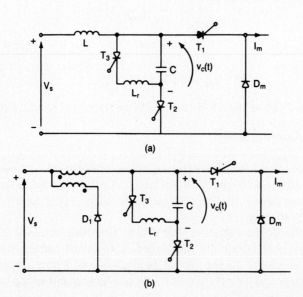

Figure 7-17 Line-side commutated circuit.

voltage of capacitor to $-2V_s$ and T_3 is also self-commutated. Assuming that thyristor T_1 is conducting and carries a load current of I_m, thyristor T_2 is fired to turn off T_1. The firing of thyristor T_2 will forward bias the diode D_m and apply a reverse voltage of $2V_s$ across T_1; and T_1 will be turned off. The discharging and recharging of the capacitor will be done through the supply. The connection of the load is not required to test the commutation circuit.

The inductor L carries the load current I_m and the equivalent circuit during the commutation period is shown in Fig. 7-18. The capacitor current is expressed (from Appendix D) as

$$V_s = L\frac{di}{dt} + \frac{1}{C}\int i\,dt + v_c(t = 0) \tag{7-17}$$

with initial conditions $i(t = 0) = I_m$ and $v_c(t = 0) = -2V_s$. The solution of Eq. (7-17) gives the capacitor current and voltage as

$$i(t) = I_m \cos \omega_m t + 3V_s \sqrt{\frac{C}{L}} \sin \omega_m t \tag{7-18}$$

and

$$v_c(t) = I_m \sqrt{\frac{L}{C}} \sin \omega_m t - 3V_s \cos \omega_m t + V_s \tag{7-19}$$

where

$$\omega_m = \frac{1}{\sqrt{LC}}$$

The circuit turn-off time, t_{off}, is obtained from the condition $v_c(t = t_{\text{off}}) = 0$ of Eq. (7-19) and solved after simplification as

$$t_{\text{off}} = \sqrt{LC} \left(\tan^{-1} 3x - \sin^{-1} \frac{x}{\sqrt{9x^2 + 1}} \right) \tag{7-20}$$

where

$$x = \frac{V_s}{I_m} \sqrt{\frac{C}{L}} \tag{7-21}$$

The conduction time of thyristor T_2, which can be found from the condition $i(t = t_1) = 0$ of Eq. (7-18) is given by

$$t_1 = \sqrt{LC} \tan^{-1} \frac{-1}{3x} = \sqrt{LC} \left(\pi - \tan^{-1} \frac{1}{3x} \right) \tag{7-22}$$

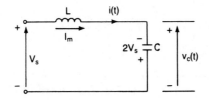

Figure 7-18 Equivalent circuit during commutation period.

Under no-load conditions, $I_m = 0$ and x is infinite. Equation (7-19) gives t_{off} as

$$t_{\text{off}} = \sqrt{LC}\,\cos^{-1}\tfrac{1}{3} = 1.231\sqrt{LC}$$

and

$$t_1 = \pi\sqrt{LC} \qquad (7\text{-}23)$$

Note. If $I_m = 0$ and $t_1 = \pi\sqrt{LC}$, the capacitor voltage in Eq. (7-19) becomes $v_c(t = t_1) = V_0 = 4V_s$ and there will be continuous buildup of the capacitor voltage. To limit the overcharging of the capacitor, the inductor L is normally replaced by an energy recovery transformer and a diode as shown in Fig. 7-17b.

7-4 COMMUTATION CIRCUIT DESIGN

The design of commutation circuits requires determining the values of capacitor C and inductor L. For the impulse commutation circuit in Fig. 7-6, the value of capacitor C is calculated from Eq. (7-8) and the reversal inductor L_r is determined from the peak permissible reversal current of Eq. (7-5). For the circuit in Fig. 7-8, the turn-off time requirement t_{off} of Eq. (7-10) can be satisfied by choosing either C or L_1.

For the resonant pulse commutation circuit in Fig. 7-12, the values of L and C can be calculated from Eqs. (7-15) and (7-16). In Eqs. (7-14) and (7-15) V_0 and V_1 also depend on L and C as in Eq. (7-7).

Example 7-7

For the impulse commutated circuit in Fig. 7-6, determine the values of capacitor C and reversing inductor L_r if the supply voltage, $V_s = 200$ V, load current, $I_m = 100$ A, the turn-off time, $t_{\text{off}} = 20$ μs, and the peak reversal current is limited to 140% of I_m. **Solution** $V_0 = V_s = 200$ V. From Eq. (7-8), $C = 100 \times 20/200 = 10$ μF. From Eq. (7-5), the peak resonant current is $1.4 \times 100 = 140 = V_s\sqrt{C/L_r} = 200\sqrt{10/L_r}$, which gives $L_r = 20.4$ μH.

Example 7-8

For the resonant commutation circuit in Fig. 7-13, determine the optimum values of C and L so that the minimum energy losses would occur during the commutation period if $I_m = 350$ A, $V_0 = 200$ V, and $t_{\text{off}} = 20$ μs. **Solution** Substituting Eq. (7-16) into Eq. (7-13), the time required for the capacitor current to rise to the level of peak load current I_m is given by

$$t_1 = \sqrt{LC}\,\sin^{-1}\frac{1}{x} \qquad (7\text{-}24)$$

where $x = I_p/I_m = (V_0/I_m)\sqrt{C/L}$. From Fig 7-13b, the available reverse bias time or turn-off time t_{off} is

$$t_{\text{off}} = t_2 - t_1 = \pi\sqrt{LC} - 2t_1 = \sqrt{LC}\left(\pi - 2\sin^{-1}\frac{1}{x}\right) \qquad (7\text{-}25)$$

From Eq. (7-11), the peak resonant current is

$$I_p = V_0\sqrt{\frac{C}{L}} \qquad (7\text{-}26)$$

Let us define a function $F_1(x)$ such that

$$F_1(x) = \frac{t_{off}}{\sqrt{LC}} = \pi - 2\sin^{-1}\frac{1}{x}$$ (7-27)

The commutation energy can be expressed as

$$W = 0.5CV_0^2 = 0.5LI_p^2$$ (7-28)

Substituting the value of I_p from Eq. (7-26) gives us

$$W = 0.5\sqrt{LC}\,V_0 I_p$$

Substituting the value of $\sqrt{LC}$ from Eq.(7-27), we have

$$W = 0.5V_0 I_m\frac{xt_{off}}{F_1(x)} = 0.5V_0 x I_m\sqrt{LC}$$ (7-29)

Let us define another function $F_2(x)$ such that

$$F_2(x) = \frac{W}{V_0 I_m t_{off}} = \frac{x}{2F_1(x)} = \frac{x}{2[\pi - 2\sin(1/x)]}$$ (7-30)

It can be shown mathematically or by plotting $F_2(x)$ against x that $F_2(x)$ becomes minimum when $x = 1.5$. Table 7-1 shows the values of $F_2(x)$ against x. For $x = 1.5$, Table 7-1 gives $F_1(x) = 1.6821375$ and $F_2(x) = 0.4458613$.

TABLE 7-1 $F_2(x)$ AGAINST x

x	$F_2(x)$	$F_1(x)$
1.2	0.5122202	1.1713712
1.3	0.4688672	1.3863201
1.4	0.4515002	1.5384548
1.5	0.4458613	1.6821375
1.6	0.4465956	1.7913298
1.7	0.4512053	1.8838431
1.8	0.4583579	1.9635311

Substituting $x = 1.5$ in Eqs. (7-25) and (7-16), we obtain

$$t_{off} = 1.682\sqrt{LC}$$

and

$$1.5 = \frac{V_0}{I_m}\sqrt{\frac{C}{L}}$$

Solving these, the optimum values of L and C are

$$L = 0.398\frac{t_{off}V_0}{I_m}$$ (7-31)

$$C = 0.8917\frac{t_{off}I_m}{V_0}$$ (7-32)

For $I_m = 350$ A, $V_0 = 200$ V, and $t_{off} = 20$ μs,

$$L = 0.398 \times 20 \times \frac{200}{350} = 6.4\ \mu\text{H}$$

---

and

$$C = 0.8917 \times 20 \times \frac{350}{200} = 31.2 \ \mu F$$

Note. Due to the freewheeling diode across the load as shown in Figs. 7-12a and 7-13a, the capacitor will get overcharged by the energy stored in inductor L. The capacitor voltage V_0, which will depend on the values of L and C, can be determined from Eq. (7-7). In this case, Eqs. (7-31) and (7-32) should be solved for the values of L and C.

Example 7-9

A freewheeling diode is connected across the output as shown in Fig. 7-12a and the capacitor gets overcharged due to the energy stored in inductor L. Determine the values of L and C. The particulars are: $V_s = 200$ V, $I_m = 350$ A, $t_{off} = 20 \ \mu s$, and $x = 1.5$.

Solution Substituting Eq. (7-7) into Eq. (7-16) yields $x = (V_s/I_m) \sqrt{C/L} + 1$. Substituting Eqs. (7-7), (7-13), and (7-14), into Eq. (7-15) gives us

$$t_{off} = \left[\frac{V_s C}{I_m} + \sqrt{LC} \right] \cos \left(\sin^{-1} \frac{1}{x} \right)$$

The values of C and L can be found from these two equations for known values of x and t_{off}. The results are $L = 20.4 \ \mu H$ and $C = 15.65 \ \mu F$.

Example 7-10

Repeat Example 7-9 for the circuit in Fig. 7-13.
Solution From Eq. (7-25), $t_{off} = \sqrt{LC} \ [\pi - 2 \sin^{-1}(1/x)]$. From Eqs. (7-7) and (7-16), $x = (V_s/I_m) \sqrt{C/L} + 1$. The values of L and C are found from these two equations for known values of x and t_{off} as $C = 10.4 \ \mu F$ and $L = 13.59 \ \mu H$.

7-5 DC THYRISTOR SPICE MODEL

A dc thyristor can be modeled by a diode and a voltage-controlled switch as shown in Fig. 7-19. The switch is controlled by the gate voltage v_g. The diode parameters can be adjusted to give required voltage drop and the reverse recovery time of the thyristor. Let us assume that the PSpice model parameters of the diode are IS=1E−25, BV=1000V, and the switch parameters are RON=0.1, ROFF=10E+6, VON=10V, VOFF=5V.

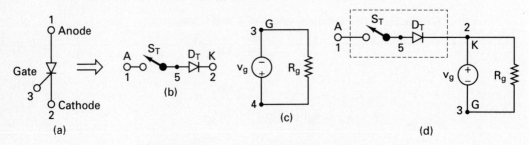

Figure 7-19 Dc thyristor SPICE model.

This model can be used as a subcircuit. The subcircuit definition for the dc thyristor model DCSCR can be described as follows:

```
*      Subcircuit for dc thyristor model
.SUBCKT    DCSCR       1        2        3         4
*           model    anode   cathode  +control  -control
*           name                      voltage   voltage
DT     5    2    DMOD              ; Switch diode
ST     1    5    3    4    SMOD    ; Switch
.MODEL    DMOD    D(IS=1E-25  BV=1000V)   ; Diode model parameters
.MODEL    SMOD    VSWITCH  (RON=0.1 ROFF=10E+6 VON=10V VOFF=5V)
.ENDS DCSCR                       ; Ends subcircuit definition
```

Example 7-11

The circuit parameters of the resonant pulse commutation circuit in Fig. 7-13a are: supply voltage V_s = 200 V, commutation capacitor C = 31.2 μF, commutation inductance L = 6.4 μH, load resistance R_m = 0.5 Ω, and load inductance L_m = 5 mH. If the thyristor is modeled by the circuit of Fig. 7-19, use PSpice to plot (a) the capacitor voltage v_c, (b) the capacitor current i_c, and (c) the load current i_L. The switching frequency is f_c = 1 kHz and the on-time of thyristor T_1 is 40%.

Solution The resonant pulse commutation circuit for PSpice simulation is shown in Fig. 7-20a. The controlling voltages V_{g1}, V_{g2}, and V_{g3} for thyristors are shown in Fig. 7-20b. The list of the circuit file is as follows:

```
Example 7-11    Resonant Pulse Chopper
VS     1    0    DC    200V
Vg1    7    0    PULSE (0V   100V   0       1US   1US   0.4MS   1MS)
Vg2    8    0    PULSE (0V   100V   0.4MS   1US   1US   0.6MS   1MS)
Vg3    9    0    PULSE (0V   100V   0       1US   1US   0.2MS   1MS)
Rg1    7    0    10MEG
Rg2    8    0    10MEG
Rg3    9    0    10MEG
CS     11   11   0.1UF
RS     11   4    750
C      1    2    31.2UF   IC=200V  ; With initial capacitor voltage
L      2    3    6.4UH
D1     4    1    DMOD
DM     0    4    DMOD
.MODEL    DMOD    D(IS=1E-25  BV=1000V)   ; Diode model parameters
RM     4    5    0.5
LM     5    6    5.0MH
VX     6    0    DC    0V           ; Measures load current
VY     1    10   DC    0V           ; Measures current of T1
*    Subcircuit calls for DC thyristor model
XT1    10   4    7    0    DCSCR    ; Thyristor T1
XT2    3    4    8    0    DCSCR    ; Thyristor T2
XT3    1    3    9    0    DCSCR    ; Thyristor T3
*    Subcircuit DCSCR which is missing must be inserted
.TRAN    0.5US   3MS   1.5MS   0.5US   ; Transient analysis
.PROBE                        ; Graphics postprocessor
.options abstol = 1.000u reltol = .01 vntol = 0.1   ITL5=20000
.END
```

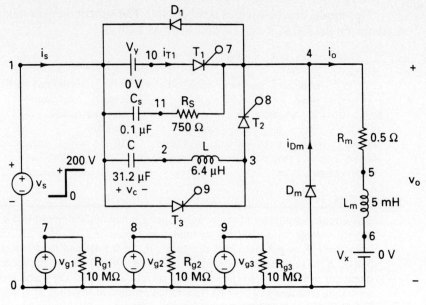

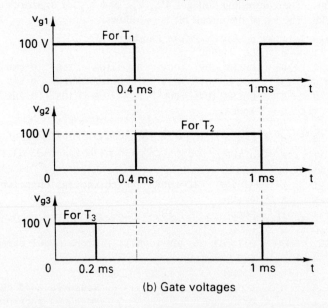

(b) Gate voltages

Figure 7-20 Resonant pulse commutation circuit for PSpice simulation.

The PSpice plots are shown in Fig. 7-21, where I(VX) = load current, I(C) = capacitor current, and V(1, 2) = capacitor voltage. From Fig. 7-21, the available turn-off time is $t_{off} = 2441.4 - 2402.1 = 39.3$ μs at a load current of $I_m = 49.474$ A. It should be noted that the instantaneous load current I(VX) has not reached the steady-state condition.

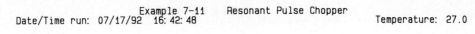

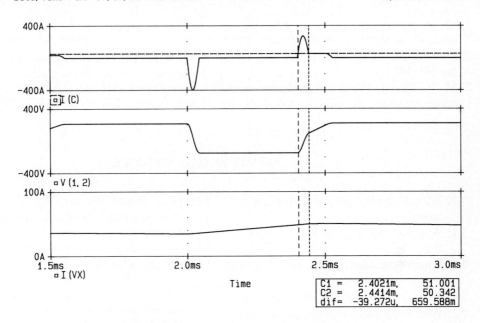

C1 =	2.4021m,	51.001
C2 =	2.4414m,	50.342
dif=	-39.272u,	659.588m

Figure 7-21   PSpice plots for Example 7-11.

7-6 COMMUTATION CAPACITORS

If the switching frequencies are below 1 kHz, the thyristor commutation time can be considered short compared with the switching period. Although the peak current through the capacitor is high, the average current could be relatively low. If the switching frequencies are above 5 kHz, the capacitor carries current for a significant part of the switching period and the capacitor must therefore be chosen for a continuous current rating.

In selecting a commutation capacitor, the specifications of peak, rms, and average current and peak-to-peak voltage must be satisfied.

SUMMARY

We have seen in this chapter that a conducting thyristor can be turned off by a natural or forced commutation. In the natural commutation, the thyristor current is reduced to zero due to the natural characteristics of the input voltage. In the forced commutation, the thyristor current is reduced to zero by an additional circuitry called a commutation circuit, and the turn-off process depends on the load current. To guarantee the turn-off of a thyristor, the circuit (or available) turn-off must be greater than the turn-off time of the thyristor, which is normally specified by the thyristor manufacturer.

REFERENCES

1. M. H. Rashid, "Commutation limits of dc chopper on output voltage control." *Electronic Engineering*, Vol. 51, No. 620, 1979, pp. 103–105.
2. M. H. Rashid, "A thyristor chopper with minimum limits on voltage control of dc drives." *International Journal of Electronics*, Vol. 53, No. 1, 1982, pp. 71–89.
3. W. McMurry, "Thyristor commutation in dc chopper: a comparative study." *IEEE Industry Applications Society Conference Record*, October 2–6, 1977, pp. 385–397.

REVIEW QUESTIONS

7-1. What are the two general types of commutation?

7-2. What are the types of forced commutation?

7-3. What is the difference between self- and natural commutation?

7-4. What is the principle of self-commutation?

7-5. What is the principle of impulse commutation?

7-6. What is the principle of resonant pulse commutation?

7-7. What is the principle of complementary commutation?

7-8. What is the principle of external pulse commutation?

7-9. What are the differences between load-side and line-side commutation?

7-10. What are the differences between voltage and current commutation?

7-11. What are the purposes of a commutation circuit?

7-12. Why should the available reverse-bias time be greater than the turn-off time of a thyristor?

7-13. What is the purpose of connecting an anti-parallel diode across the main thyristor, with or without a series inductor?

7-14. What is the ratio of peak resonant to load current for resonant pulse commutation that would minimize the commutation losses?

7-15. What are the expressions for the optimum value of commutation capacitor and inductor in a resonant pulse commutation?

7-16. Why does the commutation capacitor in a resonant pulse commutation get overcharged?

7-17. How is the voltage of the commutation capacitor reversed in a commutation circuit?

7-18. What type of capacitor is normally used in high switching frequencies?

PROBLEMS

7-1. In Fig. 7-3a, the initial capacitor voltage $V_0 = 600$ V, capacitance $C = 40$ μF, and inductance $L = 10$ μH. Determine the peak value of resonant current and the conduction time of thyristor T_1.

7-2. Repeat Prob. 7-1 if the inductor in the resonant reversal circuit has a resistance $R = 0.015$ Ω. (*Hint:* Determine the roots of a second-order system and then find the solution.)

7-3. The circuit in Fig. 7-4 has $V_s = 600$ V, $V_0 = 0$ V, $L = 20$ μH, $C = 50$ μF, and $I_m = 350$ A. Determine (a) the peak capacitor voltage and current, and (b) the conduction time of thyristor T_1.

7-4. In the commutation circuit in Fig. 7-6, capacitance $C = 20 \mu F$, input voltage V_s varies between 180 and 220 V, and load current I_m varies between 50 and 200 A. Determine the minimum and maximum values of available turn-off time t_{off}.

7-5. For the circuit in Fig. 7-6, determine the values of capacitor C and reversing inductor L_r if the supply voltage $V_s = 220$ V, load current $I_m = 150$ A, turn-off time $t_{off} = 15 \mu s$, and the reversal current is limited to 150% of I_m.

7-6. The circuit in Fig. 7-8 has $V_s = 220$ V, $C = 20 \mu F$, and $I_m = 150$ A. Determine the value of recharging inductance L_1 which will provide turn-off time $t_{off} = 15 \mu s$.

7-7. For the circuit in Fig. 7-8, determine the values of L_1 and C. The supply voltage $V_s = 200$ V, load current $I_m = 350$ A, turn-off time $t_{off} = 20 \mu s$, and the peak current through diode D_1 is limited to 2.5 times I_m. If the thyristor is modeled by the circuit of Fig. 7-19, use PSpice to plot the capacitor voltage v_c, and the capacitor current i_c, and to verify the available turn-off time t_{off}. The switching frequency is $f_c = 1$ kHz, and the on-time of thyristor T_1 is 40%.

7-8. For the impulse commutation circuit in Fig. 7-9, the supply voltage $V_s = 220$ V, capacitance $C = 20 \mu F$, and load current $R = 10 \Omega$. Determine the turn-off time t_{off}.

7-9. In the resonant pulse circuit in Fig. 7-12a, supply voltage $V_s = 200$ V, load current $I_m = 150$ A, commutation inductance $L = 4 \mu H$, and commutation capacitance $C = 20 \mu F$. Determine the peak resonant reversing current of thyristor T_3, I_k, and the turn-off time t_{off}.

7-10. Repeat Prob. 7-9 if an antiparallel diode is connected across thyristor T_1 as shown in Fig. 7-13a.

7-11. If a diode is connected across thyristor T_1 in Fig. 7-12a and the capacitor gets overcharged, determine the values of L and C. The supply voltage $V_s = 200$ V, load current $I_m = 350$ A, turn-off time $t_{off} = 20 \mu s$, and the ratio of peak resonant to load current $x = 1.5$.

7-12. Repeat Prob. 7-11 for the circuit in Fig. 7-13a.

7-13. In the circuit in Fig. 7-13a, load current $I_m = 200$ A, capacitor voltage $V_0 = 220$ V, and turn-off time $t_{off} = 15 \mu s$. Determine the optimum values of C and L so that the minimum energy losses would occur during the commutation period. If the thyristor is modeled by the circuit of Fig. 7-19, use PSpice to plot the capacitor voltage v_c and the capacitor current i_c and to verify the available turn-off time t_{off}. The switching frequency is $f_c = 1$ kHz, and the on-time of thyristor T_1 is 40%.

7-14. In the circuit in Fig. 7-18, the supply voltage $V_s = 220$ V, capacitance $C = 30 \mu F$, commutation inductance $L = 10 \mu H$, and load current $I_m = 100$ A. Determine the turn-off time t_{off} of the circuit.

7-15. Explain the operation of the circuit in Fig. 7-17a and identify the types of commutation involved in this circuit.

8

Power transistors

8-1 INTRODUCTION

Power transistors have controlled turn-on and turn-off characteristics. The transistors, which are used as switching elements, are operated in the saturation region, resulting in a low on-state voltage drop. The switching speed of modern transistors is much higher than that of thyristors and they are extensively employed in dc–dc and dc–ac converters, with inverse parallel-connected diodes to provide bidirectional current flow. However, their voltage and current ratings are lower than those of thyristors and transistors are normally used in low to medium power applications. The power transistors can be classified broadly into four categories:

1. Bipolar junction transistors (BJTs)
2. Metal-oxide-semiconductor field-effect transistors (MOSFETs)
3. Static induction transistors (SITs)
4. Insulated-gate bipolar transistors (IGBTs)

BJTs or MOSFETs, SITs or IGBTs, can be assumed as ideal switches to explain the power conversion techniques. A transistor switch is much simpler than a forced-commutated thyristor switch. However, the choice between a BJT and a MOSFET in the converter circuits is not obvious, but either of them can replace a thyristor, provided that their voltage and current ratings meet the output requirements of the converter. Practical transistors differ from ideal devices. The transistors have certain limitations and are restricted to some applications. The characteristics and ratings of each type should be examined to determine its suitability to a particular application.

A bipolar transistor is formed by adding a second *p*- or *n*-region to a *pn*-junction diode. With two *n*-regions and one *p*-region, two junctions are formed and it is known as an *NPN-transistor*, as shown in Fig. 8-1a. With two *p*-regions and one *n*-region, it is called as a *PNP-transistor,* as shown in Fig. 8-1b. The three terminals are named as *collector, emitter,* and *base.* A bipolar transistor has two junctions, collector–base junction (CBJ) and base–emitter junction (BEJ). *NPN*-transistors of various sizes are shown in Fig. 8-2.

8-2.1 Steady-State Characteristics

Although there are three possible configurations—common-collector, common-base, and common-emitter, the common-emitter configuration, which is shown in Fig. 8-3a for an *NPN*-transistor, is generally used in switching applications. The typical input characteristics of base current, I_B, against base–emitter voltage, V_{BE}, are shown in Fig. 8-3b. Figure 8-3c shows the typical output characteristics of collector current, I_C, against collector–emitter voltage, V_{CE}. For a *PNP*-transistor, the polarities of all currents and voltages are reversed.

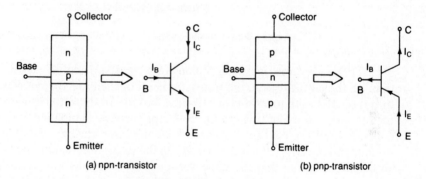

Figure 8-1 Bipolar transistors.

Figure 8-2 *NPN*-transistors.
(Courtesy of Powerex, Inc.)

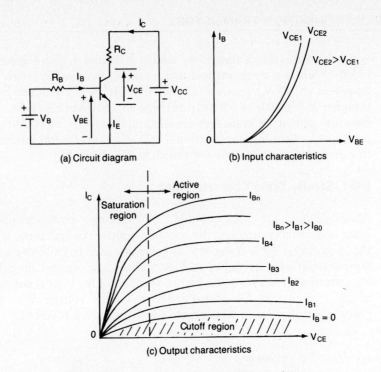

(a) Circuit diagram

(b) Input characteristics

(c) Output characteristics

Figure 8-3 Characteristics of *NPN*-transistors.

There are three operating regions of a transistor: cutoff, active, and saturation. In the cutoff region, the transistor is off or the base current is not enough to turn it on and both junctions are reverse biased. In the active region, the transistor acts as an amplifier, where the collector current is amplified by a gain and the collector–emitter voltage decreases with the base current. The CBJ is reverse biased, and the BEJ is forward biased. In the saturation region, the base current is sufficiently high so that the collector–emitter voltage is low, and the transistor acts as a switch. Both junctions (CBJ and BEJ) are forward biased. The transfer characteristic, which is a plot of V_{CE} against I_B, is shown in Fig. 8-4.

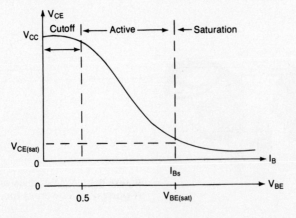

Figure 8-4 Transfer characteristics.

The model of an *NPN*-transistor is shown in Fig. 8-5 under large-signal dc operation. The equation relating the currents is

$$I_E = I_C + I_B \tag{8-1}$$

The base current is effectively the input current and the collector current is the output current. The ratio of the collector current, I_C, to base current, I_B, is known as the *current gain*, β:

$$\beta = h_{FE} = \frac{I_c}{I_B} \tag{8-2}$$

The collector current has two components: one due to the base current and the other is the leakage current of the CBJ.

$$I_C = \beta I_B + I_{CEO} \tag{8-3}$$

where I_{CEO} is the collector-to-emitter leakage current with base open circuit and can be considered negligible compared to βI_B.

From Eqs. (8-1) and (8-3),

$$I_E = I_B(1 + \beta) + I_{CEO} \tag{8-4}$$

$$\approx I_B(1 + \beta) \tag{8-4a}$$

$$I_E \approx I_C \left(1 + \frac{1}{\beta}\right) = I_C \frac{\beta + 1}{\beta} \tag{8-5}$$

The collector current can be expressed as

$$I_C \approx \alpha I_E \tag{8-6}$$

where the constant α is related to β by

$$\alpha = \frac{\beta}{\beta + 1} \tag{8-7}$$

or

$$\beta = \frac{\alpha}{1 - \alpha} \tag{8-8}$$

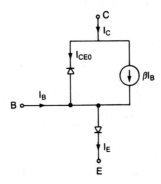

Figure 8-5 Model of *NPN*-transistors.

Let us consider the circuit of Fig. 8-6, where the transistor is operated as a switch.

$$I_B = \frac{V_B - V_{BE}}{R_B} \tag{8-9}$$

$$V_C = V_{CE} = V_{CC} - I_C R_C = V_{CC} - \frac{\beta R_C}{R_B}(V_B - V_{BE}) \tag{8-10}$$

$$V_{CE} = V_{CB} + V_{BE}$$

or

$$V_{CB} = V_{CE} - V_{BE} \tag{8-11}$$

Equation (8-11) indicates that as long as $V_{CE} \geq V_{BE}$, the CBJ will be reverse biased and the transistor will be in the active region. The maximum collector current in the active region, which can be obtained by setting $V_{CB} = 0$ and $V_{BE} = V_{CE}$, is

$$I_{CM} = \frac{V_{CC} - V_{CE}}{R_C} = \frac{V_{CC} - V_{BE}}{R_C} \tag{8-12}$$

and the corresponding value of base current

$$I_{BM} = \frac{I_{CM}}{\beta} \tag{8-13}$$

If the base current is increased above I_{BM}, V_{BE} increases and the collector current will increase and the V_{CE} will fall below V_{BE}. This will continue until the CBJ is forward biased with V_{BC} of about 0.4 to 0.5 V. The transistor then goes into saturation. The *transistor saturation* may be defined as the point above which any increase in the base current does not increase the collector current significantly.

In the saturation, the collector current remains almost constant. If the collector–emitter saturation voltage is $V_{CE(sat)}$, the collector current is

$$I_{CS} = \frac{V_{CC} - V_{CE(sat)}}{R_C} \tag{8-14}$$

and the corresponding value of base current is

$$I_{BS} = \frac{I_{CS}}{\beta} \tag{8-15}$$

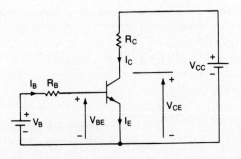

Figure 8-6 Transistor switch.

Normally, the circuit is designed so that I_B is higher than I_{BS}. The ratio of I_B to I_{BS} is called the *overdrive factor,* ODF:

$$\text{ODF} = \frac{I_B}{I_{BS}} \tag{8-16}$$

and the ratio of I_{CS} to I_B is called as *forced* β, β_f where

$$\beta_f = \frac{I_{CS}}{I_B} \tag{8-17}$$

The total power loss in the two junctions is

$$P_T = V_{BE}I_B + V_{CE}I_C \tag{8-18}$$

A high value of overdrive factor will not reduce the collector–emitter voltage significantly. However, V_{BE} will increase due to increased base current, resulting in increased power loss in the BEJ.

Example 8-1

The bipolar transistor in Fig. 8-6 is specified to have β in the range 8 to 40. The load resistance is $R_C = 11 \ \Omega$. The dc supply voltage is $V_{CC} = 200$ V and the input voltage to the base circuit is $V_B = 10$ V. If $V_{CE(\text{sat})} = 1.0$ V and $V_{BE(\text{sat})} = 1.5$ V, find (a) the value of R_B that results in saturation with an overdrive factor of 5, (b) the forced β_f, and (c) the power loss P_T in the transistor.

Solution $V_{CC} = 200$ V, $\beta_{\min} = 8$, $\beta_{\max} = 40$, $R_C = 11 \ \Omega$, ODF = 5, $V_B = 10$ V, $V_{CE(\text{sat})} = 1.0$ V, and $V_{BE(\text{sat})} = 1.5$ V. From Eq. (8-14), $I_{CS} = (200 - 1.0)/11 = 18.1$ A. From Eq. (8-15), $I_{BS} = 18.1/\beta_{\min} = 18.1/8 = 2.2625$ A. Equation (8-16) gives the base current for an overdrive factor of 5,

$$I_B = 5 \times 2.2625 = 11.3125 \text{ A}$$

(a) Equation (8-9) gives the required value of R_B,

$$R_B = \frac{V_B - V_{BE(\text{sat})}}{I_B} = \frac{10 - 1.5}{11.3125} = 0.7514 \ \Omega$$

(b) From Eq. (8-17), $\beta_f = 18.1/11.3125 = 1.6$.
(c) Equation (8-18) yields the total power loss as

$$P_T = 1.5 \times 11.3125 + 1.0 \times 18.1 = 16.97 + 18.1 = 35.07 \text{ W}$$

Note. For an overdrive factor of 10, $I_B = 22.265$ A and the power loss will be $P_T = 1.5 \times 22.265 + 18.1 = 51.5$ W. Once the transistor is saturated, the collector–emitter voltage is not reduced in relation to the increase in base current. However, the power loss is increased. At a high value of overdrive factor, the transistor may be damaged due to thermal runaway. On the other hand, if the transistor is underdriven ($I_B < I_{CB}$), it may operate in the active region and V_{CE} will increase, resulting in increased power loss.

8-2.2 Switching Characteristics

A forward-biased *pn*-junction exhibits two parallel capacitances: a depletion-layer capacitance and a diffusion capacitance. On the other hand, a reverse-biased *pn*-

junction has only depletion capacitance. Under steady-state conditions, these capacitances do not play any role. However, under transient conditions, they influence the turn-on and turn-off behavior of the transistor.

The model of a transistor under transient conditions is shown in Fig. 8-7, where C_{cb} and C_{be} are the effective capacitances of the CBJ and BEJ, respectively. The *transconductance, g_m,* of a BJT is defined as the ratio of ΔI_C to ΔV_{BE}. These capacitances are dependent on junction voltages and physical construction of the transistor. C_{cb} affects the input capacitance significantly due to the Miller multiplication effect [6]. r_{ce} and r_{be} are the resistances of collector to emitter and base to emitter, respectively.

Due to internal capacitances, the transistor does not turn on instantly. Figure 8-8 illustrates the waveforms and switching times. As the input voltage v_B rises from zero to V_1 and the base current rises to I_{B1}, the collector current does not respond immediately. There is a delay, known as *delay time, t_d* before any collector current flows. This delay is required to charge up the capacitance of the BEJ to the forward-bias voltage V_{BE} (approximately 0.7 V). After this delay, the collector current rises to the steady-state value of I_{CS}. The rise time, t_r, depends on the time constant determined by BEJ capacitance.

The base current is normally more than that required to saturate the transistor. As a result, the excess minority carrier charge is stored in the base region. The higher the overdrive factor, ODF, the greater is the amount of extra charge stored in the base. This extra charge, which is called the *saturating charge,* is proportional to the excess base drive and the corresponding current, I_e:

$$I_e = I_B - \frac{I_{CS}}{\beta} = \text{ODF} \cdot I_{BS} - I_{BS} = I_{BS}(\text{ODF} - 1) \qquad (8\text{-}19)$$

and the saturating charge is given by

$$Q_s = \tau_s I_e = \tau_s I_{BS}(\text{ODF} - 1) \qquad (8\text{-}20)$$

where τ_s is known as the *storage time constant* of the transistor.

When the input voltage is reversed from V_1 to $-V_2$ and the base current is also changed to $-I_{B2}$, the collector current does not change for a time, t_s, called the *storage time.* t_s is required to remove the saturating charge from the base. Since v_{BE} is still positive with approximately 0.7 V only, the base current reverses its direction due to the change in the polarity of v_B from V_1 to $-V_2$. The reverse

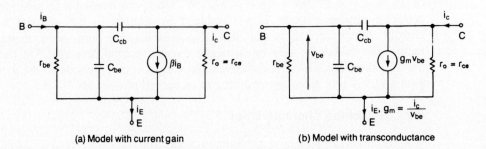

(a) Model with current gain   (b) Model with transconductance

Figure 8-7   Transient model of BJT.

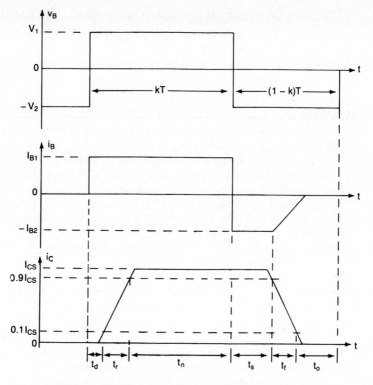

Figure 8-8 Switching times of bipolar transistors.

current, $-I_{B2}$, helps to discharge the base and remove the extra charge from the base. Without $-I_{B2}$, the saturating charge has to be removed entirely by recombination and the storage time would be longer.

Once the extra charge is removed, the BEJ capacitance charges to the input voltage $-V_2$, and the base current falls to zero. The fall time t_f depends on the time constant, which is determined by the capacitance of the reverse-biased BEJ.

Figure 8-9a shows the extra storage charge in the base of a saturated transistor. During turn-off, this extra charge is removed first in time t_s and the charge profile is changed from a to c as shown in Fig. 8-9b. During fall time, the charge profile decreases from profile c until all charges are removed.

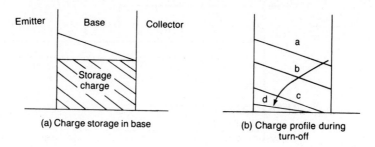

Figure 8-9 Charge storage in saturated bipolar transistors.

The turn-on time t_{on} is the sum of delay time t_d and rise time t_r:

$$t_{\text{on}} = t_d + t_r$$

and the turn-off time t_{off} is the sum of storage time t_s and fall time t_f:

$$t_{\text{off}} = t_s + t_f$$

Example 8-2

The waveforms of the transistor switch in Fig. 8-6 are shown in Fig. 8-10. The parameters are $V_{CC} = 250$ V, $V_{BE(\text{sat})} = 3$ V, $I_B = 8$ A, $V_{CE(\text{sat})} = 2$ V, $I_{CE} = 100$ A, $t_d = 0.5$ μs, $t_r = 1$ μs, $t_s = 5$ μs, $t_f = 3$ μs, and $f_s = 10$ kHz. The duty cycle is $k = 50\%$. The collector-to-emitter leakage current is $I_{CEO} = 3$ mA. Determine the power loss due to collector current (a) during turn-on $t_{\text{on}} = t_d + t_r$, (b) during conduction period t_n, (c) during turn-off $t_{\text{off}} = t_s + t_f$, (d) during off-time t_o, and (e) total average power losses P_T. (f) Plot the instantaneous power due to collector current, $P_c(t)$.

Solution $T = 1/f_s = 100$ μs, $k = 0.5$, $kT = t_d + t_r + t_n = 50$ μs, $t_n = 50 - 0.5 - 1 = 48.5$ μs, $(1 - k)T = t_s + t_f + t_o = 50$ μs, and $t_o = 50 - 5 - 3 = 42$ μs.

(a) During delay time, $0 \le t \le t_d$:

$$i_c(t) = I_{CEO}$$

$$v_{CE}(t) = V_{CC}$$

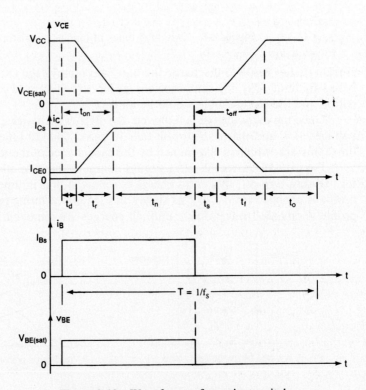

Figure 8-10 Waveforms of transient switch.

The instantaneous power due to the collector current is

$$P_c(t) = i_c v_{CE} = I_{CEO} V_{CC}$$
$$= 3 \times 10^{-3} \times 250 = 0.75 \text{ W}$$

The average power loss during the delay time is

$$P_d = \frac{1}{T} \int_0^{t_d} P_c(t) \, dt = I_{CEO} V_{CC} t_d f_s$$
$$= 3 \times 10^{-3} \times 250 \times 0.5 \times 10^{-6} \times 10 \times 10^3 = 3.75 \text{ mW}$$

$$(8\text{-}21)$$

During rise time, $0 \le t \le t_r$:

$$i_c(t) = \frac{I_{CS}}{t_r} t$$

$$v_{CE}(t) = V_{CC} + (V_{CE(sat)} - V_{CC}) \frac{t}{t_r}$$

$$(8\text{-}22)$$

$$P_c(t) = i_c v_{CE} = I_{CS} \frac{t}{t_r} \left[V_{CC} + (V_{CE(sat)} - V_{CC}) \frac{t}{t_r} \right]$$

The power $P_c(t)$ will be maximum when $t = t_m$, where

$$t_m = \frac{t_r V_{CC}}{2[V_{CC} - V_{CE(sat)}]}$$

$$= 1 \times \frac{250}{2(250 - 2)} = 0.504 \; \mu s$$

$$(8\text{-}23)$$

and Eq. (8-22) yields the peak power

$$P_p = \frac{V_{CC}^2 I_{CS}}{4[V_{CC} - V_{CE(sat)}]}$$

$$= 250^2 \times \frac{100}{4(250 - 2)} = 6300 \text{ W}$$

$$(8\text{-}24)$$

$$P_r = \frac{1}{T} \int_0^{t_r} P_c(t) \, dt = f_s I_{CS} t_r \left[\frac{V_{CC}}{2} + \frac{V_{CE(sat)} - V_{CC}}{3} \right]$$

$$= 10 \times 10^3 \times 100 \times 1 \times 10^{-6} \left[\frac{250}{2} + \frac{2 - 250}{3} \right] = 42.33 \text{ W}$$

$$(8\text{-}25)$$

The total power loss during the turn-on is

$$P_{on} = P_d + P_r$$
$$= 0.00375 + 42.33 = 42.33 \text{ W}$$

$$(8\text{-}26)$$

(b) The conduction period, $0 \le t \le t_n$:

$$i_c(t) = I_{CS}$$
$$v_{CE}(t) = V_{CE(sat)}$$
$$P_c(t) = i_c v_{CE} = V_{CE(sat)} I_{CS}$$
$$= 2 \times 100 = 200 \text{ W}$$

$$(8\text{-}27)$$

$$P_n = \frac{1}{T} \int_0^{t_n} P_c(t) \, dt = V_{CE(\text{sat})} I_{CS} t_n f_s$$

$$= 2 \times 100 \times 48.5 \times 10^{-6} \times 10 \times 10^3 = 97 \text{ W}$$

(c) The storage period, $0 \leq t \leq t_s$:

$$i_c(t) = I_{CS}$$

$$v_{CE}(t) = V_{CE(\text{sat})}$$

$$P_c(t) = i_c v_{CE} = V_{CE(\text{sat})} I_{CS}$$

$$= 2 \times 100 = 200 \text{ W}$$

(8-28)

$$P_s = \frac{1}{T} \int_0^{t_s} P_c(t) \, dt = V_{CE(\text{sat})} I_{CS} t_s f_s$$

$$= 2 \times 100 \times 5 \times 10^{-6} \times 10 \times 10^3 = 10 \text{ W}$$

The fall time, $0 \leq t \leq t_f$:

$$i_c(t) = I_{CS} \left(1 - \frac{t}{t_f} \right), \text{ neglecting } I_{CEO}$$

$$v_{CE}(t) = \frac{V_{CC}}{t_f} t, \text{ neglecting } I_{CEO}$$

(8-29)

$$P_c(t) = i_c v_{CE} = V_{CC} I_{CS} \left[\left(1 - \frac{t}{t_f} \right) \frac{t}{t_f} \right]$$

This power loss during fall time will be maximum when $t = t_f/2 = 1.5 \ \mu s$ and Eq. (8-29) gives the peak power,

$$P_m = \frac{V_{CC} I_{CS}}{4}$$

(8-30)

$$= 250 \times \frac{100}{4} = 6250 \text{ W}$$

$$P_f = \frac{1}{T} \int_0^{t_f} P_c(t) \, dt = \frac{V_{CC} I_{CS} t_f f_s}{6}$$

(8-31)

$$= \frac{250 \times 100 \times 3 \times 10^{-6} \times 10 \times 10^3}{6} = 125 \text{ W}$$

The power loss during turn-off is

$$P_{\text{off}} = P_s + P_f = I_{CS} f_s \left(t_s V_{CE(\text{Sat})} + \frac{V_{CC} \, t_f}{6} \right)$$

(8-32)

$$= 10 + 125 = 135 \text{ W}$$

(d) Off-period, $0 \leq t \leq t_o$:

$$i_c(t) = I_{CEO}$$

$$v_{CE}(t) = V_{CC}$$

$$P_c(t) = i_c v_{CE} = I_{CEO} V_{CC}$$

$$= 3 \times 10^{-3} \times 250 = 0.75 \text{ W}$$

(8-33)

$$P_0 = \frac{1}{T} \int_0^{t_o} P_c(t) \, dt = I_{CEO} V_{CC} t_o f_s$$

$$= 3 \times 10^{-3} \times 250 \times 42 \times 10^{-6} \times 10 \times 10^3 = 0.315 \text{ W}$$

(e) The total power loss in the transistor due to collector current is

$$P_T = P_{on} + P_n + P_{off} + P_0$$

$$= 42.33 + 97 + 135 + 0.315 = 274.65 \text{ W} \tag{8-34}$$

(f) The plot of the instantaneous power is shown in Fig. 8-11.

Example 8-3

For the parameters in Example 8-2, calculate the average power loss due to the base current.

Solution $V_{BE(sat)} = 3$ V, $I_B = 8$ A, $T = 1/f_s = 100 \, \mu s$, $k = 0.5$, $kT = 50 \, \mu s$, $t_d = 0.5 \, \mu s$, $t_r = 1 \, \mu s$, $t_n = 50 - 1.5 = 48.5 \, \mu s$, $t_s = 5 \, \mu s$, $t_f = 3 \, \mu s$, $t_{on} = t_d + t_r = 1.5 \, \mu s$, and $t_{off} = t_s + t_f = 5 + 3 = 8 \, \mu s$.

During the period, $0 \le t \le (t_{on} + t_n)$:

$$i_b(t) = I_{BS}$$

$$v_{BE}(t) = V_{BE(sat)}$$

The instantaneous power due to the base current is

$$P_b(t) = i_b v_{BE} = I_{BS} V_{BS(sat)}$$

$$= 8 \times 3 = 24 \text{ W}$$

During the period, $0 \le t \le t_o = (T - t_{on} - t_n - t_s - t_f)$: $P_b(t) = 0$. The average power loss is

$$P_B = I_{BS} V_{BE(sat)} (t_{on} + t_n + t_s + t_f) f_s$$

$$= 8 \times 3 \times (1.5 + 48.5 + 5 + 3) \times 10^{-6} \times 10 \times 10^3 = 13.92 \text{ W} \tag{8-35}$$

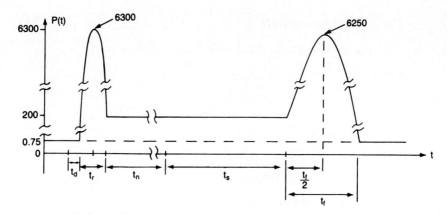

Figure 8-11 Plot of instantaneous power for Example 8-2.

8-2.3 Switching Limits

Second breakdown, SB. The secondary breakdown (SB), which is a destructive phenomenon, results from the current flow to a small portion of the base, producing localized hot spots. If the energy in these hot spots is sufficient, the excessive localized heating may damage the transistor. Thus secondary breakdown is caused by a localized thermal runaway, resulting from high current concentrations. The current concentration may be caused by defects in the transistor structure. The SB occurs at certain combinations of voltage, current, and time. Since the time is involved, the secondary breakdown is basically an energy dependent phenomenon.

Forward-biased safe operating area, FBSOA. During turn-on and on-state conditions, the average junction temperature and second breakdown limit the power-handling capability of a transistor. The manufacturers usually provide the FBSOA curves under specified test conditions. FBSOA indicates the i_c–v_{CE} limits of the transistor; and for reliable operation the transistor must not be subjected to greater power dissipation than that shown by the FBSOA curve.

Reverse-biased safe operating area, RBSOA. During turn-off, a high current and high voltage must be sustained by the transistor, in most cases with the base-to-emitter junction reverse biased. The collector–emitter voltage must be held to a safe level at or below a specified value of collector current. The manufacturers provide the I_C–V_{CE} limits during reverse-biased turn-off as reverse-biased safe operating area (RBSOA).

Power derating. The thermal equivalent circuit is shown in Fig. 8-12. If the total average power loss is P_T, the case temperature is

$$T_C = T_J - P_T R_{JC}$$

The sink temperature is

$$T_S = T_C - P_T R_{CS}$$

The ambient temperature is

$$T_A = T_S - P_T R_{SA}$$

and

$$T_J - T_A = P_T(R_{JC} + R_{CS} + R_{SA}) \tag{8-36}$$

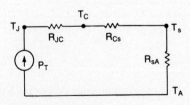

Figure 8-12 Thermal equivalent circuit of a transistor.

where R_{JC} = thermal resistance from junction to case, °C/W
  R_{CS} = thermal resistance from case to sink, °C/W
  R_{SA} = thermal resistance from sink to ambient, °C/W

The maximum power dissipation P_T is normally specified at $T_C = 25°C$. If the ambient temperature is increased to $T_A = T_{J(max)} = 150°C$, the transistor can dissipate zero power. On the other hand, if the junction temperature is $T_C = 0°C$, the device can dissipate maximum power and this is not practical. Therefore, the ambient temperature and thermal resistances must be considered when interpreting the ratings of devices. Manufacturers show the derating curves for the thermal derating and second breakdown derating.

Breakdown voltages. A *breakdown voltage* is defined as the absolute maximum voltage between two terminals with the third terminal open, shorted, or biased in either forward or reverse direction. At breakdown the voltage remains relatively constant, where the current rises rapidly. The following breakdown voltages are quoted by the manufacturers:

V_{EBO}: the maximum voltage between the emitter terminal and base terminal with collector terminal open circuited.

V_{CEV} *or* V_{CEX}: the maximum voltage between the collector terminal and emitter terminal at a specified negative voltage applied between base and emitter.

$V_{CEO(SUS)}$: the maximum sustaining voltage between the collector terminal and emitter terminal with the base open circuited. This rating is specified at the maximum collector current and voltage, appearing simultaneously across the device with a specified value of load inductance.

Let us consider the circuit in Fig. 8-13a. When the switch SW is closed, the collector current increases, and after a transient, the steady-state collector current is $I_{CS} = (V_{CC} - V_{CE(sat)})/R_C$. For an inductive load, the load line would be the path *ABC* shown in Fig. 8-13b. If the switch is opened to remove the base current, the collector current will begin to fall and a voltage of $L(di/dt)$ will be induced across the inductor to oppose the current reduction. The transistor will be subjected to a

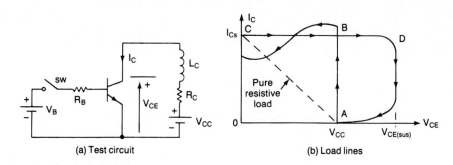

(a) Test circuit   (b) Load lines

Figure 8-13 Turn-on and turn-off load lines.

transient voltage. If this voltage reaches the sustaining voltage level, the collector voltage remains approximately constant and the collector current will fall. After a small time, the transistor will be in the off-state and the turn-off load line is shown in Fig. 8-13b by the path *CDA*.

Example 8-4

The maximum junction temperature of a transistor is $T_J = 150°C$ and the ambient temperature is $T_A = 25°C$. If the thermal impedances are $R_{JC} = 0.4°C/W$, $R_{CS} = 0.1°C/W$, and $R_{SA} = 0.5°C/W$, calculate (a) the maximum power dissipation, and (b) the case temperature.

Solution (a) $T_J - T_A = P_T(R_{JC} + R_{CS} + R_{SA}) = P_T R_{JA}$, $R_{JA} = 0.4 + 0.1 + 0.5 = 1.0$, and $150 - 25 = 1.0 P_T$, which gives the maximum power dissipation as $P_T = 125$ W.

(b) $T_C = T_J - P_T R_{JC} = 150 - 125 \times 0.4 = 100°C$.

8-2.4 Base Drive Control

The switching speed can be increased by reducing turn-on time t_{on} and turn-off time t_{off}. t_{on} can be reduced by allowing base current peaking during turn-on, resulting in low forced $\beta(\beta_F)$ at the beginning. After turn-on, β_f can be increased to a sufficiently high value to maintain the transistor in the quasi-saturation region. t_{off} can be reduced by reversing base current and allowing base current peaking during turn-off. Increasing the value of reverse base current I_{B2} decreases the storage time. A typical waveform for base current is shown in Fig. 8-14.

Apart from a fixed shape of base current as in Fig. 8-14, the forced β may be controlled continuously to match the collector current variations. The commonly used techniques for optimizing the base drive of a transistor are:

1. Turn-on control
2. Turn-off control
3. Proportional base control
4. Antisaturation control

Turn-on control. The base current peaking can be provided by the circuit of Fig. 8-15. When the input voltage is turned on, the base current is limited by resistor R_1 and the initial value of base current is

$$I_{B0} = \frac{V_1 - V_{BE}}{R_1} \tag{8-37}$$

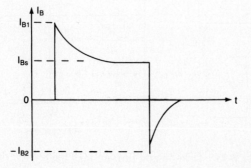

Figure 8-14 Base drive current waveform.

and the final value of the base current is

$$I_{B1} = \frac{V_1 - V_{BE}}{R_1 + R_2} \tag{8-38}$$

The capacitor C_1 charges up to a final value of

$$V_c \cong V_1 \frac{R_2}{R_1 + R_2} \tag{8-39}$$

The charging time constant of the capacitor is approximately

$$\tau_1 = \frac{R_1 R_2 C_1}{R_1 + R_2} \tag{8-40}$$

Once the input voltage v_B becomes zero, the base–emitter junction is reverse biased and C_1 discharges through R_2. The discharging time constant is $\tau_2 = R_2 C_1$. To allow sufficient charging and discharging times, the width of the base pulse must be $t_1 \geq 5\tau_1$ and the off-period of the pulse must be $t_2 \geq 5\tau_2$. The maximum switching frequency is $f_s = 1/T = 1/(t_1 + t_2) = 0.2/(\tau_1 + \tau_2)$.

Turn-off control. If the input voltage in Fig. 8-15 is changed to $-V_2$ during turn-off, the capacitor voltage V_c in Eq. (8-39) is added to V_2 as a reverse voltage across the transistor. There will be base-current peaking during turn-off. As the capacitor C_1 discharges, the reverse voltage will be reduced to a steady-state value, V_2. If different turn-on and turn-off characteristics are required, a turn-off circuit (using C_2, R_3, and R_4) as shown in Fig. 8-16 may be added. The diode D_1 isolates the forward base drive circuit from the reverse base drive circuit during turn-off.

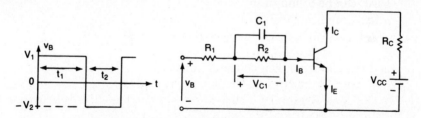

Figure 8-15 Base current peaking during turn-on.

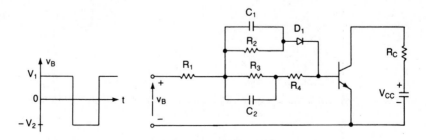

Figure 8-16 Base current peaking during turn-on and turn-off.

Proportional base control. This type of control has advantages over the constant drive circuit. If the collector current changes due to change in load demand, the base drive current is changed in proportion to the collector current. An arrangement is shown in Fig. 8-17. When switch S_1 is turned on, a pulse current of short duration would flow through the base of transistor Q_1; and Q_1 is turned on into saturation. Once the collector current starts to flow, a corresponding base current is induced due to the transformer action. The transistor would latch on itself, and S_1 can be turned off. The turns ratio is $N_2/N_1 = I_C/I_B = \beta$. For proper operation of the circuit, the magnetizing current, which must be much smaller than the collector current, should be as small as possible. Switch S_1 can be implemented by a small-signal transistor, and an additional circuitry is necessary to discharge capacitor C_1 and to reset the transformer core during turn-off of the power transistor

Antisaturation control. If the transistor is driven hard, the storage time, which is proportional to the base current, increases and the switching speed is reduced. The storage time can be reduced by operating the transistor in soft saturation rather than hard saturation. This can be accomplished by clamping the collector–emitter voltage to a predetermined level and the collector current is given by

$$I_C = \frac{V_{CC} - V_{cm}}{R_C} \tag{8-41}$$

where V_{cm} is the clamping voltage and $V_{cm} > V_{CE(\text{sat})}$. A circuit with clamping action (also known as Baker's clamp) is shown in Fig. 8-18.

The base current without clamping, which is adequate to drive the transistor hard, can be found from

$$I_B = I_1 = \frac{V_B - V_{d1} - V_{BE}}{R_B} \tag{8-42}$$

and the corresponding collector current is

$$I_C = \beta I_B \tag{8-43}$$

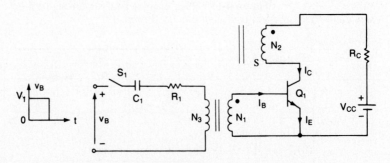

Figure 8-17 Proportional base drive circuit.

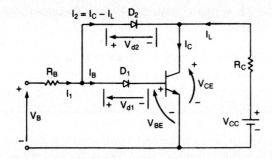

Figure 8-18 Collector clamping circuit.

After the collector current rises, the transistor is turned on, and the clamping takes place (due to the fact that D_2 gets forward biased and conducts), then

$$V_{CE} = V_{BE} + V_{d1} - V_{d2} \qquad (8\text{-}44)$$

The load current is

$$I_L = \frac{V_{CC} - V_{CE}}{R_C} = \frac{V_{CC} - V_{BE} - V_{d1} + V_{d2}}{R_C} \qquad (8\text{-}45)$$

and the collector current with clamping is

$$I_C = \beta I_B = \beta(I_1 - I_C + I_L)$$

$$= \frac{\beta}{1 + \beta}(I_1 + I_L) \qquad (8\text{-}46)$$

For clamping, $V_{d1} > V_{d2}$ and this can be accomplished by connecting two or more diodes in place of D_1. The load resistance R_C should satisfy the condition

$$\beta I_B > I_L$$

From Eq. (8-45),

$$\beta I_B R_C > (V_{CC} - V_{BE} - V_{d1} + V_{d2}) \qquad (8\text{-}47)$$

The clamping action results in a reduced collector current and almost elimination of the storage time. At the same time, a fast turn-on is accomplished. However, due to increased V_{CE}, the on-state power dissipation in the transistor is increased, whereas the switching power loss is decreased.

Example 8-5

The base drive circuit in Fig. 8-18 has $V_{CC} = 100$ V, $R_C = 1.5\ \Omega$, $V_{d1} = 2.1$ V, $V_{d2} = 0.9$ V, $V_{BE} = 0.7$ V, $V_B = 15$ V, and $R_B = 2.5\ \Omega$, and $\beta = 16$. Calculate (a) the collector current without clamping, (b) the collector–emitter clamping voltage V_{CE}, and (c) the collector current with clamping.
Solution (a) From Eq. (8-42), $I_1 = (15 - 2.1 - 0.7)/2.5 = 4.88$ A. Without clamping, $I_C = 16 \times 4.88 = 78.08$ A.
(b) From Eq. (8-44), the clamping voltage is

$$V_{CE} = 0.7 + 2.1 - 0.9 = 1.9 \text{ V}$$

(c) From Eq. (8-45), $I_L = (100 - 1.9)/1.5 = 65.4$ A. Equation (8-46) gives the collector current with clamping:

$$I_C = 16 \times \frac{4.88 + 65.4}{16 + 1} = 66.15 \text{ A}$$

8-3 POWER MOSFETs

A bipolar junction transistor (BJT) is a current-controlled device and requires base current for current flow in the collector. Since the collector current is dependent on the input (or base) current, the current gain is highly dependent on the junction temperature.

A power MOSFET is a voltage-controlled device and requires only a small input current. The switching speed is very high and the switching times are of the order of nanoseconds. Power MOSFETS are finding increasing applications in low-power high-frequency converters. MOSFETs do not have the problems of second breakdown phenomena as do BJTs. However, MOSFETs have the problems of electrostatic discharge and require special care in handling. In addition, it is relatively difficult to protect them under short-circuited fault conditions.

MOSFETs are two types: (1) depletion MOSFETs, and (2) enhancement MOSFETs. An n-channel depletion-type MOSFET is formed on a p-type silicon substrate as shown in Fig. 8-19a, with two heavily doped n^+ silicon for low-resistance connections. The gate is isolated from the channel by a thin oxide layer. The three terminals are called *gate, drain,* and *source.* The substrate is normally connected to the source. The gate-to-source voltage, V_{GS}, could be either positive or negative. If V_{GS} is negative, some of the electrons in the n-channel area will be repelled and a depletion region will be created below the oxide layer, resulting in a narrower effective channel and a high resistance from the drain to source, R_{DS}. If V_{GS} is made negative enough, the channel will be completely depleted, offering a high value of R_{DS}, and there will be no current flow from the drain to source, $I_{DS} = 0$. The value of V_{GS} when this happens is called *pinch-off voltage, V_p*. On the other hand, V_{GS} is made positive, the channel becomes wider, and I_{DS} increases due to reduction in R_{DS}. With a p-channel depletion-type MOSFET, the polarities of V_{DS}, I_{DS}, and V_{GS} are reversed.

An n-channel enhancement-type MOSFET has no physical channel, as shown in Fig. 8-20. If V_{GS} is positive, an induced voltage will attract the electrons from the p-substrate and accumulate them at the surface beneath the oxide layer. If V_{GS} is greater than or equal to a value known as *threshold voltage, V_T*, a sufficient number of electrons are accumulated to form a virtual n-channel and the current flows from the drain to source. The polarities of V_{DS}, I_{DS}, and V_{GS} are reversed for a p-channel enhancement-type MOSFET. Power MOSFETs of various sizes are shown in Fig. 8-21.

8-3.1 Steady-State Characteristics

The MOSFETs are voltage-controlled devices and have a very high input impedance. The gate draws a very small leakage current, in the order of nanoam-

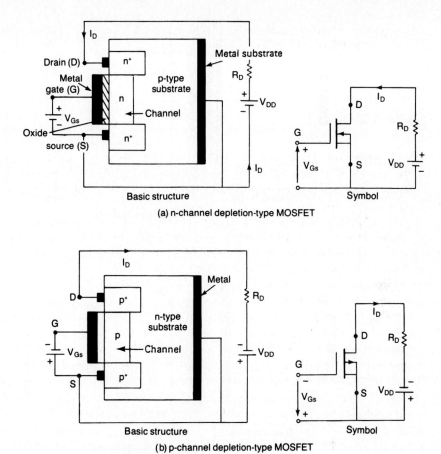

Figure 8-19 Depletion-type MOSFETs.

peres. The current gain, which is the ratio of drain current, I_D, to input gate current, I_G, is typically on the order of 10^9. However, the current gain is not an important parameter. The *transconductance,* which is the ratio of drain current to gate voltage, defines the transfer characteristics and is a very important parameter.

The transfer characteristics of *n*-channel and *p*-channel MOSFETs are shown in Fig. 8-22. Figure 8-23 shows the output characteristics of an *n*-channel enhancement MOSFET. There are three regions of operation: (1) cutoff region, where $V_{GS} \leq V_T$; (2) pinch-off or saturation region, where $V_{DS} \geq V_{GS} - V_T$; and (3) linear region, where $V_{DS} \leq V_{GS} - V_T$. The pinch-off occurs at $V_{DS} = V_{GS} - V_T$. In the linear region, the drain current, I_D varies in proportion to the drain–source voltage, V_{DS}. Due to high drain current and low drain voltage, the power MOSFETs are operated in the linear region for switching actions. In the saturation region, the drain current remains almost constant for any increase in the value of V_{DS} and the transistors are used in this region for voltage amplification. It should be noted that saturation has the opposite meaning to that for bipolar transistors.

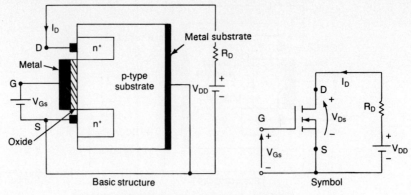

(a) n-channel enhancement-type MOSFET

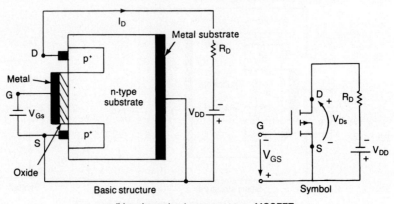

(b) p-channel enhancement-type MOSFET

Figure 8-20   Enhancement-type MOSFETs.

Figure 8-21   Power MOSFETs. (Courtesy of International Rectifier.)

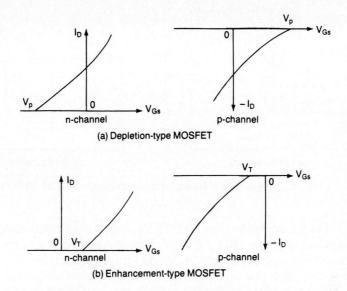

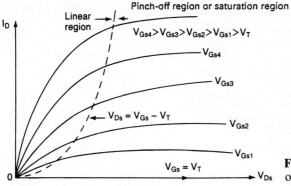

Figure 8-22 Transfer characteristics of MOSFETs.

The steady-state model, which is the same for both depletion-type and enhancement-type MOSFETs, is shown in Fig. 8-24. The transconductance, g_m, is defined as

$$g_m = \frac{\Delta I_D}{\Delta V_{GS}}\bigg|_{V_{DS}=\text{constant}} \tag{8-48}$$

The output resistance, $r_o = R_{DS}$, which is defined as

$$R_{DS} = \frac{\Delta V_{DS}}{\Delta I_D} \tag{8-49}$$

is normally very high in the pinch-off region, typically on the order of megohms and is very small in the linear region, typically on the order of milliohms.

Figure 8-23 Output characteristics of enhancement-type MOSFET.

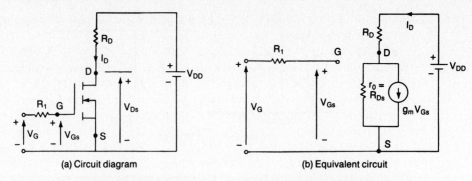

| (a) Circuit diagram | (b) Equivalent circuit |

Figure 8-24 Steady-state switching model of MOSFETs.

For the depletion-type MOSFETs, the gate (or input) voltage could be either positive or negative. But the enhancement-type MOSFETs respond to a positive gate voltage only. The power MOSFETs are generally of enhancement type. However, depletion-type MOSFETs would be advantageous and simplify the logic design in some applications which require some form of logic-compatible ac or dc switch that would remain on when the logic supply falls and V_{GS} becomes zero. The characteristics of depletion-type MOSFETs will not be discussed further.

8-3.2 Switching Characteristics

Without any gate signal, an enhancement-type MOSFET may be considered as two diodes connected back to back or as an *NPN*-transistor. The gate structure has parasitic capacitances to the source, C_{gs}, and to the drain, C_{gd}. The *npn*-transistor has a reverse-bias junction from the drain to the source and offers a capacitance, C_{ds}. Figure 8-25a shows the equivalent circuit of a parasitic bipolar transistor in parallel with a MOSFET. The base-to-emitter region of *NPN*-transistor is shorted at the chip by metalizing the source terminal and the resistance from the base to emitter due to bulk resistance of *n*- and *p*-regions, R_{be}, is small. Hence a MOSFET may be considered as having an internal diode and the equiva-

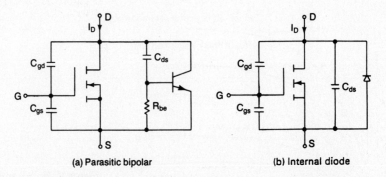

| (a) Parasitic bipolar | (b) Internal diode |

Figure 8-25 Parasitic model of enhancement of MOSFETs.

lent circuit is shown in Fig. 8-25b. The parasitic capacitances are dependent on their respective voltages.

The switching model of MOSFETs is shown in Fig. 8-26. The typical switching waveforms and times are shown in Fig. 8-27. the *turn-on delay* $t_{d(on)}$ is the time that is required to charge the input capacitance to threshold voltage level. The *rise time* t_r is the gate charging time from the threshold level to the full-gate voltage V_{GSP}, which is required to drive the transistor into the linear region. The *turn-off delay time* $t_{d(off)}$ is the time required for the input capacitance to discharge from the overdrive gate voltage V_1 to the pinch-off region. V_{GS} must decrease significantly before V_{DS} begins to rise. The *fall time* t_f is the time that is required for the input capacitance to discharge from the pinch-off region to threshold voltage. If $V_{GS} \leq V_T$, the transistor turns off.

8-3.3 Gate Drive

The turn-on time of a MOSFET depends on the charging time of the input or gate capacitance. The turn-on time can be reduced by connecting an *RC* circuit as shown in Fig. 8-28 to charge the gate capacitance faster. When the gate voltage is turned on, the initial charging current of the capacitance is

$$I_G = \frac{V_G}{R_S} \tag{8-50}$$

and the steady-state value of gate voltage is

$$V_{GS} = \frac{R_G V_G}{R_s + R_1 + R_G} \tag{8-51}$$

where R_s is the internal resistance of gate drive source.

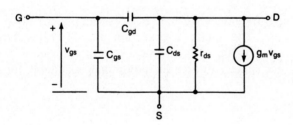

Figure 8-26 Switching model of MOSFETs.

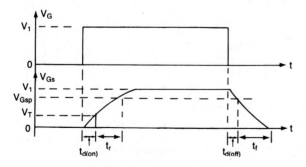

Figure 8-27 Switching waveforms and times.

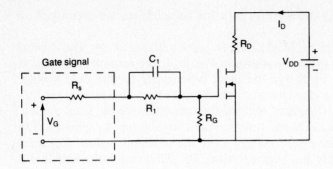

Figure 8-28 Fast-turn-on gate circuit.

In order to achieve switching speeds of the order of 100 ns or less, the gate-drive circuit should have a low output impedance and the ability to sink and source relatively large currents. A totem-pole arrangement that is capable of sourcing and sinking a large current is shown in Fig. 8-29. The *PNP*- and *NPN*-transistors act as emitter followers and offer a low output impedance. These transistors operate in the linear region rather than in the saturation mode, thereby minimizing the delay time. The gate signal for the power MOSFET may be generated by an op-amp. Feedback via the capacitor C regulates the rate of rise and fall of the gate voltage, thereby controlling the rate of rise and fall of the MOSFET drain current. A diode across the capacitor C allows the gate voltage to change rapidly in one direction only. There are a number of integrated drive circuits on the market that are designed to drive transistors and are capable of sourcing and sinking large currents for most converters.

8-4 SITs

A SIT is a high-power, high-frequency device. It is essentially the solid-state version of triode vacuum tube. The silicon cross section of a SIT [15] is shown in Fig. 8-30a and its symbol in Fig. 8-30b. It is a vertical structure device with short multichannels. Thus it is not subject to area limitation and is suitable for high-

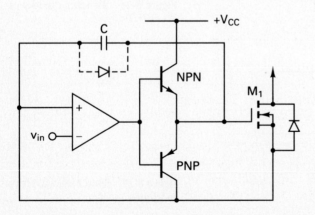

Figure 8-29 Totem-pole arrangement gate drive with pulse-edge shaping.

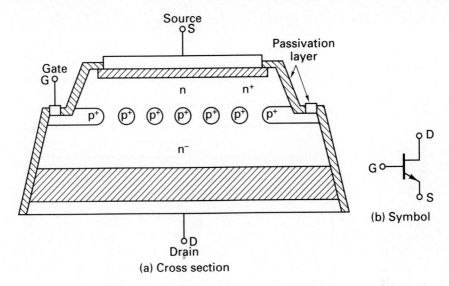

Figure 8-30 Cross section and symbol for SITs.

speed, high-power operation. The gate electrodes are buried within the drain and source n-epsi layers. A SIT is identical to a JFET except for vertical and buried gate construction, which gives a lower channel resistance, causing a lower drop. A SIT has a short channel length, low gate series resistance, low gate–source capacitance, and small thermal resistance. It has a low noise, low distortion, and high audio-frequency power capability. The turn-on and turn-off times are very small, typically 0.25 μs.

The on-state drop is high, typically 90 V for a 180-A device and 18 V for an 18-A device. A SIT is a normally-on device, and a negative gate voltage holds it off. The normally on-characteristic and the high on-state drop limit its applications for general power conversions. The current rating of SITs can be up to 300 A, 1200 V, and the switching speed can be as high as 100 kHz. It is most suitable for high-power, high-frequency applications (e.g., audio, VHF/UHF, and microwave amplifiers).

8-5 IGBTs

An IGBT combines the advantages of BJTs and MOSFETS. An IGBT has high input impedance, like MOSFETs, and low on-state conduction losses, like BJTs. But there is no second breakdown problem, like BJTs. By chip design and structure, the equivalent drain-to-source resistance, R_{DS}, is controlled to behave like that of a BJT.

The silicon cross section of an IGBT is shown in Fig. 8-31a, which is identical to that of a MOSFET except the p^+ substrate. However, the performance of an IGBT is closer to that of a BJT than a MOSFET. This is due to the p^+ substrate, which is responsible for the minority carrier injection into the n-

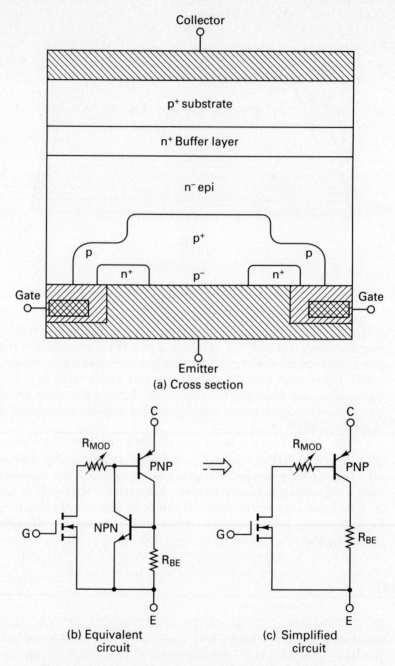

Figure 8-31 Cross section and equivalent circuit for IGBTs.

region. The equivalent circuit is shown in Fig. 8-31b, which can be simplified to Fig. 8-31c. An IGBT is made of four alternate *PNPN* layers, and could latch like a thyristor given the necessary condition: $(\alpha_{npn} + \alpha_{pnp}) > 1$. The n^+ buffer layer and the wide epi base reduce the gain of the *NPN* terminal by internal design, thereby avoiding latching. An IGBT is a voltage-controlled device similar to a power MOSFET. It has lower switching and conducting losses while sharing many of the appealing features of power MOSFETS, such as ease of gate drive, peak current, capability, and ruggedness. An IGBT is inherently faster than a BJT. However, the switching speed of IGBTs is inferior to that of MOSFETs.

The symbol and circuit of an IGBT switch are shown in Fig. 8-32. The three terminals are gate, collector, and emitter instead of gate, drain, and source for a MOSFET. The parameters and their symbols are similar to that of MOSFETs, except that the subscripts for source and drain are changed to emitter and collector, respectively. The current rating of a single IGBT can be up to 400 A, 1200 V, and the switching frequency can be up to 20 kHz. IBGTs are finding increasing applications in medium-power applications such as dc and ac motor drives, power supplies, solid-state relays, and contractors.

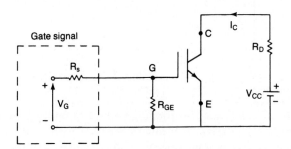

Figure 8-32 Symbol and circuit for a MOSIGT.

8-6 SERIES AND PARALLEL OPERATION

Transistors may be operated in series to increase their voltage-handling capability. It is very important that the series-connected transistors are turned on and off simultaneously. Otherwise, the slowest device at turn-on and the fastest device at turn-off will be subjected to the full voltage of the collector–emitter (or drain–source) circuit and that particular device may be destroyed due to a high voltage. The devices should be matched for gain, transconductance, threshold voltage, on-state voltage, turn-on time, and turn-off time. Even the gate or base drive characteristics should be identical. Voltage-sharing networks similar to diodes could be used.

Transistors are connected in parallel if one device cannot handle the load current demand. For equal current sharings, the transistors should be matched for gain, transconductance, saturation voltage, and turn-on time and turn-off time. But, in practice, it is not always possible to meet these requirements. A reasonable amount of current sharing (45 to 55% with two transistors) can be obtained by connecting resistors in series with the emitter (or source) terminals, as shown in Fig. 8-33.

The resistors in Fig. 8-33 will help current sharing under steady-state conditions. Current sharing under dynamic conditions can be accomplished by connecting coupled inductors as shown in Fig. 8-34. If the current through Q_1 rises,

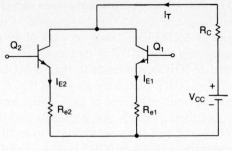

Figure 8-33 Parallel connection of transistors.

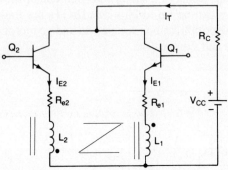

Figure 8-34 Dynamic current sharing.

the $L(di/dt)$ across L_1 increases, and a corresponding voltage of opposite polarity is induced across inductor L_2. The result is a low-impedance path, and the current is shifted to Q_2. The inductors would generate voltage spikes and they may be expensive and bulky, especially at high currents.

BJTs have a negative temperature coefficient. During current sharing, if one BJT carries more current, its on-state resistance decreases and its current increases further, whereas MOSFETS have positive temperature coefficient and parallel operation is relatively easy. The MOSFET that initially draws higher current heats up faster and its on-state resistance increases, resulting in current shifting to the other devices. IGBTs require special care to match the characteristics due to the variations of the temperature coefficients with the collector current.

Example 8-6

Two MOSFETS which are connected in parallel similar to Fig. 8-33 carry a total current of $I_T = 20$ A. The drain-to-source voltage of MOSFET M_1 is $V_{DS1} = 2.5$ V and that of MOSFET M_2 is $V_{DS2} = 3$ V. Determine the drain current of each transistor and difference in current sharing if the current sharing series resistances are (a) $R_{s1} = 0.3$ Ω and $R_{s2} = 0.2$ Ω, and (b) $R_{s1} = R_{s2} = 0.5$ Ω.

Solution (a) $I_{D1} + I_{D2} = I_T$ and $V_{DS1} + I_{D1}R_{S1} = V_{DS2} + I_{D2}R_{S2} = V_{DS2} = R_{S2}(I_T - I_{D1})$.

$$I_{D1} = \frac{V_{DS2} - V_{DS1} + I_T R_{s2}}{R_{s1} + R_{s2}} \tag{8-52}$$

$$= \frac{3 - 2.5 + 20 \times 0.2}{0.3 + 0.2} = 9 \text{ A} \quad \text{or} \quad 45\%$$

$$I_{D2} = 20 - 9 = 11 \text{ A} \quad \text{or} \quad 55\%$$

$$\Delta I = 55 - 45 = 10\%$$

(b) $I_{D1} = \dfrac{3 - 2.5 + 20 \times 0.5}{0.5 + 0.5} = 10.5$ A   or   52.5%

$I_{D2} = 20 - 10.5 = 9.5$ A   or   47.5%

$\Delta I = 52.5 - 47.5 = 5\%$

8-7 *di/dt* AND *dv/dt* LIMITATIONS

Transistors require certain turn-on and turn-off times. Neglecting the delay time t_d and the storage time t_s, the typical voltage and current waveforms of a BJT switch are shown in Fig. 8-35. During turn-on, the collector current rises and the *di/dt* is

$$\frac{di}{dt} = \frac{I_L}{t_r} = \frac{I_{cs}}{t_r} \tag{8-53}$$

During turn-off, the collector–emitter voltage must rise in relation to the fall of the collector current, and *dv/dt* is

$$\frac{dv}{dt} = \frac{V_s}{t_f} = \frac{V_{cc}}{t_f} \tag{8-54}$$

The conditions *di/dt* and *dv/dt* in Eqs. (8-53) and (8-54) are set by the transistor switching characteristics and must be satisfied during turn-on and turn-off. Protection circuits are normally required to keep the operating *di/dt* and *dv/dt* within the allowable limits of the transistor. A typical transistor switch with *di/dt* and *dv/dt* protection is shown in Fig. 8-36a, with the operating waveforms in Fig. 8-36b. The *RC* network across the transistor is known as the *snubber circuit* or *snubber* and limits the *dv/dt*. The inductor L_s, which limits the *di/dt*, is sometimes called a *series snubber*.

Let us assume that under steady-state conditions the load current I_L is freewheeling through diode D_m, which has negligible reverse recovery time. When

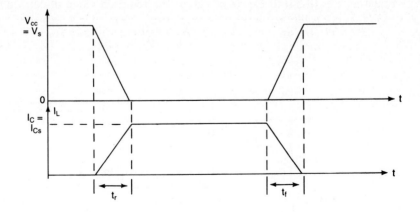

Figure 8-35   Voltage and current waveforms.

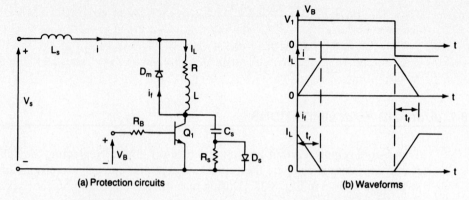

(a) Protection circuits

(b) Waveforms

Figure 8-36 Transistor switch with *di/dt* and *dv/dt* protection.

transistor Q_1 is turned on, the collector current rises and current of diode D_m falls, because D_m will behave as short-circuited. The equivalent circuit during turn-on is shown in Fig. 8-37a and turn-on *di/dt* is

$$\frac{di}{dt} = \frac{V_s}{L_s} \qquad (8-55)$$

Equating Eq. (8-53) to Eq. (8-55) gives the value of L_s,

$$L_s = \frac{V_s t_r}{I_L} \qquad (8-56)$$

During turn-off, the capacitor C_s will charge by the load current and the equivalent circuit is shown in Fig. 8-37b. The capacitor voltage will appear across the transistor and the *dv/dt* is

$$\frac{dv}{dt} = \frac{I_L}{C_s} \qquad (8-57)$$

Equating Eq. (8-54) to Eq. (8-57) gives the required value of capacitance,

$$C_s = \frac{I_L t_f}{V_s} \qquad (8-58)$$

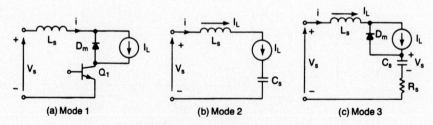

(a) Mode 1

(b) Mode 2

(c) Mode 3

Figure 8-37 Equivalent circuits.

Power Transistors   Chap. 8

Once the capacitor is charged to V_s, the freewheeling diode will turn on. Due to the energy stored in L_s, there will be a damped resonant circuit as shown in Fig. 8-37c. The transient analysis of RLC circuit is discussed in Section 16-4. The RLC circuit is normally made critically damped to avoid oscillations. For unity critical damping, $\delta = 1$, and Eq. (16-11) yields

$$R_s = 2\sqrt{\frac{L_s}{C_s}} \tag{8-59}$$

The capacitor C_s has to discharge through the transistor and this increases the peak current rating of the transistor. The discharge through the transistor can be avoided by placing resistor R_s across C_s instead of placing R_s across D_s.

The discharge current is shown in Fig. 8-38. When choosing the value of R_s, the discharge time, $R_sC_s = \tau_s$ should also be considered. A discharge time of one-third the switching period, T_s, is usually adequate.

$$3R_sC_s = T_s = \frac{1}{f_s}$$

or

$$R_s = \frac{1}{3f_sC_s} \tag{8-60}$$

Example 8-7

A bipolar transistor is operated as a chopper switch at a frequency of $f_s = 10$ kHz. The circuit arrangement is shown in Fig. 8-36a. The dc voltage of the chopper is $V_s = 220$ V and the load current is $I_L = 100$ A. $V_{CE(\text{sat})} = 0$ V. The switching times are $t_d = 0$, $t_r = 3$ μ, and $t_f = 1.2$ μs. Determine the values of (a) L_s; (b) C_s; (c) R_s for critically damped condition; (d) R_s, if the discharge time is limited to one-third of switching period; (e) R_s, if the peak discharge current is limited to 10% of load current; and (f) power loss due to RC snubber, P_s, neglecting the effect of inductor L_s on the voltage of the snubber capacitor C_s.

Solution $I_L = 100$ A, $V_s = 220$ V, $f_s = 10$ kHz, $t_r = 3$ μ, and $t_f = 1.2$ μs.
(a) From Eq. (8-56), $L_s = V_st_r/I_L = 220 \times 3/100 = 6.6$ μH.
(b) From Eq. (8-58), $C_s = I_Lt_f/V_s = 100 \times 1.2/220 = 0.55$ μF.
(c) From Eq. (8-59), $R_s = 2\sqrt{L_s/C_s} = 2\sqrt{6.6/0.55} = 6.93$ Ω.
(d) From Eq. (8-60), $R_s = 1/(3f_sC_s) = 10^3/(3 \times 10 \times 0.55) = 60.6$ Ω.
(e) $V_s/R_s = 0.1 \times I_L$ or $220/R_s = 0.1 \times 100$ or $R_s = 22$ Ω.
(f) The snubber loss, neglecting the loss in diode D_s, is

$$P_s \cong 0.5C_sV_s^2f_s$$

$$= 0.5 \times 0.55 \times 10^{-6} \times 220^2 \times 10 \times 10^3 = 133.1 \text{ W} \tag{8-61}$$

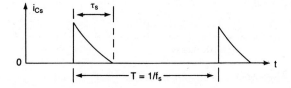

Figure 8-38 Discharge current of snubber capacitor.

For operating power transistors as switches, an appropriate gate voltage or base current must be applied to drive the transistors into the saturation mode for low on-state voltage. The control voltage should be applied between the gate and source terminals or between the base and emitter terminals. The power converters generally require multiple transistors and each transistor must be gated individually. Figure 8-39a shows the topology of a single-phase bridge inverter. The main dc voltage is V_s with ground terminal G.

The logic circuit in Fig. 8-39b generates four pulses. These pulses as shown in Fig. 8-39c are shifted in time to perform the required logic sequence for power conversion from dc to ac. However, all four logic pulses have a common terminal C. The common terminal of the logic circuit may be connected to the ground terminal G of the main dc supply as shown by dashed lines.

The terminal g_1, which has a voltage of V_{g1} with respect to terminal C, cannot be connected directly to gate terminal G_1. The signal V_{g1} should be applied between the gate terminal G_1 and source terminal S_1 of transistor M_1. There is a need for isolation and interfacing circuits between the logic circuit and power transistors. However, transistors M_2 and M_4 can be gated directly without isolation or interfacing circuits if the logic signals are compatible with the gate drive requirements of the transistors.

The importance of gating a transistor between its gate and source rather than applying gating voltage between the gate and common ground can be demonstrated with Fig. 8-40, where the load resistance is connected between the source

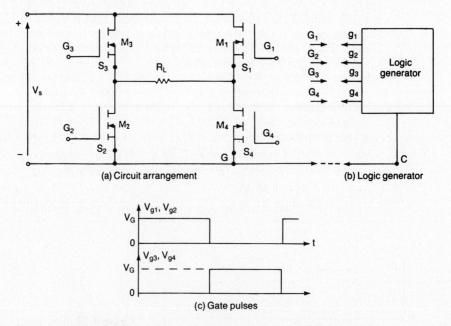

(a) Circuit arrangement

(b) Logic generator

(c) Gate pulses

Figure 8-39   Single-phase bridge-inverter and gating signals.

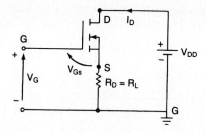

Figure 8-40 Gate voltage between gate and ground.

and ground. The effective gate-source voltage is

$$V_{GS} = V_G - R_L I_D(V_{GS})$$

where $I_D(V_{GS})$ varies with V_{GS}. The effective value of V_{GS} decreases as the transistor turns on and V_{GS} reaches a steady-state value, which is required to balance the load or drain current. The effective value of V_{GS} is unpredictable and such an arrangement is not suitable. There are basically two ways of floating or isolating the control or gate signal with respect to ground.

1. Pulse transformers
2. Optocouplers

8-8.1 Pulse Transformers

Pulse transformers have one primary winding and can have one or more secondary windings. Multiple secondary windings allow simultaneous gating signals to series- and parallel-connected transistors. Figure 8-41 shows a transformer-isolated gate drive arrangement. The transformer should have a very small leakage inductance and the rise time of the output pulse should be very small. At a relatively long pulse and low switching frequency, the transformer would saturate and its output would be distorted.

8-8.2 Optocouplers

Optocouplers combine an infrared light-emitting diode (ILED) and a silicon phototransistor. The input signal is applied to the ILED and the output is taken from the phototransistor. The rise and fall times of phototransistors are very small, with typical values of turn-on time $t_{on} = 2$ to 5 μs and turn-off time $t_{off} = 300$ ns. These turn-on and turn-off times limit the high-frequency applications. A gate

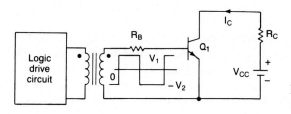

Figure 8-41 Transformer-isolated gate drive.

---

isolation circuit using a phototransistor is shown in Fig. 8-42. The phototransistor could be a Darlington pair. The phototransistors require separate power supply and add to the complexity and cost and weight of the drive circuits.

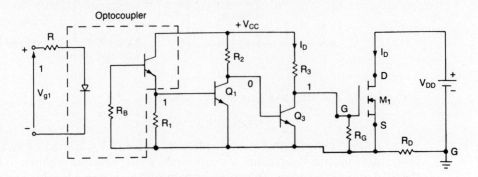

Figure 8-42   Optocoupler gate isolation.

8-9 SPICE MODELS

The PSpice model, which is based on the integral charge-control model of Gummel and Poon [16], is shown in Fig. 8-43a. The static (dc) model that is generated by PSpice is shown in Fig. 8-43b. If certain parameters are not specified, PSpice assumes the simple model of Ebers–Moll as shown in Fig. 8-43c.

The model statement for *NPN*-transistors has the general form

```
.MODEL QNAME NPN  (P1=V1 P2=V2 P3=V3 . . . . . . . PN=VN)
```

and the general form for *PNP*-transistors is

```
.MODEL QNAME PNP  (P1=V1 P2=V2 P3=V3 . . . . . . . PN=VN)
```

where QNAME is the name of the BJT model. NPN and PNP are the type symbols for *NPN*- and *PNP*-transistors, respectively. P1, P2, . . . and V1, V2, . . . are the parameters and their values, respectively. The parameters that affect the switching behavior of a BJT in power electronics are IS, BF, CJE, CJC, TR, TF. The symbol for a BJT is Q, and its name must start with Q. The general form is

```
Q⟨name⟩  NC  NB  NE  NS  QNAME [⟨area⟩ value]
```

where NC, NB, NE, and NS are the collector, base, emitter, and substrate nodes, respectively. The substrate node is optional: If not specified, it defaults to ground. Positive current is the current that flows into a terminal. That is, the current flows from the collector node, through the device, to the emitter node for an *NPN*-BJT.

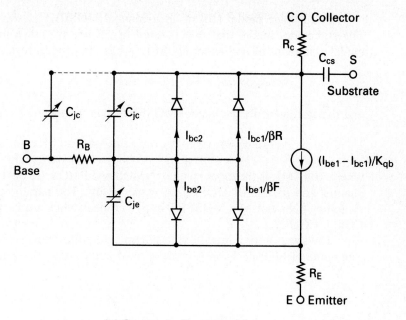

(a) Gummel – Poon model

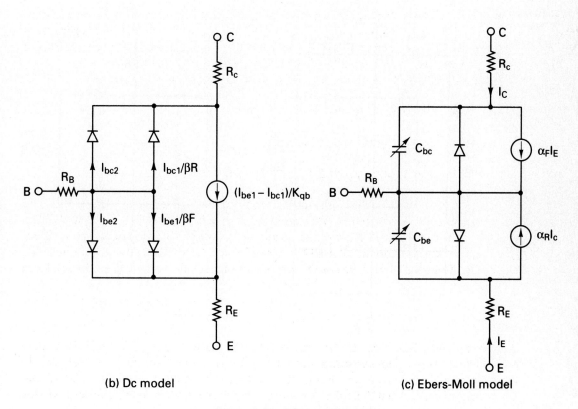

(b) Dc model

(c) Ebers-Moll model

Figure 8-43 PSpice BJT model.

The PSpice model [16] of an *n*-channel MOSFET is shown in Fig. 8-44a. The static (dc) model that is generated by PSpice is shown in Fig. 8-44b. The model statement of *n*-channel MOSFETs has the general form

```
.MODEL MNAME NMOS (P1=V1 P2=V2 P3=V3 ........PN=VN)
```

and the statement for *p*-channel MOSFETs has the form

```
.MODEL MNAME PMOS (P1=V1 P2=V2 P3=V3 ........PN=VN)
```

where MNAME is the model name. NMOS and PMOS are the type symbols of *n*-channel and *p*-channel MOSFETs, respectively. The parameters that affect the switching behavior of a MOSFET in power electronics are L, W, VTO, KP, IS, CGSO, CGDO.

The symbol for a metal-oxide silicon field-effect transistor (MOSFET) is M. The name of MOSFETs must start with M and it takes the general form

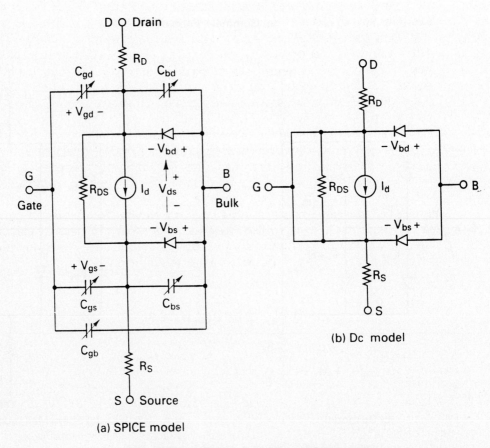

(a) SPICE model

(b) Dc model

Figure 8-44 PSpice *n*-channel MOSFET model.

```
M⟨name⟩ ND   NG   NS   NB   MNAME
+        [L=⟨value]  [W=⟨value⟩)]
+        [AD=⟨value⟩)]  [AS=⟨value⟩)]
+        [PD=⟨value⟩)]  [PS=⟨value⟩)]
+        [NRD=⟨value⟩)]  [NRS=⟨value⟩)]
+        [NRG=⟨value⟩)]  [NRB=⟨value⟩)]
```

where ND, NG, NS, and NB are the drain, gate, source, and bulk (or substrate) nodes, respectively.

SUMMARY

Power transistors are generally of four types: BJTs, MOSFETs, SITs, and IGBTs. BJTs are current-controlled devices and their parameters are sensitive to junction temperature. BJTs suffer from second breakdown and require reverse base current during turn-off to reduce the storage time. But they have low on-state or saturation voltage.

MOSFETs are voltage-controlled devices and require very low gating power and their parameters are less sensitive to junction temperature. There is no second breakdown problem and no need for negative gate voltage during turn-off. IGBTs, which combine the advantages of BJTs and MOSFETs, are voltage-controlled devices and have low on-state voltage similar to BJTs. IGBTs have no second breakdown phenomena. SITs are high-power, high-frequency devices. They are most suitable for audio, VHF/UHF, and microwave amplifiers. They have a normally on characteristic and a high on-state drop.

Transistors can be connected in series or parallel. Parallel operation usually requires current-sharing elements. Series operation requires matching of parameters, especially during turn-on and turn-off. To maintain the voltage and current relationship of transistors during turn-on and turn-off, it is generally necessary to use snubber circuits to limit the di/dt and dv/dt.

The gate signals can be isolated from the power circuit by pulse transformers or optocouplers. The pulse transformers are simple, but the leakage inductance should be very small. The transformers may be saturated at a low frequency and long pulse. Optocouplers require separate power supply.

REFERENCES

1. E. S. Oxner, *Power FETs and Their Applications.* Englewood Cliffs, N.J.: Prentice Hall, 1982.

2. B. J. Baliga and D. Y. Chen, *Power Transistors: Device Design and Applications.* New York: IEEE Press, 1984.

3. Westinghouse Electric, *Silicon Power Transistor Handbook.* Pittsburgh, Pa.: Westinghouse Electric Corporation, 1967.

4. B. R. Pelly, "Power MOSFETs: a status review." *International Power Electronics Conference,* 1983, pp. 10–32.

5. A. Ferraro, "An overview of low cost snubber technology for transistor converters."

IEEE Power Electronics Specialist Conference, 1982, pp. 466–477.

6. A. S. Sedra and K. C. Smith, *Microelectronics.* New York: CBS College Publishing, 1986.

7. R. Severns and J. Armijos, *MOSPOWER Application Handbook.* Santa Clara, Calif.: Siliconix Corporation, 1984.

8. B. R. Pelly and S. M. Clemente, *Applying International Rectifier's HEXFET Power MOSFETs,* Application Note 930A. El Segundo, Calif.: International Rectifier, 1985.

9. T. A. Radomski, "Protection of power transistors in electric vehicle drives." *IEEE Power Electronics Specialist Conference,* 1982, pp. 455–465.

10. B. J. Baliga, M. Cheng, P. Shafer, and M. W. Smith, "The insulated gate transistor (IGT): a new power switching device."

IEEE Industry Applications Society Conference Record, 1983, pp. 354–363.

11. S. Clemente and B. R. Pelly, "Understanding power MOSFET switching performance," *Solid-State Electronics,* Vol. 12, No. 12, 1982, pp. 1133–1141.

12. D. A. Grant and J. Gower, *Power MOSFETS: Theory and Applications.* New York: John Wiley & Sons, Inc., 1988.

13. B. J. Baliga, *Modern Power Devices.* New York: John Wiley & Sons, Inc., 1987.

14. *IGBT Designer's Manual.* El Segundo, Calif: International Rectifier, 1991.

15. J. Nishizawa and K. Yamamoto, "High-frequency high-power static induction transistor." *IEEE Transactions on Electron Devices,* Vol. ED25, No. 3, 1978, pp. 314–322.

16. M. H. Rashid, *SPICE for Circuits and Electronics Using PSpice.* Englewood Cliffs, N.J.: Prentice Hall, 1990.

REVIEW QUESTIONS

8-1. What is a bipolar transistor (BJT)?

8-2. What are the types of BJTs?

8-3. What are the differences between *NPN*-transistors and *PNP*-transistors?

8-4. What are the input characteristics of *NPN*-transistors?

8-5. What are the output characteristics of *NPN*-transistors?

8-6. What are the three regions of operation for BJTs?

8-7. What is a beta (β) of BJTs?

8-8. What is the difference between beta, β, and forced beta, β_F of BJTs?

8-9. What is a transductance of BJTs?

8-10. What is an overdrive factor of BJTs?

8-11. What is the switching model of BJTs?

8-12. What is the cause of delay time in BJTs?

8-13. What is the cause of storage time in BJTs?

8-14. What is the cause of rise time in BJTs?

8-15. What is the cause of fall time in BJTs?

8-16. What is a saturation mode of BJTs?

8-17. What is a turn-on time of BJTs?

8-18. What is a turn-off time of BJTs?

8-19. What is a FBSOA of BJTs?

8-20. What is a RBSOA of BJTs?

8-21. Why is it necessary to reverse bias BJTs during turn-off?

8-22. What is a second breakdown of BJTs?

8-23. What are the base drive techniques for increasing switching speeds of BJTs?

8-24. What is an antisaturation control of BJTs?

8-25. What are the advantages and disadvantages of BJTs?

8-26. What is a MOSFET?

8-27. What are the types of MOSFETs?

8-28. What are the differences between enhancement-type MOSFETs and depletion-type MOSFETs?

8-29. What is a pinch-off voltage of MOSFETs?

8-30. What is a threshold voltage of MOSFETs?

8-31. What is a transconductance of MOSFETs?

8-32. What is the switching model of *n*-channel MOSFETs?

8-33. What are the transfer characteristics of MOSFETs?

8-34. What are the output characteristics of MOSFETs?

8-35. What are the advantages and disadvantages of MOSFETs?

8-36. Why do the MOSFETs not require negative gate voltage during turn-off?

8-37. Why does the concept of saturation differ in BJTs and MOSFETs?

8-38. What is a turn-on time of MOSFETs?

8-39. What is a turn-off time of MOSFETs?

8-40. What is a SIT?

8-41. What are the advantages of SITs?

8-42. What are the disadvantages of SITs?

8-43. What is an IGBT?

8-44. What are the transfer characteristics of IGBTs?

8-45. What are the output characteristics of IGBTs?

8-46. What are the advantages and disadvantages of IGBTs?

8-47. What are the main differences between MOSFETs and BJTs?

8-48. What are the problems of parallel operation of BJTs?

8-49. What are the problems of parallel operation of MOSFETs?

8-50. What are the problems of parallel operation of IGBTs?

8-51. What are the problems of series operation of BJTs?

8-52. What are the problems of series operations of MOSFETs?

8-53. What are the problems of series operations of IGBTs?

8-54. What are the purposes of shunt snubber in transistors?

8-55. What is the purpose of series snubber in transistors?

8-56. What are the advantages and disadvantages of transformer gate isolation?

8-57. What are the advantages and disadvantages of optocoupled gate isolation?

PROBLEMS

8-1. The beta (β) of bipolar transistor in Fig. 8-6 varies from 10 to 60. The load resistance is $R_C = 5\ \Omega$. The dc supply voltage is $V_{CC} = 100$ V and the input voltage to the base circuit is $V_B = 8$ V. If $V_{CE(\text{sat})} = 2.5$ V and $V_{BE(\text{sat})} = 1.75$ V, find **(a)** the value of R_B that will result in saturation with an overdrive factor of 20; **(b)** the forced β, and **(c)** the power loss in the transistor P_T.

8-2. The beta (β) of bipolar transistor in Fig. 8-6 varies from 12 to 75. The load resistance is $R_C = 1.5\ \Omega$. The dc supply voltage is $V_{CC} = 40$ V and the input voltage to the base circuit is $V_B = 6$ V. If $V_{CE(\text{sat})} = 1.2$ V, $V_{BE(\text{sat})} = 1.6$ V, and $R_B = 0.7\ \Omega$, determine **(a)** the overdrive factor ODF, **(b)** the forced β, and **(c)** the power loss in the transistor P_T.

8-3. A transistor is used as a switch and the waveforms are shown in Fig. 8-10. The parameters are $V_{CC} = 200$ V, $V_{BE(\text{sat})} = $ 3 V, $I_B = 8$ A, $V_{CE(\text{sat})} = 2$ V, $I_{CS} = 100$ A, $t_d = 0.5\ \mu$s, $t_r = 1\ \mu$s, $t_s = 5\ \mu$s, $t_f = 3\ \mu$s, and $f_s = 10$ kHz. The duty cycle is $k = 50\%$. The collector–emitter leakage current is $I_{CEO} = 3$ mA. Determine the power loss due to the collector current **(a)** during turn-on $t_{\text{on}} = t_d + t_r$; **(b)** during conduction period t_n, **(c)** during turn-off $t_{\text{off}} = t_s + t_f$, **(d)** during off-time t_o, and **(e)** the total average power losses P_T. **(f)** Plot the instantaneous power due to the collector current $P_c(t)$.

8-4. The maximum junction temperature of the bipolar transistor in Prob. 8-3 is $T_j = 150°$C and the ambient temperature is $T_A = 25°$C. If the thermal resistances are $R_{JC} = 0.4°$C/W and $R_{CS} = 0.05°$C/W, calculate the thermal resistance of heat sink, R_{SA}. (*Hint:* Neglect the power loss due to base drive.)

8-5. For the parameters in Prob. 8-3, calculate

the average power loss due to the base current, P_B.

8-6. Repeat Prob. 8-3 if $V_{BE(sat)} = 2.3$ V, $I_B = 8$ A, $V_{CE(sat)} = 1.4$ V, $t_d = 0.1$ μs, $t_r = 0.45$ μs, $t_s = 3.2$ μs, and $t_f = 1.1$ μs.

8-7. A MOSFET is used as a switch. The parameters are $V_{DD} = 40$ V, $I_D = 35$ A, $R_{DS} = 28$ mΩ, $V_{GS} = 10$ V, $t_{d(on)} = 25$ ns, $t_r = 60$ ns, $t_{d(off)} = 70$ ns, $t_f = 25$ ns, $f_s = 20$ kHz. The drain-source leakage current is $I_{DSS} = 250$ μA. The duty cycle is $k = 60\%$. Determine the power loss due to the drain current **(a)** during turn-on $t_{on} = t_{d(n)} + t_r$; **(b)** during conduction period t_n, **(c)** during turn-off $t_{off} = t_{d(off)} + t_f$, **(d)** during off-time t_o, and **(e)** the total average power losses P_T.

8-8. The maximum junction temperature of the MOSFET in Prob. 8-7 is $T_j = 150°$C and the ambient temperature is $T_A = 30°$C. If the thermal resistances are $R_{JC} = 1$K/W and $R_{CS} = 1$K/W, calculate the thermal resistance of the heat sink, R_{SA}. (*Note:* K = °C + 273.)

The base drive circuit in Fig. 8-18 has $V_{CC} = 400$ V, $R_C = 4$ Ω, $V_{d1} = 3.6$ V, $V_{d2} = 0.9$ V, $V_{BE(sat)} = 0.7$ V, $V_B = 15$ V, $R_B = 1.1$ Ω, and $\beta = 12$. Calculate **(a)** the collector current without clamping, **(b)** the collector clamping voltage V_{CE}, and **(c)** the collector current with clamping.

8-10. Two BJTs are connected in parallel similar to Fig. 8-33. The total load current of $I_T = 200$ A. The collector–emitter voltage of transistor Q_1 is $V_{CE1} = 1.5$ V and that of transistor Q_2 is $V_{CE2} = 1.1$ V. Determine the collector current of each transistor and difference in current sharing if the current sharing series resistances are **(a)** $R_{e1} = 10$ mΩ and $R_{e2} = 20$ mΩ, and **(b)** $R_{e1} = R_{e2} = 20$ mΩ.

8-11. A bipolar transistor is operated as chopper switch at a frequency of $f_s = 20$ kHz. The circuit arrangement is shown in Fig. 8-36a. The dc input voltage of the chopper is $V_s = 400$ V and the load current is $I_L = 100$ A. The switching times are $t_r = 1$ μs and $t_f = 3$ μs. Determine the values of **(a)** L_s; **(b)** C_s; **(c)** R_s for critically damped condition; **(d)** R_s if the discharge time is limited to one-third of switching period; **(e)** R_s if peak discharge current is limited to 5% of load current; and **(f)** power loss due to RC snubber, P_s, neglecting the effect of inductor L_s on the voltage of snubber capacitor, C_s. Assume that $V_{CE(sat)} = 0$.

8-12. A MOSFET is operated as chopper switch at a frequency of $f_s = 50$ kHz. The circuit arrangement is shown in Fig. 8-36a. The dc input voltage of the chopper is $V_s = 30$ V and the load current is $I_L = 40$ A. The switching times are $t_r = 60$ ns and $t_f = 25$ ns. Determine the values of **(a)** L_s; **(b)** C_s; **(c)** R_s for critically damped condition; **(d)** R_s if the discharge time is limited to one-third of switching period; **(e)** R_s if peak discharge current is limited to 5% of load current; and **(f)** power loss due to RC snubber, P_s, neglecting the effect of inductor L_s on the voltage of snubber capacitor, C_s. Assume that $V_{CE(sat)} = 0$.

Dc choppers

9-1 INTRODUCTION

In many industrial applications, it is required to convert a fixed-voltage dc source into a variable-voltage dc source. A dc chopper converts directly from dc to dc and is also known as a *dc-to-dc converter*. A chopper can be considered as dc equivalent to an ac transformer with a continuously variable turns ratio. Like a transformer, it can be used to step-down or step-up a dc voltage source.

Choppers are widely used for traction motor control in electric automobiles, trolley cars, marine hoists, forklift trucks, and mine haulers. They provide smooth acceleration control, high efficiency, and fast dynamic response. Choppers can be used in regenerative braking of dc motors to return energy back into the supply, and this feature results in energy savings for transportation systems with frequent stops. Choppers are used in dc voltage regulators, and also used, in conjunction with an inductor, to generate a dc current source, especially for the current source inverter.

9-2 PRINCIPLE OF STEP-DOWN OPERATION

The principle of operation can be explained by Fig. 9-1a. When switch SW is closed for a time t_1, the input voltage V_s appears across the load. If the switch remains off for a time t_2, the voltage across the load is zero. The waveforms for the output voltage and load current are also shown in Fig. 9-1b. The chopper switch can be implemented by using a (1) power BJT, (2) power MOSFET, (3) GTO, or (4) forced-commutated thyristor. The practical devices have a finite voltage drop ranging from 0.5 to 2 V, and for the sake of simplicity we shall neglect the voltage drops of these power semiconductor devices.

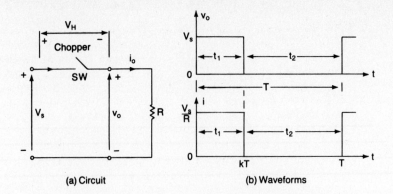

(a) Circuit　　　　　　　　**(b) Waveforms**

Figure 9-1　Step-down chopper with resistive load.

The average output voltage is given by

$$V_a = \frac{1}{T} \int_0^{t_1} v_0 \, dt = \frac{t_1}{T} V_s = ft_1 V_s = kV_s \tag{9-1}$$

and the average load current, $I_a = V_a/R = kV_s/R$, where T is the chopping period, $k = t_1/T$ is the duty cycle of chopper, and f is the chopping frequency. The rms value of output voltage is found from

$$V_o = \left(\frac{1}{T} \int_0^{kT} v_0^2 \, dt\right)^{1/2} = \sqrt{k} \, V_s \tag{9-2}$$

Assuming a lossless chopper, the input power to the chopper is the same as the output power and is given by

$$P_i = \frac{1}{T} \int_0^{kT} v_0 i \, dt = \frac{1}{T} \int_0^{kT} \frac{v_0^2}{R} \, dt = k \frac{V_s^2}{R} \tag{9-3}$$

The effective input resistance seen by the source is

$$R_i = \frac{V_s}{I_a} = \frac{V_s}{kV_s/R} = \frac{R}{k} \tag{9-4}$$

The duty cycle k can be varied from 0 to 1 by varying t_1, T, or f. Therefore, the output voltage V_o can be varied from 0 to V_s by controlling k, and the power flow can be controlled.

1. *Constant-frequency operation.*　The chopping frequency f (or chopping period T) is kept constant and the on-time t_1 is varied. The width of the pulse is varied and this type of control is known as *pulse-width-modulation* (PWM) control.

2. *Variable-frequency operation.*　The chopping frequency f is varied. Either on-time t_1 or off-time t_2 is kept constant. This is called *frequency modulation.* The frequency has to be varied over a wide range to obtain the full output voltage range. This type of control would generate harmonics at unpredictable frequencies and the filter design would be difficult.

Example 9-1

The dc chopper in Fig. 9-1a has a resistive load of $R = 10 \, \Omega$ and the input voltage is $V_s = 220$ V. When the chopper switch remains on, its voltage drop is $v_{ch} = 2$ V and the chopping frequency is $f = 1$ kHz. If the duty cycle is 50%, determine (a) the average output voltage V_a, (b) the rms output voltage V_o, (c) the chopper efficiency, (d) the effective input resistance R_i of the chopper, and (e) the rms value of the fundamental component of output harmonic voltage.

Solution $V_s = 220$ V, $k = 0.5$, $R = 10 \, \Omega$, and $v_{ch} = 2$ V.

(a) From Eq. (9-1), $V_a = 0.5 \times (220 - 2) = 109$ V.

(b) From Eq. (9-2), $V_o = \sqrt{0.5} \times (220 - 2) = 154.15$ V.

(c) The output power can be found from

$$P_o = \frac{1}{T} \int_0^{kT} \frac{v_o^2}{R} \, dt = \frac{1}{T} \int_0^{kT} \frac{(V_s - v_{ch})^2}{R} \, dt = k \frac{(V_s - v_{ch})^2}{R}$$

$$= 0.5 \times \frac{(220 - 2)^2}{10} = 2376.2 \text{ W} \tag{9-5}$$

The input power to the chopper can be found from

$$P_i = \frac{1}{T} \int_0^{kT} V_s i \, dt = \frac{1}{T} \int_0^{kT} \frac{V_s(V_s - v_{ch})}{R} \, dt = k \frac{V_s(V_s - v_{ch})}{R}$$

$$= 0.5 \times 220 \times \frac{220 - 2}{10} = 2398 \text{ W} \tag{9-6}$$

The chopper efficiency is

$$\frac{P_o}{P_i} = \frac{2376.2}{2398} = 99.09\%$$

(d) From Eq. (9-4), $R_i = 10/0.5 = 20 \, \Omega$.

(e) The output voltage as shown in Fig. 9-1b can be expressed in a Fourier series as

$$v_o(t) = kV_s + \frac{V_s}{n\pi} \sum_{n=1}^{\infty} \sin 2n\pi k \cos 2n\pi ft$$

$$+ \frac{V_s}{n\pi} \sum_{n=1}^{\infty} (1 - \cos 2n\pi k) \sin 2n\pi ft \tag{9-7}$$

The fundamental component (for $n = 1$) of output voltage harmonic can be determined from Eq. (9-7) as

$$v_1(t) = \frac{V_s}{\pi} [\sin 2\pi k \cos 2\pi ft + (1 - \cos 2\pi k) \sin 2\pi ft]$$

$$= \frac{220 \times 2}{\pi} \sin(2\pi \times 1000t) = 140.06 \sin(6283.2t) \tag{9-8}$$

and its rms value is $V_1 = 140.06/\sqrt{2} = 99.04$ V.

Note. The efficiency calculation, which includes the conduction loss of the chopper, does not take into account the switching loss due to turn-on and turn-off of practical choppers. The efficiency of a practical chopper varies between 92 and 99%.

A chopper with an *RL* load is shown in Fig. 9-2. The operation of the chopper can be divided into two modes. During mode 1, the chopper is switched on and the current flows from the supply to the load. During mode 2, the chopper is switched off and the load current continues to flow through freewheeling diode D_m. The equivalent circuits for these modes are shown in Fig. 9-3a. The load current and output voltage waveforms are shown in Fig. 9-3b.

The load current for mode 1 can be found from

$$V_s = Ri_1 + L\frac{di_1}{dt} + E \tag{9-9}$$

The solution of Eq. (9-9) with initial current $i_1(t = 0) = I_1$ gives the load current as

$$i_1(t) = I_1 e^{-tR/L} + \frac{V_s - E}{R}(1 - e^{-tR/L}) \tag{9-10}$$

This mode is valid $0 \le t \le t_1 (= kT)$; and at the end of this mode, the load current becomes

$$i_1(t = t_1 = kT) = I_2 \tag{9-11}$$

The load current for mode 2 can be found from

$$0 = Ri_2 + L\frac{di_2}{dt} + E \tag{9-12}$$

With initial current $i_2(t = 0) = I_2$ and redefining the time origin (i.e., $t = 0$) at the beginning of mode 2, we have

$$i_2(t) = I_2 e^{-tR/L} - \frac{E}{R}(1 - e^{-tR/L}) \tag{9-13}$$

This mode is valid for $0 \le t \le t_2 [= (1 - k)T]$. At the end of this mode, the load current becomes

$$i_2(t = t_2) = I_3 \tag{9-14}$$

At the end of mode 2, the chopper is turned on again in the next cycle after time, $T = 1/f = t_1 + t_2$.

Under steady-state conditions, $I_1 = I_3$. The peak-to-peak load ripple current can be determined from Eqs. (9-10), (9-11), (9-13), and (9-14). From Eqs. (9-10)

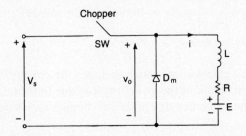

Figure 9-2 Chopper with *RL* loads.

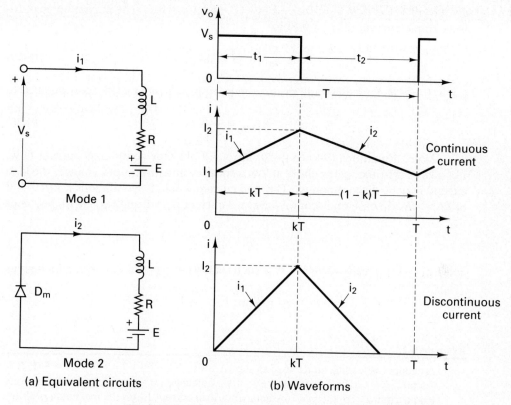

Figure 9-3 Equivalent circuits and waveforms for *RL* loads.

and (9-11), I_2 is given by

$$I_2 = I_1 e^{-kTR/L} + \frac{V_s - E}{R} (1 - e^{-kTR/L})$$ (9-15)

From Eqs. (9-13) and (9-14), I_3 is given by

$$I_3 = I_1 = I_2 e^{-(1-k)TR/L} - \frac{E}{R} (1 - e^{-(1-k)TR/L})$$ (9-16)

The peak-to-peak ripple current is

$$\Delta I = I_2 - I_1$$

which after simplifications becomes

$$\Delta I = \frac{V_s}{R} \frac{1 - e^{-kTR/L} + e^{-TR/L} - e^{-(1-k)TR/L}}{1 - e^{-TR/L}}$$ (9-17)

The condition for maximum ripple,

$$\frac{d(\Delta I)}{dk} = 0$$ (9-18)

---

gives $e^{-kTR/L} - e^{-(1-k)TR/L} = 0$ or $-k = -(1 - k)$ or $k = 0.5$. The maximum peak-to-peak ripple current (at $k = 0.5$) is

$$\Delta I_{max} = \frac{V_s}{R} \tanh \frac{R}{4fL} \qquad (9-19)$$

For $4fL >> R$, $\tanh \theta \approx \theta$ and the maximum ripple current can be approximated to

$$\Delta I_{max} = \frac{V_s}{4fL} \qquad (9-20)$$

Note. Equations (9-9) to (9-20) are valid only for continuous current flow. For a large off-time, particularly at low frequency and low output voltage, the load current may be discontinuous. The load current would be continuous if $L/R >> T$ or $Lf >> R$. In case of discontinuous load current, $I_1 = 0$ and Eq. (9-10) becomes

$$i_1(t) = \frac{V_s - E}{R} (1 - e^{-tR/L})$$

and Eq. (9-13) is valid for $0 \le t \le t_2$ such that $i_2(t = t_2) = I_3 = I_1 = 0$, which gives

$$t_2 = \frac{L}{R} \ln\left(1 + \frac{RI_2}{E}\right)$$

Example 9-2

A chopper is feeding an *RL* load as shown in Fig. 9-2 with $V_s = 220$ V, $R = 5\ \Omega$, $L = 7.5$ mH, $f = 1$ kHz, k = 0.5 and $E = 0$ V. Calculate (a) the minimum instantaneous load current I_1, (b) the peak instantaneous load current I_2, (c) the maximum peak-to-peak load ripple current, (d) the average value of load current I_a, (e) the rms load current I_o, (f) the effective input resistance R_i seen by the source, and (g) the rms chopper current I_R.

Solution $V_s = 220$ V, $R = 5\ \Omega$, $L = 7.5$ mH, $E = 0$ V, $k = 0.5$, and $f = 1000$ Hz. From Eq. (9-15), $I_2 = 0.7165I_1 + 12.473$ and from Eq. (9-16), $I_1 = 0.7165I_2 + 0$.

(a) Solving these two equations yields $I_1 = 18.37$ A.

(b) $I_2 = 25.63$ A.

(c) $\Delta I = I_2 - I_1 = 25.63 - 18.37 = 7.26$ A. From Eq. (9-19), $\Delta I_{max} = 7.26$ A and Eq. (9-20) gives the approximate value, $\Delta I_{max} = 7.33$ A.

(d) The average load current is, approximately,

$$I_a = \frac{I_2 + I_1}{2} = \frac{25.63 + 18.37}{2} = 22 \text{ A}$$

(e) Assuming that the load current rises linearly from I_1 to I_2, the instantaneous load current can be expressed as

$$i_1 = I_1 + \frac{\Delta I\, t}{kT} \qquad \text{for } 0 < t < kT$$

The rms value of load current can be found from

$$I_o = \left(\frac{1}{kT} \int_0^{kT} i_1^2\, dt\right)^{1/2} = \left[I_1^2 + \frac{(I_2 - I_1)^2}{3} + I_1(I_2 - I_1)\right]^{1/2} \qquad (9-21)$$

$$= 22.1 \text{ A}$$

(f) The average source current

$$I_s = kI_a = 0.5 \times 22 = 11 \text{ A}$$

and the effective input resistance $R_i = V_s/I_s = 220/11 = 20 \ \Omega$.

(g) The rms chopper current can be found from

$$I_R = \left(\frac{1}{T} \int_0^{kT} i_1^2 \, dt\right)^{1/2} = \sqrt{k}\left[I_1^2 + \frac{(I_2 - I_1)^2}{3} + I_1(I_2 - I_1)\right]^{1/2}$$

$$= \sqrt{k}I_o = \sqrt{0.5} \times 22.1 = 15.63 \text{ A}$$

(9-22)

Example 9-3
_____

The chopper in Fig. 9-2 has a load resistance $R = 0.25 \ \Omega$, input voltage $V_s = 550$ V, and battery voltage $E = 0$ V. The average load current $I_a = 200$ A, and chopping frequency $f = 250$ Hz. Use the average output voltage to calculate the load inductance L, which would limit the maximum load ripple current to 10% of I_a.

Solution $V_s = 550$ V, $R = 0.25 \ \Omega$, $E = 0$ V, $f = 250$ Hz, $T = 1/f = 0.004$ s, and $\Delta i = 200 \times 0.1 = 20$ A. The average output voltage $V_a = kV_s = RI_a$. The voltage across the inductor is given by

$$L\frac{di}{dt} = V_s - RI_a = V_s - kV_s = V_s(1 - k)$$

If the load current is assumed to rise linearly, $dt = t_1 = kT$ and $di = \Delta i$:

$$\Delta i = \frac{V_s(1 - k)}{L} kT$$

For the worst-case ripple conditions,

$$\frac{d(\Delta i)}{dk} = 0$$

This gives $k = 0.5$ and

$$\Delta i \, L = 20 \times L = 550(1 - 0.5) \times 0.5 \times 0.004$$

and the required value of inductance is $L = 27.5$ mH.

9-4 PRINCIPLE OF STEP-UP OPERATION

A chopper can be used to step up a dc voltage and an arrangement for step-up operation is shown in Fig. 9-4a. When switch SW is closed for time t_1, the inductor current rises and energy is stored in the inductor, L. If the switch is opened for time t_2, the energy stored in the inductor is transferred to load through diode D_1 and the inductor current falls. Assuming a continuous current flow, the waveform for the inductor current is shown in Fig. 9-4b.

When the chopper is turned on, the voltage across the inductor is

$$v_L = L\frac{di}{dt}$$

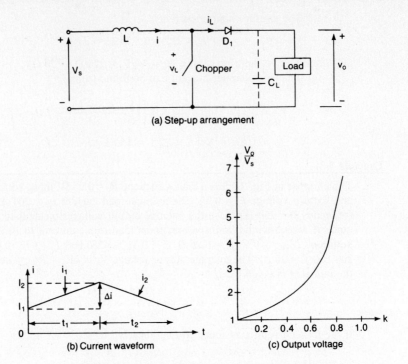

Figure 9-4 Arrangement for step-up operation.

and this gives the peak-to-peak ripple current in the inductor as

$$\Delta I = \frac{V_s}{L} t_1 \tag{9-23}$$

The instantaneous output voltage is

$$v_o = V_s + L \frac{\Delta I}{t_2} = V_s \left(1 + \frac{t_1}{t_2}\right) = V_s \frac{1}{1-k} \tag{9-24}$$

If a large capacitor C_L is connected across the load as shown by dashed lines in Fig. 9-4a, the output voltage will be continuous and v_o would become the average value V_a. We can notice from Eq. (9-24) that the voltage across the load can be stepped up by varying the duty cycle, k, and the minimum output voltage is V_s when $k = 0$. However, the chopper cannot be switched on continuously such that $k = 1$. For values of k tending to unity, the output voltage becomes very large and is very sensitive to changes in k, as shown in Fig. 9-4c.

This principle can be applied to transfer energy from one voltage source to another as shown in Fig. 9-5a. The equivalent circuits for the modes of operation are shown in Fig. 9-5b and the current waveforms in Fig. 9-5c. The inductor current for mode 1 is given by

$$V_s = L \frac{di_1}{dt}$$

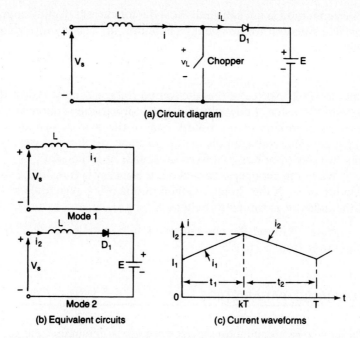

(a) Circuit diagram

Mode 1

Mode 2

(b) Equivalent circuits

(c) Current waveforms

Figure 9-5 Arrangement for transfer of energy.

and is expressed as

$$i_1(t) = \frac{V_s}{L} t + I_1 \qquad (9\text{-}25)$$

where I_1 is the initial current for mode 1. During mode 1, the current must rise and the necessary condition,

$$\frac{di_1}{dt} > 0 \qquad \text{or} \qquad V_s > 0$$

The current for mode 2 is given by

$$V_s = L \frac{di_2}{dt} + E$$

and is solved as

$$i_2(t) = \frac{V_s - E}{L} t + I_2 \qquad (9\text{-}26)$$

where I_2 is initial current for mode 2. For a stable system, the current must fall and the condition is

$$\frac{di_2}{dt} < 0 \qquad \text{or} \qquad V_s < E$$

If this condition is not satisfied, the inductor current would continue to rise and an unstable situation would occur. Therefore, the conditions for controllable power transfer are

$$0 < V_s < E \qquad (9\text{-}27)$$

Equation (9-27) indicates that the source voltage V_s must be less than the voltage E to permit transfer of power from a fixed (or variable) source to a fixed dc voltage. In electric braking of dc motors, where the motors operate as dc generators, whose terminal voltage falls as the machine speed decreases. The chopper permits transfer of power to a fixed dc source or a rheostat.

When the chopper is turned on, the energy is transferred from the source V_s to inductor L. If the chopper is then turned off, a magnitude of the energy stored in the inductor is forced to battery E.

Note. Without the chopping action, v_s must be greater than E for transferring power from V_s to E.

9-5 PERFORMANCE PARAMETERS

The power semiconductor devices require a minimum time to turn on and turn off. Therefore, the duty cycle k can only be controlled between a minimum value k_{min} and a maximum value k_{max}, thereby limiting the minimum and maximum value of output voltage. The switching frequency of the chopper is also limited. It can be noticed from Eq. (9-20) that the load ripple current depends inversely on the chopping frequency f. The frequency should be as high as possible to reduce the load ripple current and to minimize the size of any additional series inductor in the load circuit.

9-6 CHOPPER CLASSIFICATION

The step-down chopper in Fig. 9-1a only allows power to flow from the supply to the load, and is referred to as class A chopper. Depending on the directions of current and voltage flows, choppers can be classified into five types:

Class A chopper
Class B chopper
Class C chopper
Class D chopper
Class E chopper

Class A chopper. The load current flows into the load. Both the load voltage and the load current are positive, as shown in Fig. 9-6a. This is a single-quadrant chopper and is said to be operated as a rectifier. Equations in Sections 9-2 and 9-3 can be applied to evaluate the performance of an A chopper.

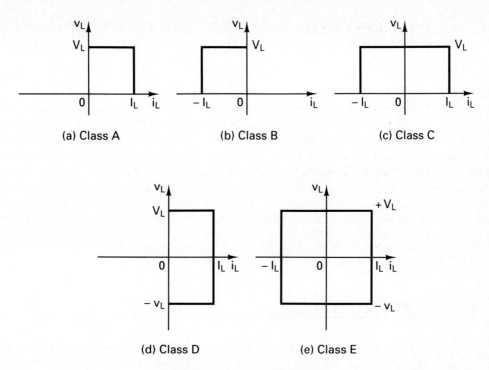

(a) Class A   (b) Class B   (c) Class C

(d) Class D   (e) Class E

Figure 9-6 Chopper classification.

Class B chopper.   The load current flows out of the load. The load voltage is positive, but the load current is negative, as shown in Fig. 9-6b. This is also a single-quadrant chopper, but operates in the second quadrant and is said to be operated as an inverter. A class B chopper is shown in Fig. 9-7a, where the battery E is a part of the load and may be the back emf of a dc motor.

When switch S_1 is turned on, the voltage E drives current through inductor L and load voltage v_L becomes zero. The instantaneous load voltage v_L and load

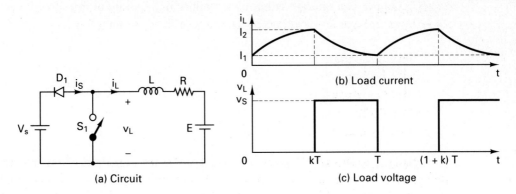

(a) Circuit   (b) Load current   (c) Load voltage

Figure 9-7 Class B chopper.

current i_L are shown in Fig. 9-7b and c, respectively. The current i_L, which rises, is described by

$$0 = L \frac{di_L}{dt} + Ri_L + E$$

which, with initial condition $i_L(t = 0) = I_1$, gives

$$i_L = I_1 e^{-(R/L)t} - \frac{E}{R}(1 - e^{-(R/L)t}) \qquad \text{for } 0 \le t \le kT \qquad (9\text{-}28)$$

At $t = t_1$,

$$i_L(t = t_1 = kT) = I_2$$

When switch S_1 is turned off, a magnitude of the energy stored in inductor L is returned to the supply V_s via diode D_1. The load current i_L falls. Redefining the time origin $t = 0$, the load current i_L is described by

$$V_s = L \frac{di_L}{dt} + Ri_L + E$$

which, with initial condition $i(t = t_2) = I_2$, gives

$$i_L = I_2 e^{-(R/L)t} + \frac{V_s - E}{R}(1 - e^{-(R/L)t}) \qquad \text{for } 0 \le t \le t_2 \qquad (9\text{-}29)$$

where $t_2 = (1 - k)T$. At $t = t_2$,

$$i_L(t = t_2) = I_1 \qquad \text{for steady-state continuous current}$$

$$= 0 \qquad \text{for steady-state discontinuous current}$$

Class C chopper. The load current is either positive or negative, as shown in Fig. 9-6c. The load voltage is always positive. This is known as a *two-quadrant chopper*. The class A and class B choppers can be combined to form a class C chopper as shown in Fig. 9-8. S_1 and D_2 operate as as a class A chopper. S_2 and D_1 operate as a class B chopper. Care must be taken to ensure that the two switches are not fired together; otherwise, the supply V_s will be short-circuited. A class C chopper can operate either as a rectifier or as an inverter.

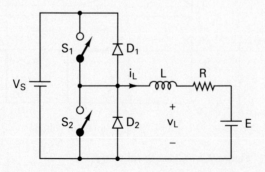

Figure 9-8 Class C chopper.

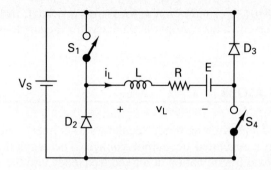

Figure 9-9 Class D chopper.

Class D chopper.   The load current is always positive. The load voltage is either positive or negative, as shown in Fig. 9-6d. A class D chopper can also operate either as a rectifier or as an inverter, as shown in Fig. 9-9. If S_1 and S_4 are turned on, v_L and i_L becomes positive. If S_1 and S_4 are turned off, load current i_L will be positive and continue to flow for a highly inductive load. Diodes D_2 and D_3 provide a path for the load current and v_L will be reversed.

Class E chopper.   The load current is either positive or negative, as shown in Fig. 9-6e. The load voltage is also either positive or negative. This is known as a *four-quadrant chopper*. Two class C choppers can be combined to form a class E chopper, as shown in Fig. 9-10a. The polarities of the load voltage and load current are shown in Fig. 9-10b. The devices that are operative in different quad-

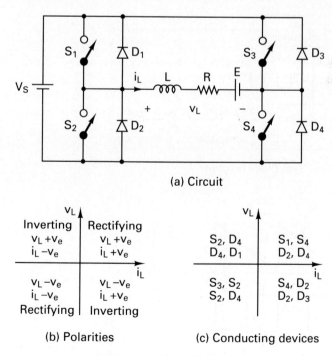

(a) Circuit

	v_L				v_L	
Inverting	Rectifying			S_2, D_4	S_1, S_4	
v_L +ve	v_L +ve			D_4, D_1	D_2, D_4	
i_L −ve	i_L +ve	i_L				i_L
v_L −ve	v_L −ve			S_3, S_2	S_4, D_2	
i_L −ve	i_L +ve			S_2, D_4	D_2, D_3	
Rectifying	Inverting					

(b) Polarities (c) Conducting devices

Figure 9-10 Class E chopper.

rants are shown in Fig. 9-10c. For operation in the fourth quadrant, the direction of battery E must be reversed. This chopper is the basis for the single-phase full-bridge inverter in Section 10-4.

9-7 SWITCHING-MODE REGULATORS

Dc choppers can be used as switching-mode regulators to convert a dc voltage, normally unregulated, to a regulated dc output voltage. The regulation is normally achieved by pulse-width modulation at a fixed frequency and the switching device is normally a ower BJT, MOSFET, or IGBT. The elements of switching-mode regulators are shown in Fig. 9-11a. We can notice from Fig. 9-1b that the output of dc choppers with resistive load is discontinuous and contains harmonics. The ripple content is normally reduced by an *LC* filter.

Switching regulators are commercially available as integrated circuits. The designer can select the switching frequency by choosing the values of *R* and *C* of frequency oscillator. As a rule of thumb, to maximize efficiency, the minimum oscillator period should be about 100 times longer than the transistor switching time; for example, if a transistor has a switching time of 0.5 μs, the oscillator period would be 50 μs, which gives the maximum oscillator frequency of 20 kHz. This limitation is due to switching loss in the transistor. The transistor switching loss increases with the switching frequency and as a result the efficiency decreases. In addition, the core loss of inductors limits the high-frequency opera-

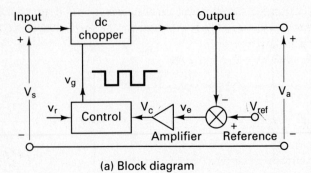

(a) Block diagram

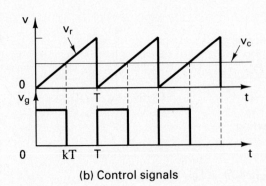

(b) Control signals

Figure 9-11 Elements of switching-mode regulators.

tion. Control voltage v_c is obtained by comparing the output voltage with its desired value. v_c can be compared with a sawtooth voltage v_r to generate the PWM control signal for the dc chopper. This is shown in Fig. 9-11b. There are four basic topologies of switching regulators:

1. Buck regulators
2. Boost regulators
3. Buck-Boost regulators
4. Cúk regulators

9-7.1 Buck Regulators

In a buck regulator, the average output voltage V_a, is less than the input voltage, V_s—hence the name "buck," a very popular regulator. The circuit diagram of a buck regulator using a power BJT is shown in Fig. 9-12a, and this is like a stepdown chopper. The circuit operation can be divided into two modes. Mode 1 begins when transistor Q_1 is switched on at $t = 0$. The input current, which rises, flows through filter inductor L, filter capacitor C, and load resistor R. Mode 2 begins when transistor Q_1 is switched off at $t = t_1$. The freewheeling diode D_m conducts due to energy stored in the inductor and the inductor current continues to flow through L, C, load, and diode D_m. The inductor current falls until transistor Q_1 is switched on again in the next cycle. The equivalent circuits for the modes of operation are shown in Fig. 9-12b. The waveforms for the voltages and currents are shown in Fig. 9-12c for a continuous current flow in the inductor L. Depending on the switching frequency, filter inductance, and capacitance, the inductor current could be discontinuous.

The voltage across the inductor L is, in general,

$$e_L = L \frac{di}{dt}$$

Assuming that the inductor current rises linearly from I_1 to I_2 in time t_1,

$$V_s - V_a = L \frac{I_2 - I_1}{t_1} = L \frac{\Delta I}{t_1} \tag{9-30}$$

or

$$t_1 = \frac{\Delta I\, L}{V_s - V_a} \tag{9-31}$$

and the inductor current falls linearly from I_2 to I_1 in time t_2,

$$-V_a = -L \frac{\Delta I}{t_2} \tag{9-32}$$

or

$$t_2 = \frac{\Delta I\, L}{V_a} \tag{9-33}$$

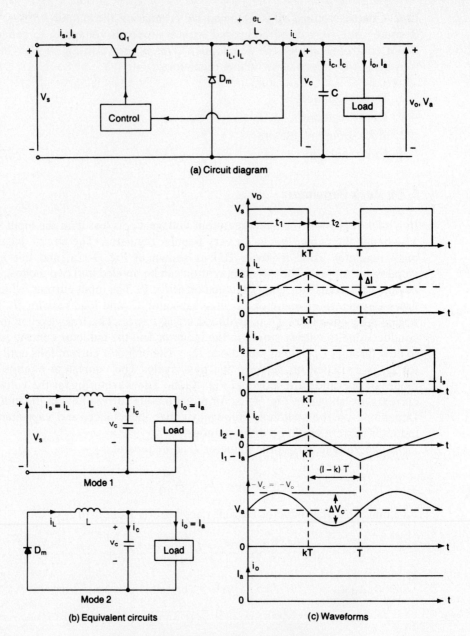

Figure 9-12 Buck regulator with continuous i_L.

where $\Delta I = I_2 - I_1$ is the peak-to-peak ripple current of the inductor L. Equating the value of ΔI in Eqs. (9-30) and (9-32) gives

$$\Delta I = \frac{(V_s - V_a)t_1}{L} = \frac{V_a t_2}{L}$$

Substituting $t_1 = kT$ and $t_2 = (1 - k)T$ yields the average output voltage as

$$V_a = V_s \frac{t_1}{T} = kV_s \qquad (9\text{-}34)$$

Assuming a lossless circuit, $V_s I_s = V_a I_a = kV_s I_a$ and the average input current

$$I_s = kI_a \qquad (9\text{-}35)$$

The switching period T can be expressed as

$$T = \frac{1}{f} = t_1 + t_2 = \frac{\Delta I\, L}{V_s - V_a} + \frac{\Delta I\, L}{V_a} = \frac{\Delta I\, L V_s}{V_a(V_s - V_a)} \qquad (9\text{-}36)$$

which gives the peak-to-peak ripple current as

$$\Delta I = \frac{V_a(V_s - V_a)}{fLV_s} \qquad (9\text{-}37)$$

or

$$\Delta I = \frac{V_s k(1 - k)}{fL} \qquad (9\text{-}38)$$

Using Kirchhoff's current law, we can write the inductor current i_L as

$$i_L = i_c + i_o$$

If we assume that the load ripple current Δi_o is very small and negligible, $\Delta i_L = \Delta i_c$. The average capacitor current, which flows into for $t_1/2 + t_2/2 = T/2$, is

$$I_c = \frac{\Delta I}{4}$$

The capacitor voltage is expressed as

$$v_c = \frac{1}{C} \int i_c\, dt + v_c(t = 0)$$

and the peak-to-peak ripple voltage of the capacitor is

$$\Delta V_c = v_c - v_c(t = 0) = \frac{1}{C} \int_0^{T/2} \frac{\Delta I}{4}\, dt = \frac{\Delta I\, T}{8C} = \frac{\Delta I}{8fC} \qquad (9\text{-}39)$$

Substituting the value of ΔI from Eq. (9-37) or (9-38) in Eq. (9-39) yields

$$\Delta V_c = \frac{V_a(V_s - V_a)}{8LCf^2V_s} \qquad (9\text{-}40)$$

or

$$\Delta V_c = \frac{V_s k(1 - k)}{8LCf^2} \qquad (9\text{-}41)$$

The buck regulator requires only one transistor, is simple, and has high efficiency greater than 90%. The di/dt of the load current is limited by inductor L. However, the input current is discontinuous and a smoothing input filter is

---

normally required. It provides one polarity of output voltage and unidirectional output current. It requires a protection circuit in case of possible short circuit across the diode path.

Example 9-4

The buck regulator in Fig. 9-12a has an input voltage of $V_s = 12$ V. The required average output voltage is $V_a = 5$ V and the peak-to-peak output ripple voltage is 20 mV. The switching frequency is 25 kHz. If the peak-to-peak ripple current of inductor is limited to 0.8 A, determine (a) the duty cycle k, (b) the filter inductance L, and (c) the filter capacitor C.

Solution $V_s = 12$ V, $\Delta V_c = 20$ mV, $\Delta I = 0.8$ A, $f = 25$ kHz, and $V_a = 5$ V.

(a) From Eq. (9-34), $V_a = kV_s$ and $k = V_a/V_s = 5/12 = 0.4167 = 41.67\%$.

(b) From Eq. (9-37),

$$L = \frac{5(12 - 5)}{0.8 \times 25{,}000 \times 12} = 145.83 \ \mu H$$

(c) From Eq. (9-39),

$$C = \frac{0.8}{8 \times 20 \times 10^{-3} \times 25{,}000} = 200 \ \mu F$$

9-7.2 Boost Regulators

In a boost regulator, the output voltage is greater than the input voltage—hence the name "boost." A boost regulator using a power MOSFET is shown in Fig. 9-13a. The circuit operation can be divided into two modes. Mode 1 begins when transistor M_1 is switched on at $t = 0$. The input current, which rises, flows through inductor L and transistor Q_1. Mode 2 begins when transistor M_1 is switched off at $t = t_1$. The current which was flowing through the transistor would now flow through L, C, load, and diode D_m. The inductor current falls until transistor M_1 is turned on again in the next cycle. The energy stored in inductor L is transferred to the load. The equivalent circuits for the modes of operation are shown in Fig. 9-13b. The waveforms for voltages and currents are shown in Fig. 13c for continuous load current.

Assuming that the inductor current rises linearly from I_1 to I_2 in time t_1,

$$V_s = L \frac{I_2 - I_1}{t_1} = L \frac{\Delta I}{t_1} \tag{9-42}$$

or

$$t_1 = \frac{\Delta I \, L}{V_s} \tag{9-43}$$

and the inductor current falls linearly from I_2 to I_1 in time t_2,

$$V_s - V_a = -L \frac{\Delta I}{t_2} \tag{9-44}$$

or

$$t_2 = \frac{\Delta I \, L}{V_a - V_s} \tag{9-45}$$

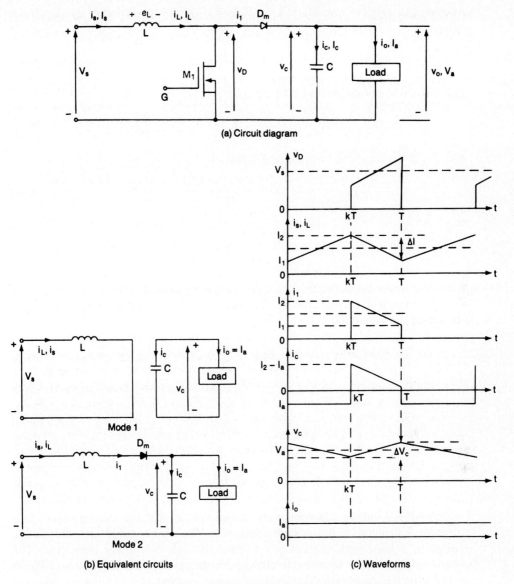

(a) Circuit diagram

Mode 1

Mode 2

(b) Equivalent circuits

(c) Waveforms

Figure 9-13 Boost regulator with continuous i_L.

where $\Delta I = I_2 - I_1$ is the peak-to-peak ripple current of inductor L. From Eqs. (9-42) and (9-44),

$$\Delta I = \frac{V_s t_1}{L} = \frac{(V_a - V_s)t_2}{L}$$

Substituting $t_1 = kT$ and $t_2 = (1 - k)T$ yields the average output voltage,

$$V_a = V_s \frac{T}{t_2} = \frac{V_s}{1 - k} \tag{9-46}$$

Assuming a lossless circuit, $V_s I_s = V_a I_a = V_s I_a/(1-k)$ and the average input current is

$$I_s = \frac{I_a}{1-k} \tag{9-47}$$

The switching period T can be found from

$$T = \frac{1}{f} = t_1 + t_2 = \frac{\Delta I \, L}{V_s} + \frac{\Delta I \, L}{V_a - V_s} = \frac{\Delta I \, L V_a}{V_s(V_a - V_s)} \tag{9-48}$$

and this gives the peak-to-peak ripple current.

$$\Delta I = \frac{V_s(V_a - V_s)}{fLV_a} \tag{9-49}$$

or

$$\Delta I = \frac{V_s k}{fL} \tag{9-50}$$

When the transistor is on, the capacitor supplies the load current for $t = t_1$. The average capacitor current during time t_1 is $I_c = I_a$ and the peak-to-peak ripple voltage of the capacitor is

$$\Delta V_c = v_c - v_c(t = 0) = \frac{1}{C}\int_0^{t_1} I_c \, dt = \frac{1}{C}\int_0^{t_1} I_a = \frac{I_a t_1}{C} \tag{9-51}$$

Equation (9-46) gives $t_1 = (V_a - V_s)/(V_a f)$ and substituting t_1 in Eq. (9-51) gives

$$\Delta V_c = \frac{I_a(V_a - V_s)}{V_a fC} \tag{9-52}$$

or

$$\Delta V_c = \frac{I_a k}{fC} \tag{9-53}$$

A boost regulator can step up the output voltage without a transformer. Due to single transistor, it has a high efficiency. The input current is continuous. However, a high peak current has to flow through the power transistor. The output voltage is very sensitive to changes in duty cycle k and it might be difficult to stabilize the regulator. The average output current is less than the average inductor current by a factor of $(1-k)$, and a much higher rms current would flow through the filter capacitor, resulting in the use of a larger filter capacitor and a larger inductor than those of a buck regulator.

Example 9-5

A boost regulator in Fig. 9-13a has an input voltage of $V_s = 5$ V. The average output voltage $V_a = 15$ V and the average load current $I_a = 0.5$ A. The switching frequency is 25 kHz. If $L = 150$ μH and $C = 220$ μF, determine (a) the duty cycle k, (b) the ripple current of inductor ΔI, (c) the peak current of inductor I_2, and (d) the ripple voltage of filter capacitor ΔV_c.

Solution $V_s = 5$ V, $V_a = 15$ V, $f = 25$ kHz, $L = 150$ μH, and $C = 220$ μF.
(a) From Eq. (9-46), $15 = 5/(1 - k)$ or $k = 2/3 = 0.6667 = 66.67\%$.
(b) From Eq. (9-49),

$$\Delta I = \frac{5 \times (15 - 5)}{25,000 \times 150 \times 10^{-6} \times 15} = 0.89 \text{ A}$$

(c) From Eq. (9-47), $I_s = 0.5/(1 - 0.667) = 1.5$ A and peak inductor current,

$$I_2 = I_s + \frac{\Delta I}{2} = 1.5 + \frac{0.89}{2} = 1.945 \text{ A}$$

(d) From Eq. (9-53),

$$\Delta V_c = \frac{0.5 \times 0.6667}{25,000 \times 220 \times 10^{-6}} = 60.61 \text{ mV}$$

9-7.3 Buck–Boost Regulators

A buck–boost regulator provides an output voltage which may be less than or greater than the input voltage—hence the name "buck–boost"; the output voltage polarity is opposite to that of the input voltage. This regulator is also known as an *inverting regulator*. The circuit arrangement of a buck–boost regulator is shown in Fig. 9-14a.

The circuit operation can be divided into two modes. During mode 1, transistor Q_1 is turned on and diode D_m is reversed biased. The input current, which rises, flows through inductor L and transistor Q_1. During mode 2, transistor Q_1 is switched off and the current, which was flowing through inductor L, would flow through L, C, D_m, and the load. The energy stored in inductor L would be transferred to the load and the inductor current would fall until transistor Q_1 is switched on again in the next cycle. The equivalent circuits for the modes are shown in Fig. 9-14b. The waveforms for steady-state voltages and currents of the buck–boost regulator are shown in Fig. 9-14c for a continuous load current.

Assuming that the inductor current rises linearly from I_1 to I_2 in time t_1,

$$V_s = L\frac{I_2 - I_1}{t_1} = L\frac{\Delta I}{t_1} \tag{9-54}$$

or

$$t_1 = \frac{\Delta I L}{V_s} \tag{9-55}$$

and the inductor current falls linearly from I_2 to I_1 in time t_2,

$$V_a = -L\frac{\Delta I}{t_2} \tag{9-56}$$

or

$$t_2 = \frac{-\Delta I L}{V_a} \tag{9-57}$$

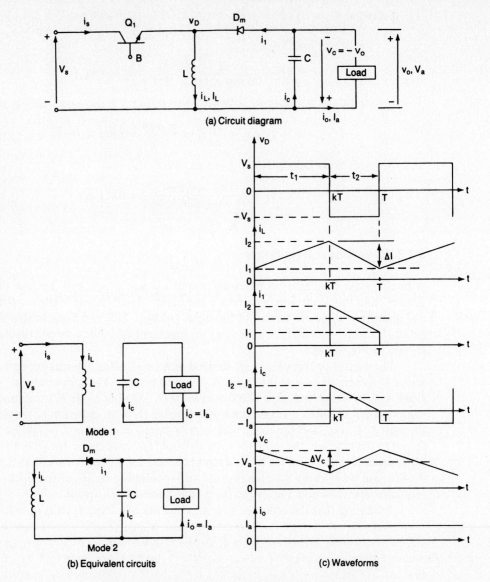

(a) Circuit diagram

Mode 1

Mode 2

(b) Equivalent circuits

(c) Waveforms

Figure 9-14 Buck–boost regulator with continuous i_L.

where $\Delta I = I_2 - I_1$ is the peak-to-peak ripple current of inductor L. From Eqs. (9-54) and (9-56),

$$\Delta I = \frac{V_s t_1}{L} = \frac{-V_a t_2}{L}$$

Substituting $t_1 = kT$ and $t_2 = (1 - k)T$, the average output voltage is

$$V_a = -\frac{V_s k}{1 - k} \tag{9-58}$$

Assuming a lossless circuit, $V_s I_s = -V_a I_a = V_s I_a k/(1 - k)$ and the average input current I_s is related to the average output current I_a by

$$I_s = \frac{I_a k}{1 - k} \qquad (9\text{-}59)$$

The switching period T can be found from

$$T = \frac{1}{f} = t_1 + t_2 = \frac{\Delta I L}{V_s} - \frac{\Delta I L}{V_a} = \frac{\Delta I L (V_a - V_s)}{V_s V_a} \qquad (9\text{-}60)$$

and this gives the peak-to-peak ripple current,

$$\Delta I = \frac{V_s V_a}{f L (V_a - V_s)} \qquad (9\text{-}61)$$

or

$$\Delta I = \frac{V_s k}{f L} \qquad (9\text{-}62)$$

When transistor Q_1 is on, the filter capacitor supplies the load current for $t = t_1$. The average discharging current of the capacitor is $I_c = I_a$ and the peak-to-peak ripple voltage of the capacitor is

$$\Delta V_c = \frac{1}{C} \int_0^{t_1} I_c \, dt = \frac{1}{C} \int_0^{t_1} I_a \, dt = \frac{I_a t_1}{C} \qquad (9\text{-}63)$$

Equation (9-58) gives $t_1 = V_a/[(V_a - V_s)f]$ and Eq. (9-63) becomes

$$\Delta V_c = \frac{I_a V_a}{(V_a - V_s)fC} \qquad (9\text{-}64)$$

or

$$\Delta V_c = \frac{I_a k}{fC} \qquad (9\text{-}65)$$

A buck–boost regulator provides output voltage polarity reversal without a transformer. It has high efficiency. Under a fault condition of the transistor, the di/dt of the fault current is limited by the inductor L and will be V_s/L. Output short-circuit protection would be easy to implement. However, the input current is discontinuous and a high peak current flows through transistor Q_1.

Example 9-6

The buck–boost regulator in Fig. 9-14a has an input voltage of $V_s = 12$ V. The duty cycle $k = 0.25$ and the switching frequency is 25 kHz. The inductance $L = 150$ μH and filter capacitance $C = 220$ μF. The average load current $I_a = 1.25$ A. Determine (a) the average output voltage, V_a; (b) the peak-to-peak output voltage ripple, ΔV_c; (c) the peak-to-peak ripple current of inductor, ΔI; and (d) the peak current of the transistor, I_p.

Solution $V_s = 12$ V, $k = 0.25$, $I_a = 1.25$ A, $f = 25$ kHz, $L = 150$ μH, and $C = 220$ μF.

(a) From Eq. (9-58), $V_a = -12 \times 0.25/(1 - 0.25) = -4$V.

(b) From Eq. (9-65), the peak-to-peak output ripple voltage is

$$\Delta V_c = \frac{1.25 \times 0.25}{25,000 \times 220 \times 10^{-6}} = 56.8 \text{ mV}$$

(c) From Eq. (9-62), the peak-to-peak inductor ripple is

$$\Delta I = \frac{12 \times 0.25}{25,000 \times 150 \times 10^{-6}} = 0.8 \text{ A}$$

(d) From Eq. (9-59), $I_s = 1.25 \times 0.25/(1 - 0.25) = 0.4167$ A. Since I_s is the average of duration kT, the peak-to-peak current of the transistor,

$$I_p = \frac{I_s}{k} + \frac{\Delta I}{2} = \frac{0.4167}{0.25} + \frac{0.8}{2} = 2.067 \text{ A}$$

9-7.4 Cúk Regulators

The circuit arrangement of the Cúk regulator using a power BJT is shown in Fig. 9-15a. Similar to the buck–boost regulator, the Cúk regulator provides an output voltage which is less than or greater than the input voltage, but the output voltage polarity is opposite to that of the input voltage. It is named after its inventor [1]. When the input voltage is turned on and transistor Q_1 is switched off, diode D_m is forward biased and capacitor C_1 is charged through L_1, D_m, and the input supply, V_s.

The circuit operation can be divided into two modes. Mode 1 begins when transistor Q_1 is turned on at $t = 0$. The current through inductor L_1 rises. At the same time, the voltage of capacitor C_1 reverse biases diode D_m and turns it off. The capacitor C_1 discharges its energy to the circuit formed by C_1, C_2, the load, and L_2. Mode 2 begins when transistor Q_1 is turned off at $t = t_1$. The capacitor C_1 is charged from the input supply and the energy stored in the inductor L_2 is transferred to the load. The diode D_m and transistor Q_1 provide a synchronous switching action. The capacitor C_1 is the medium for transferring energy from the source to the load. The equivalent circuits for the modes are shown in Fig. 9-15b and the waveforms for steady-state voltages and currents are shown in Fig. 9-15c for a continuous load current.

Assuming that the current of inductor L_1 rises linearly from I_{L11} to I_{L12} in time t_1,

$$V_s = L_1 \frac{I_{L12} - I_{L11}}{t_1} = L_1 \frac{\Delta I_1}{t_1} \tag{9-66}$$

or

$$t_1 = \frac{\Delta I_1 L_1}{V_s} \tag{9-67}$$

and due to the charged capacitor C_1, the current of inductor L_1 falls linearly from I_{L12} to I_{L11} in time t_2,

$$V_s - V_{c1} = -L_1 \frac{\Delta I_1}{t_2} \tag{9-68}$$

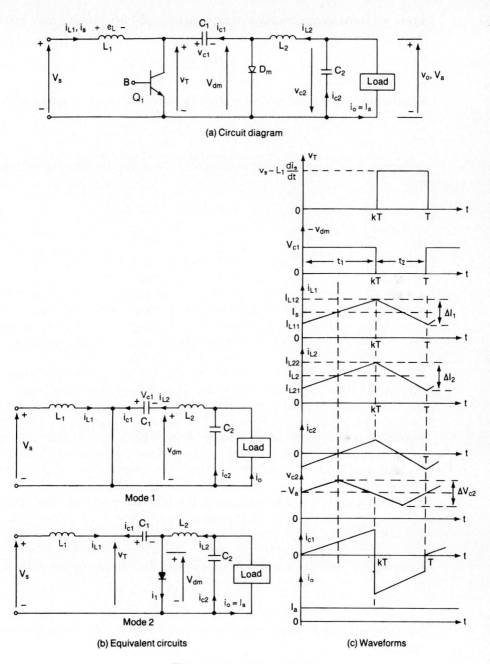

(a) Circuit diagram

Mode 1

Mode 2

(b) Equivalent circuits

(c) Waveforms

Figure 9-15 Cúk regulator.

or

$$t_2 = \frac{-\Delta I_1 L_1}{V_s - V_{c1}} \tag{9-69}$$

where V_{c1} is the average voltage of capacitor C_1, and $\Delta I_1 = I_{L12} - I_{L11}$. From Eqs. (9-66) and (9-68).

$$\Delta I_1 = \frac{V_s t_1}{L_1} = \frac{-(V_s - V_{c1})t_2}{L_1}$$

Substituting $t_1 = kT$ and $t_2 = (1 - k)T$, the average voltage of capacitor C_1 is

$$V_{c1} = \frac{V_s}{1 - k} \qquad (9\text{-}70)$$

Assuming that the current of filter inductor L_2 rises linearly from I_{L21} to I_{L22} in time t_1,

$$V_{c1} + V_a = L_2 \frac{I_{L22} - I_{L21}}{t_1} = L_2 \frac{\Delta I_2}{t_1} \qquad (9\text{-}71)$$

or

$$t_1 = \frac{\Delta I_2 L_2}{V_{c1} + V_a} \qquad (9\text{-}72)$$

and the current of inductor L_2 falls linearly from I_{L22} to I_{L21} in time t_2,

$$V_a = -L_2 \frac{\Delta I_2}{t_2} \qquad (9\text{-}73)$$

or

$$t_2 = -\frac{\Delta I_2 L_2}{V_a} \qquad (9\text{-}74)$$

where $\Delta I_2 = I_{L22} - I_{L21}$. From Eqs. (9-71) and (9-73),

$$\Delta I_2 = \frac{(V_{c1} + V_a)t_1}{L_2} = -\frac{V_a t_2}{L_2}$$

Substituting $t_1 = kT$ and $t_2 = (1 - k)T$, the average voltage of capacitor C_1 is

$$V_{c1} = -\frac{V_a}{k} \qquad (9\text{-}75)$$

Equating Eq. (9-70) to Eq. (9-75), we can find the average output voltage as

$$V_a = -\frac{kV_s}{1 - k} \qquad (9\text{-}76)$$

Assuming a lossless circuit, $V_s I_s = -V_a I_a = V_s I_a k/(1 - k)$ and the average input current,

$$I_s = \frac{kI_a}{1 - k} \qquad (9\text{-}77)$$

The switching period T can be found from Eqs. (9-67) and (9-69):

$$T = \frac{1}{f} = t_1 + t_2 = \frac{\Delta I_1 L_1}{V_s} - \frac{\Delta I_1 L_1}{V_s - V_{c1}} = \frac{-\Delta I_1 L_1 V_{c1}}{V_s(V_s - V_{c1})} \qquad (9\text{-}78)$$

which gives the peak-to-peak ripple current of inductor L_1 as

$$\Delta I_1 = \frac{-V_s(V_s - V_{c1})}{fL_1 V_{c1}} \tag{9-79}$$

or

$$\Delta I_1 = \frac{V_s k}{fL_1} \tag{9-80}$$

The switching period T can also be found from Eqs. (9-72) and (9-74):

$$T = \frac{1}{f} = t_1 + t_2 = \frac{\Delta I_2 L_2}{V_{c1} + V_a} - \frac{\Delta I_2 L_2}{V_a} = \frac{-\Delta I_2 L_2 V_{c1}}{V_a(V_{c1} + V_a)} \tag{9-81}$$

and this gives the peak-to-peak ripple current of inductor L_2 as

$$\Delta I_2 = \frac{-V_a(V_{c1} + V_a)}{fL_2 V_{c1}} \tag{9-82}$$

or

$$\Delta I_2 = -\frac{V_a(1 - k)}{fL_2} = \frac{kV_s}{fL_2} \tag{9-83}$$

When transistor Q_1 is off, the energy transfer capacitor C_1 is charged by the input current for time $t = t_2$. The average charging current for C_1 is $I_{c1} = I_s$ and the peak-to-peak ripple voltage of the capacitor C_1 is

$$\Delta V_{c1} = \frac{1}{C_1} \int_0^{t_2} I_{c1} \, dt = \frac{1}{C_1} \int_0^{t_2} I_s = \frac{I_s t_2}{C_1} \tag{9-84}$$

Equation (9-76) gives $t_2 = V_s/[(V_s - V_a)f]$ and Eq. (9-84) becomes

$$\Delta V_{c1} = \frac{I_s V_s}{(V_s - V_a)fC_1} \tag{9-85}$$

or

$$\Delta V_{c1} = \frac{I_s(1 - k)}{fC_1} \tag{9-86}$$

If we assume that the load current ripple Δi_o is negligible, $\Delta i_{L2} = \Delta i_{c2}$. The average charging current of C_2, which flows for time $T/2$, is $I_{c2} = \Delta I_2/4$ and the peak-to-peak ripple voltage of capacitor C_2 is

$$\Delta V_{c2} = \frac{1}{C_2} \int_0^{T/2} I_{c2} \, dt = \frac{1}{C_2} \int_0^{T/2} \frac{\Delta I_2}{4} = \frac{\Delta I_2}{8fC_2} \tag{9-87}$$

or

$$\Delta V_{c2} = -\frac{V_a(1 - k)}{8C_2 L_2 f^2} = \frac{kV_s}{8C_2 L_2 f^2} \tag{9-88}$$

The Cúk regulator is based on the capacitor energy transfer. As a result, the input current is continuous. The circuit has low switching losses and has high

---

efficiency. When transistor Q_1 is turned on, it has to carry the currents of inductors L_1 and L_2. As a result a high peak current flows through transistor Q_1. Since the capacitor provides the energy transfer, the ripple current of the capacitor C_1 is also high. This circuit also requires an additional capacitor and inductor.

Example 9-7

The input voltage of a Cúk converter in Fig. 9-15a, $V_s = 12$ V. The duty cycle $k = 0.25$ and the switching frequency is 25 kHz. The filter inductance is $L_2 = 150$ μH and filter capacitance is $C_2 = 220$ μF. The energy transfer capacitance is $C_1 = 200$ μF and inductance $L_1 = 180$ μH. The average load current is $I_a = 1.25$ A. Determine (a) the average output voltage V_a; (b) the average input current I_s; (c) the peak-to-peak ripple current of inductor L_1, ΔI_1; (d) the peak-to-peak ripple voltage of capacitor C_1, ΔV_{c1}; (e) the peak-to-peak ripple current of inductor L_2, ΔI_2; (f) the peak-to-peak ripple voltage of capacitor C_2, ΔV_{c2}; and (g) the peak current of the transistor I_p.

Solution $V_s = 12$ V, $k = 0.25$, $I_a = 1.25$ A, $f = 25$ kHz, $L_1 = 180$ μH, $C_1 = 200$ μF, $L_2 = 150$ μH, and $C_2 = 220$ μF.

(a) From Eq. (9-76), $V_a = -0.25 \times 12/(1 - 0.25) = -4$V.

(b) From Eq. (9-77), $I_s = 1.25 \times 0.25/(1 - 0.25) = 0.42$ A.

(c) From Eq. (9-80), $\Delta I_1 = 12 \times 0.25/(25,000 \times 180 \times 10^{-6}) = 0.67$ A.

(d) From Eq. (9-86), $\Delta V_{c1} = 0.42 \times (1 - 0.25)/(25,000 \times 200 \times 10^{-6}) = 63$ mV.

(e) From Eq. (9-83), $\Delta I_2 = 0.25 \times 12/(25,000 \times 150 \times 10^{-6}) = 0.8$ A.

(f) From Eq. (9-87), $\Delta V_{c2} = 0.8/(8 \times 25,000 \times 220 \times 10^{-6}) = 18.18$ mV.

(g) The average voltage across the diode can be found from

$$V_{dm} = -kV_{c1} = -V_a k \frac{1}{-k} = V_a \qquad (9\text{-}89)$$

For a lossless circuit, $I_{L2}V_{dm} = V_a I_a$ and the average value of the current in inductor L_2 is

$$I_{L2} = \frac{I_a V_a}{V_{dm}} = I_a \qquad (9\text{-}90)$$
$$= 1.25 \text{ A}$$

Therefore, the peak current of transistor is

$$I_p = I_s + \frac{\Delta I_1}{2} + I_{L2} + \frac{\Delta I_2}{2} = 0.42 + \frac{0.67}{2} + 1.25 + \frac{0.8}{2} = 2.405 \text{ A}$$

9-7.5 Limitations of Single-Stage Conversion

The four regulators use only one transistor, employing only one stage conversion, and require inductors or capacitors for energy transfer. Due to the current-handling limitation of a single transistor, the output power of these regulators is small, typically tens of watts. At a higher current, the size of these components increases, with increased component losses, and the efficiency decreases. In addition, there is no isolation between the input and output voltage, which is a highly desirable criterion in most applications. For high-power applications, multistage conversions are used, where a dc voltage is converted to ac by an inverter. The ac output is isolated by a transformer and then converted to dc by rectifiers. The multistage conversions are discussed in Section 13-4.

A thyristor chopper circuit uses a fast turn-off thyristor as a switch and requires commutation circuitry to turn it off. There are various techniques by which a thyristor can be turned off and these are described in detail in Chapter 7. During the early development stage of fast-turn-off thyristors, a number of chopper circuits have been published. The various circuits are the outcome of meeting certain criteria: (1) reduction of minimum on-time limit, (2) high frequency of operation, and (3) reliable operation. However, with the development of alternative switching devices (e.g., power transistors, GTOs), the applications of thyristor chopper circuits are limited to high power levels and especially, to traction motor control. Some of the chopper circuits used by traction equipment manufacturers are discussed in this section.

9-8.1 Impulse-Commutated Choppers

The impulse-commutated chopper is a very common circuit with two thyristors as shown in Fig. 9-16 and is also known as a *classical chopper*. At the beginning of operation, thyristor T_2 is fired and this causes the commutation capacitor C to charge through the load to voltage V_c, which should be the supply voltage V_s in the first cycle. The plate A becomes positive with respect to plate B. The circuit operation can be divided into five modes, and the equivalent circuits under steady-state conditions are shown in Fig. 9-17. We shall assume that the load current remains constant at a peak value I_m during the commutation process. We shall also redefine the time origin, $t = 0$, at the beginning of each mode.

Mode 1 begins with T_1 is fired. The load is connected to the supply. The commutation capacitor C reverses also its charge through the resonant reversing circuit formed by T_1, D_1, and L_m. The resonant current is given by

$$i_r = V_c \sqrt{\frac{C}{L_m}} \sin \omega_m t \tag{9-91}$$

The peak value of resonant reversal current is

$$I_p = V_c \sqrt{\frac{C}{L_m}} \tag{9-92}$$

The capacitor voltage is found from

$$v_c(t) = V_c \cos \omega_m t \tag{9-93}$$

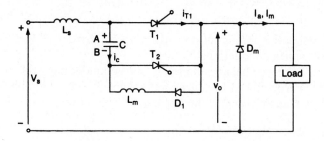

Figure 9-16 Impulse-commutated chopper.

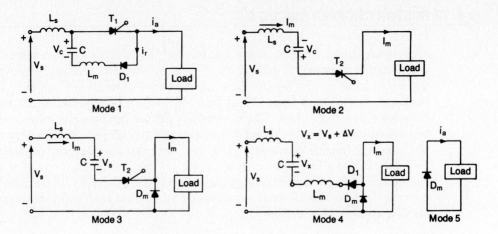

Figure 9-17 Mode equivalent circuits.

where $\omega_m = 1/\sqrt{L_m C}$. After time $t = t_r = \pi\sqrt{L_m C}$, the capacitor voltage is reversed to $-V_c$. This is sometimes called *commutation readiness* of the chopper.

Mode 2 begins when the commutation thyristor T_2 is fired. A reverse voltage of V_c is applied across main thyristor T_1 and it is turned off. The capacitor C discharges through the load from $-V_c$ to zero and this discharging time, which is also called the *circuit (or available) turn-off time*, is given by

$$t_{\text{off}} = \frac{V_c C}{I_m} \tag{9-94}$$

where I_m is the peak load current. The circuit turn-off time t_{off} must be greater than the turn-off time of the thyristor, t_q. t_{off} varies with the load current and must be designed for the worst-case condition, which occurs at the maximum value of load current and the minimum value of capacitor voltage.

The time required for the capacitor to recharge back to the supply voltage is called the *recharging time* and is given by

$$t_d = \frac{V_s C}{I_m} \tag{9-95}$$

Thus the total time necessary for the capacitor to discharge and recharge is called the *commutation time*, which is

$$t_c = t_{\text{off}} + t_d \tag{9-96}$$

This mode ends at $t = t_c$ when the commutation capacitor C recharges to V_s and the freewheeling diode D_m starts conducting.

Mode 3 begins when the freewheeling diode D_m starts conducting and the load current decays. The energy stored in source inductance L_s (plus any stray inductance in the circuit) is transferred into the capacitor. The current is

$$i_s(t) = I_m \cos \omega_s t \tag{9-97}$$

and the instantaneous capacitor voltage is

$$v_c(t) = V_s + I_m \sqrt{\frac{L_s}{C}} \sin \omega_s t \tag{9-98}$$

where $\omega_s = 1/\sqrt{L_s C}$. After time $t = t_s = 0.5 \pi \sqrt{L_s C}$, this overcharging current becomes zero and the capacitor is recharged to

$$V_x = V_s + \Delta V \tag{9-99}$$

where ΔV and V_x are the overvoltage and peak voltage of commutation capacitor, respectively. Equation (9-98) gives the overcharging voltage as

$$\Delta V = I_m \sqrt{\frac{L_s}{C}} \tag{9-100}$$

Mode 4 begins when the overcharging is complete and the load current continues to decay. It is important to note that this mode exists due to diode D_1, because it allows the resonant oscillation in mode 3 to continue through the circuit formed by D_m, D_1, C, and the supply. This will undercharge the commutation capacitor C and the undercharging current through the capacitor is given by

$$i_c(t) = -\Delta V \sqrt{\frac{C}{(L_s + L_m)}} \sin \omega_u t \tag{9-101}$$

The commutation capacitor voltage is

$$v_c(t) = V_x - \Delta V(1 - \cos \omega_u t) \tag{9-102}$$

where $\omega_u = 1/\sqrt{C(L_s + L_m)}$. After time $t = t_u = \pi \sqrt{C(L_s + L_m)}$, the undercharging current becomes zero and diode D_1 stops conducting. Equation (9-102) gives the available commutation voltage of the capacitor as

$$V_c = V_x - 2\Delta V = V_s - \Delta V \tag{9-103}$$

If there is no overcharge, there will not be any undercharge.

Mode 5 begins when the commutation process is complete and the load current continues to decay through diode D_m. This mode ends when the main thyristor is refired at the beginning of next cycle. The different waveforms for the currents and voltages are shown in Fig. 9-18.

The average output voltage of the chopper is

$$V_o = \frac{1}{T}\left[V_s kT + t_c \frac{1}{2}(V_c + V_s)\right] \tag{9-104}$$

It could be noticed from Eq. (9-104) that even at $k = 0$, the output voltage becomes

$$V_o(k = 0) = 0.5 f t_c (V_c + V_s) \tag{9-105}$$

This limits the minimum output voltage of the chopper. However, the thyristor T_1 must be on for a minimum time of $t_r = \pi \sqrt{L_m C}$ to allow the charge reversal of the capacitor and t_r is fixed for a particular circuit design. Therefore, the minimum duty cycle and minimum output voltage are also set.

$$t_r = k_{\min} T = \pi \sqrt{L_m C} \tag{9-106}$$

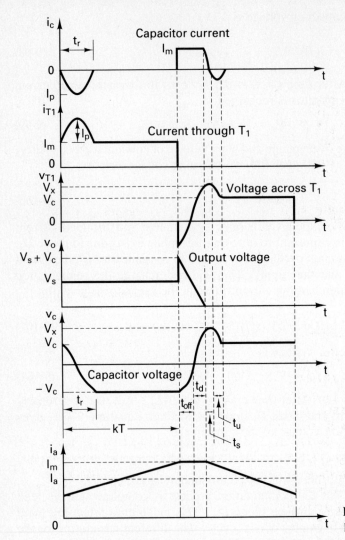

Figure 9-18 Waveforms for impulse-commutated chopper.

The minimum duty cycle

$$k_{\min} = t_r f = \pi f \sqrt{L_m C} \tag{9-107}$$

The minimum average output voltage

$$V_{o(\min)} = k_{\min} V_s + 0.5 t_c (V_c + V_s) f$$
$$= f[V_s t_r + 0.5 t_c (V_c + V_s)] \tag{9-108}$$

The minimum output voltage, $V_{o(\min)}$, can be varied by controlling the chopping frequency. Normally, $V_{o(\min)}$ is fixed by the design requirement to a permissible value.

The maximum value of the duty cycle is also limited in order to allow the commutation capacitor to discharge and recharge. The maximum value of this duty cycle is given by

$$k_{max}T = T - t_c - t_s - t_u$$

and

$$k_{max} = 1 - \frac{t_c + t_s + t_u}{T}$$

(9-109)

The maximum output voltage

$$V_{o(max)} = k_{max}V_s + 0.5t_c(V_c + V_s)f$$

(9-110)

An ideal thyristor chopper should have no limit on (1) the minimum on-time, (2) the maximum on-time, (3) the minimum output voltage, and (4) the maximum chopping frequency. The turn-off time t_{off} should be independent of the load current. At higher frequency, the load ripple currents and the supply harmonic current become smaller. In addition, the size of the input filter is reduced.

This chopper circuit is very simple and requires two thyristors and one diode. However, the main thyristor T_1 has to carry the resonant reversal current, thereby increasing its peak current rating and limiting the mininum output voltage. The discharging and charging time of commutation capacitor are dependent on the load current, and this limits the high-frequency operation, especially at a low load current. This chopper cannot be tested without connecting the load. This circuit has many disadvantages. However, it points out the problems of thyristor commutation.

Note. The available turn-off time t_{off}, commutation time t_c, and the overvoltage depend on the peak load current I_m rather than on the average value I_a.

Example 9-8

A highly inductive load controlled by the chopper in Fig. 9-16 requires an average current of $I_a = 425$ A with a peak value of $I_m = 450$ A. The input supply voltage is $V_s = 220$ V. The chopping frequency is $f = 400$ Hz and the turn-off time of the main thyristor is $t_q = 18$ μs. If the peak current through the main thyristor is limited to 180% of I_m and the source inductance (including the stray inductance) is negligible ($L_s = 0$), determine (a) the commutation capacitance C, (b) the inductance L_m, and (c) the minimum and maximum output voltage.

Solution $I_a = 425$ A, $I_m = 450$ A, $f = 400$ Hz, $t_q = 18$ μs, and $L_s = 0$.

(a) From Eqs. (9-99), (9-100), and (9-103), the overvoltage is $\Delta V = 0$ and $V_c = V_x = V_s = 220$ V. From Eq. (9-94), the turn-off requirement gives

$$t_{off} = \frac{V_c C}{I_m} > t_q$$

and $C > I_m t_q / V_c = (450 \times 18/220) = 36.8$ μF. Let $C = 40$ μF.

(b) From Eq. (9-92), the peak resonant current

$$I_p = 1.8 \times 450 - 450 = 220 \sqrt{\frac{40\ \mu F}{L_m}}$$

which gives inductance, $L_m = 14.94$ μH.

(c) From Eq. (9-94), discharging time, $t_{off} = (220 \times 40)/450 = 19.56$ μs. From Eq. (9-95), recharging time, $t_d = (220 \times 40)/450 = 19.56$ μs. From Eq. (9-96), total

---

time, $t_c = 19.56 \times 2 = 39.12 \; \mu s$. From Eq. (9-106), resonant reversal time,

$$t_r = \pi[(14.94 \times 40) \times 10^{-12}]^{1/2} = 76.8 \; \mu s$$

From Eq. (9-107), the minimum duty cycle, $k_{min} = t_r f = 0.0307 = 3.07\%$. From Eq. (9-108), the minimum output voltage.

$$V_{o(min)} = 0.0307 \times 220 + 0.5 \times 39.12 \times 10^{-6} \times 2 \times 220 \times 400$$
$$= 6.75 + 3.44 = 10.19 \; V$$

Since there is no overcharging, there will be no overcharging period, and the overcharging and undercharging time, $t_u = t_s = 0$. From Eq. (9-109), the maximum duty cycle, $k_{max} = 1 - (t_c + t_u + t_s)f = 0.984$; and from Eq. (9-110), the maximum output voltage,

$$V_{o(max)} = 0.984 \times 220 + 0.5 \times 39.12 \times 10^{-6} \times 2 \times 220 \times 400$$
$$= 216.48 + 3.44 = 219.92 \; V$$

9-8.2 Effects of Source and Load Inductance

The source inductance plays a significant role on the operation of the chopper, and this inductance should be as small as possible to limit the transient voltage within an acceptable level. It is evident from Eq. (9-100) that the commutation capacitor is overcharged due to the source inductance L_s, and the semiconductor devices will be subjected to this capacitor voltage. If the minimum value of the source inductance cannot be guaranteed, an input filter is required. In practical systems, the stray inductance always exists and its value depends on the type of wiring and the layout of the components. Therefore, L_s in Eq. (9-100) has a finite value and the capacitor gets always overcharged.

Due to inductance L_s and diode D_1 in Fig. 9-16, the capacitor also gets undercharged and this may cause commutation problem of the chopper. Equation (9-20) indicates that the load ripple current is an inverse function of load inductance and chopping frequency. Hence the peak load current is dependent on the load inductance. Therefore, the performances of the chopper are also influenced by the load inductance. A smoothing choke is normally connected in series with the load to limit the load ripple current.

Example 9-9

If the supply in Example 9-8 has an inductance of $L_s = 4 \; \mu H$, determine (a) the peak capacitor voltage V_x, (b) the available turn-off time t_{off}, and (c) the commutation time t_c.

Solution $I_a = 425 \; A$, $I_m = 450 \; A$, $V_s = 220 \; V$, $f = 400 \; Hz$, $t_q = 18 \; \mu s$, $L_s = 4 \; \mu H$, and $C = 40 \; \mu F$.

(a) From Eq. (9-100), overvoltage, $\Delta V = 450 \times \sqrt{4/40} = 142.3 \; V$. From Eq. (9-99), peak capacitor, $V_x = 220 + 142.3 = 362.3 \; V$ and from Eq. (9-103), available commutation voltage, $V_c = 220 - 142.3 = 77.7$.

(b) From Eq. (9-94), available turn-off time, $t_{off} = (77.7 \times 40)/450 = 6.9 \; \mu s$.

(c) From Eq. (9-95), recharging time, $t_d = (220 \times 40)/450 = 19.56 \; \mu s$ and from Eq. (9-96), commutation time, $t_c = 6.0 + 19.56 = 26.46 \; \mu s$.

Note. The turn-off requirement of the main thyristor is 18 μs, whereas the available turn-off time is only 6.9 μs. Therefore, a commutation failure will occur.

9-8.3 Impulse-Commutated Three-Thyristor Choppers

This problem of undercharging can be remedied by replacing diode D_1 with thyristor T_3, as shown in Fig. 9-19. In a good chopper, the commutation time, t_c, should ideally be independent of the load current. t_c could be made less dependent on the load current by adding an antiparallel diode D_f across the main thyristor as shown in Fig. 9-19 by dashed lines. A modified version of the circuit is shown in Fig. 9-20, where the charge reversal of the capacitor is done independently of main thyristor T_1 by firing thyristor T_3. There are four possible modes and their equivalent circuits are shown in Fig. 9-21.

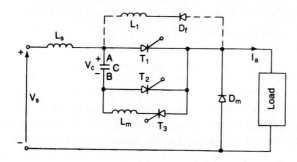

Figure 9-19 Impulse-commutated three-thyristor chopper.

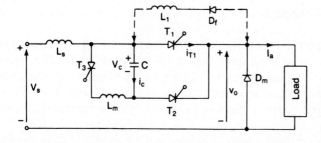

Figure 9-20 Impulse-commutated chopper with independent charge reversal.

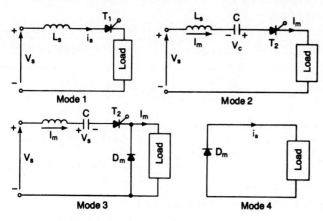

Figure 9-21 Equivalent circuits.

Mode 1 begins when the main thyristor T_1 is fired and the load is connected to the supply. Thyristor T_3 may be fired at the same time as T_1 to reverse the charge on capacitor C. If this charge reversal is done independently, the minimum output voltage would not be limited due to the resonant reversal as in the case of classical chopper in Fig. 9-16.

Mode 2 begins when commutation thyristor T_2 is fired and capacitor C discharges and recharges through the load at a rate determined by the load current.

Mode 3 begins when the capacitor is recharged to the supply voltage and the freewheeling diode D_m starts conducting. During this mode the capacitor overcharges due to the energy stored in the source inductance, L_s, and the load current decays through D_m. This mode ends when the overcharging current reduces to zero.

Mode 4 begins when the thyristor T_2 stops conducting. The freewheeling diode D_m continues to conduct and the load current continues to decay.

All the equations for the classical chopper except Eqs. (9-101), (9-102), and (9-103) are valid for this chopper, and mode 4 of the classical chopper is not applicable. The available commutation voltage

$$V_c = V_x = V_s + \Delta V \qquad (9\text{-}111)$$

For the chopper in Fig. 9-20, the resonant reversal is independent of the main thyristor and the minimum on-time of the chopper is not limited. However, the commutation time is dependent on the load current and the high-frequency operation is limited. The chopper circuit cannot be tested without connecting the load.

9-8.4 Resonant Pulse Choppers

A resonant pulse chopper is shown in Fig. 9-22. As soon as the supply is switched on, the capacitor is charged to a voltage V_c through L_m, D_1, and load. The circuit operation can be divided into six modes and the equivalent circuits are shown in Fig. 9-23. The waveforms for currents and voltages are shown in Fig. 9-24. In the following analysis, we shall redefine the time origin $t = 0$ at the beginning of each mode.

Mode 1 begins when main thyristor T_1 is fired and the supply is connected to the load. This mode is valid for $t = kT$.

Mode 2 begins when commutation thyristor T_2 is fired. The commutation capacitor reverses its charge through C, L_m, and T_2. The reversal current is given

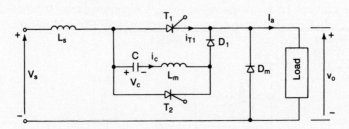

Figure 9-22 Resonant pulse chopper.

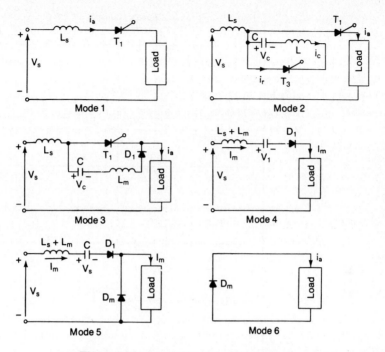

Figure 9-23 Equivalent circuits for modes.

by

$$i_r = -i_c = V_c \sqrt{\frac{C}{L_m}} \sin \omega_m t = I_p \sin \omega_m t \qquad (9\text{-}112)$$

and the capacitor voltage is

$$v_c(t) = V_c \cos \omega_m t \qquad (9\text{-}113)$$

where $\omega_m = 1/\sqrt{L_m C}$. After time, $t = t_r = \pi\sqrt{L_m c}$, the capacitor voltage is reversed to $-V_c$. However, the resonant oscillation continues through diode D_1 and T_1. The peak resonant current, I_p, must be greater than load current I_m and the circuit is normally designed for a ratio of $I_p/I_m = 1.5$.

Mode 3 begins when T_2 is self-commutated and the capacitor discharges due to resonant oscillation through diode D_1 and T_1. This mode ends when the capacitor current rises to the level of I_m. Assuming that the capacitor current rises linearly from 0 to I_m and that the current of thyristor T_1 falls from I_m to 0 in time t_x, the time duration for this mode is

$$t_x = \frac{L_m I_m}{V_c} \qquad (9\text{-}114)$$

and the capacitor voltage falls to

$$V_1 = V_c - \frac{t_x I_m}{2C} = V_c - \frac{L_m I_m^2}{2CV_c} \qquad (9\text{-}115)$$

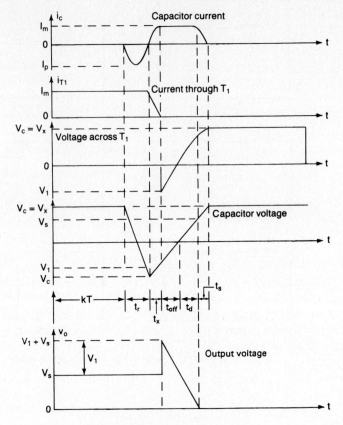

Figure 9-24 Chopper waveforms.

Mode 4 begins when the current through T_1 falls to zero. The capacitor continues to discharge through the load at a rate determined by the peak load current. The available turn-off time,

$$t_{\text{off}} = \frac{V_1 C}{I_m} \qquad (9\text{-}116)$$

The time required for the capacitor to recharge to the supply voltage,

$$t_d = \frac{V_s C}{I_m} \qquad (9\text{-}117)$$

The total time to discharge and recharge to the supply V_s is $t_c = t_{\text{off}} + t_d$.

Mode 5 begins when the freewheeling diode D_m starts conducting and the load current decays through D_m. The energy stored in commutation inductance L_m and source inductance L_s is transferred to capacitor C. After time $t_s = \pi \sqrt{(L_s + L_m)C}$, the overcharging current becomes zero and the capacitor is recharged to

$$V_x = V_s + \Delta V \qquad (9\text{-}118)$$

where

$$\Delta V = I_m \sqrt{\frac{L_m + L_s}{C}} \tag{9-119}$$

Mode 6 begins when the overcharging is complete and diode D_1 turns off. The load current continues to decay until the main thyristor is refired in the next cycle. In the steady-state condition $V_c = V_x$. The average output voltage is given by

$$V_o = \frac{1}{T} [V_s kT + V_s(t_r + t_x) + 0.5t_c(V_1 + V_s)] \tag{9-120}$$

$$= V_s k + f[(t_r + t_x)V_s + 0.5t_c(V_1 + V_s)]$$

Although the circuit does not have any constraint on the minimum value of duty cycle k, in practice the value of k cannot be zero. The maximum value of k is

$$k_{\max} = 1 - (t_r + t_x + t_c)f \tag{9-121}$$

Due to commutation by a resonant pulse, the reverse di/dt of thyristor T_1 is limited by inductor L_m, and this is also known as *soft commutation*. The resonant reversal is independent of thyristor T_1. However, the inductance, L_m, overcharges capacitor C and this will increase the voltage ratings of the components. After thyristor T_2 is fired, the capacitor has to reverse its charge before turning off thyristor T_1. There is an inherent delay in commutation and this limits the minimum on-time of the chopper. The commutation time t_c is dependent on the load current.

Example 9-10

A highly inductive load that is controlled by the chopper in Fig. 9-22 requires an average current $I_a = 425$ A with a peak value, $I_m = 450$ A. The supply voltage $V_s = 220$ V. The chopping frequency $f = 400$ Hz, commutation inductance $L_m = 8$ μH, and commutation capacitance $C = 40$ μF. If the source inductance (including the stray inductance) is $L_s = 4$ μH, determine (a) the peak resonant current I_p; (b) the peak load voltage V_x, (c) the turn-off time t_{off}, and (d) the minimum and maximum output voltage.

Solution The reversing time, $t_r = \pi \sqrt{8 \times 40} = 56.2$ μs. From Eq. (9-119), overvoltage, $\Delta V = 450 \sqrt{(8 + 4)/40} = 246.5$ V and from Eq. (9-118), peak capacitor voltage, $V_c = V_x = 220 + 246.5 = 466.5$ V.

(a) From Eq. (9-112), $I_p = 466.5 \sqrt{40/8} = 1043.1$ A.

(b) From Eq. (9-114), $t_x = 8 \times 450/466.5 = 7.72$ μs and from Eq. (9-115), the peak load voltage

$$V_1 = 466.5 - \frac{8 \times 450 \times 450}{2 \times 40 \times 466.5} = 423.1 \text{ V}$$

(c) From Eq. (9-116), the turn-off time $t_{\text{off}} = 423.1 \times 40/450 = 37.6$ μs.

(d) From Eq. (9-117), $t_d = 220 \times 40/450 = 19.6$ μs and $t_c = 37.6 + 19.6 = 57.2$ μs. From Eq. (9-121), the maximum duty cycle

$$k_{\max} = 1 - (56.2 + 7.72 + 57.2) \times 400 \times 10^{-6} = 0.952$$

For $k = k_{max}$, Eq. (9-120) gives the maximum output voltage

$$V_{o(max)} = 220 \times 0.952 + 400 \times [(56.2 + 7.72) \times 220 + 0.5 \times 57.2$$
$$\times (423.1 + 220)] \times 10^{-6} = 209.4 + 12.98 = 222.4 \text{ V}$$

The minimum output voltage (for $k = 0$) $V_{o(min)} = 12.98$ V.

9-9 CHOPPER CIRCUIT DESIGN

The major requirement for the design of commutation circuit is to provide an adequate turn-off time to switch the main thyristor off. The analyses of the mode equations for the classical chopper in Section 9-8.1 and the resonant pulse chopper in Section 9-8.4 show that the turn-off time depends on the voltage of commutation capacitor V_c.

It is much simpler to design the commutation circuit if the source inductance can be neglected or the load current is not high. But in the case of higher load current, stray inductances, which are always present in practical systems, play a significant part in the design of the commutation circuit because the energy stored in the circuit inductance increases as the square of the peak load current. The source inductance makes the design equations nonlinear and an iterative method of solution is required to determine the commutation components. The voltage stresses on the power devices depend on the source inductance and load current.

There are no fixed rules for the design of a chopper circuit and the design varies with the types of circuits used. The designer has a wide range of choice and the values of L_mC components are influenced by the designer's choice of peak resonant reversal current, and peak allowable voltage of the circuit. The voltage and current ratings of L_mC components and devices give the minimum limits, but the actual selection of components and devices is left to the designer based on the considerations of price, availability, and safety margin. In general, the following steps are involved in the design:

1. Identify the modes of operation for the chopper circuit.
2. Determine the equivalent circuits for the various modes.
3. Determine the currents and voltages for the modes and their waveforms.
4. Evaluate the values of commutation components L_mC that would satisfy the design limits.
5. Determine the voltage and current rating requirements of all components and devices.

We can notice from Eq. (9-7) that the output voltage contains harmonics. An output filter of C, LC, L type may be connected to the output in order to reduce the output harmonics. The techniques for filter design are similar to that of Examples 3-21 and 5-14.

A chopper with a highly inductive load is shown in Fig. 9-25a. The load current ripple is negligible ($\Delta I = 0$). If the average load current is I_a, the peak load

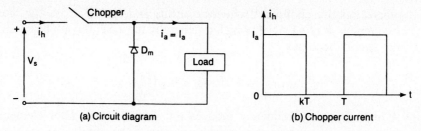

| (a) Circuit diagram | (b) Chopper current |

Figure 9-25   Input current waveform of chopper.

current is $I_m = I_a + \Delta I = I_a$. The input current, which is of pulsed shape as shown in Fig. 9-25b, contains harmonics and can be expressed in Fourier series as

$$i_{nh}(t) = kI_a + \frac{I_a}{n\pi} \sum_{n=1}^{\infty} \sin 2n\pi k \cos 2n\pi ft$$

$$+ \frac{I_a}{n\pi} \sum_{n=1}^{\infty} (1 - \cos 2n\pi k) \sin 2n\pi ft \qquad (9\text{-}122)$$

The fundamental component ($n = 1$) of the chopper-generated harmonic current at the input side is given by

$$i_{1h}(t) = \frac{I_a}{\pi} \sin 2\pi k \cos 2\pi ft + \frac{I_a}{\pi} (1 - \cos 2\pi k) \sin 2\pi ft \qquad (9\text{-}123)$$

In practice, an input filter is shown in Fig. 9-26 is normally connected to filter out the chopper generated harmonics from the supply line. The equivalent circuit for the chopper-generated harmonic currents is shown in Fig. 9-27, and the rms value of the nth harmonic component in the supply can be calculated from

$$I_{ns} = \frac{1}{1 + (2n\pi f)^2 L_e C_e} I_{nh} = \frac{1}{1 + (nf/f_0)^2} I_{nh} \qquad (9\text{-}124)$$

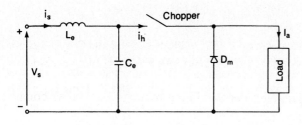

Figure 9-26   Chopper with input filter.

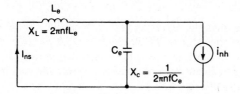

Figure 9-27   Equivalent circuit for harmonic currents.

where f is the chopping frequency and $f_0 = 1/(2\pi\sqrt{L_eC_e})$ is the filter resonant frequency. If $(f/f_0) >> 1$, which is generally the case, the nth harmonic current in the supply becomes

$$I_{ns} = I_{nh}\left(\frac{f_0}{nf}\right)^2 \tag{9-125}$$

A high chopping frequency reduces the sizes of input filter elements. But the frequencies of chopper-generated harmonics in the supply line are also increased, and this may cause interference problems with control and communication signals.

If the source has some inductances, L_s, and the chopper switch as in Fig. 9-1a is turned on, an amount of energy will be stored in the source inductance. If an attempt is made to turn off the chopper switch, the power semiconductor devices might be damaged due to an induced voltage resulting from this stored energy. The LC input filter provides a low-impedance source for the chopper action.

Example 9-11

It is required to design the impulse-commutated chopper in the circuit of Fig. 9-19. It operates from a supply voltage of $V_s = 220$ V and the peak load current is $I_m = 440$ A. The minimum output voltage should be less than 5% of V_s, the peak resonant current should be limited to 80% of I_m, the turn-off time requirement is $t_{off} = 25$ μs, and the source inductance is $L_s = 4$ μH. Determine (a) the values of L_mC components, (b) the maximum allowable chopping frequency, and (c) the ratings of all components and devices. Assume negligible load ripple current.

Solution $V_s = 220$ V, $I_m = 440$ A, $t_{off} = 25$ μs, $L_s = 4$ μH, and $V_{o(min)} = 0.05 \times 220 = 11$ V. The waveforms for the various currents and capacitor voltage are shown in Fig. 9-28.

(a) From Eqs. (9-94), (9-99), and (9-100), the turn-off time is

$$t_{off} = \frac{V_cC}{I_m} = \left(V_s + I_m\sqrt{\frac{L_s}{C}}\right)\frac{C}{I_m} = \frac{V_sC}{I_m} + \sqrt{L_sC}$$

or

$$\left(t_{off} - \frac{V_sC}{I_m}\right)^2 = t_{off}^2 + \left(\frac{V_sC}{I_m}\right)^2 - \frac{2V_sCt_{off}}{I_m} = L_sC$$

Substituting the numerical values, $0.25C^2 - 29C + 625 = 0$ and $C = 87.4$ μF or 28.6 μF. Choose the lowest value, $C = 28.6$ μF, and let $C = 30$ μF.

(b) From Eq. (9-100), the overvoltage is $\Delta V = 440\sqrt{4/30} = 160$ V, and from Eq. (9-99), the capacitor voltage, $V_c = V_x = 220 + 160 = 380$ V. From Eq. (9-92), the peak resonant current,

$$I_p = 380\sqrt{\frac{30}{L_m}} = 0.8 \times 440 = 352 \qquad \text{or} \qquad L_m = 34.96 \ \mu\text{H}$$

Let $L_m = 35$ μH; then the reversing time, $t_r = \pi\sqrt{35 \times 30} = 101.8$ μs. From Eq. (9-94), the turn-off time, $t_{off} = 380 \times 30/440 = 25.9$ μs, from Eq. (9-95), $t_d = 220 \times 30/440 = 15$ μs. From Eq. (9-96), the commutation time, $t_c = 25.9 + 15 = 40.9$ μs. The chopping frequency can be determined from the condition of minimum voltage to satisfy Eq. (9-108):

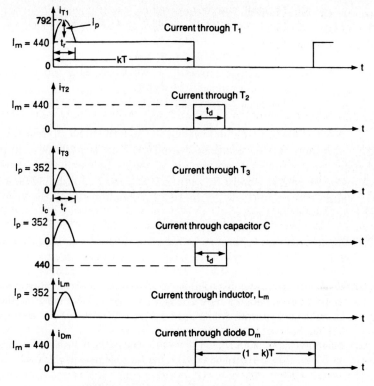

Figure 9-28   Waveforms for Example 9-11.

$$11 = f[220 \times 101.8 + 0.5 \times 40.9 \times (380 + 220)] \times 10^{-6} \text{ or } f = 317 \text{ Hz}$$

The maximum chopping frequency is $f = 317$ Hz; let $f = 300$ Hz.

(c) At this stage we have all the particulars to determine the ratings.

T_1: The average current $I_{av} = 440$ A (assuming duty cycle, $k \cong 1$.)
The peak current $I_p = 440 + 0.8 \times 440 = 792$ A.
The maximum rms current due to the load, $I_{r1} = 440$ A.
The rms current due to resonant reversal,

$$I_{r2} = 0.8 \times 440 \sqrt{t_r f/2} = 0.352 \sqrt{101.8 \times 300/2} = 43.5 \text{ A}$$

The effective rms current $I_{rms} = (440^2 + 43.5^2)^{1/2} = 442.14$ A.

T_s: The peak current $I_p = 440$ A.
The rms current $I_{rms} = 440\sqrt{ft_c} = 0.44\sqrt{300 \times 40.9} = 48.7$ A.
The average current $I_{av} = I_p t_c f = 440 \times 40.9 \times 300 \times 10^{-6} = 5.4$ A.

T_3: The peak current $I_p = 0.8 \times 440 = 352$ A.
The rms current $I_{rms} = I_p\sqrt{ft_r/2} = 0.352\sqrt{101.8 \times 300/2} = 43.5$ A.
The average current

$$I_{av} = 2I_p ft_r/\pi = 2 \times 352 \times 300 \times 101.8 \times 10^{-6}/\pi = 6.84 \text{ A}$$

C: The value of capacitance $C = 30$ μF.
The peak-to-peak voltage $V_{pp} = 2 \times 380 = 760$ V.

The peak current I_p = 440 A.
The rms current I_{rms} = $(48.7^2 + 43.5^2)^{1/2}$ = 65.3 A.

L_m: The peak current I_p = 352 A.
The rms current I_{rms} = 43.5 A.

D_m: The average current I_{av} = 440 A (assuming duty cycle, $k \cong 0$).
The rms current I_{rms} = 440 A.
The peak current I_p = 440 A.

Note. Due to resonant reversal through the main thyristor, its effective rms current ratings and losses are increased. The main thyristor as in Fig. 9-20 can be avoided in the reversal process. If V_s varies between $V_{s(min)}$ and $V_{s(max)}$ and L_s varies between $L_{s(min)}$ and $L_{s(max)}$, then $V_{s(min)}$ and $L_{s(min)}$ should be used to calculate the values of L_m and C. $V_{s(max)}$ and $L_{s(max)}$ should be used to determine the ratings of components and devices.

Example 9-12

It is required to design the resonant pulse chopper circuit in Fig. 9-22. It operates from a supply voltage of V_s = 220 V and a peak load current, I_m = 440 A. The peak resonant current should be limited to 150% of I_m; the turn-off time requirement, t_{off} = 25 μs, and the source inductance, L_s = 4 μH. Determine (a) the values of L_mC components, (b) the overcharged voltage ΔV, and (c) the available commutation voltage V_c.

Solution I_m = 440 A, I_p = 1.5 × 440 = 660 A, L_s = 4 μH, t_{off} = 25 μs, and V_s = 220 V. From Eqs. (9-115) and (9-116), the turn-off time is given as

$$t_{off} = \frac{V_c C}{I_m} - \frac{L_m I_m}{2V_c}$$

From Eq. (9-112), the peak resonant current, $I_p = V_c\sqrt{C/L_m}$. From Eqs. (9-118) and (9-119), the capacitor voltage,

$$V_c = V_x = V_s + I_m \sqrt{\frac{L_s + L_m}{C}}$$

Substituting for $V_c = I_p\sqrt{L_m/C}$ gives t_{off} as

$$t_{off} = \sqrt{CL_m}\left(\frac{I_p}{I_m} - \frac{I_m}{2I_p}\right) = \sqrt{CL_m}\left(x - \frac{1}{2x}\right) \qquad (9\text{-}126)$$

where $x = I_p/I_m$. Substituting for V_c in $I_p = V_c\sqrt{C/L_m}$ gives

$$I_p = \sqrt{\frac{C}{L_m}}\left(V_s + I_m\sqrt{\frac{L_s + L_m}{C}}\right) = V_s\sqrt{\frac{C}{L_m}} + I_m\sqrt{1 + \frac{L_s}{L_m}} \qquad (9\text{-}127)$$

Solving $\sqrt{CL_m}$ from Eq. (9-127) and substituting it in Eq. (9-126), we get

$$t_{off} = \frac{L_m I_m}{V_s}\left(x - \sqrt{1 + \frac{L_s}{L_m}}\right)\left(x - \frac{1}{2x}\right) \qquad (9\text{-}128)$$

which can be solved for L_m by iteration, where L_m is increased by a small amount until the desired value of t_{off} is obtained. Once L_m is found, C can be determined from Eq. (9-126).

Find the values of L_m and C that would satisfy the conditions of t_{off} and I_p. An iterative method of solution yields:

(a) $L_m = 25.29$ μH, $C = 18.16$ μF.
(b) $\Delta V = 558.86$ V.
(c) $V_c = 220 + 558.86 = 778.86$ V, and Eq. (9-115) gives $V_1 = 605.63$ V.

Note. For $L_s = 0$, $L_m = 21.43$ μH, $C = 21.43$ μF, $\Delta V = 440$ V, $V_c = 660$ V, and $V_1 = 513.33$ V.

Example 9-13

A highly inductive load is supplied by a chopper. The average load current is $I_a = 100$ A and the load ripple current can be considered negligible ($\Delta I = 0$). A simple LC input filter with $L_e = 0.3$ mH and $C_e = 4500$ μF is used. If the chopper is operated at a frequency of 350 Hz and a duty cycle of 0.5, determine the maximum rms value of the fundamental component of chopper-generated harmonic current in the supply line.

Solution For $I_a = 100$ A, $f = 350$ Hz, $k = 0.50$, $C_e = 4500$ μF, and $L_e = 0.3$ mH, $f_0 = 1/(2\pi\sqrt{C_eL_e}) = 136.98$ Hz. Equation (9-123) can be written as

$$I_{1h}(t) = A_1 \cos 2\pi ft + B_1 \sin 2\pi ft$$

where $A_1 = (I_a/\pi) \sin 2\pi k$ and $B_1 = (I_a/\pi)(1 - \cos 2\pi k)$. The peak magnitude of this current is calculated from

$$I_{ph} = (A_1^2 + B_1^2)^{1/2} = \frac{\sqrt{2}\, I_a}{\pi}(1 - \cos 2\pi k)^{1/2}$$

The rms value of this current is

$$I_{1h} = \frac{I_a}{\pi}(1 - \cos 2\pi k)^{1/2} = 45.02 \text{ A}$$

and this becomes maximum at $k = 0.5$. The fundamental component of chopper-generated harmonic current in the supply can be calculated from Eq. (9-124) and is given by

$$I_{1s} = \frac{1}{1 + (f/f_0)^2} I_{1h} = \frac{45.02}{1 + (350/136.98)^2} = 5.98 \text{ A}$$

If $f/f_0 \gg 1$, the harmonic current in the supply becomes approximately

$$I_{1s} = I_{1h}\left(\frac{f_0}{f}\right)^2$$

Example 9-14

A buck chopper is shown in Fig. 9-29. The input voltage is $V_s = 110$ V, the average load voltage is $V_a = 60$ V, and the average load current is $I_a = 20$ A. The chopping frequency is $f = 20$ kHz. The peak-to-peak ripples are 2.5% for load voltage, 5% for load current, and 10% for filter L_e current. (a) Determine the values of L_e, L, and C_e. Use PSpice (b) to verify the results by plotting the instantaneous capacitor voltage v_C, and instantaneous load current i_L, and (c) to calculate the Fourier coefficients and the input current i_S. The SPICE model parameters of the transistor are IS=6.734f, BF=416.4, BR=.7371, CJC=3.638P, CJE=4.493P, TR=239.5N, TF=301.2P, and that of the diode are IS=2.2E−15, BV=1800V, TT=0.

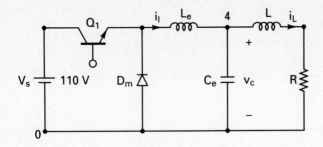

Figure 9-29 Buck chopper.

Solution $V_s = 110$ V, $V_a = 60$ V, $I_a = 20$ A.

$$\Delta V_c = 0.025 \times V_a = 0.025 \times 60 = 1.5 \text{ V}$$

$$R = \frac{V_a}{I_a} = \frac{60}{20} = 3 \text{ } \Omega$$

From Eq. (9-34),

$$k = \frac{V_a}{V_s} = \frac{60}{110} = 0.5455$$

From Eq. (9-35),

$$I_s = kI_a = 0.5455 \times 20 = 10.91 \text{ A}$$

$$\Delta I_L = 0.05 \times I_a = 0.05 \times 20 = 1 \text{ A}$$

$$\Delta I = 0.1 \times I_a = 0.1 \times 20 = 2 \text{ A}$$

(a) From Eq. (9-37), we get the value of L_e:

$$L_e = \frac{V_a(V_s - V_a)}{\Delta I f V_s} = \frac{60 \times (110 - 60)}{2 \times 20 \text{ kHz} \times 110} = 681.82 \text{ } \mu\text{H}$$

From Eq. (9-39) we get the value of C_e:

$$C_e = \frac{\Delta I}{V_c \times 8f} = \frac{2}{1.5 \times 8 \times 20 \text{ kHz}} = 8.33 \text{ } \mu\text{F}$$

Assuming a linear rise of load current i_L during the time from $t = 0$ to $t_1 = kT$, we can write approximately

$$L \frac{\Delta I_L}{t_1} = L \frac{\Delta I_L}{kT} = \Delta V_C$$

which gives the approximate value of L:

$$L = \frac{kT \Delta V_c}{\Delta I_L} = \frac{k \Delta V_c}{\Delta I_L f} \tag{9-129}$$

$$= \frac{0.5454 \times 1.5}{1 \times 20 \text{ kHz}} = 40.91 \text{ } \mu\text{H}$$

(b) $k = 0.5455$, $f = 20$ kHz, $T = 1/f = 50 \text{ } \mu\text{s}$, and $t_{on} = k \times T = 27.28 \text{ } \mu\text{s}$. The buck chopper for PSpice simulation is shown in Fig. 9-30a. The control voltage V_g is shown in Fig. 9-30b. The list of the circuit file is as follows:

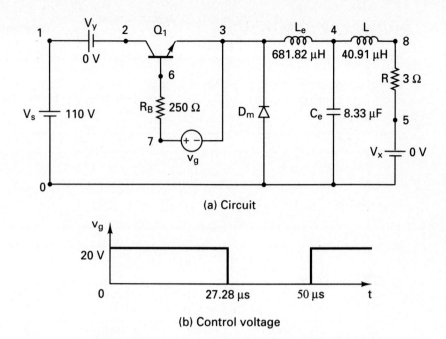

(a) Circuit

(b) Control voltage

Figure 9-30 Buck chopper for PSpice simulation.

```
Example 9-14    Buck Chopper
VS     1    0    DC      110V
VY     1    2    DC      0V   ; Voltage source to measure input current
Vg     7    3    PULSE (0V  20V   0    0.1NS  0.1NS   27.28US    50US)
RB     7    6    250                       ; Transistor base resistance
LE     3    4    681.82UH
CE     4    0    8.33UF    IC=60V ; initial voltage
L      4    8    40.91UH
R      8    5    3
VX     5    0    DC      0V          ; Voltage source to measure load current
DM     0    3    DMOD                      ; Freewheeling diode
.MODEL    DMOD    D(IS=2.2E-15 BV=1800V TT=0)  ; Diode model parameters
Q1     2    6    3    QMOD                 ; BJT switch
.MODEL   QMOD   NPN (IS=6.734F BF=416.4 BR=.7371 CJC=3.638P
+ CJE=4.493P TR=239.5N TF=301.2P)         ; BJT model parameters
.TRAN    1US    1.6MS    1.5MS    1US    UIC   ; Transient analysis
.PROBE                                    ; Graphics postprocessor
.options abstol = 1.00n reltol = 0.01 vntol = 0.1 ITL5=50000 ; convergence
.FOUR    20KHZ    I(VY)                    ; Fourier analysis
.END
```

The PSpice plots are shown in Fig. 9–31, where I(VX) = load current, I(Le) = inductor L_e current, and V(4) = capacitor voltage. Using the PSpice cursor in Fig. 9–31 gives $V_a = V_c = 59.462$ V, $\Delta V_c = 1.782$ V, $\Delta I = 2.029$ A, $I_{(av)} = 19.813$ A, $\Delta I_L = 0.3278$ A, and $I_a = 19.8249$ A. This verifies the design; however, ΔI_L gives a better result than expected.

Example 9-14   A Buck Chopper

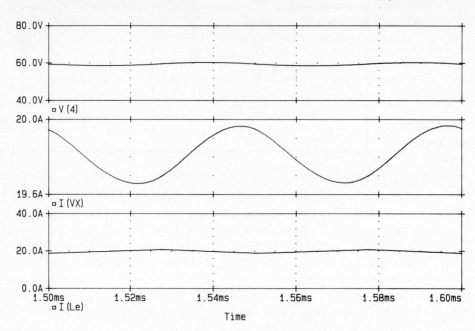

Figure 9-31   PSpice plots for Example 9-14.

(c) The Fourier coefficients of the input current are

FOURIER COMPONENTS OF TRANSIENT RESPONSE I(VY)
DC COMPONENT =   1.079535E+01

HARMONIC NO	FREQUENCY (HZ)	FOURIER COMPONENT	NORMALIZED COMPONENT	PHASE (DEG)	NORMALIZED PHASE (DEG)
1	2.000E+04	1.251E+01	1.000E+00	−1.195E+01	0.000E+00
2	4.000E+04	1.769E+00	1.415E−01	7.969E+01	9.163E+01
3	6.000E+04	3.848E+00	3.076E−01	−3.131E+01	−1.937E+01
4	8.000E+04	1.686E+00	1.348E−01	5.500E+01	6.695E+01
5	1.000E+05	1.939E+00	1.551E−01	−5.187E+01	−3.992E+01
6	1.200E+05	1.577E+00	1.261E−01	3.347E+01	4.542E+01
7	1.400E+05	1.014E+00	8.107E−02	−7.328E+01	−6.133E+01
8	1.600E+05	1.435E+00	1.147E−01	1.271E+01	2.466E+01
9	1.800E+05	4.385E−01	3.506E−02	−9.751E+01	−8.556E+01

TOTAL HARMONIC DISTORTION =   4.401661E+01 PERCENT

9-10 MAGNETIC CONSIDERATIONS

Inductances, which are used to create resonant oscillation for voltage reversal of commutation capacitor and turning off thyristors, act as energy storage elements in switched-mode regulators, and as filter elements to smooth out the current harmonics. We can notice from Eqs. (B-17) and (B-18) in Appendix B that the magnetic loss increases with the square of frequency. On the other hand, a higher frequency reduces the size of inductors for the same value of ripple current and filtering requirement. The design of dc–dc converters requires a compromise among switching frequency, inductor sizes, and switching losses.

SUMMARY

A dc chopper can be used as a dc transformer to step up or step down a fixed dc voltage. The chopper can also be used for switching-mode voltage regulators and for transferring energy between two dc sources. However, harmonics are generated at the input and load side of the chopper, and these harmonics can be reduced by input and output filters. A chopper can operate on either fixed frequency or variable frequency. A variable-frequency chopper generates harmonics of variable frequencies and a filter design becomes difficult. A fixed-frequency chopper is normally used. To reduce the sizes of filters and to lower the load ripple current, the chopping frequency should be high. Thyristor choppers require extra circuitry to turn off the main thyristor, and as a result the chopping frequency and minimum on-time are limited.

REFERENCES

1. S. Cúk and R. D. Middlebrook, "Advances in switched mode power conversion." *IEEE Transactions on Industrial Electronics,* Vol. IE30, No. 1, 1983, pp. 10–29.

2. C. E. Band and D. W. Venemans, "Chopper control on a 1600-V dc traction supply." *IRCA, Cybernatics and Electronics on the Railways,* Vol. 5, No. 12, 1968, pp. 473–478.

3. F. Nouvion, "Use of power semiconductors to control locomotive traction motors in the French National Railways." *Proceedings, IEE,* Vol. 55, No. 3, 1967.

4. Westinghouse Electric, "Choppers for São Paulo metro follow BART pattern." *Railway Gazette International,* Vol. 129, No. 8, 1973, pp. 309–310.

5. T. Tsuboi, S. Izawa, K. Wajima, T. Ogawa, and T. Katta, "Newly developed thyristor chopper equipment for electric railcars." *IEEE Transactions on Industry and General Applications,* Vol. IA9, No. 3, 1973.

6. J. Gouthiere, J. Gregoire, and H. Hologne, "Thyristor choppers in electric tractions." *ACEC Review,* No. 2, 1970, pp. 46–47.

7. M. H. Rashid, "A thyristor chopper with minimum limits on voltage control of dc drives." *International Journal of Electronics,* Vol. 53, No. 1, 1982, pp. 71–81.

8. R. P. Severns and G. E. Bloom, *Modern DC-to-DC Switchmode Power Converter Circuits.* New York: Van Nostrand Reinhold Company, Inc., 1983.

9. P. Wood, *Switching Power Converters*. New York: Van Nostrand Reinhold Company, Inc., 1981.

10. S. Cúk, "Survey of switched mode power supplies." *IEEE International Conference on Power Electronics and Variable Speed Drives,* London, 1985, pp. 83–94.

11. S. A. Chin, D. Y. Chen, and F. C. Lee, "Optimization of the energy storage inductors for dc to dc converters." *IEEE Transactions on Aerospace and Electronic Systems,* Vol. AES19, No. 2, 1983, pp. 203–214.

12. M. Ehsani, R. L. Kustom, and R. E. Fuja, "Microprocessor control of a current source dc–dc converter." *IEEE Transactions on Industry Applications,* Vol. IA19, No. 5, 1983, pp. 690–698.

13. M. H. Rashid, *SPICE for Power Electronics Using PSpice.* Englewood Cliffs, N.J.: Prentice Hall, 1993, Chapters 10 and 11.

REVIEW QUESTIONS

9-1. What is a dc chopper or dc–dc converter?

9-2. What is the principle of operation of a step-down chopper?

9-3. What is the principle of operation of a step-up chopper?

9-4. What is pulse-width-modulation control of a chopper?

9-5. What is frequency-modulation control of a chopper?

9-6. What are the advantages and disadvantages of a variable-frequency chopper?

9-7. What is the effect of load inductance on the load ripple current?

9-8. What is the effect of chopping frequency on the load ripple current?

9-9. What are the constraints for controllable transfer of energy between two dc voltage sources?

9-10. What are the performance parameters of a chopper?

9-11. What is a switching-mode regulator?

9-12. What are the four basic types of switching-mode regulators?

9-13. What are the advantages and disadvantages of a buck regulator?

9-14. What are the advantages and disadvantages of a boost regulator?

9-15. What are the advantages and disadvantages of a buck–boost regulator?

9-16. What are the advantages and disadvantages of a Cúk regulator?

9-17. What is the purpose of the commutation circuit of a chopper?

9-18. What is the difference between the circuit turn-off time and the turn-off time of a thyristor?

9-19. Why does a commutation capacitor get overcharged?

9-20. Why is the minimum output voltage of the classical chopper limited?

9-21. What are the advantages and disadvantages of the classical chopper?

9-22. What are the effects of source inductance?

9-23. Why should the resonant reversal be independent of the main thyristor?

9-24. Why should the peak resonant current of the resonant pulse chopper be greater than the peak load current?

9-25. What are the advantages and disadvantages of a resonant pulse chopper?

9-26. At what duty cycle does the load ripple current become maximum?

9-27. Why may the design of commutation circuits require an iterative method of solution?

9-28. What are the general steps for the design of chopper circuits?

9-29. Why is the peak load current rather than the average load current used in the design of thyristor choppers?

9-30. What are the effects of chopping frequency on filter sizes?

PROBLEMS

9-1. The dc chopper in Fig. 9-1a has a resistive load, $R = 20 \, \Omega$ and input voltage, $V_s = 220$ V. When the chopper remains on, its voltage drop is $V_{ch} = 1.5$ V and chopping frequency is $f = 10$ kHz. If the duty cycle is 80%, determine **(a)** the average output voltage V_a, **(b)** the rms output voltage V_o, **(c)** the chopper efficiency, **(d)** the effective input resistance R_i, and **(e)** the rms value of the fundamental component of harmonics on the output voltage.

9-2. A chopper is feeding an RL load as shown in Fig. 9-2 with $V_s = 220$ V, $R = 10 \, \Omega$, $L = 15.5$ mH, $f = 5$ kHz, and $E = 20$ V. Calculate **(a)** the minimum instantaneous load current I_1, **(b)** the peak instantaneous load current I_2, **(c)** the maximum peak-to-peak ripple current in the load, **(d)** the average load current I_a, **(e)** the rms load current I_o, **(f)** the effective input resistance R_i, and **(g)** the rms value of chopper current I_R.

9-3. The chopper in Fig. 9-2 has load resistance, $R = 0.2 \, \Omega$, input voltage, $V_s = 220$ V, and battery voltage is $E = 10$ V. The average load current, $I_a = 200$ A, and the chopping frequency is $f = 200$ Hz ($T = 5$ ms). Use the average output voltage to calculate the value of load inductance, L, which would limit the maximum load ripple current to 5% of I_a.

9-4. The dc chopper shown in Fig. 9-5a is used to control power flow from a dc voltage, $V_s = 110$ V to a battery voltage, $E = 220$ V. The power transferred to the battery is 30 kW. The current ripple of the inductor is negligible. Determine **(a)** the duty cycle K, **(b)** the effective load resistance R_{eq}, and **(c)** the average input current I_s.

9-5. For Prob 9-4, plot the instantaneous inductor current and current through the battery E if inductor L has a finite value of $L = 7.5$ mH, $f = 250$ Hz, and $k = 0.5$.

9-6. An RL load as shown in Fig. 9-2 is controlled by a chopper. If load resistance $R = 0.25 \, \Omega$, inductance $L = 20$ mH, supply voltage $V_s = 600$, battery voltage $E = 150$ V, and chopping frequency $f = 250$ Hz, determine the minimum and maximum load current, the peak-to-peak load ripple current, and average load current for $k = 0.1$ to 0.9 with a step of 0.1.

9-7. Determine the maximum peak-to-peak ripple current of Prob. 9-6 by using Eqs. (9-19) and (9-20), and compare the results.

9-8. The buck regulator in Fig. 9-12a has an input voltage, $V_s = 15$ V. The required average output voltage $V_a = 5$ V and the peak-to-peak output ripple voltage is 10 mV. The switching frequency is 20 kHz. The peak-to-peak ripple current of inductor is limited to 0.5 A. Determine **(a)** the duty cycle k, **(b)** the filter inductance L, and **(c)** the filter capacitor C.

9-9. The boost regulator in Fig. 9-13a has an input voltage, $V_s = 6$ V. The average output voltage, $V_a = 15$ V and average load current, $I_a = 0.5$ A. The switching frequency is 20 kHz. If $L = 250 \, \mu H$ and $C = 440 \, \mu F$, determine **(a)** the duty cycle k **(b)** the ripple current of inductor, ΔI, **(c)** the peak current of inductor, I_2, and **(d)** the ripple voltage of filter capacitor, ΔV_c.

9-10. The buck–boost regulator in Fig. 9-14a has an input voltage, $V_s = 12$ V. The duty cycle, $k = 0.6$ and the switching frequency is 25 kHz. The inductance, $L = 250 \, \mu H$ and filter capacitance, $C = 220 \, \mu F$. The average load current, $I_a = 1.5$ A. Determine **(a)** the average output voltage V_a, **(b)** the peak-to-peak output ripple voltage ΔV_c, **(c)** the peak-to-peak ripple current of inductor, ΔI, and **(d)** the peak current of the transistor, I_p.

9-11. The Cúk regulator in Fig. 9-15a has an input voltage, $V_s = 15$ V. The duty cycle $k = 0.4$ and the switching frequency is 25 kHz. The filter inductance $L_2 = 350 \, \mu H$ and filter capacitance, $C_2 = 220 \, \mu F$. The energy transfer capacitance $C_1 = 400 \, \mu F$ and inductance $L_1 = 250 \, \mu H$. The average load current $I_a = 1.25$ A. Determine **(a)** the average output voltage V_a, **(b)** the average input current I_s, **(c)** the peak-to-peak ripple current of inductor L_1, ΔI_1, **(d)**

---

the peak-to-peak ripple voltage of capacitor C_1, ΔV_{c1}, **(e)** the peak-to-peak ripple current of inductor L_2, ΔI_2, **(f)** the peak-to-peak ripple voltage of capacitor C_2, ΔV_{c2}, and **(g)** the peak current of the transistor, I_p.

9-12. An inductive load is controlled by an impulse commutation chopper in Fig. 9-16 and peak load current, $I_m = 450$ A at a supply voltage of 220 V. The chopping frequency $f = 275$ Hz, commutation capacitor $C = 60$ μF, and reversing inductance $L_m = 20$ μH. The source inductance $L_s = 8$ μH. Determine the circuit turn-off time and minimum and maximum output voltage limits.

9-13. Repeat Prob. 9-12 for the case when the source inductance is negligible ($L_s = 0$).

9-14. An inductive load is controlled by the chopper in Fig. 9-20 and peak load current $I_m = 350$ A at a supply voltage $V_s = 750$ V. The chopping frequency $f = 250$ Hz, commutation capacitance $C = 15$ μF, and commutation inductance $L_m = 70$ μH. If the source inductance $L_s = 10$ μH, determine the circuit turn-off time t_{off}, minimum and maximum output voltage, and output voltage for a duty cycle of $k = 0.5$.

9-15. Repeat Prob. 9-12 for the resonant pulse chopper circuit in Fig. 9-22 if commutation capacitance $C = 30$ μF and commutation inductance $L_m = 35$ μH.

9-16. Design the values of commutation components L_m and C to provide a circuit turn-off time, $t_{off} = 20$ μs for the circuit in Fig. 9-16. The specifications for the circuit are $V_s = 600$ V, $I_m = 350$ A, and $L_s = 6$ μH. The peak current through T_1 is not to exceed $2I_m$.

9-17. Repeat Prob. 9-16 for the chopper circuit in Fig. 9-19 if the peak current of diode D_1 is limited to $2I_m$. Determine C and L_1.

9-18. Repeat Prob. 9-16 for the chopper circuit in Fig. 9-20 if the peak resonant reversal current is limited to I_m.

9-19. Repeat Prob. 9-16 for the circuit in Fig. 9-22 if the resonant reversal current through T_2 is limited to $2I_m$.

9-20. Design the value of the commutation capacitor C to provide a turn-off-time requirement of $t_{off} = 20$ μs for the circuit in Fig. 9-20 if $V_s = 600$ V, $I_m = 350$ A, and $L_s = 8$ μH.

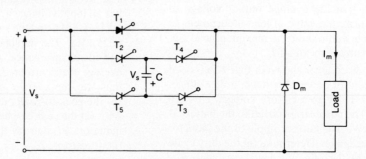

Figure P9-20

9-21. A highly inductive load is controlled by a chopper as shown in Fig. P9-20. The average load current is 250 A, which has negligible ripple current. A simple LC input filter with $L_e = 0.4$ mH and $C_e = 5000$ μF is used. If the chopper is operated at frequency, $f = 250$ Hz, determine the total chopper generated harmonic current in

the supply for $k = 0.5$. (Hint: Consider up to the seventh harmonic, and refer to Fig. 9–26.)

9-22. The chopper circuit in Example 9-11 uses the simple RC snubber network, as shown in Fig. 4-8b, for thyristors T_1, T_2, and T_3. If the dv/dt of all thyristors is limited to 200 V/μs and the discharging currents are limited to 10% of their respective peak values, determine **(a)** the values of snubber resistors and capacitors, and **(b)** the power ratings of resistors. The effects of load circuit and source inductance L_s can be neglected.

9-23. The holding current of thyristor T_1 in the chopper circuit in Fig. 9-20 is $I_H = 200$ mA and the delay time of T_1 is 1.5 μs. The dc input voltage is 220 V and source inductance L_s is negligible. It has a load of $L = 10$ mH and $R = 2$ Ω. Determine the minimum width t_G of gate pulse width.

9-24. The buck chopper in Fig. 9-29 has a dc input voltage $V_s = 110$ V, average load voltage $V_a = 80$ V, and average load current $I_a = 20$ A. The chopping frequency is $f = 10$ kHz. The peak-to-peak ripples are 5% for load voltage, 2.5% for load current, and 10% for filter L_e current. **(a)** Determine the values of L_e, L, and C_e. Use PSpice **(b)** to verify the results by plotting the instantaneous capacitor voltage v_C and instantaneous load current i_L, and **(c)** to calculate the Fourier coefficients of the input current i_s. Use SPICE model parameters of Example 9-14.

9-25. The boost chopper in Fig. 9-13a has a dc input voltage $V_S = 5$ V. The load resistance R is 100 Ω. The inductance is $L = 150$ μH, and the filter capacitance is $C = 220$ μF. The chopping frequency is $f = 20$ kHz and the duty cycle of the chopper is $k = 60\%$. Use PSpice **(a)** to plot the output voltage v_C, the input current i_s, and the MOSFET voltage v_T, and **(b)** to calculate the Fourier coefficients of the input current i_s. The SPICE model parameters of the MOSFET are L=2U, W=.3, VTO=2.831, KP=20.53U, IS=194E−18, CGSO=9.027N, CGDO=1.679N.

9-26. The circuit parameters of the impulse commutated circuit in Fig. 9-19 are: supply voltage $V_s = 200$ V, commutation capacitor $C = 20$ μF, commutation inductance $L_m = 20$ μH, discharging inductance $L_1 = 25$ μH, load resistance $R_R = 1\Omega$, and load inductance $L_L = 5$ mH. If the thyristor is modeled by the circuit of Fig. 7-19, use PSpice to plot capacitor voltage v_c, capacitor current, i_c, and load current i_L. The switching frequency is $f = 1$ kHz and the on-time of thyristor T_1 is 40%.

10

Pulse-width-modulated inverters

10-1 INTRODUCTION

DC-to-ac converters are known as *inverters*. The function of an inverter is to change a dc input voltage to a symmetrical ac output voltage of desired magnitude and frequency. The output voltage could be fixed or variable at a fixed or variable frequency. A variable output voltage can be obtained by varying the input dc voltage and maintaining the gain of the inverter constant. On the other hand, if the dc input voltage is fixed and it is not controllable, a variable output voltage can be obtained by varying the gain of the inverter, which is normally accomplished by pulse-width-modulation (PWM) control within the inverter. The *inverter gain* may be defined as the ratio of the ac output voltage to dc input voltage.

The output voltage waveforms of ideal inverters should be sinusoidal. However, the waveforms of practical inverters are nonsinusoidal and contain certain harmonics. For low- and medium-power applications, square-wave or quasi-square-wave voltages may be acceptable; and for high-power applications, low distorted sinusoidal waveforms are required. With the availability of high-speed power semiconductor devices, the harmonic contents of output voltage can be minimized or reduced significantly by switching techniques.

Inverters are widely used in industrial applications (e.g., variable-speed ac motor drives, induction heating, standby power supplies, uninterruptible power supplies). The input may be a battery, fuel cell, solar cell, or other dc source. The typical single-phase outputs are (1) 120 V at 60 Hz, (2) 220 V at 50 Hz, and (3) 115 V at 400 Hz. For high-power three-phase systems, typical outputs are (1) 220/380 V at 50 Hz, (2) 120/208 V at 60 Hz, and (3) 115/200 V at 400 Hz.

Inverters can be broadly classified into two types: (1) single-phase inverters, and (2) three-phase inverters. Each type can use controlled turn-on and turn-off devices (e.g., BJTs, MOSFETs, IGBTs, MCTs, SITs, GTOs) or forced-commutated thyristors depending on applications. These inverters generally use PWM control signals for producing an ac output voltage. An inverter is called a *voltage-fed inverter* (VFI) if the input voltage remains constant, a *current-fed inverter* (CFI) if the input current is maintained constant, and a *variable dc linked inverter* if the input voltage is controllable.

10-2 PRINCIPLE OF OPERATION

The principle of single-phase inverters can be explained with Fig. 10-1a. The inverter circuit consists of two choppers. When only transistor Q_1 is turned on for a time $T_0/2$, the instantaneous voltage across the load v_0 is $V_s/2$. If transistor Q_2 only is turned on for a time $T_0/2$, $-V_s/2$ appears across the load. The logic circuit should be designed such that Q_1 and Q_2 are not turned on at the same time. Figure 10-1b shows the waveforms for the output voltage and transistor currents with a resistive load. This inverter requires a three-wire dc source, and when a transistor is off, its reverse voltage is V_s instead of $V_s/2$. This inverter is known as a *half-bridge inverter*.

The rms output voltage can be found from

$$V_o = \left(\frac{2}{T_0} \int_0^{T_0/2} \frac{V_s^2}{4} \, dt \right)^{1/2} = \frac{V_s}{2} \tag{10-1}$$

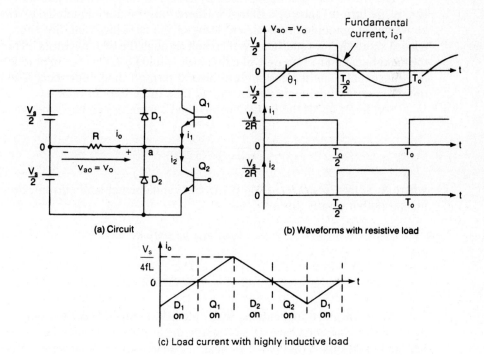

(a) Circuit

(b) Waveforms with resistive load

(c) Load current with highly inductive load

Figure 10-1 Single-phase half-bridge inverter.

The instantaneous output voltage can be expressed in Fourier series as

$$v_0 = \sum_{n=1,3,5,\ldots}^{\infty} \frac{2V_s}{n\pi} \sin n\omega t \tag{10-2}$$

$$= 0 \qquad \text{for } n = 2, 4, \ldots$$

where $\omega = 2\pi f_0$ is the frequency of output voltage in rad/s. For $n = 1$, Eq. (10-2) gives the rms value of fundamental component as

$$V_1 = \frac{2V_s}{\sqrt{2}\,\pi} = 0.45V_s \tag{10-3}$$

For an inductive load, the load current cannot change immediately with the output voltage. If Q_1 is turned off at $t = T_0/2$, the load current would continue to flow through D_2, load, and the lower half of the dc source until the current falls to zero. Similary, when Q_2 is turned off at $t = T_0$, the load current flows through D_1, load, and the upper half of the dc source. When diode D_1 or D_2 conducts, energy is fed back to the dc source and these diodes are known as *feedback diodes*. Figure 10-1c shows the load current and conduction intervals of devices for a purely inductive load. It can be noticed that for a purely inductive load, a transistor conducts only for $T_0/2$ (or 90°). Depending on the load power factor, the conduction period of a transistor would vary from 90 to 180°.

The transistors can be replaced by GTOs or forced-commutated thyristors. If t_q is the turn-off time of a thyristor, there must be a minimum delay time of t_q between the outgoing thyristor and firing of the next incoming thyristor. Otherwise, a short-circuit condition would result through the two thyristors. Therefore, the maximum conduction time of a thyristor would be $T_0/2 - t_q$. In practice, even the transistors require a certain turn-on and turn-off time. For successful operation of inverters, the logic circuit should take these into account.

For an RL load, the instantaneous load current i_0 can be found from

$$i_0 = \sum_{n=1,3,5,\ldots}^{\infty} \frac{2V_s}{n\pi\sqrt{R^2 + (n\omega L)^2}} \sin(n\omega t - \theta_n) \tag{10-4}$$

where $\theta_n = \tan^{-1}(n\omega L/R)$. If I_{01} is the rms fundamental load current, the fundamental output power (for $n = 1$) is

$$P_{01} = V_1 I_{01} \cos \theta_1 = I_{01}^2 R \tag{10-5}$$

$$= \left[\frac{2V_s}{\sqrt{2}\pi\sqrt{R^2 + (\omega L)^2}} \right]^2 R \tag{10-5a}$$

Note. In most applications (e.g., electric motor drives) the output power due to the fundamental current is generally the useful power, and the power due to harmonic currents is dissipated as heat and increases the load temperature.

The output of practical inverters contain harmonics and the quality of an inverter is normally evaluated in terms of the following performance parameters.

Harmonic factor of nth harmonic, HF_n. The harmonic factor (of the nth harmonic), which is a measure of individual harmonic contribution, is defined as

$$HF_n = \frac{V_n}{V_1} \qquad (10\text{-}6)$$

where V_1 is the rms value of the fundamental component and V_n is the rms value of the nth harmonic component.

Total harmonic distortion THD. The total harmonic distortion, which is a measure of closeness in shape between a waveform and its fundamental component, is defined as

$$THD = \frac{1}{V_1} \left(\sum_{n=2,3,\ldots}^{\infty} V_n^2 \right)^{1/2} \qquad (10\text{-}7)$$

Distortion factor DF. THD gives the total harmonic content, but it does not indicate the level of each harmonic component. If a filter is used at the output of inverters, the higher-order harmonics would be attenuated more effectively. Therefore, a knowledge of both the frequency and magnitude of each harmonic is important. The distortion factor indicates the amount of harmonic distortion that remains in a particular waveform after the harmonics of that waveform have been subjected to a second-order attenuation (i.e., divided by n^2). Thus DF is a measure of effectiveness in reducing unwanted harmonics without having to specify the values of a second-order load filter and is defined as

$$DF = \frac{1}{V_1} \left[\sum_{n=2,3,\ldots}^{\infty} \left(\frac{V_n}{n^2} \right)^2 \right]^{1/2} \qquad (10\text{-}8)$$

The distortion factor of an individual (or nth) harmonic component is defined as

$$DF_n = \frac{V_n}{V_1 n^2} \qquad (10\text{-}9)$$

Lowest-order harmonic LOH. The lowest-order harmonic is that harmonic component whose frequency is closest to the fundamental one, and its amplitude is greater than or equal to 3% of the fundamental component.

Example 10-1

The single-phase half-bridge inverter in Fig. 10-1a has a resistive load of $R = 2.4 \ \Omega$ and the dc input voltage is $V_s = 48$ V. Determine (a) the rms output voltage at the fundamental frequency V_1, (b) the output power P_o, (c) the average and peak currents of each transistor, (d) the peak reverse blocking voltage V_{BR} of each transistor, (e) the total harmonic distortion THD, (f) the distortion factor DF, and (g) the harmonic factor and distortion factor of the lowest-order harmonic.

Solution $V_s = 48$ V and $R = 2.4 \ \Omega$.

(a) From Eq. (10-3), $V_1 = 0.45 \times 48 = 21.6$ V.

(b) From Eq. (10-1), $V_o = V_s/2 = 48/2 = 24$ V. The output power, $P_o = V_o^2/R = 24^2/2.4 = 240$ W.

(c) The peak transistor current $I_p = 24/2.4 = 10$ A. Since each transistor conducts for a 50% duty cycle, the average current of each transistor is $I_D = 0.5 \times 10 = 5$ A.

(d) The peak reverse blocking voltage $V_{BR} = 2 \times 24 = 48$ V.

(e) From Eq. (10-3), $V_1 = 0.45V_s$ and the rms harmonic voltage V_h

$$V_h = \left(\sum_{n=3,5,7,\ldots}^{\infty} V_n^2 \right)^{1/2} = (V_0^2 - V_1^2)^{1/2} = 0.2176V_s$$

From Eq. (10-7), THD $= (0.2176V_s/(0.45V_s) = 48.34\%$.

(f) From Eq. (10-2), we can find V_n and then find,

$$\left[\sum_{n=3,5,\ldots}^{\infty} \left(\frac{V_n}{n^2} \right)^2 \right]^{1/2} = \left[\left(\frac{V_3}{3^2} \right)^2 + \left(\frac{V_5}{5^2} \right)^2 + \left(\frac{V_7}{7^2} \right)^2 + \cdots \right]^{1/2} = 0.01712V_s$$

From Eq. (10-8), DF $= 0.01712V_s/(0.45V_s) = 3.804\%$.

(g) The lowest-order harmonic is the third, $V_3 = V_1/3$. From Eq. (10-6), HF$_3 = V_3/V_1 = 1/3 = 33.33\%$, and from Eq. (10-9), DF$_3 = (V_3/3^2)/V_1 = 1/27 = 3.704\%$. Since $V_3/V_1 = 33.33\%$, which is greater than 3%, LOH $= V_3$.

10-4 SINGLE-PHASE BRIDGE INVERTERS

A single-phase bridge inverter is shown in Fig. 10-2a. It consists of four choppers. When transistors Q_1 and Q_2 are turned on simultaneously, the input voltage V_s appears across the load. If transistors Q_3 and Q_4 are turned on at the same time, the voltage across the load is reversed and is $-V_s$. The waveform for the output voltage is shown in Fig. 10-2b.

The rms output voltage can be found from

$$V_o = \left(\frac{2}{T_0} \int_0^{T_0/2} V_s^2 \, dt \right)^{1/2} = V_s \qquad (10\text{-}10)$$

Equation (10-2) can be extended to express the instantaneous output voltage in a Fourier series as

$$v_o = \sum_{n=1,3,5,\ldots}^{\infty} \frac{4V_s}{n\pi} \sin n\omega t \qquad (10\text{-}11)$$

and for $n = 1$, Eq. (10-11) gives the rms value of fundamental component as

$$V_1 = \frac{4V_s}{\sqrt{2}\,\pi} = 0.90V_s \qquad (10\text{-}12)$$

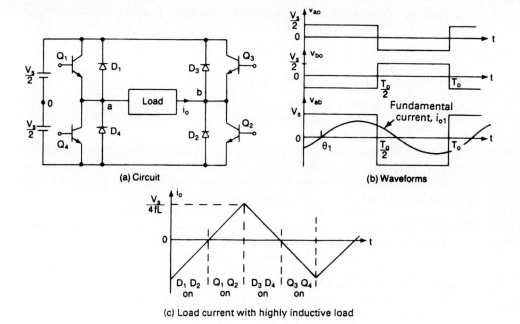

(a) Circuit

(b) Waveforms

(c) Load current with highly inductive load

Figure 10-2   Single-phase full-bridge inverter.

Using Eq. (10-4), the instantaneous load current i_0 for an RL load becomes

$$i_0 = \sum_{n=1,3,5,\ldots}^{\infty} \frac{4V_s}{n\pi \sqrt{R^2 + (n\omega L)^2}} \sin(n\omega t - \theta_n) \tag{10-13}$$

where $\theta_n = \tan^{-1}(n\omega L/R)$.

When diodes D_1 and D_2 conduct, the energy is fed back to the dc source and they are known as *feedback diodes*. Figure 10-1c shows the waveform of load current for an inductive load.

Example 10-2

Repeat Example 10-1 for a single-phase bridge inverter in Fig. 10-2a.
Solution $V_s = 48$ V and $R = 2.4\ \Omega$.

(a) From Eq. (10-12), $V_1 = 0.90 \times 48 = 43.2$ V.

(b) From Eq. (10-10), $V_o = V_s = 48$ V. The output power, $P_o = V_s^2/R = 48^2/2.4 = 960$ W.

(c) The peak transistor current, $I_p = 48/2.4 = 20$ A. Since each transistor conducts for a 50% duty cycle, the average current of each transistor is $I_D = 0.5 \times 20 = 10$ A.

(d) The peak reverse blocking voltage, $V_{BR} = 48$ V.

(e) From Eq. (10-12), $V_1 = 0.9V_S$. The rms harmonic voltage V_h is

$$V_h = \left(\sum_{n=3,5,7,\ldots}^{\infty} V_n^2 \right)^{1/2} = (V_0^2 - V_1^2)^{1/2} = 0.4352V_s$$

From Eq. (10-7), THD $= 0.4359V_s/(0.9V_s) = 48.34\%$.

(f) $\left[\displaystyle\sum_{n=3,5,7,\ldots}^{\infty} \left(\dfrac{V_n}{n^2}\right)^2\right]^{1/2} = 0.03424V_s$

From Eq. (10-8), DF $= 0.03424V_s/(0.9V_s) = 3.804\%$.

(g) The lowest-order harmonic is the third, $V_3 = V_1/3$. From Eq. (10-6), $HF_3 = V_3/V_1 = 1/3 = 33.33\%$ and from Eq. (10-9), $DF_3 = (V_3/3^2)/V_1 = 1/27 = 3.704\%$.

Note. The peak reverse blocking voltage of each transistor and the quality of output voltage for half-bridge and full-bridge inverters are the same. However, for full-bridge inverters, the output power is four times higher and the fundamental component is twice that of half-bridge inverters.

Example 10-3*

The bridge inverter in Fig. 10-2a has an *RLC* load with $R = 10\ \Omega$, $L = 31.5$ mH, and $C = 112\ \mu F$. The inverter frequency, $f_0 = 60$ Hz and dc input voltage, $V_s = 220$ V. (a) Express the instantaneous load current in Fourier series. Calculate (b) the rms load current at the fundamental frequency I_1; (c) the THD of the load current; (d) the power absorbed by the load P_0 and the fundamental power P_{01}; (e) the average current of dc supply I_s; and (f) the rms and peak current of each transistor. (g) Draw the waveform of fundamental load current and show the conduction intervals of transistors and diodes. Calculate the conduction time of (h) the transistors, and (i) the diodes.

Solution $V_s = 220$ V, $f_0 = 60$ Hz, $R = 10\ \Omega$, $L = 31.5$ mH, $C = 112\ \mu F$, and $\omega = 2\pi \times 60 = 377$ rad/s. The inductive reactance for the nth harmonic voltage is

$$X_L = j2n\pi \times 60 \times 31.5 \times 10^{-3} = j11.87n\ \Omega$$

The capacitive reactance for the nth harmonic voltage is

$$X_c = -\frac{j10^6}{2n\pi \times 60 \times 112} = \frac{-j23.68}{n}\ \Omega$$

The impedance for the nth harmonic voltage is

$$|Z_n| = [10^2 + (11.87n - 23.68/n)^2]^{1/2}$$

and the power factor angle for the nth harmonic voltage is

$$\theta_n = \tan^{-1}\frac{11.87n - 23.68/n}{10} = \tan^{-1}\left(1.187n - \frac{2.368}{n}\right)$$

(a) From Eq. (10-11), the instantaneous output voltage can be expressed as

$$v_o(t) = 280.1 \sin(377t) + 93.4 \sin(3 \times 377t) + 56.02 \sin(5 \times 377t)$$
$$+ 40.02 \sin(7 \times 377t) + 31.12 \sin(9 \times 377t) + \cdots$$

Dividing the output voltage by the load impedance and considering the appropriate delay due to the power factor angles, we can obtain the instantaneous load current as

$$i_o(t) = 18.1 \sin(377t + 49.72°) + 3.17 \sin(3 \times 377t - 70.17°)$$
$$+ \sin(5 \times 377t - 79.63°) + 0.5 \sin(7 \times 377t - 82.85°)$$
$$+ 0.3 \sin(9 \times 377t - 84.52°) + \cdots$$

(b) The peak fundamental load current, $I_{m1} = 18.1$ A. The rms load current at fundamental frequency, $I_{o1} = 18.1/\sqrt{2} = 12.8$ A.

(c) Considering up to the ninth harmonic, the peak load current,

$$I_m = (18.1^2 + 3.17^2 + 1.0^2 + 0.5^2 + 0.3^2)^{1/2} = 18.41 \text{ A}$$

The rms harmonic load current is

$$I_h = (I_m^2 - I_{m1}^2)^{1/2} = \frac{18.41^2 - 18.1^2}{\sqrt{2}} = 2.3789 \text{ A}$$

Using Eq. (10-7), the THD of the load current,

$$\text{THD} = \frac{(I_m^2 - I_{m1}^2)^{1/2}}{I_{m1}} = \left[\left(\frac{18.41}{18.1}\right)^2 - 1\right]^{1/2} = 18.59\%$$

(d) The rms load current $I_o \cong I_m/\sqrt{2} = 18.41/\sqrt{2} = 13.02$ A, and the load power, $P_o = 13.02^2 \times 10 = 1695$ W. Using Eq. (10-5), the fundamental output power is

$$P_{o1} = I_{o1}^2 R = 12.8^2 \times 10 = 1638.4 \text{ W}$$

(e) The average supply current $I_s = 1695/220 = 7.7$ A.

(f) The peak transistor current $I_p \cong I_m = 18.41$ A. The maximum permissible rms current of each transistor, $I_R = I_o/\sqrt{2} = I_p/2 = 18.41/2 = 9.2$ A.

(g) The waveform for fundamental load current, $i_1(t)$, is shown in Fig. 10.3.

(h) From Fig. 10-3, the conduction time of each transistor is found approximately from $\omega t_0 = 180 - 49.72 = 130.28°$ or $t_0 = 130.28 \times \pi/(180 \times 377) = 6031$ μs.

(i) The conduction time of each diode is approximately

$$t_d = (180 - 130.28) \times \frac{\pi}{180 \times 377} = 2302 \text{ } \mu\text{s}$$

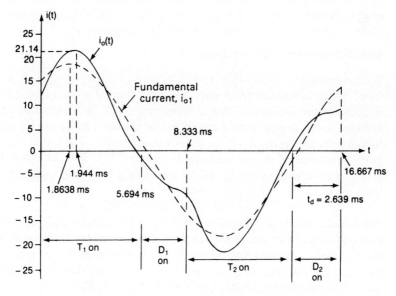

Figure 10-3 Waveforms for Example 10-3.

Notes:

1. To calculate the exact values of the peak current, the conduction time of transistors and diodes, the instantaneous load current $i_o(t)$ should be plotted as shown in Fig. 10-3. The conduction time of a transistor must satisfy the condition $i_o(t = t_0) = 0$, and a plot of $i_o(t)$ by a computer program gives $I_p = 21.14$ A, $t_0 = 5694$ μs, and $t_d = 2639$ μs.

2. This example can be repeated to evaluate the performance of an inverter with R, RL, or RLC load with an appropriate change in load impedance Z_L and load angle θ_n.

10-5 THREE-PHASE INVERTERS

Three-phase inverters are normally used for high-power applications. Three single-phase half (or full)-bridge inverters can be connected in parallel as shown in Fig. 10-4a to form the configuration of a three-phase inverter. The gating signals of single-phase inverters should be advanced or delayed by 120° with respect to each other in order to obtain three-phase balanced (fundamental) voltages. The transformer primary windings must be isolated from each other, while the secondary windings may be connected in wye or delta. The transformer secondary is normally connected in wye to eliminate triplen harmonics ($n = 3, 6, 9, \ldots$) appearing on the output voltages and the circuit arrangement is shown in Fig. 10-4b. This arrangement requires three single-phase transformers, 12 transistors, and 12 diodes. If the output voltages of single-phase inverters are not perfectly balanced in magnitudes and phases, the three-phase output voltages will be unbalanced.

A three-phase output can be obtained from a configuration of six transistors and six diodes as shown in Fig. 10-5a. Two types of control signals can be applied to the transistors: 180° conduction or 120° conduction.

10-5.1 180-Degree Conduction

Each transistor conducts for 180°. Three transistors remain on at any instant of time. When transistor Q_1 is switched on, terminal *a* is connected to the positive terminal of the dc input voltage. When transistor Q_4 is switched on, terminal *a* is brought to the negative terminal of the dc source. There are six modes of operation in a cycle and the duration of each mode is 60°. The transistors are numbered in the sequence of gating the transistors (e.g., 123, 234, 345, 456, 561, 612). The gating signals shown in Fig. 10-5b are shifted from each other by 60° to obtain three-phase balanced (fundamental) voltages.

The load may be connected in wye or delta as shown in Fig. 10-6. For a delta-connected load, the phase currents can be obtained directly from the line-to-line voltages. Once the phase currents are known, the line currents can be determined. For a wye-connected load, the line-to-neutral voltages must be determined to find the line (or phase) currents. There are three modes of operation in a half-cycle and the equivalent circuits are shown in Fig. 10-7a for a wye-connected load.

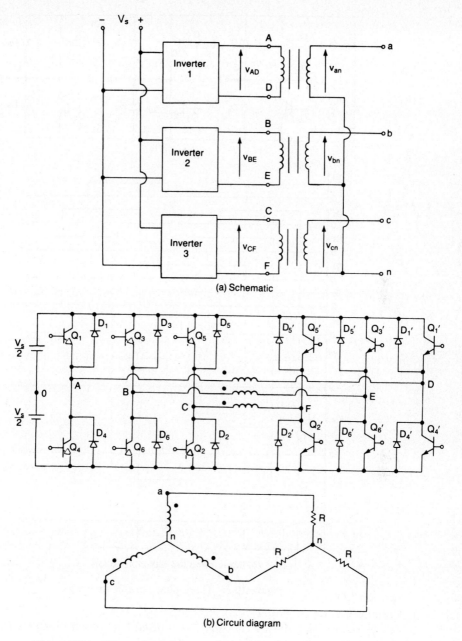

(a) Schematic

(b) Circuit diagram

Figure 10-4 Three-phase inverter formed by three single-phase inverters.

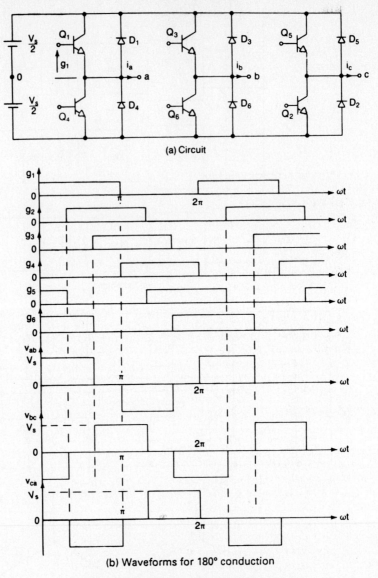

(a) Circuit

(b) Waveforms for 180° conduction

Figure 10-5 Three-phase bridge inverter.

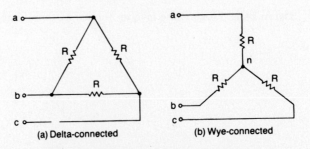

(a) Delta-connected   (b) Wye-connected

Figure 10-6 Delta/wye-connected load.

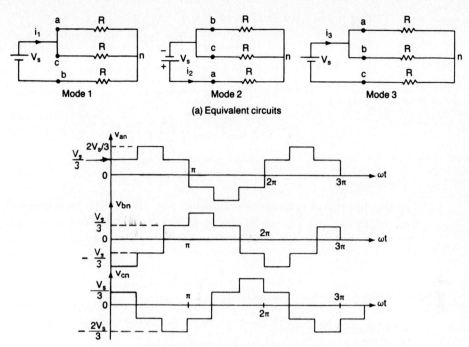

(a) Equivalent circuits

(b) Phase voltages for 180° conduction

Figure 10-7 Equivalent circuits for wye-connected resistive load.

During mode 1 for $0 \leq \omega t < \pi/3$,

$$R_{eq} = R + \frac{R}{2} = \frac{3R}{2}$$

$$i_1 = \frac{V_s}{R_{eq}} = \frac{2V_s}{3R}$$

$$v_{an} = v_{cn} = \frac{i_1 R}{2} = \frac{V_s}{3}$$

$$v_{bn} = -i_1 R = \frac{-2V_s}{3}$$

During mode 2 for $\pi/3 \leq \omega t < 2\pi/3$,

$$R_{eq} = R + \frac{R}{2} = \frac{3R}{2}$$

$$i_2 = \frac{V_s}{R_{eq}} = \frac{2V_s}{3R}$$

$$v_{an} = i_2 R = \frac{2V_s}{3}$$

$$v_{bn} = v_{cn} = \frac{-i_2 R}{2} = \frac{-V_s}{3}$$

During mode 3 for $2\pi/3 \le \omega t < \pi$,

$$R_{eq} = R + \frac{R}{2} = \frac{3R}{2}$$

$$i_3 = \frac{V_s}{R_{eq}} = \frac{2V_s}{3R}$$

$$v_{an} = v_{bn} = \frac{i_3 R}{2} = \frac{V_s}{3}$$

$$v_{cn} = -i_3 R = \frac{-2V_s}{3}$$

The line-to-neutral voltages are shown in Fig. 10-7b. The instantaneous line-to-line voltage, v_{ab}, in Fig. 10-5b can be expressed in a Fourier series, recognizing that v_{ab} is shifted by $\pi/6$ and the even harmonics are zero,

$$v_{ab} = \sum_{n=1,3,5,\ldots}^{\infty} \frac{4V_s}{n\pi} \cos \frac{n\pi}{6} \sin n \left(\omega t + \frac{\pi}{6} \right) \tag{10-14}$$

v_{bc} and v_{ca} can be found from Eq. (10-14) by phase shifting v_{ab} by 120° and 240°, respectively,

$$v_{bc} = \sum_{n=1,3,5,\ldots}^{\infty} \frac{4V_s}{n\pi} \cos \frac{n\pi}{6} \sin n \left(\omega t - \frac{\pi}{2} \right) \tag{10-15}$$

$$v_{ca} = \sum_{n=1,3,5,\ldots}^{\infty} \frac{4V_s}{n\pi} \cos \frac{n\pi}{6} \sin n \left(\omega t - \frac{7\pi}{6} \right) \tag{10-16}$$

We can notice from Eqs. (10-14), (10-15), and (10-16) that the triplen harmonics ($n = 3, 9, 15, \ldots$) would be zero in the line-to-line voltages.

The line-to-line rms voltage can be found from

$$V_L = \left[\frac{2}{2\pi} \int_0^{2\pi/3} V_s^2 \, d(\omega t) \right]^{1/2} = \sqrt{\frac{2}{3}} \, V_s = 0.8165 V_s \tag{10-17}$$

From Eq. (10-14), the rms nth component of the line voltage is

$$V_{Ln} = \frac{4V_s}{\sqrt{2} \, n\pi} \cos \frac{n\pi}{6} \tag{10-18}$$

which, for $n = 1$, gives the fundamental line voltage.

$$V_{L1} = \frac{4V_s \cos 30°}{\sqrt{2} \, \pi} = 0.7797 V_s \tag{10-19}$$

The rms value of line-to-neutral voltages can be found from the line voltage,

$$V_p = \frac{V_L}{\sqrt{3}} = \frac{\sqrt{2} \, V_s}{3} = 0.4714 V_s \tag{10-20}$$

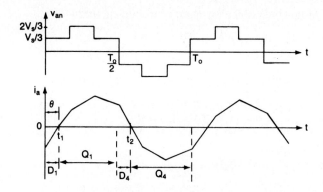

Figure 10-8   Three-phase inverter with *RL* load.

With resistive loads, the diodes across the transistors have no functions. If the load is inductive, the current in each arm of the inverter would be delayed to its voltage as shown in Fig. 10-8. When transistor Q_4 in Fig. 10-5a is off, the only path for the negative line current i_a is through D_1. Hence the load terminal a is connected to the dc source through D_1 until the load current reverses its polarity at $t = t_1$. During the period for $0 \leq t \leq t_1$, transistor Q_1 will not conduct. Similarly, transistor Q_4 will only start to conduct at $t = t_2$. The transistors must be continuously gated, since the conduction time of transistors and diodes depends on the load power factor.

For a wye-connected load, the phase voltage is $v_{an} = v_{ab}/\sqrt{3}$ with a delay of 30°. Using Eq. (10-14), the line current i_a for an *RL* load is given by

$$i_a = \sum_{n=1,3,5,\ldots}^{\infty} \left[\frac{4V_s}{\sqrt{3} \, n\pi \, \sqrt{R^2 + (n\omega L)^2}} \cos \frac{n\pi}{6} \right] \sin(n\omega t - \theta_n) \qquad (10\text{-}21)$$

where $\theta_n = \tan^{-1}(n\omega L/R)$.

Example 10-4

The three-phase inverter in Fig. 10-5a has a wye-connected resistive load of $R = 5 \ \Omega$ and $L = 23$ mH. The inverter frequency is $f_0 = 60$ Hz and the dc input voltage is $V_s = 220$ V. (a) Express the instantaneous line-to-line voltage $v_{ab}(t)$ and line current $i_a(t)$ in a Fourier series. Determine (b) the rms line voltage V_L; (c) the rms phase voltage V_p; (d) the rms line voltage V_{L1} at the fundamental frequency; (e) the rms phase voltage at the fundamental frequency, V_{p1}; (f) the total harmonic distortion THD; (g) the distortion factor DF; (h) the harmonic factor and distortion factor of the lowest-order harmonic; (i) the load power P_o; (j) the average transistor current I_D; and (k) the rms transistor current I_R.

Solution $V_s = 220$ V, $R = 5 \ \Omega$, $f_0 = 60$ Hz, and $\omega = 2\pi \times 60 = 377$ rad/s.

(a) Using Eq. (10-14), the instantaneous line-to-line voltage $v_{ab}(t)$ can be written as

$$v_{ab}(t) = 242.58 \sin(377t + 30°) - 48.52 \sin 5(377t + 30°)$$

$$- 34.66 \sin 7(377t + 30°) + 22.05 \sin 11(377t + 30°)$$

$$+ 18.66 \sin 13(377t + 30°) - 14.27 \sin 17(377t + 30°) + \cdots$$

$$Z_L = \sqrt{R^2 + (n\omega L)^2} \; \underline{/\tan^{-1}(n\omega L/R)} = \sqrt{5^2 + (8.67n)^2} \; \underline{/\tan^{-1}(8.67n/5)}$$

Using Eq. (10-21), the instantaneous line (or phase) current is given by

$$i_{a(t)} = 14 \sin(377t - 60°) - 0.64 \sin(5 \times 377t - 83.4°)$$

$$- 0.33 \sin(7 \times 377t - 85.3°) + 0.13 \sin(11 \times 377t - 87°)$$

$$+ 0.10 \sin(13 \times 377t - 87.5°) - 0.06 \sin(17 \times 377t - 88°) - \cdots$$

(b) From Eq. (10-17), $V_L = 0.8165 \times 220 = 179.63$ V.
(c) From Eq. (10-20), $V_p = 0.4714 \times 220 = 103.7$ V.
(d) From Eq. (10-19), $V_{L1} = 0.7797 \times 220 = 171.53$ V.
(e) $V_{p1} = V_{L1}/\sqrt{3} = 99.03$ V.
(f) From Eq. (10-19), $V_{L1} = 0.7797V_S$

$$\left(\sum_{n=5,7,11,\ldots}^{\infty} V_{Ln}^2 \right)^{1/2} = (V_L^2 - V_{L1}^2)^{1/2} = 0.24236V_s$$

From Eq. (10-7), THD $= 0.24236V_s/(0.7797V_s) = 31.08\%$. The rms harmonic line voltage is

$$\text{(g) } V_{Lh} = \left[\sum_{n=5,7,11,\ldots}^{\infty} \left(\frac{V_{Ln}}{n^2} \right)^2 \right]^{1/2} = 0.00666V_s$$

From Eq. (10-8), DF $= 0.00666V_s/(0.7797V_s) = 0.854\%$.

(h) The lowest-order harmonic is the fifth, $V_{L5} = V_{L1}/5$. From Eq. (10-6), $\text{HF}_5 = V_{L5}/V_{L1} = 1/5 = 20\%$, and from Eq. (10-9), $\text{DF}_5 = (V_{L5}/5^2)/V_{L1} = 1/125 = 0.8\%$.

(i) For wye-connected loads, the line current is the same as the phase current and the rms line current,

$$I_L = \frac{(14^2 + 0.64^2 + 0.33^2 + 0.13^2 + 0.10^2 + 0.06^2)^{1/2}}{\sqrt{2}} = 9.91 \text{ A}$$

The load power $P_0 = 3I_L^2 R = 3 \times 9.91^2 \times 5 = 1473$ W.

(j) The average supply current $I_s = P_o/220 = 1473/220 = 6.7$ A and the average transistor current $I_D = 6.7/3 = 2.23$ A.

(k) Since the line current is shared by two transistors, the rms value of a transistor current is $I_R = I_L/\sqrt{2} = 9.91/\sqrt{2} = 5.72$ A.

10-5.2 120-Degree Conduction

In this type of control, each transistor conducts for 120°. Only two transistors remain on at any instant of time. The gating signals are shown in Fig. 10-9. The conduction sequence of transistors is 61, 12, 23, 34, 45, 56, 61. There are three modes of operation in one half-cycle and the equivalent circuits for a wye-connected load are shown in Fig. 10-10. During mode 1 for $\le \omega t \le \pi/3$, transistors 1 and 6 conduct.

$$v_{an} = \frac{V_s}{2} \qquad v_{bn} = -\frac{V_s}{2} \qquad v_{cn} = 0$$

During mode 2 for $\pi/3 \le \omega t \le 2\pi/3$, transistors 1 and 2 conduct.

$$v_{an} = \frac{V_s}{2} \qquad v_{bn} = 0 \qquad v_{cn} = -\frac{V_s}{2}$$

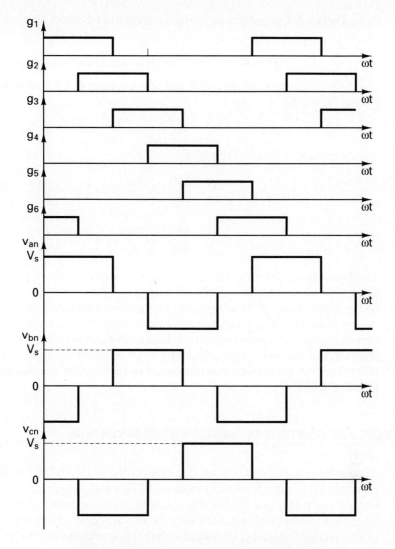

Figure 10-9   Gating signals for 120° conduction.

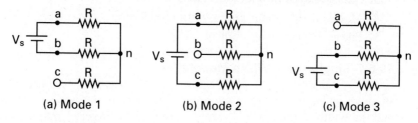

Figure 10-10   Equivalent circuits for wye-connected resistive load.

During mode 3 for $2\pi/3 \le \omega t \le 3\pi/3$, transistors 2 and 3 conduct.

$$v_{an} = 0 \qquad v_{bn} = \frac{V_s}{2} \qquad v_{cn} = -\frac{V_s}{2}$$

The line-to-neutral voltages that are shown in Fig. 10-9 can be expressed in Fourier series as

$$v_{an} = \sum_{n=1,3,5,\ldots}^{\infty} \frac{2V_s}{n\pi} \cos\frac{n\pi}{6} \sin n\left(\omega t + \frac{\pi}{6}\right) \qquad (10\text{-}22)$$

$$v_{bn} = \sum_{n=1,3,5,\ldots}^{\infty} \frac{2V_s}{n\pi} \cos\frac{n\pi}{6} \sin n\left(\omega t - \frac{\pi}{2}\right) \qquad (10\text{-}23)$$

$$v_{cn} = \sum_{n=1,3,5,\ldots}^{\infty} \frac{2V_s}{n\pi} \cos\frac{n\pi}{6} \sin n\left(\omega t - \frac{7\pi}{6}\right) \qquad (10\text{-}24)$$

The line a-to-b voltage is $v_{ab} = \sqrt{3}\, v_{an}$ with a phase advance of 30°. There is a delay of $\pi/6$ between the turning off of Q_1 and turning on of Q_4. Thus there should be no short circuit of the dc supply through one upper and one lower transistors. At any time, two load terminals are connected to the dc supply and the third one remains open. The potential of this open terminal will depend on the load characteristics and would be unpredictable. Since one transistor conducts for 120°, the transistors are less utilized as compared to that of 180° conduction for the same load condition.

10-6 VOLTAGE CONTROL OF SINGLE-PHASE INVERTERS

In many industrial applications, it is often required to control the output voltage of inverters (1) to cope with the variations of dc input voltage, (2) for voltage regulation of inverters, and (3) for the constant volts/frequency control requirement. There are various techniques to vary the inverter gain. The most efficient method of controlling the gain (and output voltage) is to incorporate pulse-width-modulation (PWM) control within the inverters. The commonly used techniques are:

1. Single-pulse-width modulation
2. Multiple-pulse-width modulation
3. Sinusoidal pulse-width modulation
4. Modified sinusoidal pulse-width modulation
5. Phase-displacement control

10-6.1 Single-Pulse-Width Modulation

In single-pulse-width modulation control, there is only one pulse per half-cycle and the width of the pulse is varied to control the inverter output voltage. Figure 10-11 shows the generation of gating signals and output voltage of single-phase

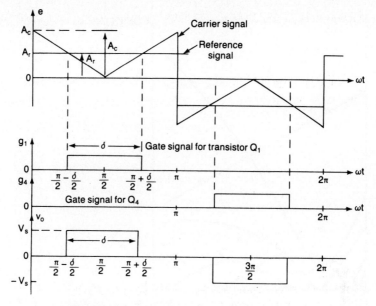

Figure 10-11   Single pulse-width modulation.

full-bridge inverters. The gating signals are generated by comparing a rectangular reference signal of amplitude, A_r, with a triangular carrier wave of amplitude, A_c. The frequency of the reference signal determines the fundamental frequency of output voltage. By varying A_r from 0 to A_c, the pulse width, δ, can be varied from 0 to 180°. The ratio of A_r to A_c is the control variable and defined as the amplitude *modulation index*. The amplitude modulation index, or simply modulation index

$$M = \frac{A_r}{A_c} \qquad (10\text{-}25)$$

The rms output voltage can be found from

$$V_o = \left[\frac{2}{2\pi} \int_{(\pi-\delta)/2}^{(\pi+\delta)/2} V_s^2 \, d(\omega t) \right]^{1/2} = V_s \sqrt{\frac{\delta}{\pi}} \qquad (10\text{-}26)$$

The Fourier series of output voltage yields

$$v_o(t) = \sum_{n=1,3,5,\ldots}^{\infty} \frac{4V_s}{n\pi} \sin \frac{n\delta}{2} \sin n\omega t \qquad (10\text{-}27)$$

A computer program named PROG-5 is developed to evaluate the performance of single-pulse modulation for single-phase full-bridge inverters and the program is listed in Appendix F. Figure 10-12 shows the harmonic profile with the variation of modulation index, M. The dominant harmonic is the third, and the distortion factor increases significantly at a low output voltage.

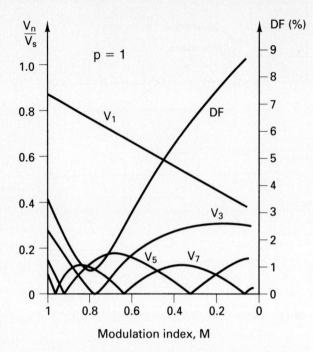

Figure 10-12 Harmonic profile of single-pulse-width modulation.

10-6.2 Multiple-Pulse-Width Modulation

The harmonic content can be reduced by using several pulses in each half-cycle of output voltage. The generation of gating signals for turning on and off of transistors is shown in Fig. 10-13a by comparing a reference signal with a triangular carrier wave. The frequency of reference signal sets the output frequency, f_o, and the carrier frequency, f_c, determines the number of pulses per half-cycle, p. The modulation index controls the output voltage. This type of modulation is also known as *uniform pulse-width modulation* (UPWM). The number of pulses per half-cycle is found from

$$p = \frac{f_c}{2f_o} = \frac{m_f}{2} \tag{10-28}$$

where $m_f = f_c/f_o$ is defined as the *frequency modulation ratio*.

The variation of modulation index M from 0 to 1 varies the pulse width from 0 to π/p and the output voltage from 0 to V_s. The output voltage for single-phase bridge inverters is shown in Fig. 10-13b for UPWM.

If δ is the width of each pulse, the rms output voltage can be found from

$$V_o = \left[\frac{2p}{2\pi} \int_{(\pi/p-\delta)/2}^{(\pi/p+\delta)/2} V_s^2 \, d(\omega t) \right]^{1/2} = V_s \sqrt{\frac{p\delta}{\pi}} \tag{10-29}$$

The general form of a Fourier series for the instantaneous output voltage is

$$v_o(t) = \sum_{n=1,3,5,\ldots}^{\infty} B_n \sin n\omega t \tag{10-30}$$

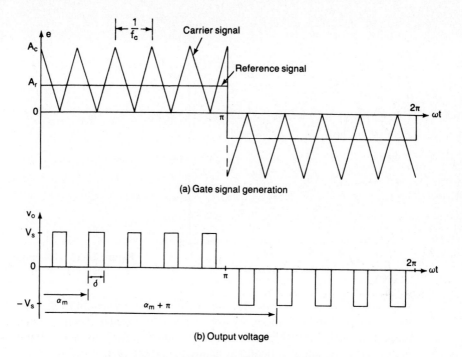

Figure 10-13 Multiple-pulse-width modulation.

The coefficient B_n in Eq. (10-30) can be determined by considering a pair of pulses such that the positive pulse of duration δ starts at $\omega t = \alpha$ and the negative one of the same width starts at $\omega t = \pi + \alpha$. This is shown in Fig. 10-13b. The effects of all pulses can be combined together to obtain the effective output voltage.

If the positive pulse of mth pair starts at $\omega t = \alpha_m$ and ends at $\omega t = \alpha_m + \pi$, the Fourier coefficient for a pair of pulses is

$$b_n = \frac{1}{\pi}\left[\int_{\alpha_m}^{\alpha_m+\delta}\cos n\omega t\ d(\omega t) - \int_{\pi+\alpha_m}^{\pi+\alpha_m+\delta}\cos n\omega t\ d(\omega t)\right]$$

$$= \frac{2V_s}{n\pi}\sin\frac{n\delta}{2}\left[\sin n\left(\alpha_m+\frac{\delta}{2}\right) - \sin n\left(\pi+\alpha_m+\frac{\delta}{2}\right)\right] \quad (10\text{-}31)$$

The coefficient B_n of Eq. (10-30) can be found by adding the effects of all pulses,

$$B_n = \sum_{m=1}^{p}\frac{2V_s}{n\pi}\sin\frac{n\delta}{2}\left[\sin n\left(\alpha_m+\frac{\delta}{2}\right) - \sin n\left(\pi+\alpha_m+\frac{\delta}{2}\right)\right] \quad (10\text{-}32)$$

A computer program named PROG-5 is used to evaluate the performance of multiple pulse modulation, and the program is listed in Appendix F. Figure 10-14 shows the harmonic profile against the variation of modulation index for five pulses per half-cycle. The order of harmonics is the same as that of single-pulse modulation. The distortion factor is reduced significantly compared to that of single-pulse modulation. However, due to larger number of switching on and off

---

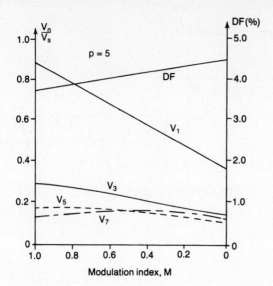

Figure 10-14 Harmonic profile of multiple-pulse-width modulation.

processes of power transistors, the switching losses would increase. With larger values of p, the amplitudes of lower-order harmonics would be lower, but the amplitudes of some higher-order harmonics would increase. However, such higher-order harmonics produce negligible ripple or can easily be filtered out.

10-6.3 Sinusoidal Pulse-Width Modulation

Instead of maintaining the width of all pulses the same as in the case of multiple-pulse modulation, the width of each pulse is varied in proportion to the amplitude of a sine wave evaluated at the center of the same pulse. The distortion factor and lower-order harmonics are reduced significantly. The gating signals as shown in Fig. 10-15a are generated by comparing a sinusoidal reference signal with a triangular carrier wave of frequency, f_c. This type of modulation is commonly used in industrial applications and abbreviated as SPWM. The frequency of reference signal, f_r, determines the inverter output frequency, f_o, and its peak amplitude, A_r, controls the modulation index, M, and then in turn the rms output voltage, V_o. The number of pulses per half-cycle depends on the carrier frequency. Within the constraint that two transistors of the same arm (Q_1 and Q_4) cannot conduct at the same time, the intantaneous output voltage is shown in Fig. 10-15a. The same gating signals can be generated by using unidirectional triangular carrier wave as shown in Fig. 10-15b.

The rms output voltage can be varied by varying the modulation index M. It can be observed that the area of each pulse corresponds approximately to the area under the sine wave between the adjacent midpoints of off periods on the gating signals. If δ_m is the width of mth pulse, Eq. (10-29) can be extended to find the rms output voltage

$$V_o = V_s \left(\sum_{m=1}^{p} \frac{\delta_m}{\pi} \right)^{1/2} \tag{10-33}$$

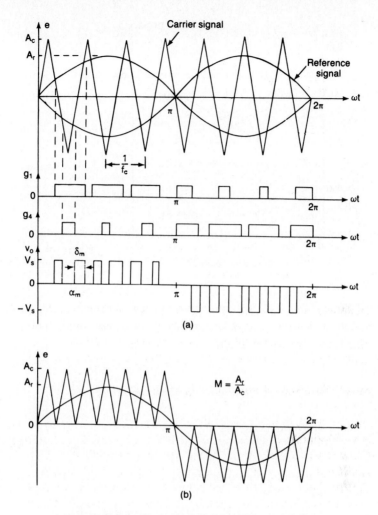

Figure 10-15 Sinusoidal pulse-width modulation.

Equation (10-32) can also be applied to determine the Fourier coefficient of output voltage as

$$B_n = \sum_{m=1}^{p} \frac{2V_s}{n\pi} \sin \frac{n\delta_m}{2} \left[\sin n \left(\alpha_m + \frac{\delta_m}{2} \right) - \sin n \left(\pi + \alpha_m + \frac{\delta_m}{2} \right) \right]$$

for $n = 1, 3, 5, \ldots$ (10-34)

A computer program named PROG-6 is developed to determine the width of pulses and to evaluate the harmonic profile of sinusoidal modulation. The harmonic profile is shown in Fig. 10-16 for five pulses per half-cycle. The distortion factor is significantly reduced compared to that of multiple-pulse modulation. This type of modulation eliminates all harmonics less than or equal to $2p - 1$. For $p = 5$, the lowest-order harmonic is ninth.

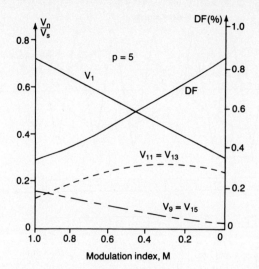

Figure 10-16 Harmonic profile of sinusoidal pulse-width modulation.

The output voltage of an inverter contains harmonics. The PWM pushes the harmonics into a high-frequency range around the switching frequency f_c and its multiples, that is, around harmonics m_f, $2m_f$, $3m_f$, and so on. The frequencies at which the voltage harmonics occur can be related by

$$f_n = (jm_f \pm k)f_c \tag{10-35}$$

where the nth harmonic equals the kth sideband of jth times the frequency-modulation ratio m_f.

$$
\begin{aligned}
n &= jm_f \pm k \\
&= 2jp \pm k \qquad \text{for } j = 1, 2, 3, \ldots \text{ and } k = 1, 3, 5, \ldots
\end{aligned}
\tag{10-36}
$$

The peak fundamental output voltage for PWM and SPWM control can be found approximately from

$$V_{m1} = dV_s \qquad \text{for } 0 \le d \le 1.0 \tag{10-37}$$

For $d = 1$, Eq. (10-37) gives the maximum peak amplitude of the fundamental output voltage as $V_{m1(\max)} = V_s$. But according to Eq. (10-11), $V_{m1(\max)}$ could be as high as $4V_s/\pi = 1.273V_s$ for a square-wave output. In order to increase the fundamental output voltage, d must be increased beyond 1.0. the operation beyond $d = 1.0$ is called *overmodulation*. The value of d at which $V_{m1(\max)}$ equals $1.273V_s$ is dependent on the number of pulses per half-cycle p and is approximately 3 for $p = 7$, as shown in Fig. 10-17. Overmodulation basically leads to a square-wave operation and adds more harmonics as compared to operation in the linear range (with $d \le 1.0$). Overmodulation is normally avoided in applications requiring low distortion [e.g., uninterruptible power supplies (UPSs)].

10-6.4 Modified Sinusoidal Pulse-Width Modulation

Figure 10-15 indicates that the widths of pulses that are nearer the peak of the sine wave do not change significantly with the variation of modulation index. This is

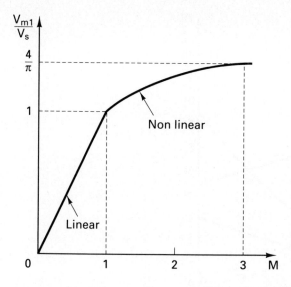

Figure 10-17 Peak fundamental output voltage versus modulation index M.

due to the characteristics of a sine wave, and the SPWM technique can be modified so that the carrier wave is applied during the first and last 60° intervals per half-cycle (e.g., 0 to 60° and 120 to 180°). This type of modulation is known as MSPWM and shown in Fig. 10-18. The fundamental component is increased and its harmonic characteristics are improved. It reduces the number of switching of power devices and also reduces switching losses.

A computer program named PROG-7, which is listed in Appendix F, determines the pulse widths and evaluates the performance of modified SPWM. The harmonic profile is shown in Fig. 10-19 for five pulses per half-cycle. The number

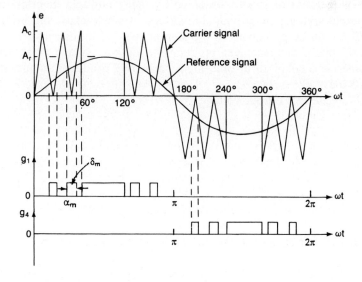

Figure 10-18 Modified sinusoidal pulse-width modulation.

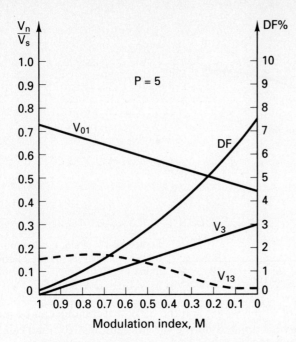

Figure 10-19 Harmonic profile of modified sinusoidal pulse-width modulation.

of pulses, q, in the 60° period is normally related to the frequency ratio, particularly in three-phase inverters, by

$$\frac{f_c}{f_o} = 6q + 3 \tag{10-38}$$

10-6.5 Phase-Displacement Control

Voltage control can be obtained by using multiple inverters and summing the output voltages of individual inverters. A single-phase full-bridge inverter in Fig. 10-2a can be perceived as the sum of two half-bridge inverters in Fig. 10-1a. A 180° phase displacement produces an output voltage as shown in Fig. 10-20c, whereas a delay (or displacement) angle of β produces an output as shown in Fig. 10-20e.

The rms output voltage,

$$V_o = V_s \sqrt{\frac{\beta}{\pi}} \tag{10-39}$$

If

$$v_{ao} = \sum_{n=1,3,5,\ldots}^{\infty} \frac{2V_s}{n\pi} \sin n\omega t$$

then

$$v_{bo} = \sum_{n=1,3,5,\ldots}^{\infty} \frac{2V_s}{n\pi} \sin n(\omega t - \beta)$$

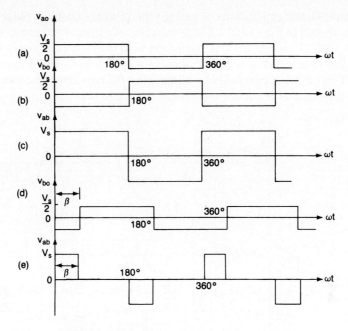

Figure 10-20 Phase-displacement control.

The instantaneous output voltage,

$$v_{ab} = v_{ao} - v_{bo} = \sum_{n=1,3,5,\ldots}^{\infty} \frac{2V_s}{n\pi} [\sin n\omega t - \sin n(\omega t - \beta)] \qquad (10\text{-}40)$$

since $\sin A - \sin B = 2 \sin[(A - B)/2] \cos[(A + B)/2]$, Eq. (10-40) can be simplified to

$$v_{ab} = \sum_{n=1,3,5,\ldots}^{\infty} \frac{4V_s}{n\pi} \sin \frac{n\beta}{2} \cos n \left(\omega t - \frac{\beta}{2} \right) \qquad (10\text{-}41)$$

The rms value of the fundamental output voltage is

$$V_1 = \frac{4V_s}{\sqrt{2}} \sin \frac{\beta}{2} \qquad (10\text{-}42)$$

Equation (10-42) indicates that the output voltage can be varied by varying the delay angle. This type of control is especially useful for high-power applications, requiring a large number of transistors in parallel.

10-7 VOLTAGE CONTROL OF THREE-PHASE INVERTERS

A three-phase inverter may be considered as three single-phase inverters and the output of each single-phase inverter is shifted by 120°. The voltage control techniques discussed in Section 10-6 are applicable to three-phase inverters. As an

---

example, the generations of gating signals with sinusoidal pulse-width modulation are shown in Fig. 10-21. There are three sinusoidal reference waves each shifted by 120°. A carrier wave is compared with the reference signal corresponding to a phase to generate the gating signals for that phase. The output voltage as shown in Fig. 10-21, is generated by eliminating the condition that two switching devices in the same arm cannot conduct at the same time.

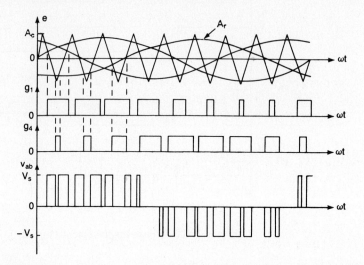

Figure 10-21 Sinusoidal pulse-width modulation for three-phase inverter.

Example 10-5

A single-phase full-bridge inverter controls the power in a resistive load. The nominal value of input dc voltage is $V_s = 220$ V and a uniform pulse-width modulation with five pulses per half-cycle is used. For the required control, the width of each pulse is 30°. (a) Determine the rms voltage of the load. (b) If the dc supply increases by 10%, determine the pulse width to maintain the same load power. If the maximum possible pulse width is 35°, determine the minimum allowable limit of the dc input source.

Solution   (a) $V_s = 220$ V, $p = 5$, and $\delta = 30°$. From Eq. (10-29), $V_o = 220 \sqrt{5 \times 30/180} = 200.8$ V.

(b) $V_s = 1.1 \times 220 = 242$ V. Using Eq. (10-29), $242\sqrt{5\delta/180} = 200.8$ and this gives the required value of pulse width, $\delta = 24.75°$.

To maintain the output voltage of 200.8 V at the maximum possible pulse width of $\delta = 35°$, the input voltage can be found from $200.8 = V_s\sqrt{5 \times 35/180}$, and this yields the minimum allowable input voltage, $V_s = 203.64$ V.

10-8 ADVANCED MODULATION TECHNIQUES

The SPWM, which is most commonly used, suffers from drawbacks (e.g., low fundamental output voltage). The other techniques that offer improved performances are:

Trapezoidal modulation
Staircase modulation
Stepped modulation
Harmonic injection modulation
Delta modulation

For the sake of simplicity, we shall show the output voltage, v_{ao}, for a half-bridge inverter. For a full-bridge inverter, $v_o = v_{ao} - v_{bo}$, where v_{bo} is the inverse of v_{ao}.

Trapezoidal modulation.   The gating signals are generated by comparing a triangular carrier wave with a modulating trapezoidal wave [6] as shown in Fig. 10-22. The trapezoidal wave can be obtained from a triangular wave by limiting its magnitude to $\pm A_r$, which is related to the peak value $A_{r(max)}$ by

$$A_r = \sigma A_{r(max)}$$

where σ is called the *triangular factor*, because the waveform becomes a triangular wave when $\sigma = 1$. The modulation index M is

$$M = \frac{A_r}{A_c} = \frac{\sigma A_{r(max)}}{A_c} \qquad \text{for } 0 \le M \le 1 \qquad (10\text{-}43)$$

The angle of the flat portion of the trapezoidal wave is given by

$$2\phi = (1 - \sigma)\pi \qquad (10\text{-}44)$$

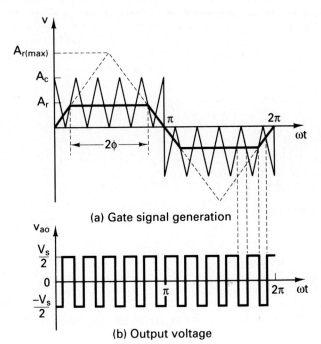

(a) Gate signal generation

(b) Output voltage

Figure 10-22 Trapezoidal modulation.

For fixed values of $A_{r(\text{max})}$ and A_c, M that varies with the output voltage can be varied by changing the triangular factor, σ. This type of modulation increases the peak fundamental output voltage up to $1.05V_s$, but the output contains lower-order harmonics.

Staircase modulation. The modulating signal is a staircase wave, as shown in Fig. 10-23. The staircase is not a sampled approximation to the sine wave. The levels of the stairs are calculated to eliminate specific harmonics. The modulation frequency ratio m_f and the number of steps are chosen to obtain the desired quality of output voltage. This is an optimized PWM and is not recommended for fewer than 15 pulses in one cycle. It has been shown [7] that for high fundamental output voltage and low distortion factor, the optimum number of pulses in one cycle is 15 for two levels, 21 for three levels, and 27 for four levels. This type of control provides a high-quality output voltage with a fundamental value of up to $0.94V_s$.

Stepped modulation. The modulating signal is a stepped wave [8] as shown in Fig. 10-24. The stepped wave is not a sampled approximation to the sine wave. It is divided into specified intervals, say 20°, with each interval being controlled individually to control the magnitude of the fundamental component and to eliminate specific harmonics. This type of control gives low distortion, but a higher fundamental amplitude compared to that of normal PWM control.

Harmonic injected modulation. The modulating signal is generated by injecting selected harmonics to the sine wave. This results in flat-topped waveform and reduces the amount of overmodulation. It provides a higher fundamental amplitude and low distortion of the output voltage. The modulating signal [9] is generally composed of

$$v_r = 1.15 \sin \omega t + 0.27 \sin 3\omega t - 0.029 \sin 9\omega t$$

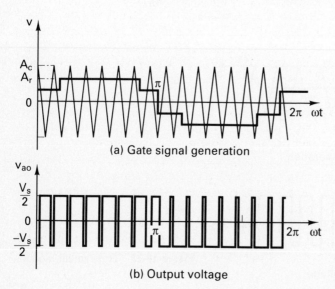

(a) Gate signal generation

(b) Output voltage

Figure 10-23 Staircase modulation.

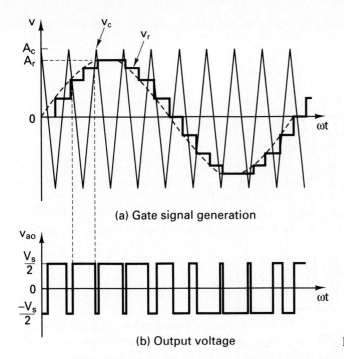

(a) Gate signal generation

(b) Output voltage

Figure 10-24 Stepped modulation.

The modulating signal with third and ninth harmonic injections is shown in Fig. 10-25. It should be noted that the injection of $3n$th harmonics will not affect the quality of the output voltage, because the output of a three-phase inverter does not contain triplen harmonics. If only third harmonic is injected, v_r is given by

$$v_r = 1.15 \sin \omega t + 0.19 \sin 3\omega t$$

The modulating signal [10] can be generated from $2\pi/3$ segments of a sine wave as shown in Fig. 10-26. This is the same as injecting $3n$th harmonics to a sine wave. The line-to-line voltage is sinusoidal PWM and the amplitude of the fundamental component is approximately 15% more than that of a normal sinusoidal PWM. Since each arm is switched off for one-third of the period, the heating of the switching devices is reduced.

Delta modulation. In delta modulation [11], a triangular wave is allowed to oscillate within a defined window ΔV above and below the reference sine wave v_r. The inverter switching function, which is identical to the output voltage v_o is generated from the vertices of the triangular wave v_c as shown in Fig. 10-27. It is also known as *hysteresis modulation*. If the frequency of the modulating wave is changed keeping the slope of the triangular wave constant, the number of pulses and pulses widths of the modulated wave would change.

The fundamental output voltage can be up to $1V_s$ and is dependent on the peak amplitude A_r and frequency f_r of the reference voltage. The delta modulation can control the ratio of voltage to frequency, which is a desirable feature in ac motor control.

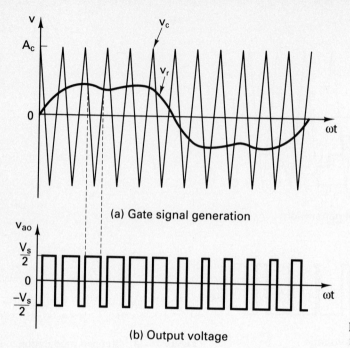

(a) Gate signal generation

(b) Output voltage

Figure 10-25 Selected harmonic injection modulation.

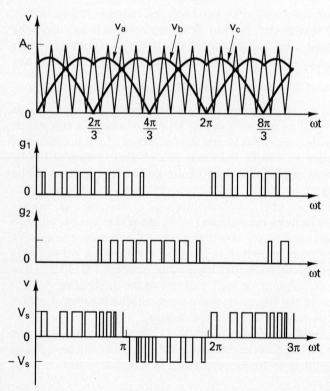

Figure 10-26 Harmonic injection modulation.

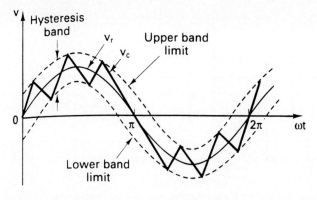

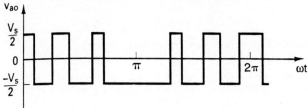

Figure 10-27 Delta modulation.

10-9 HARMONIC REDUCTIONS

Equation (10-41) indicates that the nth harmonic can be eliminated by a proper choice of displacement angle, β, if

$$\sin \frac{n\beta}{2} = 0$$

or

$$\beta = \frac{360°}{n} \tag{10-45}$$

and the third harmonic will be eliminated if $\beta = 360/3 = 120°$. A pair of unwanted harmonics at the output of single-phase inverters can be eliminated by introducing a pair of symmetrically placed bipolar voltage *notches* as shown in Fig. 10-28.

The Fourier series of output voltage can be expressed as

$$v_o = \sum_{n=1,3,5,\ldots}^{\infty} B_n \sin n\omega t \tag{10-46}$$

where

$$B_n = \frac{4V_s}{\pi} \left[\int_0^{\alpha_1} \sin n\omega t \, d(\omega t) - \int_{\alpha_1}^{\alpha_2} \sin n\omega t \, d(\omega t) + \int_{\alpha_2}^{\pi/2} \sin n\omega t \, d(\omega t) \right]$$

$$= \frac{4V_s}{\pi} \frac{1 - 2 \cos n\alpha_1 + 2 \cos n\alpha_2}{n} \tag{10-47}$$

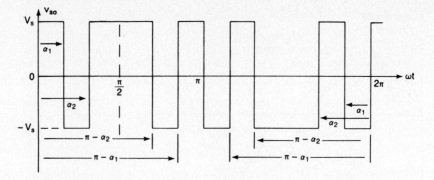

Figure 10-28 Output voltage with two bipolar notches per half-wave.

Equation (10-47) can be extended to m notches per quarter-wave:

$$B_n = \frac{4V_s}{n\pi} (1 - 2\cos n\alpha_1 + 2\cos n\alpha_2 - 2\cos n\alpha_3 + 2\cos n\alpha_4 - \cdots) \quad (10\text{-}48)$$

The third and fifth harmonics would be eliminated if $B_3 = B_5 = 0$ and Eq. (10-47) gives the necessary equations to be solved.

$$1 - 2\cos 3\alpha_1 + 2\cos 3\alpha_2 = 0 \quad \text{or} \quad \alpha_2 = \tfrac{1}{3}\cos^{-1}(\cos 3\alpha_1 - 0.5)$$

$$1 - 2\cos 5\alpha_1 + 2\cos 5\alpha_2 = 0 \quad \text{or} \quad \alpha_1 = \tfrac{1}{5}\cos^{-1}(\cos 5\alpha_2 + 0.5)$$

These equations can be solved iteratively by initially assuming that $\alpha_1 = 0$ and repeating the calculations for α_1 and α_2. The result is $\alpha_1 = 23.62°$ and $\alpha_2 = 33.3°$.

With unipolar voltage notches as shown in Fig. 10-29, the coefficient B_n is given by

$$B_n = \frac{4V_s}{\pi} \left[\int_0^{\alpha_1} \sin n\omega t \, d(\omega t) + \int_{\alpha_1}^{\pi/2} \sin n\omega t \, d(\omega t) \right]$$

$$= \frac{4V_s}{\pi} \frac{1 - \cos n\alpha_1 + \cos n\alpha_2}{n} \quad (10\text{-}49)$$

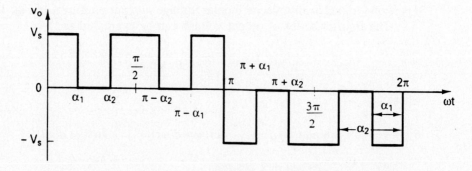

Figure 10-29 Unipolar output voltage with two notches per half-cycle.

The third and fifth harmonics would be eliminated if

$$1 - \cos 3\alpha_1 + \cos 3\alpha_2 = 0$$

$$1 - \cos 5\alpha_1 + \cos 5\alpha_2 = 0$$

Solving these equations by iterations, we get $\alpha_1 = 17.83°$ and $\alpha_2 = 37.97°$.

The modified sinusoidal pulse-width-modulation techniques can be applied to generate the notches which would eliminate certain harmonics effectively in the output voltage, as shown in Fig. 10-30.

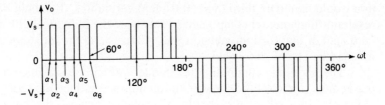

Figure 10-30 Output voltage for modified sinusoidal pulse-width modulation.

The output voltages of two or more inverters may be connected in series through a transformer to reduce or eliminate certain unwanted harmonics. The arrangement for combining two inverter output voltages is shown in Fig. 10-31a. The waveforms for the output of each inverter and the resultant output voltage are shown in Fig. 10-31b. The second inverter is phase shifted by $\pi/3$.

From Eq. (10-11), the output of first inverter can be expressed as

$$v_{o1} = A_1 \sin \omega t + A_3 \sin 3\omega t + A_5 \sin 5\omega t + \cdots$$

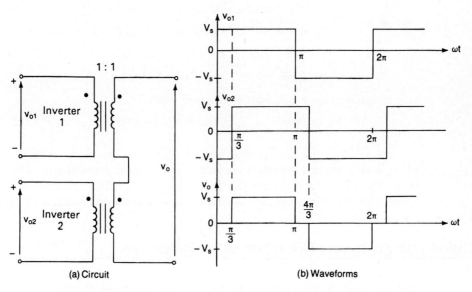

(a) Circuit　　　　(b) Waveforms

Figure 10-31 Elimination of harmonics by transformer connection.

Since the output of second inverter, v_{o2}, is delayed by $\pi/3$,

$$v_{o2} = A_1 \sin\left(\omega t - \frac{\pi}{3}\right) + A_3 \sin 3\left(\omega t - \frac{\pi}{3}\right) + A_5 \sin 5\left(\omega t - \frac{\pi}{3}\right) + \cdots$$

The resultant voltage v_o is obtained by vector addition.

$$v_o = v_{o1} + v_{o2} = \sqrt{3}\left[A_1 \sin\left(\omega t - \frac{\pi}{6}\right) + A_5 \sin 5\left(\omega t + \frac{\pi}{6}\right) + \cdots\right]$$

Therefore, a phase shifting of $\pi/3$ and combining voltages by transformer connection would eliminate third (and all triplen) harmonics. It should be noted that the resultant fundamental component is not twice the individual voltage, but is $\sqrt{3}/2$ ($= 0.866$) of that for individual output voltages and the effective output has been reduced by ($1 - 0.866 =$) 13.4%.

The harmonic elimination techniques, which are suitable only for fixed output voltage, increase the order of harmonics and reduce the sizes of output filter. However, this advantage should be weighed against the increased switching losses of power devices and increased iron (or magnetic losses) in the transformer due to higher harmonic frequencies.

Example 10-6

A single-phase full-wave inverter uses multiple notches to give bipolar voltage as shown in Fig. 10-28, and is required to eliminate the fifth, seventh, eleventh, and thirteenth harmonics from the output wave. Determine the number of notches and their angles.

Solution For elimination of the fifth, seventh, eleventh, and thirteenth harmonics, $A_5 = A_7 = A_{11} = A_{13} = 0$; that is, $m = 4$. Four notches per quarter-wave would be required. Equation (10-48) gives the following set of nonlinear simultaneous equations to solve for the angles.

$$1 - 2 \cos 5\alpha_1 + 2 \cos 5\alpha_2 - 2 \cos 5\alpha_3 + 2 \cos 5\alpha_4 = 0$$

$$1 - 2 \cos 7\alpha_1 + 2 \cos 7\alpha_2 - 2 \cos 7\alpha_3 + 2 \cos 7\alpha_4 = 0$$

$$1 - 2 \cos 11\alpha_1 + 2 \cos 11\alpha_2 - 2 \cos 11\alpha_3 + 2 \cos 11\alpha_4 = 0$$

$$1 - 2 \cos 13\alpha_1 + 2 \cos 13\alpha_2 - 2 \cos 13\alpha_3 + 2 \cos 13\alpha_4 = 0$$

Solution of these equations by iteration yields

$$\alpha_1 = 10.55° \qquad \alpha_2 = 16.09° \qquad \alpha_3 = 30.91° \qquad \alpha_4 = 32.87°$$

Note. It is not always necessary to eliminate the third harmonic (and triplen), which are not normally present in three-phase connections. Therefore, in three-phase inverters, it is preferable to eliminate the fifth, seventh, and eleventh harmonics of output voltages, so that the lowest-order harmonic is the thirteenth.

10-10 FORCED-COMMUTATED THYRISTOR INVERTERS

Although transistors or other devices can be employed as switching devices for inverters, they are used mostly in low- and medium-power applications. Transistors, GTOs, and IBGTs are becoming more competitive and taking over thyristors. For high-voltage and high-current applications, it is necessary to connect them in series and/or parallel combinations; and this results in an increased complexity of the circuit. Fast-switching thyristors, which are available in high voltage and current ratings, are more suitable for high-power applications. However, thyristors require extra commutation circuits for turn-off and the various techniques for thyristor commutation are discussed in Chapter 7. At the earlier stage of power electronics, many thyristor commutation circuits for inverters were developed. Two types of commutation circuits commonly used in inverter applications are:

1. Auxiliary commutated inverters
2. Complementary commutated inverters

10-10.1 Auxiliary-Commutated Inverters

A single-phase full-bridge thyristor inverter using auxiliary commutation is shown in Fig. 10-32a. A commutation circuit is shared by two thyristors. Let us assume that thyristor T_1 is conducting and supplies the peak load current, I_m; and the capacitor C_m is charged to V_o with polarity as shown. The waveforms for the capacitor voltage and current are shown in Fig. 10-32b. The commutation process is similar to that of the resonant pulse circuit in Fig. 7-13a. The commutation process of a thyristor can be divided into four modes.

Mode 1. This mode begins when thyristor T_{11} is fired to turn off thyristor T_1 which was conducting. Firing of T_{11} causes a resonant current flow through the capacitor and forces the current of T_1 to fall. This can be considered as a reverse current through the circuit formed by L_m, C_m, T_1, and T_{11}. This mode ends when the forward current of T_1 falls to zero and the capacitor current rises to the load current I_m at $t = t_1$.

Mode 2. This mode begins when diode D_1 starts to conduct and the resonant oscillation continues through L_m, C_m, D_1, and T_{11}. This mode ends when the capacitor current falls back to the load current at $t = t_2$ and diode D_1 stops conducting.

Mode 3. This mode starts when D_1 stops conducting. The capacitor recharges through the load at an approximately constant current of I_m. This mode ends when the capacitor voltage becomes equal to the dc supply voltage V_s at $t = t_3$ and tends to overcharge due to the energy stored in inductor L_m.

Mode 4. This mode begins when the capacitor voltage tends to be greater than V_s, and D_4 is forward biased. The energy stored in inductor L_m is transferred to the capacitor, causing it to be overcharged with respect to supply voltage, V_s. This mode ends when the capacitor current falls again to zero and the capacitor voltage is reversed to that of original polarity. The capacitor is now ready to turn off T_4 if T_{44} is fired.

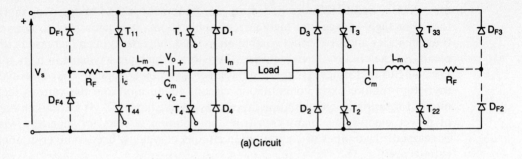

(a) Circuit

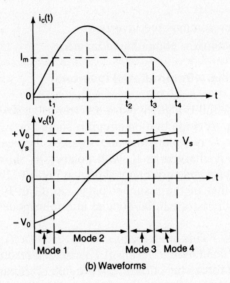

(b) Waveforms

Figure 10-32 Single-phase auxiliary commutated inverter.

This inverter is commonly known as the *McMurray inverter*. The circuit operation is similar to that in Fig. 7-13a. Equations (7-24) to (7-32) for the available turn-off time and the design conditions are applicable for this inverter circuit. From Eq. (7-24), the available turn-off time or reverse bias time is

$$t_{\text{off}} = \sqrt{L_m C_m} \left(\pi - 2 \sin^{-1} \frac{1}{x} \right) \tag{10-50}$$

where

$$x = \frac{V_o}{I_m} \sqrt{\frac{C_m}{L_m}} \tag{10-51}$$

$$V_o = V_s + I_m \sqrt{\frac{L_m}{C_m}} \tag{10-52}$$

In an inverter, the load current varies as a function of time and the commutation circuit should be designed for the peak load current. The capacitor voltage

V_o, which depends on the load current at the instant of commutation, increases the voltage and current ratings of devices and components. By connecting diodes, the excess energy can be returned to the dc source as shown in Fig. 10-32a by dashed lines. A part of the energy would be dissipated in resistor R, which may be replaced by a feedback winding discussed in Section 3-5.

10-10.2 Complementary-Commutated Inverters

If two inductors are tightly coupled, firing of one thyristor turns off another thyristor in the same arm. This type of commutation is known as *complementary commutation*. This principle can be applied to forced commutated inverter circuits, and Fig. 10-33a shows one arm of a single-phase full-bridge inverter. This circuit is also known as a *McMurray–Bedford* inverter. The circuit operation can be divided into three modes and the equivalent circuits for modes are shown in Fig. 10-33b. The waveforms for voltages and currents are shown in Fig. 10-33c under the assumption that the load current remains constant during the commutation period. In the following analysis, we shall redefine the time origin, $t = 0$, at the beginning of each mode.

Mode 1. This mode begins when T_2 is fired to turn off T_1 which was conducting. The equivalent circuit is shown in Fig. 10-33b. At the start of this mode, capacitor C_2 is charged to V_s. C_1 was shorted previously by T_1 and has no voltage. The voltage across L_2 is $v_{L2} = V_s$ and the current through L_2 induces a voltage of $v_{L1} = V_s$ across L_1. A reverse voltage of $v_{ak} = V_s - v_{L1} - v_{L2} = -V_s$ is applied across T_1 and the forward current of T_1 is forced to zero. i_{T1} falls to zero and i_{T2} rises to the level of the instantaneous load current, $i_{T2} = I_m$.

Assuming that $C_1 = C_2 = C_m$ and by completing the loop around C_1, C_2, and the dc source, the capacitor currents are described by

$$\frac{1}{C_m} \int i_{c1} \, dt + v_{c1}(t = 0) - \frac{1}{C_m} \int i_{c2} \, dt + v_{c2}(t = 0) = V_s \qquad (10\text{-}53)$$

Since $v_{c1}(t = 0) = 0$ and $v_{c2}(t = 0) = V_s$, Eq. (10-53) yields

$$i_{c1} = i_{c2} \qquad (10\text{-}54)$$

Using Kirchhoff's current law at the node B,

$$I_m - i_{c1} + i_1 - i_{c2} = 0 \qquad \text{or} \qquad I_m + i_1 = i_{c1} + i_{c2} = 2i_{c1}$$

or

$$i_{c1} = i_{c2} = \frac{I_m + i_1}{2} \qquad (10\text{-}55)$$

Assuming that $L_1 = L_2 = L_m$ and completing the loop formed by L_2, T_2, and C_2 gives

$$L_m \frac{di_1}{dt} + \frac{1}{C_m} \int i_{c2} \, dt - v_{c2}(t = 0) = 0 \qquad (10\text{-}56)$$

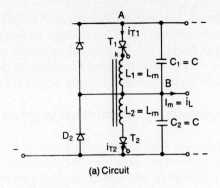

(a) Circuit

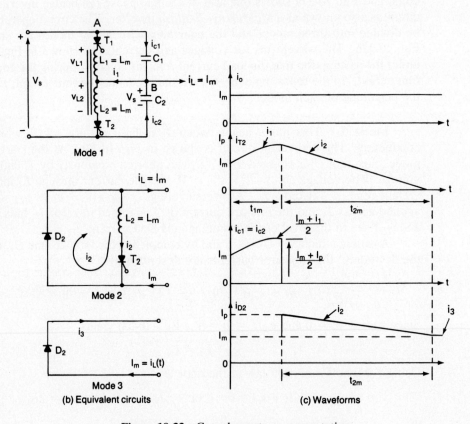

(b) Equivalent circuits

(c) Waveforms

Figure 10-33 Complementary commutation.

with initial conditions $i_1(t = 0) = I_m$ and $v_{c2}(t = 0) = V_s$. The solution of Eq. (10-56) with initial conditions yields

$$i_1(t) = 2I_m \cos \omega t + V_s \sqrt{\frac{2C_m}{L_m}} \sin \omega t - I_m \qquad (10\text{-}57)$$

where

$$\omega = \frac{1}{\sqrt{2L_m C_m}} \tag{10-58}$$

The voltage across inductor L_2,

$$v_{L2}(t) = v_{L1}(t) = v_{c2}(t) = L_m \frac{di_1}{dt} = V_s \cos \omega t - 2I_m \sqrt{\frac{L_m}{2C_m}} \sin \omega t \tag{10-59}$$

The reverse bias voltage across T_1 is

$$v_{ak}(t) = V_s - 2v_{L2} = V_s - 2V_s \cos \omega t + 4I_m \sqrt{\frac{L_m}{2C_m}} \sin \omega t \tag{10-60}$$

The available (or circuit) turn-off time can be determined from the condition $v_{ak}(t = t_{off}) = 0$ in Eq. (10-60), which after simplification gives

$$t_{off} = \sqrt{2L_m C_m} \left[\cos^{-1} \frac{1}{2(1 + x^2)^{1/2}} - \tan^{-1} x \right] \tag{10-61}$$

where

$$x = \frac{I_m}{V_s} \sqrt{\frac{2L_m}{C_m}} \tag{10-62}$$

The circuit turn-off time is dependent on the load current I_m and will be maximum when $I_m = 0$. The maximum value of t_{off} is

$$t_{off(max)} = \frac{\pi}{3} \sqrt{2L_m C_m} \tag{10-63}$$

This mode ends when the voltage on capacitor C_2 becomes zero and $v_{c2}(t)$ tends to charge in opposite direction. The time duration for this mode can be found from the condition $v_{L2}(t = t_{1m}) = v_{c2}(t = t_{1m}) = 0$, which is also the condition for peak thyristor current. From Eq. (10-59),

$$V_s \cos \omega t_{1m} - 2I_m \sqrt{\frac{L_m}{2C_m}} \sin \omega t_{1m} = 0$$

or

$$t_m = t_{1m} = \sqrt{2L_m C_m} \tan^{-1} \frac{1}{x} \tag{10-64}$$

The thyristor current i_{T2} becomes maximum at $t = t_m = t_{1m}$ and at the end of this mode,

$$i_{T2} = i_1(t = t_{1m}) = I_1 = I_p \tag{10-65}$$

Mode 2. This mode begins when diode D_2 starts conducting. The equivalent circuit is shown in Fig. 10-33b. The energy stored in inductor L_2 is lost in the circuit formed by T_2, D_2, and L_2. The load current $i_L(t) (= I_m)$ also flows through diode D_2. If V_d is the forward voltage drop of diode D_2 and thyristor T_2, the instantaneous current, $i_2(t)$, for mode 2 is given by

$$L_m \frac{di_2}{dt} + V_d = 0 \tag{10-66}$$

With initial condition $i_2(t = 0) = I_p$, the solution of Eq. (10-66) is

$$i_2(t) = I_p - \frac{V_d}{L_m} t \tag{10-67}$$

This mode ends when $i_2(t)$ falls to zero and thyristor T_2 is turned off due to self-commutation. The duration of this mode is approximately

$$t_{2m} = \frac{I_p L_m}{V_d} \tag{10-68}$$

Mode 3. This mode begins when T_2 is turned off. The equivalent circuit is shown in Fig. 10-33b. Diode D_2 continues to carry the load current until the load current falls to zero. The reverse-bias voltage for T_2 is provided by the forward voltage drop of diode D_2.

Example 10-7*

The single-phase complementary inverter in Fig. 10-33a has $L_1 = L_2 = L_m = 30\ \mu\text{H}$, $C_m = 50\ \mu\text{F}$, and the peak load current is $I_m = 175$ A. The dc input voltage is $V_s = 220$ V and the inverter frequency is $f_0 = 60$ Hz. The voltage drop of the circuit formed by thyristor T_2 and diode D_2 is, approximately, $V_d = 2$ V. Determine (a) the circuit turn-off time t_{off}; (b) the maximum circuit turn-off time $t_{\text{off(max)}}$ if $I_m = 0$; (c) the peak current of thyristors I_p; (d) the duration of the commutation process, $t_c = t_{1m} + t_{2m}$; and (e) the energy tapped in inductor L_2 at the end of mode 1.

Solution $V_s = 220$ V, $L_m = 30\ \mu\text{H}$, $C_m = 50\ \mu\text{F}$, and $I_m = 175$ A.

(a) From Eq. (10-62), $x = (175/220)\sqrt{2 \times 30/50} = 0.8714$. From Eq. (10-61),

$$t_{\text{off}} = \sqrt{2 \times 30 \times 50} \times \left[\cos^{-1} \frac{1}{2(1 + 0.8714^2)^{1/2}} - \tan^{-1}(0.8714) \right] = 25.6\ \mu\text{s}$$

(b) From Eq. (10-63), $t_{\text{off(max)}} = (\pi/3)\sqrt{2 \times 30 \times 50} = 57.36\ \mu\text{s}$.

(c) From Eq. (10-64), the time for the peak current is

$$t_m = t_{1m} = \sqrt{2 \times 30 \times 50}\ \tan^{-1} \frac{1}{0.8714} = 46.78\ \mu\text{s}$$

$$\omega = \frac{10^6}{\sqrt{2 \times 30 \times 50}} = 18{,}257\ \text{rad/s}$$

From Eqs. (10-57) and (10-65), the peak thyristor current is

$$I_p = 2 \times 175 \cos(1.8257 \times 0.4678) + 220 \sqrt{2 \times \frac{50}{30}} \sin(1.8257 \times 0.4678)$$

$$- 175 = 357.76\ \text{A}$$

(d) From Eq. (10-68), $t_{2m} = 175 \times 30\ \mu\text{s}/2 = 2625\ \mu\text{s}$ and the commutation time is

$$t_c = t_{1m} + t_{2m} = 46.78 + 2625 = 2671.78\ \mu\text{s}$$

(e) At the end of mode 1, the energy stored in inductor L_2 is

$$W = 0.5 L_m I_p^2 = 0.5 \times 30 \times 10^{-6} \times 357.76^2 = 1.92 \text{ J}$$

Note. It takes a relatively long time to dissipate the stored energy and reduces the efficiency and output frequency of the inverter. Due to this energy dissipation in the power devices, there could be a thermal problem. This trapped energy can be returned to the source by connecting a feedback transformer and diodes as shown in Fig. 10-34.

Example 10-8*

If the turns ratio of feedback transformer in Fig. 10-34 is $N_1/N_2 = a = 0.1$, determine (a) the duration of commutation process, $t_c = t_{1m} + t_{2m}$; (b) the energy trapped in inductor L_2 at the end of mode 1; and (c) the peak thyristor current I_p.

Solution $V_s = 220 \text{ V}$, $L_1 = L_2 = L_m = 30 \ \mu\text{H}$, $C = 50 \ \mu\text{F}$, and $I_m = 175 \text{ A}$. From Eq. (10-62),

$$x = \frac{175}{220} \sqrt{\frac{2 \times 30}{50}} = 0.8714$$

$$\omega = \frac{10^6}{\sqrt{2 \times 30 \times 50}} = 18{,}257 \text{ rad/s}$$

(a) The commutation interval can be divided into two modes. In the following analysis, we shall redefine the time origin, $t = 0$, at the beginning of each mode. The equivalent circuit for mode 1 is shown in Fig. 10-35a, which is the same as that in Fig. 10-33a. If v_1 and v_2 are the primary and secondary voltage of the feedback transformer, respectively, D_2 and D_{11} would conduct if

$$v_2 \le V_s \quad \text{or} \quad v_1 \le aV_s$$

$$v_1 = av_2 \tag{10-69}$$

where a is the turns ratio of the transformer and $a \le 1$. For D_2 and D_{11} to conduct,

$$v_1 = v_{L1} = v_{L2} = -aV_s \tag{10-70}$$

The duration of mode 1 can be found from Eq. (10-59):

$$v_{L2}(t = t_{1m}) = V_s \cos \omega t_{1m} - 2I_m \sqrt{\frac{L_m}{2C_m}} \sin \omega t_{1m} = -aV_s$$

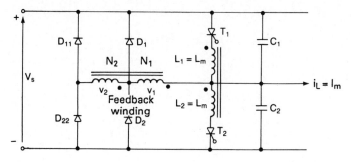

Figure 10-34 Complementary commutation with feedback windings.

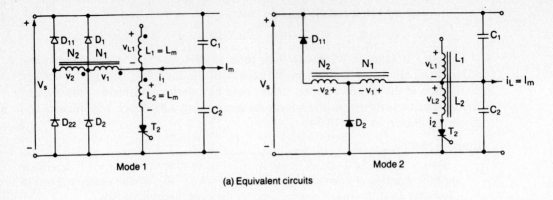

Mode 1

Mode 2

(a) Equivalent circuits

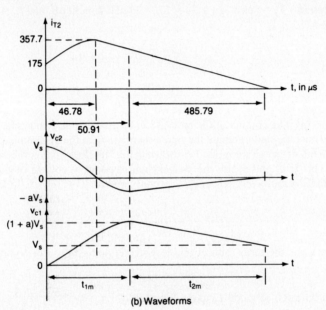

(b) Waveforms

Figure 10-35 Equivalent circuits and waveforms for Example 10-8.

and solving for t_{1m} yields

$$t_{1m} = \sqrt{2L_m C_m} \left[\sin^{-1} \frac{a}{(1 + x^2)^{1/2}} + \tan^{-1} \frac{1}{x} \right]$$

$$= 50.91 \ \mu s \tag{10-71}$$

This mode ends when D_2 and D_{11} conducts. At the end of mode 1, Eq. (10-57) gives the current in thyristor T_2:

$$i_{T2} = i_1(t = t_{1m}) = I_1 \tag{10-72}$$

From Eqs. (10-57) and (10-72),

$$I_1 = 2 \times 175 \cos(1.8257 \times 0.5091) + 220 \sqrt{2 \times \frac{50}{30}} \sin(1.8257 \times 0.5091)$$

$$- 175 = 356.24 \text{ A}$$

The voltage on capacitor C_2

$$V_{c2} = V_{L2} = -aV_s$$

$$= -0.1 \times 220 = -22 \text{ V} \tag{10-73}$$

The voltage on capacitor C_1

$$V_{c1} = V_s - V_{c2} = (1 + a)V_s$$

$$= 1.1 \times 220 = 242 \text{ V} \tag{10-74}$$

Mode 2 begins when D_2 and D_{11} conduct and the voltage of inductor L_2 is clamped to $-aV_s$. The equivalent circuit is shown in Fig. 10-35a.

$$v_{L2} = L_m \frac{di_2}{dt} = -av_2 = -aV_s \tag{10-75}$$

with initial condition $i_2(t = 0) = I_1$ and the solution of Eq. (10-75) yields

$$i_2(t) = I_1 - \frac{aV_s}{L_m} t \tag{10-76}$$

Mode 2 ends when $i_2(t)$ becomes zero at $t = t_{2m}$ and

$$t_{2m} = \frac{L_m I_1}{aV_s}$$

$$= 30 \times 10^{-6} \times \frac{356.24}{0.1 \times 220} = 485.78 \ \mu s \tag{10-77}$$

The commutation time is

$$t_c = t_{1m} + t_{2m} = 50.91 + 485.79 = 536.7 \ \mu s$$

(b) At the end of mode 1, the energy stored in inductor L_2 is

$$W = 0.5 L_m I_1^2 = 0.5 \times 30 \times 10^{-6} \times 356.24^2 = 1.904 \text{ J}$$

(c) From Eq. (10-64), $t_m = 46.78 \ \mu s$, and from Eq. (10-57), $I_p = 357.76 \text{ A}$.

Note. The trapped energy is returned to the supply. The commutation time can be reduced by lowering the turns ratio a; this would increase the voltage ratings of feedback diodes.

In the previous sections the inverters are fed from a voltage source and the load current is forced to fluctuate from positive to negative, and vice versa. To cope with inductive loads, the power switches with freewheeling diodes are required, whereas in a current-source inverter (CSI), the input behaves as a current source. The output current is maintained constant irrespective of load on the inverter and the output voltage is forced to change. The circuit diagram of a single-phase transistorized inverter is shown in Fig. 10-36a. Since there must be a continuous current flow from the source, two switches must always conduct—one from the upper and one from the lower switches. The conduction sequence is 12, 23, 34, and 41 as shown in Fig. 10-36b. The output current waveform is shown in Fig. 10-36c. The diodes in series with the transistors are required to block the reverse voltages on the transistors.

When two devices in different arms conduct, the source current I_L flows through the load. When two devices in the same arm conduct, the source current is bypassed from the load. The design to the current source is similar to Example 9-14. From Eq. (10-19), the load current can be expressed as

$$i_0 = \sum_{n=1,3,5,...}^{\infty} \frac{4I_L}{n\pi} \sin \frac{n\delta}{2} \sin n(\omega t) \tag{10-78}$$

With a current-source inverter, the commutation circuits for thyristors require only capacitors and are simpler, as shown in Fig. 10-37c. Let us assume that T_1 and T_2 are conducting, and capacitors C_1 and C_2 are charged with polarity as shown. Firing of thyristors T_3 and T_4 reverse biases thyristors T_1 and T_2. T_1 and T_2 are turned off by impulse commutation. The current now flows through $T_3C_1D_1$, load, and $D_2C_2T_4$. The capacitors C_1 and C_2 are discharged and recharged at a constant rate determined by load current, $I_m = I_L$. When C_1 and C_2 are charged to the load voltage and their currents fall to zero, the load current will be transferred from diode D_1 to D_3 and D_2 to D_4. D_1 and D_2 will be turned off when the load current is completely reversed. The capacitor is now ready to turn off T_3 and T_4 if thyristors T_1 and T_2 are fired in the next half-cycle. The commutation time will depend on the magnitude of load current and load voltage. The diodes in Fig. 10-37c isolate the capacitors from the load voltage.

Figure 10-38a shows the circuit diagram of a three-phase current-source inverter. The waveforms for gating signals and line currents for a wye-connected load are shown in Fig. 10-38b. At any instant, only two thyristors conduct at the same time. Each device conducts for 120°. From Eq. (10-14), the current for phase a can be expressed as

$$i_a = \sum_{n=1,3,5,...}^{\infty} \frac{4I_L}{n\pi} \cos \frac{n\pi}{6} \sin n\left(\omega t + \frac{\pi}{6}\right) \tag{10-79}$$

The PWM, SPWM, MSPWM, or MSPWN technique can be applied to vary the load current and to improve the quality of its waveform.

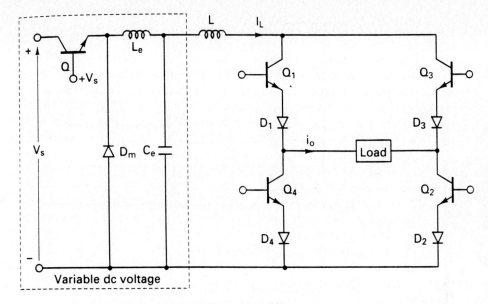

(a) Transistor CSI

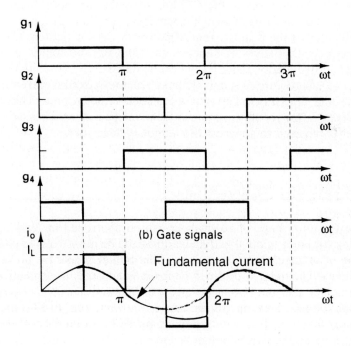

(b) Gate signals

(c) Load current

Figure 10-36 Single-phase current source.

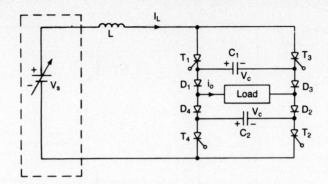

Figure 10-37 Single-phase thyristor current-source inverter.

The current-source inverter (CSI) is a dual of a voltage-source inverter (VSI). The line-to-line voltage of a VSI is similar in shape to the line current of a CSI. The advantages of the CSI are: (1) since the input dc current is controlled and limited, misfiring of switching devices, or a short circuit, would not be serious problems; (2) the peak current of power devices is limited; (3) the commutation circuits for thyristors are simpler; and (4) it has the ability to handle reactive or regenerative load without freewheeling diodes.

A CSI requires a relatively large reactor to exhibit current-source characteristics and an extra converter stage to control the current. The dynamic response is slower. Due to current transfer from one pair of switches to another, an output filter is required to suppress the output voltage spikes.

10-12 VARIABLE DC-LINK INVERTER

The output voltage of an inverter can be controlled by varying the modulation index (or pulse widths) and maintaining the dc input voltage constant; but in this type of voltage control, a range of harmonics would be present on the output voltage. The pulse widths can be maintained fixed to eliminate or reduce certain harmonics and the output voltage can be controlled by varying the level of dc input voltage. Such an arrangement as shown in Fig. 10-39 is known as a *variable dc-link inverter*. This arrangement requires an additional converter stage; and if it is a chopper, the power cannot be fed back to the dc source. In order to obtain the desired quality and harmonics of the output voltage, the shape of the output voltage can be predetermined, as shown in Fig. 10-2b or Fig. 10-36. The dc supply is varied to give variable ac output.

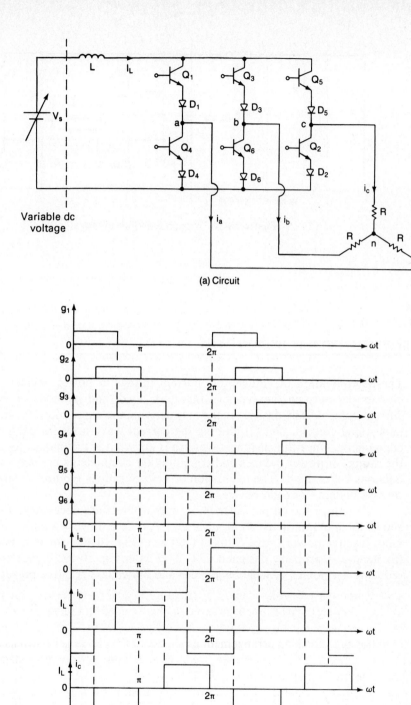

(a) Circuit

(b) Waveforms

Figure 10-38 Three-phase current source transistor inverter.

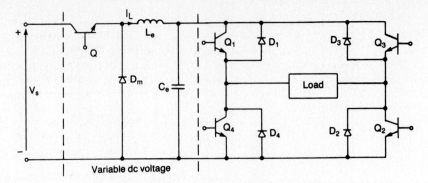

Figure 10-39 Variable dc-link inverter.

10-13 INVERTER CIRCUIT DESIGN

The determination of voltage and current ratings of power devices in inverter circuits depends on the types of inverters, load, and methods of voltage and current control. The design requires (1) deriving the expressions for the instantaneous load current, and (2) plotting the current waveforms for each device and component. Once the current waveform is known, the techniques for calculating the ratings of power devices and commutation components are similar to that in Sections 7-4 and 9-8. The evaluation of voltage ratings requires establishing the reverse voltages of each device.

To reduce the output harmonics, output filters are necessary. Figure 10-40 shows the commonly used output filters. A C-filter is very simple, but it draws more reactive power. An LC-tuned filter as in Fig. 10-40b can eliminate only one frequency. A properly designed CLC-filter as in Fig. 10-40c is more effective in reducing harmonics of wide bandwidth and draws less reactive power.

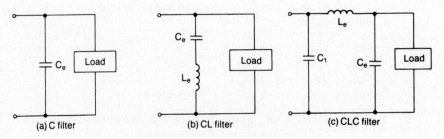

Figure 10-40 Output filters.

Example 10-9*

The single-phase full-bridge inverter in Fig. 10-2a supplies a load of $R = 10\ \Omega$, $L = 31.5$ mH, and $C = 112\ \mu$F. The dc input voltage is $V_s = 220$ V and the inverter frequency is $f_o = 60$ Hz. The output voltage has two notches such that third and fifth harmonics are eliminated. (a) Determine the expression for the load current $i_o(t)$. (b) If an output C-filter is used to eliminate seventh and higher-order harmonics, determine the filter capacitance C_e.

Solution The output voltage waveform is shown in Fig. 10-28. $V_S = 220$ V, $f_o = 60$ Hz, $R = 10\ \Omega$, $L = 31.5$ mH, and $C = 112\ \mu$F. $\omega_o = 2\pi \times 60 = 377$ rad/s.

The inductive reactance for the nth harmonic voltage is

$$X_L = j2n\pi \times 60 \times 31.5 \times 10^{-3} = j11.87n\ \Omega$$

The capacitive reactance for the nth harmonic voltage is

$$X_c = -\frac{j10^6}{2n\pi \times 60 \times 112} = -\frac{j23.68}{n}\ \Omega$$

The impedance for the nth harmonic voltage is

$$|Z_n| = \left[10^2 + \left(11.87n - \frac{23.68}{n}\right)^2\right]^{1/2}$$

and the power factor angle for the nth harmonic voltage is

$$\theta_n = \tan^{-1}\frac{11.87n - 23.68/n}{10} = \tan^{-1}\left(1.187n - \frac{2.368}{n}\right)$$

(a) Equation (10-47) gives the coefficients of the Fourier series,

$$B_n = \frac{4V_s}{\pi}\frac{1 - 2\cos n\alpha_1 + 2\cos n\alpha_2}{n}$$

For $\alpha_1 = 23.62°$ and $\alpha_2 = 33.3°$, the third and fifth harmonics would be absent. From Eq. (10-46) the instantaneous output voltage can be expressed as

$$v_o(t) = 235.1\sin 377t + 69.4\sin(7 \times 377t) + 85.1\sin(9 \times 377t) + \cdots$$

Dividing the output voltage by the load impedance and considering the appropriate delay due to the power factor angles give the load current as

$$i_o(t) = 15.19\sin(377t + 49.74°) + 0.86\sin(7 \times 377t - 82.85°)$$
$$+ 1.09\sin(9 \times 377t - 84.52°) + 0.66\sin(11 \times 377t - 85.55°) + \cdots$$

(b) The nth and higher-order harmonics would be reduced significantly if the filter impedance is much smaller than that of the load, and a ratio of $1:10$ is normally adequate,

$$|Z_n| = 10X_e$$

where the filter impedance is $|X_e| = 1/(377nC_e)$. The value of filter capacitance C_e can be found from

$$\left[10^2 + \left(11.87n - \frac{23.68}{n}\right)^2\right]^{1/2} = \frac{10}{377nC_e}$$

For the seventh harmonic, $n = 7$ and $C_e = 47.3\ \mu$F.

Example 10-10

The inverter in Example 10-9 uses auxiliary commutated thyristors, shown in Fig. 10-32a. The turn-off time of thyristors is $t_q = t_{off} = 25$ μs. Determine (a) the optimum values of commutation capacitor C_m and inductor L_m to minimize the energy, and (b) the peak currents of thyristors T_1 and diode D_1.

Solution (a) Neglecting the harmonics, the peak load current, $I_m = I_p = 15.19$ A. Equations (7-31) and (7-32) give the optimum values of L_m and C_m.

$$L_m = \frac{0.398t_{off}V_o}{I_p} = \frac{0.398 \times 25 \times 10^{-6}}{15.19} V_o$$

$$C_m = \frac{0.8917t_{off}I_p}{V_o} = \frac{0.8917 \times 25 \times 10^{-6} \times 15.19}{V_o}$$

From Eq. (10-52),

$$V_o = V_s + I_p \sqrt{\frac{L_m}{C_m}} = 220 + 15.19 \sqrt{\frac{L_m}{C_m}}$$

Substituting for V_o yields relations between C_m and L_m. Solving for C_m and L_m yields $C_m = 0.51$ μF, $L_m = 434$ μH, and $V_o = 662.6$ V.

(b) The peak thyristor current $I_p = 15.19$ A, and the peak resonant current

$$I_{pk} = V_o \sqrt{\frac{C_m}{L_m}} = 662.6 \sqrt{\frac{0.51}{434}} = 22.77 \text{ A}$$

From Fig. 10-32b the peak diode current due to resonant oscillation is

$$I_{pd} = I_{pk} - I_p = 22.77 - 15.19 = 7.58 \text{ A}$$

Diode D_1 will carry both the inductive load current and resonant current. Although these two components would determine the rms and average currents of the diode, the peak diode current in this example should be the same as that of thyristor current, namely, 15.19 A.

Example 10-11

The single-phase inverter of Fig. 10-2a uses the PWM control as shown in Fig. 10-13a with five pulses per half-cycle. The dc supply voltage is $V_s = 100$. The modulation index M is 0.6. The output frequency is $f_o = 60$ Hz. The load is resistive with $R = 2.5$ Ω. Use PSpice (a) to plot the output voltage v_o, and (b) to calculate its Fourier coefficients. The SPICE model parameters of the transistor are IS=6.734F, BF=416.4, CJC=3.638P, CJE=4.493P, and that of diodes are IS=2.2E−15, BV=1800V, TT=0.

Solution (a) $M = 0.6$, $f_o = 60$ Hz, $T = 1/f_o = 16.667$ ms. The inverter for PSpice simulation is shown in Fig. 10-41a. An op-amp as shown in Fig. 10-41b is used as a comparator and produces the PWM control signals. The carrier and reference signals are shown in Fig. 10-41c. The list of the circuit file is as follows:

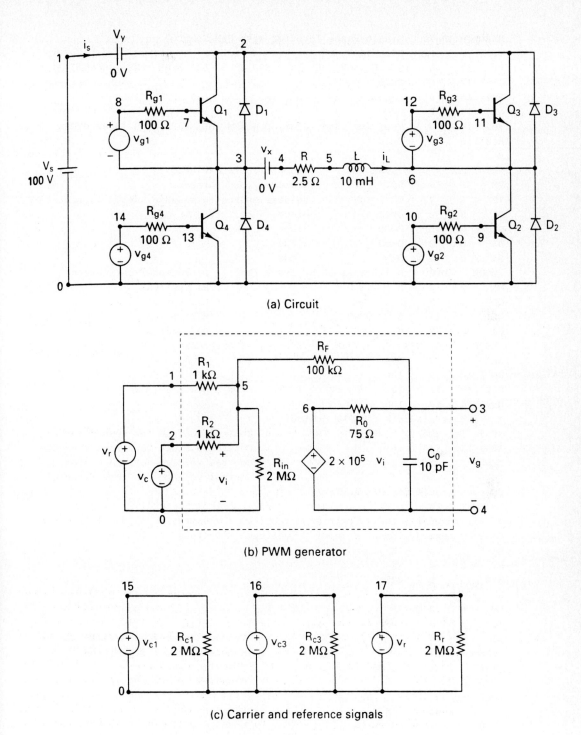

(a) Circuit

(b) PWM generator

(c) Carrier and reference signals

Figure 10-41   Single-phase inverter for PSpice simulation.

```
Example 10-11    Single-Phase Inverter with PWM Control
VS    1    0    DC    100V
Vr    17   0    PULSE (50V  0V   0    833.33US  833.33US  INS  16666.67US)
Rr    17   0    2MEG
Vcl   15   0    PULSE (0  −30V  0    INS   INS   8333.33US  16666.67US)
Rcl   15   0    2MEG
Vc3   16   0    PULSE (0  −30V  8333.33US  INS  INS  8333.33US  16666.67US)
Rc3   16   0    2MEG
R     4    5    2.5
*L    5    6    10MH          ; Inductor L is excluded
VX    3    4    DC    0V      ; Measures load current
VY    1    2    DC    0V      ; Voltage source to measure supply current
D1    3    2    DMOD          ; Diode
D2    0    6    DMOD          ; Diode
D3    6    2    DMOD          ; Diode
D4    0    3    DMOD          ; Diode
.MODEL    DMOD    D (IS=2.2E−15 BV=1800V TT=0) ; Diode model parameters
Q1    2    7    3    QMOD                    ; BJT switch
Q2    6    9    0    QMOD                    ; BJT switch
Q3    2    11   6    QMOD                    ; BJT switch
Q4    3    13   0    QMOD                    ; BJT switch
.MODEL   QMOD NPN (IS=6.734F BF=416.4 CJC=3.638P CJE=4.493P) ; BJT parameters
Rg1   8    7    100
Rg2   10   9    100
Rg3   12   11   100
Rg4   14   13   100
*    Subcircuit call for PWM control
XPW1  17   15   8    3    PWM      ; Control voltage for transistor Q1
XPW2  17   15   10   0    PWM      ; Control voltage for transistor Q2
XPW3  17   16   12   6    PWM      ; Control voltage for transistor Q3
XPW4  17   16   14   0    PWM      ; Control voltage for transistor Q4
*    Subcircuit for PWM control
.SUBCKT   PWM      1        2        3         4
*         model    ref.     carrier  +control  −control
*         name     input    input    voltage   voltage
R1    1    5    1K
R2    2    5    1K
RIN   5    0    2MEG
RF    5    3    100K
RO    6    3    75
CO    3    4    10PF
E1    6    4    0    5    2E+5        ; Voltage-controlled voltage source
.ENDS    PWM                         ; Ends subcircuit definition
.TRAN    10US  16.67MS   0    10US   ; Transient analysis
.PROBE                               ; Graphics postprocessor
.options abstol = 1.00n  reltol = 0.01  vntol = 0.1 ITL5=20000 ; convergence
.FOUR    60HZ    V(3,6)              ; Fourier analysis
.END
```

        The PSpice plots are shown in Fig. 10-42, where V(17) = reference signal and
V(3,6) = output voltage.

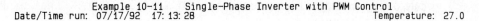

Example 10-11   Single-Phase Inverter with PWM Control
Date/Time run: 07/17/92 17: 13: 28   Temperature: 27.0

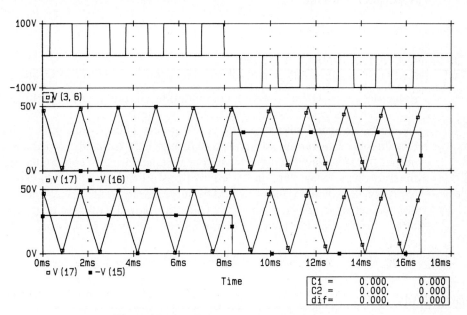

Figure 10-42 PSpice plots for Example 10-11.

(b) FOURIER COMPONENTS OF TRANSIENT RESPONSE V(3,6)
DC COMPONENT = 6.335275E−03

HARMONIC NO	FREQUENCY (HZ)	FOURIER COMPONENT	NORMALIZED COMPONENT	PHASE (DEG)	NORMALIZED PHASE (DEG)
1	6.000E+01	7.553E+01	1.000E+00	6.275E−02	0.000E+00
2	1.200E+02	1.329E−02	1.759E−04	5.651E+01	5.645E+01
3	1.800E+02	2.756E+01	3.649E−01	1.342E−01	7.141E−02
4	2.400E+02	1.216E−02	1.609E−04	6.914E+00	6.852E+00
5	3.000E+02	2.027E+01	2.683E−01	4.379E−01	3.752E−01
6	3.600E+02	7.502E−03	9.933E−05	−4.924E+01	−4.930E+01
7	4.200E+02	2.159E+01	2.858E−01	4.841E−01	4.213E−01
8	4.800E+02	2.435E−03	3.224E−05	−1.343E+02	−1.343E+02
9	5.400E+02	4.553E+01	6.028E−01	6.479E−01	5.852E−01

TOTAL HARMONIC DISTORTION = 8.063548E+01 PERCENT

Note. For $M = 0.6$ and $p = 5$, the computer program FIG10-11.BAS for uniform PWM gives $V_1 = 54.59$ V (rms) and THD = 100.65% as compared to values of $V_1 = 75.53/\sqrt{2} = 53.41$ V (rms) and THD = 80.65% from PSpice. In calculating the THD, PSpice uses only up to ninth harmonics by default, instead of all harmonics. Thus if the harmonics higher than ninth have significant values as compared to the fundamental component, PSpice will give a low and erroneous value of THD. However the PSpice version 5.1 (or higher) allows an argument to specify the number of harmonics to be calculated. For example, the statement for

---

Sec. 10-13   Inverter Circuit Design   **409**

calculating up to 30th harmonic will be .FOUR 60HZ 30 V(3,6). The default value is the ninth harmonic.

10-14 MAGNETIC CONSIDERATIONS

Inductors are used in thyristor commutation circuits and for input and output filters. The magnetic loss is dependent on the frequency and these inductors should be designed with magnetic cores of very high permeability to reduce the core losses. The output of inverters are normally isolated from the load by an output transformer. The inverter output voltage normally contain harmonics and the transformer losses are increased. A transformer, which is designed to operate at purely sinusoidal voltages, would be subjected to higher losses and it should be derated when it is being operated from the output voltages of inverters. The output voltage should have no dc component; otherwise, the core may be saturated.

SUMMARY

Inverters can provide single-phase and three-phase ac voltages from a fixed or variable dc voltage. There are various voltage control techniques and they produce a range of harmonics on the output voltage. The sinusoidal pulse-width modulation (SPWM) is more effective in reducing the lower-order harmonics. With a proper choice of the switching patterns for power devices, certain harmonics can be eliminated. Due to the development of fast-switching power devices such as transistors, GTOs, IGBTs, and MCTs the applications of forced-commutated thyristor inverters are limited to high power inverters.

REFERENCES

1. B. D. Bedford and R. G. Hoft, *Principle of Inverter Circuits*. New York: John Wiley & Sons, Inc., 1964.

2. H. S. Patel and R. G. Hoft, "Generalized techniques of harmonic elimination and voltage control in thyristor converter." *IEEE Transactions on Industry Applications*, Vol. IA9, No. 3, 1973, pp. 310–317, and Vol. IA10, No. 5, 1974, pp. 666–673.

3. T. Ohnishi and H. Okitsu, "A novel PWM technique for three-phase inverter/con-verter." *International Power Electronics Conference*, 1983, pp. 384–395.

4. M. F. Schlecht, "Novel topologies alternatives to the design of a harmonic-free utility/dc interface." *Power Electronic Specialist Conference*, 1983, pp. 206–216.

5. P. D. Ziogas, V. T. Ranganathan, and V. R. Stefanovic, "A four-quadrant current regulated converter with a high frequency link." *IEEE Transactions on Industry Applications*, Vol. IA18, No. 5, 1982, pp. 499–505.

6. K. Taniguchi and H. Irie, "Trapezoidal modulating signal for three-phase PWM inverter." *IEEE Transactions on Industrial Electronics,* Vol. IE3, No. 2, 1986, pp. 193–200.

7. K. Thorborg and A. Nystorm, "Staircase PWM: an uncomplicated and efficient modulation technique for ac motor drives." *IEEE Transactions on Power Electronics,* Vol. PE3, No. 4, 1988, pp. 391–398.

8. J. C. Salmon, S. Olsen, and N. Durdle, "A three-phase PWM strategy using a stepped reference waveform." *IEEE Transactions on Industry Applications,* Vol. IA27, No. 5, 1991, pp. 914–920.

9. M. A. Boost and P. D. Ziogas, "State-of-the-art carrier PWM techniques: a critical evaluation." *IEEE Transactions on Industry Applications,* Vol. IA24, No. 2, 1988, pp. 271–279.

10. K. Taniguchi and H. Irie, "PWM technique for power MOSFET inverter." *IEEE Transactions on Power Electronics,* Vol. PE3, No. 3, 1988, pp. 328–334.

11. P. D. Ziogas, "The delta modulation techniques in static PWM inverters." *IEEE Transactions on Industry Applications,* March/April 1981, pp. 199–204.

12. A. A. Rahman, J. E. Quaicoe, and M. A. Chowdhury, "Performance analysis of delta modulated PWM inverters." *IEEE Transactions on Power Electronics,* Vol. PE2, No. 3, 1987, pp. 227–232.

REVIEW QUESTIONS

10-1. What is an inverter?

10-2. What is the principle of operation of an inverter?

10-3. What are the types of inverters?

10-4. What are the differences between half-bridge and full-bridge inverters?

10-5. What are the performance parameters of inverters?

10-6. What are the purposes of feedback diodes in inverters?

10-7. What are the arrangements for obtaining three-phase output voltages?

10-8. What are the methods for voltage control within the inverters?

10-9. What are the advantages and disadvantages of displacement-angle control?

10-10. What are the techniques for harmonic reductions?

10-11. What are the effects of eliminating lower-order harmonics?

10-12. What is the effect of thyristor turn-off time on inverter frequency?

10-13. What are the advantages and disadvantages of transistorized inverters compared to thyristor inverters?

10-14. What is the principle of auxiliary commutated inverters?

10-15. What is the principle of complimentary commutated inverters?

10-16. What is the purpose of feedback transformer in complementary commutated inverters?

10-17. What are the advantages and disadvantages of current-source inverters?

10-18. What are the main differences between voltage-source and current-source inverters?

10-19. What are the main advantages and disadvantages of variable dc link inverters?

10-20. What are the reasons for adding a filter on the inverter output?

10-21. What are the differences between ac and dc filters?

PROBLEMS

10-1. The single-phase half-bridge inverter in Fig. 10-1a has a resistive load of $R = 10$ Ω and the dc input voltage is $V_s = 220$ V. Determine **(a)** the rms output voltage at the fundamental frequency, V_1; **(b)** the output power P_o; **(c)** the average, rms, and peak currents of each transistor; **(d)** the peak off-state voltage V_{BR} of each transistor; **(e)** the total harmonic distortion THD; **(f)** the distortion factor DF; and **(g)** the harmonic factor and distortion factor of the lowest-order harmonic.

10-2. Repeat Prob. 10-1 for the single-phase full-bridge inverter in Fig. 10-2a.

10-3. The full-bridge inverter in Fig. 10-2a has an RLC load with $R = 5$ Ω, $L = 10$ mH, and $C = 26$ μF. The inverter frequency, $f_o = 400$ Hz, and the dc input voltage, $V_s = 220$ V. **(a)** Express the instantaneous load current in a Fourier series. Calculate **(b)** the rms load current at the fundamental frequency, I_1; **(c)** the THD of the load current; **(d)** the average supply current I_s; and **(e)** the average, rms, and peak currents of each transistor.

10-4. Repeat Prob. 10-3 for $f_o = 60$ Hz, $R = 4$ Ω, $L = 25$ mH, and $C = 10$ μF.

10-5. Repeat Prob. 10-3 for $f_o = 60$ Hz, $R = 5$ Ω, and $L = 20$ mH.

10-6. The three-phase full-bridge inverter in Fig. 10-5a has a wye-connected resistive load of $R = 5$ Ω. The inverter frequency is $f_o = 400$ Hz and the dc input voltage is $V_s = 220$ V. Express the instantaneous phase voltages and phase currents in a Fourier series.

10-7. Repeat Prob. 10-6 for the line-to-line voltages and line currents.

10-8. Repeat Prob. 10-6 for a delta-connected load.

10-9. Repeat Prob. 10-7 for a delta-connected load.

10-10. The three-phase full-bridge inverter in Fig. 10-5a has a wye-connected load and each phase consists of $R = 5$ Ω, $L = 10$ mH, and $C = 25$ μF. The inverter frequency is $f_o = 60$ Hz and the dc input voltage, $V_s = 220$ V. Determine the rms, average, and peak currents of the transistors.

10-11. The output voltage of a single-phase full-bridge inverter is controlled by pulse-width modulation with one pulse per half-cycle. Determine the required pulse width so that the fundamental rms component is 70% of dc input voltage.

10-12. A single-phase full-bridge inverter uses a uniform PWM with two pulses per half-cycle for voltage control. Plot the distortion factor, fundamental component, and lower-order harmonics against modulation index.

10-13. A single-phase full-bridge inverter, which uses a uniform PWM with two pulses per half-cycle, has a load of $R = 5$ Ω, $L = 15$ mH, and $C = 25$ μF. The dc input voltage is $V_s = 220$ V. Express the instantaneous load current $i_o(t)$ in a Fourier series for $M = 0.8$, $f_o = 60$ Hz.

10-14. A single-phase full-bridge inverter uses a uniform PWM with seven pulses per half-cycle for voltage control. Plot the distortion factor, fundamental component, and lower-order harmonics against the modulation index.

10-15. A single-phase full-bridge inverter uses an SPWM with seven pulses per half-cycle for voltage control. Plot the distortion factor, fundamental component, and lower-order harmonics against the modulation index.

10-16. Repeat Prob. 10-15 for the modified SPWM with two pulses per quarter-cycle.

10-17. A single-phase full-bridge inverter uses a uniform PWM with five pulses per half-cycle. Determine the pulse width if the rms output voltage is 80% of the dc input voltage.

10-18. A single-phase full-bridge inverter uses displacement-angle control to vary the output voltage and has one pulse per half-cycle, as shown in Fig. 10-20a. Determine the delay (or displacement) angle if the fundamental component of output voltage is 70% of dc input voltage.

10-19. A single-phase full-bridge inverter uses multiple bipolar notches and it is required to eliminate third, fifth, seventh, and eleventh harmonics from the output waveform. Determine the number of notches and their angles.

10-20. Repeat Prob. 10-19 to eliminate third, fifth, seventh, and ninth harmonics.

10-21. The single-phase full-bridge auxiliary-commutated inverter in Fig. 10-32a has a load of $R = 5\ \Omega$, $L = 10$ mH, and $C = 25$ μF. The input dc voltage is $V_s = 220$ V and the inverter frequency is $f_0 = 60$ Hz. If $t_{off} = 18\ \mu$s, determine the optimum values of commutation components C_m and L_m.

10-22. Repeat Prob. 10-21 if the peak resonant current of the commutation circuit is limited to two times the peak load current.

10-23. A single-phase full-bridge inverter in Fig. 10-33a, which uses complementary commutation, has $L_1 = L_2 = L_m = 40\ \mu$H and $C_m = 60\ \mu$F. The peak load current is $I_m = 200$ A. The dc input voltage, $V_s = 220$ V and the inverter frequency, $f_0 = 60$ Hz. The voltage drop of the circuit formed by thyristor T_2 and diode D_2 is, approximately, $V_d = 2$ V. Determine **(a)** the circuit turn-off time t_{off}; **(b)** the maximum available turn-off time $t_{off(max)}$ if $I_m = 0$; **(c)** the peak current of thyristors; **(d)** the duration of commutation process, $t_c = t_{1m} + t_{2m}$; and **(e)** the energy trapped in inductor L_2 at the end of mode. 1.

10-24. The inverter in Prob. 10-23 has a feedback transformer as shown in Fig. 10-34. The turns ratio of the feedback windings is $N_1/N_2 = a = 0.1$. Determine **(a)** the duration of commutation process, $t_c = t_{1m} + t_{2m}$; **(b)** the energy trapped in inductor L_2 at the end of mode 1; and **(c)** the peak thyristor current I_p.

10-25. Repeat Prob. 10-24 if $a = 1.0$.

10-26. Repeat Prob. 10-24 if $a = 0.01$.

10-27. A single-phase auxiliary-commutated full-bridge inverter in Fig. 10-32a supplies a load of $R = 5\ \Omega$, $L = 15$ mH, and $C = 30\ \mu$F. The dc input voltage is $V_s = 220$ V and the inverter frequency is $f_0 = 400$ Hz. The turn-off time of thyristors is $t_q = t_{off} = 20\ \mu$s. The output voltage has two notches such that third and fifth harmonics are eliminated. Determine **(a)** the expression for the instantaneous load current, $i_0(t)$; **(b)** the optimum values of commutation capacitor C_m and inductor L_m to minimize the energy; and **(c)** the average, rms, and peak currents of thyristor T_1 and diode D_1.

10-28. If in Prob. 10-27, a tuned LC filter is used to eliminate the seventh harmonic from the output voltage, determine the suitable values of filter components.

10-29. The single-phase auxiliary-commutated full-bridge inverter in Fig. 10-32a supplies a load of $R = 2\ \Omega$, $L = 25$ mH, and $C = 40\ \mu$F. The dc input voltage is $V_s = 220$ V and the inverter frequency, $f_0 = 60$ Hz. The turn-off time of thyristors, $t_q = t_{off} = 15\ \mu$s. The output voltage has three notches such that the third, fifth, and seventh harmonics are eliminated. Determfine **(a)** the expression for the instantaneous load current $I_0(t)$, and **(b)** the values of commutation capacitor C_m and inductor L_m if the peak resonant current is limited to 2.5 times the peak load current.

10-30. If in Prob. 10-29, an output C filter is used to eliminate ninth and higher-order harmonics, determine the value of filter capacitor C_e.

10-31. Repeat Example 10-11 for a SPWM.

Resonant pulse converters

11-1 INTRODUCTION

The switching devices in converters with a PWM control can be gated to synthesize the desired shape of the output voltage and/or current. However, the devices are turned "on" and "off" at the load current with a high di/dt value. The switches are subjected to a high-voltage stress, and the switching power loss of a device increases linearly with the switching frequency. The turn-on and turn-off loss could be a significant portion of the total power loss. The electromagnetic interference is also produced due to high di/dt and dv/dt in the converter waveforms.

The disadvantages of PWM control can be eliminated or minimized if the switching devices are turned "on" and "off" when the voltage across a device and/or its current become zero. The voltage and current are forced to pass through zero crossing by creating an LC-resonant circuit, thereby calling a *resonant pulse converter*. The resonant converters can be classified into eight types:

Series-resonant inverters
Parallel-resonant inverters
Class E resonant converter
Class E resonant rectifier
Zero-voltage-switching (ZVS) resonant converters
Zero-current-switching (ZCS) resonant converters
Two-quadrant zero-voltage-switching (ZVS) resonant converters
Resonant dc-link inverters

The series resonant inverters are based on resonant current oscillation. The commutating components and switching device are placed in series with the load to form an underdamped circuit. The current through the switching devices falls to zero due to the natural characteristics of the circuit. If the switching element is a thyristor, it is said to be self-commutated. This type of inverter produces an approximately sinusoidal waveform at a high output frequency, ranging from 200 Hz to 100 kHz, and is commonly used in relatively fixed output applications (e.g., induction heating, sonar transmitter, fluorescent lighting, or ultrasonic generators). Due to the high switching frequency, the size of commutating components is small.

There are various configurations of series resonant inverters, depending on the connections of the switching devices and load. The series inverters may be classified into two categories:

1. Series resonant inverters with unidirectional switches
2. Series resonant inverters with bidirectional switches

11-2.1 Series Resonant Inverters with Unidirectional Switches

Figure 11-1a shows the circuit diagram of a simple series inverter using two unidirectional thyristor switches. When thyristor T_1 is fired, a resonant pulse of current flows through the load and the current falls to zero at $t = t_{1m}$ and T_1 is self-commutated. Firing of thyristor T_2 causes a reverse resonant current through the load and T_2 is also self-commutated. The circuit operation can be divided into three modes and the equivalent circuits are shown in Fig. 11-1b. The gating signals for thyristors and the waveforms for the load current and capacitor voltage are shown in Fig. 11-1c.

The series resonant circuit formed by L, C, and load (assumed resistive) must be underdamped. That is,

$$R^2 < \frac{4L}{C} \tag{11-1}$$

Mode 1. This mode begins when T_1 is fired and a resonant pulse of current flows through T_1 and the load. The instantaneous load current for this mode is described by

$$L \frac{di_1}{dt} + Ri_1 + \frac{1}{C} \int i_1 \, dt + v_{c1}(t = 0) = V_s \tag{11-2}$$

with initial conditions $i_1(t = 0) = 0$ and $v_{c1}(t = 0) = -V_c$. Since the circuit is underdamped, the solution of Eq. (11-2) yields

$$i_1(t) = A_1 e^{-tR/2L} \sin \omega_r t \tag{11-3}$$

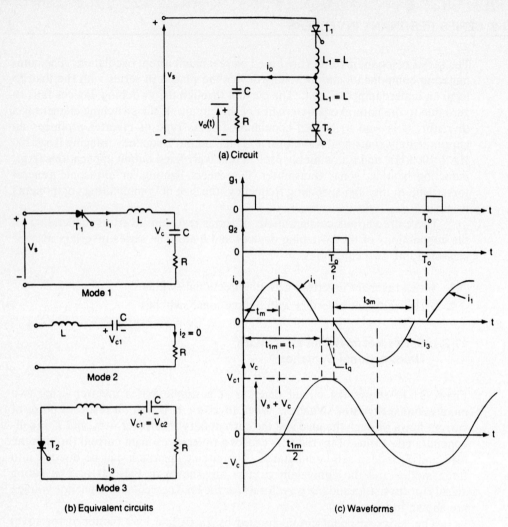

Figure 11-1 Basic series resonant inverter.

where ω_r is the resonant frequency and

$$\omega_r = \left(\frac{1}{LC} - \frac{R^2}{4L^2}\right)^{1/2} \tag{11-4}$$

The constant, A_1, in Eq. (11-3) can be evaluated from the initial condition:

$$\frac{di_1}{dt}\bigg|_{t=0} = \frac{V_s + V_c}{\omega_r L} = A_1$$

and

$$i_1(t) = \frac{V_s + V_c}{\omega_r L} e^{-\alpha t} \sin \omega_r t \tag{11-5}$$

where

$$\alpha = \frac{R}{2L} \tag{11-6}$$

The time t_m when the current $i_1(t)$ in Eq. (11-5) becomes maximum can be found from the condition

$$\frac{di_1}{dt} = 0 \qquad \text{or} \qquad \omega_r e^{-\alpha t_m} \cos \omega_r t_m - \alpha e^{-\alpha t_m} \sin \omega_r t_m = 0$$

and this gives

$$t_m = \frac{1}{\omega_r} \tan^{-1} \frac{\omega_r}{\alpha} \tag{11-7}$$

The capacitor voltage can be found from

$$v_{c1}(t) = \frac{1}{C} \int_0^t i_1(t) - V_c$$
$$= -(V_s + V_c)e^{-\alpha t} (\alpha \sin \omega_r t + \omega_r \cos \omega_r t)/\omega_r + V_s \tag{11-8}$$

This mode is valid for $0 \le t \le t_{1m}(= \pi/\omega_r)$ and ends when $i_1(t)$ becomes zero at t_{1m}. At the end of this mode,

$$i_1(t = t_{1m}) = 0$$

and

$$v_{c1}(t = t_{1m}) = V_{c1} = (V_s + V_c)e^{-\alpha \pi/\omega_r} + V_s \tag{11-9}$$

Mode 2. During this mode, thyristors T_1 and T_2 are off. Redefining the time origin, $t = 0$, at the beginning of this mode, this mode is valid for $0 \le t \le t_{2m}$.

$$i_2(t) = 0, \qquad v_{c2}(t) = V_{c1} \qquad v_{c2}(t = t_{2m}) = V_{c2} = V_{c1}$$

Mode 3. This mode begins when T_2 is switched on and a reverse resonant current flows through the load. Let us redefine the time origin, $t = 0$, at the beginning of this mode. The load current can be found from

$$L \frac{di_3}{dt} + Ri_3 + \frac{1}{C} \int i_3 \, dt + v_{c3}(t = 0) = 0 \tag{11-10}$$

with initial conditions $i_3(t = 0) = 0$ and $v_{c3}(t = 0) = -V_{c2} = -V_{c1}$. The solution of Eq. (11-10) gives

$$i_3(t) = \frac{V_{c1}}{\omega_r L} e^{-\alpha t} \sin \omega_r t \tag{11-11}$$

Sec. 11-2   Series Resonant Inverters   **417**

The capacitor voltage can be found from

$$v_{c3}(t) = \frac{1}{C} \int_0^t i_3(t) - V_{c1}$$

$$= V_{c1}e^{-\alpha t}(\alpha \sin \omega_r t + \omega_r \cos \omega_r t)/\omega_r \tag{11-12}$$

This mode is valid for $0 \le t \le t_{3m} = \pi/\omega_r$ and ends when $i_3(t)$ becomes zero. At the end of this mode,

$$i_3(t = t_{3m}) = 0$$

and in the steady state,

$$v_{c3}(t = t_{3m}) = V_{c3} = V_c = V_{c1}e^{-\alpha\pi/\omega_r} \tag{11-13}$$

Equations (11-9) and (11-13) yield

$$V_c = V_s \frac{1 + e^{-z}}{e^z - e^{-z}} = V_s \frac{e^z + 1}{e^{2z} - 1} = \frac{V_s}{e^z - 1} \tag{11-14}$$

$$V_{c1} = V_s \frac{1 + e^z}{e^z - e^{-z}} = V_s \frac{e^z(1 + e^z)}{e^{2z} - 1} = \frac{V_s e^z}{e^z - 1} \tag{11-15}$$

where $z = \alpha\pi/\omega_r$. Adding V_c from Eq. (11-14) to V_s gives

$$V_s + V_c = V_{c1} \tag{11-16}$$

Equation (11-16) indicates that under steady-state conditions, the peak values of positive current in Eq. (11-5) and of negative current in Eq. (11-11) through the load are the same.

The load current $i_1(t)$ must be zero and T_1 must be turned off before T_2 is fired. Otherwise, a short-circuit condition will result through the thyristors and dc supply. Therefore, the available off-time $t_{2m}(= t_{\text{off}})$, known as the *dead zone*, must be greater than the turn-off time of thyristors, t_q.

$$\frac{\pi}{\omega_o} - \frac{\pi}{\omega_r} = t_{\text{off}} > t_q \tag{11-17}$$

where ω_o is the frequency of the output voltage in rad/s. Equation (11-17) indicates that the maximum possible output frequency is limited to

$$f_o \le f_{\max} = \frac{1}{2(t_q + \pi/\omega_r)} \tag{11-18}$$

The resonant inverter circuit in Fig. 11-1a is very simple. However, the power flow from the dc supply is discontinuous. The dc supply will have a high peak current and would contain harmonics. An improvement of the basic inverter in Fig. 11-1a can be made if inductors are closely coupled, as shown in Fig. 11-2. When T_1 is fired and current $i_1(t)$ begins to rise, the voltage across L_1 will be positive with polarity as shown. The induced voltage on L_2 will now add to the voltage of C in reverse biasing T_2; and T_2 will be turned off. The result is that firing of one thyristor will turn off the other, even before the load current reaches to zero.

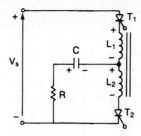

Figure 11-2 Series resonant inverter with coupled inductors.

The drawback of high pulsed current from the dc supply can be overcome in a half-bridge configuration as shown in Fig. 11-3, where $L_1 = L_2$ and $C_1 = C_2$. The power is drawn from the dc source during both half-cycles of output voltage. One-half of the load current is supplied by capacitor C_1 or C_2 and the other-half by the dc source.

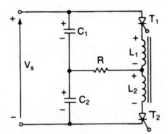

Figure 11-3 Half-bridge series resonant inverter.

A full-bridge inverter, which allows higher output power, is shown in Fig. 11-4. When T_1 and T_2 are fired, a positive resonant current flows through the load; and when T_3 and T_4 are fired, a negative load current flows. The supply current is continuous, but pulsating.

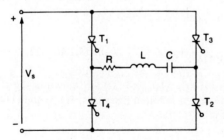

Figure 11-4 Full-bridge series resonant inverter.

The resonant frequency and available dead-zone depend on the load and for this reason, resonant inverters are most suitable for fixed-load applications. The inverter load (or resistor R) could also be connected in parallel with the capacitor. The thyristors can be replaced by BJTs, MOSFETs, IGBTs, and GTOs.

Example 11-1*

The series resonant inverter in Fig. 11-2 has $L_1 = L_2 = L = 50 \ \mu H$, $C = 6 \ \mu F$, and $R = 2\Omega$. The dc input voltage is $V_s = 220$ V and the frequency of output voltage is $f_o = 7$ kHz. The turn-off time of thyristors is $t_q = 10 \ \mu s$. Determine (a) the available (or circuit) turn-off time t_{off}, (b) the maximum permissible frequency f_{max}, (c) the peak-to-peak capacitor voltage V_{pp}, and (d) the peak load current I_p. (e) Sketch the instantaneous load current $i_o(t)$, capacitor voltage $v_c(t)$, and dc supply current $i_s(t)$. Calculate (f) the rms load current I_o, (g) the output power P_o, (h) the average supply current I_s, and (i) the average, peak, and rms thyristor currents.

Solution $V_s = 220$ V, $C = 6$ μF, $L = 50$ μH, $R = 2$ Ω, $f_o = 7$ kHz, $t_q = 10$ μs, and $\omega_o = 2\pi \times 7000 = 43,982$ rad/s. From Eq. (11-4),

$$\omega_r = \left(\frac{1}{LC} - \frac{R^2}{4L^2}\right)^{1/2} = \left(\frac{10^{12}}{50 \times 6} - \frac{2^2 \times 10^{12}}{4 \times 50^2}\right)^{1/2} = 54,160 \text{ rad/s}$$

The resonant frequency is $f_r = \omega_r/2\pi = 8619.8$ Hz, $T_r = 1/f_r = 116$ μs. From Eq. (11-6), $\alpha = 2/(2 \times 50 \times 10^{-6}) = 20,000$.

(a) From Eq. (11-17),

$$t_{off} = \frac{\pi}{43,982} - \frac{\pi}{54,160} = 13.42 \text{ } \mu s$$

(b) From Eq. (11-18), the maximum possible frequency is

$$f_{max} = \frac{1}{2(10 \times 10^{-6} + \pi/54,160)} = 7352 \text{ Hz}$$

(c) From Eq. (11-14),

$$V_c = \frac{V_s}{e^{\alpha\pi/\omega r} - 1} = \frac{220}{e^{20\pi/54.16} - 1} = 100.4 \text{ V}$$

From Eq. (11-16), $V_{c1} = 220 + 100.4 = 320.4$ V. The peak-to-peak capacitor voltage is $V_{pp} = 100.4 + 320.4 = 420.8$ V.

(d) From Eq. (11-7), the peak load current, which is the same as the peak supply current, occurs at

$$t_m = \frac{1}{\omega_r} \tan^{-1} \frac{\omega_r}{\alpha} = \frac{1}{54,160} \tan^{-1} \frac{54.16}{20} = 22.47 \text{ } \mu s$$

and Eq. (11-5) gives the peak load current as

$$i_1(t = t_m) = I_p = \frac{320.4}{0.05416 \times 50} e^{-0.02 \times 22.47} \sin(54,160 \times 22.47 \times 10^{-6}) = 70.82 \text{ A}$$

(e) The sketches for $i(t)$, $v_c(t)$, and $i_s(t)$ are shown in Fig. 11-5.

(f) The rms load current is found from Eqs. (11-5) and (11-11) by a numerical method and the result is

$$I_o = \left[2f_o \int_0^{T_r/2} i_0^2(t) \, dt\right]^{1/2} = 44.1 \text{ A}$$

(g) The output power $P_o = 44.1^2 \times 2 = 3889$ W.

(h) The average supply current $I_s = 3889/220 = 17.68$ A.

(i) The average thyristor current

$$I_A = f_o \int_0^{T_r/2} i_0(t) \, dt = 17.68 \text{ A}$$

The peak thyristor current $I_{pk} = I_p = 70.82$ A, and the rms thyristor current $I_R = I_o/\sqrt{2} = 44.1/\sqrt{2} = 31.18$ A.

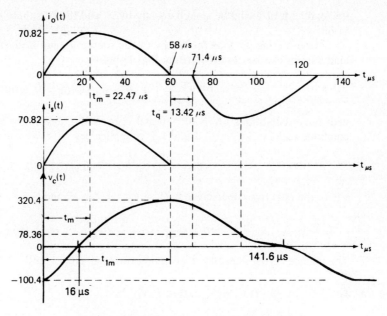

Figure 11-5 Waveforms for Example 11-1.

Example 11-2*

The half-bridge resonant inverter in Fig. 11-3 is operated at an output frequency, $f_o = 7$ kHz. If $C_1 = C_2 = C = 3$ μF, $L_1 = L_2 = L = 50$ μH, $R = 2$ Ω, and $V_s = 220$ V, determine (a) the peak supply current, (b) the average thyristor current I_A, and (c) the rms thyristor current I_R.

Solution $V_s = 220$ V, $C = 3$ μF, $L = 50$ μH, $R = 2$ Ω, and $f_o = 7$ kHz. Figure 11-6a shows the equivalent circuit when thyristor T_1 is conducting and T_2 is off. The capacitors C_1 and C_2 would initially be charged to $V_{c1}(= V_s + V_c)$ and V_c, respectively, with the polarities as shown, under steady-state conditions. Since $C_1 = C_2$,

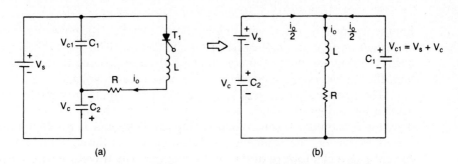

Figure 11-6 Equivalent circuits for Example 11-2.

the load current would be shared equally by C_1 and the dc supply as shown in Fig. 11-6b.

Considering the loop formed by C_2, dc source, L, and load, the instantaneous load current can be described (from Fig. 11-6b) by

$$L \frac{di_o}{dt} + Ri_0 + \frac{1}{2C_2} \int i_o \, dt + v_{c2}(t = 0) - V_s = 0 \tag{11-19}$$

with initial conditions $i_0(t = 0) = 0$ and $v_{c2}(t = 0) = -V_c$. For an underdamped condition and $C_1 = C_2 = C$, Eq. (11-5) is applicable:

$$i_0(t) = \frac{V_s + V_c}{\omega_r L} e^{-\alpha t} \sin \omega_r t \tag{11-20}$$

where the effective capacitance is $C_e = C_1 + C_2 = 2C$ and

$$\omega_r = \left(\frac{1}{2LC_2} - \frac{R^2}{4L^2} \right)^{1/2} = \left(\frac{10^{12}}{2 \times 50 \times 3} - \frac{2^2 \times 10^{12}}{4 \times 50^2} \right)^{1/2} = 54{,}160 \text{ rad/s} \tag{11-21}$$

The voltage across capacitor C_2 can be expressed as

$$v_{c2}(t) = \frac{1}{2C_2} \int_0^t i_0(t) \, dt - V_c$$
$$= -(V_s + V_c)e^{-\alpha t}(\alpha \sin \omega_r t + \omega_r \cos \omega_r t)/\omega_r + V_s \tag{11-22}$$

(a) Since the resonant frequency is the same as that of Example 11-1, the results of Example 11-1 are valid, provided that the equivalent capacitance is $C_e = C_1 + C_2 = 6 \, \mu\text{F}$. From Example 11-1, $V_c = 100.4$ V, $t_m = 22.47 \, \mu\text{s}$, and $I_o = 44.1$ A. From Eq. (11-20), the peak load current is $I_p = 70.82$ A. The peak supply current, which is half of the peak load current, is $I_{ps} = 70.82/2 = 35.41$ A.

(b) The average thyristor current $I_A = 17.68$ A.

(c) The rms thyristor current $I_R = I_o/\sqrt{2} = 31.18$ A.

Note. For the same power output and resonant frequency, the capacitances of C_1 and C_2 in Fig. 11-3 should be half that in Figs. 11-1 and 11-2. The peak supply current becomes half. The analysis of full-bridge series inverters is similar to that of the basic series inverter in Fig. 11-1a.

11-2.2 Series Resonant Inverters with Bidirectional Switches

For the resonant inverters with unidirectional switches, the power devices have to be turned on in every half-cycle of output voltage. This limits the inverter frequency and the amount of energy transfer from the source to the load. In addition, the thyristors are subjected to high peak reverse voltage.

The performance of series inverters can be significantly improved by connecting an antiparallel diode across a thyristor as shown in Fig. 11-7a. When thyristor T_1 is fired, a resonant pulse of current flows and T_1 is self-commutated at $t = t_1$. However, the resonant oscillation continues through diode D_1 until the current falls again to zero at the end of a cycle. The waveforms for currents and capacitor voltage are shown in Fig. 11-7b.

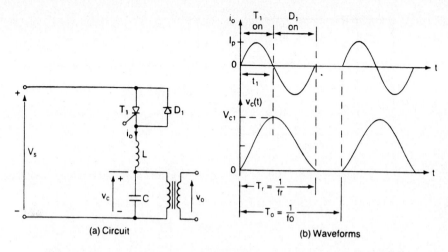

(a) Circuit   (b) Waveforms

Figure 11-7 Basic series resonant inverter with bidirectional switches.

The reverse voltage of the thyristor is limited to the forward voltage drop of a diode, typically 1 V. If the conduction time of the diode is greater than the turn-off time of the thyristor, there is no need of a dead zone and the output frequency, f_o, is the same as the resonant frequency, f_r,

$$f_o = f_r = \frac{\omega_r}{2\pi} \tag{11-23}$$

where f_r is the resonant frequency of the series circuit in hertz. If t_q is the turn-off time of a thyristor, the maximum inverter frequency is given by

$$f_{max} = \frac{1}{2t_q} \tag{11-24}$$

and f_o should be less than f_{max}.

The diode D_1 should be connected as close as possible to the thyristor and the connecting leads should be minimum to reduce any stray inductance in the loop formed by T_1 and D_1. As the reverse voltage during the recovery time of thyristor T_1 is already low, typically 1 V, any inductance in the diode path would reduce the net reverse voltage across the terminals of T_1, and thyristor T_1 may not turn off. To overcome this problem, a *reverse conducting thyristor* (RCT) is normally used. An RCT is made by integration of an asymmetrical thyristor and a fast-recovery diode into a single silicon chip, and RCTs are ideal for series resonant inverters.

The circuit diagram for the half-bridge version is shown in Fig. 11-8a and the waveform for the load current and the conduction intervals of the power devices are shown in Fig. 11-8b. The full-bridge configuration is shown in Fig. 11-9a. The inverters can be operated in two different modes: nonoverlapping and overlapping. In a nonoverlapping mode, the firing of a thyristor is delayed until the last current oscillation through a diode has been completed as in Fig. 11-8b. In an overlapping mode, a thyristor is fired, while the current in the diode of the other

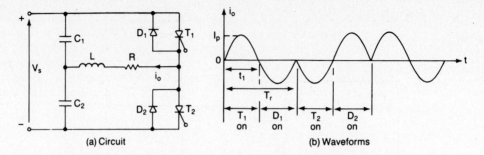

(a) Circuit

(b) Waveforms

Figure 11-8   Half-bridge series inverters with bidirectional switches.

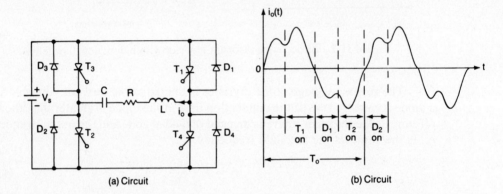

(a) Circuit

(b) Circuit

Figure 11-9   Full-bridge series inverters with bidirectional switches.

part is still conducting, as shown in Fig. 11-9b. Although overlapping operation increases the output frequency, the output power is increased.

The maximum frequency of resonant inverters are limited due to the turn-off or commutation requirements of thyristors, typically 12 to 20 μs, whereas transistors, which require only a microsecond or less, can replace the thyristors. The inverter can operate at the resonant frequency. A transistorized half-bridge inverter is shown in Fig. 11-10 with a transformer-connected load. Transistor Q_2 can be turned on almost instantaneously after transistor Q_1 is turned off.

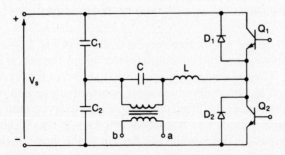

Figure 11-10   Half-bridge transistorized resonant inverter.

Example 11-3*

The resonant inverter in Fig. 11-7a has $C = 2 \ \mu\text{F}$, $L = 20 \ \mu\text{H}$, $R = 0$, and $V_s = 220$ V. The turn-off time of the thyristor, $t_{\text{off}} = 12 \ \mu\text{s}$. The output frequency, $f_o = 20$ kHz. Determine (a) the peak supply current I_p, (b) the average thyristor current I_A, (c) the rms thyristor current I_R, (d) the peak-to-peak capacitor voltage V_{pp}, (e) the maximum permissible output frequency f_{max}, and (f) the average supply current I_s.

Solution When thyristor T_1 is turned on, the current is described by

$$L \frac{di_0}{dt} + \frac{1}{C} \int i_0 \ dt + v_c(t = 0) = V_s \tag{11-25}$$

with initial conditions $i_0(t = 0) = 0$, $v_c(t = 0) = V_c = 0$. Solving for the current gives

$$i_0(t) = V_s \sqrt{\frac{C}{L}} \sin \omega_r t \tag{11-26}$$

and the capacitor voltage is

$$v_c(t) = V_s(1 - \cos \omega_r t) \tag{11-27}$$

where

$$\omega_r = 1/\sqrt{LC}$$

$$\omega_r = \frac{10^6}{\sqrt{20 \times 2}} = 158{,}114 \ \text{rad/s} \quad \text{and} \quad f_r = \frac{158{,}114}{2\pi} = 25{,}165 \ \text{Hz}$$

$$T_r = \frac{1}{f_r} = \frac{1}{25{,}165} = 39.74 \ \mu\text{s} \qquad t_1 = \frac{T_r}{2} = \frac{39.74}{2} = 19.87 \ \mu\text{s}$$

At $\omega_r t = \pi$,

$$v_c(\omega_r t = \pi) = V_{c1} = 2V_s = 2 \times 220 = 440 \ \text{V}$$

(a) $I_p = V_s\sqrt{C/L} = 220\sqrt{2/20} = 69.57$ A.

(b) $I_A = f_o \int_0^\pi I_p \sin \theta \ d\theta = I_p f_o/(\pi f_r) = 69.57 \times 20{,}000/(\pi \times 25{,}165) =$

    17.6 A

(c) $I_R = I_p\sqrt{f_o t_1/2} = 69.57\sqrt{20{,}000 \times 19.87 \times 10^{-6}/2} = 31.01$ A.

(d) The peak-to-peak capacitor voltage $V_{pp} = V_{c1} - V_c = 440$ V.

(e) From Eq. (11-24), $f_{\text{max}} = 10^6/(2 \times 12) = 41.67$ kHz.

(f) Since there is no power loss in the circuit, $I_s = 0$.

Example 11-4*

The half-bridge resonant inverter in Fig. 11-8a is operated at a frequency of $f_o = 3.5$ kHz. If $C_1 = C_2 = C = 3 \ \mu\text{F}$, $L_1 = L_2 = L = 50 \ \mu\text{H}$, $R = 2 \ \Omega$, and $V_s = 220$ V, determine (a) the peak supply current I_p, (b) the average thyristor current I_A, (c) the rms thyristor current I_R, (d) the rms load current I_o, and (e) the average supply current I_s.

Solution $V_s = 220$ V, $C_e = C_1 + C_2 = 6 \ \mu\text{F}$, $L = 50 \ \mu\text{H}$, $R = 2 \ \Omega$, and $f_o = 3500$ Hz. The analysis of this inverter is similar to that of inverter in Fig. 11-3. Instead of two current pulses, there are four pulses in a full cycle of the output voltage with one pulse through each of devices T_1, D_1, T_2, and D_2. Equation (11-20) is applicable. During the positive half-cycle, the current flows through T_1; and during the negative

---

half-cycle the current flows through D_1. In a nonoverlap control, there are two resonant cycles during the entire period of output frequency, f_o. From Eq. (11-21),

$$\omega_r = 54,160 \text{ rad/s} \qquad f_r = \frac{54,160}{2\pi} = 8619.9 \text{ Hz}$$

$$T_r = \frac{1}{8619.9} = 116 \ \mu s \qquad t_1 = \frac{116}{2} = 58 \ \mu s$$

$$T_0 = \frac{1}{3500} = 285.72 \ \mu s$$

The off-period of load current

$$t_d = T_0 - T_r = 285.72 - 116 = 169.72 \ \mu s$$

Since t_d is greater than zero, the inverter would operate in the nonoverlap mode. From Eq. (11-14), $V_c = 100.4$ V and $V_{c1} = 220 + 100.4 = 320.4$ V.

(a) From Eq. (11-7),

$$t_m = \frac{1}{54,160} \tan^{-1} \frac{54,160}{20,000} = 22.47 \ \mu s$$

$$i_0(t) = \frac{V_s + V_c}{\omega_r L} e^{-\alpha t} \sin \omega_r t$$

and the peak load current becomes $I_p = i_0(t = t_m) = 70.82$ A.

(b) A thyristor conducts from a time of t_1. The average thyristor current can be found from

$$I_A = f_o \int_0^{t_1} i_0(t) \ dt = 8.84 \text{ A}$$

(c) The rms thyristor current is

$$I_R = \left[f_o \int_0^{t_1} i_0^2(t) \ dt \right]^{1/2} = 22.05 \text{ A}$$

(d) The rms load current $I_o = 2I_R = 2 \times 22.05 = 44.1$ A.

(e) $P_o = 44.1^2 \times 2 = 3889$ W and the average supply current, $I_s = 3889/220 = 17.68$ A.

Note. With bidirectional switches, the current ratings of the devices are reduced. For the same output power, the average device current is half and the rms current is $1/\sqrt{2}$ of that for an inverter with unidirectional switches.

Example 11-5*

The full-bridge resonant inverter in Fig. 11-9a is operated at a frequency, $f_o = 3.5$ kHz. If $C = 6 \ \mu F$, $L = 50 \ \mu H$, $R = 2 \ \Omega$, and $V_s = 220$ V, determine (a) the peak supply current I_p, (b) the average thyristor current I_A, (c) the rms thyristor current I_R, (d) the rms load current I_o, and (e) the average supply current I_s.

Solution $V_s = 220$ V, $C = 6 \ \mu F$, $L = 50 \ \mu H$, $R = 2 \ \Omega$, and $f_o = 3500$ kHz. From Eq. (11-21), $\omega_r = 54,160$ rad/s and $f_r = 54,160/(2\pi) = 8619.9$ Hz. $\alpha = 20,000$, $T_r = 1/8619.9 = 116 \ \mu s$, $t_1 = 116/2 = 58 \ \mu s$, and $T_0 = 1/3500 = 285.72 \ \mu s$. The off-period of load current is $t_d = T_0 - T_r = 285.72 - 116 = 169.72 \ \mu s$, and the inverter would operate in the nonoverlap mode.

Mode 1. This mode begins when T_1 and T_2 are fired. A resonant current flows through T_1, T_2, load, and supply. The instantaneous current is described by

$$L \frac{di_0}{dt} + Ri_0 + \frac{1}{C} \int i_0 \, dt + v_c(t = 0) = V_s$$

with initial conditions $i_0(t = 0) = 0$, $v_{c1}(t = 0) = -V_c$ and the solution for the current gives

$$i_0(t) = \frac{V_s + V_c}{\omega_r L} e^{-\alpha t} \sin \omega_r t \tag{11-28}$$

$$v_c(t) = -(V_s + V_c)e^{-\alpha t}(\alpha \sin \omega_r t + \omega_r \cos \omega_r t) + V_s \tag{11-29}$$

Thyristors T_1 and T_2 are turned off at $t_1 = \pi/\omega_r$, when $i_1(t)$ becomes zero.

$$V_{c1} = v_c(t = t_1) = (V_s + V_c)e^{-\alpha \pi/\omega_r} + V_s \tag{11-30}$$

Mode 2. This mode begins when T_3 and T_4 are fired. A reverse resonant current flows through T_3, T_4, load, and supply. The instantaneous load current is described by

$$L \frac{di_0}{dt} + Ri_0 + \frac{1}{C} \int i_0 \, dt + v_c(t = 0) = -V_s$$

with initial conditions $i_2(t = 0) = 0$ and $v_c(t = 0) = V_{c1}$, and the solution for the current gives

$$i_0(t) = - \frac{V_s + V_{c1}}{\omega_r L} e^{-\alpha t} \sin \omega_r t \tag{11-31}$$

$$v_c(t) = (V_s + V_{c1})e^{-\alpha t}(\alpha \sin \omega_r t + \omega_r \cos \omega_r t)/\omega_r - V_s \tag{11-32}$$

Thyristors T_3 and T_4 are turned off at $t_1 = \pi/\omega_r$, when $i_0(t)$ becomes zero.

$$V_c = -v_c(t = t_1) = (V_s + V_{c1})e^{-\alpha \pi/\omega_r} + V_s \tag{11-33}$$

Solving for V_c and V_{c1} from Eqs. (11-30) and (11-33) gives

$$V_c = V_{c1} = V_s \frac{e^z + 1}{e^z - 1} \tag{11-34}$$

where $z = \alpha \pi/\omega_r$. For $z = 20,000\pi/54,160 = 1.1601$, Eq. (11-34) gives $V_c = V_{c1} = 420.9$ V.

(a) From Eq. (11-7),

$$t_m = \frac{1}{54,160} \tan^{-1} \frac{54,160}{20,000} = 22.47 \ \mu s.$$

From Eq. (11-28), the peak load current $I_p = i_0(t = t_m) = 141.64$ A.

(b) A thyristor conducts from a time of t_1. The average thyristor current can be found from Eq. (11-28):

$$I_A = f_o \int_0^{t_1} i_0(t) \, dt = 17.68 \text{ A}$$

(c) The rms thyristor current can be found from Eq. (11-28):

$$I_R = \left[f_o \int_0^{t_1} i_0^2(t) \, dt \right]^{1/2} = 44.1 \text{ A}$$

(d) The rms load current is $I_o = 2I_R = 2 \times 44.1 = 88.2$ A.

(e) $P_o = 88.2^2 \times 2 = 15,556$ W and the average supply current, $I_s = 15,556/220 = 70.71$ A.

Note. For the same circuit parameters, the output power is four times, and the device currents are twice, that for a half-bridge inverter.

11-2.3 Frequency Response for Series-Loaded

It can be noted from the waveforms of Figs. 11-8b and 11-9b that the output voltage can be varied by varying the switching frequency, f_s. In Figs. 11-4, 11-8, and 11-9a, the load resistance R forms a series circuit with the resonating components L and C. The equivalent circuit is shown in Fig. 11-11a. The input voltage is a square wave whose peak fundamental component is $V_{i(\text{pk})} = 4V_s/\pi$, and its rms value is $V_i = 4V_s/\sqrt{2}\,\pi$. Using the voltage-divider rule in frequency domain, the voltage gain is given by

$$G(j\omega) = \frac{V_o}{V_i}(j\omega) = \frac{1}{1 + j\omega L/R - j/(\omega CR)}$$

Let $\omega_0 = 1/\sqrt{LC}$ be the resonant frequency, and $Q_s = \omega_0 L/R$ be the quality factor. Substituting L, C, and R in terms of Q_s and ω_0, we get

$$G(j\omega) = \frac{v_o}{v_i}(j\omega) = \frac{1}{1 + jQ_s(\omega/\omega_0 - \omega_0/\omega)} = \frac{1}{1 + jQ_s(u - 1/u)}$$

where $u = \omega/\omega_o$. The magnitude of $G(j\omega)$ can be found from

$$|G(j\omega)| = \frac{1}{[1 + Q_s^2(u - 1/u)^2]^{1/2}} \qquad (11\text{-}35)$$

Figure 11-11b shows the magnitude plot of Eq. (11-35) for $Q_s = 1$ to 5. For a continuous output voltage, the switching frequency should be greater than the resonant frequency, f_0. If the inverter operates near resonance, and a short circuit occurs at the load, the current will rise to a high value, especially at a high load current. However, the output current can be controlled by raising the switching frequency. The current through the switching devices decrease as the load current decreases, thereby having lower on-state conduction losses and a high efficiency at a partial load. The series inverter is most suitable for high-voltage, low-current applications. The maximum output occurs at resonance, and the maximum gain for $u = 1$ is $|G(j\omega)|_{\text{max}} = 1$.

Under no-load conditions, $R = \infty$ and $Q_s = 0$. Thus the curve would simply be a horizontal line. That is, for $Q_s = 1$, the characteristic has a poor "selectivity" and the output voltage will change significantly from no-load to full-load conditions, thereby yielding poor regulation. The resonant inverter is normally

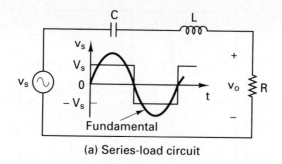

(a) Series-load circuit

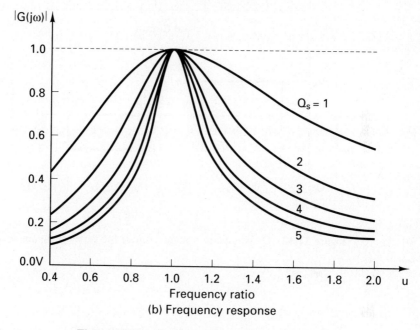

(b) Frequency response

Figure 11-11 Frequency response for series loaded.

used in applications requiring only a fixed output voltage. However, some no-load regulations can be obtained by time ratio control at frequencies lower than the resonant frequency (e.g., in Fig. 11-8b). This type of control has two disadvantages: (1) it limits how far the operating frequency can be varied up and down from the resonant frequency, and (2) due to a low Q-factor, it requires a large change in frequency in order to realize a wide range of output voltage control.

A bridge topology as shown in Fig. 11-12a can be applied to achieve the output voltage control. The switching frequency f_s is kept constant at the resonant frequency f_o. By switching two devices simultaneously a *quasi-square wave* as shown in Fig. 11-12b can be obtained. The rms fundamental input voltage is given by

$$V_i = \frac{4V_s}{\sqrt{2}\,\pi}\cos\alpha \qquad (11\text{-}36)$$

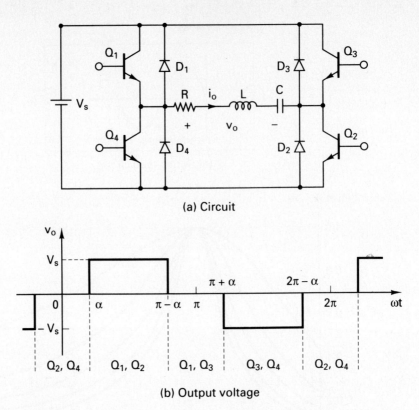

(a) Circuit

(b) Output voltage

Figure 11-12 Quasi-square voltage control for series resonant inverter.

where α is the control angle. By varying α from 0 to $\pi/2$ at a constant frequency, the voltage V_i can be controlled from $4V_s/(\pi\sqrt{2})$ to 0.

Example 11-6

A series resonance inverter with series loaded delivers a load power of $P_L = 1$ kW at resonance. The load resistance is $R = 10$ Ω. The resonant frequency is $f_0 = 20$ kHz. Determine (a) the dc input voltage V_s, (b) the quality factor Q_s if it is required to reduce the load power to 250 W by frequency control so that $u = 0.8$, (c) the inductor L, and (d) the capacitor C.

Solution (a) Since at resonance $u = 1$ and $|G(j\omega)|_{\max} = 1$, the peak fundamental load voltage is $V_p = V_{i(\text{pk})} = 4V_s/\pi$.

$$P_L = \frac{V_p^2}{2R} = \frac{4^2 V_s^2}{2R\pi} \qquad \text{or} \qquad 1000 = \frac{4^2 V_s^2}{2\pi \times 10}$$

which gives $V_s = 110$ V.

(b) To reduce the load power by (1000/250 =) 4, the voltage gain must be reduced by 2 at $u = 0.8$. That is, from Eq. (11-35), we get $1 + Q_s^2(u - 1/u)^2 = 2^2$, which gives $Q_s = 3.85$.

(c) Q_s is defined by

$$Q_s = \frac{\omega_0 L}{R} \quad \text{or} \quad 3.85 = \frac{2\pi \times 20 \text{ kHz} \times L}{10} \quad \text{which gives } L = 306.37 \ \mu H$$

(d) $f_0 = 1/2\pi\sqrt{LC}$ or 20 kHz $= 1/[2\pi\sqrt{(306.37 \ \mu H \times C)}]$, which gives $C = 0.2067 \ \mu F$.

11-2.4 Frequency Response for Parallel-Loaded

With the load connected across the capacitor C directly (or through a transformer) as shown in Fig. 11-7, the equivalent circuit is shown in Fig. 11-13a. Using the voltage-divider rule in frequency domain, the voltage gain is given by

$$G(j\omega) = \frac{V_o}{V_i}(j\omega) = \frac{1}{1 - \omega^2 LC + j\omega L/R}$$

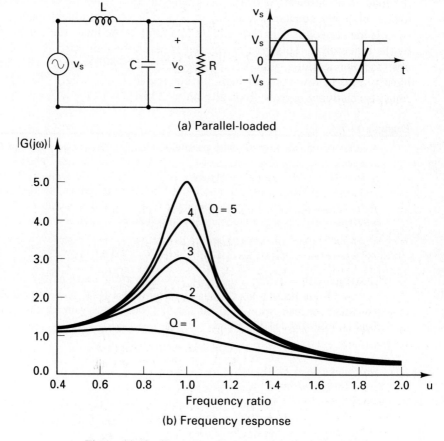

(a) Parallel-loaded

(b) Frequency response

Figure 11-13 Frequency response for parallel-loaded.

Let $\omega_0 = 1/\sqrt{LC}$ be the resonant frequency, and $Q = 1/Q_s = R/\omega_o L$ be the quality factor. Substituting L, C, and R in terms of Q and ω_o, we get

$$G(j\omega) = \frac{V_o}{V_i}(j\omega) = \frac{1}{[1 - (\omega/\omega_o)^2] + j(\omega/\omega_o)/Q} = \frac{1}{(1 - u^2) + ju/Q}$$

where $u = \omega/\omega_o$. The magnitude of $G(j\omega)$ can be found from

$$|G(j\omega)| = \frac{1}{[(1 - u^2)^2 + (u/Q)^2]^{1/2}} \tag{11-37}$$

Figure 11-13b shows the magnitude plot of the voltage gain in Eq. (11-37) for $Q = 1$ to 5. The maximum gain occurs near resonance for $Q > 2$, and its value for $u = 1$ is

$$|G(j\omega)|_{max} = Q \tag{11-38}$$

At no-load, $R = \infty$ and $Q = \infty$. Thus the output voltage at resonance is a function of load and can be very high at no-load if the operating frequency is not raised. But the output voltage is normally controlled at no-load by varying the frequency above resonance. The current carried by the switching devices is independent to the load, but increases with the dc input voltage. Thus the conduction loss remains relatively constant, resulting in poor efficiency at a light load.

If the capacitor C is shorted due to a fault in the load, the current is limited by the inductor L. This type of inverter is naturally short-circuit proof and desirable for applications with severe short-circuit requirements. This inverter is mostly used in low-voltage, high-current applications, where the input voltage range is relatively narrow, typically up to $\pm15\%$.

Example 11-7

A series resonance inverter with parallel-loaded delivers a load power of $P_L = 1$ kW at a peak sinusoidal load voltage of $V_p = 330$ V and at resonance. The load resistance is $R = 10\Omega$. The resonant frequency is $f_0 = 20$ kHz. Determine (a) the dc input voltage V_s, (b) the frequency ratio u if it is required to reduce the load power to 250 W by frequency control, (c) the inductor L, and (d) the capacitor C.

Solution (a) The peak fundamental component of a square voltage is $V_p = 4V_s/\pi$.

$$P_L = \frac{V_p^2}{2R} = \frac{4^2 V_s^2}{2\pi R} \qquad \text{or} \qquad 1000 = \frac{4^2 V_s^2}{2\pi \times 10}$$

which gives $V_s = 110$ V. $V_{i(pk)} = 4V_s/\pi = 4 \times 110/\pi = 140.06$ V.

(b) From Eq. (11-38), the quality factor is $Q = V_p/V_{i(pk)} = 330/140.06 = 2.356$. To reduce the load power by $(1000/250 =) 4$, the voltage gain must be reduced by 2. That is, from Eq. (11-37), we get

$$(1 - u^2)^2 + (u/2.356)^2 = 2^2$$

which gives $u = 1.693$.

(c) Q is defined by

$$Q = \frac{R}{\omega_0 L} \qquad \text{or} \qquad 2.356 = \frac{R}{2\pi \times 20 \text{ kHz } L}$$

which gives $L = 33.78$ μH.

(d) $f_0 = 1/2\pi\sqrt{LC}$ or $20\,\text{kHz} = 1/2\pi\sqrt{(33.78\ \mu H \times C)}$, which gives $C = 1.875\ \mu F$.

11-2.5 Frequency Response for Series–Parallel-Loaded

In Fig. 11-10 the capacitor $C_1 = C_2 = C_s$ forms a series circuit and the capacitor C is in parallel with the load. This circuit is a compromise between the characteristics of a series load and a parallel load. The equivalent circuit is shown in Fig. 11-14a. Using the voltage-divider rule in frequency domain, the voltage gain is given by

$$G(j\omega) = \frac{V_o}{V_i}(j\omega) = \frac{1}{1 + C_p/C_s - \omega^2 LC_p + j\omega L/R - j/(\omega C_s R)}$$

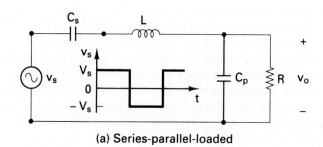

(a) Series-parallel-loaded

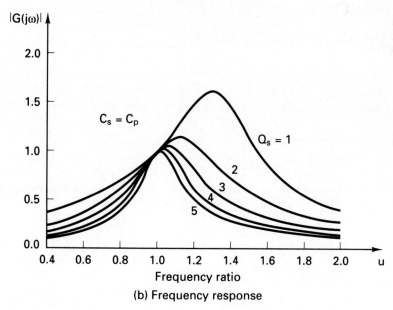

(b) Frequency response

Figure 11-14 Frequency response for series–parallel-loaded.

Let $\omega_0 = 1/\sqrt{LC_s}$ be the resonant frequency, and $Q_s = \omega_0 L/R$ be the quality factor. Substituting L, C, and R in terms of Q_s and ω_0, we get

$$G(j\omega) = \frac{V_o}{V_i}(j\omega) = \frac{1}{1 + C_p/C_s - \omega^2 LC_p + jQ_s(\omega/\omega_0 - \omega_0/\omega)}$$

$$= \frac{1}{1 + (C_p/C_s)(1 - u^2) + jQ_s(u - 1/u)}$$

where $u = \omega/\omega_o$. The magnitude of $G(j\omega)$ can be found from

$$|G(j\omega)| = \frac{1}{\{[1 + (C_p/C_s)(1 - u^2)]^2 + Q_s^2(u - 1/u)^2\}^{1/2}} \qquad (11\text{-}39)$$

Figure 11-14b shows the magnitude plot of the voltage gain in Eq. (11-39) for $Q_s = 1$ to 5 and $C_p/C_s = 1$. This inverter combines the best characteristics of the series load and parallel load, while eliminating the weak points such as lack of regulation for series load and load current independent for parallel load.

As C_p gets smaller and smaller, the inverter exhibits the characteristics of series load. With a reasonable value of C_p, the inverter exhibits some of the characteristics of parallel load and can operate under no-load. As C_p becomes smaller, the upper frequency needed for a specified output voltage increases. Choosing $C_p = C_s$ is generally a good compromise between the efficiency at partial load and the regulation at no-load with a reasonable upper frequency. For making the current to decrease with the load in order to maintain a high efficiency at partial load, the full-load Q is chosen between 4 and 5. An inverter with series–parallel-loaded can run over a wider input voltage and load ranges from no-load to full load, while maintaining excellent efficiency.

11-3 PARALLEL RESONANT INVERTERS

A parallel resonant inverter is the dual of a series resonant inverter. It is supplied from a current source so that the circuit offers a high impedance to the switching current. A parallel resonant circuit is shown in Fig. 11-15. Since the current is

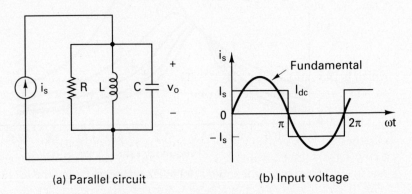

(a) Parallel circuit   (b) Input voltage

Figure 11-15 Parallel resonant circuit.

continuously controlled, this inverter gives a better short-circuit protection under fault conditions. Summing the currents through R, L, and C, gives

$$C \frac{dv}{dt} + \frac{v}{R} + \frac{1}{L} \int v \, dt = I_s$$

with initial condition $v(t = 0) = 0$ and $i_L(t = 0) = 0$. This equation is similar to Eq. (11-2) if i is replaced by v, R by $1/R$, L by C, C by L, and V_s by I_s. Using Eq. (11-5), the voltage v is given by

$$v = \frac{I_s}{\omega_r C} e^{-\alpha t} \sin \omega_r t \tag{11-40}$$

where $\alpha = 1/2RC$. The damped-resonant frequency ω_r is given by

$$\omega_r = \left(\frac{1}{LC} - \frac{1}{4R^2 C^2} \right)^{1/2} \tag{11-41}$$

Using Eq. (11-7), the voltage v in Eq. (11-40) becomes maximum at t_m given by

$$t_m = \frac{1}{\omega_r} \tan^{-1} \frac{\omega_r}{\alpha} \tag{11-42}$$

which can be approximated to π/ω_r. The input impedance is given by

$$Z(j\omega) = \frac{V_o}{I_i}(j\omega) = R \frac{1}{1 + jR/\omega L + j\omega CR}$$

where I_i is the rms ac input current, and $I_i = 4 \, I_s/\sqrt{2}\pi$. The quality factor Q_p is

$$Q_p = \omega_0 CR = \frac{R}{\omega_0 L} = R \sqrt{\frac{C}{L}} = 2\delta \tag{11-43}$$

when δ is the damping factor and $\delta = \alpha/\omega_0 = (R/2)\sqrt{C/L}$. Substituting L, C, and R in terms of Q_p and ω_0, we get

$$Z(j\omega) = \frac{V_o}{I_i}(j\omega) = \frac{1}{1 + jQ_p(\omega/\omega_0 - \omega_0/\omega)} = \frac{1}{1 + jQ_p(u - 1/u)}$$

where $u = \omega/\omega_o$. The magnitude of $Z(j\omega)$ can be found from

$$|Z(j\omega)| = \frac{1}{[1 + Q_p^2(u - 1/u)^2]^{1/2}} \tag{11-44}$$

which is the same as Eq. (11-35). The magnitude plot is shown in Fig. 11-11b. A parallel resonant inverter is shown in Fig. 11-16a. Inductor L_e acts as a current source and capacitor C is the resonating element. L_m is the mutual inductance of the transformer and acts as the resonating inductor. A constant current is switched alternatively into the resonant circuit by transistors Q_1 and Q_2. The gating signals are shown in Fig. 11-16c. Referring the load resistance R_L into the primary side and neglecting the leakage inductances of the transformer, the equivalent circuit is shown in Fig. 11-16b A practical resonant inverter that supplies a fluorescent lamp is shown in Fig. 11-17.

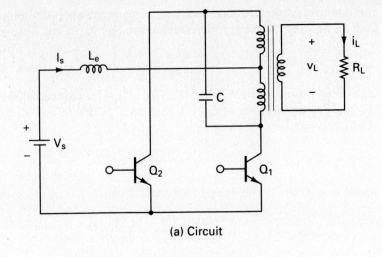

(a) Circuit

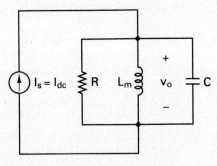

(b) Equivalent circuit

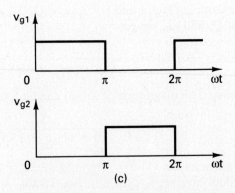

(c)

Figure 11-16 Parallel resonant inverter.

Figure 11-17 Practical resonant inverter.

The bridge topology in Fig. 11-18a can control the output voltage. The switching frequency f_s is kept constant at the resonant frequency f_o. By switching two devices simultaneously a *quasi-square wave* as shown in Fig. 11-18b can be obtained. The rms fundamental input current is given by

$$I_i = \frac{4I_s}{\sqrt{2}\,\pi} \cos \alpha \qquad (11\text{-}45)$$

By varying α from 0 to $\pi/2$ at a constant frequency, the current I_i can be controlled from $4I_s/(\sqrt{2}\pi)$ to 0.

This concept can be extended to HVDC applications in which the ac voltage is converted to the dc voltage and then converted back to ac. The transmission is normally done at a constant dc current I_{dc}. A single-phase version is shown in Fig. 11-18c. The output stage could be either a current-source inverter or a thyristor-controlled rectifier.

Example 11-8
_____

The parallel resonant inverter of Fig. 11-16a delivers a load power of $P_L = 1$ kW at a peak sinusoidal load voltage of $V_p = 170$ V and at resonance. The load resistance is $R = 10 \ \Omega$. The resonant frequency is $f_0 = 20$ kHz. Determine (a) the dc input current I_s, (b) the quality factor Q_p if it is required to reduce the load power to 250 W by frequency control so that $u = 1.25$, (c) the inductor L, and (d) the capacitor C.

Solution (a) Since at resonance $u = 1$ and $|Z(j\omega)|_{max} = 1$, the peak fundamental load current is $I_p = 4I_s/\pi$.

$$P_L = \frac{I_p^2 R}{2} = \frac{4^2 I_s^2 R}{2\pi^2} \qquad \text{or} \qquad 1000 = \frac{4^2 I_s^2 10}{2\pi^2}$$

which gives $I_s = 11.1$ A.

(b) To reduce the load power by $(1000/250 =)$ 4, the impedance must be reduced by 2 at $u = 1.25$. That is, from Eq. (11-44),we get $1 + Q_p^2(u - 1/u)^2 = 2^2$ which gives $Q_p = 3.85$.

(c) Q_p is defined by $Q_p = \omega_0 CR$ or $3.85 = 2\pi \times 20$ kHz $\times C \times 10$, which gives $C = 3.06 \ \mu$F.

(d) $f_o = 1/2\pi\sqrt{LC}$ or 20 kHz $= 1/[2\pi\sqrt{(3.06 \ \mu\text{F} \times L)}]$, which gives $L = 20.67 \ \mu$H.

_____

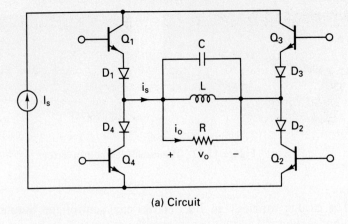

(a) Circuit

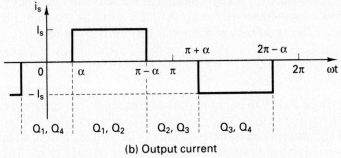

(b) Output current

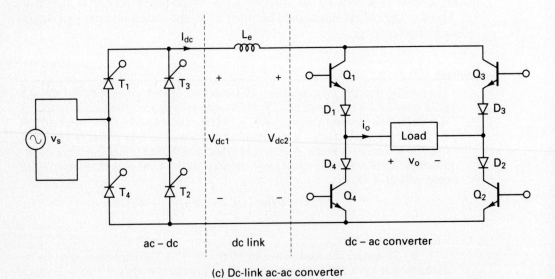

(c) Dc-link ac-ac converter

Figure 11-18 Quasi-square current control for parallel resonant inverter.

A class E resonant inverter uses only one transistor and has low switching losses, yielding a high efficiency of more than 95%. The circuit is shown in Fig. 11-19a. It is normally used for low-power applications requiring less than 100 W, particularly in high-frequency electronic lamp ballasts. The switching device has to withstand a high voltage. This inverter is normally used for fixed output voltage. However, the output voltage can be varied by varying the switching frequency. The circuit operation can be divided into two modes: mode 1 and mode 2.

Mode 1. During this mode, the transistor Q_1 is turned on. The equivalent circuit is shown in Fig. 11-19b. The switch current i_T consists of source current i_s and load current i_o. In order to obtain an almost sinusoidal output current, the values of L and C are chosen to have a high-quality factor, $Q \geq 7$, and a low damping ratio, usually $\delta \leq 0.072$. The switch is turned off at zero voltage. When the switch is turned off, its current is immediately diverted through capacitor C_e.

Mode 2. During this mode, transistor Q_1 is turned off. The equivalent circuit is shown in Fig. 11-19b. The capacitor current i_e becomes the sum of i_s and i_o. The switch voltage rises from zero to a maximum value and falls to zero again. When the switch voltage falls to zero, $i_e = C_e \, dv_T/dt$ will normally be negative. Thus the switch voltage would tend to be negative. To limit this negative voltage, an antiparallel diode as shown in Fig. 11-19a by dashed lines is connected. If the switch is a MOSFET, its negative voltage is limited by its built-in diode to a diode drop.

Mode 3. This mode will exist only if the switch voltage falls to zero with a finite negative slope. The equivalent circuit is similar to that for mode 1, except the initial conditions. The load current falls to zero at the end of mode 3. However, if the circuit parameters are such that the switch voltage falls to zero with a zero slope, there will be no need for a diode and this mode would not exist. That is, $v_T = 0$, and $dv_T/dt = 0$. The optimum parameters that usually satisfy these conditions and give the maximum efficiency are given by [7 to 0]:

$$L_e = 0.4001R/\omega_s$$

$$C_e = \frac{2.165}{R\omega_s}$$

$$\omega_s L - \frac{1}{\omega_s C} = 0.3533R$$

where ω_s is the switching frequency. The duty cycle is $k = t_{on}/T_s = 30.4\%$. The waveforms of the output current, switch current, and switch voltage are shown in Fig. 11-19c.

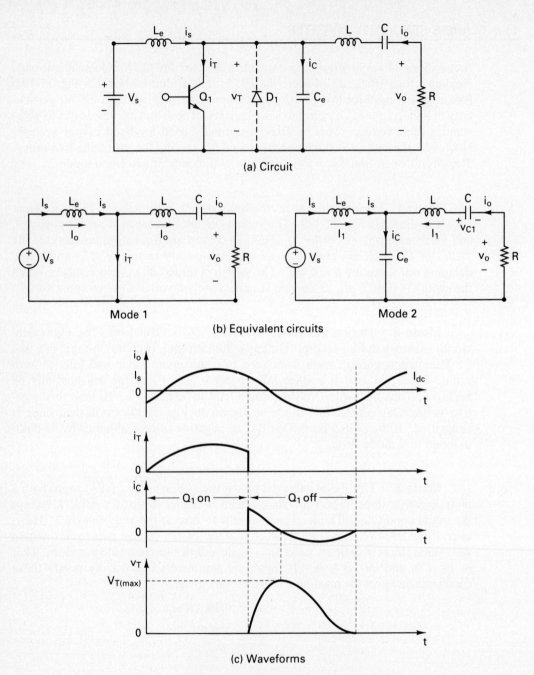

(a) Circuit

(b) Equivalent circuits

(c) Waveforms

Figure 11-19   Class E resonant inverter.

Example 11-9

The class E inverter of Fig. 11-19a operates at resonance and has $V_s = 12$ V and $R = 10 \Omega$. The switching frequency is $f_s = 25$ kHz. (a) Determine the optimum values of $L, C, C_e,$ and L_e. (b) Use PSpice to plot the output voltage v_o and the switch voltage v_T for $k = 0.304$. Assume that $Q = 7$.

Solution $V_s = 12$ V, $R = 10 \Omega$, and $\omega_s = 2\pi f_s = 2\pi \times 25$ kHz $= 157.1$ krad/s.

(a) $L_e = \dfrac{0.4001 R}{\omega_s} = 0.4001 \times \dfrac{10}{157.1 \text{ krad/s}} = 25.47 \ \mu\text{H}$

$C_e = \dfrac{2.165}{R\omega_s} = \dfrac{2.165}{10 \times 157.1 \text{ krad/s}} = 1.38 \ \mu\text{F}$

$L = \dfrac{QR}{\omega_s} = \dfrac{7 \times 10}{157.1 \text{ krad/s}} = 445.63 \ \mu\text{H}$

$\omega_s L - 1/\omega_s C = 0.3533 \, R$ or $7 \times 10 - 1/\omega_s C = 0.3533 \times 10$, which gives $C = 0.0958$ μF. The damping factor is

$$\delta = (R/2)\sqrt{C/L} = (10/2)\sqrt{0.0958/445.63} = 0.0733$$

which is very small, and the output current should essentially be sinusoidal. The resonant frequency is

$$f_0 = \frac{1}{2\pi\sqrt{LC}} = \frac{1}{2\pi\sqrt{(445.63 \ \mu\text{H} \times 0.0958 \ \mu\text{F})}} = 24.36 \text{ kHz}$$

(b) $T_s = 1/f_s = 1/25$ kHz $= 40 \ \mu$s, and $t_{on} = kT_s = 0.304 \times 40 = 12.24 \ \mu$s. The circuit for PSpice simulation is shown in Fig. 11-20a and the control voltage in Fig. 11-20b. The list of the circuit file is as follows:

```
Example 11-9     Class-E Resonant Inverter
VS     1    0    DC     12V
VY     1    2    DC     0V    ; Voltage source to measure input current
VG     8    0    PULSE (0V    20V   0    1NS    1NS   12.24US    40US)
RB     8    7    250         ; Transistor base-drive resistance
R      6    0    10
LE     2    3    25.47UH
CE     3    0    1.38UF
C      3    4    0.0958UF
L      5    6    445.63UH
VX     4    5    DC     0V   ; Voltage source to measure load current of L2
Q1     3    7    0    MODQ1                ; BJT switch
.MODEL   MODQ1   NPN (IS=6.734F BF=416.4 ISE=6.734F BR=.7371
+      CJE=3.638P MJC=.3085 VJC=.75 CJE=4.493P MJE=.2593 VJE=.75
+      TR=239.5N TF=301.2P)              ; Transistor model parameters
.TRAN    2US    300US    180US    1US    UIC      ; Transient analysis
.PROBE                                   ; Graphics postprocessor
.OPTIONS ABSTOL = 1.00N RELTOL = 0.01 VNTOL = 0.1 ITL5=20000 ; convergence
.END
```

The PSpice plots are shown in Fig. 11-21, where V(3) = switch voltage and V(6) = output voltage. Using the PSpice cursor in Fig. 11-21 gives $V_{o(pp)} = 29.18$ V and $V_{T(peak)} = 31.481$ V.

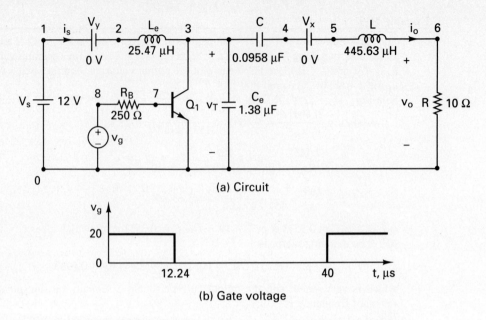

(a) Circuit

(b) Gate voltage

Figure 11-20 Class E resonant inverter for PSpice simulation.

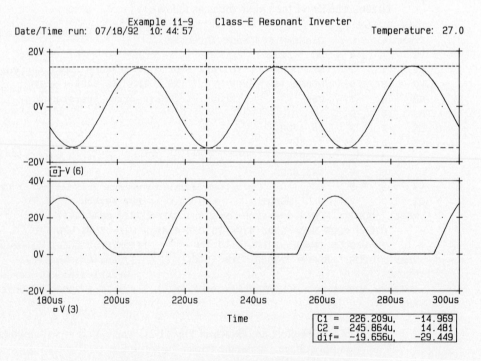

Figure 11-21 PSpice plots for Example 11-9.

Since dc/dc converters generally consist of a dc/ac resonant inverter and an ac/dc rectifier, a high-frequency diode rectifier suffers from disadvantages such as conduction and switching losses, parasitic oscillations, and high harmonic content of the input current. A class E resonant rectifier as shown in Fig. 11-22a overcomes these limitations. It uses the principle of zero-voltage switching of the diode. That is, the diode turns off at zero voltage. The diode junction capacitance is included in the resonant capacitance C and therefore does not adversely affect the circuit operation. The circuit operation can be divided into two modes: mode 1 and mode 2. Let us assume that C_f is sufficiently large so that the dc output voltage V_o is constant. Let the input voltage $v_s = V_m \sin \omega t$.

Mode 1. During this mode, the diode is off. The equivalent circuit is shown in Fig. 11-22b. The values of L and C are such that $\omega L = 1/\omega C$ at the operating frequency, f. The voltage appearing across L and C is $v_{(LC)} = V_s \sin \omega t - V_o$.

Mode 2. During this mode, the diode is on. The equivalent circuit is shown in Fig. 11-22b. The voltage appearing across L is $v_L = V_s \sin \omega t - V_o$. When the diode current i_D, which is the same as the inductor current i_L, reaches zero, the diode turns off. At turn-off, $i_D = i_L = 0$ and $v_D = v_C = 0$. That is, $i_c = C \, dv_c/dt = 0$, which gives $dv_c/dt = 0$. Therefore, the diode voltage is zero at turn-off, thereby reducing the switching losses. The inductor current can be expressed approximately by

$$i_L = I_m \sin(\omega t - \phi) - I_o \tag{11-46}$$

where $I_m = V_m/R$ and $I_o = V_o/R$. When the diode is on, the phase shift ϕ will be 90°. When the diode is off, it will be 0°, provided that $\omega L = 1/\omega C$. Therefore, ϕ will have a value between 0 and 90°, and its value depends on the load resistance R. The peak-to-peak current will be $2V_m/R$. The input current has a dc component I_o and a phase delay ϕ. In order to improve the input power factor, an input capacitor as shown in Fig. 11-22a by dashed lines is normally connected.

Example 11-10

The class E rectifier of Fig. 11-22a supplies a load power of $P_L = 400$ mW at $V_o = 4$V. The peak supply voltage is $V_m = 10$ V. The supply frequency is $f = 250$ kHz. The peak-to-peak ripple on the dc output voltage is $\Delta V_o = 40$ mV. (a) Determine the values of L, C, and C_f, and (b) the rms and dc currents of L and C. (c) Use PSpice to plot the output voltage v_o and the inductor current i_L.

Solution $V_m = 10$ V, $V_o = 4$V, $\Delta V_o = 40$ mV, and $f = 250$ kHz.

(a) Choose a suitable value of C. Let $C = 10$ nF. Let the resonant frequency be $f_o = f = 250$ kHz. 250 kHz $= f_o = 1/[2\pi\sqrt{(L \times 10 \text{ nF})}]$, which gives $L = 40.5 \ \mu$H. $P_L = V_o^2/R$ or 400 mW $= 4^2/R$, which gives $R = 40 \ \Omega$. $I_o = V_o/R = 4/40 = 100$ mA. The value of capacitance C_f is given by

$$C_f = \frac{I_o}{2f \, \Delta V_o} = \frac{100 \text{ mA}}{2 \times 250 \text{ kHz} \times 40 \text{ mV}} = 5 \ \mu\text{F}$$

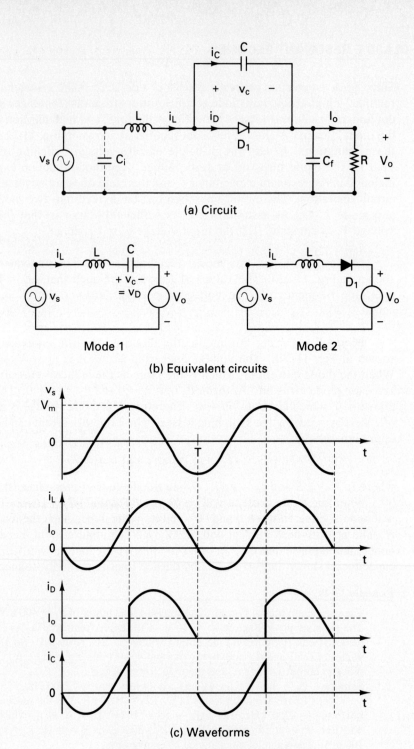

(a) Circuit

(b) Equivalent circuits

(c) Waveforms

Figure 11-22 Class E resonant rectifier.

(b) $I_m = V_m/R = 10/40 = 250$ mA. The rms inductor current I_L is

$$I_{L(\text{rms})} = \sqrt{100^2 + \frac{250^2}{2}} = 203.1 \text{ mA}$$

$$I_{L(\text{dc})} = 100 \text{ mA}$$

The rms current of the capacitor C is

$$I_{C(\text{rms})} = \frac{250}{\sqrt{2}} = 176.78 \text{ mA}$$

$$I_{C(\text{dc})} = 0$$

(c) $T = 1/f = 1/250$ kHz $= 4$ μs. The circuit for PSpice simulation is shown in Fig. 11-23. The list of the circuit file is as follows:

```
Example  11-10    Class E Resonant Rectifier
VS     1    0    SIN (0    10V    250KHZ)
VY     1    2    DC    0V   ; Voltage source to measure input current
R      4    5    40
L      2    3    40.5UH
C      3    4    10NF
CF     4    0    5UF
VX     5    0    DC    0V   ; Voltage source to measure current through R
D1     3    4    DMOD            ; Rectifier diode
.MODEL    DMOD    D              ; Diode default parameters
.TRAN    0.1US    1220US   1200US   0.1US   UIC  ; Transient analysis
.PROBE                          ; Graphics postprocessor
.OPTIONS ABSTOL = 1.00N RETOL1 = 0.01 VNTOL = 0.1 ITL5=40000 ; convergence
.END
```

The PSpice plot is shown in Fig. 11-24, where I(L) = inductor current and V(4) = output voltage. Using the PSpice cursor in Fig. 11-24 gives $V_o = 3.98$ V, $\Delta V_o = 63.04$ mV, and $i_{L(\text{pp})} = 489.36$ mA.

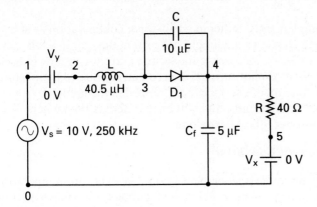

Figure 11-23 Class E resonant rectifier for PSpice simulation.

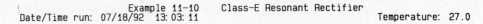

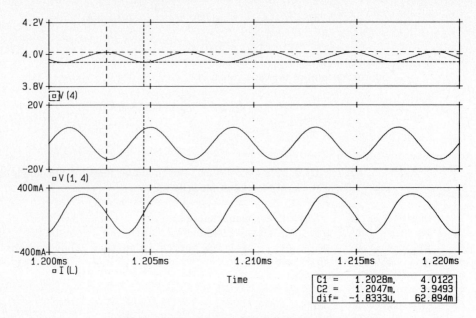

C1 =	1.2028m,	4.0122
C2 =	1.2047m,	3.9493
dif=	−1.8333u,	62.894m

Figure 11-24   PSpice plots for Example 11-10.

11-6 ZERO-CURRENT-SWITCHING RESONANT CONVERTERS

The switches of a zero-current-switching (ZCS) resonant converter turn "on" and "off" at zero current. The resonant circuit that consists of switch S_1, inductor L, and capacitor C is shown in Fig. 11-25a. It is classified by Liu et al. [11] into two types: L type and M type. In both types, the inductor L limits the di/dt of the switch current, and L and C constitute a series-resonant circuit. When the switch current is zero, there will be a current $i = C_j \, dv_T/dt$ flowing through the internal capacitance C_j due to a finite slope of the switch voltage at turn-off. This current flow will cause power dissipation in the switch and limits the high-switching frequency.

The switch can be implemented either in a half-wave configuration as shown in Fig. 11-25b, where diode D_1 allows unidirectional current flow, or in a full-wave configuration as shown in Fig. 11-25c, where the switch current can flow bidirectionally. The practical devices do not turn off at zero current due to their recovery times. As a result, an amount of energy will be trapped in the inductor L of the L-type configuration, and voltage transients will appear across the switch. This favors L-type configuration over the M-type one.

11-6.1 L-Type ZCS Resonant Converter

An L-type ZCS resonant converter is shown in Fig. 11-26a. The circuit operation can be divided into five modes, whose equivalent circuits are shown in Fig.

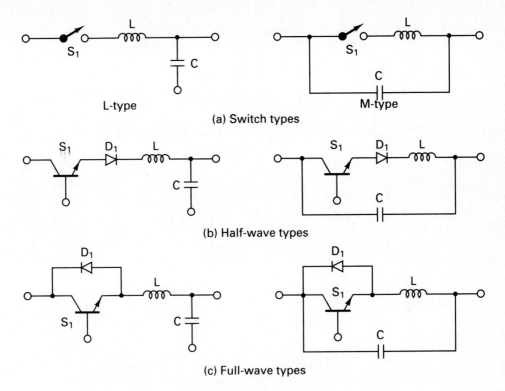

L-type   **M-type**

(a) Switch types

(b) Half-wave types

(c) Full-wave types

Figure 11-25   Switch configurations for ZCS resonant converters.

11-26b. We shall redefine the time origin, $t = 0$, at the beginning of each mode.

Mode 1.   This mode is valid for $0 \leq t \leq t_1$. Switch S_1 is turned on and diode D_m conducts. The inductor current i_L, which rises linearly, is given by

$$i_L = \frac{V_s}{L} t \qquad (11\text{-}47)$$

This mode ends at time $t = t_1$ when $i_L(t = t_1) = I_o$. That is, $t_1 = I_o L / V_s$.

Mode 2.   This mode is valid for $0 \leq t \leq t_2$. Switch S_1 remains on, but diode D_m is off. The inductor current i_L is given by

$$i_L = I_m \sin \omega_0 t + I_o \qquad (11\text{-}48)$$

where $I_m = V_s \sqrt{C/L}$, and $\omega_0 = 1/\sqrt{LC}$. The capacitor voltage v_c is given by

$$v_c = V_s(1 - \cos \omega_0 t)$$

The peak switch current, which occurs at $t = (\pi/2)\sqrt{LC}$, is

$$I_p = I_m + I_o$$

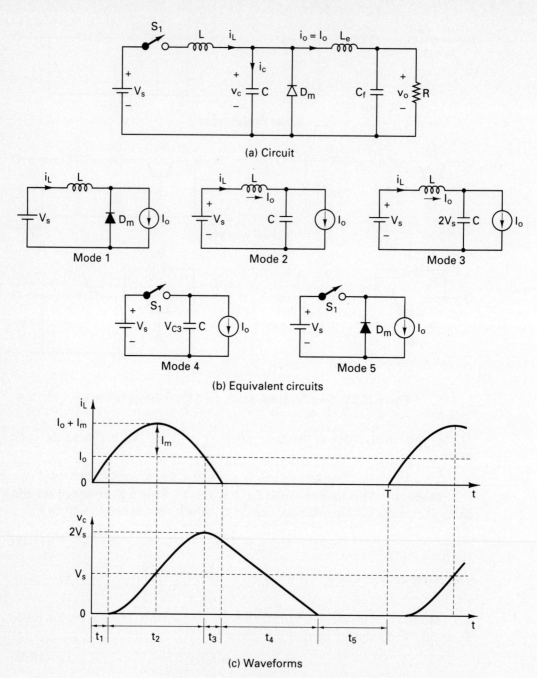

(a) Circuit

Mode 1   Mode 2   Mode 3

Mode 4   Mode 5

(b) Equivalent circuits

(c) Waveforms

Figure 11-26   L-type ZCS resonant converter.

The peak capacitor voltage is

$$V_{c(pk)} = 2V_s$$

This mode ends at $t = t_2$ when $i_L(t = t_2) = I_o$, and $v_c(t = t_2) = V_{c2} = 2V_s$. Therefore, $t_2 = \pi\sqrt{LC}$.

Mode 3. This mode is valid for $0 \le t \le t_3$. The inductor current that falls from I_o to zero is given by

$$i_L = I_o - I_m \sin \omega_0 t \qquad (11\text{-}49)$$

The capacitor voltage is given by

$$v_c = 2V_s \cos \omega_0 t \qquad (11\text{-}50)$$

This mode ends at $t = t_3$ when $i_L(t = t_3) = 0$ and $v_c(t = t_3) = V_{c3}$. Thus $t_3 = \sqrt{LC}$ $\sin^{-1}(1/x)$ where $x = I_m/I_o = (V_s/I_o)\sqrt{C/L}$.

Mode 4. This mode is valid for $0 \le t \le t_4$. The capacitor supplies the load current I_o, and its voltage is given by

$$v_c = V_{c3} - \frac{I_o}{C}t \qquad (11\text{-}51)$$

This mode ends at time $t = t_4$ when $v_c(t = t_4) = 0$. Thus $t_4 = V_{c3}C/I_o$.

Mode 5. This mode is valid for $0 \le t \le t_5$. When the capacitor voltage tends to be negative, the diode D_m conducts. The load current I_o flows through the diode D_m. This mode ends at time $t = t_5$ when the switch S_1 is turned on again, and the cycle is repeated. That is, $t_5 = T - (t_1 + t_2 + t_3 + t_4)$.

The waveforms for i_L and v_c are shown in Fig. 11-26c. The peak switch voltage equals to the dc supply voltage, V_s. Since the switch current is zero at turn-on and turn-off, the switching loss, which is the product of v and i, becomes very small. The peak resonant current I_m must be higher than the load current I_o, and this sets a limit on the minimum value of load resistance, R. However, by placing an antiparallel diode across the switch, the output voltage can be made insensitive to load variations.

Example 11-11

The ZCS resonant converter of Fig. 11-26a delivers a maximum power of $P_L = 400$ mW at $V_o = 4$ V. The supply voltage is $V_s = 12$ V. The maximum operating frequency is $f_{max} = 50$ kHz. Determine the values of L and C. Assume that the intervals t_1 and t_3 are very small, and $x = 1.5$.

Solution $V_s = 12$ V, $f = f_{max} = 50$ kHz, and $T = 1/50$ kHz $= 20$ μs. $P_L = V_oI_o$ or 400 mW $= 4I_o$, which gives $I_o = 100$ mA. The maximum frequency will occur when $t_5 = 0$. Since $t_1 = t_3 = t_5 = 0$, $t_2 + t_4 = T$. Substituting $t_4 = 2V_sC/I_m$ and using $x = (V_s/I_o)$ $\sqrt{C/L}$ gives

$$\pi\sqrt{LC} + \frac{2V_sC}{I_o} = T \qquad \text{or} \qquad \frac{\pi V_s}{xI_o}C + \frac{2V_s}{I_o}C = T$$

which gives $C = 0.0407$ μF. Thus $L = (V_s/xI_o)^2C = 260.52$ μH.

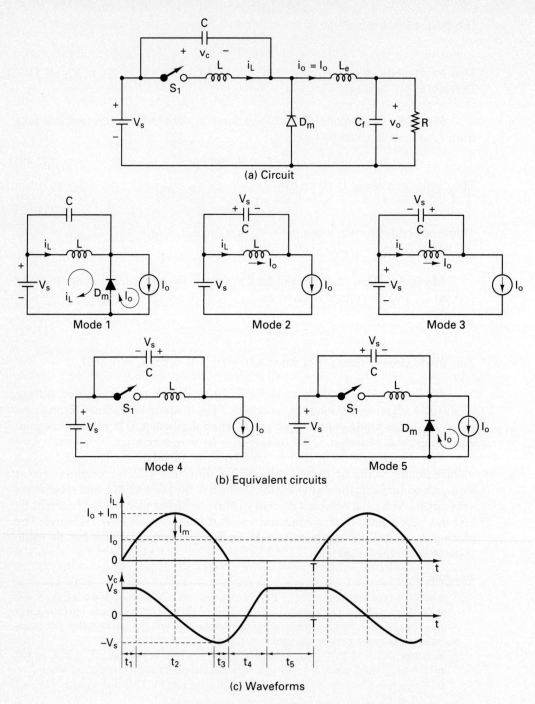

Figure 11-27 M-type ZCS resonant converter.

11-6.2 M-Type ZCS Resonant Converter

An M-type ZCS resonant converter is shown in Fig. 11-27a. The circuit operation can be divided into five modes, whose equivalent circuits are shown in Fig. 11-27b. We shall redefine the time origin, $t = 0$, at the beginning of each mode. The mode equations are similar to that of an L-type converter, except the following.

Mode 2. The capacitor voltage v_c is given by

$$v_c = V_s \cos \omega_0 t \qquad (11\text{-}52)$$

The peak capacitor voltage is $V_{c(pk)} = V_s$. At the end of this mode at $t = t_2$, $v_c(t = t_2) = V_{c2} = -V_s$.

Mode 3. The capacitor voltage is given by

$$v_c = -V_s \cos \omega_0 t \qquad (11\text{-}53)$$

At the end of this mode at $t = t_3$, $v_c(t = t_3) = V_{c3}$. It should be noted that V_{c3} will have a negative value.

Mode 4. This mode ends at $t = t_4$ when $v_c(t = t_4) = V_s$. Thus $t_4 = (V_s - V_{c3})C/I_o$. The waveforms for i_L and v_c are shown in Fig. 11-27c.

11-7 ZERO-VOLTAGE-SWITCHING RESONANT CONVERTERS

The switches of a zero-voltage-switching (ZVS) resonant converter turn "on" and "off" at zero voltage. The resonant circuit is shown in Fig. 11-28a. The capacitor C is connected in parallel with the switch S_1 to achieve zero-voltage switching. The internal switch capacitance C_j is added with the capacitor C, and it affects the resonant frequency only, thereby contributing no power dissipation in the switch. If the switch is implemented with a transistor Q_1 and an antiparallel diode D_1 as shown in Fig. 11-28b, the voltage across C is clamped by D_1, and the switch is operated in a half-wave configuration. If the diode D_1 is connected in series with Q_1 as shown in Fig. 11-28c, the voltage across C can oscillate freely, and the switch is operated in a full-wave configuration. A ZVS resonant converter is shown in Fig. 11-29a. A ZVS resonant converter is the dual of the ZCS resonant converter in Fig. 11-27a. Equations for the M-type ZCS resonant converter can be applied if i_L is replaced by v_c and vice versa, L by C and vice versa, and V_s by I_o and vice versa. The circuit operation can be divided into five modes whose equivalent circuits are shown in Fig. 11-29b. We shall redefine the time origin, $t = 0$, at the beginning of each mode.

Mode 1. This mode is valid for $0 \le t \le t_1$. Both switch S_1 and diode D_m are off. Capacitor C charges at a constant rate of load current I_o. The capacitor

---

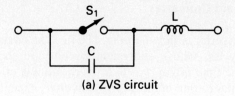

(a) ZVS circuit

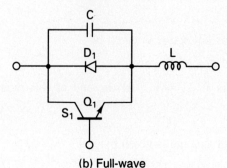

(b) Full-wave

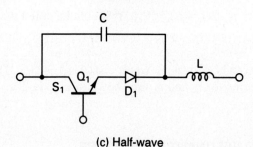

(c) Half-wave

Figure 11-28 Switch configurations for ZVS resonant converters.

voltage v_c, which rises, is given by

$$v_c = \frac{I_o}{C} t \tag{11-54}$$

This mode ends at time $t = t_1$ when $v_c(t = t_1) = V_s$. That is, $t_1 = V_s C / I_o$.

Mode 2. This mode is valid for $0 \le t \le t_2$. The switch S_1 is still off, but diode D_m turns on. The capacitor voltage v_c is given by

$$v_c = V_m \sin \omega_0 t + V_s \tag{11-55}$$

where $V_m = I_o \sqrt{L/C}$. The peak switch voltage, which occurs at $t = (\pi/2)\sqrt{LC}$, is

$$V_{T(pk)} = V_{C(pk)} = I_o \sqrt{\frac{L}{C}} + V_s \tag{11-56}$$

The inductor current i_L is given by

$$i_L = I_o \cos \omega_0 t \tag{11-57}$$

This mode ends at $t = t_2$ when $v_c(t = t_2) = V_s$, and $i_L(t = t_2) = -I_o$. Therefore, $t_2 = \pi \sqrt{LC}$.

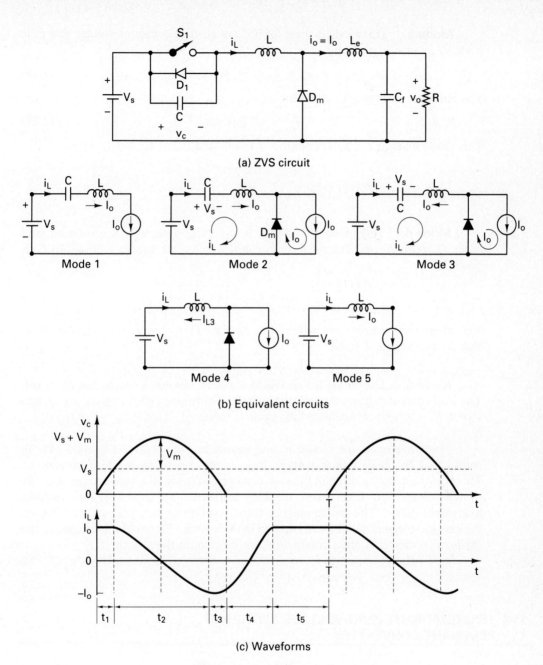

(a) ZVS circuit

(b) Equivalent circuits

(c) Waveforms

Figure 11-29   ZVS resonant converter.

Mode 3. This mode is valid for $0 \leq t \leq t_3$. The capacitor voltage that falls from V_s to zero is given by

$$v_c = V_s - V_m \sin \omega_0 t \qquad (11\text{-}58)$$

The inductor current i_L is given by

$$i_L = -I_o \cos \omega_0 t \qquad (11\text{-}59)$$

This mode ends at $t = t_3$ when $v_c(t = t_3) = 0$, and $i_L(t = t_3) = I_{L3}$. Thus

$$t_3 = \sqrt{LC} \sin^{-1} x$$

where $x = V_s/V_m = (V_s/I_o)\sqrt{C/L}$.

Mode 4. This mode is valid for $0 \leq t \leq t_4$. Switch S_1 is turned on, and diode D_m remains on. The inductor current, which rises linearly from I_{L3} to I_o, is given by

$$i_L = I_{L3} + \frac{V_s}{L} t \qquad (11\text{-}60)$$

This mode ends at time $t = t_4$ when $i_L(t = t_4) = 0$. Thus $t_4 = (I_o - I_{L3})(L/V_s)$. Note that I_{L3} is a negative value.

Mode 5. This mode is valid for $0 \leq t \leq t_5$. Switch S_1 is on, but D_m is off. The load current I_o flows through the switch. This mode ends at time $t = t_5$, when switch S_1 is turned off again and the cycle is repeated. That is, $t_5 = T - (t_1 + t_2 + t_3 + t_4)$.

The waveforms for i_L and v_c are shown in Fig. 11-29c. Equation (11-56) shows that the peak switch voltage $V_{T(\text{pk})}$ is dependent on the load current I_o. Therefore, a wide variation in the load current will result in a wide variation of the switch voltage. For this reason, the ZVS converters are used only for constant-load applications. The switch must be turned on only at zero voltage. Otherwise, the energy stored in C will be dissipated in the switch. To avoid this situation, the antiparallel diode D_1 must conduct before turning on the switch.

11-8 TWO-QUADRANT ZERO-VOLTAGE-SWITCHING RESONANT CONVERTERS

The ZVS concept can be extended to a two-quadrant class A chopper as shown in Fig. 11-30a, where the capacitors $C_+ = C_- = C/2$. The inductor L has such a value, so that it forms a resonant circuit. The resonant frequency is $f_o = 1/(2\pi\sqrt{LC})$, and it is much larger than the switching frequency f_s. Assuming the filter capacitance C_e to be large, the load is replaced by a dc voltage V_{dc} as shown in Fig. 11-30b. The circuit operations can be divided into six modes. The equivalent circuits for various modes are shown in Fig. 11-30d.

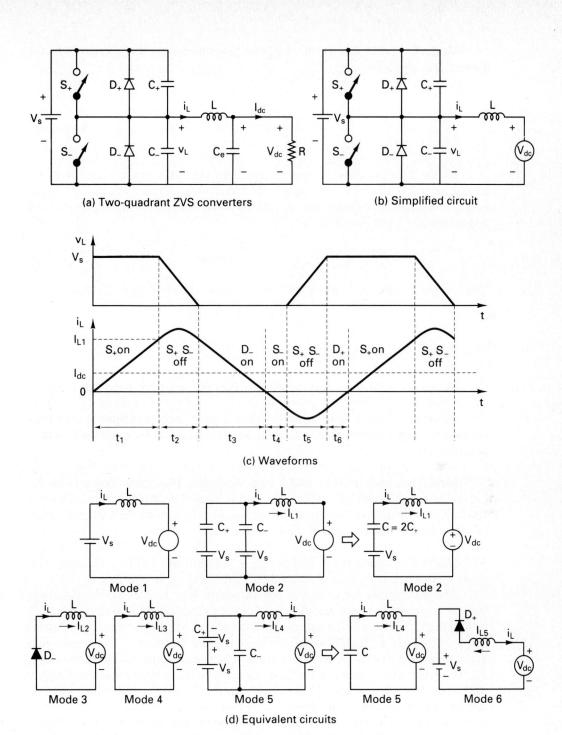

(a) Two-quadrant ZVS converters

(b) Simplified circuit

(c) Waveforms

(d) Equivalent circuits

Figure 11-30 Two-quadrant ZVS resonant converter.

Mode 1. Switch S_+ is on. Assuming an initial current of I_{L0}, the inductor current i_L is given by

$$i_L = \frac{V_s}{L} t \qquad (11\text{-}61)$$

This mode ends when the voltage on capacitor C_+ is zero and S_+ is turned off. The voltage on C_- is V_s.

Mode 2. Switches S_+ and S_- are both off. This mode begins with C_+ having zero voltage and C_- having V_s. The equivalent of this mode can be simplified to a resonant circuit of C and L with an initial inductor current I_{L1}. i_L can be approximately represented by

$$i_L = (V_s - V_{dc}) \sqrt{\frac{L}{C}} \sin \omega_0 t + I_{L1} \qquad (11\text{-}62)$$

The voltage v_o can be approximated to fall linearly from V_s to 0. That is,

$$v_o = V_s - \frac{V_s C}{I_{L1}} t \qquad (11\text{-}63)$$

This mode ends when v_o becomes zero and diode D_- turns on.

Mode 3. Diode D_- is turned on. i_L falls linearly from $I_{L2}(= I_{L1})$ to 0.

Mode 4. Switch S_- is turned on when i_L and v_o becomes zero. i_L continues to fall in the negative direction to I_{L4} until the switch voltage becomes zero, and S_- is turned off.

Mode 5. Switches S_+ and S_- are both off. This mode begins with C_- having zero voltage and C_+ having V_s, and is similar to mode 2. The voltage v_o can be approximated to rise linearly from 0 to V_s. This mode ends when v_o tends to become more than V_s and diode D_+ turns on.

Mode 6. Diode D_+ is turned on. i_L falls linearly from I_{L5} to zero. This mode ends when $i_L = 0$. S_+ is turned on, and the cycle is repeated.

The waveforms for i_L and v_o are shown in Fig. 11-30c. For zero-voltage switching i_L must flow in either direction so that a diode conducts before its switch is turned on. The output voltage can be made almost square wave by choosing the resonant frequency f_o much larger than the switching frequency. The output voltage can be regulated by frequency control. The switch voltage is clamped to only V_s. However, the switches have to carry i_L, which has high ripples and higher peak than the load current I_o. The converter can be operated under current-regulated mode to obtain the desired waveform of i_L.

The circuit in Fig. 11-30a can be extended to a single-phase half-bridge inverter as shown in Fig. 11-31. A three-phase version is shown in Fig. 11-32a, where the load inductance L constitutes the resonant circuit. One arm of a three-phase circuit in which a separate resonant inductor [10] is used is shown in Fig. 11-32b.

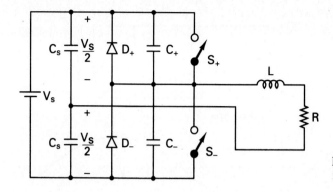

Figure 11-31 Single-phase ZVS resonant inverter.

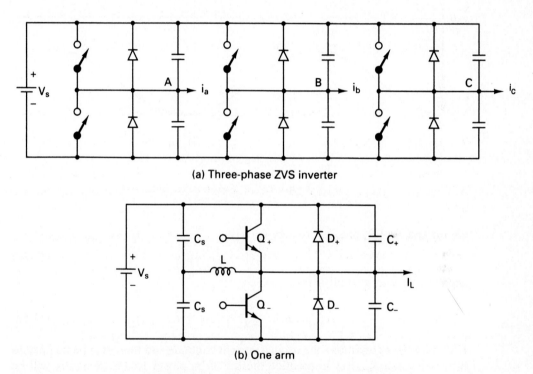

(a) Three-phase ZVS inverter

(b) One arm

Figure 11-32 Three-phase ZVS resonant inverter.

11-9 RESONANT DC-LINK INVERTERS

In resonant dc-link inverters, a resonant circuit is connected between the dc input voltage and the PWM inverter, so that the input voltage to the inverter oscillates between zero and a value slightly greater than twice the dc input voltage. The resonant link, which is similar to the class E inverter in Fig. 11-19a, is shown in Fig. 11-33a, where I_o is the current drawn by the inverter. Assuming a lossless

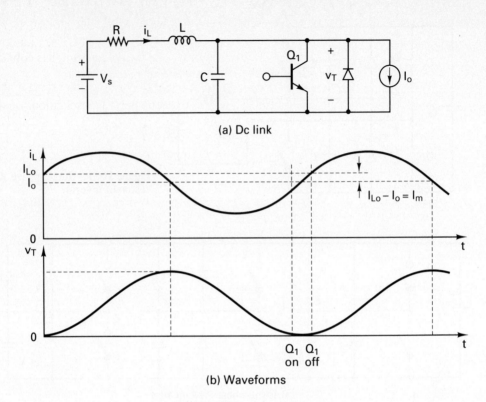

(a) Dc link

(b) Waveforms

Figure 11-33 Resonant dc link.

circuit and $R = 0$, the link voltage is

$$v_c = V_s(1 - \cos \omega_0 t) \tag{11-64}$$

and the inductor current i_L is

$$i_L = V_s \sqrt{\frac{C}{L}} \sin \omega_0 t + I_o \tag{11-65}$$

Under lossless conditions the oscillation will continue and there will be no need to turn on switch S_1. But in practice there will be power loss in R and i_L will be damped sinusoidal and S_1 is turned on to bring the current to the initial level. The value of R is small and the circuit is underdamped. Under this condition, i_L and v_c can be shown [13] as

$$i_L \approx I_o + e^{-\alpha t}\left[\frac{V_s}{\omega L} \sin \omega t + (I_{Lo} - I_o) \cos \omega t\right] \tag{11-66}$$

and the capacitor voltage v_c is

$$v_c \approx V_s + e^{-\alpha t}[\omega L(I_{Lo} - I_o) \sin \omega t - V_s \cos \omega t] \tag{11-67}$$

The waveforms for v_c and i_L are shown in Fig. 11-33b. Switch S_1 is turned on when the capacitor voltage falls to zero and is turned off when the current i_L

reaches the level of the initial current I_{Lo}. It can be noted that the capacitor voltage depends only on the difference $I_m(= I_{Lo} - I_o)$ rather than the load current, I_o. Thus the control circuit should monitor $(i_L - I_o)$ when the switch is conducting and turn off the switch when the desired value of I_m is reached.

A three-phase resonant dc-link inverter [14] is shown in Fig. 11-34a. The six inverter devices are gated in such a way as to set up periodic oscillations on the dc link LC circuit. The devices are turned on and off at zero-link voltages, thereby accomplishing lossless turn-on and turn-off of all devices. The waveforms for the link voltage and the inverter line-to-line voltage are shown in Fig. 11-34b.

The dc-link resonant cycle is normally started with a fixed value of initial capacitor current. This causes the voltage across the resonant dc link to exceed $2V_s$, and all the inverter devices are subjected to this high-voltage stress. An active clamp [14] as shown in Fig. 11-35a can limit the link voltage shown in Fig.

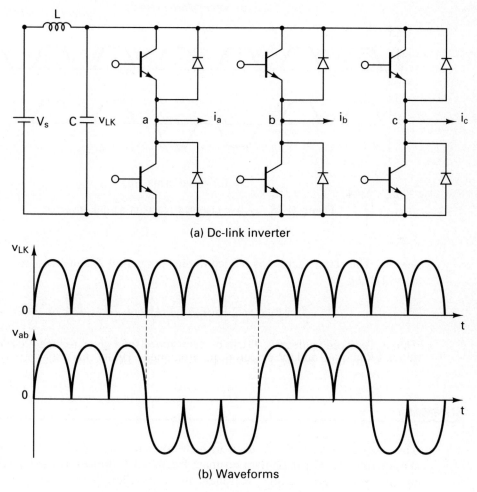

(a) Dc-link inverter

(b) Waveforms

Figure 11-34 Three-phase resonant dc-link inverter.

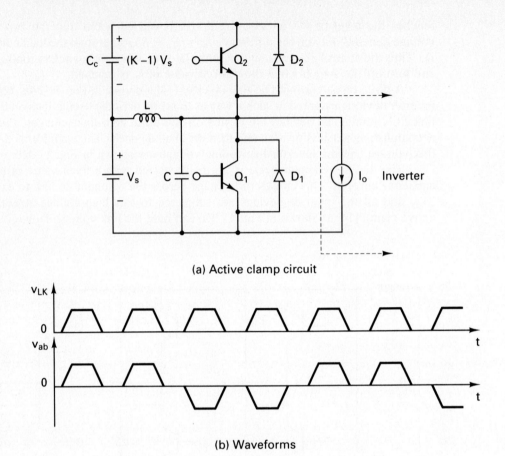

(a) Active clamp circuit

(b) Waveforms

Figure 11-35 Active clamped resonant dc-link inverter.

11-35b. The clamp factor k is related to the tank period T_k and resonant frequency $\omega_0 = 1/\sqrt{LC}$ by

$$\frac{f_o}{f_k} = T_k\omega_0 = 2\left[\cos^{-1}(1 - k) + \frac{\sqrt{k(2 - k)}}{k - 1}\right] \qquad (11\text{-}68)$$

That is, for fixed value of k, T_k can be determined for a given resonant circuit. For $k = 1.5$, the tank period T_k should be $T_k = 7.65\sqrt{LC}$.

SUMMARY

The resonant inverters are used in high-frequency applications requiring fixed output voltage. The maximum resonant frequency is limited by the turn-off times of thyristors or transistors. Resonant inverters allow limited regulation of the output voltage. Parallel-resonant inverters are supplied from a constant dc source and give a sinusoidal output voltage. Class E inverters and rectifiers are simple and are used primarily for low-power, high-frequency applications. The zero-voltage-switching (ZVS) and zero-current-switching (ZCS) converters are getting increasingly popular because they are turned on and off at zero current/voltage, thereby eliminating switching losses. In resonant dc-link inverters, a resonant circuit is connected between the inverter and the dc supply. The resonant voltage pulses are produced at the input of the inverter, and the inverter devices are turned "on" and "off" at zero voltages.

REFERENCES

1. F. C. Schwarz, "An improved method of resonant current pulse modulation for power converters." *IEEE Transactions on Industrial Electronics and Control Instrumentation,* Vol. IECI23, No. 2, 1976, pp. 133–141.

2. J. Vitnis, A. Schweizer, and J. L. Steiner, "Reverse conducting thyristors for high power series resonant circuits." *IEEE Industry Applications Society Conference Record,* 1985, pp. 715–722.

3. D. M. Divan, "Design considerations for very high frequency resonant mode dc/dc converters." *IEEE Industry Applications Society Conference Record,* 1986, pp. 640–647.

4. A. K. S. Bhat and S. B. Dewan, "A generalized approach for the steady state analysis of resonant inverters." *IEEE Industry Applications Society Conference Record,* 1986, pp. 664–671.

5. P. D. Ziogas, V. T. Ranganathan, and V. R. Stefanovic, "A four-quadrant current regulated converter with a high frequency link." *IEEE Transactions on Industry Applications,* Vol. IA18, No. 5, 1982, pp. 499–505.

6. R. L. Steigerwald, "A compromise of half-bridge resonance converter topologies." *IEEE Transactions on Power Electronics,* Vol. PE3, No. 2, 1988, pp. 174–182.

7. N. O. Sokal and A. D. Sokal, "Class E: a new class of high-efficiency tuned single-ended switching power amplifiers." *IEEE Journal of Solid-State Circuits,* Vol. 10, No. 3, 1975, pp. 168–176.

8. J. Ebert and M. K. Kazimierczuk, "Class-E high-efficiency tuned power oscillator." *IEEE Journal of Solid-State Circuits,* Vol. 16, No. 2, 1981, pp. 62–66.

9. R. E. Zuliski, "A high-efficiency self-regulated class-E power inverter/converter." *IEEE Transactions on Industrial Electronics,* Vol. IE33, No. 3, 1986, pp. 340–342.

10. M. K. Kazimierczuk and J. Jozwik, "Class-E zero voltage switching and zero-current switching rectifiers." *IEEE Transactions on Circuits and Systems,* Vol. CS37, No. 3, 1990, pp. 436–444.

11. K. Liu, R. Oruganti, and F. C. Y. Lee, "Quasi-resonant converters: topologies and characteristics." *IEEE Transactions on Power Electronics,* Vol. PE2, No. 1, 1987, pp. 62–71.

12. J. A. Ferreira, P. C. Theron, and J. D. van Wyk, *Conference Proceedings of the IEEE-IAS Annual Meeting,* 1991, pp. 1462–1468.

13. D. M. Devan, "The resonant DC link converter: a new concept in static power conversion." *IEEE Transactions on Industry Applications,* Vol. IA25, No. 2, 1989, pp. 317–325.

14. D. M. Devan and G. Skibinski, "Zero-switching loss inverters for high power applications." *IEEE Transactions on Industry Applications,* Vol. IA25, No. 4, 1989, pp. 634–643.

11-1. What is the principle of series resonant inverters?

11-2. What is the dead zone of a resonant inverter?

11-3. What are the advantages and disadvantages of resonant inverters with bidirectional switches?

11-4. What are the advantages and disadvantages of resonant inverters with unidirectional switches?

11-5. What is the necessary condition for series resonant oscillation?

11-6. What is the purpose of coupled inductors in half-bridge resonant inverters?

11-7. What are the advantages of reverse-conducting thyristors in resonant inverters?

11-8. What is an overlap control of resonant inverters?

11-9. What is an nonoverlap control of inverters?

11-10. What are the effects of series loading in a series-resonant inverter?

11-11. What are the effects of parallel loading in a series-resonant inverter?

11-12. What are the effects of both series and parallel loading in a series-resonant inverter?

11-13. What are the methods for voltage control of series-resonant inverters?

11-14. What are the advantages of parallel-resonant inverters?

11-15. What is the class E resonant inverter?

11-16. What are the advantages and limitations of class E resonant inverters?

11-17. What is a class E resonant rectifier?

11-18. What are the advantages and limitations of class E resonant rectifiers?

11-19. What is the principle of zero-current-switching (ZCS) resonant converters?

11-20. What is the principle of zero-voltage-switching (ZVS) resonant converters?

11-21. What are the advantages and limitations of ZCS converters?

11-22. What are the advantages and limitations of ZVS converters?

PROBLEMS

11-1. The basic series resonant inverter in Fig. 11-1a has $L_1 = L_2 = L = 25$ μH, $C = 2$ μF, and $R = 5$ Ω. The dc input voltage, $V_s = 220$ V and the output frequency, $f_o = 6.5$ kHz. The turn-off time of thyristors, $t_q = 15$ μs. Determine **(a)** the available (or circuit) turn-off time t_{off}, **(b)** the maximum permissible frequency f_{max}, **(c)** the peak-to-peak capacitor voltage V_{pp}, and **(d)** the peak load current I_p. **(e)** Sketch the instantaneous load current $i_0(t)$; capacitor voltage $v_c(t)$; and dc supply current $I_s(t)$. Calculate **(f)** the rms load current I_o, **(g)** the output power P_o, **(h)** the average supply current I_s, and **(i)** the average, peak, and rms thyristor currents.

11-2. The half-bridge resonant inverter in Fig. 11-3 uses nonoverlapping control. The inverter frequency is $f_0 = 8.5$ kHz. If $C_1 = C_2 = C = 2$ μF, $L_1 = L_2 = L = 40$ μH, $R = 2$ Ω, and $V_s = 220$ V. Determine **(a)** the peak supply current, **(b)** the average thyristor current I_A, and **(c)** the rms thyristor current I_R.

11-3. The resonant inverter in Fig. 11-7a has $C = 2$ μF, $L = 30$ μH, $R = 0$, and $V_s = 220$ V. The turn-off time of thyristor, $t_q = 12$ μs. The output frequency, $f_o = 15$ kHz. Determine **(a)** the peak supply current I_{ps}, **(b)** the average thyristor current I_A, **(c)** the rms thyristor current I_R, **(d)** the peak-to-peak capacitor voltage V_c, **(e)** the maximum permissible output frequency f_{max}, and **(f)** the average supply current I_s.

11-4. The half-bridge resonant inverter in Fig. 11-8a is operated at frequency, $f_0 = 3.5$ kHz in the nonoverlap mode. If $C_1 = C_2 = C = 2$ μF, $L = 20$ μH, $R = 1.5$ Ω, and $V_s = 220$ V, determine **(a)** the peak supply current I_{ps}, **(b)** the average thyristor current I_A, **(c)** the rms thyristor current I_R, **(d)** the rms load current I_o, and **(e)** the average supply current I_s.

11-5. Repeat Prob. 11-4 with an overlapping control so that the firing of T_1 and T_2 are advanced by 50% of the resonant frequency.

11-6. The full-bridge resonant inverter in Fig. 11-9a is operated at a frequency of $f_0 = 3.5$ kHz. If $C = 2$ μF, $L = 20$ μH, $R = 1.5$ Ω, and $V_s = 220$ V, determine **(a)** the peak supply current I_s, **(b)** the average thyristor current I_A, **(c)** the rms thyristor current I_R, **(d)** the rms load current I_o, and **(e)** the average supply current I_s.

11-7. A series resonant inverter with series-loaded delivers a load power of $P_L = 2$ kW at resonance. The load resistance is $R = 10$ Ω. The resonant frequency is $f_0 = 25$ kHz. Determine **(a)** the dc input voltage V_s, **(b)** the quality factor Q_s if it is required to reduce the load power to 500 W by frequency control so that $u = 0.8$, **(c)** the inductor L, and **(d)** the capacitor C.

11-8. A series resonant inverter with parallel-loaded delivers a load power of $P_L = 2$ kW at a peak sinusoidal load voltage of $V_p = 330$ V and at resonance. The load resistance is $R = 10$ Ω. The resonant frequency is $f_0 = 25$ kHz. Determine **(a)** the dc input voltage V_s, **(b)** the frequency ratio u if it is required to reduce the load power to 500 W by frequency control, **(c)** the inductor L, and **(d)** the capacitor C.

11-9. A parallel resonant inverter delivers a load power of $P_L = 2$ kW at a peak sinusoidal load voltage of $V_p = 170$ V and at resonance. The load resistance is $R = 10$ Ω. The resonant frequency is $f_0 = 25$ kHz. Determine **(a)** the dc input current I_s, **(b)** the quality factor Q_p if it is required to reduce the load power to 500 W by frequency control so that $u = 1.25$, **(c)** the inductor L, and **(d)** the capacitor C.

11-10. The class E inverter of Fig. 11-19a operates at resonance and has $V_s = 18$ V and $R = 10$ Ω. The switching frequency is $f_s = 50$ kHz. **(a)** Determine the optimum values of L, C, C_e, and L_e. **(b)** Use PSpice to plot the output voltage v_o and the switch voltage v_T for $k = 0.304$. Assume that $Q = 7$.

11-11. The class E rectifier of Fig. 11-22a supplies a load power of $P_L = 1$ kW at $V_o = 5$ V. The peak supply voltage is $V_m = 12$ V. The supply frequency is $f = 350$ kHz. The peak-to-peak ripple on the output voltage is $\Delta V_o = 20$ mV. **(a)** Determine the values of L, C, and C_f, and **(b)** the rms and dc currents of L and C. **(c)** Use PSpice to plot the output voltage v_o and the inductor current i_L.

11-12. The ZCS resonant converter of Fig. 11-26a delivers a maximum power of $P_L = 1$ kW at $V_o = 5$ V. The supply voltage is $V_s = 15$ V. The maximum operating frequency is $f_{max} = 40$ kHz. Determine the values of L and C. Assume that the intervals t_1 and t_3 are very small, and $x = I_m/I_o = 1.5$.

11-13. The ZVS resonant converter of Fig. 11-29a supplies a load power of $P_L = 1$ kW at $V_o = 5$ V. The supply voltage is $V_s = 15$ V. The operating frequency is $f = 40$ kHz. The values of L and C are $L = 150$ μH and $C = 0.05$ μF. **(a)** Determine the peak switch voltage V_p and current I_p, and **(b)** the time durations of each mode.

11-14. For the active clamped of Fig. 11-35, plot the f_0/f_k ratio for $1 < k \le 2$.

Static switches

12-1 INTRODUCTION

Thyristors that can be turned on and off within a few microseconds may be operated as fast-acting switches to replace mechanical and electromechanical circuit breakers. For low-power dc applications, power transistors can also be used as switches. The static switches have many advantages (e.g., very high switching speeds, no moving parts, and no contact bounce upon closing).

In addition to applications as static switches, the thyristor (or transistor) circuits can be designed to provide time-delay, latching, over- and under-current, and voltage detections. The transducers, for detecting mechanical, electrical, position, proximity, and so on, can generate the gating or control signals for the thyristors (or transistors).

The static switches can be classified into two types: (1) ac switches, and (2) dc switches. The ac switches can be subdivided into (a) single-phase, and (b) three-phase. For ac switches, the thyristors are line or natural commutated and the switching speed is limited by the frequency of the ac supply and the turn-off time of thyristors. The dc switches are forced commutated and the switching speed depends on the commutation circuitry and the turn-off time of fast thyristors.

12-2 SINGLE-PHASE AC SWITCHES

The circuit diagram of a single-phase full-wave switch is shown in Fig. 12-1a, where the two thyristors are connected in inverse parallel. Thyristor T_1 is fired at $\omega t = 0$ and thyristor T_2 is fired at $\omega t = \pi$. The output voltage is the same as the input voltage. The thyristors act like switches and are line commutated. The

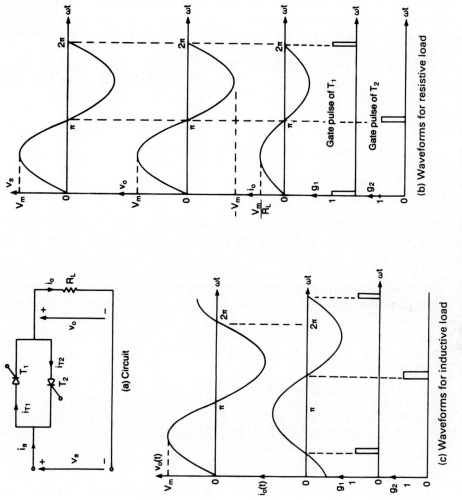

(a) Circuit

(b) Waveforms for resistive load

(c) Waveforms for inductive load

Figure 12-1   Single-phase thyristor ac switch.

waveforms for the input voltage, output voltage, and output current are shown in Fig. 12-1b.

With an inductive load, thyristor T_1 should be fired when the current passes through the zero crossing after the positive half-cycle of input voltage and thyristor T_2 should be fired when the current passes through the zero crossing after the negative half-cycle of input voltage. The triggering pulses for T_1 and T_2 are shown in Fig. 12-1c. A TRIAC may be used instead of two thyristors as shown in Fig. 12-2.

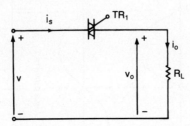

Figure 12-2   Single-phase TRIAC ac switch.

If the instantaneous line current is $i_s(t) = I_m \sin \omega t$, the rms line current is

$$I_s = \left[\frac{2}{2\pi} \int_0^\pi I_m^2 \sin^2 \omega t \; d(\omega t) \right]^{1/2} = \frac{I_m}{\sqrt{2}} \qquad (12\text{-}1)$$

Since each thyristor carries current for only one-half cycle, the average current through each thyristor is

$$I_A = \frac{1}{2\pi} \int_0^\pi I_m \sin \omega t \; d(\omega t) = \frac{I_m}{\pi} \qquad (12\text{-}2)$$

and the rms current of each thyristor is

$$I_R = \left[\frac{1}{2\pi} \int_0^\pi I_m^2 \sin^2 \omega t \; d(\omega t) \right]^{1/2} = \frac{I_m}{2} \qquad (12\text{-}3)$$

The circuit in Fig. 12-1a can be modified as shown in Fig. 12-3a, where the two thyristors have a common cathode and the gating signals have a common terminal. Thyristor T_1 and diode D_1 conduct during the positive half-cycle and thyristor T_2 and diode D_2 conduct during the negative half-cycle.

A diode bridge rectifier and a thyristor T_1 as shown in Fig. 12-4a can perform the same function as that of Fig. 12-1a. The current though the load is ac and that through thyristor T_1 is dc. A transistor can replace thyristor T_1. The unit comprising the transistor (or thyristor or GTO) and bridge rectifier is known as a *bidirectional switch*.

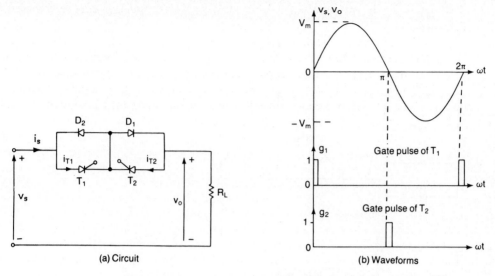

Figure 12-3 Single-phase bridge diode and thyristor ac switch.

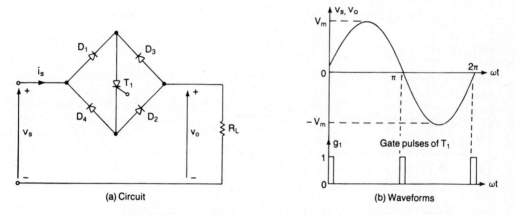

Figure 12-4 Single-phase bridge rectifier and thyristor ac switch.

12-3 THREE-PHASE AC SWITCHES

The concept of single-phase ac switching can be extended to three-phase applications. Three single-phase switches in Fig. 12-1a can be connected to form a three-phase switch as shown in Fig. 12-5a. The gating signals for thyristors and the current through T_1 are shown in Fig. 12-5b. The load can be connected in either wye or delta.

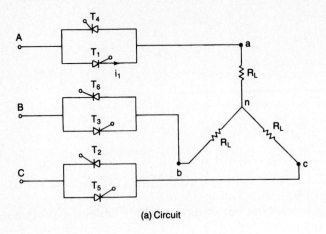

(a) Circuit

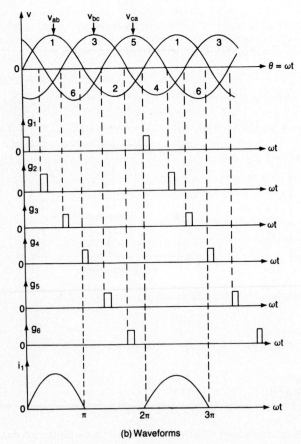

(b) Waveforms

Figure 12-5 Three-phase thyristor ac switch.

To reduce the number of thyristors and costs, a diode and a thyristor can also be used to form a three-phase switch as shown in Fig. 12-6. In case of two thristors connected in back to back, there is the possibility to stop the current flow in every half-cycle. But with a diode and a thyristor, the current flow can only be

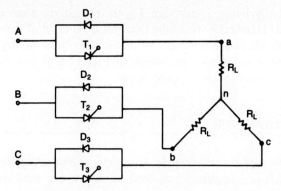

Figure 12-6 Three-phase diode and thyristor ac switch.

stopped in every cycle of input voltage and the reaction time becomes slow (e.g., 16.67 ms for a 60-Hz supply).

12-4 THREE-PHASE REVERSING SWITCHES

The reversal of three-phase power supplied to a load can be achieved by adding two more single-phase switches to the three-phase switch in Fig. 12-5a. This is shown in Fig. 12-7. Under normal operation, thyristors T_7 through T_{10} are turned off by gate pulse inhibiting (or suppression) and thyristors T_1 through T_6 are turned on. Line A feeds terminal a, line B feeds terminal b, and line C feeds terminal c. Under phase-reversing operation, thyristors T_2, T_3, T_5, and T_6 are turned off by

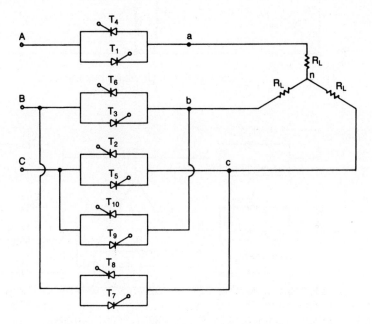

Figure 12-7 Three-phase reversing thyristor ac switch.

gate pulse inhibiting and thyristors T_7 through T_{10} are operative. Line B feeds terminal c and line C feeds terminal b, resulting in a phase reversal of the voltage applied to the load. To obtain phase reversal, all the devices must be thyristors. A combination of thyristors and diodes as shown in Fig. 12-6 cannot be used; otherwise, phase-to-phase short circuits would occur.

12-5 AC SWITCHES FOR BUS TRANSFER

The static switches can be used for bus transfer from one source to another. In a practical supply system, it is sometimes required to switch the load from the normal source to an alternative source in case of (1) unavailability of normal source, and (2) undervoltage or overvoltage condition of normal source. Figure 12-8 shows a single-phase bus transfer switch. When thyristors T_1 and T_2 are operative, the load is connected to the normal source; and for transfer to an alternative source, thyristors T_1' and T_2' are operative, while T_1 and T_2 are turned off by gate signal inhibiting. The extension of single-phase bus transfer to three-phase transfer is shown in Fig. 12-9.

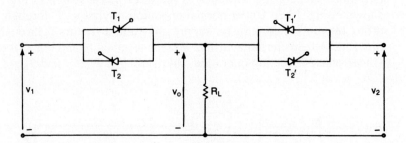

Figure 12-8 Single-phase bus transfer.

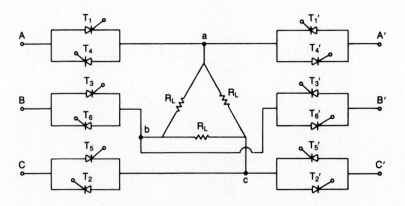

Figure 12-9 Three-phase bus transfer.

In the case of dc switches, the input voltage is dc and power transistors or fast-switching thyristors or GTOs can be used. Once a thyristor is turned on, it must be turned off by forced commutation and the techniques for forced commutation are discussed in Chapter 7. A single-pole transistor switch is shown in Fig. 12-10, with a resistive load; and in case of an inductive load, a diode (as shown by dashed lines) must be connected across the load to protect the transistor from the transient voltage during switch-off. The single-pole switches can also be applied to bus transfer from one source to another.

If forced commutated thyristors are used, the commutation circuit is an integral part of the switch and a dc switch for high-power applications is shown in Fig. 12-11. If thyristor T_3 is fired, the capacitor C is charged through the supply V_s, L, and T_3. From Eqs. (7-2) and (7-3), the charging current i and the capacitor voltage v_c are expressed as

$$i(t) = V_s \sqrt{\frac{C}{L}} \sin \omega t \tag{12-4}$$

and

$$v_c(t) = V_s(1 - \cos \omega t) \tag{12-5}$$

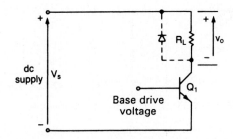

Figure 12-10 Single-pole transistor dc switch.

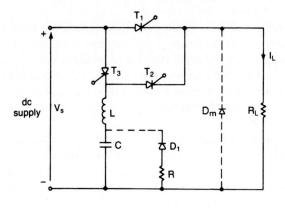

Figure 12-11 Single-pole thyristor dc switch.

where $\omega = 1/\sqrt{LC}$. After time $t = t_0 = \pi\sqrt{LC}$, the charging current becomes zero and the capacitor is charged to $2V_s$. If thyristor T_1 is conducting and supplies power to the load, thyristor T_2 is fired to switch T_1 off. T_3 is self-commutated. Firing T_2 causes a resonant pulse of current of the capacitor C through capacitor C, inductor L, and thyristor T_2. As the resonant current increases, the current through thyristor T_1 reduces. When the resonant current rises to the load current, I_L, the current of thyristor T_1 falls to zero and thyristor T_1 is turned off. The capacitor discharges its remaining charge through the load resistance, R_L. T_2 is self-commutated. A freewheeling diode, D_m, across the load is necessary for an inductive load. The capacitor must be discharged completely in every switching action; and a negative voltage on the capacitor can be prevented by connecting a resistor and a diode as shown in Fig. 12-11 by dashed lines. It is not very easy to turn-off dc circuits, and dc static switches require additional circuits for turning off.

Dc switches can be applied for control of power flow in very high voltage and high current applications (e.g., fusion reactor) [1] and these can also be used as fast-acting current breaker [2]. Instead of transistors, gate-turn-off thyristors (GTOs) can be used. A GTO is turned on by the application of a short positive pulse to its gate similar to normal thyristors; however, a GTO can be turned off by applying a short negative pulse to its gate and it does not require any commutation circuitry. A single-pole GTO switch is shown in Fig. 12-12.

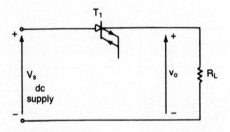

Figure 12-12 Single-pole GTO dc switch.

12-7 SOLID-STATE RELAYS

Static switches can be used as solid-state relays (SSRs), which are used for the control of ac and dc power. SSRs find many applications in industrial control (e.g., control of motor loads, transformers, resistance heating, etc.) to replace electromechanical relays. For ac applications, thyristors or TRIACs can be used; and for dc applications, transistors are used. The SSRs are normally isolated electrically between the control circuit and the load circuit by reed relay, transformer, or optocoupler.

Figure 12-13 shows two basic circuits for dc SSRs, one with reed relay isolation and the other with an optocoupler. Although the single-phase circuit in Fig. 12-1a can be operated as an SSR, the circuit in Fig. 12-2 with a TRIAC is normally used for ac power, because of the requirement of only one gating circuit

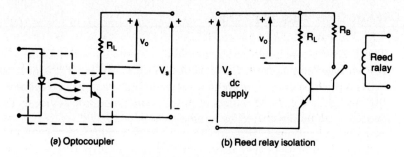

(a) Optocoupler

(b) Reed relay isolation

Figure 12-13   Dc solid-state relays.

for a TRIAC. Figure 12-14 shows SSRs with reed relay, a transformer isolation, and an optocoupler. If the application requirements demand thyristors for high power levels, the circuit in Fig. 12-1a can also be used to operate as an SSR even though the complexity of the gating circuit would increase.

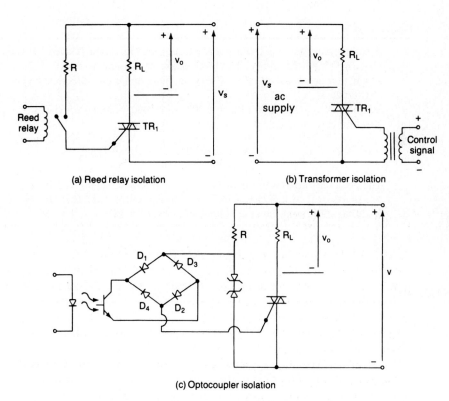

(a) Reed relay isolation

(b) Transformer isolation

(c) Optocoupler isolation

Figure 12-14   Ac solid-state relays.

The solid-state switches are available commercially with limited voltage and current ratings ranging from 1 A to 50 A and up to 440 V. If it is necessary to design SSRs to meet specific requirements, the design is simple and requires determining the voltage and current ratings of power semiconductor devices. The design procedures can be illustrated by examples.

Example 12-1

A single-phase ac switch with configuration in Fig. 12-1a is used between a 120-V 60-Hz supply and an inductive load. The load power is 5 kW at a power factor of 0.88 lagging. Determine (a) the voltage and current ratings of thyristors, and (b) the firing angles of thyristors.

Solution $P_o = 5000$ W, PF = 0.88, and $V_s = 120$ V.

(a) The peak load current $I_m = \sqrt{2} \times 5000/(120 \times 0.88) = 66.96$ A. From Eq. (12-2) the average current $I_A = 66.96/\pi = 21.31$ A and from Eq. (12-3) the rms current, $I_R = 66.96/2 = 33.48$ A. The peak inverse voltage PIV $= \sqrt{2} \times 120 = 169.7$ V.

(b) $\cos \theta = 0.88$ or $\theta = 28.36°$. Thus the firing angle of T_1 is $\alpha_1 = 28.36°$ and for thyristor T_2, $\alpha_2 = 180° + 28.36° = 208.36°$.

Example 12-2

A three-phase ac switch with configuration in Fig. 12-5a is used between a three-phase 440-V 60-Hz supply and a three-phase wye-connected load. The load power is 20 kW at a power factor of 0.707 lagging. Determine the voltage and current ratings of thyristors.

Solution $P_o = 20,000$ W, PF = 0.707, $V_L = 440$ V, and $V_s = 440/\sqrt{3} = 254.03$ V. The line current is calculated from the power as

$$I_s = \frac{20,000}{\sqrt{3} \times 400 \times 0.707} = 37.119 \text{ A}$$

The peak current of a thyristor $I_m = \sqrt{2} \times 37.119 = 52.494$ A. The average current of a thyristor $I_A = 52.494/\pi = 16.71$ A. The rms current of a thyristor $I_R = 52.494/2 = 26.247$ A. The peak inverse voltage of a thyristor PIV $= \sqrt{2} \times 440 = 622.3$ V.

SUMMARY

Solid-state ac and dc switches have a number of advantages over conventional electromechanical switches and relays. With the developments of power semiconductor devices and integrated circuits, static switches are finding a wide range of applications in industrial control. Static switches can be interfaced with digital or computer control systems.

REFERENCES

1. W. F. Praeg, "Detailed design of a 13-kA, 13-kV DC solid state turn-off switch." *IEEE Industry Applications Conference Record,* 1985, pp. 1221–1226.

2. P. F. Dawson, L. E. Lansing, and S. B. Dewan, "A fast dc current breaker." *IEEE Transactions on Industry Applications,* Vol. IA21, No. 5, 1985, pp. 1176–1181.

REVIEW QUESTIONS

12-1. What is a static switch?

12-2. What are the differences between ac and dc switches?

12.3. What are the advantages of static switches over mechanical or electromechanical switches?

12-4. What are the advantages and disadvantages of inverse-parallel thyristor ac switches?

12-5. What the the advantages and disadvantages of TRIAC ac switches?

12-6. What are the advantages and disadvantages of diode and thyristor ac switches?

12-7. What are the advantages and disadvantages of bridge rectifier and thyristor ac switches?

12-8. What are the effects of load inductance on the gating requirements of ac switches?

12-9. What is the principle of operation of SSRs?

12-10. What are the methods of isolating the control circuit from the load circuit of SSRs?

12-11. What are the factors involved in the design of dc switches?

12-12. What are the factors involved in the design of ac switches?

12-13. What type of commutation is required for dc switches?

12-14. What type of commutation is required for ac switches?

PROBLEMS

12-1. A single-phase ac switch with configuration in Fig. 12-1a is used between a 120-V 60-Hz supply and an inductive load. The load power is 15 kW at a power factor of 0.90 lagging. Determine the voltage and current ratings of thyristors.

12-2. Determine the firing angles of thyristors T_1 and T_2 in Prob. 12-1.

12-3. A single-phase ac switch with configuration in Fig. 12-3a is used between a 120-V 60-Hz supply and an inductive load. The load power is 15 kW at a power factor of 0.90 lagging. Determine the voltage and current ratings of diodes and thyristors.

12-4. A single-phase ac switch with configuration in Fig. 12-4a is used between a 120-V 60-Hz supply and an inductive load. The load power is 15 kW at a power factor of 0.90 lagging. Determine the voltage and current ratings of the thyristor and diodes in the bridge rectifier.

12-5. Determine the firing angles of thyristor T_1 in Prob. 12-4.

12-6. A three-phase ac switch with configuration in Fig. 12-5a is used between a three-phase 440-V 60-Hz supply and a three-phase wye-connected load. The load power is 20 kW at a power factor of 0.86

lagging. Determine the voltage and current ratings of thyristors.

12-7. Determine the firing angles of thyristors in Prob. 12-6.

12-8. Repeat Prob. 12-6 for a delta-connected load.

12-9. A three-phase ac switch with configuration in Fig. 12-6 has a three-phase 440-V 60-Hz supply and a three-phase wye-connected load. The load power is 20 kW at a power factor of 0.86 lagging. Determine the voltage and current ratings of diodes and thyristors.

12-10. The thyristor dc switch in Fig. 12-11 has a load resistance $R_L = 5\ \Omega$, dc supply voltage $V_s = 220$ V, inductance $L = 40$ μH, and capacitance $C = 40$ μF. Determine **(a)** the peak current through thyristor T_3, and **(b)** the time required to re-

duce the current of thyristor T_1 from the steady-state value to zero.

12-11. For Prob. 12-10, determine the time required for the capacitor to discharge from $2V_s$ to zero after the firing of thyristor T_2.

12-12. The thyristor dc switch in Fig. 12-11 has a load resistance $R_L = 0.5\ \Omega$, supply voltage $V_s = 220$ V, inductance $L = 40$ μH, and capacitance $C = 80$ μF. If the switch is operated at a frequency of 60 Hz, determine **(a)** the peak, rms, and average currents of thyristors T_1, T_2, and T_3; and **(b)** the rms current rating of capacitor C.

12-13. For Prob. 12-12, determine the time required for the capacitor to discharge from $2V_s$ to zero after the firing of thyristor T_2.

Power supplies

13-1 INTRODUCTION

Power supplies, which are used extensively in industrial applications, are often required to meet all or most of the following specifications:

1. Isolation between the source and the load
2. High power density for reduction of size and weight
3. Controlled direction of power flow
4. High conversion efficiency
5. Input and output waveforms with a low total harmonic distortion for small filters
6. Controlled power factor if the source is an ac voltage

The single-stage ac–dc or ac–ac or dc–dc or dc–ac converters discussed in Chapters 5, 6, 9, and 10, respectively, do not meet most of these specifications, and multistage conversions are normally required. There are various possible conversion topologies, depending on the permissible complexity and the design requirements. Only the basic topologies are discussed in this chapter. Depending on the type of output voltages, the power supplies can be categorized into two types:

1. Dc power supplies
2. Ac power supplies

The ac–dc converters in Chapter 5 can provide the isolation between the input and output through an input transformer, but the harmonic contents are high. The switched-mode regulators in Section 9-7 do not provide the necessary isolation and the output power is low. The common practice is to use two-stage conversions, dc–ac and ac–dc. In the case of ac input, it is three-stage conversions, ac–dc, dc–ac, and ac–dc. The isolation is provided by an interstage transformer. The dc–ac conversion can be accomplished by PWM or resonant inverter. Based on the type of conversion techniques and the direction of power control, the dc power supplies can be subdivided into three types:

1. Switched-mode power supplies
2. Resonant power supplies
3. Bidirectional power supplies

13-2.1 Switched-Mode DC Power Supplies

There are four common configurations for the switched-mode or PWM operation of the inverter (or dc–ac converter) stage: flyback, push-pull, half-bridge, and full-bridge. The output of the inverter, which is varied by a PWM technique, is converted to a dc voltage by a diode rectifier. Since the inverter can operate at a very high frequency, the ripples on the dc output voltage can easily be filtered out with small filters.

The circuit topology for the *flyback* converter is shown in Fig. 13-1a. When transistor Q_1 is turned on, the supply voltage appears across the transformer primary and a corresponding voltage is induced in the secondary. When Q_1 is off, a voltage of opposite polarity is induced in the primary by the secondary due to the transformer action. The minimum open-circuit voltage of the transistor is $V_{oc} = 2V_s$. If I_s is the average input current with negligible ripple and the duty cycle is $k = 50\%$, the peak transistor current is $I_p = I_s/k = 2I_s$. The input current is pulsating and discontinuous. Without diode D_2, a dc current would flow through the transformer. When Q_1 is off, diode D_2 and capacitor C_1 reset the transformer core. C_1 discharges through R_1, when D_2 is off and energy is lost in every cycle. This circuit is very simple and is restricted to applications below 500 W. This is a forward converter and it requires a voltage control feedback loop.

The transformer core can also be reset by having a reset winding as shown in Fig. 13-1b, where the energy stored in the transformer core is returned to the supply and the efficiency is increased. The open-circuit voltage of the transistor in Fig. 13-1b is

$$V_{oc} = V_s \left(1 + \frac{N_p}{N_r} \right) \tag{13-1}$$

where N_p and N_r are the number of turns on the primary and reset windings, respectively. The reset turns ratio is related to the duty cycle as

$$a_r = \frac{N_r}{N_p} = \frac{1-k}{k} \tag{13-2}$$

For a duty cycle of $k = 0.8$, $N_p/N_r = 0.8/(1 - 0.8) = 4$ and the open-circuit voltage becomes $V_{oc} = V_s(1 + 4) = 5V_s$. The open-circuit voltage of the transistor is much higher than the supply voltage.

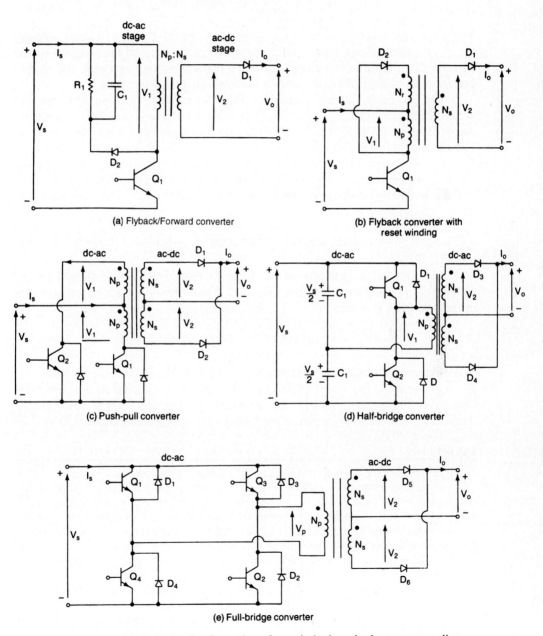

Figure 13-1 Configurations for switched-mode dc power supplies.

The push-pull configuration is shown in Fig. 13-1c. When Q_1 is turned on, V_s appears across one-half of the primary. When Q_2 is turned on, V_s is applied across the other half of the transformer. The voltage of a primary winding swings from $-V_s$ to V_s. The average current through the transformer should ideally be zero. The average output voltage is

$$V_o = V_2 = \frac{N_s}{N_p} V_1 = aV_1 = aV_s \qquad (13\text{-}3)$$

Transistors Q_1 and Q_2 operate with a 50% duty cycle. The open-circuit voltage, $V_{oc} = 2V_s$, the average current of a transistor, $I_A = I_s/2$, and the peak transistor current, $I_p = I_s$. Since the open-circuit transistor voltage is twice the supply voltage, this configuration is suitable for low-voltage applications.

The half-bridge circuit is shown in Fig. 13-1d. When Q_1 is on, $V_s/2$ appears across the transformer primary. When Q_2 is on, a reverse voltage of $V_s/2$ appears across the transformer primary. The primary voltage swings from $-V_s/2$ to $V_s/2$. The open-circuit transistor voltage is $V_{oc} = V_s$ and the peak transistor current is $I_p = 2I_s$. The average transistor current is $I_A = I_s$. In high-voltage applications, the half-bridge circuit is preferable to the push-pull circuit. Whereas for low-voltage applications, the push-pull circuit is preferable due to low transistor currents. The average output voltage is

$$V_o = V_2 = \frac{N_s}{N_p} V_1 = aV_1 = 0.5aV_s \qquad (13\text{-}4)$$

The full-bridge arrangement is shown in Fig. 13-1e. When Q_1 and Q_2 are turned on, V_s appears across the primary. When Q_3 and Q_4 are turned on, the primary voltage is reversed to $-V_s$. The average output voltage is

$$V_o = V_2 = \frac{N_s}{N_p} V_1 = aV_1 = aV_s \qquad (13\text{-}5)$$

The open-circuit transistor voltage is $V_{oc} = V_s$ and the peak transistor current is $I_p = I_s$. The average current of a transistor is only $I_A = I_s/2$. Of all the configurations, this circuit operates with the minimum voltage and current stress of the transistors, and it is very popular for high-power applications above 750 W.

Example 13-1

The average (or dc) output voltage of the push-pull circuit in Fig. 13-1c is $V_o = 24$ V at a resistive load of $R = 0.8\ \Omega$. The on-state voltage drops of transistors and diodes are $V_t = 1.2$ V and $V_d = 0.7$ V, respectively. The turns ratio of the transformer is $a = N_s/N_p = 0.25$. Determine (a) the average input current I_s, (b) the efficiency η, (c) the average transistor current I_A, (d) the peak transistor current I_p, (e) the rms transistor current I_R, and (f) the open-circuit transistor voltage V_{oc}. Neglect the losses in the transformer, and the ripple current of the load and input supply is negligible.

Solution $a = N_s/N_p = 0.25$ and $I_o = V_o/R = 24/0.8 = 30$ A.

(a) The output power $P_o = V_o I_o = 24 \times 30 = 720$ W. The secondary voltage $V_2 = V_o + V_d = 24 + 0.7 = 24.7$ V. The primary voltage $V_1 = V_2/a = 24.7/0.25 = 98.8$ V. The input voltage $V_s = V_1 + V_t = 98.8 + 1.2 = 100$ and the input power is

$$P_i = V_s I_s = 1.2I_A + 1.2I_A + V_d I_o + P_o$$

Substituting $I_A = I_s/2$ gives

$$I_s(100 - 1.2) = 0.7 \times 30 + 720$$

$$I_s = \frac{741}{98.8} = 7.5 \text{ A}$$

(b) $P_i = V_s I_s = 100 \times 7.5 = 750$ W. The efficiency $\eta = 7.5/750 = 96.0\%$.

(c) $I_A = I_s/2 = 7.5/2 = 3.75$ A.

(d) $I_p = {}_s = 7.5$A.

(e) $I_R = \sqrt{k} I_p = \sqrt{0.5} \times 7.5 = 5.30$ A, for 50% duty cycle

(f) $V_{oc} = 2V_s = 2 \times 100 = 200$ V.

13-2.2 Resonant DC Power Supplies

If the variation of the dc output voltage is not wide, resonant pulse inverters can be used. The inverter frequency, which could be the same as the resonant frequency, is very high, and the inverter output voltage is almost sinusoidal. Due to resonant oscillation, the transformer core is always reset and there are no dc saturation problems. The half-bridge and full-bridge configurations of the resonant inverters are shown in Fig. 13-2. The sizes of the transformer and output filter are reduced due to high inverter frequency.

Example 13-2

The average output voltage of the half-bridge circuit in Fig. 13-2a is $V_0 = 24$ V at a resistive load of $R_L = 0.8$ Ω. The inverter operates at the resonant frequency. The circuit parameters are $C_1 = C_2 = C = 1$ μF, $L = 20$ μH and $R = 0$. The dc input voltage is $V_s = 100$ V. The on-state voltage drops of transistors and diodes are negligible. The turns radio of the transformer is $a = N_s/N_p = 0.25$. Determine (a) the average input current I_s, (b) The average transistor current I_A, (c) the peak transistor current I_p, (d) the rms transistor current I_s, and (e) the open-circuit transistor voltage V_{oc}. Neglect the losses in the transformer, and the effect of the load on the resonant frequency is negligible.

Solution $C_e = C_1 + C_2 = 2C$. The resonant frequency $\omega_r = 10^6/\sqrt{2 \times 20} = 158$, 158,113.8 rad/s or $f_r = 25,164.6$ Hz, $a = N_s/N_p = 0.25$, and $I_o = V_o/R = 24/0.8 = 30$A.

(a) The output power $P_o = V_o I_o = 24 \times 30 = 720$ W. From Eq. (3-62), the rms secondary voltage $V_2 = \pi V_o/(2\sqrt{2}) = 1.1107V_o = 26.66$ V. The average input current $I_s = 720/100 = 7.2$ A.

(b) The average transistor current $I_A = I_s = 7.2$ A.

(c) For a sinusoidal pulse of current through the transistor, $I_A = I_p/\pi$ and the peak transistor current $I_p = 7.2\pi = 22.62$ A.

(d) With a sinusoidal pulse of current with 180° conduction, the rms transistor current $I_R = I_p/2 = 11.31$ A.

(e) $V_{oc} = V_s = 100$ V.

13-2.3 Bidirectional Power Supplies

In some applications, such as battery charging and discharging, it is desirable to have bidirectional power flow capability. A bidirectional power supply is shown in Fig. 13-3. The direction of the power flow will depend on the values of V_o, V_s,

---

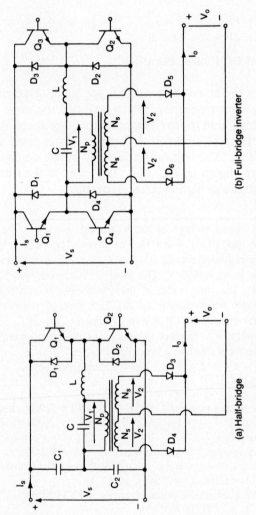

Figure 13-2 Configurations for resonant dc power supplies.

(a) Half-bridge

(b) Full-bridge inverter

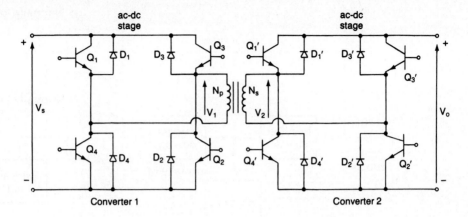

Figure 13-3 Bidirectional dc power supply.

and turns ratio ($a = N_s/N_p$). For power flow from the source to the load, the inverter operates in the inversion mode if

$$V_o < aV_s \qquad (13\text{-}6)$$

For power flow from the output to the input, the inverter operates as a rectifier if

$$V_o > aV_s \qquad (13\text{-}7)$$

The bidirectional converters allow the inductive current to flow in either direction and the current flow becomes continuous.

13-3 AC POWER SUPPLIES

The ac power supplies are commonly used as standby sources for critical loads and in applications where normal ac supplies are not available. The standby power supplies are also known as uninterruptible power supply (UPS) systems. The two configurations for commonly used in UPSs are shown in Fig. 13-4. The load in the configuration of Fig. 13-4a is normally supplied from the ac main supply and the rectifier maintains the full charge of the battery. If the supply fails, the load is switched to the output of the inverter, which then takes over the main supply. This configuration requires breaking the circuit momentarily and the transfer by a solid-state switch usually takes 4 to 5 ms. The switchover by a mechanical contactor may take 30 to 50 ms. The inverter runs only during the time when the supply failure occurs.

The inverter in the configuration of Fig. 13-4b operates continuously and its output is connected to the load. There is no need for breaking the supply in the event of supply failure. The rectifier supplies the inverter and maintains the charge on the standby battery. The inverter can be used to condition the supply to the load, to protect the load from the transients in the main supply, and to maintain the load frequency at the desired value. In case of inverter failure, the load is switched to the main supply.

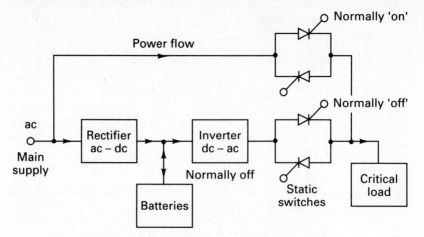

(a) Load normally connected to ac main supply

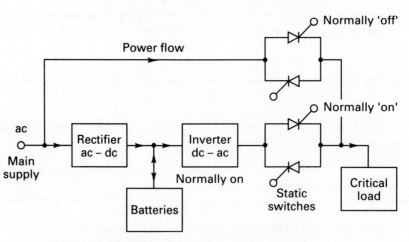

(b) Load normally connected to inverter

Figure 13-4 UPS configurations.

The standby battery is normally either nickel–cadmium or lead–acid type. A nickel–cadmium battery is preferable to a lead–acid battery. Because the electrolyte of a nickel–cadmium battery is noncorrosive and does not emit explosive gas. It has a longer life due to its ability to withstand overheating or discharging. However, its cost is at least three times that of a lead–acid battery. An alternative arrangement of an UPS system is shown in Fig. 13-5, which consists of a battery, inverter, and static switch. In case of power failure, the battery supplies the inverter. When the main supply is on, the inverter operates as a rectifier and charges the battery. In this arrangement the inverter has to operate at the fundamental output frequency. Consequently, the high-frequency capability of the inverter is not utilized in reducing the size of the transformer. Similar to the dc power supplies, the ac power supplies can be categorized into three types:

1. Switched-mode ac power supplies
2. Resonant ac power supplies
3. Bidirectional ac power supplies

13-3.1 Switched-Mode AC Power Supplies

The size of the transformer in Fig. 13-5 can be reduced by the addition of a high-frequency dc link as shown in Fig. 13-6. There are two inverters. The input-side inverter operates with a PWM control at a very high frequency to reduce the size

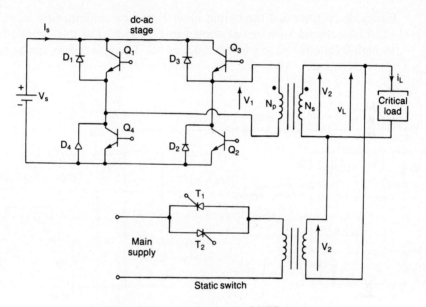

Figure 13-5   Arrangement of UPS systems.

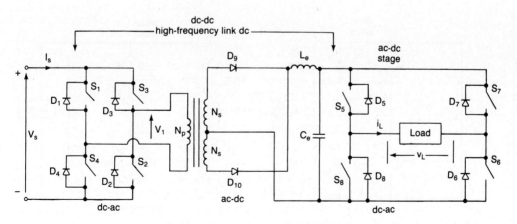

Figure 13-6   Switched-mode ac power supplies.

of the transformer and the dc filter at the input of the output-side inverter. The output-side inverter operates at the output frequency.

13-3.2 Resonant AC Power Supplies

The input-stage inverter in Fig. 13-6 can be replaced by a resonant inverter as shown in Fig. 13-7. The output-side inverter operates with a PWM control at the output frequency.

13-3.3 Bidirectional AC Power Supplies

The diode rectifier and the output inverter can be combined by a cycloconverter with bidirectional switches as shown in Fig. 13-8. The cycloconverter converts the high-frequency ac to a low-frequency ac. The power flow can be controlled in either direction.

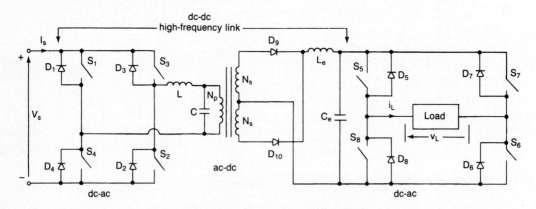

Figure 13-7   Resonant ac power supply.

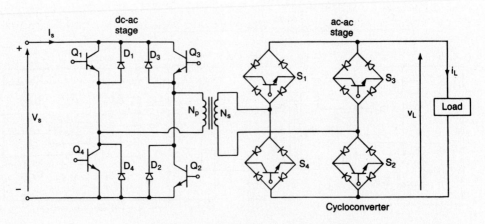

Figure 13-8   Bidirectional ac power supplies.

Example 13-3

The load resistance of the ac power supply in Fig. 13-6 is $R = 2.5\ \Omega$. The dc input voltage is $V_s = 100$ V. The input inverter operates at a frequency of 20 kHz with one pulse per half-cycle. The on-state voltage drops of transistor switches and diodes are negligible. The turns ratio of the transformer is $a = N_s/N_p = 0.5$ The output inverter operates with a uniform PWM of four pulses per half-cycle. The width of each pulse is $\delta = 18°$. Determine the rms load current. The ripple voltage on the output of the rectifier is negligible. Neglect the losses in the transformer, and the effect of the load on the resonant frequency is negligible.

Solution   The rms output voltage of the input inverter is $V_1 = V_s = 100$ V. The rms transformer secondary voltage, $V_2 = aV_1 = 0.5 \times 100 = 50$ V. The dc voltage of the rectifier, $V_o = V_2 = 50$ V. With the pulse width of $\delta = 18°$, Eq. (10-26) gives the rms load voltage, $V_L = V_o\sqrt{(p\delta/\pi)} = 50\sqrt{4 \times 18/180} = 31.6$ V. The rms load current, $I_L = V_L/R = 31.6/2.5 = 12.64$ A.

13-4 MULTISTAGE CONVERSIONS

If the input is an ac source, an input-stage rectifier is required as shown in Fig. 13-9 and there are four conversions: ac–dc–ac–dc–ac. The rectifier and inverter pair can be replaced by a converter with bidirectional ac switches as shown in Fig. 13-10. The switching functions of this converter can be synthesized to combine the functions of the rectifier and inverter. This converter, which converts ac–ac directly, is called the forced commutated cycloconverter [3]. The ac–dc–ac–dc–ac conversions in Fig. 13-9 can be performed by two forced-commutated cycloconverters as shown in Fig. 13-10.

13-5 POWER FACTOR CONDITIONING

Diode rectifiers are the most commonly used circuits for applications where the input is the ac supply (e.g., in computers, telecommunications, fluorescent lighting, and air conditioning). The power factor of diode rectifiers with a resistive load can be as high as 0.9, and it is lower with a reactive load. With the aid of

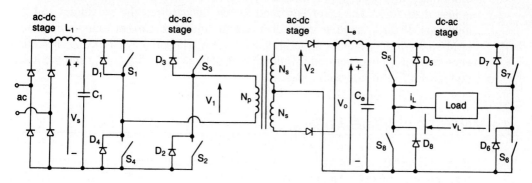

Figure 13-9   Multistage conversions.

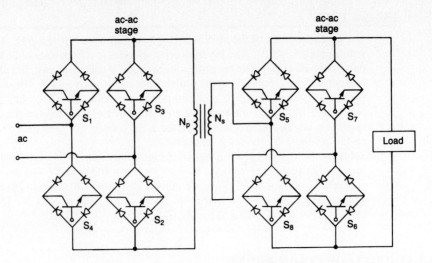

Figure 13-10 Cycloconverters with bilateral switches.

modern control technique, the input current of the rectifiers can be made sinusoidal and in phase with the input voltage, thereby having an input power factor approximately unity. A unity power factor circuit that combines a full-bridge rectifier and a boost chopper is shown in Fig. 13-11a. The input current of the chopper is controlled to follow the full-rectified waveform of the sinusoidal input voltage by PWM control [7,8]. The PWM control signals can be generated by using the bang-bang hysteresis (BBH) technique, similar to the delta modulation in Fig. 10-27. This technique, which is shown in Fig. 13-11b, has the advantage of yielding instantaneous current control, resulting in a fast response. However, the switching frequency is not constant and varies over a wide range during each half-cycle of the ac input voltage. The frequency is also sensitive to the values of the circuit components.

The switching frequency can be maintained constant by using the reference current I_{ref} and feedback current I_{fb} averaged over the per-switching period. This is shown in Fig. 13-11c. I_{ref} is compared with I_{fb}. If $I_{ref} > I_{fb}$, the duty cycle is more than 50%. For $I_{ref} = I_{fb}$, the duty cycle is 50%. For $I_{ref} < I_{fb}$, the duty cycle is less than 50%. The error is forced to remain between the maximum and the minimum of the triangular waveform and the inductor current follows the reference sine wave, which is superimposed with a triangular waveform. The reference current I_{ref} is generated from the error voltage V_e $(= V_{ref} - V_o)$ and the input voltage V_{in} to the boost chopper.

13-6 MAGNETIC CONSIDERATIONS

If there is any dc imbalance, the transformer core may saturate, resulting in a high magnetizing current. An ideal core should exhibit a very high relative permeability in the normal operating region and under dc imbalance conditions, it should not

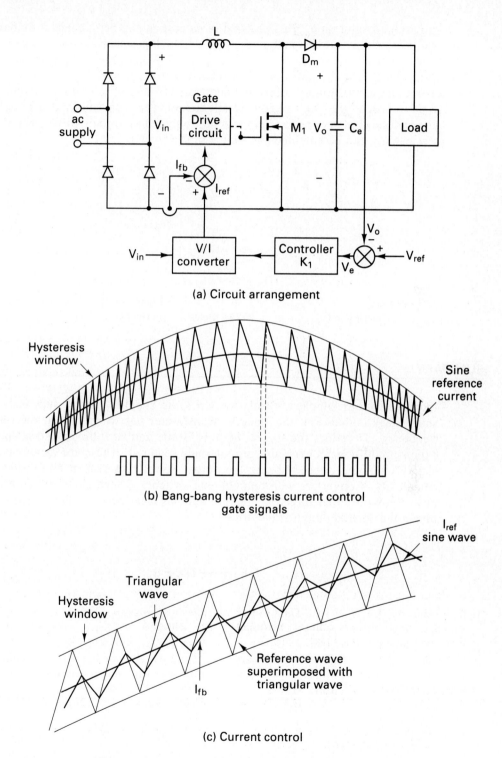

Figure 13-11 Power factor conditioning of diode rectifiers.

go into hard saturation. This problem of saturation can be minimized by having two permeability regions in the core, low and high permeability. An air gap may be inserted as shown in a toroid in Fig. 13-12, where the inner part has high permeability and the outer part has relatively low permeability. Under normal operation the flux flows through the inner part. In case of saturation, the flux has to flow through the outer region, which has a lower permeability due to the air gap, and the core does not go into hard saturation. Two toroids with high and low permeability can be combined as shown in Fig. 13-12b.

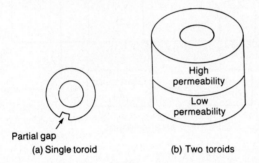

Partial gap
(a) Single toroid

(b) Two toroids

Figure 13-12 Cores with two permeability regions.

SUMMARY

Industrial power supplies are of two types: dc supplies and ac supplies. In a single-stage conversion, the isolating transformer has to operate at the output frequency. To reduce the size of the transformer and meet the industrial specifications, multistage conversions are normally required. There are various power supply topologies, depending on the output power requirement and acceptable complexity. Converters with bidirectional switches, which allow the control of power flow in either direction, require synthesizing the switching functions to obtain the desired output waveforms.

REFERENCES

1. R. E. Hnatek, *Design of Solid-State Power Supplies*. New York: Van Nostrand Reinhold Company, Inc., 1981.

2. S. Manias and P. D. Ziogas, "A novel sine wave in ac–dc converter with high frequency transformer isolation." *IEEE Transactions on Industrial Electronics*. Vol. IE32, No. 4, 1985, pp. 430–438.

3. K. A. Haddad, T. Krishnan, and V. Rajago-paloan, "Dc to dc converters with high frequency ac link." *IEEE Transactions on In-*

dustry Applications, Vol. IA22, No. 2, 1986, pp. 244–254.

4. P. D. Ziogas, S. I. Khan, and M. H. Rashid, "Analysis and design of cycloconverter structures with improved transfer characteristics." *IEEE Transactions on Industrial Electronics,* Vol. IE33, No. 3, 1986, pp. 271–280.

5. E. D. Weichman, P. D. Ziogas, and V. R. Stefanovic, "A novel bilateral power conversion scheme for variable frequency static

power supplies." *IEEE Transactions on Industry Applications,* Vol. IA21, No. 5, 1985, pp. 1226–1233.

6. I. J. Pitel, "Phase-modulated resonant power conversion techniques for high frequency link inverters." *IEEE Industry Applications Society Conference Record,* 1985, pp. 1163–1172.

7. M. Kazerani, P. D. Ziogas, and G. Joos, "A novel active current waveshaping technique for solid-state input power factor conditioners." *IEEE Transactions on Industrial Electronics,* Vol. IE38, No. 1, 1991, pp. 72–78.

8. I. Takahashi, "Power factor improvements of a diode rectifier circuit by dither signals." *Conference Proceedings of the IEEE–IAS Annual Meeting,* Seattle, Wash., October 1990, pp. 1279–1294.

REVIEW QUESTIONS

13-1. What are the normal specifications of power supplies?

13-2. What the types of power supplies in general?

13-3. Name three types of dc power supplies.

13-4. Name three types of ac power supplies.

13-5. What the advantages and disadvantages of single-stage conversion?

13-6. What are the advantages and disadvantages of switched-mode power supplies?

13-7. What are the advantages and disadvantages of resonant power supplies?

13-8. What are the advantages and disadvantages of bidirectional power supplies?

13-9. What are the advantages and disadvantages of flyback converters?

13-10. What are the advantages and disadvantages of push-pull converters?

13-11. What are the advantages and disadvantages of half-bridge converters?

13-12. What are the various configurations of resonant dc power supplies?

13-13. What are the advantages and disadvantages of high-frequency link power supplies?

13-14. What is the general arrangement of UPS systems?

13-15. What are the problems of the transformer core?

13-16. What is the principle of power factor conditioning in diode rectifiers?

PROBLEMS

13-1. The dc output voltage of the push-pull circuit in Fig. 13-1c is $V_o = 24$ V at a resistive load of $R = 0.4 \ \Omega$. The on-state voltage drops of transistors and diodes are $V_t = 1.2$ V and $V_d = 0.7$ V, respectively. The turns ratio of the transformer, $a = N_s/N_p = 0.5$. Determine **(a)** the average input current I_s, **(b)** the efficiency η, **(c)** the average transistor current I_A, **(d)** the peak transistor current I_p, **(e)** the rms transistor current I_R, and **(f)** the open-circuit transistor voltage V_{oc}. Neglect the losses in the transformer, and the ripple current of the load and input supply is negligible.

13-2. Repeat Prob. 13-1 for the circuit in Fig. 13-1b, for $k = 0.5$.

13-3. Repeat Prob. 13-1 for the circuit in Fig. 13-1d.

13-4. Repeat Prob. 13-1 for the circuit in Fig. 13-1e.

13-5. The dc output voltage of the half-bridge circuit in Fig. 13-2a is $V_o = 24$ V at a load resistance of $R = 0.8 \ \Omega$. The inverter operates at the resonant frequency. The circuit parameters are $C_1 = C_2 = C = 2\ \mu F$, $L = 5\ \mu H$, and $R = 0$. The dc input voltage, $V_s = 50$ V. The on-state voltage drops of transistors and diodes are negligible. The turns ratio of the transformer, $a = N_s/N_p = 0.5$. Determine **(a)** the average input current I_s, **(b)** the average transistor current I_A, **(c)** the peak transistor current I_p, **(d)** the rms transistor current I_R, and **(e)** the open-circuit transistor voltage V_{oc}. Neglect the losses in the transformer, and the effect of the load on the resonant frequency is negligible.

13-6. Repeat Prob. 13-5 for the full-bridge circuit in Fig. 13-2b.

13-7. The load resistance of the ac power supplied in Fig. 13-5 is $R = 1.5\ \Omega$. The dc input voltage is $V_s = 24$ V. The input inverter operates at a frequency of 400 Hz with a uniform PWM of eight pulses per half-cycle and the width of each pulse is $\delta = 20°$. The on-state voltage drops of transistor switches and diodes are negligi-

ble. The turns ratio of the transformer, $a = N_s/N_p = 4$. Determine the rms load current. Neglect the losses in the transformer, and the effect of the load on the resonant frequency is negligible.

13-8. The load resistance of the ac power supplies in Fig. 13-6 is $R = 1.5\ \Omega$. The dc input voltage, $V_s = 24$ V. The input inverter operates at a frequency of 20 kHz with a uniform PWM of four pulses per half-cycle and the width of each pulse is $\delta_i = 40°$. The on-state voltage drops of transistor switches and diodes are negligible. The turns ratio of the transformer, $a = N_s/N_p = 0.5$. The output inverter operates with a uniform PWM of eight pulses per half-cycle and the width of each pulse is $\delta_0 = 20°$. Determine the rms load current. The ripple voltage on the output of the rectifier is negligible. Neglect the losses in the transformer, and the effect of the load on the resonant frequency is negligible.

Dc drives

14-1 INTRODUCTION

Direct-current (dc) motors have variable characteristics and are used extensively in variable-speed drives. Dc motors can provide a high starting torque and it is also possible to obtain speed control over a wide range. The methods of speed control are normally simpler and less expensive than those of ac drives. Dc motors play a significant role in modern industrial drives. Both series and separately excited dc motors are normally used in variable-speed drives, but series motors are traditionally employed for traction applications. Due to commutators, dc motors are not suitable for very high-speed applications and require more maintenance than do ac motors. With the recent advancements in power conversions, control techniques, and microcomputers, the ac motor drives are becoming increasingly competitive with dc motor drives. Although the future trend is toward ac drives, dc drives are currently used in many industries. It might be a few decades before the dc drives are completely replaced by ac drives.

Controlled rectifiers provide a variable dc output voltage from a fixed ac voltage, whereas choppers can provide a variable dc voltage from a fixed dc voltage. Due to their ability to supply a continuously variable dc voltage, controlled rectifiers and dc choppers made a revolution in modern industrial control equipment and variable-speed drives, with power levels ranging from fractional horsepower to several megawatts. Controlled rectifiers are generally used for the speed control of dc motors as shown in Fig. 14-1a. The alternative form would be a diode rectifier followed by a chopper as shown in Fig. 14-1b. Dc drives can be classified, in general, into three types:

1. Single-phase drives
2. Three-phase drives
3. Chopper drives

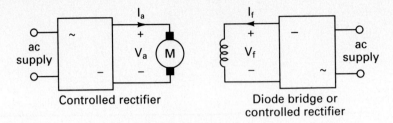

Controlled rectifier

Diode bridge or
controlled rectifier

(a) Controlled rectifier-fed drive

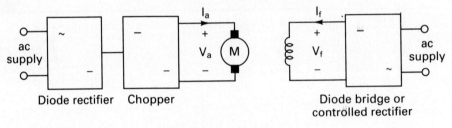

Diode rectifier   Chopper

Diode bridge or
controlled rectifier

(b) Chopper-fed drive

Figure 14-1   Controller rectifier- and chopped-fed drives.

14-2 BASIC CHARACTERISTICS OF DC MOTORS

The equivalent circuit for a separately excited dc motor is shown in Fig. 14-2. When a separately excited motor is excited by a field current of i_f and an armature current of i_a flows in the armature circuit, the motor develops a back emf and a torque to balance the load torque at a particular speed. The field current i_f of a separately excited motor is independent of the armature current i_a and any change in the armature current has no effect on the field current. The field current is normally much less than the armature current.

The equations describing the characteristics of a separately excited motor can be determined from Fig. 14-2. The instantaneous field current i_f is described as

$$v_f = R_f i_f + L_f \frac{di_f}{dt}$$

The instantaneous armature current can be found from

$$v_a = R_a i_a + L_a \frac{di_a}{dt} + e_g$$

The motor back emf, which is also known as *speed voltage*, is expressed as

$$e_g = K_v \omega i_f$$

The torque developed by the motor is

$$T_d = K_t i_f i_a$$

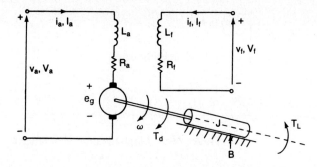

Figure 14-2 Equivalent circuit of separately excited dc motors.

The developed torque must be equal to the load torque:

$$T_d = J\frac{d\omega}{dt} + B\omega + T_L$$

where ω = motor speed, rad/s
  B = viscous friction constant, N·m/rad/s
  K_v = voltage constant, V/A-rad/s
  $K_t = K_v$ = torque constant
  L_a = armature circuit inductance, H
  L_f = field circuit inductance, H
  R_a = armature circuit resistance, Ω
  R_f = field circuit resistance, Ω
  T_L = load torque, N·m

Under steady-state conditions, the time derivatives in these equations are zero and the steady-state average quantities are

$$V_f = R_f I_f \tag{14-1}$$

$$E_g = K_v \omega I_f \tag{14-2}$$

$$V_a = R_a I_a + E_g$$

$$= R_a I_a + K_v \omega I_f \tag{14-3}$$

$$T_d = K_t I_f I_a \tag{14-4}$$

$$= B\omega + T_L \tag{14-5}$$

The developed power is

$$P_d = T_d \omega \tag{14-6}$$

The relationship between the field current I_f and the back emf E_g is nonlinear due to magnetic saturation. The relationship, which is shown in Fig. 14-3, is known as *magnetization characteristic* of the motor. From Eq. (14-3), the speed of a sepa-

---

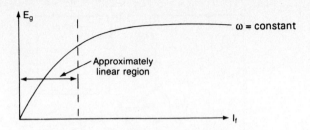

Figure 14-3 Magnetization characteristic.

rately excited motor can be found from

$$\omega = \frac{V_a - R_a I_a}{K_v I_f} = \frac{V_a - R_a I_a}{K_v V_f / R_f} \tag{14-7}$$

We can notice from Eq. (14-7) that the motor speed can be varied by (1) controlling the armature voltage, V_a, known as *voltage control*; (2) controlling the field current, I_f, known as *field control*; or (3) torque demand, which corresponds to an armature current, I_a, for a fixed field current, I_f. The speed, which corresponds to the rated armature voltage, rated field current and rated armature current, is known as the *base speed*.

In practice, for a speed less than the base speed, the armature current and field currents are maintained constant to meet the torque demand, and the armature voltage, V_a, is varied to control the speed. For speed higher than the base speed, the armature voltage is maintained at the rated value and the field current is varied to control the speed. However, the power developed by the motor (= torque × speed) remains constant. Figure 14-4 shows the characteristics of torque, power, armature current, and field current against the speed.

The field of a dc motor may be connected in series with the armature circuit as shown in Fig. 14-5, and this type of motor is called a *series motor*. The field circuit is designed to carry the armature current. The steady-state average quantities are

$$E_g = K_v \omega I_a \tag{14-8}$$

$$V_a = (R_a + R_f) I_a + E_g \tag{14-9}$$

$$= (R_a + R_f) I_a + K_v \omega I_f \tag{14-10}$$

$$T_d = K_t I_a I_f \tag{14-11}$$

$$= B\omega + T_L$$

The speed of a series motor can be determined from Eq. (14-10):

$$\omega = \frac{V_a - (R_a + R_f) I_a}{K_v I_f} \tag{14-12}$$

The speed can be varied by controlling the (1) armature voltage, V_a, or (2) armature current, which is a measure of the torque demand. Equation (14-11) indicates

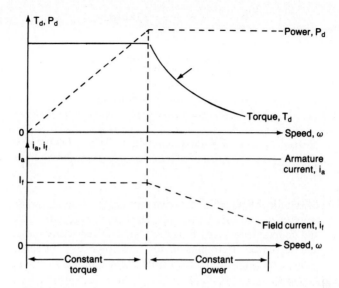

Figure 14-4 Characteristics of separately excited motors.

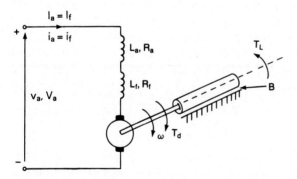

Figure 14-5 Equivalent circuit of dc series motors.

that a series motor can provide a high torque, especially at starting; and for this reason, series motors are commonly used in traction applications.

For a speed up to the base speed, the armature voltage is varied and the torque is maintained constant. Once the rated armature voltage is applied, the speed–torque relationship follows the natural characteristic of the motor and the power (= torque × speed) remains constant. As the torque demand is reduced, the speed increases. At a very light load, the speed could be very high and it is not advisable to run a dc series motor without a load. Figure 14-6 shows the characteristics of dc series motors.

Example 14-1

A 15-hp 220-V 2000-rpm separately excited dc motor controls a load requiring a torque of $T_L = 45$ N·m at a speed of 1200 rpm. The field circuit resistance is $R_f = $

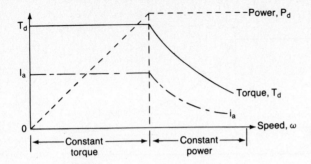

Figure 14-6 Characteristics of dc series motors.

147 Ω, the armature circuit resistance is $R_a = 0.25\ \Omega$, and the voltage constant of the motor is $K_v = 0.7032$ V/A-rad/s. The field voltage is $V_f = 220$ V. The viscous friction and no-load losses are negligible. The armature current may be assumed continuous and ripple free. Determine (a) the back emf E_g, (b) the required armature voltage V_a, and (c) the rated armature current of the motor.

Solution $R_f = 147\ \Omega$, $R_a = 0.25\ \Omega$, $K_v = K_t = 0.7032$ V/A-rad/s, $V_f = 220$ V, $T_d = T_L = 45$ N·m, $\omega = 1200\ \pi/30 = 125.66$ rad/s, and $I_f = 220/147 = 1.497$ A.

(a) From Eq. (14-4), $I_a = 45/(0.7032 \times 1.497) = 42.75$ A. From Eq. (14-2), $E_g = 0.7032 \times 125.66 \times 1.497 = 132.28$ V.

(b) From Eq. (14-3), $V_a = 0.25 \times 42.75 + 132.28 = 142.97$ V.

(c) Since 1 hp is equal to 746 W, $I_{\text{rated}} = 15 \times 746/220 = 50.87$ A.

14-3 OPERATING MODES

In variable-speed applications, a dc motor may be operating in one or more modes: motoring, regenerative braking, dynamic braking, plugging, and four quadrants.

Motoring. The arrangements for motoring are shown in Fig. 14-7a. Back-emf E_g is less than supply voltage V_a. Both armature and field currents are positive. The motor develops torque to meet the load demand.

Regenerative braking. The arrangements for regenerative braking are shown in Fig. 14-7b. The motor acts as a generator and develops an induced voltage E_g. E_g must be greater than supply voltage V_a. The armature current is negative, but the field current is positive. The kinetic energy of the motor is returned to the supply. A series motor is usually connected as a self-excited generator. For self-excitation, it is necessary that the field current aids the residual flux. This is normally accomplished by reversing the armature terminals or the field terminals.

Dynamic braking. The arrangements shown in Fig. 14-7c are similar to those of regenerative braking, except the supply voltage V_a is replaced by a braking resistance R_b. The kinetic energy of the motor is dissipated in R_b.

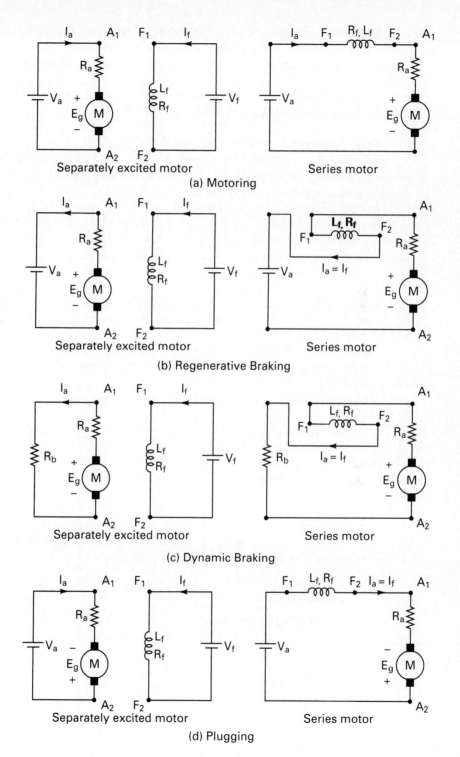

Figure 14-7 Operating modes.

Plugging.   Plugging is a type of braking. The connections for plugging are shown in Fig. 14-7d. The armature terminals are reversed while running. The supply voltage V_a and the induced voltage E_g act in the same direction. The armature current is reversed, thereby producing a braking torque. The field current is positive. For a series motor, either the armature terminals or field terminals should be reversed, but not both.

Four quadrants.   Figure 14-8 shows the polarities of the supply voltage V_a, back emf E_g, and armature current I_a for a separately excited motor. In forward motoring (quadrant I), V_a, E_g, and I_a are all positive. The torque and speed are also positive in this quadrant.

During forward braking (quadrant II), the motor runs in the forward direction and the induced emf E_g will continue to be positive. For the torque to be negative and the direction of energy flow to reverse, the armature current must be negative. The supply voltage V_a should be kept less than E_g.

In reverse motoring (quadrant III), V_a, E_g, and I_a are all negative. The torque and speed are also negative in this quadrant. To keep the torque negative and the energy flow from the source to the motor, the back emf E_g must satisfy the condition $|V_a| > |E_g|$. The polarity of E_g can be reversed by changing the direction of field current or by reversing the armature terminals.

During reverse braking (quadrant IV), the motor runs in the reverse direction. V_a and E_g will continue to be negative. For the torque to be positive and the energy to flow from the motor to the source, the armature current must be positive. The induced emf E_g must satisfy the condition $|V_a| < |E_g|$.

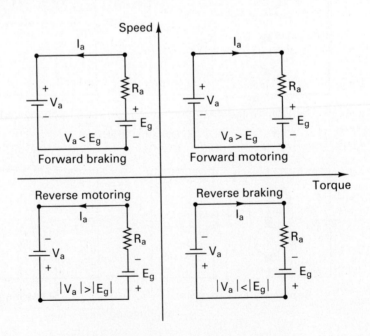

Figure 14-8   Conditions for four quadrants.

If the armature circuit of a dc motor is connected to the output of a single-phase controlled rectifier, the armature voltage can be varied by varying the delay angle of the converter, α_a. The forced-commutated ac–dc converters can also be used to improve the power factor and to reduce the harmonics. The basic circuit agreement for a single-phase converter-fed separately excited motor is shown in Fig. 14-9. At a low delay angle, the armature current may be discontinuous, and this would increase the losses in the motor. A smoothing inductor, L_m, is normally connected in series with the armature circuit to reduce the ripple current to an acceptable magnitude. A converter is also applied in the field circuit to control the field current by varying the delay angle, α_f. For operating the motor in a particular mode, it is often necessary to use contactors for reversing the armature circuit as shown in Fig. 14-10a or the field circuit as shown in Fig. 14-10b. To avoid inductive voltage surges, the field or the armature reversing is performed at a zero armature current. The delay (or firing) angle is normally adjusted to give a zero current; and additionally, a dead time of typically 2 to 10 ms is provided to ensure that the armature current becomes zero. Because of a relatively large time constant of the field winding, the field reversal takes a longer time. A semi- or full converter can be used to vary the field voltage. But a full converter is preferable. Due to the ability to reverse the voltage, a full converter can reduce the field current much faster than a semiconverter. Depending on the type of single-phase converters, single-phase drives may be subdivided into:

1. Single-phase half-wave-converter drives
2. Single-phase semiconverter drives
3. Single-phase full-converter drives
4. Single-phase dual-converter drives

14-4.1 Single-Phase Half-Wave-Converter Drives

A single-phase half-wave converter feeds a dc motor as shown in Fig. 14-11a. The armature current is normally discontinuous unless a very large inductor is connected in the armature circuit. A freewheeling diode is always required for a dc

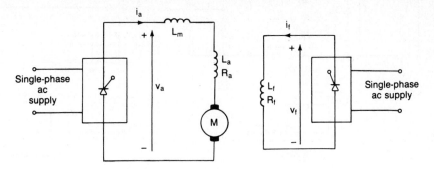

Figure 14-9 Basic circuit arrangement of a single-phase dc drive.

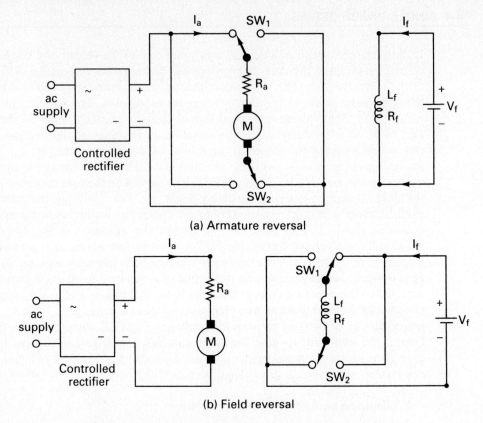

(a) Armature reversal

(b) Field reversal

Figure 14-10   Field and armature reversals using contactors.

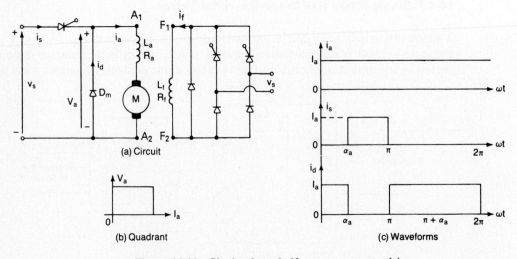

(a) Circuit

(b) Quadrant

(c) Waveforms

Figure 14-11   Single-phase half-wave converter drive.

motor load and it is a one-quadrant drive, as shown in Fig. 14-11b. The applications of this drive are limited to the $\frac{1}{2}$-kW power level. Figure 14-11c shows the waveforms for a highly inductive load. The converter in the field circuit can be a semiconverter. A half-wave converter in the field circuit would increase the magnetic losses of the motor due to a high ripple content on the field excitation current.

With a single-phase half-wave converter in the armature circuit, Eq. (5-1) gives the average armature voltage as

$$V_a = \frac{V_m}{2\pi}(1 + \cos \alpha_a) \qquad \text{for } 0 \le \alpha_a \le \pi \qquad (14\text{-}13)$$

where V_m is the peak voltage of the ac supply. With a semiconverter in the field circuit, Eq. (5-5) gives the average field voltage as

$$V_f = \frac{V_m}{\pi}(1 + \cos \alpha_f) \qquad \text{for } 0 \le \alpha_f \le \pi \qquad (14\text{-}14)$$

14-4.2 Single-Phase Semiconverter Drives

A single-phase semiconverter feeds the armature circuit as shown in Fig. 14-12a. It is a one-quadrant drive as shown in Fig. 14-12b and is limited to applications up to 15 kW. The converter in the field circuit can be a semiconverter. The current waveforms for a highly inductive load are shown in Fig. 14-12c.

With a single-phase semiconverter in the armature circuit, Eq. (5-5) gives the average armature voltage as

$$V_a = \frac{V_m}{\pi}(1 + \cos \alpha_a) \qquad \text{for } 0 \le \alpha_a \le \pi \qquad (14\text{-}15)$$

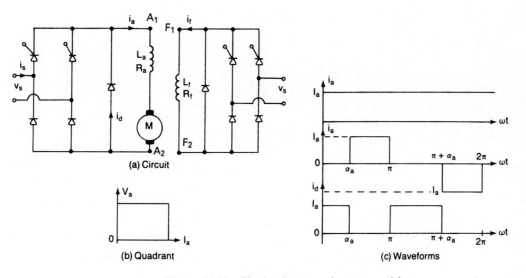

Figure 14-12 Single-phase semiconverter drive.

With a semiconverter in the field circuit, Eq. (5-5) gives the average field voltage as

$$V_f = \frac{V_m}{\pi}(1 + \cos \alpha_f) \qquad \text{for } 0 \le \alpha_f \le \pi \qquad (14\text{-}16)$$

14-4.3 Single-Phase Full-Converter Drives

The armature voltage is varied by a single-phase full-wave converter as shown in Fig. 14-13a. It is a two-quadrant drive as shown in Fig. 14-13b and is limited to applications up to 15 kW. The armature converter gives $+V_a$ or $-V_a$, and allows operation in the first and fourth quadrants. During regeneration for reversing the direction of power flow, the back emf of the motor can be reversed by reversing the field excitation. The converter in the field circuit could be a semi-, full-, or even dual converter. The reversal of the armature or field will allow operation in the second and third quadrants. The current waveforms for a highly inductive load are shown in Fig. 14-13c for powering action. A 9.5-kW 40-A single-phase full-converter drive is shown in Fig. 14-14, where the power stack is on the back of the panel and the control signals are implemented by analog electronics.

With a single-phase full-wave converter in the armature circuit, Eq. (5-21) gives the average armature voltage as

$$V_a = \frac{2V_m}{\pi} \cos \alpha_a \qquad \text{for } 0 \le \alpha_a \le \pi \qquad (14\text{-}17)$$

With a single-phase full-converter in the field circuit, Eq. (5-21) gives the field voltage as

$$V_f = \frac{2V_m}{\pi} \cos \alpha_f \qquad \text{for } 0 \le \alpha_f \le \pi \qquad (14\text{-}18)$$

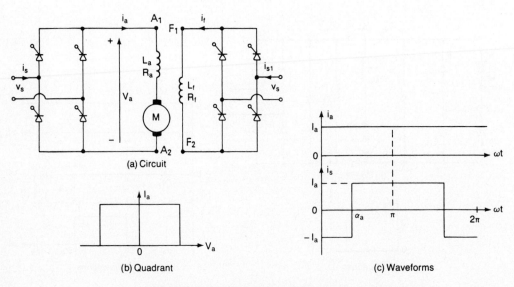

(a) Circuit

(b) Quadrant

(c) Waveforms

Figure 14-13 Single-phase full-converter drive.

Figure 14-14 A 9.5-kW analog-based single-phase full-wave drive. (Reproduced by permission of Brush Electrical Machines Ltd., England.)

14-4.4 Single-Phase Dual-Converter Drives

Two single-phase full-wave converters are connected as shown in Fig. 14-15. Either converter 1 operates to supply a positive armature voltage, V_a, or converter 2 operates to supply a negative armature voltage, $-V_a$. Converter 1 provides operation in the first and fourth quadrants, and converter 2 in the second and third quadrants. It is a four-quadrant drive and permits four modes of operation: forward powering, forward braking (regeneration), reverse powering, and reverse braking (regeneration). It is limited to applications up to 15 kW. The field converter could be a full-wave or semi- or dual converter.

If converter 1 operates with a delay angle of α_{a1}, Eq. (5-31) gives the armature voltage as

$$V_a = \frac{2V_m}{\pi} \cos \alpha_{a1} \qquad \text{for } 0 \le \alpha_{a1} \le \pi \tag{14-19}$$

If converter 2 operates with a delay angle of α_{a2}, Eq. (5-32) gives the armature voltage as

$$V_a = \frac{2V_m}{\pi} \cos \alpha_{a2} \qquad \text{for } 0 \le \alpha_{a2} \le \pi \tag{14-20}$$

where $\alpha_{a2} = \pi - \alpha_{a1}$. With a full converter in the field circuit, Eq. (5-21) gives the field voltage as

$$V_f = \frac{2V_m}{\pi} \cos \alpha_f \qquad \text{for } 0 \le \alpha_f \le \pi \tag{14-21}$$

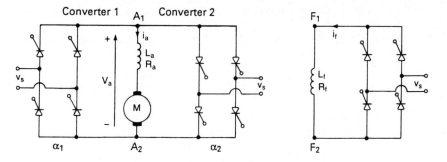

Figure 14-15 Single-phase dual-converter drive.

Example 14-2

The speed of a separately excited motor is controlled by a single-phase semiconverter in Fig.14-12a).The field current, which is also controlled by a semiconverter, is set to the maximum possible value. The ac supply voltage to the armature and field converters is one-phase, 208 V, 60 Hz. The armature resistance is $R_a = 0.25$ Ω, the field resistance is $R_f = 147$ Ω, and the motor voltage constant is $K_v = 0.7032$ V/A-rad/s. The load torque is $T_L = 45$ N·m at 1000 rpm. The viscous friction and no-load losses are negligible. The inductances of the armature and field circuits are sufficient enough to make the armature and field currents continuous and ripple-free. Determine (a) the field current I_f; (b) the delay angle of the converter in the armature circuit, α_a; and (c) the input power factor PF of the armature circuit converter.

Solution $V_s = 208$ V, $V_m = \sqrt{2} \times 208 = 294.16$ V, $R_a = 0.25$ Ω, $R_f = 147$ Ω, $T_d = T_L = 45$ N·m, $K_v = 0.7032$ V/A-rad/s, and $\omega = 1000 \, \pi/30 = 104.72$ rad/s.

(a) From Eq. (14-16), the maximum field voltage (and current) is obtained for a delay angle of $\alpha_f = 0$ and

$$V_f = \frac{2V_m}{\pi} = \frac{2 \times 294.16}{\pi} = 187.27 \text{ V}$$

The field current is

$$I_f = \frac{V_f}{R_f} = \frac{187.27}{147} = 1.274 \text{ A}$$

(b) From Eq. (14-4),

$$I_a = \frac{T_d}{K_v I_f} = \frac{45}{0.7032 \times 1.274} = 50.23 \text{ A}$$

From Eq. (14-2),

$$E_g = K_v \omega I_f = 0.7032 \times 104.72 \times 1.274 = 93.82 \text{ V}$$

From Eq. (14-3), the armature voltage is

$$V_a = 93.82 + I_a R_a = 93.82 + 50.23 \times 0.25 = 93.82 + 12.56 = 106.38 \text{ V}$$

From Eq. (14-15), $V_a = 106.38 = (294.16/\pi) \times (1 + \cos \alpha_a)$ and this gives the delay angle as $\alpha_a = 82.2°$.

(c) If the armature current is constant and ripple-free, the output power is $P_o = V_a I_a = 106.38 \times 50.23 = 5343.5$ W. If the losses in the armature converter are neglected, the power from the supply is $P_a = P_o = 5343.5$ W. The rms input current of the armature converter as shown in Fig. 14-12 is

$$I_{sa} = \left(\frac{2}{2\pi} \int_{\alpha_a}^{\pi} I_a^2 \, d\theta \right)^{1/2} = I_a \left(\frac{\pi - \alpha_a}{\pi} \right)^{1/2}$$

$$= 50.23 \left(\frac{180 - 82.2}{180} \right)^{1/2} = 37.03 \text{ A}$$

and the input volt-ampere rating, VI $= V_s I_{sa} = 208 \times 37.03 = 7702.24$. Assuming negligible harmonics, the input power factor is approximately

$$\text{PF} = \frac{P_o}{\text{VI}} = \frac{5343.5}{7702.24} = 0.694 \text{ (lagging)}$$

From Eq. (5-14),

$$PF = \frac{\sqrt{2}\,(1 + \cos 82.2°)}{[\pi(\pi - 82.2°)]^{1/2}} = 0.694 \text{ (lagging)}$$

Example 14-3

The speed of a separately excited dc motor is controlled by a single-phase full-wave converter in Fig. 14-13a. The field circuit is also controlled by a full converter and the field current is set to the maximum possible value. The ac supply voltage to the armature and field converters is one-phase, 440 V, 60 Hz. The armature resistance is $R_a = 0.25\ \Omega$, the field circuit resistance is $R_f = 175\ \Omega$, and the motor voltage constant is $K_v = 1.4$ V/A-rad/s. The armature current corresponding to the load demand is $I_a = 45$ A. The viscous friction and no-load losses are negligible. The inductances of the armature and field circuits are sufficient to make the armature and field currents continuous and ripple-free. If the delay angle of the armature converter is $\alpha_a = 60°$ and the armature current is $I_a = 45$ A, determine (a) the torque developed by the motor T_d, (b) the speed ω, and (c) the input power factor PF of the drive.

Solution $V_s = 440$ V, $V_m = \sqrt{2} \times 440 = 622.25$ V, $R_a = 0.25\ \Omega$, $R_f = 175\ \Omega$, $\alpha_a = 60°$, and $K_v = 1.4$ V/A-rad/s.

(a) From Eq. (14-18), the maximum field voltage (and current) would be obtained for a delay angle of $\alpha_f = 0$ and

$$V_f = \frac{2V_m}{\pi} = \frac{2 \times 622.25}{\pi} = 396.14 \text{ V}$$

The field current is

$$I_f = \frac{V_f}{R_f} = \frac{396.14}{175} = 2.26 \text{ A}$$

From Eq. (14-4), the developed torque is

$$T_d = T_L = K_v I_f I_a = 1.4 \times 2.26 \times 45 = 142.4 \text{ N·m}$$

From Eq. (14-17), the armature voltage is

$$V_a = \frac{2V_m}{\pi} \cos 60° = \frac{2 \times 622.25}{\pi} \cos 60° = 198.07 \text{ V}$$

The back emf is

$$E_g = V_a - I_a R_a = 198.07 - 45 \times 0.25 = 186.82 \text{ V}$$

From Eq. (14-2), the speed is

$$\omega = \frac{E_g}{K_v I_f} = \frac{186.82}{1.4 \times 2.26} = 59.05 \text{ rad/s or } 564 \text{ rpm}$$

(c) Assuming lossless converters, the total input power from the supply is

$$P_i = V_a I_a + V_f I_f = 198.07 \times 45 + 396.14 \times 2.26 = 9808.4 \text{ W}$$

The input current of the armature converter for a highly inductive load is shown in Fig. 14-9b and its rms value is $I_{sa} = I_a = 45$ A. The rms value of the input current of field converter is $I_{sf} = I_f = 2.26$ A. The effective rms supply current can be found from

$$I_s = (I_{sa}^2 + I_{sf}^2)^{1/2}$$

$$= (45^2 + 2.26^2)^{1/2} = 45.06 \text{ A}$$

and the input volt-ampere rating, $VI = V_s I_s = 440 \times 45.06 = 19{,}826.4$. Neglecting the ripples, the input power is approximately

$$PF = \frac{P_i}{VI} = \frac{9808.4}{19{,}826.4} = 0.495 \text{ (lagging)}$$

From Eq. (5-27),

$$PF = \left(\frac{2\sqrt{2}}{\pi}\right) \cos \alpha_a = \left(\frac{2\sqrt{2}}{\pi}\right) \cos 60° = 0.45 \text{ (lagging)}$$

Example 14-4

If the polarity of the motor back emf in Example 14-3 is reversed by reversing the polarity of the field current, determine (a) the delay angle of the armature circuit converter, α_a, to maintain the armature current constant at the same value of $I_a = 45$ A; and (b) the power fed back to the supply due to regenerative braking of the motor.

Solution (a) From part (b) of Example 14-3, the back emf at the time of polarity reversal is $E_g = 186.82$ V and after polarity reversal $E_g = -186.82$ V. From Eq. (14-3),

$$V_a = E_{gs} + I_a R_a = -186.82 + 45 \times 0.25 = -175.57 \text{ V}$$

From Eq. (14-17),

$$V_a = \frac{2V_m}{\pi} \cos \alpha_a = \frac{2 \times 622.25}{\pi} \cos \alpha_a = -175.57 \text{ V}$$

and this yields the delay angle of the armature converter as $\alpha_a = 116.31°$.

(b) The power fed back to the supply, $P_a = V_a I_a = 175.57 \times 45 = 7900.7$ W.

Note. The speed and back emf of the motor will decrease with time. If the armature current is to be maintained constant at $I_a = 45$ A during regeneration, the delay angle of the armature converter has to be reduced. This would require a closed-loop control to maintain the armature current constant and to adjust the delay angle continuously.

14-5 THREE-PHASE DRIVES

The armature circuit is connected to the output of a three-phase controlled rectifier or a forced-commutated three-phase ac–dc converter. Three-phase drives are used for high-power applications up to megawatts power level. The ripple frequency of the armature voltage is higher than that of single-phase drives and it requires less inductance in the armature circuit to reduce the armature ripple current. The armature current is mostly continuous, and therefore the motor performance is better compared to that of single-phase drives. Similar to the single-phase drives, three-phase drives may also be subdivided into:

1. Three-phase half-wave-converter drives
2. Three-phase semiconverter drives

3. Three-phase full-converter drives

4. Three-phase dual-converter drives

14-5.1 Three-Phase Half-Wave-Converter Drives

A three-phase half-wave converter-fed dc motor drive operates in one quadrant and could be used in applications up to a 40-kW power level. The field converter could be a single-phase or three-phase semiconverter. This drive is not normally used in industrial applications because the ac supply contains dc components.

With a three-phase half-wave converter in the armature circuit, Eq. (5-51) gives the armature voltage as

$$V_a = \frac{3\sqrt{3}\,V_m}{2\pi} \cos \alpha_a \qquad \text{for } 0 \le \alpha_a \le \pi \qquad (14\text{-}22)$$

where V_m is the peak phase voltage of a wye-connected three-phase ac supply. With a three-phase semiconverter in the field circuit, Eq. (5-54) gives the field voltage as

$$V_f = \frac{3\sqrt{3}\,V_m}{2\pi} (1 + \cos \alpha_f) \qquad \text{for } 0 \le \alpha_f \le \pi \qquad (14\text{-}23)$$

14-5.2 Three-Phase Semiconverter Drives

A three-phase semiconverter-fed drive is a one-quadrant drive without field reversal, and is limited to applications up to 115 kW. The field converter should also be a single-phase or three-phase semiconverter.

With a three-phase semiconverter in the armature circuit, Eq. (5-54) gives the armature voltage as

$$V_a = \frac{3\sqrt{3}\,V_m}{2\pi} (1 + \cos \alpha_a) \qquad \text{for } 0 \le \alpha_a \le \pi \qquad (14\text{-}24)$$

With a three-phase semiconverter in the field circuit, Eq. (5-54) gives the field voltage as

$$V_f = \frac{3\sqrt{3}\,V_m}{2\pi} (1 + \cos \alpha_f) \qquad \text{for } 0 \le \alpha_f \le \pi \qquad (14\text{-}25)$$

14-5.3 Three-Phase Full-Converter Drives

A three-phase full-wave converter drive is a two-quadrant drive without any field reversal, and is limited to applications up to 1500 kW. During regeneration for reversing the direction of power flow, the back emf of the motor is reversed by reversing the field excitation. The converter in the field circuit should be a single- or three-phase full converter. A 68-kW 170-A microprocessor-based three-phase full-converter dc drive is shown in Fig. 14-16, where the power semiconductor stacks are at the back of the panel.

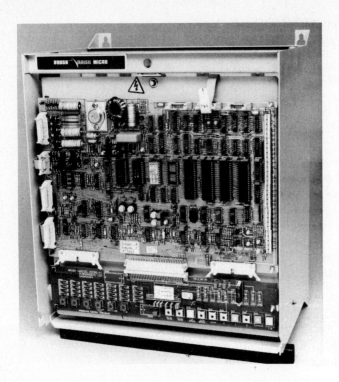

Figure 14-16 A 68-kW microprocessor-based three-phase full converter. (Reproduced by permission of Brush Electrical Machines Ltd., England.)

With a three-phase full-wave converter in the armature circuit, Eq. (5-57) gives the armature voltage as

$$V_a = \frac{3\sqrt{3}\ V_m}{\pi} \cos \alpha_a \qquad \text{for } 0 \le \alpha_a \le \pi \qquad (14\text{-}26)$$

With a three-phase full converter in the field circuit, Eq. (5-57) gives the field voltage as

$$V_f = \frac{3\sqrt{3}\ V_m}{\pi} \cos \alpha_f \qquad \text{for } 0 \le \alpha_f \le \pi \qquad (14\text{-}27)$$

14-5.4 Three-Phase Dual-Converter Drives

Two three-phase full-wave converters are connected in an arrangement similar to Fig. 14-15a. Either converter 1 operates to supply a positive armature voltage, V_a, or converter 2 operates to supply a negative armature voltage, $-V_a$. It is a four-quadrant drive and is limited to applications up to 1500 kW. Similar to single-phase drives, the field converter can be a full-wave or semiconverter.

A 12-pulse ac–dc converter for a 360-kW motor for driving a cement kiln is shown in Fig. 14-17, where the control electronics are mounted on the cubicle door and the pulse drive boards are mounted on the front of the thyristor stacks. The cooling fans are mounted at the top of each stack. If converter 1 operates

Figure 14-17 A 360-kW 12-pulse ac–dc converter for dc drives. (Reproduced by permission of Brush Electrical Machines Ltd., England.)

with a delay angle of α_{a1}, Eq. (5-57) gives the average armature voltage as

$$V_a = \frac{3\sqrt{3}\,V_m}{\pi} \cos \alpha_{a1} \qquad \text{for } 0 \leq \alpha_{a1} \leq \pi \qquad (14\text{-}28)$$

If converter 2 operates with a delay angle of α_{a2}, Eq. (5-57) gives the average armature voltage as

$$V_a = \frac{3\sqrt{3}\,V_m}{\pi} \cos \alpha_{a2} \qquad \text{for } 0 \leq \alpha_{a2} \leq \pi \qquad (14\text{-}29)$$

With a three-phase full converter in the field circuit, Eq. (5-57) gives the average field voltage as

$$V_f = \frac{3\sqrt{3}\,V_m}{\pi} \cos \alpha_f \qquad \text{for } 0 \leq \alpha_f \leq \pi \qquad (14\text{-}30)$$

Example 14-5

The speed of a 20-hp 300-V 1800-rpm separately excited dc motor is controlled by a three-phase full-converter drive. The field current is also controlled by a three-phase full-converter and is set to the maximum possible value. The ac input is a three-phase wye-connected 208-V 60-Hz supply. The armature resistance is $R_a = 0.25\ \Omega$, the field resistance is $R_f = 245\ \Omega$, and the motor voltage constant is $K_v = 1.2$ V/A-rad/s. The armature and field currents can be assumed to be continuous and ripple-free. The viscous friction is negligible. Determine (a) the delay angle of the armature converter, α_a, if the motor supplies the rated power at the rated speed; (b) the no-load speed if the delay angles are the same as in part (a) and the armature current at no-load is 10% of the rated value; and (c) the speed regulation.

Solution $R_a = 0.25\ \Omega$, $R_f = 245\ \Omega$, $K_v = 1.2$ V/A-rad/s, $V_L = 208$ V, and $\omega = 1800\ \pi/30 = 188.5$ rad/s. The phase voltage is $V_p = V_L/\sqrt{3} = 208/\sqrt{3} = 120$ V and $V_m = 120 \times \sqrt{2} = 169.7$ V. Since 1 hp is equal to 746 W, the rated armature current is $I_{rated} = 20 \times 746/300 = 49.73$ A; and for maximum possible field current, $\alpha_f = 0$. From Eq. (14-27),

$$V_f = 3\sqrt{3} \times \frac{169.7}{\pi} = 280.7\ \text{V}$$

$$I_f = \frac{V_f}{R_f} = \frac{280.7}{245} = 1.146\ \text{A}$$

(a) $I_a = I_{rated} = 49.73$ A and

$$E_g = K_v I_f \omega = 1.2 \times 1.146 \times 188.5 = 259.2\ \text{V}$$

$$V_a = 259.2 + I_a R_a = 259.2 + 49.73 \times 0.25 = 271.63\ \text{V}$$

From Eq. (14-26),

$$V_a = 271.63 = \frac{3\sqrt{3}\ V_m}{\pi} \cos\alpha_a = \frac{3\sqrt{3} \times 169.7}{\pi} \cos\alpha_a$$

and this gives the delay angle as $\alpha_a = 14.59°$.

(b) $I_a = 10\%$ of $49.73 = 4.973$ A and

$$E_{go} = V_a - R_a I_a = 271.63 - 0.25 \times 4.973 = 270.39\ \text{V}$$

From Eq. (14-2), the no-load speed is

$$\omega_0 = \frac{E_{go}}{K_v I_f} = \frac{270.39}{1.2 \times 1.146} = 196.62\ \text{rad/s} \qquad \text{or} \qquad 196.62 \times \frac{30}{\pi} = 1877.58\ \text{rpm.}$$

(c) The speed regulation is defined as

$$\frac{\text{no-load speed} - \text{full-load speed}}{\text{full-load speed}} = \frac{1877.58 - 1800}{1800} = 0.043 \quad \text{or} \quad 4.3\%$$

Example 14-6

The speed of a 20-hp 300-V 900-rpm separately excited dc motor is controlled by a three-phase full converter. The field circuit is also controlled by a three-phase full converter. The ac input to the armature and field converters is three-phase, wye-connected, 208 V, 60 Hz. The armature resistance is $R_a = 0.25\ \Omega$, the field circuit resistance is $R_f = 145\ \Omega$, and the motor voltage constant is $K_v = 1.2$ V/A-rad/s. The viscous friction and no-load losses can be considered negligible. The armature and field currents are continuous and ripple-free. (a) If the free converter is operated at the maximum field current and the developed torque is $T_d = 116$ N·m at 900 rpm, determine the delay angle of the armature converter, α_a. (b) If the field circuit converter is set for the maximum field current, the developed torque is $T_d = 116$ N·m, and the delay angle of the armature converter is $\alpha_a = 0$, determine the speed of the motor. (c) For the same load demand as in part (b), determine the delay angle of the field converter if the speed has to be increased to 1800 rpm.

Solution $R_a = 0.25\ \Omega$, $R_f = 145\ \Omega$, $K_v = 1.2$ V/A-rad/s, and $V_L = 208$ V. The phase voltage is $V_p = 208/\sqrt{3} = 120$ V and $V_m = \sqrt{2} \times 120 = 169.7$ V.

(a) $T_d = 116$ N·m and $\omega = 900\,\pi/30 = 94.25$ rad/s. For maximum field current, $\alpha_f = 0$. From Eq. (14-27),

$$V_f = \frac{3 \times \sqrt{3} \times 169.7}{\pi} = 280.7 \text{ V}$$

$$I_f = \frac{280.7}{145} = 1.936 \text{ A}$$

From Eq. (14-4),

$$I_a = \frac{T_d}{K_v I_f} = \frac{116}{1.2 \times 1.936} = 49.93 \text{ A}$$

$$E_g = K_v I_f \omega = 1.2 \times 1.936 \times 94.25 = 218.96 \text{ V}$$

$$V_a = E_g + I_a R_a = 218.96 + 49.93 \times 0.25 = 231.44 \text{ V}$$

From Eq. (14-26),

$$V_a = 231.44 = \frac{3 \times \sqrt{3} \times 169.7}{\pi} \cos \alpha_a$$

which gives the delay angle as $\alpha_a = 34.46°$.

(b) $\alpha_a = 0$ and

$$V_a = \frac{3 \times \sqrt{3} \times 169.7}{\pi} = 280.7 \text{ V}$$

$$E_g = 280.7 - 49.93 \times 0.25 = 268.22 \text{ V}$$

and the speed

$$\omega = \frac{E_g}{K_v I_f} = \frac{268.22}{1.2 \times 1.936} = 115.45 \text{ rad/s} \qquad \text{or} \qquad 1102.5 \text{ rpm}$$

(c) $\omega = 1800\,\pi/30 = 188.5$ rad/s

$$E_g = 268.22 \text{ V} = 1.2 \times 188.5 \times I_f \qquad \text{or} \qquad I_f = 1.186 \text{ A}$$

$$V_f = 1.186 \times 145 = 171.97 \text{ V}$$

From Eq. (14-27),

$$V_f = 171.97 = \frac{3 \times \sqrt{3} \times 169.7}{\pi} \cos \alpha_f$$

which gives the delay angle as $\alpha_f = 52.2°$.

14-6 CHOPPER DRIVES

Chopper drives are widely used in traction applications all over the world. A dc chopper is connected between a fixed-voltage dc source and a dc motor to vary the armature voltage. In addition to armature voltage control, a dc chopper can provide regenerative braking of the motors and can return energy back to the supply. This energy-saving feature is particulary attractive to transportation systems with frequent stops such as mass rapid transit (MRT). Chopper drives are

---

also used in battery electric vehicles (BEVs). A dc motor can be operated in one of the four quadrants by controlling the armature and/or field voltages (or currents). It is often required to reverse the armature or field terminals in order to operate the motor in the desired quadrant.

If the supply is nonreceptive during the regenerative braking, the line voltage would increase and regenerative braking may not be possible. In this case, an alternative form of braking is necessary, such as rheostatic braking. The possible control modes of a dc chopper drive are:

1. Power (or acceleration) control
2. Regenerative brake control
3. Rheostatic brake control
4. Combined regenerative and rheostatic brake control

14-6.1 Principle of Power Control

The chopper is used to control the armature voltage of a dc motor. The circuit arrangement of a chopper-fed dc separately excited motor is shown in Fig. 14-18a. The chopper switch could be a transistor or forced-commutated thyristor chopper, as discussed in Section 9-8. This is a one-quadrant drive, as shown in Fig. 14-18b. The waveforms for the armature voltage, load current, and input current are shown in Fig. 14-18c, assuming a highly inductive load.

The average armature voltage is

$$V_a = kV_s \qquad (14\text{-}31)$$

where k is the duty cycle of the chopper. The power supplied to the motor is

$$P_0 = V_a I_a = kV_s I_a \qquad (14\text{-}32)$$

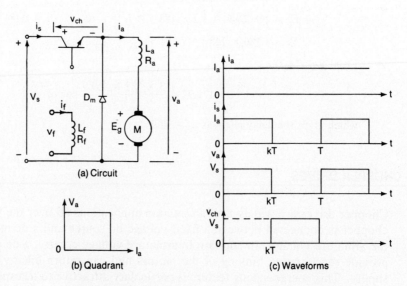

(a) Circuit

(b) Quadrant

(c) Waveforms

Figure 14-18 Chopper-fed dc drive in power control.

where I_a is the average armature current of the motor and it is ripple-free. Assuming a lossless chopper, the input power is $P_i = P_0 = kV_sI_s$. The average value of the input current is

$$I_s = kI_a \qquad (14\text{-}33)$$

The equivalent input resistance of the chopper drive seen by the source is

$$R_{eq} = \frac{V_s}{I_s} = \frac{V_s}{I_a}\frac{1}{k} \qquad (14\text{-}34)$$

By varying the duty cycle, k, the power flow to the motor (and speed) can be controlled. For a finite armature circuit inductance, Eq. (9-19) can be applied to find the maximum peak-to-peak ripple current as

$$\Delta I_{max} = \frac{V_s}{R_m}\tanh\frac{R_m}{4fL_m} \qquad (14\text{-}35)$$

where R_m and L_m are the total armature circuit resistance and inductance respectively. For a separately excited motor, $R_m = R_a +$ any series resistance, and $L_m = L_a +$ any series inductance. For a series motor, $R_m = R_a + R_f +$ any series resistance, and $L_m = L_a + L_f +$ any series inductance.

Example 14-7

A dc separately excited motor is powered by a dc chopper (as shown in Fig. 14-18a) from a 600-V dc source. The armature resistance is $R_a = 0.05\ \Omega$. The back emf constant of the motor is $K_v = 1.527$ V/A-rad/s. The average armature current is $I_a = 250$ A. The field current is $I_f = 2.5$ A. The armature current is continuous and has negligible ripple. If the duty cycle of the chopper is 60%, determine (a) the input power from the source, (b) the equivalent input resistance of the chopper drive, (c) the motor speed, and (d) the developed torque.

Solution $V_s = 600$ V, $I_a = 250$ A, and $k = 0.6$. The total armature circuit resitance is $R_m = R_a = 0.05\ \Omega$.

(a) From Eq. (14-32),

$$P_i = kV_sI_a = 0.6 \times 600 \times 250 = 90 \text{ kW}$$

(b) From Eq. (14-34), $R_{eq} = 600/(250 \times 0.6) = 4\ \Omega$.
(c) From Eq. (14-31), $V_a = 0.6 \times 600 = 360$ V. The back emf is

$$E_g = V_a - R_mI_m = 360 - 0.05 \times 250 = 347.5 \text{ V}$$

From Eq. (14-2), the motor speed is

$$\omega = \frac{347.5}{1.527 \times 2.5} = 91.03 \text{ rad/s} \qquad \text{or} \qquad 91.03 \times \frac{30}{\pi} = 869.3 \text{ rpm}$$

(d) From Eq. (14-4),

$$T_d = 1.527 \times 250 \times 2.5 = 954.38 \text{ N·m}$$

14-6.2 Principle of Regenerative Brake Control

In regenerative braking, the motor acts as a generator and the kinetic energy of the motor and load is returned back to the supply. The principle of energy transfer from one dc source to another of higher voltage is discussed in Section 9-5, and this can be applied in regenerative braking of dc motors.

The application of dc choppers in regenerative braking can be explained with Fig. 14-19a. It requires rearranging the switch from powering mode to regenerative braking. Let us assume that the armature of a separately excited motor is rotating due to the inertia of the motor (and load); and in case of a transportation system, the kinetic energy of the vehicle or train would rotate the armature shaft. Then if the transistor is switched on, the armature current will rise due to the short-circuiting of the motor terminals. If the chopper is turned off, diode D_m would be turned on and the energy stored in the armature circuit inductances would be transferred to the supply, provided that the supply is receptive. It is a one-quadrant drive and operates in the second quadrant as shown in Fig. 14-19b. Figure 14-19c shows the voltage and current waveforms assuming that the armature current is continuous and ripple-free.

The average voltage across the chopper is

$$V_{ch} = (1 - k)V_s \tag{14-36}$$

If I_a is the average armature current, the regenerated power can be found from

$$P_g = I_a V_s (1 - k) \tag{14-37}$$

The voltage generated by the motor acting as a generator is

$$\begin{aligned} E_g &= K_v I_f \omega \\ &= V_{ch} + R_m I_a = (1 - k)V_s + R_m I_a \end{aligned} \tag{14-38}$$

where K_v is machine constant and ω is the machine speed in rad/s. Therefore, the equivalent load resistance of the motor acting as a generator is

$$R_{eq} = \frac{E_g}{I_a} = \frac{V_s}{I_a}(1 - k) + R_m \tag{14-39}$$

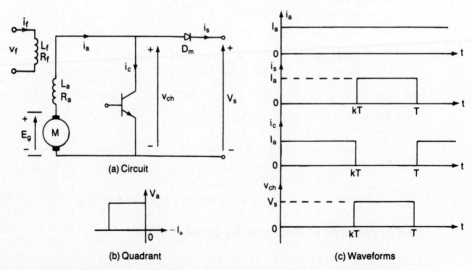

(a) Circuit

(b) Quadrant

(c) Waveforms

Figure 14-19 Regenerative braking of dc separately excited motors.

By varying the duty cycle, k, the equivalent load resistance seen by the motor can be varied from R_m to $(V_s/I_a + R_m)$ and the regenerative power can be controlled.

From Eq. (9-27), the conditions for permissible potentials and polarity of the two voltages are

$$0 \leq (E_g - R_m I_a) \leq V_s \qquad (14\text{-}40)$$

which gives the minimum braking speed of the motor as

$$E_g = K_v \omega_{\min} I_f = R_m I_a$$

or

$$\omega_{\min} = \frac{R_m}{K_v} \frac{I_a}{I_f} \qquad (14\text{-}41)$$

and $\omega \geq \omega_{\min}$. The maximum braking speed of a series motor can be found from Eq. (14-40):

$$K_v \omega_{\max} I_f - R_m I_a = V_s$$

or

$$\omega_{\max} = \frac{V_s}{K_v I_f} + \frac{R_m}{K_v} \frac{I_a}{I_f} \qquad (14\text{-}42)$$

and $\omega \leq \omega_{\max}$.

The regenerative braking would be effective only if the motor speed is between these two speed limits (e.g., $\omega_{\min} < \omega < \omega_{\max}$). At any speed less than $\omega_{\min}$, an alternative braking arrangement would be required.

Although dc series motors are traditionally used for traction applications due to their high starting torque, a series-excited generator is unstable when working into a fixed voltage supply. Thus for running on the traction supply, a separate excitation control is required and such an arrangement of series motor is, normally, sensitive to supply voltage fluctuations and a fast dynamic response is required to provide an adequate brake control. The application of a dc chopper allows the regenerative braking of dc series motors due to its fast dynamic response.

A separately excited dc motor is stable in regenerative braking. The armature and field can be controlled independently to provide the required torque during starting. A chopper-fed series and separately excited dc motors are both suitable for traction applications.

Example 14-8

A dc chopper is used in regenerative braking of a dc series motor similar to the arrangement shown in Fig. 14-19a. The dc supply voltage is 600 V. The armature resistance is $R_a = 0.02\ \Omega$ and the field resistance is $R_f = 0.03\ \Omega$. The back emf constant is $K_v = 15.27$ mV/A-rad/s. The average armature current is maintained constant at $I_a = 250$ A. The armature current is continuous and has negligible ripple. If the duty cycle of the chopper is 60%, determine (a) the average voltage across the chopper, V_{ch}; (b) the power regenerated to the dc supply, P_g; (c) the equivalent load resistance of the motor acting as a generator, R_{eq}; (d) the minimum permissible braking speed, ω_{min}; (e) the maximum permissible braking speed, ω_{max}; and (f) the motor speed.

Solution $V_s = 600$ V, $I_a = 250$ A, $K_v = 0.01527$ V/A-rad/s, $k = 0.6$. For a series motor $R_m = R_a + R_f = 0.02 + 0.03 = 0.05 \, \Omega$.

(a) From Eq. (14-36), $V_{ch} = (1 - 0.6) \times 600 = 240$ V.

(b) From Eq. (14-37), $P_g = 250 \times 600 \times (1 - 0.6) = 60$ kW.

(c) From Eq. (14-39), $R_{eq} = (600/250)(1 - 0.6) + 0.05 = 1.01 \, \Omega$.

(d) From Eq. (14-41), the minimum permissible braking speed,

$$\omega_{min} = \frac{0.05}{0.01527} = 3.274 \text{ rad/s} \qquad \text{or} \qquad 3.274 \times \frac{30}{\pi} = 31.26 \text{ rpm}$$

(e) From Eq. (14-42), the maximum permissible braking speed,

$$\omega_{max} = \frac{600}{0.01527 \times 250} + \frac{0.05}{0.01527} = 160.445 \text{ rad/s} \qquad \text{or} \qquad 1532.14 \text{ rpm}$$

(f) From Eq. (14-38), $E_g = 240 + 0.05 \times 250 = 252.5$ V and the motor speed,

$$\omega = \frac{252.5}{0.01527 \times 250} = 66.14 \text{ rad/s or } 631.6 \text{ rpm}$$

Note. The motor speed would decrease with time. To maintain the armature current at the same level, the effective load resistance of the series generator should be adjusted by varying the duty cycle of the chopper.

14-6.3 Principle of Rheostatic Brake Control

In a rheostatic braking, the energy is dissipated in a rheostat and it may not be a desirable feature. In mass rapid transit (MRT) systems, the energy may be used in heating the trains. The rheostatic braking is also known as *dynamic braking*. An arrangement for the rheostatic braking of dc separately excited motor is shown in Fig. 14-20a. This is a one-quadrant drive and operates in the second quadrant as shown in Fig. 14-20b. Figure 14-20c shows the waveforms for the current and voltage, assuming that the armature current is continuous and ripple free.

The average current of the braking resistor,

$$I_b = I_a(1 - k) \tag{14-43}$$

and the average voltage across the braking resistor,

$$V_b = R_b I_a(1 - k) \tag{14-44}$$

The equivalent load resistance of the generator,

$$R_{eq} = \frac{V_b}{I_a} = R_b(1 - k) + R_m \tag{14-45}$$

The power dissipated in the resistor, R_b, is

$$P_b = I_a^2 R_b(1 - k) \tag{14-46}$$

By controlling the duty cycle k, the effective load resistance can be varied from R_m to $R_m + R_b$; and the braking power can be controlled. The braking resistance R_b determines the maximum voltage rating of the chopper.

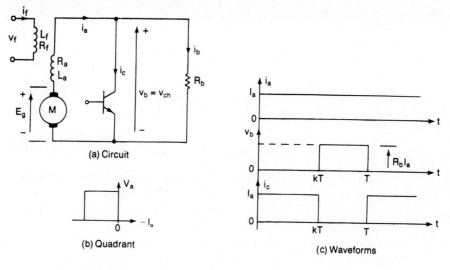

(a) Circuit

(b) Quadrant

(c) Waveforms

Figure 14-20   Rheostatic braking of dc separately excited motors.

Example 14-9

A dc chopper is used in rheostatic braking of a dc separately excited motor as shown in Fig. 14-20a. The armature resistance is $R_a = 0.05$ Ω. The braking resistor, $R_b = 5$ Ω. The back emf constant is $K_v = 1.527$ V/A-rad/s. The average armature current is maintained constant at $I_a = 150$ A. The armature current is continuous and has negligible ripple. The field current is $I_f = 1.5$ A. If the duty cycle of the chopper is 40%, determine (a) the average voltage across the chopper, V_{ch}; (b) the power dissipated in the braking resistor, P_b; (c) the equivalent load resistance of the motor acting as a generator, R_{eq}; (d) the motor speed; and (e) the peak chopper voltage, V_p.

Solution $I_a = 150$ A, $K_v = 1.527$ V/A-rad/s, $k = 0.4$, and $R_m = R_a = 0.05$ Ω.

(a) From Eq. (14-44), $V_{ch} = V_b = 5 \times 150 \times (1 - 0.4) = 450$ V.

(b) From Eq. (14-46), $P_b = 150 \times 150 \times 5 \times (1 - 0.4) = 67.5$ kW.

(c) From Eq. (14-45), $R_{eq} = 5 \times (1 - 0.4) + 0.05 = 3.05$ Ω.

(d) The generated emf $E_g = 450 + 0.05 \times 150 = 457.5$ V and the braking speed,

$$\omega = \frac{E_g}{K_v I_f} = \frac{457.5}{1.527 \times 1.5} = 199.74 \text{ rad/s} \qquad \text{or} \qquad 1907.4 \text{ rpm}$$

(e) The peak chopper voltage $V_p = I_a R_b = 150 \times 5 = 750$ V.

14-6.4 Principle of Combined Regenerative and Rheostatic Brake Control

Regenerative braking is energy-efficient braking. On the other hand, the energy is dissipated as heat in rheostatic braking. If the supply is partly receptive, which is normally the case in practical traction systems, a combined regenerative and rheostatic brake control would be the most energy efficient. Figure 14-21 shows an arrangement in which rheostatic braking is combined with regenerative braking.

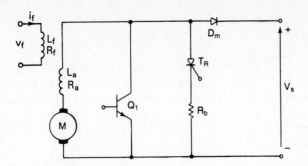

Figure 14-21 Combined regenerative and rheostatic braking.

During regenerative brakings, the line voltage is sensed continuously. If it exceeds a certain preset value, normally 20% above the line voltage, the regenerative braking is removed and a rheostatic braking is applied. It allows an almost instantaneous transfer from regenerative to rheostatic braking if the line becomes nonreceptive, even momentarily. In every cycle, the logic circuit determines the receptivity of the supply. If it is nonreceptive, thyristor T_R is turned "on" to divert the motor current to the resistor R_b. Thyristor T_R is self-commutated when transistor Q_1 is turned "on" in the next cycle.

14-6.5 Two/Four-Quadrant Chopper Drives

During power control, a chopper-fed drive operates in the first quadrant, where the armature voltage and armature current are positive as shown in Fig. 14-18b. In a regenerative braking, the chopper drive operates in the second quadrant, where the armature voltage is positive and the armature current is negative, as shown in Fig. 14-19b. Two-quadrant operation, as shown in Fig. 14-22a, is required to allow power and regenerative braking control. The circuit arrangement of a transistorized two-quadrant drive is shown in Fig. 14-22b.

Power control. Transistor Q_1 and diode D_2 operate. When Q_1 is turned on, the supply voltage V_s is connected to the motor terminals. When Q_1 is turned

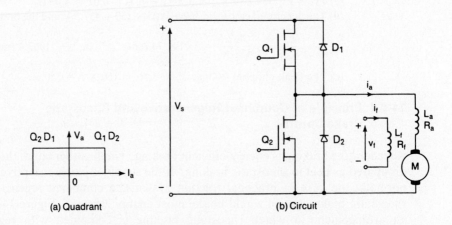

(a) Quadrant   (b) Circuit

Figure 14-22 Two-quadrant transistorized chopper drive.

off, the armature current which flows through the freewheeling diode D_2 decays.

Regenerative control. Transistor Q_2 and diode D_1 operate. When Q_2 is turned on, the motor acts as a generator and the armature current rises. When Q_2 is turned off, the motor, acting as a generator, returns energy to the supply through the regenerative diode D_1. In industrial applications, four-quadrant operation, as shown in Fig. 14-23a, is required. A transistorized four-quadrant drive is shown in Fig. 14-23b.

Forward power control. Transistors Q_1 and Q_2 operate. Transistors Q_3 and Q_4 are off. When Q_1 and Q_2 are turned on together, the supply voltage appears across the motor terminals and the armature current rises. When Q_1 is turned off and Q_2 is still turned on, the armature current decays through Q_2 and D_4. Alternatively, both Q_1 and Q_2 can be turned off, while the armature current is forced to decay through D_3 and D_4.

Forward regeneration. Transistors Q_1, Q_2, and Q_3 are turned off. When transistor Q_4 is turned on, the armature current, which rises, flows through Q_4 and D_2. When Q_4 is turned off, the motor, acting as a generator, returns energy to the supply through D_1 and D_2.

Reverse power control. Transistors Q_3 and Q_4 operate. Transistors Q_1 and Q_2 are off. When Q_3 and Q_4 are turned on together, the armature current rises and flows in the reverse direction. When Q_3 is turned off and Q_4 is turned on, the armature current falls through Q_4 and D_2. Alternatively, both Q_3 and Q_4 can be turned off, while forcing the armature current to decay through D_1 and D_2.

Reverse regeneration. Transistors Q_1, Q_3, and Q_4 are off. When Q_2 is turned on, the armature current rises through Q_2 and D_4. When Q_2 is turned off, the armature current falls and the motor returns energy to the supply through D_3 and D_4.

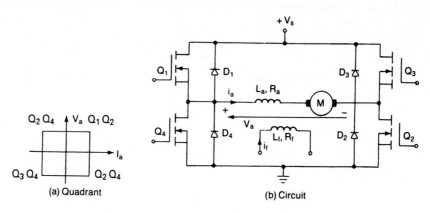

(a) Quadrant

(b) Circuit

Figure 14-23 Four-quadrant transistorized chopper drive.

14-6.6 Multiphase Choppers

If two or more choppers are operated in parallel and are phase shifted from each other by π/u as shown in Fig. 14-24a, the amplitude of the load current ripples decreases and the ripple frequency increases. As a result, the chopper-generated harmonic currents in the supply are reduced. The sizes of the input filter are also reduced. The multiphase operation allows the reduction of smoothing chokes, which are normally connected in the armature circuit of dc motors. Separate inductors in each phase are used for current sharing. Figure 14-24b shows the waveforms for the currents with u choppers.

For u choppers in multiphase operation, it can be proved that Eq. (9-19) is satisfied when $k = 1/2u$ and the maximum peak-to-peak load ripple current becomes

$$\Delta I_{max} = \frac{V_s}{R_m} \tanh \frac{R_m}{4ufL_m} \tag{14-47}$$

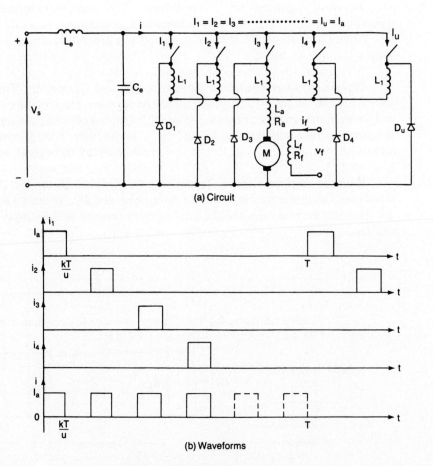

(a) Circuit

(b) Waveforms

Figure 14-24 Multiphase choppers.

where L_m and R_m are the total armature inductance and resistance, respectively. For $4ufL_m >> R_m$, the maximum peak-to-peak load ripple current can be approximated to

$$\Delta I_{\max} = \frac{V_s}{4ufL_m} \qquad (14\text{-}48)$$

If an *LC*-input filter is used, Eq. (9-124) can be applied to find the rms *n*th harmonic component of chopper-generated harmonics in the supply

$$I_{ns} = \frac{1}{1 + (2n\pi uf)^2 L_e C_e} I_{nh}$$
$$= \frac{1}{1 + (nuf/f_0)^2} I_{nh} \qquad (14\text{-}49)$$

where I_{nh} is the rms value of the nth harmonic component of the chopper current and $f_0[= 1/2 \pi \,) \, (L_e C_e)]$ is the resonant frequency of the input filter. If $(nuf/f_o) >> 1$, the *n*th harmonic current in the supply becomes

$$I_{ns} = I_{nh} \left(\frac{f_0}{nuf}\right)^2 \qquad (14\text{-}50)$$

Multiphase operations are advantageous for large motor drives, especially if the load current requirement is large. However, considering the additional complexity involved in increasing the number of choppers, there is not much reduction in the chopper-generated harmonics in the supply line if more than two choppers are used [7]. In practice, both the magnitude and frequency of the line current harmonics are important factors to determine the level of interferences into the signaling circuits. In many rapid transit systems, the power and signaling lines are very close; in three-line systems, they even share a common line. The signaling circuits are sensitive to particular frequencies and the reduction in the magnitude of harmonics by using a multiphase operation of choppers could generate frequencies within the sensitivity band—which might cause more problems than it solves.

Example 14-10

Two choppers control a dc separately excited motor and they are phase shifted in operation by $\pi/2$. The supply voltage to the chopper drive $V_s = 220$ V, the total armature circuit resistance $R_m = 4\ \Omega$, the total armature circuit inductance $L_m = 15$ mH, and the frequency of each chopper, $f = 350$ Hz. Calculate the maximum peak-to-peak load ripple current.

Solution The effective chopping frequency is $f_e = 2 \times 350 = 700$ Hz, $R_m = 4\ \Omega$, $L_m = 15$ mH, $u = 2$, and $V_s = 220$ V. $4ufL_m = 4 \times 2 \times 350 \times 15 \times 10^{-3} = 42$. Since $42 >> 4$, the approximate equation (14-48) can be used, and this gives the maximum peak-to-peak load ripple current, $\Delta I_{\max} = 220/42 = 5.24$ A.

Example 14-11

A dc separately excited motor is controlled by two multiphase choppers. The average armature current, $I_a = 100$ A. A simple *LC*-input filter with $L_e = 0.3$ mH and

C_e = 4500 μF is used. Each chopper is operated at a frequency of f = 350 Hz. Determine the rms fundamental component of the chopper-generated harmonic current in the supply.

Solution I_a = 100 A, u = 2, L_e = 0.3 mH, C_e = 4500 μF, and f_0 = $1/(2\pi\sqrt{L_eC_e})$ = 136.98 Hz. The effective chopper frequency is f_e = 2 $\times$ 350 = 700 Hz. From the results of Example 9-13, the rms value of the fundamental component of the chopper current is I_{1h} = 45.02 A. From Eq. (14-49), the fundamental component of the chopper-generated harmonic current is

$$I_{1s} = \frac{45.02}{1 + (2 \times 350/136.98)^2} = 1.66 \text{ A}$$

14-7 CLOSED-LOOP CONTROL OF DC DRIVES

The speed of dc motors changes with the load torque. To maintain a constant speed, the armature (and or field) voltage should be varied continuously by varying the delay angle of ac–dc converters or duty cycle of dc choppers. In practical drive systems it is required to operate the drive at a constant torque or constant power; in addition, controlled acceleration and deceleration are required. Most industrial drives operate as closed-loop feedback systems. A closed-loop control system has the advantages of improved accuracy, fast dynamic response, and reduced effects of load disturbances and system nonlinearities.

The block diagram of a closed-loop converter-fed separately excited dc drive is shown in Fig. 14-25. If the speed of the motor decreases due to the application of additional load torque, the speed error V_e increases. The speed controller responses with an increased control signal V_c, change the delay angle or duty cycle of the converter, and increase the armature voltage of the motor. An increased armature voltage develops more torque to restore the motor speed to the original value. The drive normally passes through a transient period until the developed torque is equal to the load torque.

14-7.1 Open-Loop Transfer Function

The steady-state characteristics of dc drives, which are discussed in the preceding sections, are of major importance in the selection of dc drives and are not suffi-

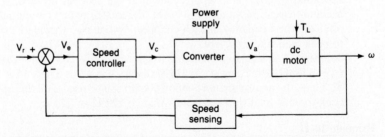

Figure 14-25 Block diagram of a closed-loop converter-fed dc motor drive.

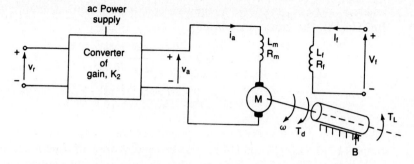

Figure 14-26 Converter-fed separately excited dc motor drive.

cient when the drive is in closed-loop control. Knowledge of the dynamic behavior, which is normally expressed in the form of a transfer function, is also important.

The circuit arrangement of a converter-fed separately excited dc motor drive with open-loop control is shown in Fig. 14-26. The motor speed is adjusted by setting reference (or control) voltage v_r. Assuming a linear power converter of gain K_2, the armature voltage of the motor is

$$v_a = K_2 v_r \tag{14-51}$$

Assuming that the motor field current I_f and the back emf constant K_v remain constant during any transient disturbances, the system equations are

$$e_g = K_v I_f \omega \tag{14-52}$$

$$v_a = R_m i_a + L_m \frac{di_a}{dt} + e_g = R_m i_a + L_m \frac{di_a}{dt} + K_v I_f \omega \tag{14-53}$$

$$T_d = K_t I_f i_a \tag{14-54}$$

$$T_d = K_t I_f i_a = J \frac{d\omega}{dt} + B\omega + T_L \tag{14-55}$$

The transient behavior may be analyzed by changing the system equations into Laplace's transforms with zero initial conditions. Transforming Eqs. (14-51), (14-53), and (14-55) yields

$$V_a(s) = K_2 V_r(s) \tag{14-56}$$

$$V_a(s) = R_m I_a(s) + s L_m I_a(s) + K_v I_f \omega(s) \tag{14-57}$$

$$T_d(s) = K_t I_f I_a(s) = s J \omega(s) + B\omega(s) + T_L(s) \tag{14-58}$$

From Eq. (14-57), the armature current is

$$I_a(s) = \frac{V_a(s) - K_v I_f \omega(s)}{s L_m + R_m} \tag{14-59}$$

$$= \frac{V_a(s) - K_v I_f \omega(s)}{R_m(s \tau_a + 1)} \tag{14-60}$$

where $\tau_a = L_m/R_m$ is known as the *time constant* of motor armature circuit. From Eq. (14-58), the motor speed is

$$\omega(s) = \frac{T_d(s) - T_L(s)}{sJ + B} \tag{14-61}$$

$$= \frac{T_d(s) - T_L(s)}{B(s\tau_m + 1)} \tag{14-62}$$

where $\tau_m = J/B$ is known as the *mechanical time constant* of the motor. Equations (14-56), (14-60), and (14-62) can be used to draw the open-loop block diagram as shown in Fig. 14-27. Two possible disturbances are control voltage, V_r, and load torque, T_L. The steady-state responses can be determined by combining the individual response due to V_r and T_L.

The response due to a step change in the reference voltage is obtained by setting T_L to zero. From Fig. 14-27, we can obtain the speed response due to reference voltage as

$$\frac{\omega(s)}{V_r(s)} = \frac{K_2 K_v I_f/(R_m B)}{s^2(\tau_a \tau_m) + s(\tau_a + \tau_m) + 1 + (K_v I_f)^2/R_m B} \tag{14-63}$$

The response due to a change in load torque, T_L, can be obtained by setting V_r to zero. The block diagram for a step change in load torque disturbance is shown in Fig. 14-28.

$$\frac{\omega(s)}{T_L(s)} = -\frac{(1/B)(s\tau_a + 1)}{s^2(\tau_a \tau_m) + s(\tau_a + \tau_m) + 1 + (K_v I_f)^2/R_m B} \tag{14-64}$$

Using the final value theorem, the steady-state relationship of a change in speed, $\Delta\omega$, due to a step change in control voltage, ΔV_r, and a step change in load torque, ΔT_L, can be found from Eqs. (14-63) and (14-64), respectively, by substituting $s = 0$.

$$\Delta\omega = \frac{K_2 K_v I_f}{R_m B + (K_v I_f)^2} \Delta V_r \tag{14-65}$$

$$\Delta\omega = -\frac{R_m}{R_m B + (K_v I_f)^2} \Delta T_L \tag{14-66}$$

The dc series motors are used extensively in traction applications where the steady-state speed is determined by the friction and gradient forces. By adjusting

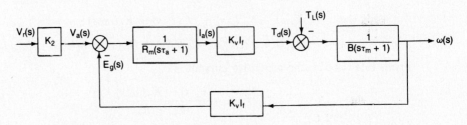

Figure 14-27 Open-loop block diagram of separately excited dc motor drive.

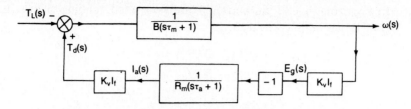

Figure 14-28 Open-loop block diagram for torque disturbance input.

the armature voltage, the motor may be operated at a constant torque (or current) up to the base speed, which corresponds to the maximum armature voltage. A chopper-controlled dc series motor drive is shown in Fig. 14-29.

The armature voltage is related to the control (or reference) voltage by a linear gain of the chopper, K_2. Assuming that the back emf constant K_v does not change with the armature current and remains constant, the system equations are

$$v_a = K_2 v_r \tag{14-67}$$

$$e_g = K_v i_a \omega \tag{14-68}$$

$$v_a = R_m i_a + L_m \frac{di_a}{dt} + e_g \tag{14-69}$$

$$T_d = K_t i_a^2 \tag{14-70}$$

$$T_d = J \frac{d\omega}{dt} + B\omega + T_L \tag{14-71}$$

Equation (14-70) contains a product of variable-type nonlinearities, and as a result the application of transfer function techniques would no longer be valid. However, these equations can be linearized by considering a small perturbation at the operating point. Let us define the system parameters around the operating point as

$$e_g = E_{g0} + \Delta e_g \qquad i_a = I_{a0} + \Delta i_a \qquad v_a = V_{a0} + \Delta v_a \qquad T_d = T_{d0} + \Delta T_d$$
$$\omega = \omega_0 + \Delta \omega \qquad v_r = V_{r0} + \Delta v_r \qquad T_L = T_{L0} + \Delta T_L$$

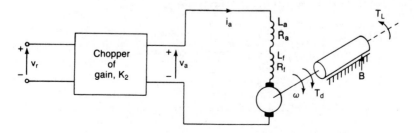

Figure 14-29 Chopper-fed dc series motor drive.

Recognizing that $\Delta i_a \, \Delta\omega$ and $(\Delta i_a)^2$ are very small, tending to zero, Eqs. (14-67) to (14-71) can be linearized to

$$\Delta v_a = K_2 \, \Delta v_r$$

$$\Delta e_g = K_v(I_{a0} \, \Delta\omega + \omega_0 \, \Delta i_a)$$

$$\Delta v_a = R_m \, \Delta i_a + L_m \frac{d(\Delta i_a)}{dt} + \Delta e_g$$

$$\Delta T_d = 2K_v I_{a0} \, \Delta i_a$$

$$\Delta T_d = J \frac{d(\Delta\omega)}{dt} + B \, \Delta\omega + \Delta T_L$$

Transforming these equations in Laplace's domain gives us

$$\Delta V_a(s) = K_2 \Delta V_r(s) \tag{14-72}$$

$$\Delta E_g(s) = K_v[I_{a0} \, \Delta\omega(s) + \omega_0 \, \Delta I_a(s)] \tag{14-73}$$

$$\Delta V_a(s) = R_m \, \Delta I_a(s) + sL_m \, \Delta I_a(s) + \Delta E_g(s) \tag{14-74}$$

$$\Delta T_d(s) = 2K_v I_{a0} \, \Delta I_a(s) \tag{14-75}$$

$$\Delta T_d(s) = sJ \, \Delta\omega(s) + B \, \Delta\omega(s) + \Delta T_L(s) \tag{14-76}$$

These five equations are sufficient to establish the block diagram of a dc series motor drive as shown in Fig. 14-30. It is evident from Fig. 14-30 that any change in either reference voltage or load torque will result in a change of speed. The block diagram for a change in reference voltage is shown in Fig. 14-31a and that for a change in load torque is shown in Fig. 14-31b.

14-7.2 Closed-Loop Transfer Function

Once the models for motors are known, feedback paths can be added to obtain the desired output response. To change the open-loop arrangement in Fig. 14-26 into a closed-loop system, a speed sensor is connected to the output shaft. The output of the sensor, which is proportional to the speed, is amplified by a factor of K_1 and is compared with the reference voltage V_r to form the error voltage V_e. The complete block diagram is shown in Fig. 14-32.

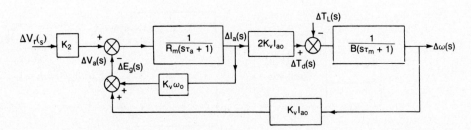

Figure 14-30 Open-loop block diagram of chopper-fed dc series drive.

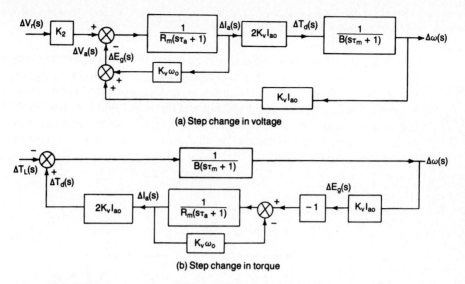

(a) Step change in voltage

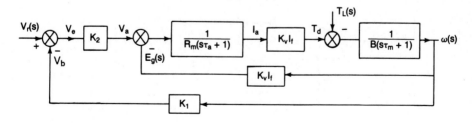

(b) Step change in torque

Figure 14-31 Block diagram for reference voltage and load torque distributions.

The closed-loop step response due to a change in reference voltage can be found from Fig. 14-28 with $T_L = 0$. The transfer function becomes

$$\frac{\omega(s)}{V_r(s)} = \frac{K_2 K_v I_f / (R_m B)}{s^2(\tau_a \tau_m) + s(\tau_a + \tau_m) + 1 + [(K_v I_f)^2 + K_1 K_2 K_v I_f]/R_m B} \quad (14\text{-}77)$$

The response due to a change in the load torque T_L can also be obtained from Fig. 14-32 by setting V_r to zero. The transfer function becomes

$$\frac{\omega(s)}{T_L(s)} = -\frac{(1/B)(s\tau_a + 1)}{s^2(\tau_a \tau_m) + s(\tau_a + \tau_m) + 1 + [(K_v I_f)^2 + K_1 K_2 K_v I_f]/R_m B} \quad (14\text{-}78)$$

Using the final value theorem, the steady-state change in speed, $\Delta\omega$, due to a step change in control voltage, ΔV_r, and a step change in load torque, ΔT_L, can be found from Eqs. (14-77) and (14-78), respectively, by substituting $s = 0$.

$$\Delta\omega = \frac{K_2 K_v I_f}{R_m B + (K_v I_f)^2 + K_1 K_2 K_v I_f} \Delta V_r \quad (14\text{-}79)$$

Figure 14-32 Block diagram for closed-loop control of separately excited dc motor.

$$\Delta\omega = -\frac{R_m}{R_mB + (K_vI_f)^2 + K_1K_2K_vI_f}\Delta T_L \qquad (14\text{-}80)$$

Figure 14-32 uses a speed feedback only. In practice, the motor is required to operate at a desired speed, but it has to meet the load torque, which depends on the armature current. While the motor is operating at a particular speed, if a load is applied suddenly, the speed will fall and the motor will take time to come up to the desired speed. A speed feedback with an inner current loop as shown in Fig. 14-33 provides faster response to any disturbances in speed command, load torque, and supply voltage.

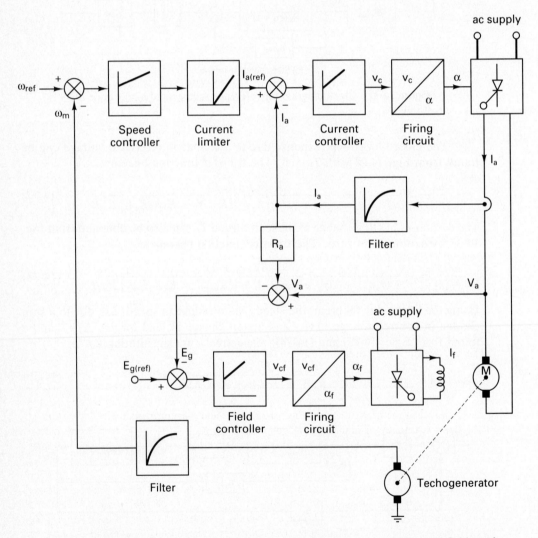

Figure 14-33 Closed-loop speed control with inner current loop and field weakening.

The current loop is used to cope with a sudden torque demand under transient condition. The output of the speed controller e_c is applied to the current limiter, which sets the current reference $I_{a(ref)}$ for the current loop. The armature current I_a is sensed by a current sensor, filtered normally by an active filter to remove ripple, and compared with the current reference $I_{a(ref)}$. The error current is processed through a current controller whose output v_c adjusts the firing angle of the converter and brings the motor speed to the desired value.

Any positive speed error caused by an increase in either speed command or load torque demand will produce a high reference current $I_{a(ref)}$. The motor will accelerate to correct the speed error, and finally settles at a new $I_{a(ref)}$, which makes the motor torque equal to the load torque, resulting in a speed error close to zero. For any large positive speed error, the current limiter saturates and limits the reference current $I_{a(ref)}$ to a maximum value $I_{a(max)}$. The speed error is then corrected at the maximum permissible armature current $I_{a(max)}$ until the speed error becomes small and the current limiter comes out of saturation. Normally, the speed error is corrected with I_a less than the permissible value $I_{a(max)}$.

The speed control from zero to base speed is normally done at the maximum field by armature voltage control, and control above the base speed should be done by field weakening at the rated armature voltage. In the field control loop, the back emf E_g ($= V_a - R_a I_a$) is compared with a reference voltage $E_{g(ref)}$, which is generally between 0.85 to 0.95 of the rated armature voltage. For speeds below the base speed, the field error e_f is large and the field controller saturates, thereby applying the maximum field voltage and current.

When the speed is close to the base speed, V_a is almost near to the rated value and the field controller comes out of saturation. For a speed command above the base speed, the speed error causes a higher value of V_a. The motor accelerates, the back emf E_g increases, and the field error e_f decreases. The field current then decreases and the motor speed continues to increase until the motor speed reaches the desired speed. Thus the speed control above the base speed is obtained by the field weakening while the armature terminal voltage is maintained at near the rated value. In the field weakening mode, the drive responds very slowly due to the large field time constant. A full converter is normally used in the field, because it has the ability to reverse the voltage, thereby reducing the field current much faster that a semi-converter.

Example 14-12

A 50-kW 240-V 1700-rpm separately excited dc motor is controlled by a converter as shown in the block diagram Fig. 14-32. The field current is maintained constant at $I_f = 1.4$ A and the machine back emf constant is $K_v = 0.91$ V/A-rad/s. The armature resistance is $R_m = 0.1$ Ω and the viscous friction constant is $B = 0.3$ N·m/rad/s. The amplification of the speed sensor is $K_1 = 95$ mV/rad/s and the gain of the power controller is $K_2 = 100$. (a) Determine the rated torque of the motor. (b) Determine the reference voltage V_r to drive the motor at the rated speed. (c) If the reference voltage is kept unchanged, determine the speed at which the motor develops the rated torque. (d) If the load torque is increased by 10% of the rated value, determine the motor speed. (e) If the reference voltage is reduced by 10%, determine the motor speed. (f) If the load torque is increased by 10% of the rated value and the reference voltage is reduced by 10%, determine the motor speed. (g) If there was no feedback in an open-loop control, determine the speed regulation for a reference voltage of $V_r = 2.31$ V. (h) Determine the speed regulation with a closed-loop control.

Solution $I_f = 1.4$ A, $K_v = 0.91$ V/A-rad/s, $K_1 = 95$ mV/rad/s, $K_2 = 100$, $R_m = 0.1$ Ω, $B = 0.3$ N·m/rad/s, and $\omega_{\text{rated}} = 1700\ \pi/30 = 178.02$ rad/s.

(a) The rated torque, $T_L = 50,000/178.02 = 280.87$ N·m.

(b) Since $V_a = K_2 V_r$, for open-loop control Eq. (14-65) gives

$$\frac{\omega}{V_a} = \frac{\omega}{K_2 V_r} = \frac{K_v I_f}{R_m B + (K_v I_f)^2} = \frac{0.91 \times 1.4}{0.1 \times 0.3 + (0.91 \times 1.4)^2} = 0.7707$$

At rated speed,

$$V_a = \frac{\omega}{0.7707} = \frac{178.02}{0.7707} = 230.98\ \text{V}$$

and feedback voltage,

$$V_b = K_1 \omega = 95 \times 10^{-3} \times 178.02 = 16.912\ \text{V}$$

With closed-loop control, $(V_r - V_b)K_2 = V_a$ or $(V_r - 16.912) \times 100 = 230.98$, which gives the reference voltage, $V_r = 19.222$ V.

(c) For $V_r = 19.222$ V and $\Delta T_L = 280.87$ N·m, Eq. (14-80) gives

$$\Delta\omega = -\frac{0.1 \times 280.86}{0.1 \times 0.3 + (0.91 \times 1.4)^2 + 95 \times 10^{-3} \times 100 \times 0.91 \times 1.4}$$
$$= -2.04\ \text{rad/s}$$

The speed at rated torque,

$$\omega = 178.02 - 2.04 = 175.98\ \text{rad/s} \quad \text{or} \quad 1680.5\ \text{rpm}$$

(d) $\Delta T_L = 1.1 \times 280.87 = 308.96$ N·m and Eq. (14-80) gives

$$\Delta\omega = -\frac{0.1 \times 308.96}{0.1 \times 0.3 + (0.91 \times 1.4)^2 + 95 \times 10^{-3} \times 100 \times 0.91 \times 1.4}$$

$$= -2.246\ \text{rad/s}$$

The motor speed

$$\omega = 178.02 - 2.246 = 175.774\ \text{rad/s} \quad \text{or} \quad 1678.5\ \text{rpm}$$

(e) $\Delta V_r = -0.1 \times 19.222 = -1.9222$ V and Eq. (14-79) gives the change in speed,

$$\Delta\omega = -\frac{100 \times 0.91 \times 1.4 \times 1.9222}{0.1 \times 0.3 + (0.91 \times 1.4)^2 + 95 \times 10^{-3} \times 100 \times 0.91 \times 1.4}$$

$$= -17.8\ \text{rad/s}$$

The motor speed is

$$\omega = 178.02 - 17.8 = 160.22\ \text{rad/s} \quad \text{or} \quad 1530\ \text{rpm}$$

(f) The motor speed can be obtained by using superposition:

$$\omega = 178.02 - 2.246 - 17.8 = 158\ \text{rad/s} \quad \text{or} \quad 1508.5\ \text{rpm}$$

(g) $\Delta V_r = 2.31$ V and Eq. (14-65) gives

$$\Delta\omega = \frac{100 \times 0.91 \times 1.4 \times 2.31}{0.1 \times 0.3 + (0.91 \times 1.4)^2} = 178.02 \text{ rad/s} \quad \text{or} \quad 1700 \text{ rpm}$$

and the no-load speed is $\omega = 178.02$ rad/s or 1700 rpm. For full load, $\Delta T_L = 280.87$ N·m, Eq. (14-66) gives

$$\Delta\omega = -\frac{0.1 \times 280.87}{0.1 \times 0.3 + (0.91 \times 1.4)^2} = -16.99 \text{ rad/s}$$

and the full-load speed

$$\omega = 178.02 - 16.99 = 161.03 \text{ rad/s} \quad \text{or} \quad 1537.7 \text{ rpm}$$

The speed regulation with open-loop control is

$$\frac{1700 - 1537.7}{1537.7} = 10.55\%$$

(h) Using the speed from part (c), the speed regulation with closed-loop control is

$$\frac{1700 - 1680.5}{1680.5} = 1.16\%$$

Note. By closed-loop control, the speed regulation is reduced by a factor of approximately 10, from 10.55% to 1.16%.

14-7.3 Phase-Locked-Loop Control

For precise speed control of servo systems, closed-loop control is normally used. The speed, which is sensed by analog sensing devices (e.g., tachometer), is compared with the reference speed to generate the error signal and to vary the armature voltage of the motor. These analog devices for speed sensing and comparing signals are not ideal and the speed regulation is more than 0.2%. The speed regulator can be improved if digital phase-locked-loop (PLL) control is used. The block diagram of a converter-fed dc motor drive with phase-locked-loop control is shown in Fig. 14-34a and the transfer function block diagram is shown in Fig. 14-29b.

In a phase-locked loop control system, the motor speed is converted to a digital pulse train by using a speed encoder. The output of the encoder acts as the speed feedback signal of frequency f_0. The phase detector compares the reference pulse train (or frequency) f_r with the feedback frequency f_0 and provides a pulse-width-modulated output voltage V_e which is proportional to the difference in phases and frequencies of the reference and feedback pulse trains. The phase detector (or comparator) is available in integrated circuits. A low-pass loop filter converts the pulse train V_e to a continuous dc level V_c which varies the output of the power converter and in turn the motor speed.

When the motor runs at the same speed as the reference pulse train, the two frequencies would be synchronized (or locked) together with a phase difference. The output of the phase detector would be a constant voltage proportional to the phase difference and the steady-state motor speed would be maintained at a fixed

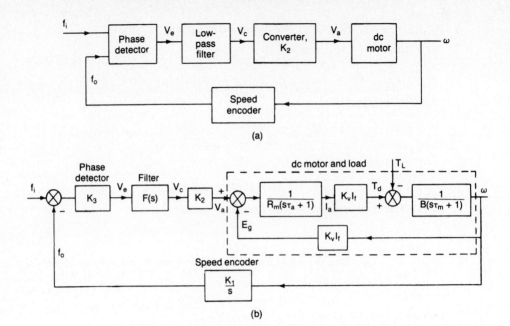

Figure 14-34 Phase-locked-loop control system.

value irrespective of the load on the motor. Any disturbances contributing to the speed change would result in a phase difference and the output of the phase detector would respond immediately to vary the speed of the motor in such a direction and magnitude as to retain the locking of the reference and feedback frequencies. The response of the phase detector is very fast. As long as the two frequencies are locked, the speed regulation should ideally be zero. However, in practice the speed regulation is limited to 0.002%, and this represents a significant improvement over the analog speed control system.

14-7.4 Microcomputer Control of DC Drives

The analog control scheme for a converter-fed dc motor drive can be implemented by hardwired electronics. An analog control scheme has several disadvantages: nonlinearity of speed sensor, temperature dependency, drift, and offset. Once a control circuit is built to meet certain performance criteria, it may require major changes in the hardwired logic circuits to meet other performance requirements.

A microcomputer control reduces the size and costs of hardwired electronics, improving reliability and control performance. This control scheme is implemented in the software and is flexible to change the control strategy to meet different performance characteristics or to add extra control features. A microcomputer control system can also perform various desirable functions: on/off of the main power supply, start/stop of the drive, speed control, current control, monitoring the control variables, initiating protection and trip circuit, diagnostics for built-in fault finding, and communication with a supervisory central computer. Figure 14-35 shows a schematic diagram for a microcomputer control of a converter-fed four-quadrant dc drive.

The speed signal is fed into the microcomputer using an A/D (analog-to-digital) converter. To limit the armature current of the motor, an inner current-control loop is used. The armature current signal can be fed into the microcomputer through an A/D converter or by sampling the armature current. The line synchronizing circuit is required to synchronize the generation of the firing pulses with the supply line frequency. Although the microcomputer can perform the functions of gate pulse generator and logic circuit, these are shown outside the microcomputer. The pulse amplifier provides the necessary isolation and produces gate pulses of required magnitude and duration. A microprocessor-controlled drive has become a norm. Analog control has become almost obsolete.

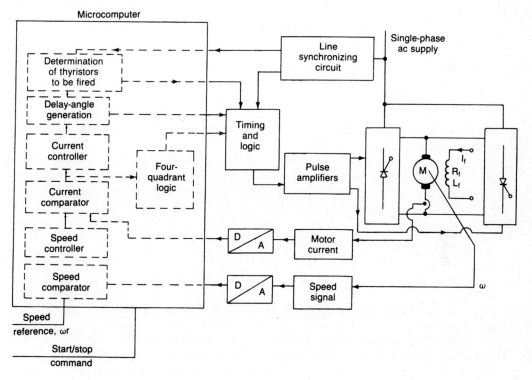

Figure 14-35 Schematic diagram of computer-controlled four-quadrant dc drive.

SUMMARY

In dc drives, the armature and field voltages of dc motors are varied by either ac–dc converters or dc choppers. The ac–dc converter-fed drives are normally used in variable-speed applications, whereas chopper-fed drives are more suited for traction applications. Dc series motors are mostly used in traction applications, due to their capability of high starting torque.

Dc drives can be classified broadly into three types depending on the input supply: (1) single-phase drives, (2) three-phase drives, and (3) chopper drives. Again each drive could be subdivided into three types depending on the modes of operation: (a) one-quadrant drives, (b) two-quadrant drives, and (c) four-quadrant drives. The energy-saving feature of chopper-fed drives is very attractive for use in transportation systems requiring frequent stops.

Closed-loop control, which has many advantages, is normally used in industrial drives. The speed regulation of dc drives can be significantly improved by using phase-locked-loop (PLL) control. The analog control schemes, which are hardwired electronics, are limited in flexibility and have certain disadvantages, whereas microcomputer control drives, which are implemented in software, are more flexible and can perform many desirable functions.

REFERENCES

1. J. F. Lindsay and M. H. Rashid, *Electromechanics and Electrical Machinery*. Englewood Cliffs, N.J.: Prentice Hall, 1986.

2. P. C. Sen, *Thyristor DC Drives*. New York: John Wiley & Sons, Inc., 1981.

3. P. C. Sen and M. L. McDonald, "Thyristorized dc drives with regenerative braking and speed reversal." *IEEE Transactions on Industrial Electronics and Control Instrumentation,* Vol. IECI25, No. 4, 1978, pp. 347–354.

4. M. H. Rashid, "Dynamic responses of dc chopper controlled series motor." *IEEE Transactions on Industrial Electronics and Control Instrumentation,* Vol. IECI28, No. 4, 1981, pp. 323–340.

5. M. H. Rashid, "Regenerative characteristics of dc chopper controlled series motor." *IEEE Transactions on Vehicular Technology,* Vol. VT33, No. 1, 1984, pp. 3–13.

6. E. Reimers, "Design analysis of multiphase dc chopper motor drive." *IEEE Transactions on Industry Applications,* Vol. IA8, No. 2, 1972, pp. 136–144.

7. M. H. Rashid, "Design of *LC* input filter for multiphase dc choppers." *Proceedings IEE,* Vol. B130, No. 1, 1983, pp. 310–44.

8. D. F. Geiger, *Phaselock Loops for DC Motor Speed Control.* New York: John Wiley & Sons, Inc., 1981.

9. S. K. Tso and P. T. Ho, "Dedicated microprocessor scheme for thyristor phase control of multiphase converters." *Proceedings IEE,* Vol. B22, 1981, pp. 101–108.

10. J. Best and P. Mutschler, "Control of armature and field current of a chopper-fed dc motor drive by a single chip microcomputer." *3rd IFAC Symposium on Control in Power Electronics and Electrical Drives,* Lausanne, Switzerland, 1983, pp. 515–522.

11. G. K. Dubey, *Power Semiconductor Controlled Drives.* Englewood Cliffs, N.J.: Prentice Hall, 1989.

12. P. C. Sen, "Electric motor drives and control: past, present and future." *IEEE Transactions on Industrial Electronics,* Vol. IE37, No. 6, 1990, pp. 562–575.

REVIEW QUESTIONS

14-1. What are the three types of dc drives based on the input supply?

14-2. What is the magnetization characteristic of dc motors?

14-3. What is the purpose of a converter in dc drives?

14-4. What is a base speed of dc motors?

14-5. What are the parameters to be varied for speed control of separately excited dc motors?

14-6. Which are the parameters to be varied for speed control of dc series motors?

14-7. Why are the dc series motors mostly used in traction applications?

14-8. What is a speed regulation of dc drives?

14-9. What is the principle of single-phase full-converter-fed dc motor drives?

14-10. What is the principle of three-phase semiconverter-fed dc motor drives?

14-11. What are the advantages and disadvantages of single-phase full-converter-fed dc motor drives?

14-12. What are the advantages and disadvantages of single-phase semiconverter-fed dc motor drives?

14-13. What are the advantages and disadvantages of three-phase full-converter-fed dc motor drives?

14-14. What are the advantages and disadvantages of three-phase semiconverter-fed dc motor drives?

14-15. What are the advantages and disadvantages of three-phase dual-converter-fed dc motor drives?

14-16. Why is it preferable to use a full con-verter for field control of separately excited motors?

14-17. What is a one-quadrant dc drive?

14-18. What is a two-quadrant dc drive?

14-19. What is a four-quadrant dc drive?

14-20. What is the principle of regenerative braking of dc chopper-fed dc motor drives?

14-21. What is the principle of rheostatic braking of dc chopper-fed dc motor drives?

14-22. What are the advantages of chopper-fed dc drives?

14-23. What are the advantages of multiphase choppers?

14-24. What is the principle of closed-loop control of dc drives?

14-25. What are the advantages of closed-loop control of dc drives?

14-26. What is the principle of phase-locked-loop control of dc drives?

14-27. What are the advantages of phase-locked-loop control of dc drives?

14-28. What is the principle of microcomputer control of dc drives?

14-29. What are the advantages of microcomputer control of dc drives?

14-30. What is a mechanical time constant of dc motors?

14-31. What is an electrical time constant of dc motors?

PROBLEMS

14-1. A separately excited dc motor is supplied from a dc source of 600 V to control the speed of a mechanical load and the field current is maintained constant. The armature resistance and losses are negligible. **(a)** If the load torque is $T_L = 550$ N·m at 1500 rpm, determine the armature current I_a. **(b)** If the armature current remains the same as that in part (a) and the field current is reduced such that the motor runs at a speed of 2800, determine the load torque.

14-2. Repeat Prob. 14-1 if the armature resistance is $R_a = 0.12 \ \Omega$. The viscous friction and the no-load losses are negligible.

14-3. A 30-hp 440-V 2000-rpm separately excited motor dc controls a load requiring a torque of $T_L = 85$ N·m at 1200 rpm. The field circuit resistance is $R_f = 294 \ \Omega$, the armature circuit resistance is $R_a = 0.12 \ \Omega$, and the motor voltage constant is $K_v = 0.7032$ V/A-rad/s. The field voltage is $V_f = 440$ V. The viscous friction and the no-load losses are negligible. The armature current may be assumed continuous and ripple-free. Determine **(a)** the back emf E_g, **(b)** the required armature voltage V_a, **(c)** the rated armature current of the motor, and **(d)** the speed regulation at full load.

14-4. A 120-hp 600-V 1200-rpm dc series motor controls a load requiring a torque of $T_L = 185$ N·m at 1100 rpm. The field circuit resistance $R_f = 0.06$ Ω, the armature circuit resistance $R_a = 0.04$ Ω, and the voltage constant $K_v = 32$ mV/A-rad/s. The viscous friction and the no-load losses are negligible. The armature current is continuous and ripple-free. Determine **(a)** the back emf E_g, **(b)** the required armature voltage V_a, **(c)** the rated armature current, and **(d)** the speed regulation at full speed.

14-5. The speed of a separately excited motor is controlled by a single-phase semiconverter in Fig. 14-12a. The field current is also controlled by a semiconverter and the field current is set to the maximum possible value. The ac supply voltage to the armature and field converter is one-phase, 208 V, 60 Hz. The armature resistance is $R_a = 0.12$ Ω, the field resistance is $R_f = 220$ Ω, and the motor voltage constant is $K_v = 1.055$ V/A-rad/s. The load torque is $T_L = 75$ N·m at a speed of 700 rpm. The viscous friction and no-load losses are negligible. The armature and field currents are continuous and ripple-free. Determine **(a)** the field current I_f; **(b)** the delay angle of the converter in the armature circuit α_a; and **(c)** the input power factor of the armature circuit.

14-6. The speed of a separately excited dc motor is controlled by a single-phase full-wave converter in Fig. 14-13a. The field circuit is also controlled by a full converter and the field current is set to the maximum possible value. The ac supply voltage to the armature and field converters is one-phase, 208 V, 60 Hz. The armature resistance is $R_a = 0.50$ Ω, the field circuit resistance is $R_f = 345$ Ω, and

the motor voltage constant is $K_v = 0.71$ V/A-rad/s. The viscous friction and no-load losses are negligible. The armature and field currents are continuous and ripple-free. If the delay angle of the armature converter is $\alpha_a = 45°$ and the armature current of the motor is $I_a = 55$ A, determine **(a)** the torque developed by the motor T_d, **(b)** the speed ω, and **(c)** the input power factor PF of the drive.

14-7. If the polarity of the motor back emf in Prob. 14-6 is reversed by reversing the polarity of the field current, determine **(a)** the delay angle of the armature circuit converter, α_a, to maintain the armature current constant at the same value of $I_a = 55$ A; and **(b)** the power fed back to the supply during regenerative braking of the motor.

14-8. The speed of a 20-hp 300-V 1800-rpm separately excited dc motor is controlled by a three-phase full-converter drive. The field current is also controlled by a three-phase full-converter and is set to the maximum possible value. The ac input is three-phase, wye-connected, 208 V, 60 Hz. The armature resistance, $R_a = 0.35$ Ω, the field resistance, $R_f = 250$ Ω, and the motor voltage constant, $K_v = 1.15$ V/A-rad/s. The armature and field currents are continuous and ripple-free. The viscous friction and no-load losses are negligible. Determine **(a)** the delay angle of the armature converter, α_a, if the motor supplies the rated power at the rated speed; **(b)** the no-load speed if the delay angles are the same as in part (a) and the armature current at no-load is 10% of the rated value; and **(c)** the speed regulation.

14-9. Repeat Prob. 14-8 if both armature and field circuits are controlled by three-phase semiconverters.

14-10. The speed of a 20-hp 300-V 900-rpm separately excited dc motor is controlled by a three-phase full converter. The field circuit is also controlled by a three-phase full converter. The ac input to armature and field converters is three-phase, wye-connected, 208 V, 60 Hz. The armature resistance $R_a = 0.15$ Ω, the field circuit resistance $R_f = 145$ Ω, and the motor voltage constant $K_v = 1.15$ V/A-rad/s. The viscous friction and no-load losses are negligible. The armature and field currents are continuous and ripple-free. **(a)** If the field converter is operated at the maximum field current and the developed torque is $T_d = 106$ N·m at 750 rpm, determine the delay angle of the armature converter, α_a. **(b)** If the field circuit converter is set to the maximum field current, the developed torque is $T_d = 108$ N·m, and the delay angle of the armature converter is $\alpha_a = 0$, determine the speed. **(c)** For the same load demand as in part (b), determine the delay angle of the field circuit converter if the speed has to be increased to 1800 rpm.

14-11. Repeat Prob. 14-10 if both the armature and field circuits are controlled by three-phase semiconverters.

14-12. A dc chopper controls the speed of a dc series motor. The armature resistance $R_a = 0.04$ Ω, field circuit resistance $R_f = 0.06$ Ω, and back emf constant $K_v = 35$ mV/rad/s. The dc input voltage of the chopper $V_s = 600$ V. If it is required to maintain a constant developed torque of $T_d = 547$ N·m, plot the motor speed against the duty cycle k of the chopper.

14-13. A dc chopper controls the speed of a separately excited motor. The armature resistance is $R_a = 0.05$ Ω. The back emf constant is $K_v = 1.527$ V/A-rad/s. The rated field current is $I_f = 2.5$ A. The dc input voltage to the chopper is $V_s = 600$ V. If it is required to maintain a constant developed torque of $T_d = 547$ N·m, plot the motor speed against the duty cycle k of the chopper.

14-14. A dc series motor is powered by a dc chopper as shown in Fig. 14-18a from a 600-V dc source. The armature resistance is $R_a = 0.03$ Ω and the field resistance is $R_f = 0.05$ Ω. The back emf constant of the motor is $K_v = 15.27$ mV/A-rad/s. The average armature current $I_a = 450$ A. The armature current is continuous and has negligible ripple. If

the duty cycle of the chopper is 75%, determine **(a)** the input power from the source, **(b)** the equivalent input resistance of the chopper drive, **(c)** the motor speed, and **(d)** the developed torque of the motor.

14-15. The drive in Fig. 14-19a is operated in regenerative braking of a dc series motor. The dc supply voltage is 600 V. The armature resistance $R_a = 0.03$ Ω and the field resistance, $R_f = 0.05$ Ω. The back emf constant of the motor $K_v = 12$ mV/A-rad/s. The average armature current is maintained constant at $I_a = 350$ A. The armature current is continuous and has negligible ripple. If the duty cycle of the chopper is 50%, determine **(a)** the average voltage across the chopper, V_h; **(b)** the power regenerated to the dc supply, P_g; **(c)** the equivalent load resistance of the motor acting as a generator, R_{eq}; **(d)** the minimum permissible braking speed, ω_{min}; **(e)** the maximum permissible braking speed, ω_{max}; and **(f)** the motor speed.

14-16. A dc chopper is used in rheostatic braking of a dc series motor as shown in Fig. 14-20. The armature resistance $R_a = 0.03$ Ω and the field resistance $R_f = 0.05$ Ω. The braking resistor $R_b = 5$ Ω. The back emf constant $K_v = 14$ mV/A-rad/s. The average armature current is maintained constant at $I_a = 250$ A. The armature current is continuous and has negligible ripple. If the duty cycle of the chopper is 60%, determine **(a)** the average voltage across the chopper, V_{ch}; **(b)** the power dissipated in the resistor, P_b; **(c)** the equivalent load resistance of the motor acting as a generator, R_{eq}; **(d)** the motor speed; and **(e)** the peak chopper voltage, V_p.

14-17. Two choppers control a dc motor as shown in Fig. 14-24a and they are phase shifted in operation by π/m, where m is the number of multiphase choppers. The supply voltage $V_s = 440$ V, total armature circuit resistance $R_m = 8$ Ω, armature circuit inductance $L_m = 12$ mH, and the frequency of each chopper $f = 250$

Hz. Calculate the maximum value of peak-to-peak load ripple current.

14-18. For Prob. 14-17 plot the maximum value of peak-to-peak load ripple current against the number of multiphase choppers.

14-19. A dc motor is controlled by two multiphase choppers. The average armature current, $I_a = 250$ A. A simple LC-input filter with $L_e = 0.35$ mH and $C_e = 5600$ μF is used. Each chopper is operated at a frequency $f = 250$ Hz. Determine the rms fundamental component of the chopper-generated harmonic current in the supply.

14-20. For Prob. 14-19, plot the rms fundamental component of the chopper-generated harmonic current in the supply against the number of multiphase choppers.

14-21. A 40-hp 230-V 3500-rpm separately excited dc motor is controlled by a linear converter of gain $K_2 = 200$. The moment of inertia of the motor load, $J = 0.156$ N·m/rad/s, viscous friction constant is negligible, total armature resistance $R_m = 0.045$ Ω, and total armature inductance $L_m = 730$ mH. The back emf constant is $K_v = 0.502$ V/A-rad/s and the field current is maintained constant at $I_f = 1.25$ A. (a) Obtain the open-loop transfer function $\omega(s)/V_r(s)$ and $\omega(s)/T_L(s)$ for the motor. (b) Calculate the motor steady-state speed if the reference voltage is $V_r = 1$ V and the load torque is 60% of the rated value.

14-22. Repeat Prob. 14-21 with a closed-loop control if the amplification of speed sensor is $K_1 = 3$ mV/rad/s.

14-23. The motor in Prob. 14-21 is controlled by a linear converter of gain K_2 with a closed-loop control. If the amplification of speed sensor is $K_1 = 3$ mV/rad/s, determine the gain of the converter K_2 to limit the speed regulation at full load to 1%.

14-24. A 60-hp 230-V 1750-rpm separately excited dc motor is controlled by a con-

verter as shown in the block diagram in Fig. 14-32. The field current is maintained constant at $I_f = 1.25$ A and the machine back emf constant, $K_v = 0.81$ V/A-rad/s. The armature resistance, $R_a = 0.02$ Ω and the viscous friction constant, $B = 0.3$ N·m/rad/s. The amplification of the speed sensor $K_1 = 96$ mV/rad/s and the gain of the power controller, $K_2 = 150$. (a) Determine the rated torque of the motor. (b) Determine the reference voltage V_r to drive the motor at the rated speed. (c) If the reference voltage is kept unchanged, determine the speed at which the motor develops the rated torque.

14-25. Repeat Prob. 14-24. (a) If the load torque is increased by 20% of the rated value, determine the motor speed. (b) If the reference voltage is reduced by 10%, determine the motor speed. (c) If the load torque is reduced by 15% of the rated value and the reference voltage is reduced by 20%, determine the motor speed. (d) If there was no feedback, as in an open-loop control, determine the speed regulation for a reference voltage $V_r = 1.24$ V. (e) Determine the speed regulation with a closed-loop control.

14-26. A 40-hp 230-V 3500-rpm series excited dc motor is controlled by a linear converter of gain $K_2 = 200$. The moment of inertia of the motor load $J = 0.156$ N·m/rad/s, viscous friction constant is negligible, total armature resistance $R_m = 0.045$ Ω, and total armature inductance, $L_m = 730$ mH. The back emf constant is $K_v = 340$ mV/A-rad/s. The field resistance $R_f = 0.035$ Ω and field inductance $L_f = 450$ mH. (a) Obtain the open-loop transfer function $\omega(s)/V_r(s)$ and $\omega(s)/T_L(s)$ for the motor. (b) Calculate the motor steady-state speed if the reference voltage, $V_r = 1$ V and the load torque is 60% of the rated value.

14-27. Repeat Prob. 14-26 with closed-loop control if the amplification of speed sensor $K_1 = 3$ mV/rad/s.

Ac drives

15-1 INTRODUCTION

The control of dc motors requires providing a variable dc voltage which can be obtained from dc choppers or controlled rectifiers. These voltage controllers are simple and less expensive. Dc motors are relatively expensive and require more maintenance, due to the brushes and commutators. However, dc drives are used in many industrial applications. Ac motors exhibit highly coupled, nonlinear, and multivariable structures as opposed to much simpler decoupled structures of separately excited dc motors. The control of ac drives generally requires complex control algorithms that can be performed by microprocessors and/or microcomputers along with fast switching power converters.

The ac motors have a number of advantages; they are lightweight (20 to 40% lighter than equivalent dc motors), inexpensive, and of low maintenance compared to dc motors. They require control of frequency, voltage, and current for variable-speed applications. The power converters, inverters and ac voltage controllers, can control the frequency, voltage, and/or current to meet the drive requirements. These power controllers, which are relatively complex and more expensive, require advanced feedback control techniques such as model reference, adaptive control, sliding mode control, and field-oriented control. However, the advantages of ac drives outweigh the disadvantages. There are two types of ac drives:

1. Induction motor drives
2. Synchronous motor drives

Three-phase induction motors are commonly used in adjustable-speed drives and they have three-phase stator and rotor windings. The stator windings are supplied with balanced three-phase ac voltages, which produce induced voltages in the rotor windings due to transformer action. It is possible to arrange the distribution of stator windings so that there is an effect of multiple poles, producing several cycles of magnetomotive force (mmf) (or field) around the air gap. This field establishes a spatially distributed sinusoidal flux density in the air gap. The speed of rotation of the field is called the *synchronous speed*, which is defined by

$$\omega_s = \frac{2\omega}{p} \tag{15-1}$$

where p is the number of poles and ω is the supply frequency in rad/s.

If a stator phase voltage, $v_s = \sqrt{2}\, V_s \sin \omega t$, produces a flux linkage (in the rotor) given by

$$\phi(t) = \phi_m \cos(\omega_m t + \delta - \omega_s t) \tag{15-2}$$

the induced voltage per phase in the rotor winding is

$$
\begin{aligned}
e_r &= N_r \frac{d\phi}{dt} = N_r \frac{d}{dt} [\phi_m \cos(\omega_m t + \delta - \omega_s t)] \\
&= -N_r \phi_m (\omega_s - \omega_m) \sin[(\omega_s - \omega_m)t - \delta] \\
&= -s E_m \sin(s\omega_s t - \delta) \\
&= -s\sqrt{2}\, E_r \sin(s\omega_s t - \delta)
\end{aligned}
\tag{15-3}
$$

where N_r = number of turns on each rotor phase
  ω_m = angular speed of the rotor
  δ = relative position of the rotor
  E_r = rms value of the rotor induced voltage per phase

and s is the slip, defined as

$$s = \frac{\omega_s - \omega_m}{\omega_s} \tag{15-4}$$

which gives the motor speed as $\omega_m = \omega_s(1 - s)$. The equivalent circuit for one phase of the rotor is shown in Fig. 15-1a, where R_r' is the resistance per phase of the rotor windings, X_r' is the leakage reactance per phase of the rotor at the supply frequency, and E_r represents the induced rms phase voltage when the speed is zero (or $s = 1$). The rotor current is given by

$$I_r' = \frac{sE_r}{R_r' + jsX_r'} \tag{15-5}$$

$$= \frac{E_r}{R'R/s + jX'_r} \tag{15-5a}$$

where R_r' and X_r' are referred to the rotor winding.

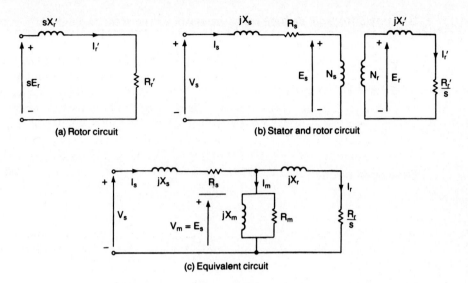

Figure 15-1 Circuit model of induction motors.

The per phase circuit model of induction motors is shown in Fig. 15-1b, where R_s and X_s are the per phase resistance and leakage reactance of the stator winding. The complete circuit model with all parameters referred to the stator is shown in Fig. 15-1c, where R_m represents the resistance for excitation (or core) loss and X_m is the magnetizing reactance. There will be stator core loss, when the supply is connected and the rotor core loss depends on the slip. The friction and windage loss, $P_{\text{no load}}$, exists when the machine rotates. The core loss, P_c, may be included as a part of rotational loss, $P_{\text{no load}}$.

15-2.1 Performance Characteristics

The rotor current, I_r, and stator current, I_s, can be found from the circuit model in Fig. 15-1c where R_r and X_r are referred to the stator windings. Once the values of I_r and I_s are known, the performance parameters of a three-phase motor can be determined as follows:

Stator copper loss

$$P_{su} = 3I_s^2 R_s \qquad (15\text{-}6)$$

Rotor copper loss

$$P_{ru} = 3I_r^2 R_r \qquad (15\text{-}7)$$

Core loss

$$P_c = \frac{3V_m^2}{R_m} \approx \frac{3V_s^2}{R_m} \qquad (15\text{-}8)$$

Gap power (power passing from the stator to the rotor through the air gap)

$$P_g = 3I_r^2 \frac{R_r}{s} \tag{15-9}$$

Developed power

$$P_d = P_g - P_{ru} = 3I_r^2 \frac{R_r}{s}(1 - s) \tag{15-10}$$

$$= P_g(1 - s) \tag{15-11}$$

Developed torque

$$T_d = \frac{P_d}{\omega_m} \tag{15-12}$$

$$= \frac{P_g(1 - s)}{\omega_s(1 - s)} = \frac{P_g}{\omega_s} \tag{15-12a}$$

Input power

$$P_i = 3V_s I_s \cos \theta_m \tag{15-13}$$

$$= P_c + P_{su} + P_g \tag{15-13a}$$

where θ_m is the angle between I_s and V_s. Output power

$$P_o = P_d - P_{\text{no load}}$$

Efficiency

$$\eta = \frac{P_o}{P_i} = \frac{P_d - P_{\text{no load}}}{P_c + P_{su} + P_g} \tag{15-14}$$

If $P_g \gg (P_c + P_{su})$ and $P_d \gg P_{\text{no load}}$, the efficiency becomes approximately

$$\eta \approx \frac{P_d}{P_g} = \frac{P_g(1 - s)}{P_g} = 1 - s \tag{15-14a}$$

The value of X_m is normally large and R_m, which is much larger, can be removed from the circuit model to simplify the calculations. If $X_m^2 \gg (R_s^2 + X_s^2)$, then $V_s \approx V_m$, and the magnetizing reactance X_m may be moved to the stator winding to simplify further; this is shown in Fig. 15-2.

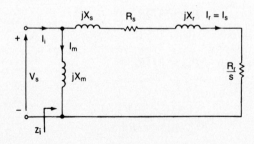

Figure 15-2 Approximate per-phase equivalent circuit.

The input impedance of the motor becomes

$$\mathbf{Z}_i = \frac{-X_m(X_s + X_r) + jX_m(R_s + R_r/s)}{R_s + R_r/s + j(X_m + X_s + X_r)} \tag{15-15}$$

and the power factor angle of the motor

$$\theta_m = \pi - \tan^{-1}\frac{R_s + R_r/s}{X_s + X_r} + \tan^{-1}\frac{X_m + X_s + X_r}{R_s + R_r/s} \tag{15-16}$$

From Fig. (15-2), the rms rotor current

$$I_r = \frac{V_s}{[(R_s + R_r/s)^2 + (X_s + X_r)^2]^{1/2}} \tag{15-17}$$

Substituting I_r from Eq. (15-17) in Eq. (15-9) and then P_g in Eq. (15-12a) yields

$$T_d = \frac{3R_r V_s^2}{s\omega_s[(R_s + R_r/s)^2 + (X_s + X_r)^2]} \tag{15-18}$$

If the motor is supplied from a fixed voltage at a constant frequency, the developed torque is a function of the slip and the torque–speed characteristics can be determined from Eq. (15-18). A typical plot of developed torque as a function of slip or speed is shown in Fig. 15-3. The operation in the reverse motoring and regenerative braking is obtained by the reversal of the phase sequence of the motor terminals. The reverse speed–torque characteristics are shown by dashed lines. There are three regions of operation: (1) motoring or powering, $0 \le s \le 1$; (2) regeneration, $s < 0$; and (3) plugging, $1 \le s \le 2$. In motoring, the motor rotates in the same direction as the field; and as the slip increases, the torque also increases while the air-gap flux remains constant. Once the torque reaches its maximum value, T_m at $s = s_m$, the torque decreases, with the increase in slip due to reduction of the air-gap flux.

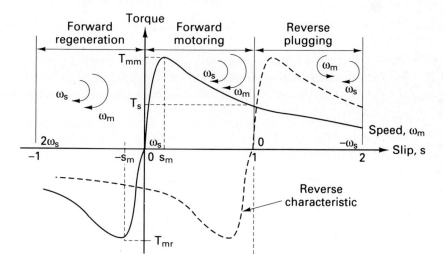

Figure 15-3 Torque–speed characteristics.

In regeneration, the speed ω_m is greater than the synchronous speed ω_s with ω_m and ω_s being in the same direction, and the slip is negative. Therefore, R_r/s is negative. This means that power is being fed back from the shaft into the rotor circuit and the motor operates as a generator. The motor returns power to the supply system. The torque–speed characteristic is similar to that of motoring, but having negative value of torque.

In reverse plugging, the speed is opposite to the direction of the field and the slip is greater than unity. This may happen if the sequence of the supply source is reversed while forward motoring, so that the direction of the field is also reversed. The developed torque, which is in the same direction as the field, opposes the motion and acts as braking torque. Since $s > 1$, the motor currents will be high, but the developed torque will be low. The energy due to a plugging brake must be dissipated within the motor and this may cause excessive heating of the motor. This type of braking is not normally recommended.

At starting, the machine speed is $\omega_m = 0$ and $s = 1$. The starting torque can be found from Eq. (15-18) by setting $s = 1$ as

$$T_s = \frac{3R_r V_s^2}{\omega_s[(R_s + R_r)^2 + (X_s + X_r)^2]} \tag{15-19}$$

The slip for maximum torque, s_m, can be determined by setting $dT_d/ds = 0$ and Eq. (15-18) yields

$$s_m = \pm \frac{R_r}{[R_s^2 + (X_s + X_r)^2]^{1/2}} \tag{15-20}$$

Substituting $s = s_m$ in Eq. (15-18) gives the maximum developed torque during motoring, which is also called *pull-out torque* or *breakdown torque*,

$$T_{mm} = \frac{3V_s^2}{2\omega_s[R_s + \sqrt{R_s^2 + (X_s + X_r)^2}]} \tag{15-21}$$

and the maximum regenerative torque can be found from Eq. (15-18) by letting

$$s = -s_m$$

$$T_{mr} = \frac{3V_s^2}{2\omega_s[-R_s + \sqrt{R_s^2 + (X_s + X_r)^2}]} \tag{15-22}$$

If R_s is considered small compared to other circuit impedances, which is usually a valid approximation for motors of more than 1 kW rating, the corresponding expressions become

$$T_d = \frac{3R_r V_s^2}{s\omega_s[(R_r/s)^2 + (X_s + X_r)^2]} \tag{15-23}$$

$$T_s = \frac{3R_r V_s^2}{\omega_s[(R_r^2 + (X_s + X_r)^2]} \tag{15-24}$$

$$s_m = \pm \frac{R_r}{X_s + X_r} \tag{15-25}$$

$$T_{mm} = -T_{mr} = \frac{3V_s^2}{2\omega_s(X_s + X_r)} \tag{15-26}$$

Normalizing Eqs. (15-23) and (15-24) with respect to Eq. (15-26) gives

$$\frac{T_d}{T_{mm}} = \frac{2R_r(X_s + X_r)}{s[(R_r/s)^2 + (X_s + X_r)^2]} = \frac{2ss_m}{s_m^2 + s^2} \tag{15-27}$$

and

$$\frac{T_s}{T_{mm}} = \frac{2R_r(X_s + X_r)}{R_r^2 + (X_s + X_r)^2} = \frac{2s_m}{s_m^2 + 1} \tag{15-28}$$

If $s < 1$, $s^2 << s_m^2$ and Eq. (15-27) can be approximated to

$$\frac{T_d}{T_{mm}} = \frac{2s}{s_m} = \frac{2(\omega_s - \omega_m)}{s_m\omega_s} \tag{15-29}$$

which gives the speed as a function of torque,

$$\omega_m = \omega_s \left(1 - \frac{s_m}{2T_{mm}} T_d\right) \tag{15-30}$$

It can be noticed from Eqs. (15-29) and (15-30) that if the motor operates with small slip, the developed torque is proportional to slip and the speed decreases with torque. The rotor current, which is zero at the synchronous speed, increases due to the decrease in R_r/s as the speed is decreased. The developed torque also increases until it becomes maximum at $s = s_m$. For $s < s_m$, the motor operates stably on the portion of the speed–torque characteristic. If the rotor resistance is low, s_m is low. That is, the change of motor speed from no-load to rated torque is only a small percentage. The motor operates essentially at a constant speed. When the load torque exceeds the breakdown torque, the motor stops and the overload protection must immediately disconnect the source to prevent damage due to overheating. It should be noted that for $s > s_m$, the torque decreases despite an increase in the rotor current and the operation is unstable for most motors. The speed and torque of induction motors can be varied by one of the following means:

1. Stator voltage control
2. Rotor voltage control
3. Frequency control
4. Stator voltage and frequency control
5. Stator current control
6. Voltage, current, and frequency control

To meet the torque-speed duty cycle of a drive, the voltage, current and frequency control are normally used.

Example 15-1*

A three-phase 460-V 60-Hz four-pole wye-connected induction motor has the following equivalent-circuit parameters: $R_s = 0.42\ \Omega$, $R_r = 0.23\ \Omega$, $X_s = X_r = 0.82\ \Omega$, and

$X_m = 22\ \Omega$. The no-load loss, which is $P_{no\ load} = 60$ W, may be assumed constant. The rotor speed is 1750 rpm. Use the approximate equivalent circuit in Fig. 15-2 to determine (a) the synchronous speed ω_s; (b) the slip s; (c) the input current I_i; (d) the input power P_i; (e) the input power factor of the supply, PF_s; (f) the gap power P_g; (g) the rotor copper loss P_{ru}; (h) the stator copper loss P_{su}; (i) the developed torque T_d; (j) the efficiency; (k) the starting current I_{rs} and starting torque T_s; (l) the slip for maximum torque s_m; (m) the maximum developed torque in motoring, T_{mm}; (n) the maximum regenerative developed torque T_{mr}; and (o) T_{mm} and T_{mr} if R_s is neglected.

Solution $f = 60$ Hz, $p = 4$, $R_s = 0.42\ \Omega$, $R_r = 0.23\ \Omega$, $X_s = X_r = 0.82\ \Omega$, $X_m = 22\ \Omega$, and $N = 1750$ rpm. The phase voltage is $V_s = 460/\sqrt{3} = 265.58$ V, $\omega = 2\pi \times 60 = 377$ rad/s, and $\omega_m = 1750\ \pi/30 = 183.26$ rad/s.

(a) From Eq. (15-1), $\omega_s = 2\omega/p = 2 \times 377/4 = 188.5$ rad/s.

(b) From Eq. (15-4), $s = (188.5 - 183.26)/188.5 = 0.028$.

(c) From Eq. (15-15),

$$\mathbf{Z}_i = \frac{-22 \times (0.82 + 0.82) + j22 \times (0.42 + 0.23/0.028)}{0.42 + 0.23/0.028 + j(22 + 0.82 + 0.82)} = 7.732\ \underline{/149.2°}$$

$$\mathbf{I}_i = \frac{V_s}{\mathbf{Z}_i} = \frac{265.58}{7.732}\ \underline{/-149.2°} = 34.35\ \underline{/-149.2°}\ \text{A}$$

(d) The power factor of the motor is

$$PF_m = \cos(-149.2°) = 0.858\ \text{(lagging)}$$

From Eq. (15-13),

$$P_i = 3 \times 265.58 \times 34.35 \times 0.858 = 23,482\ \text{W}$$

(e) The power factor of the input supply is $PF_s = PF_m = 0.858$ (lagging) which is the same as the motor power factor, PF_m, because the supply is sinusoidal.

(f) From Eq. (15-17), the rms rotor current is

$$I_r = \frac{265.58}{[(0.42 + 0.23/0.028)^2 + (0.82 + 0.82)^2]^{1/2}} = 30.1\ \text{A}$$

From Eq. (15-9),

$$P_g = \frac{3 \times 30.1^2 \times 0.23}{0.028} = 22,327\ \text{W}$$

(g) From Eq. (15-7), $P_{ru} = 3 \times 30.1^2 \times 0.23 = 625$ W.

(h) The stator copper loss, $P_{su} = 3 \times 30.1^2 \times 0.42 = 1142$ W.

(i) From Eq. (15-12a), $T_d = 22,327/188.5 = 118.4$ N·m.

(j) $P_0 = P_g - P_{ru} - P_{no\ load} = 22,327 - 625 - 60 = 21,642$ W.

(k) For $s = 1$, Eq. (15-17) gives the starting rms rotor current

$$I_{rs} = \frac{265.58}{[(0.42 + 0.23)^2 + (0.82 + 0.82)^2]^{1/2}} = 150.5\ \text{A}$$

From Eq. (15-19),

$$T_s = \frac{3 \times 0.23 \times 150.5^2}{188.5} = 82.9\ \text{N·m}$$

(l) From Eq. (15-20), the slip for maximum torque (or power)

$$s_m = \pm \frac{0.23}{[0.42^2 + (0.82 + 0.82)^2]^{1/2}} = \pm 0.1359$$

(m) From Eq. (15-21), the maximum developed torque

$$T_{mm} = \frac{3 \times 265.58^2}{2 \times 188.5 \times [0.42 + \sqrt{0.42^2 + (0.82 + 0.82)^2}]}$$

$$= 265.64 \text{ N·m}$$

(n) From Eq. (15-22), the maximum regenerative torque is

$$T_{mr} = -\frac{3 \times 265.58^2}{2 \times 188.5 \times [-0.42 + \sqrt{0.42^2 + (0.82 + 0.82)^2}]}$$

$$= -440.94 \text{ N·m}$$

(o) From Eq. (15-25),

$$s_m = \pm \frac{0.23}{0.82 + 0.82} = \pm 0.1402$$

From Eq. (15-26),

$$T_{mm} = -T_{mr} = \frac{3 \times 265.58^2}{2 \times 188.5 \times (0.82 + 0.82)} = 342.2 \text{ N·m}$$

Note. R_s spreads the difference between T_{mm} and T_{mr}. For $R_s = 0$, $T_{mm} = -T_{mr} = 342.2$ N·m, as compared to $T_{mm} = 265.64$ N·m and $T_{mr} = -440.94$ N·m.

15-2.2 Stator Voltage Control

Equation (15-18) indicates that the torque is proportional to the square of the stator supply voltage and a reduction in stator voltage will produce a reduction in speed. If the terminal voltage is reduced to bV_s, Eq. (15-18) gives the developed torque

$$T_d = \frac{3R_r(bV_s)^2}{s\omega_s[(R_s + R_r/s)^2 + (X_s + X_r)^2]}$$

where $b \leq 1$.

Figure 15-4 shows the typical torque–speed characteristics for various values of b. The points of intersection with the load line define the stable operating points. In any magnetic circuit, the induced voltage is proportional to flux and frequency, and the rms air-gap flux can be expressed as

$$V_a = bV_s = K_m\omega\phi$$

or

$$\phi = \frac{V_a}{K_m\omega} = \frac{bV_s}{K_m\omega} \tag{15-31}$$

where K_m is a constant and depends on the number of turns of the stator winding. As the stator voltage is reduced, the air-gap flux and the torque are also reduced.

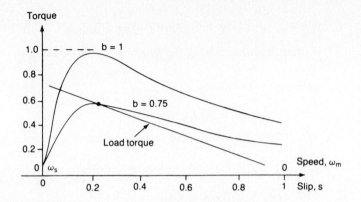

Figure 15-4 Torque–speed characteristics with variable stator voltage.

At a lower voltage, the current will be peaking at a slip of $s_a = \frac{1}{3}$. The range of speed control depends on the slip for maximum torque, s_m. For a low-slip motor, the speed range is very narrow. This type of voltage control is not suitable for a constant-torque load and is normally applied to applications requiring low starting torque and a narrow range of speed at a relatively low slip.

The stator voltage can be varied by three-phase (1) ac voltage controllers, (2) voltage-fed variable dc link inverters, or (3) PWM inverters. However, due to limited speed range requirements, the ac voltage controllers are normally used to provide the voltage control. The ac voltage controllers are very simple. However, the harmonic contents are high and the input power factor of the controllers is low. They are used mainly in low power applications, such as fans, blowers, and centrifugal pumps, where the starting torque is low. They are also used for starting high-power induction motors to limit the in-rush current.

Example 15-2*

A three-phase 460-V 60-Hz four-pole wye-connected induction motor has the following parameters: $R_s = 1.01\ \Omega$, $R_r = 0.69\ \Omega$, $X_s = 1.3\ \Omega$, $X_r = 1.94\ \Omega$, and $X_m = 43.5\ \Omega$. The no-load loss, $P_{\text{no load}}$, is negligible. The load torque, which is proportional to the speed squared, is 41 N·m at 1740 rpm. If the motor speed is 1550 rpm, determine (a) the load torque T_L; (b) the rotor current I_r; (c) the stator supply voltage V_a; (d) the motor input current I_i; (e) the motor input power P_i; (f) the slip for maximum current s_a; (g) the maximum rotor current $I_{r(\text{max})}$; (h) the speed at maximum rotor current ω_a; and (i) the torque at the maximum current T_a.

Solution $p = 4, f = 60$ Hz, $V_s = 460/\sqrt{3} = 265.58$ V, $R_s = 1.01\ \Omega$, $R_r = 0.69\ \Omega$, $X_s = 1.3\ \Omega$, $X_r = 1.94\ \Omega$, and $X_m = 43.5\ \Omega$, $\omega = 2\pi \times 60 = 377$ rad/s, and $\omega_s = 377 \times 2/4 = 188.5$ rad/s. Since torque is proportional to speed squared,

$$T_L = K_m \omega_m^2 \tag{15-32}$$

At $\omega_m = 1740\ \pi/30 = 182.2$ rad/s, $T_L = 41$ N·m and Eq. (15-32) yields $K_m = 41/182.2^2 = 1.235 \times 10^{-3}$ and $\omega_m = 1550\ \pi/30 = 162.3$ rad/s. From Eq. (15-4), $s = (188.5 - 162.3)/188.500 = 0.139$.

(a) From Eq. (15-32), $T_L = 1.235 \times 10^{-3} \times 162.3^2 = 32.5$ N·m.

(b) From Eqs. (15-10) and (15-12),

$$P_d = 3I_r^2 \frac{R_r}{s} (1 - s) = T_L \omega_m + P_{\text{no load}} \tag{15-33}$$

For negligible no-load loss,

$$I_r = \left[\frac{s T_L \omega_m}{3 R_r (1 - s)} \right]^{1/2}$$

$$= \left[\frac{0.139 \times 32.5 \times 162.3}{3 \times 0.69 (1 - 0.139)} \right]^{1/2} = 20.28 \text{ A} \tag{15-34}$$

(c) The stator supply voltage

$$V_a = I_r \left[\left(R_s + \frac{R_r}{s} \right)^2 + (X_s + X_r)^2 \right]^{1/2}$$

$$= 20.28 \times \left[\left(1.01 + \frac{0.69}{0.139} \right)^2 + (1.3 + 1.94)^2 \right]^{1/2} = 137.82 \tag{15-35}$$

(d) From Eq. (15-15),

$$\mathbf{Z}_i = \frac{-43.5 \times (1.3 + 1.94) + j43.5 \times (1.01 + 0.69/0.139)}{1.01 + 0.69/0.139 + j(43.5 + 1.3 + 1.94)} = 6.27 \underline{/144.26°}$$

$$\mathbf{I}_i = \frac{V_a}{\mathbf{Z}_i} = \frac{137.82}{6.27} \underline{/-144.26°} = 22 \underline{/-144.26°} \text{ A}$$

(e) $\text{PF}_m = \cos(-144.26°) = 0.812$ (lagging). From Eq. (15-13),

$$P_i = 3 \times 137.82 \times 22.0 \times 0.812 = 7386 \text{ W}$$

(f) Substituting $\omega_m = \omega_s (1 - s)$ and $T_L = K_m \omega_m^2$ in Eq. (15-34) yields

$$I_r = \left[\frac{s T_L \omega_m}{3 R_r (1 - s)} \right]^{1/2} = (1 - s) \omega_s \left(\frac{s K_m \omega_s}{3 R_r} \right)^{1/2} \tag{15-36}$$

The slip at which I_r becomes maximum can be obtained by setting $dI_r/ds = 0$, and this yields

$$s_a = \tfrac{1}{3} \tag{15-37}$$

(g) Substituting $s_a = \tfrac{1}{3}$ in Eq. (15-36) gives the maximum rotor current

$$I_{r(\text{max})} = \omega_s \left(\frac{4 K_m \omega_s}{81 R_r} \right)^{1/2}$$

$$= 188.5 \times \left(\frac{4 \times 1.235 \times 10^{-3} \times 188.5}{81 \times 0.69} \right)^{1/2} = 24.3 \text{ A} \tag{15-38}$$

(h) The speed at the maximum current

$$\omega_a = \omega_s (1 - s_a) = (2/3) \omega_s = 0.6667 \omega_s$$

$$= 188.5 \times 2/3 = 125.27 \text{ rad/s} \quad \text{or} \quad 1200 \text{ rpm} \tag{15-39}$$

(i) From Eqs. (15-9), (15-12a), and (15-36),

$$T_a = 9I_{r(max)}^2 \frac{R_r}{\omega_s}$$

(15-40)

$$= 9 \times 24.3^2 \times \frac{0.69}{188.5} = 19.45 \text{ N·m}$$

15-2.3 Rotor Voltage Control

In a wound-rotor motor, an external three-phase resistor may be connected to its slip rings as shown in Fig. 15-5a. The developed torque may be varied by varying the resistance, R_x. If R_x is referred to the stator winding and added to R_r, Eq. (15-18) may be applied to determine the developed torque. The typical torque–speed characteristics for variations in rotor resistance are shown in Fig. 15-5b. This method increases the starting torque while limiting the starting current. However, this is an inefficient method and there would be inbalances in voltages and currents if the resistances in the rotor circuit are not equal. A wound-rotor induction motor is designed to have a low rotor resistance so that the running efficiency is high and the full-load slip is low. The increase in the rotor resistance does not affect the value of maximum torque but increases the slip at maximum torque. The wound-rotor motors are widely used in applications requiring frequent starting and braking with large motor torques (e.g., crane hoists). Because of the availability of rotor windings for changing the rotor resistance, the wound rotor offers greater flexibility for control. But it increases the cost and needs maintenance due to slip rings and brushes. The wound-rotor motor is less widely used as compared to the squirrel-case motor.

The three-phase resistor may be replaced by a three-phase diode rectifier and a chopper as shown in Fig. 15-6a, where the GTO operates as a chopper switch. The inductor, L_d, acts as a current source, I_d, and the chopper varies the effective resistance, which can be found from Eq. (14-45):

$$R_e = R(1 - k)$$

(15-41)

where k is the duty cycle of the chopper. The speed can be controlled by varying the duty cycle. The portion of the air-gap power, which is not converted into

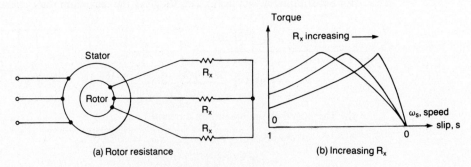

Figure 15-5   Speed control by motor resistance.

mechanical power, is called *slip power*. The slip power is dissipated in the resistance R.

The slip power in the rotor circuit may be returned to the supply by replacing the chopper and resistance, R, with a three-phase full converter as shown in Fig. 15-6b. The converter is operated in the inversion mode with delay range of

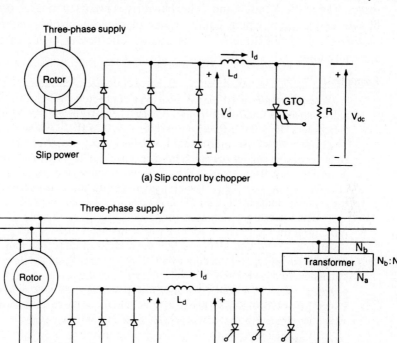

(a) Slip control by chopper

(b) Static Kramer drive

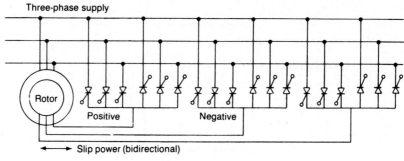

(c) Static Scherbius drive

Figure 15-6 Slip power control.

$\pi/2 \leq \alpha \leq \pi$, thereby returning energy to the source. The variation of the delay angle permits power flow and speed control. This type of drive is known as a *static Kramer* drive. Again, by replacing the bridge rectifiers by three three-phase dual converters (or cycloconverters) as shown in Fig. 15-6c, the slip power flow in either direction is possible and this arrangement is called a *static Scherbius* drive. The static Kramer and Scherbius drives are used in large power pump and blower applications where limited range of speed control is required. Since the motor is connected directly to the source, the power factor of these drives is generally high.

Example 15-3*

A three-phase 460-V 60-Hz six-pole wye-connected wound-rotor induction motor whose speed is controlled by slip power as shown in Fig. 15-6a has the following parameters: $R_s = 0.041 \, \Omega$, $R_r = 0.044 \, \Omega$, $X_s = 0.29 \, \Omega$, $X_r = 0.44 \, \Omega$, and $X_m = 6.1 \, \Omega$. The turns ratio of the rotor to stator windings is $n_m = N_r/N_s = 0.9$. The inductance L_d is very large and its current I_d has negligible ripple. The values of R_s, R_r, X_s, and X_r for the equivalent circuit in Fig. 15-2 can be considered negligible compared to the effective impedance of L_d. The no-load loss of the motor is negligible. The losses in the rectifier, inductor L_d and the GTO chopper are also negligible.

The load torque, which is proportional to speed squared, is 750 N·m at 1175 rpm. (a) If the motor has to operate with a minimum speed of 800 rpm, determine the resistance, R. With this value of R, if the desired speed is 1050 rpm, calculate (b) the inductor current I_d, (c) the duty cycle of the chopper k, (d) the dc voltage V_d, (e) the efficiency, and (f) the input power factor PF_s of the drive.

Solution $V_a = V_s = 460/\sqrt{3} = 265.58$ V, $p = 6$, $\omega = 2\pi \times 60 = 377$ rad/s, and $\omega_s = 2 \times 377/6 = 125.66$ rad/s. The equivalent circuit of the drive is shown in Fig. 15-7a, which is reduced to Fig. 15-7b provided the motor parameters are neglected. From

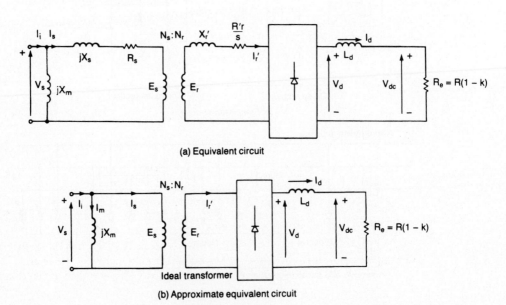

(a) Equivalent circuit

(b) Approximate equivalent circuit

Figure 15-7   Equivalent circuits for Example 15-3.

Eq. (15-41), the dc voltage at the rectifier output is

$$V_d = I_d R_e = I_d R(1 - k) \tag{15-42}$$

and

$$E_r = sV_s \frac{N_r}{N_s} = sV_s n_m \tag{15-43}$$

For a three-phase rectifier, Eq. (3-77) relates E_r and V_d as

$$V_d = 1.654 \times \sqrt{2} E_r = 2.3394 E_r$$

Using Eq. (15-43),

$$V_d = 2.3394 s V_s n_m \tag{15-44}$$

If P_r is the slip power, Eq. (15-9) gives the gap power

$$P_g = \frac{P_r}{s}$$

and Eq. (15-10) gives the developed power as

$$P_d = 3(P_g - P_r) = 3\left(\frac{P_r}{s} - P_r\right) = \frac{3P_r(1 - s)}{s} \tag{15-45}$$

Since the total slip power is $3P_r = V_d I_d$ and $P_d = T_L \omega_m$, Eq. (15-45) becomes

$$P_d = \frac{(1 - s)V_d I_d}{s} = T_L \omega_m = T_L \omega_s(1 - s) \tag{15-46}$$

Substituting V_d from Eq. (15-44) in Eq. (15-46) and solving for I_d gives

$$I_d = \frac{T_L \omega_s}{2.3394 V_s n_m} \tag{15-47}$$

which indicates that the inductor current is independent of the speed. Equating Eq. (15-42) to Eq. (15-44) gives

$$2.3394 s V_s n_m = I_d R(1 - k)$$

which gives

$$s = \frac{I_d R(1 - k)}{2.3394 V_s n_m} \tag{15-48}$$

The speed can be found from Eq. (15-48) as

$$\omega_m = \omega_s(1 - s) = \omega_s \left[1 - \frac{I_d R(1 - k)}{2.3394 V_s n_m}\right] \tag{15-49}$$

$$= \omega_s \left[1 - \frac{T_L \omega_s R(1 - k)}{(2.3394 V_s n_m)^2}\right] \tag{15-50}$$

which shows that for a fixed duty cycle, the speed decreases with load torque. By varying k from 0 to 1, the speed can be varied from a minimum value to ω_s.

(a) $\omega_m = 800 \, \pi/30 = 83.77$ rad/s. From Eq. (15-32) the torque at 900 rpm is

$$T_L = 750 \times \left(\frac{800}{1175}\right)^2 = 347.67 \text{ N·m}$$

From Eq. (15-47), the corresponding inductor current is

$$I_d = \frac{347.67 \times 125.66}{2.3394 \times 265.58 \times 0.9} = 78.13 \text{ A}$$

The speed will be minimum when the duty cycle k is zero and Eq. (15-49) gives the minimum speed,

$$83.77 = 125.66 \left(1 - \frac{78.13R}{2.3394 \times 265.58 \times 0.9}\right)$$

and this yields $R = 2.3856 \ \Omega$.

(b) At 1050 rpm

$$T_L = 750 \times \left(\frac{1050}{1175}\right)^2 = 598.91 \text{ N·m}$$

$$I_d = \frac{598.91 \times 125.66}{2.3394 \times 265.58 \times 0.9} = 134.6 \text{ A}$$

(c) $\omega_m = 1050 \ \pi/30 = 109.96$ rad/s and Eq. (15-49) gives

$$109.96 = 125.66 \left[1 - \frac{134.6 \times 2.3856(1 - k)}{2.3394 \times 265.58 \times 0.9}\right]$$

which gives $k = 0.782$.

(d) Using Eq. (15-4), the slip is

$$s = \frac{125.66 - 109.96}{125.66} = 0.125$$

From Eq. (15-44),

$$V_d = 2.3394 \times 0.125 \times 265.58 \times 0.9 = 69.9 \text{ V}$$

(e) The power loss,

$$P_1 = V_d I_d = 69.9 \times 134.6 = 9409 \text{ W}$$

The output power,

$$P_o = T_L \omega_m = 598.91 \times 109.96 = 65,856 \text{ W}$$

The rms rotor current referred to the stator is

$$I_r = \sqrt{\frac{2}{3}} I_d n_m = \sqrt{\frac{2}{3}} \times 134.6 \times 0.9 = 98.9 \text{ A}$$

The rotor copper loss, $P_{ru} = 3 \times 0.044 \times 98.9^2 = 1291$ W, and the stator copper loss, $P_{su} = 3 \times 0.041 \times 98.9^2 = 1203$ W. The input power

$$P_i = 65,856 + 9409 + 1291 + 1203 = 77,759 \text{ W}$$

The efficiency is $65,856/77,759 = 85\%$.

(f) From Eq. (5-61) for n = 1, the fundamental component of the rotor current referred to the stator is

$$I_{r1} = 0.7797 I_d \frac{N_r}{N_s} = 0.7797 I_d n_m$$

$$= 0.7797 \times 134.6 \times 0.9 = 94.45 \text{ A}$$

and the rms current through the magnetizing branch is

$$I_m = \frac{V_a}{X_m} = \frac{265.58}{6.1} = 43.54 \text{ A}$$

The rms fundamental component of the input current is

$$I_{i1} = \left[(0.7797 I_d n_m)^2 + \left(\frac{V_a}{X_m} \right)^2 \right]^{1/2}$$
$$= (94.45^2 + 43.54^2)^{1/2} = 104 \text{ A} \tag{15-51}$$

The power factor angle is given approximately by

$$\theta_m = -\tan^{-1} \frac{V_a/X_m}{0.7797 I_d n_m}$$
$$= -\tan^{-1} \frac{43.54}{94.45} = \underline{/-24.74^\circ} \tag{15-52}$$

The input power factor $PF_s = \cos(-24.74^\circ) = 0.908$ (lagging).

Example 15-4*

The induction motor in Example 15-3 is controlled by a static Kramer drive as shown in Fig. 15-6b. The turns ratio of the converter ac voltage to supply voltage is $n_c = N_a/N_b = 0.40$. The load torque is 750 N·m at 1175 rpm. If the motor is required to operate at a speed of 1050 rpm, calculate (a) the inductor current I_d; (b) the dc voltage V_d; (c) the delay angle of the converter, α; (d) the efficiency; and (e) the input power factor of the drive, PF_s. The losses in the diode rectifier, converter, transformer and inductor L_d are negligible.

Solution $V_a = V_s = 460/\sqrt{3} = 265.58$ V, $p = 6$, $\omega = 2\pi \times 60 = 377$ rad/s, $\omega_s = 2 \times 377/6 = 125.66$ rad/s, and $\omega_m = 1050 \, \pi/30 = 109.96$ rad/s. Then

$$s = \frac{125.66 - 109.96}{125.66} = 0.125$$

$$T_L = 750 \times \left(\frac{1050}{1175} \right)^2 = 598.91 \text{ N·m}$$

(a) The equivalent circuit of the drive is shown in Fig. 15-8, where the motor parameters are neglected. From Eq. (15-47), the inductor current is

$$I_d = \frac{598.91 \times 125.66}{2.3394 \times 265.58 \times 0.9} = 134.6 \text{ A}$$

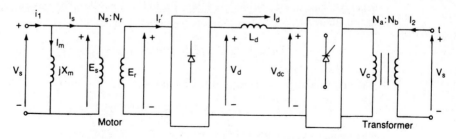

Figure 15-8 Equivalent circuit for static Kramer drive.

(b) From Eq. (15-44),

$$V_d = 2.3394 \times 0.125 \times 265.58 \times 0.9 = 69.9 \text{ V}$$

(c) Since the ac input voltage to the converter is $V_c = n_c V_s$, Eq. (5-57) gives the average voltage at the dc side of the converter as

$$V_{dc} = -\frac{3\sqrt{3}\sqrt{2}n_c V_s}{\pi} \cos \alpha = -2.3394 n_c V_s \cos \alpha \qquad (15\text{-}53)$$

Since $V_d = V_{dc}$, Eqs. (15-44) and (15-53) give

$$2.3394 s V_s n_m = -2.3394 n_c V_s \cos \alpha$$

which gives

$$s = \frac{-n_c \cos \alpha}{n_m} \qquad (15\text{-}54)$$

The speed, which is independent of torque, becomes

$$\omega_m = \omega_s(1 - s) = \omega_s \left(1 + \frac{n_c \cos \alpha}{n_m}\right)$$

$$109.96 = 125.66 \times \left(1 + \frac{0.4 \cos \alpha}{0.9}\right) \qquad (15\text{-}55)$$

which gives the delay angle, $\alpha = 106.3°$.

(d) The power fed back

$$P_1 = V_d I_d = 69.9 \times 134.6 = 9409 \text{ W}$$

The output power

$$P_o = T_L \omega_m = 598.91 \times 109.96 = 65,856 \text{ W}$$

The rms rotor current referred to the stator is

$$I_r = \sqrt{\frac{2}{3}} I_d n_m = \sqrt{\frac{2}{3}} \times 134.6 \times 0.9 = 98.9 \text{ A}$$

$$P_{ru} = 3 \times 0.044 \times 98.9^2 = 1291 \text{ W}$$

$$P_{su} = 3 \times 0.041 \times 98.9^2 = 1203 \text{ W}$$

$$P_i = 65,856 + 1291 + 1203 = 68,350 \text{ W}$$

The efficiency is $65,856/68,350 = 96\%$.

(e) From part (f) in Example 15-3, $I_{r1} = 0.7797 I_d n_m = 94.45$ A, $I_m = 265.58/6.1 = 43.54$ A, and $\mathbf{I}_{i1} = 104 \underline{/-24.74°}$. From Example 5-11, the rms current fed back to the supply is

$$\mathbf{I}_{i2} = \sqrt{\frac{2}{3}} I_d n_c \underline{/-\alpha} = \sqrt{\frac{2}{3}} \times 134.6 \times 0.4 \underline{/-\alpha} = 41.98 \underline{/-106.3°}$$

The effective input current of the drive is

$$\mathbf{I}_i = \mathbf{I}_{i1} + \mathbf{I}_{i2} = 104 \underline{/-24.74°} + 41.98 \underline{/-106.3°} = 117.7 \underline{/-45.4°} \text{ A}$$

The input power factor is $PF_s = \cos(-45.4°) = 0.702$ (lagging).

Note. The efficiency of this drive is higher than that of the rotor resistor control by a chopper. The power factor is dependent on the turns ratio of the transformer (e.g., if $n_c = 0.9$, $\alpha = 97.1°$ and $PF_s = 0.5$; if $n_c = 0.2$, $\alpha = 124.2°$, and $PF_s = 0.8$).

15-2.4 Frequency Control

The torque and speed of induction motors can be controlled by changing the supply frequency. We can notice from Eq. (15-31) that at the rated voltage and rated frequency, the flux will be the rated value. If the voltage is maintained fixed at its rated value while the frequency is reduced below its rated value, the flux will increase. This would cause saturation of the air-gap flux, and the motor parameters would not be valid in determining the torque–speed characteristics. At low frequency, the reactances will decrease and the motor current may be too high. This type of frequency control is not normally used.

If the frequency is increased above its rated value, the flux and torque would decrease. If the synchronous speed corresponding to the rated frequency is called the *base speed* ω_b, the synchronous speed at any other frequency becomes

$$\omega_s = \beta\omega_b$$

and

$$s = \frac{\beta\omega_b - \omega_m}{\beta\omega_b} = 1 - \frac{\omega_m}{\beta\omega_b} \tag{15-56}$$

The torque expression in Eq. (15-18) becomes

$$T_d = \frac{3R_r V_a^2}{s\beta\omega_b[(R_s + R_r/s)^2 + (\beta X_s + \beta X_r)^2]} \tag{15-57}$$

The typical torque–speed characteristics are shown in Fig. 15-9 for various values of β. The three-phase inverter in Fig. 10-5a can vary the frequency at a fixed voltage. If R_s is negligible, Eq. (15-26) gives the maximum torque at the base

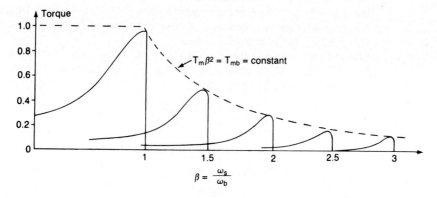

Figure 15-9 Torque characteristics with frequency control.

speed as

$$T_{mb} = \frac{3V_a^2}{2\omega_b(X_s + X_r)} \tag{15-58}$$

The maximum torque at any other frequency is

$$T_m = \frac{3}{2\omega_b(X_s + X_r)} \left(\frac{V_a}{\beta}\right)^2 \tag{15-59}$$

and from Eq. (15-25), the corresponding slip is

$$s_m = \frac{R_r}{\beta(X_s + X_r)} \tag{15-60}$$

Normalizing Eq. (15-59) with respect to Eq. (15-58) yields

$$\frac{T_m}{T_{mb}} = \frac{1}{\beta^2} \tag{15-61}$$

and

$$T_m\beta^2 = T_{mb} \tag{15-62}$$

Thus from Eqs. (15-61) and (15-62), it can be concluded that the maximum torque is inversely proportional to frequency squared, and $T_m\beta^2$ remains constant, similar to the behavior of dc series motors. In this type of control, the motor is said to be operated in a *field-weakening mode*. For $\beta > 1$, the motor is operated at a constant terminal voltage and the flux is reduced, thereby limiting the torque capability of the motor. For $1 < \beta < 1.5$, the relation between T_m and β can be considered approximately linear. For $\beta < 1$, the motor is normally operated at a constant flux by reducing the terminal voltage V_a along with the frequency so that the flux remains constant.

Example 15-5

A three-phase 11.2-kW 1750-rpm 460-V 60-Hz four-pole wye-connected induction motor has the following parameters: $R_s = 0$, $R_r = 0.38\ \Omega$, $X_s = 1.14\ \Omega$, $X_r = 1.71\ \Omega$, and $X_m = 33.2\ \Omega$. The motor is controlled by varying the supply frequency. If the breakdown torque requirement is 35 N·m, calculate (a) the supply frequency, and (b) the speed ω_m at the maximum torque.

Solution $V_a = V_s = 460/\sqrt{3} = 258 \cdot 58$ V, $\omega_b = 2\pi \times 60 = 377$ rad/s, $p = 4$, $P_0 = 11,200$ W, $T_{mb} \times 1750\ \pi/30 = 11,200$, $T_{mb} = 61.11$ N·m, and $T_m = 35$ N·m.

(a) From Eq. (15-62),

$$\beta = \sqrt{\frac{T_{mb}}{T_m}} = \sqrt{\frac{61.11}{35}} = 1.321$$

$$\omega_s = \beta\omega_b = 1.321 \times 377 = 498.01 \text{ rad/s}$$

From Eq. (15-1), the supply frequency is

$$\omega = \frac{4 \times 498.01}{2} = 996 \text{ rad/s} \quad \text{or} \quad 158.51 \text{ Hz}$$

(b) From Eq. (15-60), the slip for maximum torque is

$$s_m = \frac{R_r/\beta}{X_s + X_r} = \frac{0.38/1.321}{1.14 + 1.71} = 0.101$$

$$\omega_m = 498.01 \times (1 - 0.101) = 447.711 \text{ rad/s} \quad \text{or} \quad 4275 \text{ rpm}$$

15-2.5 Voltage and Frequency Control

If the ratio of voltage to frequency is kept constant, the flux in Eq. (15-31) remains constant. Equation (15-59) indicates that the maximum torque, which is independent of frequency, can be maintained approximately constant. However, at a low frequency, the air-gap flux is reduced due to the drop in the stator impedance and the voltage has to be increased to maintain the torque level. This type of control is usually known as *volts/hertz* control.

If $\omega_s = \beta\omega_b$, and the voltage-to-frequency ratio is constant so that

$$\frac{V_a}{\omega_s} = d \tag{15-63}$$

The ratio d, which is determined from the rated terminal voltage V_s and the base speed ω_b, is given by

$$d = \frac{V_s}{\omega_b} \tag{15-64}$$

Substituting V_a from Eq. (15-56) into Eq. (15-57) yields the torque T_d, and the slip for maximum torque is

$$s_m = \frac{R_r}{[R_s^2 + \beta^2(X_s + X_r)^2]^{1/2}} \tag{15-65}$$

The typical torque–speed characteristics are shown in Fig. 15-10. As the frequency is reduced, β decreases and the slip for maximum torque increases. For a given torque demand, the speed can be controlled according to Eq. (15-64) by changing the frequency. Therefore, by varying both the voltage and frequency, the torque and speed can be controlled. The torque is normally maintained constant while the speed is varied. The voltage at variable frequency can be obtained from three-phase inverters or cycloconverters. The cycloconverters

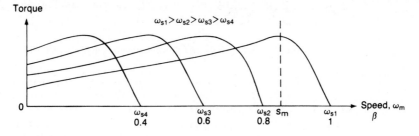

Figure 15-10 Torque–speed characteristics with volts/hertz control.

are used in very large power applications (e.g., locomotives and cement mills), where the frequency requirement is one-half or one-third of the line frequency.

Three possible circuit arrangements for obtaining variable voltage and frequency are shown in Fig. 15-11. In Fig. 15-11a, the dc voltage remains constant and the PWM techniques are applied to vary both the voltage and frequency within the inverter. Due to diode rectifier, regeneration is not possible and the inverter would generate harmonics into the ac supply. In Fig. 15-11b, the chopper varies the dc voltage to the inverter and the inverter controls the frequency. Due to the chopper, the harmonic injection into the ac supply is reduced. In Fig. 15-11c, the dc voltage is varied by the dual converter and frequency is controlled within the inverter. This arrangement permits regeneration; however, the input power factor of the converter is low, especially at a high delay angle.

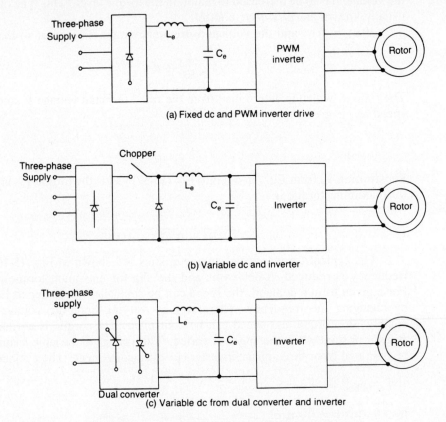

(a) Fixed dc and PWM inverter drive

(b) Variable dc and inverter

(c) Variable dc from dual converter and inverter

Figure 15-11 Voltage-source induction motor drives.

Example 15-6*

A three-phase 11.2-kW 1750-rpm 460-V 60-Hz four-pole wye-connected induction motor has the following parameters: $R_s = 0.66\ \Omega$, $R_r = 0.38\ \Omega$, $X_s = 1.14\ \Omega$, $X_r = 1.71\ \Omega$, and $X_m = 33.2\ \Omega$. The motor is controlled by varying both the voltage and frequency. The volts/hertz ratio, which corresponds to the rated voltage and rated frequency, is maintained constant. (a) Calculate the maximum torque T_m and the corresponding speed ω_m for 60 Hz and 30 Hz. (b) Repeat part (a) if R_s is negligible.

Solution $p = 4$, $V_a = V_s = 460/\sqrt{3} = 265.58$ V, $\omega = 2\pi \times 60 = 377$ rad/s, and from Eq. (15-1), $\omega_b = 2 \times 377/4 = 188.5$ rad/s. From Eq. (15-63), $d = 265.58/188.5 = 1.409$.

(a) At 60 Hz, $\omega_b = \omega_s = 188.5$ rad/s, $\beta = 1$, and $V_a = d\omega_s = 1.409 \times 188.5 = 265.58$ V. From Eq. (15-65),

$$s_m = \frac{0.38}{[0.66^2 + (1.14 + 1.71)^2]^{1/2}} = 0.1299$$

$$\omega_m = 188.5 \times (1 - 0.1299) = 164.01 \text{ rad/s} \quad \text{or} \quad 1566 \text{ rpm}$$

From Eq. (15-21), the maximum torque is

$$T_m = \frac{3 \times 265.58^2}{2 \times 188.5 \times [0.66 + \sqrt{0.66^2 + (1.14 + 1.71)^2}]} = 156.55 \text{ N·m}$$

At 30 Hz, $\omega_s = 2 \times 2 \times \pi\,30/4 = 94.25$ rad/s, $\beta = 30/60 = 0.5$, and $V_a = d\omega_s = 1.409 \times 94.25 = 132.79$ V. From Eq. (15-65), the slip for maximum torque is

$$s_m = \frac{0.38}{[0.66^2 + 0.5^2 \times (1.14 + 1.71)^2]^{1/2}} = 0.242$$

$$\omega_m = 94.25 \times (1 - 0.242) = 71.44 \text{ rad/s} \quad \text{or} \quad 682 \text{ rpm}$$

$$T_m = \frac{3 \times 132.79^2}{2 \times 94.25 \times [0.66 + \sqrt{0.66^2 + 0.5^2 \times (1.14 + 1.71)^2}]} = 125.82 \text{ N·m}$$

(b) At 60 Hz, $\omega_b = \omega_s = 188.5$ rad/s and $V_a = 265.58$ V. From Eq. (15-60),

$$s_m = \frac{0.38}{1.14 + 1.71} = 0.1333$$

$$\omega_m = 188.5 \times (1 - 0.1333) = 163.36 \text{ rad/s or } 1560 \text{ rpm}$$

From Eq. (15-59), the maximum torque is $T_m = 196.94$ N·m.

At 30 Hz, $\omega_s = 94.25$ rad/s, $\beta = 0.5$, and $V_a = 132.79$ V. From Eq. (15-60),

$$s_m = \frac{0.38/0.5}{1.14 + 1.71} = 0.2666$$

$$\omega_m = 94.25 \times (1 - 0.2666) = 69.11 \text{ rad/s} \quad \text{or} \quad 660 \text{ rpm}$$

From Eq. (15-59), the maximum torque is $T_m = 196.94$ N·m.

Note. Neglecting R_s may introduce a significant error in the torque estimation, especially at a low frequency.

15-2.6 Current Control

The torque of induction motors can be controlled by varying the rotor current. The input current, which is readily accessible, is varied instead of the rotor current. For a fixed input current, the rotor current depends on the relative values of the magnetizing and rotor circuit impedances. From Fig. 15-2 the rotor current can be found as

$$\bar{I}_r = \frac{jX_m I_i}{R_s + R_r/s + j(X_m + X_s + X_r)} = I_r \underline{/\theta_1} \tag{15-66}$$

From Eqs. (15-9) and (15-12a), the developed torque is

$$T_d = \frac{3R_r(X_m I_i)^2}{s\omega_s[(R_s + R_r/s)^2 + (X_m + X_s + X_r)^2]} \qquad (15\text{-}67)$$

and the starting torque at $s = 1$ is

$$T_s = \frac{3R_r(X_m I_i)^2}{\omega_s[(R_s + R_r)^2 + (X_m + X_s + X_r)^2]} \qquad (15\text{-}68)$$

The slip for maximum torque is

$$s_m = \pm \frac{R_r}{[R_s^2 + (X_m + X_s + X_r)^2]^{1/2}} \qquad (15\text{-}69)$$

In a real situation as shown in Fig. 15-1b and c, the stator current through R_s and X_s will be constant at I_i. Generally, X_m is much greater than X_s and R_s, which can be neglected for most applications. Neglecting the values of R_s and X_s, Eq. (15-69) becomes

$$s_m = \pm \frac{R_r}{X_m + X_r} \qquad (15\text{-}70)$$

and at $s = s_m$, Eq. (15-67) gives the maximum torque,

$$T_m = \frac{3X_m^2}{2\omega_s(X_m + X_r)} I_i^2 = \frac{3L_m^2}{2(L_m + L_r)} I_i^2 \qquad (15\text{-}71)$$

It can be noticed from Eq. (15-71) that the maximum torque depends on the square of the current and is approximately independent of the frequency. The typical torque–speed characteristics are shown in Fig. 15-12. Since X_m is large as compared to X_s and X_r, the starting torque is low. As the speed increases (or slip decreases), the stator voltage rises and the torque increases. The starting current is low due to the low values of flux (as I_m is low and X_m is large) and rotor current compared to their rated values. The torque increases with the speed due to the increase in flux. A further increase in speed toward the positive slope of the characteristics increases the terminal voltage beyond the rated value. The flux and the magnetizing current are also increased, thereby saturating the flux. The torque can be controlled by the stator current and slip. To keep the air-gap flux constant and to avoid saturation due to high voltage, the motor is normally operated on the negative slope of the equivalent torque–speed characteristics with

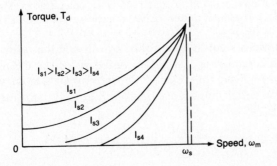

Figure 15-12 Torque–speed characteristics by current control.

voltage control. The negative slope is in the unstable region and the motor must be operated in closed-loop control. At a low slip, the terminal voltage could be excessive and the flux would saturate. Due to saturation, the torque peaking as shown in Fig. 15-12 will be less than that as shown.

The constant current can be supplied by three-phase current source inverters. The current-fed inverter has the advantages of fault current control and the current is less sensitive to the motor parameter variations. However, they generate harmonics and torque pulsation. Two possible configurations of current-fed inverter drives are shown in Fig. 15-13. In Fig. 15-13a the inductor acts as a current source and the controlled rectifier controls the current source. The input power factor of this arrangement is very low. In Fig. 15-13b the chopper controls the current source and the input power factor is higher.

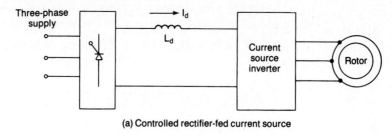

(a) Controlled rectifier-fed current source

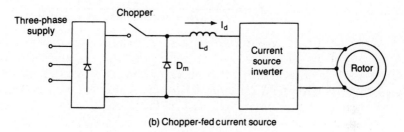

(b) Chopper-fed current source

Figure 15-13 Current-source inductor motor drive.

Example 15-7*

A three-phase 11.2-kW 1750-rpm 460-V 60-Hz four-pole wye-connected induction motor has the following parameters: $R_s = 0.66\ \Omega$, $R_r = 0.38\ \Omega$, $X_s = 1.14\ \Omega$, $X_r = 1.71\ \Omega$, and $X_m = 33.2\ \Omega$. The no-load loss is negligible. The motor is controlled by a current-source inverter and the input current is maintained constant at 20 A. If the frequency is 40 Hz and the developed torque is 55 N·m, determine (a) the slip for maximum torque s_m and maximum torque T_m; (b) the slip s; (c) the rotor speed ω_m; (d) the terminal voltage per phase, V_a; and (e) the power factor PF_m.

Solution $V_{a(\text{rated})} = 460/\sqrt{3} = 265.58$ V, $I_i = 20$ A, $T_L = T_d = 55$ N·m, and $p = 4$. At 40 Hz, $\omega = 2\pi \times 40 = 251.33$ rad/s, $\omega_s = 2 \times 251.33/4 = 125.66$ rad/s, $R_s = 0.66\ \Omega$, $R_r = 0.38\ \Omega$, $X_s = 1.14 \times 40/60 = 0.76\ \Omega$, $X_r = 1.71 \times 40/60 = 1.14\ \Omega$, and $X_m = 33.2 \times 40/60 = 22.13\ \Omega$.

(a) From Eq. (15-69),

$$s_m = \frac{0.38}{[0.66^2 + (22.13 + 0.78 + 1.14)^2]^{1/2}} = 0.0158$$

From Eq. (15-67), $T_m = 94.68$ N·m.

(b) From Eq. (15-67),

$$T_d = 55 = \frac{3(R_r/s)(22.13 \times 20)^2}{125.66 \times [(0.66 + R_r/s)^2 + (22.13 + 0.76 + 1.14)^2]}$$

which gives $(R_r/s)^2 - 83.74(R_r/s) + 578.04 = 0$, and solving for R_r/s yields

$$\frac{R_s}{s} = 76.144 \quad \text{or} \quad 7.581$$

and $s = 0.00499$ or 0.0501. Since the motor is normally operated with a large slip in the negative slope of the torque–speed characteristic,

$$s = 0.0501$$

(c) $\omega_m = 125.656 \times (1 - 0.0501) = 119.36$ rad/s or 1140 rpm.

(d) From Fig. 15-2, the input impedance can be derived as

$$\overline{Z}_i = R_i + jX_i = (R_i^2 + X_i^2)^{1/2} \; \underline{/\theta_m} = Z_i \; \underline{/\theta_m}$$

where

$$R_i = \frac{X_m^2(R_s + R_r/s)}{(R_s + R_r/s)^2 + (X_m + X_s + X_r)^2} \tag{15-72}$$

$$= 6.26 \; \Omega$$

$$X_i = \frac{X_m[(R_s + R_r/s)^2 + (X_s + X_r)(X_m + X_s + X_r)]}{(R_s + R_r/s)^2 + (X_m + X_s + X_r)^2} \tag{15-73}$$

$$= 3.899 \; \Omega$$

and

$$\theta_m = \tan^{-1}\frac{X_i}{R_i}$$

$$= 31.9°$$

$$Z_i = (6.26^2 + 3.899^2)^{1/2} = 7.38 \; \Omega \tag{15-74}$$

$$V_a = Z_iI_i = 7.38 \times 20 = 147.6 \text{ V}$$

(e) $\text{PF}_m = \cos(31.9°) = 0.849$ (lagging).

Note. If the maximum torque is calculated from Eq. (15-71), $T_m = 100.49$ and V_a (at $s = s_m$) is 313 V. For a supply frequency of 90 Hz, recalculations of the values give $\omega_s = 282.74$ rad/s, $X_s = 1.71 \; \Omega$, $X_r = 2.565 \; \Omega$, $X_m = 49.8 \; \Omega$, $s_m = 0.00726$, $T_m = 96.1$ N·m, $s = 0.0225$, $V_a = 316$ V, and V_a (at $s = s_m$) = 699.6 V. It is evident that at a high frequency and low slip, the terminal voltage would exceed the rated value and saturate the air-gap flux.

15-2.7 Voltage, Current, and Frequency Control

The torque–speed characteristics of induction motors depend on the type of control. It may be necessary to vary the voltage, frequency, and current to meet the torque–speed requirements as shown in Fig. 15-14, where there are three regions.

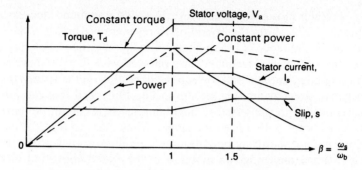

Figure 15-14 Control variables versus frequency.

In the first region, the speed can be varied by voltage (or current) control at constant torque. In the second region, the motor is operated at constant current and the slip is varied. In the third region, the speed is controlled by frequency at a reduced stator current.

The torque and power variations for a given stator current and frequencies below the rated frequency are shown by dots in Fig. 15-15. For $\beta < 1$, the motor operates at a constant flux. For $\beta > 1$, the motor is operated by frequency control, but at a constant voltage. Therefore, the flux decreases in the inverse ratio of per-unit frequency, and the motor operates in the field weakening mode.

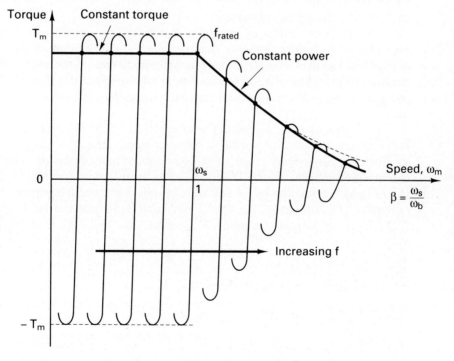

Figure 15-15 Torque–speed characteristics for variable frequency control.

When motoring, a decrease in speed command decreases the supply frequency. This shifts the operation to regenerative braking. The drive decelerates under the influence of braking torque and load torque. For speed below rate value ω_b, the voltage and frequency are reduced with speed to maintain the desired V/f ratio or constant flux and to keep the operation on the speed–torque curves with a negative slope by limiting the slip speed. For speed above ω_b, the frequency alone is reduced with the speed to maintain the operation on the portion of the speed–torque curves with a negative slope. When close to the desired speed, the operation shifts to motoring operation and the drive settles at the desired speed.

When motoring, an increase in the speed command increases the supply frequency. The motor torque exceeds the load torque and the motor decelerates. The operation is maintained on the portion of the speed–torque curves with a negative slope by limiting the slip speed. Finally, the drive settles at the desired speed.

15-2.8 Closed-Loop Control of Induction Motors

A closed-loop control is normally required in order to satisfy the steady-state and transient performance specifications of ac drives. The control strategy can be implemented by (1) *scalar control*, where the control variables are dc quantities and only their magnitudes are controlled; (2) *vector control*, where both the magnitude and phase of the control variables are controlled; or (3) *adaptive control*, where the parameters of the controller are continuously varied to adapt to the variations of the output variables.

The dynamic model of induction motors differs significantly from that of Fig. 15-1 and is more complex than dc motors. The design of feedback-loop parameters requires complete analysis and simulation of the entire drive. The control and modeling of ac drives are beyond the scope of this book [2, 5, 17, 18]; and only some of the basic scalar feedback techniques are discussed in this section.

A control system is generally characterized by the hierarchy of the control loops, where the outer loop controls the inner loops. The inner loops are designed to execute progressively faster. The loops are normally designed to have limited command excursion. Figure 15-16a shows an arrangement for stator voltage control of induction motors by ac voltage controllers at fixed frequency. The speed controller, K_1, processes the speed error and generates the reference current $I_{s(ref)}$. K_2 is the current controller. K_3 generates the delay angle of thyristor converter and the inner current-limit loop sets the torque limit indirectly. The current limiter instead of current clamping has the advantage of feeding back the short-circuit current in case of fault. The speed controller, K_1, may be a simple gain (proportional type), proportional-integral type, or a lead-lag compensator. This type of control is characterized by poor dynamic and static performance and is generally used in fans, pumps, and blower drives. A 187-kW three-phase ac voltage regulator (or controller) for a wire-drawing machine is shown in Fig. 15-17, where the control electronics are mounted on the side panel.

The arrangement in Fig. 15-16a can be extended to a volt/hertz control with the addition of a controlled rectifier and dc voltage control loop as shown in

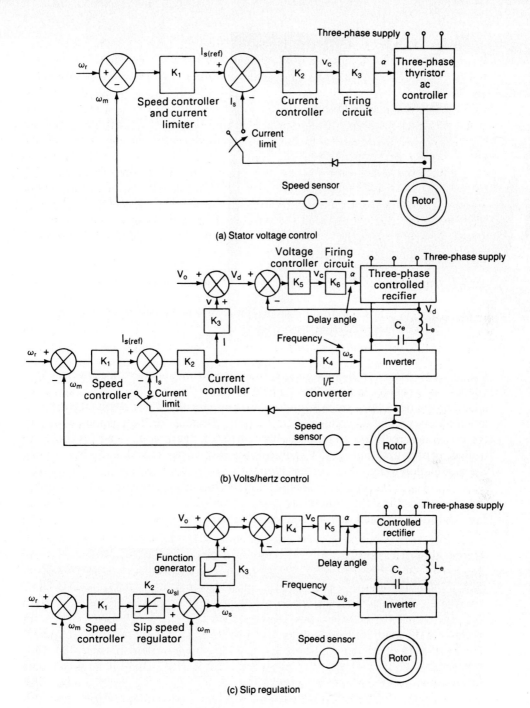

(a) Stator voltage control

(b) Volts/hertz control

(c) Slip regulation

Figure 15-16 Closed-loop control of induction motors.

Figure 15-17 A 187-kW three-phase ac voltage regulator. (Reproduced by permission of Brush Electrical Machines Ltd., England.)

Fig. 15-16b. After the current limiter, the same signal generates the inverter frequency and provides input to the dc link gain controller, K_3. A small voltage V_0 is added to the dc voltage reference to compensate for the stator resistance drop at low frequency. The dc voltage V_d acts as the reference for the voltage control of the controlled rectifier. In case of PWM inverter, there is no need for the controlled rectifier and the signal V_d controls the inverter voltage directly by varying the modulation index. For current monitoring, it requires a sensor, which introduces a delay in the system response. Fifteen forced-commutated inverter cubicles for controlling the undergrate cooler fans of a cement kiln are shown in Fig. 15-18, where each unit is rated at 100 kW.

Figure 15-18 Fifteen forced-commutated inverter cubicles for a cement kiln. (Reproduced by permission of Brush Electrical Machines Ltd., England.)

Since the torque of induction motors is proportional to the slip frequency, $\omega_{sl} = \omega_s - \omega_m = s\omega_s$, the slip frequency instead of the stator current can be controlled. The speed error generates the slip frequency command as shown in Fig. 15-16c, where the slip limits set the torque limits. The function generator, which produces the command signal for voltage control in response to the frequency, ω_s, is nonlinear and can take also into account the compensating drop V_o at a low frequency. The compensating drop V_o is shown in Fig. 15-16c. For a step change in the speed command, the motor accelerates or decelerates within the torque limits to a steady-state slip value corresponding to the load torque. This arrangement controls the torque indirectly within the speed control loop and does not require the current sensor.

A simple arrangement for current control is shown in Fig. 15-19. The speed error generates the reference signal for the dc link current. The slip frequency, $\omega_{sl} = \omega_s - \omega_m$, is fixed. With a step speed command, the machine accelerates with a high current which is proportional to the torque. In the steady state the motor current is low. However, the air-gap flux fluctuates, and due to varying flux at different operating points, the performance of this drive is poor.

A practical arrangement for current control, where the flux is maintained constant, is shown in Fig. 15-20. The speed error generates the slip frequency, which controls the inverter frequency and the dc link current source. The function generator produces the current command to maintain the air-gap flux constant, normally at the rated value.

The arrangement in Fig. 15-16a for speed control with inner current control loop can be applied to a static Kramer drive as shown in Fig. 15-21, where the torque is proportional to the dc link current I_d. The speed error generates the dc link current command. A step increase in speed clamps the current to the maximum value and the motor accelerates at a constant torque which corresponds to the maximum current. A step decrease in the speed sets the current command to zero and the motor decelerates due to the load torque. A 110-kW static Kramer drive for fans in cement works is shown in Fig. 15-22.

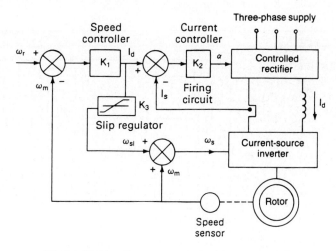

Figure 15-19 Current control with constant slip.

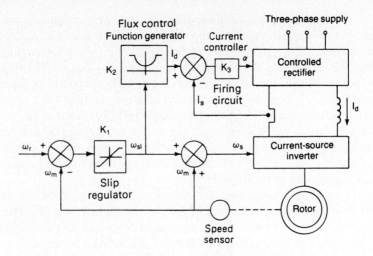

Figure 15-20 Current control with constant flux operation.

The control methods that have been discussed so far provide satisfactory steady-state performance, but their dynamic response is poor. An induction motor has a nonlinear multivariables highly coupled characteristic. *Field-oriented control* (FOC) decouples the two components of stator current: one providing the air-gap flux and the other producing the torque. It provides independent control of flux and torque, and the control characteristic is linearized. The stator currents are converted to a fictitious synchronously rotating reference frame aligned with the flux vector and are transformed back to the stator frame before feeding back to the machine. The two components are d-axis i_{ds} analogous to armature current, and q-axis i_{qs} analogous to the field current of a separately excited dc motor. The rotor flux linkage vector is aligned along the d-axis of the reference frame. The axis of rotation for various quantities is shown in Fig. 15-23a.

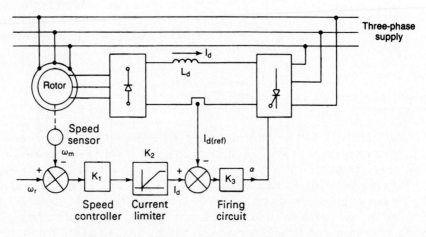

Figure 15-21 Speed control of static Kramer drive.

Figure 15-22 A 110-kW static Kramer drive. (Reproduced by permission of Brush Electrical Machines Ltd., England.)

This type of control can be implemented by either direct method or indirect method. In direct method, the flux vector is computed from the terminal quantities of the motor as shown in Fig. 15-23b. The indirect method [18] uses the motor slip frequency ω_{sl} to compute the desired flux vector as shown in Fig. 15-23c. It is simpler to implement than the direct method and is used increasingly in induction motor control. T_d is the desired motor torque, ω_r is the rotor flux linkage, τ_r is the rotor time constant, and L_m is the mutual inductance. The amount of decoupling is dependent on the motor parameters unless the flux is measured directly. Without the exact knowledge of the motor parameters, an ideal decoupling is not possible.

15-3 SYNCHRONOUS MOTOR DRIVES

Synchronous motors have a polyphase winding on the stator, also known as armature, and a field winding carrying a dc current on the rotor. There are two mmfs involved: one due to the field current and other due to the armature current. The resultant mmf produces the torque. The armature is identical to the stator of induction motors, but there is no induction in the rotor. A synchronous motor is a constant-speed machine and always rotates with zero slip at the synchronous speed, which depends on the frequency and the number of poles, as given by Eq. (15-1). A synchronous motor can be operated as a motor or genera-

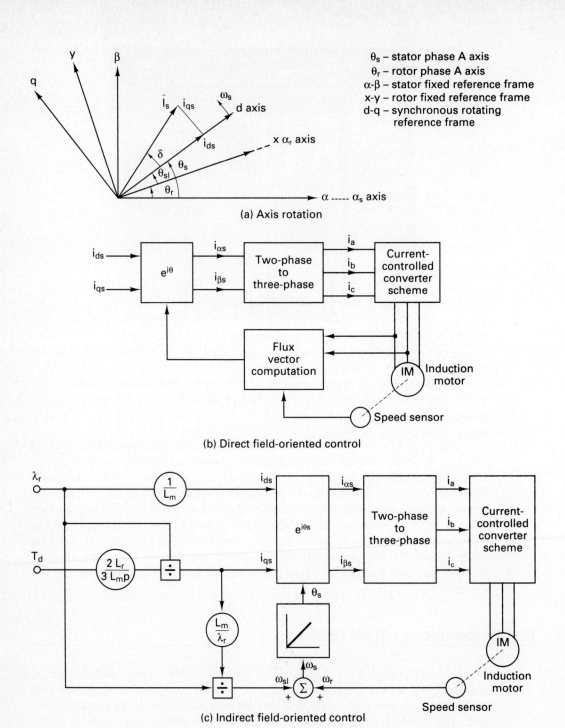

(a) Axis rotation

θ_s – stator phase A axis
θ_r – rotor phase A axis
α-β – stator fixed reference frame
x-y – rotor fixed reference frame
d-q – synchronous rotating
    reference frame

(b) Direct field-oriented control

(c) Indirect field-oriented control

Figure 15-23   Field-oriented control of induction motor.

tor. The power factor can be controlled by varying the field current. The cyclo-converters and inverters are widening the applications of synchronous motors in variable-speed drives. The synchronous motors can be classified into six types:

1. Cylindrical rotor motors
2. Salient-pole motors
3. Reluctance motors
4. Permanent-magnet motors
5. Switched reluctance motors
6. Brushless dc and ac motors

15-3.1 Cylindrical Rotor Motors

The field winding is wound on the rotor, which is cylindrical, and these motors have a uniform air gap. The reactances are independent on the rotor position. The equivalent circuit per phase, neglecting the no-load loss, is shown in Fig. 15-24a, where R_a is the armature resistance per phase and X_s is the *synchronous reactance* per phase. V_f, which is dependent on the field current, is known as *excitation* or *field* voltage.

The power factor depends on the field current. The V-curves, which show the typical variations of the armature current against the excitation current, are shown in Fig. 15-25. For the same armature current, the power factor could be lagging or leading, depending on the excitation current, I_f.

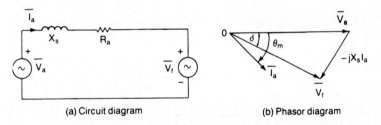

(a) Circuit diagram   (b) Phasor diagram

Figure 15-24 Equivalent circuit of synchronous motors.

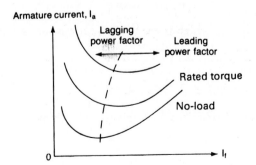

Figure 15-25 Typical V-curves of synchronous motors.

If θ_m is the lagging power factor angle of the motor, Fig. 15-24a gives

$$\overline{V}_f = V_a \underline{/0} - \overline{I}_a(R_a + jX_s) \tag{15-75}$$

$$= V_a \underline{/0} - I_a(\cos\theta_m - j\sin\theta_m)(R_a + jX_s)$$

$$= V_a - I_aX_s\sin\theta_m - I_aR_a\cos\theta_m - jI_a(X_s\cos\theta_m - R_a\sin\theta_m) \tag{15-75a}$$

$$= V_f \underline{/\delta} \tag{15-75b}$$

where

$$\delta = \tan^{-1}\frac{-(I_aX_s\cos\theta_m - I_aR_a\sin\theta_m)}{V_a - I_aX_s\sin\theta_m - I_aR_a\cos\theta_m} \tag{15-76}$$

and

$$V_f = [(V_a - I_aX_s\sin\theta_m - I_aR_a\cos\theta_m)^2$$

$$+ (I_aX_s\cos\theta_m - I_aR_a\sin\theta_m)^2]^{1/2} \tag{15-77}$$

The phasor diagram in Fig. 15-24b yields

$$\overline{V}_f = V_f(\cos\delta + j\sin\delta) \tag{15-78}$$

$$\overline{I}_a = \frac{\overline{V}_a - \overline{V}_f}{R_a + jX_s} = \frac{[V_a - V_f(\cos\delta + j\sin\delta)](R_a - jX_s)}{R_a^2 + X_s^2} \tag{15-79}$$

The real part of Eq. (15-79) becomes

$$I_a\cos\theta_m = \frac{R_a(V_a - V_f\cos\delta) - V_fX_s\sin\delta}{R_a^2 + X_s^2} \tag{15-80}$$

The input power can be determined from Eq. (15-80),

$$P_i = 3V_aI_a\cos\theta_m$$

$$= \frac{3[R_a(V_a^2 - V_aV_f\cos\delta) - V_aV_fX_s\sin\delta]}{R_a^2 + X_s^2} \tag{15-81}$$

The stator (or armature) copper loss is

$$P_{su} = 3I_a^2R_a \tag{15-82}$$

The gap power, which is the same as the developed power, is

$$P_d = P_g = P_i - P_{su} \tag{15-83}$$

If ω_s is the synchronous speed, which is the same as the rotor speed, the developed torque becomes

$$T_d = \frac{P_d}{\omega_s} \tag{15-84}$$

If the armature resistance is negligible, T_d in Eq. (15-84) becomes

$$T_d = -\frac{3V_aV_f\sin\delta}{X_s\omega_s} \tag{15-85}$$

and Eq. (15-76) becomes

$$\delta = -\tan^{-1} \frac{I_a X_s \cos \theta_m}{V_a - I_a X_s \sin \theta_m} \qquad (15\text{-}86)$$

For motoring δ is negative and torque in Eq. (15-85) becomes positive. In the case of generating, δ is positive and the power (and torque) becomes negative. The angle δ is called the *torque angle*. For a fixed voltage and frequency, the torque depends on the angle, δ, and is proportional to the excitation voltage, V_f. For fixed values of V_f and δ, the torque depends on the voltage-to-frequency ratio and a constant volts/hertz control will provide speed control at a constant torque. If V_a, V_f, and δ remain fixed, the torque decreases with the speed and the motor operates in the field weakening mode.

If $\delta = 90°$, the torque becomes maximum and the maximum developed torque, which is called the *pull-out torque*, becomes

$$T_p = T_m = -\frac{3 V_a V_f}{X_s \omega_s} \qquad (15\text{-}87)$$

The plot of developed torque against the angle, δ is shown in Fig. 15-26. For stability considerations, the motor is operated in the positive slope of T_d–δ characteristics and this limits the range of torque angle, $-90° \le \delta \le 90°$.

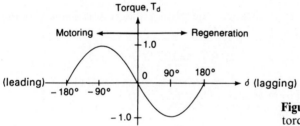

Figure 15-26 Torque versus torque angle with cylindrical rotor.

Example 15-8*

A three-phase 460-V 60-Hz six-pole wye-connected cylindrical rotor synchronous motor has a synchronous reactance of $X_s = 2.5\ \Omega$ and the armature resistance is negligible. The load torque, which is proportional to the speed squared, is $T_L = 398\ \text{N·m}$ at 1200 rpm. The power factor is maintained at unity by field control and the voltage-to-frequency ratio is kept constant at the rated value. If the inverter frequency is 36 Hz and the motor speed is 720 rpm, calculate (a) the input voltage V_a, (b) the armature current I_a, (c) the excitation voltage V_f, (d) the torque angle δ, and (e) the pull-out torque T_p.

Solution PF $= \cos \theta_m = 1.0$, $\theta_m = 0$, $V_{a(\text{rated})} = V_b = V_s = 460/\sqrt{3} = 265.58$ V, $p = 6$, $\omega = 2\pi \times 60 = 377$ rad/s, $\omega_b = \omega_s = \omega_m = 2 \times 377/6 = 125.67$ rad/s or 1200 rpm, and $d = V_b/\omega_b = 265.58/125.67 = 2.1133$. At 720 rpm,

$$T_L = 398 \times \left(\frac{720}{1200}\right)^2 = 143.28\ \text{N·m} \qquad \omega_s = \omega_m = 720 \times \frac{\pi}{30} = 75.4\ \text{rad/s}$$

$$P_0 = 143.28 \times 75.4 = 10{,}803\ \text{W}$$

(a) $V_a = d\omega_s = 2.1133 \times 75.4 = 159.34$ V.
(b) $P_0 = 3V_aI_a$ PF $= 10,803$ or $I_a = 10,803/(3 \times 159.34) = 22.6$ A.
(c) From Eq. (15-75),

$$\overline{V}_f = 159.34 - 22.6 \times (1 + j0)(j2.5) = 169.1 \; \underline{/-19.52°}$$

(d) The torque angle, $\delta = -19.52°$.
(e) From Eq. (15-87),

$$T_p = \frac{3 \times 159.34 \times 169.1}{2.5 \times 75.4} = 428.82 \text{ N·m}$$

15-3.2 Salient-Pole Motors

The armature of salient-pole motors is similar to that of cylindrical rotor motors. However, due to saliency, the air gap is not uniform and the flux is dependent on the position of the rotor. The field winding is normally wound on the pole pieces. The armature current and the reactances can be resolved into direct and quadrature axis components. I_d and I_q are the components of the armature current in the direct (or d) axis and quadrature (or q) axis, respectively. X_d and X_q are the d-axis reactance and q-axis reactance, respectively. Using Eq. (15-75), the excitation voltage becomes

$$\overline{V}_f = \overline{V}_a - jX_d\overline{I}_d - jX_q\overline{I}_q - R_a\overline{I}_a$$

For negligible armature resistance, the phasor diagram is shown in Fig. 15-27. From the phasor diagram,

$$I_d = I_a \sin(\theta_m - \delta) \tag{15-88}$$

$$I_q = I_a \cos(\theta_m - \delta) \tag{15-89}$$

$$I_dX_d = V_a \cos \delta - V_f \tag{15-90}$$

$$I_qX_q = V_a \sin \delta \tag{15-91}$$

Substituting I_q from Eq. (15-89) in Eq. (15-91), we have

$$V_a \sin \delta = X_qI_a \cos(\theta_m - \delta)$$
$$= X_qI_a(\cos \delta \cos \theta_m + \sin \delta \sin \theta_m) \tag{15-92}$$

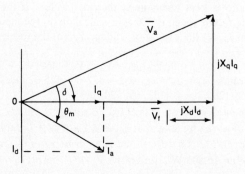

Figure 15-27 Phase diagram for salient-pole synchronous motors.

Dividing both sides by cos δ and solving for δ gives

$$\delta = -\tan^{-1}\frac{I_a X_q \cos\theta_m}{V_a - I_a X_q \sin\theta_m} \tag{15-93}$$

where the negative sign signifies that V_f lags V_a. If the terminal voltage is resolved into a d-axis and a q-axis,

$$V_{ad} = -V_a \sin\delta \text{ and } V_{aq} = V_a \cos\delta$$

The input power becomes

$$
\begin{aligned}
P &= -3(I_d V_{ad} + I_q V_{aq}) \\
&= 3I_d V_a \sin\delta - 3I_q V_a \cos\delta
\end{aligned} \tag{15-94}
$$

Substituting I_d from Eq. (15-90) and I_q from Eq. (15-91) in Eq. (15-94) yields

$$P_d = -\frac{3V_a V_f}{X_d}\sin\delta - \frac{3V_a^2}{2}\left[\frac{X_d - X_q}{X_d X_q}\sin 2\delta\right] \tag{15-95}$$

Dividing Eq. (15-95) by speed gives the developed torque as

$$T_d = -\frac{3V_a V_f}{X_d \omega_s}\sin\delta - \frac{3V_a^2}{2\omega_s}\left[\frac{X_d - X_q}{X_d X_q}\sin 2\delta\right] \tag{15-96}$$

The torque in Eq. (15-96) has two components. The first component is the same as that of cylindrical rotor if X_d is replaced by X_s and the second component is due to the rotor saliency. The typical plot of T_d against torque angle is shown in Fig. 15-28, where the torque has a maximum value at $\delta = \pm\delta_m$. For stability the torque angle is limited in the range of $-\delta_m \leq \delta \leq \delta_m$ and in this stable range, the slope of T_d–δ characteristic is higher than that of cylindrical rotor motor.

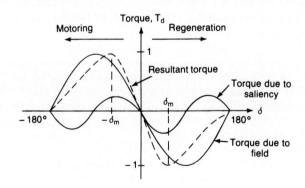

Figure 15-28 Torque versus torque angle with salient-pole rotor.

15-3.3 Reluctance Motors

The reluctance motors are similar to the salient-pole motors, except that there is no field winding on the rotor. The armature circuit, which produces rotating magnetic field in the air gap, induces a field in the rotor that has a tendency to align with the armature field. The reluctance motors are very simple and are used in applications where a number of motors are required to rotate in synchronism. These motors have low lagging power factor, typically in the range 0.65 to 0.75.

With $V_f = 0$, Eq. (15-96) can be applied to determine the reluctance torque,

$$T_d = -\frac{3V_a^2}{2\omega_s} \quad [\frac{X_d - X_q}{X_d X_q} \sin 2\delta]$$ (15-97)

where

$$\delta = -\tan^{-1} \frac{I_a X_q \cos \theta_m}{V_a - I_a X_q \sin \theta_m}$$ (15-98)

The pull-out torque for $\delta = -45°$ is

$$T_p = \frac{3V_a^2}{2\omega_s} \quad [\frac{X_d - X_q}{X_d X_q}]$$ (15-99)

15-3.4 Permanent-Magnet Motors

The permanent-magnet motors are similar to salient-pole motors, except that there is no field winding on the rotor and the field is provided by mounting permanent magnets at the rotor. The excitation voltage cannot be varied. For the same frame size, permanent-magnet motors have higher pull-out torque. The equations for the salient-pole motors may be applied to permanent-magnet motors if the excitation voltage, V_f, is assumed constant. The elimination of field coil, dc supply, and slip rings reduces the motor loss and the complexity. These motors are also known as *brushless motors* and finding increased applications in robots and machine tools. A permanent-magnet motor (PM) can be fed by either rectangular current or sinusoidal current. The rectangular current-fed motors, which have concentrated windings on the stator inducing a square or trapezoidal voltage, are normally used in low-power drives. The sinusoidal current-fed motors, which have distributed windings on the stator, provide smoother torque and are normally used in high-power drives.

Example 15-9*

A three-phase 230-V 60-Hz four-pole wye-connected reluctance motor has $X_d = 22.5\ \Omega$ and $X_q = 3.5\ \Omega$. The armature resistance is negligible. The load torque is $T_L = 12.5\ \text{N·m}$. The voltage-to-frequency ratio is maintained constant at the rated value. If the supply frequency is 60 Hz, determine (a) the torque angle δ, (b) the line current I_a, and (c) the input power factor PF.

Solution $T_L = 12.5\ \text{N·m}$, $V_{a(\text{rated})} = V_b = 230/\sqrt{3} = 132.79\ \text{V}$, $p = 4$, $\omega = 2\pi \times 60 = 377\ \text{rad/s}$, $\omega_b = \omega_s = \omega_m = 2 \times 377/4 = 188.5\ \text{rad/s}$ or 1800 rpm, and $V_a = 132.79\ \text{V}$.

(a) $\omega_s = 188.5\ \text{rad/s}$. From Eq. (15-97),

$$\sin 2\delta = -\frac{12.5 \times 2 \times 188.5 \times 22.5 \times 3.5}{3 \times 132.79^2 \times (22.5 - 3.5)}$$

and $\delta = -10.84°$.

(b) $P_0 = 12.5 \times 188.5 = 2356\ \text{W}$. From Eq. (15-98),

$$\tan(10.84°) = \frac{3.5 I_a \cos \theta_m}{132.79 - 3.5 I_a \sin \theta_m}$$

and $P_0 = 2356 = 3 \times 132.79 I_a \cos \theta_m$. From these two equations, I_a and θ_m can be determined by an iterative method of solution, which yields $I_a = 9.2$ A and $\theta_m = 49.98°$.

(c) PF = $\cos(49.98°) = 0.643$.

15-3.5 Switched Reluctance Motors

A switched reluctance motor (SRM) is a variable reluctance step motor. A cross-sectional view is shown in Fig. 15-29a. Three phases ($q = 3$) are shown with six stator teeth, $N_s = 6$, and four rotor-teeth, $N_r = 4$. N_r is related to N_s and q by $N_r =$

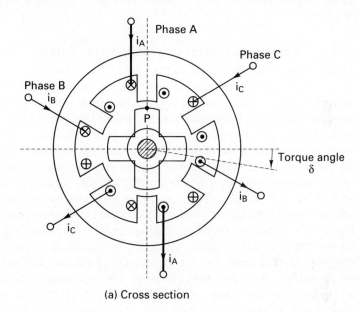

(a) Cross section

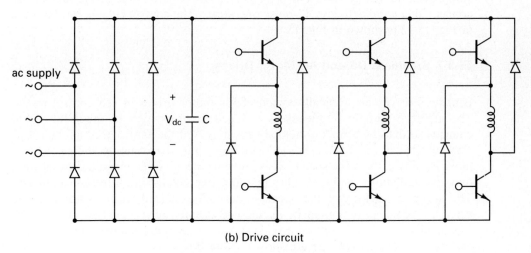

(b) Drive circuit

Figure 15-29 Switched reluctance motor.

$N_s \pm N_s/q$. Each phase winding is placed on two diametrically opposite teeth. If phase A is excited by a current i_a, a torque is developed and it causes a pair of rotor poles to be magnetically aligned with the poles of phase A. If the subsequent phase B and phase C were excited in sequence, a further rotation will take place. The motor speed can be varied by exciting in sequence phases A, B, and C. A commonly used circuit to drive an SRM is shown in Fig. 15-29b. An absolute position sensor is usually required to directly control the angles of stator excitation with respect to the rotor position. A position feedback control is provided in generating the gating signals. If the switching takes place at a fixed rotor position relative to the rotor poles, an SRM would exhibit the characteristics of a dc series motor. By varying the rotor position, a range of operating characteristics can be obtained.

15-3.6 Closed-Loop Control of Synchronous Motors

The typical characteristics of torque, current, and excitation voltage against frequency ratio β are shown in Fig. 15-30. There are two regions of operation: constant torque and constant power. In the constant-torque region, the volts/hertz is maintained constant, and in the constant-power region, the torque decreases with frequency. Speed–torque characteristics for different frequencies are shown in Fig. 15-30b. Similar to induction motors, the speed of synchronous motors can be controlled by varying the voltage, frequency, and current. There are various configurations for closed loop control of synchronous motors. A basic arrangement for constant volts/hertz control of synchronous motors is shown in Fig. 15-31, where the speed error generates the frequency and voltage command for the PWM inverter. Since the speed of synchronous motors depends on the supply frequency only, they are employed in multimotor drives requiring accurate speed tracking between motors, as in fiber spinning mills, paper mills, textile mills, and machine tools. A 3.3-kW soft-start system for synchronous motor compressor drives is shown in Fig. 15-32, where two three-phase stacks are connected in parallel to give 12-pulse operation. The control cubicle for the drive in Fig. 15-32 is shown in Fig. 15-33.

15-3.7 Brushless DC and AC Motor Drives

Brushless drives are basically synchronous motor drives in self-control mode. The armature supply frequency is changed in proportion to the rotor speed changes so that the armature field always moves at the same speed as the rotor. The self-control ensures that for all operating points the armature and rotor fields move exactly at the same speed. This prevents the motor from pulling out of step, hunting oscillations, and instability due to a step change in torque or frequency. The accurate tracking of the speed is normally realized with a rotor position sensor. The power factor can be maintained at unity by varying the field current. The block diagrams of a self-controlled synchronous motor fed from a three-phase inverter or a cycloconverter are shown in Fig. 15-34.

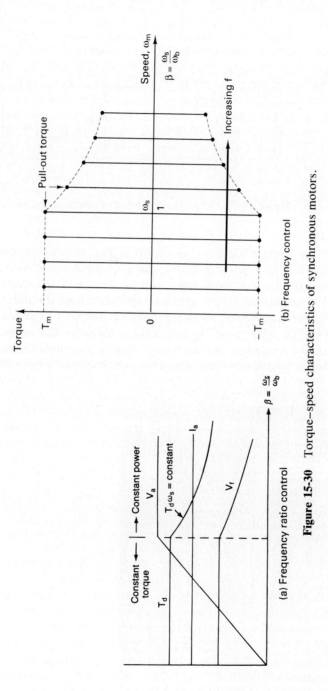

Figure 15-30 Torque–speed characteristics of synchronous motors.

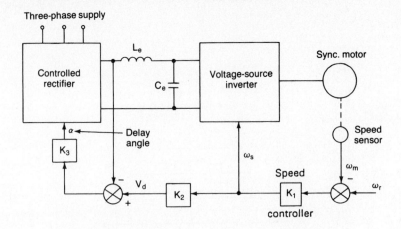

Figure 15-31 Volts/hertz control of synchronous motors.

For an inverter-fed drive as shown in Fig. 15-34a, the input source is dc. Depending on the type of inverter, the dc source could be current source, a constant current, or a controllable voltage source. The inverter's frequency is changed in proportion to the speed so that the armature and rotor mmf waves revolve at the same speed, thereby producing a steady torque at all speeds, as in a dc motor. The rotor position and inverter perform the same function as the brushes and commutator in a dc motor. Due to the similarity in operation with a dc motor, an inverter-fed self-controlled synchronous motor is known as a *com-*

Figure 15-32 A 3.3-MW synchronous motor compressor drive. (Reproduced by permission of Brush Electrical Machines Ltd., England.)

Figure 15-33 Control cubicle for the drive in Fig. 15-32. (Reproduced by permission of Brush Electrical Machines Ltd., England.)

mutatorless dc motor. If the synchronous motor is a permanent-magnet motor or a reluctance motor or a wound field motor with a brushless excitation, it is known as a *brushless* and *commutatorless dc motor* or simply a *brushless dc motor*. Connecting the field in series with the dc supply gives the characteristics of a dc series motor. The brushless dc motors offer the characteristics of dc motors and do not have limitations such as frequent maintenance and inability to operate in explosive environments. They are finding increasing applications in servo drives.

If the synchronous motor is fed from an ac source shown in Fig. 15-34b, it is called a *brushless* and *commutatorless ac motor* or simply a *brushless ac motor*. These ac motors are used for high-power applications (up to megawatts range) such as compressors, blowers, fans, conveyers, steel rolling mills, large ship steering, and cement plants. The self-control is also used for starting large synchronous motors in gas turbine and pump storage power plants.

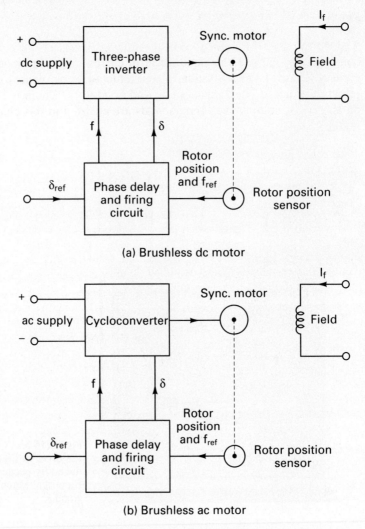

(a) Brushless dc motor

(b) Brushless ac motor

Figure 15-34 Self-controlled synchronous motors.

SUMMARY

Although ac drives require advanced control techniques for control of voltage, frequency, and current, they have advantages over dc drives. The voltage and frequency can be controlled by voltage-source inverters. The current and frequency can be controlled by current-source inverters. The slip power recovery schemes use controlled rectifiers to recover the slip power of induction motors. The most common method of closed-loop control of induction motors is volts/ hertz, flux, or slip control. Both squirrel-cage and wound-rotor motors are used in variable-speed drives. A voltage-source inverter can supply a number of motors connected in parallel, whereas a current-source inverter can supply only one motor.

Synchronous motors are constant-speed machines and their speeds can be controlled by voltage, frequency, and/or current. Synchronous motors are of six types: cylindrical rotors, salient poles, reluctance, permanent magnets, switched reluctance motors and brushless dc and ac motors. There is abundant literature on ac drives, so only the fundamentals are covered in this chapter.

REFERENCES

1 B. K. Bose, *Power Electronics and AC Drives*. Englewood Cliffs, N.J.: Prentice Hall, 1986.

2. B. K. Bose, *Adjustable AC Drives*. New York: IEEE Press, 1980.

3. S. B. Dewan, G. B. Slemon, and A. Straughen, *Power Semiconductor Drives*. New York: John Wiley & Sons, Inc., 1984.

4. W. Leonhard, "Control of ac machines with the help of electronics," *3rd IFAC Symposium on Control in Power Electronics and Electrical Drives*, Lausanne, Switzerland, Tutorial Session, September 1983, pp. 35–58.

5. W. Leonhard, *Control of Electrical Drives*. New York: Springer-Verlag, 1985.

6. Y. D. Landau, *Adaptive Control*. New York: Marcel Dekker, Inc., 1979.

7. H. Le-Huy, R. Perret, and D. Roye, "Microprocessor control of a current-fed synchronous motor drive." *IEEE Industry Applications Society Conference Record*, 1980, pp. 562–569.

8. A. Brickwedde, "Microprocessor-based adaptive speed and position control for electrical drives." *IEEE Industry Applications Society Conference Record*, 1984, pp. 411–417.

9. K. Masato, M. Yano, I. Kamiyana, and S. Yano, "Microprocessor-based vector control system for induction motor drives with rotor time constant identification." *IEEE Transactions on Industry Applications*, Vol. IA22, No. 3, 1986, pp. 453–459.

10. S. K. Biswas, S. Sahtiakumar, and J. Vi-thayathil, "High efficiency direct torque control scheme for CSI fed induction motor." *IEEE Industry Applications Society Conference Record*, 1986, pp. 216–221.

11. E. Prasad, J. F. Lindsay, and M. H. Rashid, "Parameter estimation and dynamic performance of permanent magnet synchronous motors." *IEEE Industry Applications Society Conference Record*, 1985, pp. 627–633.

12. T. J. E. Miller, "Converter volt-ampere requirements of the switched reluctance motor drive." *IEEE Industry Applications Society Conference Record*, 1984, pp. 813–819.

13. C. Wang, D. V. Novotny, and T. A. Lipo, "An automated rotor time constant measurement system for indirect field oriented drives." *IEEE Industry Applications Society Conference Record*, 1986, pp. 140–146.

14. R. Krishnan and P. Pillay, "Sensitivity analysis and comparison of parameter compensation scheme in vector control induction motor drives." *IEEE Industry Applications Society Conference Record*, 1986, pp. 155–161.

15. B. K. Bose, "Sliding mode control of induction motor." *IEEE Industry Applications Society Conference Record*, 1985, pp. 479–486.

16. A. Smith, "Static Scherbius system of induction motor speed control." *Proceedings IEE*, Vol. 124, 1977, pp. 557–565.

17. G. K. Dubey, *Power Semiconductor Controlled Drives*. Englewood Cliffs, N.J.: Prentice Hall, 1989.

18. E. Ho and P. C. Sen, "Decoupling control of induction motor drives." *IEEE Transactions on Industrial Electronics,* Vol. IE35, No. 2, 1988, pp. 253–262.

19. P. C. Sen, "Electric motor drives and control: past, present and future." *IEEE Transactions on Industrial Electronics,* Vol. IE37, No. 6, 1990, pp. 562–575.

REVIEW QUESTIONS

15-1. What are the types of induction motors?

15-2. What is a synchronous speed?

15-3. What is a slip of induction motors?

15-4. What is a slip frequency of induction motors?

15-5. What is the slip at starting of induction motors?

15-6. What are the torque–speed characteristics of induction motors?

15-7. What are various means for speed control of induction motors?

15-8. What are the advantages of volts/hertz control?

15-9. What is a base frequency of induction motors?

15-10. What are the advantages of current control?

15-11. What is a scalar control?

15-12. What is a vector control?

15-13. What is an adaptive control?

15-14. What is a static Kramer drive?

15-15. What is a static Scherbius drive?

15-16. What is a field-weakening mode of induction motor?

15-17. What are the effects of frequency control of induction motors?

15-18. What are the advantages of flux control?

15-19. What are the various types of synchronous motors?

15-20. What is the torque angle of synchronous motors?

15-21. What are the differences between salient-pole motors and reluctance motors?

15-22. What are the differences between salient-pole motors and permanent-magnet motors?

15-23. What is a pull-out torque of synchronous motors?

15-24. What is the starting torque of synchronous motors?

15-25. What are the torque–speed characteristics of synchronous motors?

15-26. What are the V-curves of synchronous motors?

15-27. What are the advantages of voltage-source inverter-fed drives?

15-28. What are the advantages and disadvantages of reluctance motor drives?

15-29. What are the advantages and disadvantages of permanent-magnet motors?

15-30. What is a switched reluctance motor?

15-31. What is a self-control mode of synchronous motors?

15-32. What is a brushless dc motor?

15-33. What is a brushless ac motor?

PROBLEMS

15-1. A three-phase 460-V 60-Hz eight-pole wye-connected induction motor has $R_s = 0.08 \ \Omega$, $R_r = 0.1 \ \Omega$, $X_s = 0.62 \ \Omega$, $X_r = 0.92 \ \Omega$, and $R_m = 6.7 \ \Omega$. The no-load loss, $P_{\text{no load}} = 300$ W. At a motor speed of 850 rpm, use the approximate equivalent circuit in Fig. 15-2 to determine **(a)** the synchronous speed ω_s; **(b)** the slip s; **(c)** the input current I_i; **(d)** the input power P_i; **(e)** the input power factor of the supply, PF$_s$; **(f)** the gap power P_g; **(g)** the rotor copper loss P_{ru}; **(h)** the stator copper loss P_{su}; **(i)** the developed torque T_d; **(j)** the efficiency; **(k)** the starting rotor current I_{rs} and the starting torque T_s; **(l)** the slip for maximum torque, s_m; **(m)** the maximum motoring developed torque, T_{mm}; and **(n)** the maximum regenerative developed torque, T_{mr}.

15-2. Repeat Prob. 15-1 if R_s is negligible.

15-3. Repeat Prob. 15-1 if the motor has two poles and the parameters are $R_s = 1.02\ \Omega$, $R_r = 0.35\ \Omega$, $X_s = 0.72\ \Omega$, $X_r = 1.08\ \Omega$, and $R_m = 60\ \Omega$. The no-load loss is $P_{\text{no load}} = 70$ W and the rotor speed is 3450 rpm.

15-4. A three phase 460-V 60-Hz six-pole wye-connected induction motor has $R_s = 0.32\ \Omega$, $R_r = 0.18\ \Omega$, $X_s = 1.04\ \Omega$, $X_r = 1.6\ \Omega$, and $X_m = 18.8\ \Omega$. The no-load loss, $P_{\text{no load}}$, is negligible. The load torque, which is proportional to speed squared, is 180 N·m at 1180 rpm. If the motor speed is 950 rpm, determine **(a)** the load torque demand T_L; **(b)** the rotor current I_r; **(c)** the stator supply voltage V_a; **(d)** the motor input current I_i; **(e)** the motor input power P_i; **(f)** the slip for maximum current s_a; **(g)** the maximum rotor current, $I_{r(max)}$; **(h)** the speed at maximum rotor current, ω_a; and **(i)** the torque at the maximum current, T_a.

15-5. Repeat Prob. 15-4 if R_s is negligible.

15-6. Repeat Prob. 15-4 if the motor has four poles and the parameters are $R_s = 0.25\ \Omega$, $R_r = 0.14\ \Omega$, $X_s = 0.7\ \Omega$, $X_r = 1.05\ \Omega$, and $X_m = 20.6\ \Omega$. The load torque is 121 N·m at 1765 rpm. The motor speed is 1525 rpm.

15-7. A three-phase 460-V 60-Hz six-pole wye-connected wound-rotor induction motor whose speed is controlled by slip power as shown in Fig. 15-6a has the following parameters: $R_s = 0.11\ \Omega$, $R_r = 0.09\ \Omega$, $X_s = 0.4\ \Omega$, $X_r = 0.6\ \Omega$, and $X_m = 11.6\ \Omega$. The turns ratio of the rotor to stator windings is $n_m = N_r/N_s = 0.9$. The inductance L_d is very large and its current, I_d, has negligible ripple. The values of R_s, R_r, X_s, and X_r for the equivalent circuit in Fig. 15-2 can be considered negligible compared to the effective impedance of L_d. The no-load loss is 275 W. The load torque, which is proportional to speed squared, is 455 N·m at 1175 rpm. **(a)** If the motor has to operate with a minimum speed of 750 rpm, determine the resistance, R. With this value of R, if the desired speed is 950 rpm, calculate **(b)** the inductor current I_d, **(c)** the duty cycle k of the chopper, **(d)** the dc voltage V_d, **(e)** the efficiency, and **(f)** the input power factor PF_s of the drive.

15-8. Repeat Prob. 15-7 if the minimum speed is 650 rpm.

15-9. Repeat Prob. 15-7 if the motor has eight poles and the motor parameters are $R_s = 0.08\ \Omega$, $R_r = 0.1\ \Omega$, $X_s = 0.62\ \Omega$, $X_r = 0.92\ \Omega$, and $R_m = 6.7\ \Omega$. The no-load loss is $P_{\text{no load}} = 300$ W. The load torque, which is proportional to speed, is 604 N·m at 885 rpm. The motor has to operate with a minimum speed of 650 rpm, and the desired speed is 750 rpm.

15-10. A three-phase 460-V 60-Hz six-pole wye-connected wound-rotor induction motor whose speed is controlled by a static Kramer drive as shown in Fig. 15-6b has: $R_s = 0.11\ \Omega$, $R_r = 0.09\ \Omega$, $X_s = 0.4\ \Omega$, $X_r = 0.6\ \Omega$, $X_m = 11.6\ \Omega$. The turns ratio of the rotor to stator windings is $n_m = N_r/N_s = 0.9$. The inductance, L_d, is very large and its current, I_d, has negligible ripple. The values of R_s, R_r, X_s, and X_r for the equivalent circuit in Fig. 15-2 can be considered negligible compared to the effective impedance of L_d. The no-load loss is 275 W. The turns ratio of the converter ac voltage to supply voltage is $n_c = N_a/N_b = 0.5$. If the motor is required to operate at a speed of 950 rpm, calculate **(a)** the inductor current I_d, **(b)** the dc voltage V_d, **(c)** the delay angle α of the converter, **(d)** the efficiency, and **(e)** the input power factor PF_s of the drive. The load torque, which is proportional to speed squared, is 455 N·m at 1175 rpm.

15-11. Repeat Prob. 15-10 for $n_c = 0.9$.

15-12. For Prob. 15-10 plot the power factor against the turns ratio n_c.

15-13. A three-phase 56-kW 3560-rpm 460-V 60-Hz two-pole wye-connected induction motor has the following parameters: $R_s = 0$, $R_r = 0.18 \ \Omega$, $X_s = 0.13 \ \Omega$, $X_r = 0.2 \ \Omega$, and $X_m = 11.4 \ \Omega$. The motor is controlled by varying the supply frequency. If the breakdown torque requirement is 160 N·m, calculate (a) the supply frequency, and (b) the speed ω_m at the maximum torque.

15-14. If $R_s = 0.07 \ \Omega$ and the frequency is changed from 60 Hz to 40 Hz in Prob. 15-13, determine the change in breakdown torque.

15-15. The motor in Prob. 15-13 is controlled by a constant volts/hertz ratio corresponding to the rated voltage and rated frequency. Calculate the maximum torque, T_m, and the corresponding speed, ω_m, for supply frequency of (a) 60 Hz, and (b) 30 Hz.

15-16. Repeat Prob. 15-15 if R_s is 0.2 Ω.

15-17. A three-phase 40-hp 880-rpm 60-Hz eight-pole wye-connected induction motor has the following parameters: $R_s = 0.19 \ \Omega$, $R_r = 0.22 \ \Omega$, $X_s = 1.2 \ \Omega$, $X_r = 1.8 \ \Omega$, and $X_m = 13 \ \Omega$. The no-load loss is negligible. The motor is controlled by a current-source inverter and the input current is maintained constant at 50 A. If the frequency is 40 Hz and the developed torque is 200 N·m, determine (a) the slip for maximum torque s_m and maximum torque T_m, (b) the slip s, (c) the rotor speed ω_m, (d) the terminal voltage per phase V_a, and (e) the power factor PF$_m$.

15-18. Repeat Prob. 15-17 if frequency is 80 Hz.

15-19. A three-phase 460-V 60-Hz 10-pole wye-connected cylindrical rotor synchronous motor has a synchronous reactance of $X_s = 0.8 \ \Omega$ per phase and the armature resistance is negligible. The load torque, which is proportional to the speed squared, is $T_L = 1250$ N·m at 720 rpm. The power factor is maintained at 0.8 lagging by field control and the voltage-to-frequency ratio is kept constant at the rated value. If the inverter frequency is 45 Hz and the motor speed is 540 rpm, calculate (a) the input voltage V_a, (b) the armature current I_a, (c) the excitation voltage V_f, (d) the torque angle δ, and (e) the pull-out torque T_p.

15-20. A three-phase 230-V 60-Hz 40-kW eight-pole wye-connected salient-pole synchronous motor has $X_d = 2.5 \ \Omega$ and $X_q = 0.4$. The armature resistance is negligible. If the motor operates with an input power of 25 kW at a leading power factor of 0.86, determine (a) the torque angle δ, (b) the excitation voltage V_f, and (c) the torque T_d.

15-21. A three-phase 230-V 60-Hz 10-pole wye-connected reluctance motor has $X_d = 18.5 \ \Omega$ and $X_q = 3 \ \Omega$. The armature resistance is negligible. The load torque, which is proportional to speed, is $T_L = 12.5$ N·m. The voltage-to-frequency ratio is maintained constant at the rated value. If the supply frequency is 60 Hz, determine (a) the torque angle δ, (b) the line current I_a, and (c) the input power factor PF$_m$.

16

Protection of devices and circuits

16-1 INTRODUCTION

Due to the reverse recovery process of power devices and switching actions in the presence of circuit inductances, voltage transients occur in the converter circuits. Even in carefully designed circuits, short-circuit fault conditions may exist, resulting in an excessive current flow through the devices. The heat produced by losses in a semiconductor device must be dissipated sufficiently and effectively to operate the device within its upper temperature limit. The reliable operation of a converter would require ensuring that at all times the circuit conditions do not exceed the ratings of the power devices, by providing protection against overvoltage, overcurrent, and overheating. In practice, the power devices are protected from (1) thermal runaway by heat sinks, (2) high dv/dt and di/dt by snubbers, (3) reverse recovery transients, (4) supply and load side transients, and (5) fault conditions by fuses.

16-2 COOLING AND HEAT SINKS

Due to on-state and switching losses, heat is generated within the power device. This heat must be transferred from the device to a cooling medium to maintain the operating junction temperature within the specified range. Although this heat transfer can be accomplished by conduction, convection, or radiation, natural or forced-air, convection cooling is commonly used in industrial applications.

The heat must flow from the device to the case and then to the heat sink in the cooling medium. If P_A is the average power loss in the device, the electrical analog of a device, which is mounted on a heat sink, is shown in Fig. 16-1. The

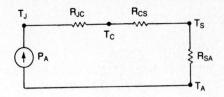

Figure 16-1 Electrical analog of heat transfer.

junction temperature of a device T_J is given by

$$T_J = P_A(R_{JC} + R_{CS} + R_{SA}) \qquad (16\text{-}1)$$

where R_{JC} = thermal resistance from junction to case, °C/W
R_{CS} = thermal resistance from case to sink, °C/W
R_{SA} = thermal resistance from sink to ambient, °C/W
T_A = ambient temperature, °C

R_{JC} and R_{CS} are normally specified by the power device manufacturers. Once the device power loss, P_A, is known, the required thermal resistance of the heat sink can be calculated for a known ambient temperature, T_A. The next step is to choose a heat sink and its size which would meet the thermal resistance requirement.

A wide variety of extruded aluminum heat sinks are commercially available and they use cooling fins to increase the heat transfer capability. The thermal resistance characteristics of a typical heat sink with natural and forced cooling are shown in Fig. 16-2, where the power dissipation against the sink temperature rise

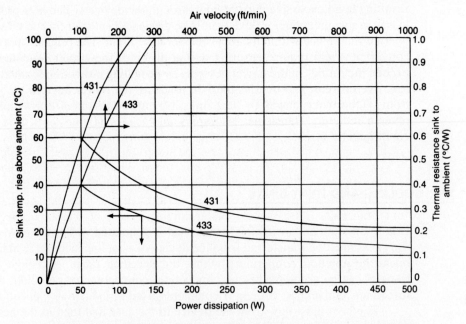

Figure 16-2 Thermal resistance characteristics. (Courtesy of EG&G Wakefield Engineering.)

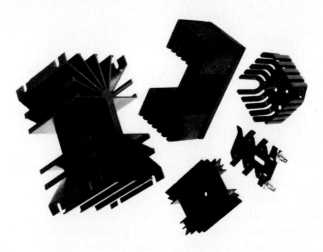

Figure 16-3   Heat sinks. (Courtesy of EG&G Wakefield Engineering.)

is depicted for natural cooling. In forced cooling, the thermal resistance decreases with the air velocity. However, above a certain velocity, the reduction in thermal resistance is not significant. Heat sinks of various types are shown in Fig. 16-3.

The contact area between the device and heat sink is extremely important to minimize the thermal resistance between the case and sink. The surfaces should be flat, smooth, and free of dirt, corrosion, and surface oxides. Silicone greases are normally applied to improve the heat transfer capability and to minimize the formation of oxides and corrosion.

The device must be mounted properly on the heat sink to obtain the correct mounting pressure between the mating surfaces. The proper installation procedures are usually recommended by the device manufacturers. In case of stud-mounted devices, excessive mounting torques may cause mechanical damage of the silicon wafer; and the stud and nut should not be greased or lubricated because the lubrication increases the tension on the stud.

The device may be cooled by heat pipes partially filled with low-vapor-pressure liquid. The device is mounted on one side of the pipe and a condensing mechanism (or heat sink) on the other side as shown in Fig. 16-4. The heat produced by the device vaporizes the liquid and the vapor, then flows to the condensing end, where it condenses and the liquid returns to the heat source. The device may be at some distance from the heat sink.

In high-power applications, the devices are more effectively cooled by liquids, normally, oil or water. Water cooling is very efficient and approximately

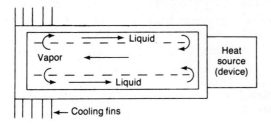

Figure 16-4   Heat pipes.

Figure 16-5 Water-cooled ac switches. (Courtesy of Powerex, Inc.)

three times more effective than oil cooling. However, it is necessary to use distilled water to minimize corrosion and antifreeze to avoid freezing. Oil is flammable. Oil cooling, which may be restricted to some applications, provides good insulation and eliminates the problems of corrosion and freezing. Heat pipes and liquid-cooled heat sinks are commercially available. Two water-cooled ac switches are shown in Fig. 16-5. Power converters are available in assembly units as shown in Fig. 16-6.

The thermal impedance of a power device is very small, and as a result the junction temperature of the device varies with the instantaneous power loss. The instantaneous junction temperature must always be maintained lower than the acceptable value. A plot of the transient thermal impedance versus square-wave pulse duration is supplied by the device manufacturers as a part of data sheet. From the knowledge of current waveform through a device, a plot of power loss against time can be determined and then the transient impedance characteristics can be used to calculate the temperature variations with time. If the cooling medum fails in practical systems, the temperature rise of the heat sinks is normally served to switch off the power converters, especially in high power applications.

The step response of a first order system can be applied to express the transient thermal impedance. If Z_0 is the steady-state junction case thermal impedance, the instantaneous thermal impedance can be expressed as

$$Z(t) = Z_0(1 - e^{-t/\tau_{th}}) \tag{16-2}$$

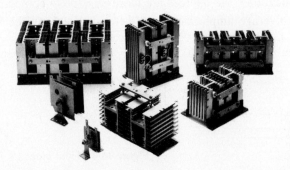

Figure 16-6 Assembly units. (Courtesy of Powerex, Inc.)

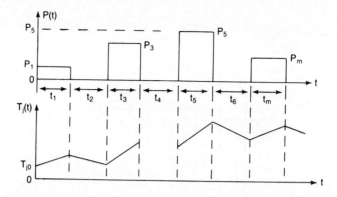

Figure 16-7 Junction temperature with rectangular power pulses.

where τ_{th} is the thermal time constant of the device. If the power loss is P_d, the instantaneous junction temperature rise above the case is

$$T_J = P_d Z(t) \tag{16-3}$$

If the power loss is a pulsed type as shown in Fig. 16-7, Eq. (16-3) can be applied to plot the step responses of the junction temperature, $T_J(t)$. If t_n is the duration of nth power pulse, the corresponding thermal impedances at the beginning and end of nth pulse are $Z_0 = Z(t = 0) = 0$ and $Z_n = Z(t = t_n)$, respectively. The thermal impedance $Z_n = Z(t = t_n)$ corresponding to the duration of t_n can be found from the transient thermal impedance characteristics. If $P_1, P_2, P_3, \ldots$ are the power pulses with $P_2 = P_4 = \cdots = 0$, the junction temperature at the end of mth pulse can be expressed as

$$T_J(t) = T_{J0} + P_1(Z_1 - Z_2) + P_3(Z_3 - Z_4) + P_5(Z_5 - Z_6) + \cdots$$
$$= T_{J0} + \sum_{n=1,3,\ldots}^{m} P_n(Z_n - Z_{n+1}) \tag{16-4}$$

where T_{J0} is the initial junction temperature. The negative signs of $Z_2, Z_4, \ldots$ signify that the junction temperature falls during the intervals $t_2, t_4, t_6, \ldots$

The step response concept of junction temperature can be extended to other power waveforms. A waveform of any shape can be represented approximately by rectangular pulses of equal or unequal duration, with the amplitude of each pulse being equal to the average amplitude of the actual pulse over the same period. The accuracy of such approximations can be improved by increasing the number of pulses and reducing the duration of each pulse. This is shown in Fig. 16-8.

The junction temperature at the end of mth pulse can be found from

$$T_J(t) = T_{J0} + Z_1 P_1 + Z_2(P_2 - P_1) + Z_3(P_3 - P_2) + \cdots$$
$$= T_{J0} + \sum_{n=1,2\ldots}^{m} Z_n(P_n - P_{n-1}) \tag{16-5}$$

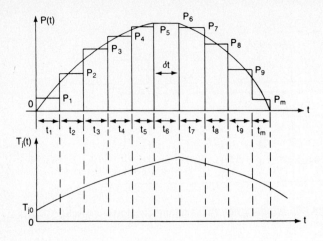

Figure 16-8 Approximation of a power pulse by rectangular pulses.

where Z_n is the impedance at the end of nth pulse of duration $t_n = \delta t$. P_n is the power loss for the nth pulse and $P_0 = 0$. t is the time interval.

Example 16-1

The power loss of a device is shown in Fig. 16-9. Plot the instantaneous junction temperature rise above the case. $P_2 = P_4 = P_6 = 0$, $P_1 = 800$ W, $P_3 = 1200$ W, and $P_5 = 600$ W. For $t_1 = t_3 = t_5 = 1$ ms, the data sheet gives

$$Z(t = t_1) = Z_1 = Z_3 = Z_5 = 0.035°C/W$$

For $t_2 = t_4 = t_6 = 0.5$ ms,

$$Z(t = t_2) = Z_2 = Z_4 = Z_6 = 0.025°C/W$$

Solution   Equation (16-4) can be applied directly to calculate the junction temperature rise.

$$\Delta T_J(t = 1 \text{ ms}) = T_J(t = 1 \text{ ms}) - T_{J0} = Z_1 P_1 = 0.035 \times 800 = 28°C$$

$$\Delta T_J(t = 1.5 \text{ ms}) = 28 - Z_2 P_1 = 28 - 0.025 \times 800 = 8°C$$

$$\Delta T_J(t = 2.5 \text{ ms}) = 8 + Z_3 P_3 = 8 + 0.035 \times 1200 = 50°C$$

$$\Delta T_J(t = 3 \text{ ms}) = 50 - Z_4 P_3 = 50 - 0.025 \times 1200 = 20°C$$

$$\Delta T_J(t = 4 \text{ ms}) = 20 + Z_5 P_5 = 20 + 0.035 \times 600 = 41°C$$

$$\Delta T_J(t = 4.5 \text{ ms}) = 41 - Z_6 P_5 = 41 - 0.025 \times 600 = 26°C$$

The junction temperature rise above case is shown in Fig. 16-10.

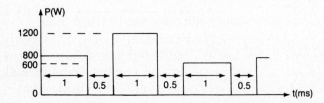

Figure 16-9   Device power loss.

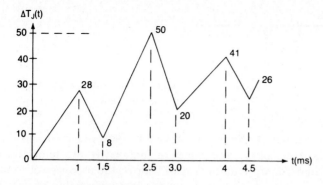

Figure 16-10 Junction temperature rise for Example 16-1.

16-3 SNUBBER CIRCUITS

An *RC* snubber is normally connected across a semiconductor device to limit the
dv/dt within the maximum allowable rating. The snubber could be polarized or
unpolarized. A forward-polarized snubber is suitable when a thyristor or transistor is connected with an antiparallel diode as shown in Fig. 16-11a. The resistor,
R, limits the forward *dv/dt*; and R_1 limits the discharge current of the capacitor
when the device is turned on.

A reverse-polarized snubber which limits the reverse *dv/dt* is shown in
Fig. 16-11b, where R_1 limits the discharge current of the capacitor. The capacitor
does not discharge through the device, resulting in reduced losses in the device.

When a pair of thyristors is connected in inverse parallel, the snubber must
be effective in either direction. An unpolarized snubber is shown in Fig. 16-11c.

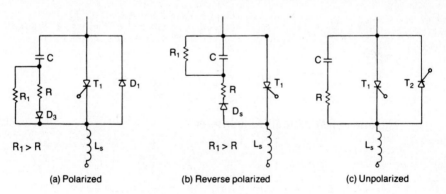

Figure 16-11 Snubber networks.

16-4 REVERSE RECOVERY TRANSIENTS

Due to the reverse recovery time t_{rr} and recovery current I_R, an amount of energy
is trapped in the circuit inductances and as a result transient voltage appears
across the device. In addition to *dv/dt* protection, the snubber limits the peak
transient voltage across the device. The equivalent circuit for a circuit arrange-

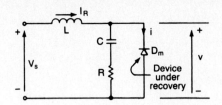

Figure 16-12 Equivalent circuit during recovery.

ment is shown in Fig. 16-12, where the initial capacitor voltage is zero and the inductor carries an initial current of I_R. The values of snubber RC are selected so that the circuit is slightly underdamped and Fig. 16-13 shows the recovery current and transient voltage. Critical damping usually results in a large value of initial reverse voltage RI_R; and insufficient damping causes large overshoot of the transient voltage. In the following analysis, it is assumed that the recovery is abrupt and the recovery current is suddenly switched to zero.

The snubber current is expressed as

$$L \frac{di}{dt} + Ri + \frac{1}{C} \int i \, dt + v_c(t=0) = V_s \tag{16-6}$$

$$v = V_s - L \frac{di}{dt} \tag{16-7}$$

with initial conditions $i(t=0) = I_R$ and $v_c(t=0) = 0$. We have seen in Section 3-3 that the form of the solution for Eq. (16-6) depends on the values of RLC. For an underdamped case, the solutions of Eqs. (16-6) and (16-7) yield the reverse voltage across the device as

$$v(t) = V_s - (V_s - RI_R)\left(\cos \omega t - \frac{\alpha}{\omega} \sin \omega t\right) e^{-\alpha t} + \frac{I_R}{\omega C} e^{-\alpha t} \sin \omega t \tag{16-8}$$

where

$$\alpha = \frac{R}{2L} \tag{16-9}$$

The undamped natural frequency is

$$\omega_0 = \frac{1}{\sqrt{LC}} \tag{16-10}$$

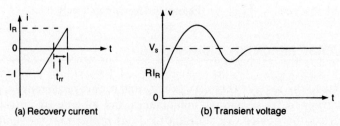

(a) Recovery current (b) Transient voltage

Figure 16-13 Recovery transient.

The damping ratio is

$$\delta = \frac{\alpha}{\omega_0} = \frac{R}{2} \sqrt{\frac{C}{L}} \tag{16-11}$$

and the damped natural frequency is

$$\omega = \sqrt{\omega_0^2 - \alpha^2} = \omega_0 \sqrt{1 - \delta^2} \tag{16-12}$$

Differentiating Eq. (16-8) yields

$$\frac{dv}{dt} = (V_s - RI_R) \left(2\alpha \cos \omega t + \frac{\omega^2 - \alpha^2}{\omega} \sin \omega t \right) e^{-\alpha t}$$

$$+ \frac{I_R}{C} \left(\cos \omega t - \frac{\alpha}{\omega} \sin \omega t \right) e^{-\alpha t} \tag{16-13}$$

The initial reverse voltage and dv/dt can be found from Eqs. (16-8) and (16-13) by setting $t = 0$:

$$v(t = 0) = RI_R \tag{16-14}$$

$$\frac{dv}{dt}\bigg|_{t=0} = (V_s - RI_R)2\alpha + \frac{I_R}{C} = \frac{(V_s - RI_R)R}{L} + \frac{I_R}{C}$$

$$= V_s \omega_0 (2\delta - 4d\delta^2 + d) \tag{16-15}$$

where the current factor (or ratio) d is given by

$$d = \frac{I_R}{V_s} \sqrt{\frac{L}{C}} = \frac{I_R}{I_p} \tag{16-16}$$

If the initial dv/dt in Eq. (16-15) is negative, the initial inverse voltage RI_R is the maximum and this may produce a destructive dv/dt. For a positive dv/dt, $V_s \omega_0 (2\delta - 4d\delta^2 + d) > 0$ or

$$\delta < \frac{1 + \sqrt{1 + 4d^2}}{4d} \tag{16-17}$$

and the reverse voltage will be maximum at $t = t_1$. The time t_1, which can be obtained by setting Eq. (16-13) equal to zero, is found as

$$\tan \omega t_1 = \frac{\omega[(V_s - RI_R) 2\alpha + IR/C]}{(V_s - RI_R)(\omega^2 - \alpha^2) - \alpha I_R/C} \tag{16-18}$$

and the peak voltage can be found from Eq. (16-8):

$$V_p = v(t = t_1) \tag{16-19}$$

The peak reverse voltage depends on the damping ratio, δ, and the current factor, d. For a given value of d, there is an optimum value of damping ratio, δ_o, which will minimize the peak voltage. However, the dv/dt varies with d and minimizing the peak voltage may not minimize the dv/dt. It is necessary to make a compromise between the peak voltage, V_p, and dv/dt. McMurray [1] proposed to minimize the product $V_p(dv/dt)$ and the optimum design curves are shown in Fig.

16-14, where dv/dt is the average value over time t_1 and d_o is the optimum value of current factor.

The energy stored in the inductor, L, which is transferred to the snubber capacitor, C, is dissipated mostly in the snubber resistor. This power loss is dependent on the switching frequency and load current. For high-power converters, where the snubber loss is significant, a nondissipative snubber which uses an energy recovery transformer as shown in Fig. 16-15 can improve the circuit efficiency. When the primary current rises, the induced voltage E_2 is positive and diode D_1 is reverse biased. If the recovery current of diode D_m starts to fall, the induced voltage E_2 becomes negative and diode D_1 conducts, returning energy to the dc supply.

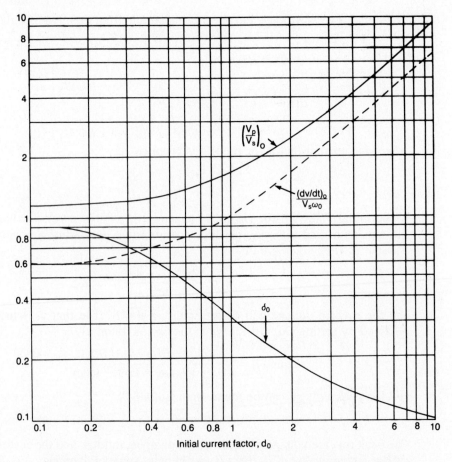

Figure 16-14 Optimum snubber parameters for compromise design. (Reproduced from W. McMurray, "Optimum snubbers for power semiconductors," *IEEE Transactions on Industry Applications*, Vol. 1A8, No. 5, 1972, pp. 503–510, Fig. 7, © 1972 by IEEE.)

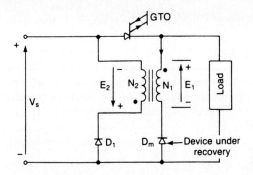

Figure 16-15 Nondissipative snubber.

Example 16-2*

The recovery current of a diode as shown in Fig. 16-12 is $I_R = 20$ A and the circuit inductance is $L = 50 \ \mu H$. The input voltage is $V_s = 220$ V. If it is necessary to limit the peak transient voltage to 1.5 times the input voltage, determine (a) the optimum value of current factor α_o, (b) the optimum damping factor δ_o, (c) the snubber capacitance C, (d) the snubber resistance R, (e) the average dv/dt, and (f) the initial reverse voltage.

Solution $I_R = 20$ A, $L = 50 \ \mu H$, $V_s = 220$ V, and $V_p = 1.5 \times 220 = 330$ V. For $V_p/V_s = 1.5$, Fig. 16-14 gives:

(a) The optimum current factor $d_o = 0.75$.

(b) The optimum damping factor $\delta_o = 0.4$.

(c) From Eq. (16-16), the snubber capacitance (with $d = d_o$) is

$$C = L \left[\frac{I_R}{dV_s} \right]^2 \qquad (16\text{-}20)$$

$$= 50 \left[\frac{20}{0.75 \times 220} \right]^2 = 0.735 \ \mu F$$

(d) From Eq. (16-11), the snubber resistance is

$$R = 2\delta \ \sqrt{\frac{L}{C}} \qquad (16\text{-}21)$$

$$= 2 \times 0.4 \ \sqrt{\frac{50}{0.735}} = 6.6 \ \Omega$$

(e) From Eq. (16-10),

$$\omega_0 = \frac{10^6}{\sqrt{50 \times 0.735}} = 164{,}957 \ \text{rad/s}$$

From Fig. 16-14,

$$\frac{dv/dt}{V_s \omega_0} = 0.88$$

or

$$\frac{dv}{dt} = 0.88 V_s \omega_0 = 0.88 \times 220 \times 164{,}957 = 31.9 \ \text{V}/\mu s$$

(f) From Eq. (16-14), the initial reverse voltage is

$$v(t = 0) = 6.6 \times 20 = 132 \text{ V}$$

Example 16-3*

An *RC*-snubber circuit as shown in Fig. 16-11c has $C = 0.75 \, \mu\text{F}$, $R = 6.6 \, \Omega$, and input voltage, $V_s = 220 \text{ V}$. The circuit inductance is $L = 50 \, \mu\text{H}$. Determine (a) the peak forward voltage V_p, (b) the initial dv/dt, and (c) the maximum dv/dt.

Solution $R = 6.6 \, \Omega$, $C = 0.75 \, \mu\text{F}$, $L = 50 \, \mu\text{H}$, and $V_s = 220 \text{ V}$. By setting $I_R = 0$, the forward voltage across the device can be determined from Eq. (16-8),

$$v(t) = V_s - V_s \left(\cos \omega t - \frac{\alpha}{\omega} \sin \omega t\right) e^{-\alpha t} \tag{16-22}$$

From Eq. (16-13) for $I_R = 0$,

$$\frac{dv}{dt} = V_s \left(2\alpha \cos \omega t + \frac{\omega^2 - \alpha^2}{\omega} \sin \omega t\right) e^{-\alpha t} \tag{16-23}$$

The initial dv/dt can be found either from Eq. (16-23) by setting $t = 0$, or from Eq. (16-15) by setting $I_R = 0$

$$\frac{dv}{dt}\bigg|_{t=0} = V_s 2\alpha = \frac{V_s R}{L} \tag{16-24}$$

The forward voltage will be maximum at $t = t_1$. The time t_1, which can be obtained either by setting Eq. (16-23) equal to zero, or by setting $I_R = 0$ in Eq. (16-18), is given by

$$\tan \omega t_1 = -\frac{2\alpha\omega}{\omega^2 - \alpha^2} \tag{16-25}$$

$$\cos \omega t_1 = -\frac{\omega^2 - \alpha^2}{\omega^2 + \alpha^2} \tag{16-26}$$

$$\sin \omega t_1 = \frac{2\alpha\omega}{\omega^2 + \alpha^2} \tag{16-27}$$

Substituting Eqs. (16-26) and (16-27) into Eq. (16-22), the peak voltage is found as

$$V_p = v(t = t_1) = V_s(1 + e^{-\alpha t_1}) \tag{16-28}$$

where

$$\alpha t_1 = \frac{\delta}{\sqrt{1 - \delta^2}} \left(\pi - \tan^{-1} \frac{-2\delta\sqrt{1 - \delta^2}}{1 - 2\delta^2}\right) \tag{16-29}$$

Differentiating Eq. (16-23) with respect to t, and setting equal to zero, dv/dt will be maximum at $t = t_m$ when

$$-\frac{\alpha(3\omega^2 - \alpha^2)}{\omega} \sin \omega t_m + (\omega^2 - 3\alpha^2) \cos \omega t_m = 0$$

or

$$\tan \omega t_m = \frac{\omega(\omega^2 - 3\alpha^2)}{\alpha(3\omega^2 - \alpha^2)} \tag{16-30}$$

Substituting the value of t_m in Eq. (16-23) and simplifying the sine and cosine terms yield the maximum value of dv/dt,

$$\frac{dv}{dt}\Big|_{max} = \sqrt{\omega^2 + \alpha^2}\, e^{-\alpha t_m} \qquad \text{for } \delta \leq 0.5 \qquad (16\text{-}31)$$

For a maximum to occur, $d(dv/dt)/dt$ must be positive if $t \leq t_m$; and Eq. (16-30) gives the necessary condition as

$$\omega^2 - 3\alpha^2 \geq 0 \qquad \text{or} \qquad \frac{\alpha}{\omega} \leq \frac{1}{\sqrt{3}} \qquad \text{or} \qquad \delta \leq 0.5$$

Equation (16-31) is valid for $\delta \leq 0.5$. For $\delta > 0.5$, dv/dt which becomes maximum when $t = 0$ is obtained from Eq. (16-23),

$$\frac{dv}{dt}\Big|_{max} = \frac{dv}{dt}\Big|_{t=0} = V_s 2\alpha = \frac{V_s R}{L} \qquad \text{for } \delta > 0.5 \qquad (16\text{-}32)$$

(a) From Eq. (16-9), $\alpha = 6.6/(2 \times 50 \times 10^{-6}) = 66,000$ and from Eq. (16-10),

$$\omega_0 = \frac{10^6}{\sqrt{50 \times 0.75}} = 163,299 \text{ rad/s}$$

From Eq. (16-11), $\delta = (6.6/2)\sqrt{0.75/50} = 0.404$, and from Eq. (16-12),

$$\omega = 163,299\sqrt{1 - 0.404^2} = 149,379 \text{ rad/s}$$

From Eq. (16-29), $t_1 = 15.46\ \mu s$; therefore, Eq. (16-28) yields the peak voltage, $V_p = 220(1 + 0.36) = 299.3$ V.

(b) Equation (16-24) gives the initial dv/dt of $(220 \times 6.6/50) = 29$ V/μs.

(c) Since $\delta < 0.5$, Eq. (16-31) should be used to calculate the maximum dv/dt. From Eq. (16-30), $t_m = 2.16\ \mu s$ and Eq. (16-31) yields the maximum dv/dt as 31.2 V/μs.

Note. $V_p = 299.3$ V and the maximum $dv/dt = 31.2\ \mu s$. The optimum snubber design in Example 16-2 gives $V_p = 330$ V, and the average $dv/dt = 31.9\ \mu s$.

16-5 SUPPLY- AND LOAD-SIDE TRANSIENTS

A transformer is normally connected to the input side of the converters. Under steady-state condition, an amount of energy is stored in the magnetizing inductance, L_m, of transformer, and switching off the supply produces a transient voltage to the input of the converter. A capacitor may be connected across the primary or secondary of the transformer to limit the transient voltage as shown in Fig. 16-16a, and in practice a resistance is also connected in series with the capacitor to limit the transient voltage oscillation.

Let us assume that the switch has been closed for a sufficiently long time. Under steady-state conditions, $v_s = V_m \sin \omega t$ and the magnetizing current is given by

$$L_m \frac{di}{dt} = V_m \sin \omega t$$

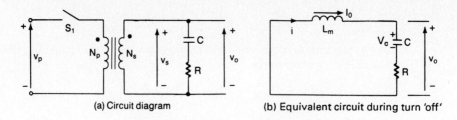

(a) Circuit diagram   (b) Equivalent circuit during turn 'off'

Figure 16-16   Switching off transient.

which gives

$$i(t) = -\frac{V_m}{\omega L_m} \cos \omega t$$

If the switch is turned off at $\omega t = \theta$, the capacitor voltage at the beginning of switch-off is

$$V_c = V_m \sin \theta \qquad (16\text{-}33)$$

and the magnetizing current is

$$I_0 = -\frac{V_m}{\omega L_m} \cos \theta \qquad (16\text{-}34)$$

The equivalent circuit during the transient condition is shown in Fig. 16-16b and the capacitor current is expressed as

$$L_m \frac{di}{dt} + Ri + \frac{1}{C} \int i\, dt + v_c(t = 0) = 0 \qquad (16\text{-}35)$$

and

$$v_0 = -L_m \frac{di}{dt} \qquad (16\text{-}36)$$

with initial conditions $i(t = 0) = -I_0$ and $v_c(t = 0) = V_c$. The transient voltage $v_0(t)$ can be determined from Eqs. (16-35) and (16-36), for underdamped conditions. A damping ratio of $\delta = 0.5$ is normally satisfactory. The analysis can be simplified by assuming small damping tending to zero, i.e., $\delta = 0$, (or, $R = 0$). Eq. (7-19), which is similar to Eq. (16-35), can be applied to determine the transient voltage, $v_o(t)$. The transient voltage, $v_o(t)$ is the same as the capacitor voltage, $v_c(t)$.

$$v_0(t) = v_c(t) = V_c \cos \omega_0 t + I_0 \sqrt{\frac{L_m}{C}} \sin \omega_0 t \qquad (16\text{-}37)$$

$$= \left(V_c^2 + I_0^2 \frac{L_m}{C} \right)^{1/2} \sin(\omega_0 t + \phi)$$

$$= V_m \left(\sin^2 \theta + \frac{1}{\omega^2 L_m C} \cos^2 \theta \right)^{1/2} \sin(\omega_0 t + \phi)$$

$$= V_m \left(1 + \frac{\omega_0^2 - \omega^2}{\omega^2} \cos^2 \theta \right)^{1/2} \sin(\omega_0 t + \phi) \tag{16-38}$$

where

$$\phi = \tan^{-1} \frac{V_c}{I_0} \sqrt{\frac{C}{L_m}} \tag{16-39}$$

and

$$\omega_0 = \frac{1}{\sqrt{CL_m}} \tag{16-40}$$

If $\omega_0 < \omega$, the transient voltage in Eq. (16-38) which will be maximum when $\cos \theta = 0$ (or $\theta = 90°$) is

$$V_p = V_m \tag{16-41}$$

In practice, $\omega_0 > \omega$ and the transient voltage, which will be maximum when $\cos \theta = 1$ (or $\theta = 0°$), is

$$V_p = V_m \frac{\omega_0}{\omega} \tag{16-42}$$

which gives the peak transient voltage due to turning off the supply. Using the voltage and current relationship in a capacitor, the required amount of capacitance to limit the transient voltage can be determined from

$$C = \frac{I_0}{V_p \omega_0} \tag{16-43}$$

Substituting ω_0 from Eq. (16-42) into Eq. (16-43) gives us

$$C = \frac{I_0 V_m}{V_p^2 \omega} \tag{16-44}$$

Now with the capacitor connected across the transformer secondary, the maximum instantaneous capacitor voltage will depend on the instantaneous ac input voltage at the instant of switching on the input voltage. The equivalent circuit during switch-on is shown in Fig. 16-17, where L is the equivalent supply inductance plus the leakage inductance of the transformer.

Under normal operation, an amount of energy is stored in the supply inductance and the leakage inductance of the transformer. When the load is disconnected, transient voltages are produced due to the energy stored in the inductances. The equivalent circuit due to load disconnection is shown in Fig. 16-18.

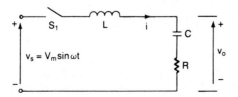

Figure 16-17 Equivalent circuit during switching on the supply.

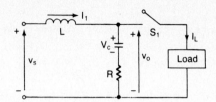

Figure 16-18 Equivalent circuit due to load disconnection.

Example 16-4*

A capacitor is connected across the secondary of an input transformer as shown in Fig. 16-16a, with zero damping resistance $R = 0$. The secondary voltage is $V_s = 120$ V, 60 Hz. If the magnetizing inductance referred to the secondary is $L_m = 2$ mH and the input supply to the transformer primary is disconnected at an angle of $\theta = 180°$ of the input ac voltage, determine (a) the initial capacitor value V_0, (b) the magnetizing current I_0, and (c) the capacitor value to limit the maximum transient capacitor voltage to $V_p = 300$ V.

Solution $V_s = 120$ V, $V_m = \sqrt{2} \times 120 = 169.7$ V, $\theta = 180°$, $f = 60$ Hz, $L_m = 2$ mH, and $\omega = 2\pi \times 60 = 377$ rad/s.

(a) From Eq. (16-33), $V_c = 169.7 \sin \theta = 0$.

(b) From Eq. (16-34),

$$I_0 = -\frac{V_m}{\omega L_m} \cos \theta = \frac{169.7}{377 \times 0.002} = 225 \text{ A}$$

(c) $V_P = 300$ V. From Eq. (16-44), the required capacitance is

$$C = 225 \times \frac{169.7}{300^2 \times 377} = 1125.3 \ \mu\text{F}$$

16-6 VOLTAGE PROTECTION BY SELENIUM DIODES AND METAL-OXIDE VARISTORS

The selenium diodes may be used for protection against transient overvoltages. These diodes have low forward voltage but well-defined reverse breakdown voltage. The characteristics of selenium diodes are shown in Fig. 16-19a. Normally, the operating point lies before the knee of the characteristic curve and draws very small current from the circuit. However, when an overvoltage appears, the knee point is crossed and the reverse current flow through the selenium increases suddenly, thereby typically limiting the transient voltage to twice the normal voltage.

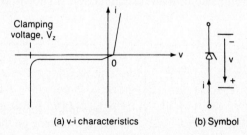

(a) v-i characteristics

(b) Symbol

Figure 16-19 Characteristics of selenium diode.

A selenium diode (or suppressor) must be capable of dissipating the surge energy without undue temperature rise. Each cell of a selenium diode is normally rated at an rms voltage of 25 V, with a clamping voltage of typically 72 V. For the protection of dc circuit, the suppression circuit is polarized as shown in Fig. 16-20a. In ac circuits as in Fig. 16-20b, the suppressors are nonpolarized, so that they can limit overvoltages in both directions. For three-phase circuits, wye-connected-polarized suppressors as shown in Fig. 16-20c can be used.

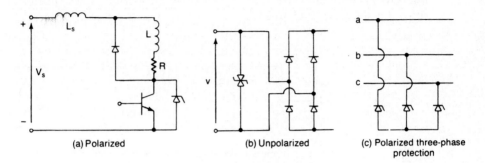

(a) Polarized   (b) Unpolarized   (c) Polarized three-phase protection

Figure 16-20   Voltage-suppression diodes.

If a dc circuit of 240 V is to be protected with 25-V selenium cells, then $240/25 \approx 10$ cells would be required and the total clamping voltage would be $10 \times 72 = 720$ V. To protect a single-phase ac circuit of 208 V, 60 Hz, with 25-V selenium cells, $208/25 \approx 9$ cells would be required in each direction and total of $2 \times 9 = 18$ cells would be necessary for nonpolarized suppression. Due to low internal capacitance, the selenium diodes do not limit the dv/dt to the same extent as compared to the RC-snubber circuits. However, they limit the transient voltages to well-defined magnitudes. In protecting a device, the reliability of an RC circuit is better than that of selenium diodes.

Varistors are nonlinear variable-impedance devices, consisting of metal-oxide particles, separated by an oxide film or insulation. As the applied voltage is increased, the film becomes conductive and the current flow is increased. The current is expressed as

$$I = KV^{\alpha} \tag{16-45}$$

where K is a constant and V is the applied voltage. The value of α varies between 30 and 40.

16-7 CURRENT PROTECTIONS

The power converters may develop short circuits or faults and the resultant fault currents must be cleared quickly. Fast-acting fuses are normally used to protect the semiconductor devices. As the fault current increases, the fuse opens and clears the fault current in few milliseconds.

16-7.1 Fusing

The semiconductor devices may be protected by carefully choosing the locations of the fuses as shown in Fig. 16-21. However, the fuse manufacturers recommend placing a fuse in series with each device as shown in Fig. 16-22. The individual protection which permits better coordination between a device and its fuse, allows superior utilization of the device capabilities and protects from short through faults (e.g., through T_1 and T_4 in Fig. 16-22a). The various sizes of semiconductor fuses are shown in Fig. 16-23.

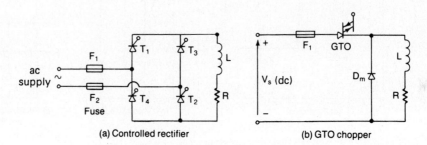

(a) Controlled rectifier   (b) GTO chopper

Figure 16-21   Protection of power devices.

When the fault current rises, the fuse temperature also rises until $t = t_m$, at which time the fuse melts and arcs are developed across the fuse. Due to the arc, the impedance of the fuse is increased, thereby reducing the current. However, an arc voltage is formed across the fuse. The generated heat vaporizes the fuse element, resulting in an increased arc length and further reduction of the current. The cumulative effect is the extinction of the arc in a very short time. When the arcing is complete in time t_a, the fault is clear. The faster the fuse clears, the higher is the arc voltage.

The clearing time t_c is the sum of melting time t_m and arc time t_a. t_m is dependent on the load current, whereas t_a is dependent on the power factor or parameters of the fault circuit. The fault is normally cleared before the fault current reaches its first peak, and the fault current, which might have blown

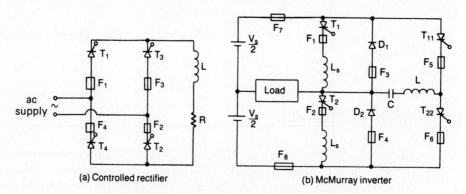

(a) Controlled rectifier   (b) McMurray inverter

Figure 16-22   Individual protection of devices.

Figure 16-23 Semiconductor fuses. (Reproduced by permission of Brush Electrical Machines Ltd., England.)

if there was no fuse, is called the *prospective fault current*. This is shown in Fig. 16-24.

The current–time curves of devices and fuses may be used for the coordination of a fuse for a device. Figure 16-25a shows the current–time characteristics of a device and its fuse, where the device will be protected over the whole range of overloads. This type of protection is normally used for low-power converters. Figure 16-25b shows the more commonly used system in which the fuse is used for short-circuit protection at the beginning of the fault; and the normal overload protection is provided by circuit breaker or other current-limiting system.

If R is the resistance of the fault circuit and i is the instantaneous fault current between the instant of fault occurring and the instant of arc extinction, the energy fed to the circuit can be expressed as

$$W_e = \int Ri^2 \, dt \tag{16-46}$$

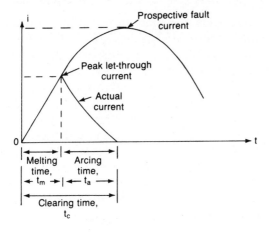

Figure 16-24 Fuse current.

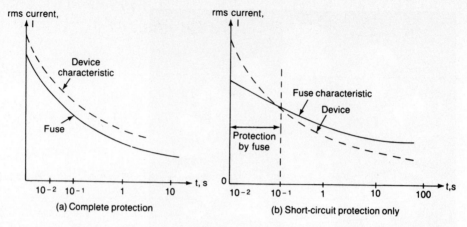

Figure 16-25 Current–time characteristics of device and fuse.

If the resistance, R, remains constant, the value of i^2t is proportional to the energy fed to the circuit. The i^2t value is termed as the *let-through energy* and is responsible for melting the fuse. The fuse manufacturers specify the i^2t characteristic of the fuse and Fig. 16-26 shows the typical characteristics of IR fuses, types TT350.

In selecting a fuse it is necessary to estimate the fault current and then to satisfy the following requirements:

1. The fuse must carry continuously the device rated current.
2. The i^2t let-through value of the fuse before the fault current is cleared must be less than the rated i^2t of the device to be protected.
3. The fuse must be able to withstand the voltage, after the arc extinction.
4. The peak arc voltage must be less than the peak voltage rating of the device.

In some applications it may be necessary to add a series inductance to limit the di/dt of the fault current and to avoid excessive di/dt stress on the device and fuse. However, this inductance may affect the normal performance of the converter.

Thyristors have more overcurrent capability than that of transistors. As a result, it is more difficult to protect transistors than thyristors. Bipolar transistors are gain-dependent and current-control devices. The maximum collector current is dependent on its base current. As the fault current rises, the transistor may go out of saturation and the collector–emitter voltage will rise with the fault current, particularly if the base current is not changed to cope with the increased collector current. This secondary effect may cause higher power loss within the transistor due to the rising collector–emitter voltage and may damage the transistor, even though the fault current is not sufficient enough to melt the fuse and clear the fault current. Thus fast-acting fuses may not be appropriate in protecting bipolar transistors under fault conditions.

Transistors can be protected by a crowbar circuit as shown in Fig. 16-27. A crowbar is used for protecting circuits or equipment under fault conditions, where

T350 SERIES

290V/175-450A r.m.s. Semiconductor Fuses

Suitable for protecting High Power Semiconductor Devices

Conforms to BS88: Part 4: 1976 and IEC 269-4. ASTA certificate of short circuit ratings and verification of I^2t cut-off and arc voltage characteristics are available.

IMPORTANT

Note 1: Thyristors/diodes are rated in average current while fuses are rated in r.m.s. current. During steady state operation the fuse must not be operated in excess of its maximum r.m.s. rating.

Note 2: The maximum cap temperature and cap temperature rise above ambient of a fuse are critical design parameters. Caution should be taken during installation to ensure that the specified ratings are not exceeded. Some form of heatsink may be necessary.

The T350 Series of semiconductor fuses are available with I700 indicator fuses already fitted, for dimensional details refer to page E-12. For electrical, thermal and mechanical specifications on I700 refer to page E-5.
To complete part number add prefix "I" e.g. IT350-450.

ELECTRICAL SPECIFICATIONS

Maximum r.m.s. voltage rating:	290V
Maximum tested peak voltage	450V
Maximum d.c. voltage rating (L/R ≤ 15ms)	160V
Maximum arcing voltage for AC Supply Voltage = 240V	490V

For variation in arcing voltage with AC Supply Voltage

$$V_A = 100 + 1.63 V_S$$

where V_A = Peak arc voltage, V_S = AC Supply Voltage

Fusing Factor	1.25
Force cooling Current uprating factor at 5 m/s	1.2

THERMAL AND MECHANICAL SPECIFICATIONS

Maximum cap temperature:	100°C
Maximum cap temperature rise above ambient	75°C
Weight:	170g (5.95 oz.)

Part number	RMS CURRENT (1) T_{amb} = 25°C	RMS CURRENT (1) T_{amb} = 25°C	MAX. POWER LOSS	PRE-ARCING (2) I^2t	TOTAL I^2t (2) at 120 V_{RMS}	TOTAL I^2t (2) at 240 V_{RMS}	NOTES
	A	A	W	A^2s	A^2s	A^2s	
T350-150	175	155	17	1600	7000	16000	1) Maximum current carrying ability, natural convection cooling using test arrangement as BS88: Part 4: 1976 conductors 1.0 to 1.6 A/mm² attachment.
T350-200	210	190	28	2100	10000	20000	
T350-250	250	230	28	4800	20000	40000	
T350-300	315	290	35	9000	34000	70000	
T350-350	355	320	35	13000	50000	100000	2) Typical values of I^2t at 20 times rated RMS current
T350-400	400	350	40	20000	75000	160000	
T350-450	450	400	42	30000	110000	220000	

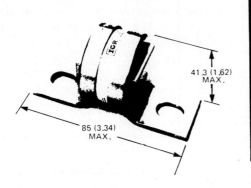

41.3 (1.62) MAX.

85 (3.34) MAX.

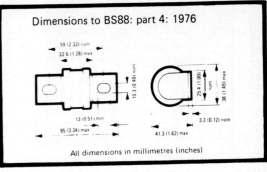

Dimensions to BS88: part 4: 1976

59 (2.32) nom
32.6 (1.28) max
10.3 (0.40) nom
13 (0.51) min
85 (3.34) max

25.4 (1.00) nom
38 (1.49) max
3.2 (0.12) nom
41.3 (1.62) max

All dimensions in millimetres (inches)

Figure 16-26 Data sheet of IR fuse, type T350. (Courtesy of International Rectifier.)

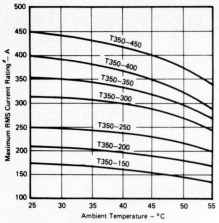

Fig. 1 – Current Rating Characteristic

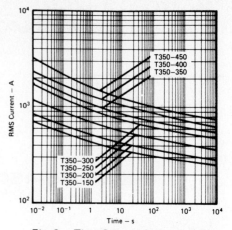

Fig. 2 – Time Current Characteristic

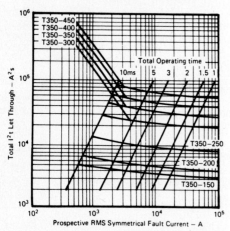

Fig. 3 – I²t Let Through Characteristic (60V~)

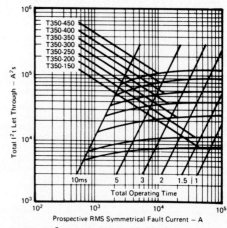

Fig. 4 – I²t Let Through Characteristic (120V~)

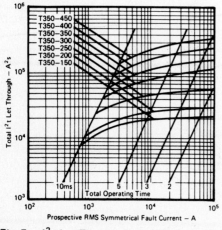

Fig. 5 – I²t Let Through Characteristic (240V~)

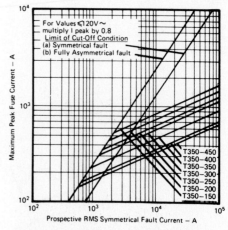

Fig. 6 – Cut-Off Characteristics (240V~)

Figure 16-26 *(continued)*

TT350 SERIES

290V/400-900A r.m.s. Semiconductor Fuses

Suitable for protecting High Power Semiconductor Devices

Conforms to BS88: Part 4: 1976 and IEC 269-4.

IMPORTANT:
Note 1: Thyristors/diodes are calibrated in average current ratings while fuses are calibrated in r.m.s. current ratings. During steady state operation the fuse must not be operated in excess of its maximum r.m.s. rating.

Note 2: The maximum cap temperature and cap temperature rise above ambient of a fuse are critical design parameters. Caution should be taken during installation to ensure that the specified ratings are not exceeded.

The TT350 Series of semiconductor fuses are available with 1700 indicator fuses already fitted, for dimensional details refer to page E-67. For electrical, thermal and mechanical specifications refer to page E-66.
To complete part number add prefix "I" e.g. ITT350-900.

ELECTRICAL SPECIFICATIONS

Maximum r.m.s. voltage rating:	290V
Maximum tested peak voltage:	450V
Maximum d.c. voltage rating (L/R≤15ms)	160V
Maximum arcing voltage for AC Supply Voltage = 240V	490V

For variation in arcing voltage with AC Supply Voltage

$$V_A = 100 + 1.63 \, V_S$$

Where V_A = Peak arc voltage, V_S = AC Supply Voltage

Fusing Factor:	1.25
Force cooling Current uprating factor at 5 m/s	1.2

THERMAL AND MECHANICAL SPECIFICATIONS

Maximum cap temperature:	100°C
Maximum cap temperature rise above ambient	75°C
Maximum gravitational withstand capability:	1500g (52.5 oz.)
(for device mounted radially to rotation.)	

Part number	RMS CURRENT (1) $T_{amb} = 25°C$	RMS CURRENT (1) $T_{amb} = 45°C$	MAX. POWER LOSS	PRE-ARCING (2) I^2t	TOTAL I^2t (2) at 120 V_{RMS}	TOTAL I^2t (L) at 240 V_{RMS}	NOTES
	A	A	W	kA^2s	kA^2s	kA^2s	
TT350-400	400	350	60	8	35	80	1) Maximum current carrying ability, natural convection cooling using test arrangement as BS88: Part 4: 1976.
TT350-500	500	430	64	19	80	170	
TT350-600	630	540	75	35	150	300	
TT350-700	710	580	77	50	200	420	2) Typical values of I^2t at 20 times rated RMS Current.
TT350-800	800	660	82	70	300	650	
TT350-900	900	740	97	100	400	850	

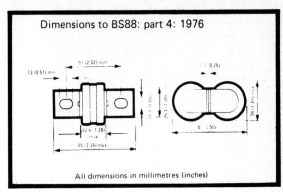

Dimensions to BS88: part 4: 1976

All dimensions in millimetres (inches)

Figure 16-26 *(continued)*

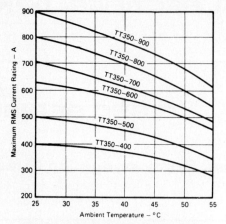

Fig. 1 — Current Rating Characteristic

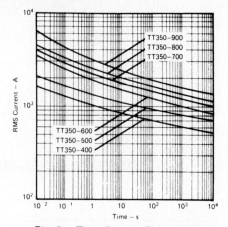

Fig. 2 — Time Current Characteristic

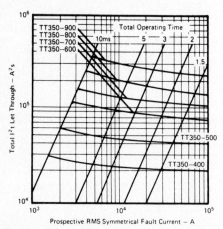

Fig. 3 — I²t Let Through Characteristic (60V∼)

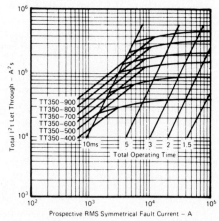

Fig. 4 — I²t Let Through Characteristic (120V∼)

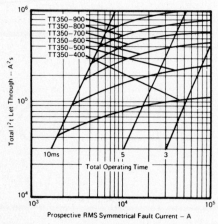

Fig. 5 — I²t Let Through Characteristics (240V∼)

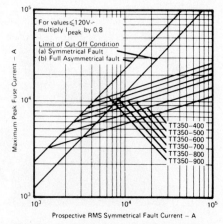

Fig. 6 — Cut-Off Characteristics (240V∼)

Figure 16-26 *(continued)*

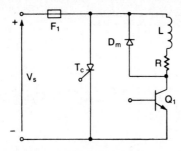

Figure 16-27 Protection by crow-bar circuit.

the amount of energy involved is too high and the normal protection circuits cannot be used. A crowbar consists of a thyristor with a voltage- or current-sensitive firing circuit. The crowbar thyristor is placed across the converter circuit to be protected. If the fault conditions are sensed and crowbar thyristor T_c is fired, a virtual short circuit is created and the fuse link F_1 is blown, thereby relieving the converter from overcurrent.

MOSFETS are voltage control devices; and as the fault current rises, the gate voltage needs not be changed. The peak current is typically three times the continuous rating. If the peak current is not exceeded and the fuse clears quickly enough, a fast-acting fuse may protect a MOSFET. However, a crowbar protection is also recommended. The fusing characteristics of IGBTs are similar to that of BJTs.

16-7.2 Fault Current with AC Source

An ac circuit is shown in Fig. 16-28, where the input voltage is $v = V_m \sin \omega t$. Let us assume that the switch is closed at $\omega t = \theta$. Redefining the time origin, $t = 0$, at the instant of closing the switch, the input voltage is described by $v_s = V_m \sin(\omega t + \theta)$ for $t \geq 0$. Eq. (6-13) gives the current as

$$i = \frac{V_m}{|Z_x|} \sin(\omega t + \theta - \phi_x) - \frac{V_m}{|Z_x|} \sin(\theta - \phi_x)e^{-Rt/L} \qquad (16.47)$$

where $|Z_x| = \sqrt{R_m^2 + (\omega L_x)^2}$, $\phi_x = \tan^{-1}(\omega L_x/R_x)$, $R_x = R + R_m$, and $L_x = L + L_m$. Figure 16-28 describes the initial current at the beginning of the fault. If there is a fault across the load as shown in Fig. 16-29, Eq. (16-47), which can be applied with an initial current of I_0 at the beginning of the fault, gives the fault current as

$$i = \frac{V_m}{|Z|} \sin(\omega t + \theta - \phi) + \left(I_0 - \frac{V_m}{|Z|}\right) \sin(\theta - \phi)e^{-Rt/L} \qquad (16-48)$$

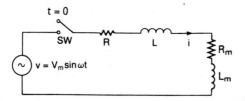

Figure 16-28 *RL* circuit.

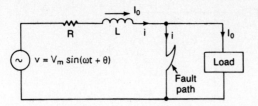

Figure 16-29 Fault in ac circuit.

where $|Z| = \sqrt{R^2 + (\omega L)^2}$ and $\phi = \tan^{-1}(\omega L/R)$. The fault current will depend on the initial current I_0, power factor angle of the short-circuit path, ϕ, and the angle of fault occurring, θ. Figure 16-30 shows the current and voltage waveforms during the fault conditions in an ac circuit. For a highly inductive fault path, $\phi = 90°$ and $e^{-Rt/L} = 1$, and Eq. (16-48) becomes

$$i = -I_0 \cos \theta + \frac{V_m}{|Z|} [\cos \theta - \cos(\omega t + \theta)] \qquad (16\text{-}49)$$

If the fault occurs at $\theta = 0$, that is, at the zero crossing of the ac input voltage, $\omega t = 2n\pi$. Eq. (16-49) becomes

$$i = -I_0 + \frac{V_m}{Z} (1 - \cos \omega t) \qquad (16\text{-}50)$$

and Eq. (16-50) gives the maximum peak fault current, $-I_0 + 2V_m/Z$, which occurs $\omega t = \pi$. But in practice, due to damping the peak current will be less than this.

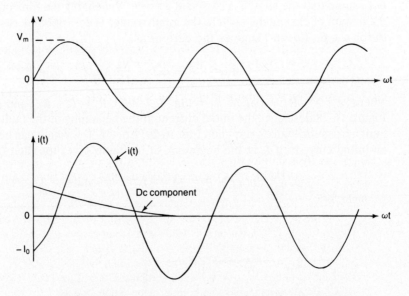

Figure 16-30 Transient voltage and current waveforms.

16-7.3 Fault Current with DC Source

The current of a dc circuit in Fig. 16-31 is given by

$$i = \frac{V_s}{R_x}(1 - e^{-R_x t/L_x})$$

(16-51)

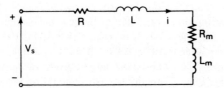

Figure 16-31 Dc circuit.

With an initial current of I_0 at the beginning of the fault current as shown in Fig. 16-32, the fault current is expressed as

$$i = I_0 e^{-Rt/L} + \frac{V_s}{R}(1 - e^{-Rt/L})$$

(16-52)

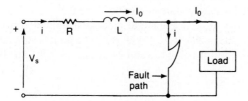

Figure 16-32 Fault in dc circuit.

The fault current and the fuse clearing time will be dependent on the time constant of the fault circuit. If the prospective current is low, the fuse may not clear the fault and a slowly rising fault current may produce arcs continuously without breaking the fault current. The fuse manufacturers specify the current–time characteristics for ac circuits and there is no equivalent curves for dc circuits. Since the dc fault currents do not have natural periodic zeros, the extinction of the arc is more difficult. For circuits operating from dc voltage, the voltage rating of the fuse should be typically 1.5 times the equivalent ac rms voltage. The fuse protection of dc circuits requires more careful design than that of ac circuits.

Example 16-5

A fuse is connected in series with each IR thyristor of type S30EF in the single-phase full converter as shown in Fig. 5-3a. The input voltage is 208 V, 60 Hz, and the average current of each thyristor is $I_a = 400$ A. The ratings of thyristors are $I_{T(AV)} = 540$ A, $I_{T(RMS)} = 850$ A, $I^2 t = 300 \, kA^2 s$ at 8.33 ms, $i^2 \sqrt{t} = 4650 \, kA^2 \sqrt{s}$, and $I_{TSM} = 10$ kA with reapplied $V_{RRM} = 0$, which will be the case if the fuse opens within a half-cycle. If the resistance of fault circuit is negligible and inductance is $L = 0.07$ mH, select the rating of a suitable fuse from Fig. 16-26.

Solution $V_s = 240$ V, $f_s = 60$ Hz. Let us try an IR fuse, type TT350-600. The short circuit current which is also known as the prospective rms symmetrical fault current is

$$I_{sc} = \frac{V_s}{Z} = 240 \times \frac{1000}{2\pi \times 60 \times 0.07} = 9094 \text{ A}$$

For the 540-A fuse type TT350-600 and $I_{sc} = 9094$ A, the maximum peak fuse current is 8500 A, which is less than the peak thyristor current of $I_{TSM} = 10$ kA. The fuse $i^2 t$ is 280 kA^2s and total clearing time is $t_c = 8$ ms. Since t_c is less than 8.33 ms, the $i^2\sqrt{t}$ rating of the thyristor must be used. If thyristor $i^2\sqrt{t} = 4650 \times 10^3$ $kA^2\sqrt{s}$, then at $t_c = 8$ ms, thyristor $i^2 t = 4650 \times 10^3 \sqrt{0.008} = 416$ kA^2s, which is 48.6% higher than the $i^2 t$ rating of the fuse. The $i^2 t$ and peak surge current ratings of the thyristor are higher than those of the fuse. Thus the thyristor should be protected by the fuse.

Note. As a general rule of thumb, a fast-acting fuse with a rms current rating equal or less than the average current rating of the thyristor or diode will normally provide adequate protection under fault conditions.

Example 16-6

The ac circuit shown in Fig. 16-33a has $R = 1.5$ Ω and $L = 15$mH. The load parameters are $R_m = 5$ Ω and $L_m = 15$ mH. The input voltage is 208 V (rms), 60 Hz. The circuit has reached a steady-state condition. The fault across the load occurs at $\omega t + \theta = 2\pi$; that is, $\theta = 0$. Use PSpice to plot the instantaneous fault current.
Solution $V_m = \sqrt{2} \times 208 = 294.16$ V, $f = 60$ Hz. The fault is simulated by a voltage control switch, whose control voltage is shown in Fig. 16-33b. The list of the circuit

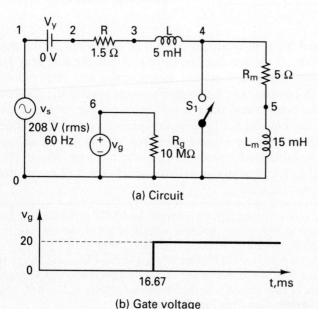

(a) Circuit

(b) Gate voltage

Figure 16-33 Fault in ac circuit for PSpice simulation.

file is as follows:

```
Example 16-6      Fault Current in AC Circuit
VS    1    0    SIN (0    294.16V  60HZ)
VY    1    2    DC   OV    ; Voltage source to measure input current
Vg    6    0    PWL (16666.67US  OV  16666.68US  20V  60MS  20V)
Rg    6    0    10MEG      ; A very high resistance for control voltage
R     2    3    1.5
L     3    4    5MH
RM    4    5    5
LM    5    0    15MH
S1    4    0    6    0    SMOD               ; Voltage-controlled switch
.MODEL    SMOD    VSWITCH  (RON=0.01  ROFF=10E+5  VON=0.2V  VOFF=OV)
.TRAN     10US    40MS    0    50US          ; Transient analysis
.PROBE                                       ; Graphics postprocessor
.options abstol = 1.00n reltol = 0.01 vntol = 0.1 ITL5=50000 ;
convergence
.END
```

The PSpice plot is shown in Fig. 16-34, where I(VY) = fault current. Using the PSpice cursor in Fig. 16-34 gives the initial current $I_o = -22.28$ A and the prospective fault current $I_p = 132.132$ A.

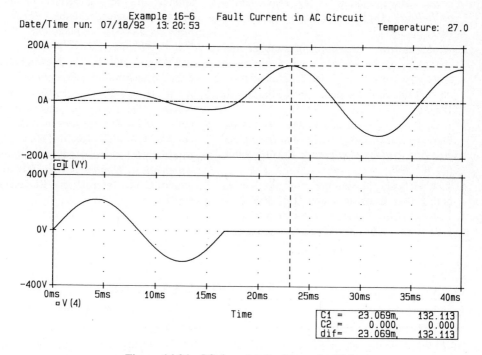

Figure 16-34 PSpice plot for Example 16-6.

SUMMARY

The power converters must be protected from overcurrents and overvoltages. The junction temperature of power semiconductor devices must be maintained within their maximum permissible values. The heat produced by the device may be transferred to the heat sinks by air and liquid cooling. Heat pipes can also be used. The reverse recovery currents and disconnection of load (and supply line) cause voltage transients due to energy stored in the line inductances.

The voltage transients are normally suppressed by the same *RC*-snubber circuit, which is used for *dv/dt* protection. The snubber design is very important to limit the *dv/dt* and peak voltage transients within the maximum ratings. Selenium diodes and varistors can be used for transient voltage suppression.

A fast-acting fuse is normally connected in series with each device for overcurrent protection under fault conditions. However, fuses may not be adequate for protecting transistors and other protection means, (e.g., a crowbar) may be required.

REFERENCES

1. W. McMurray, "Optimum snubbers for power semiconductors." *IEEE Transactions on Industry Applications,* Vol. IA8, No. 5, 1972, pp. 503–510.

2. W. McMurray, "Selection of snubber and clamps to optimize the design of transistor switching converters." *IEEE Transactions on Industry Applications,* Vol. IA16, No. 4, 1980, pp. 513–523.

3. J. B. Rice, "Design of snubber circuits for thyristor converters." *IEEE Industry General Applications Conference Record,* 1969, pp. 485–489.

4. T. Undeland, "A snubber configuration for both power transistors and GTO PWM inverter." *IEEE Power Electronics Specialist Conference,* 1984, pp. 42–53.

5. C. G. Steyn and J. D. V. Wyk, "Study and application of non-linear turn-off snubber for power electronics switches." *IEEE Transactions on Industry Applications,* Vol. IA22, No. 3, 1986, pp. 471–477.

6. A. Wright and P. G. Newbery, *Electric Fuses.* London: Peter Peregrinus Ltd., 1984.

7. A. F. Howe, P. G. Newbery, and N. P. Nurse, "Dc fusing in semiconductor circuits." *IEEE Transactions on Industry Applications,* Vol. IA22, No. 3, 1986, pp. 483–489.

8. International Rectifiers, *Semiconductor Fuse Applications Handbook* (No. HB50). El Segundo, Calif.: International Rectifiers, 1972.

REVIEW QUESTIONS

16-1. What is a heat sink?

16-2. What is the electrical analog of heat transfer from a power semiconductor device?

16-3. What are the precautions to be taken in mounting a device on a heat sink?

16-4. What is a heat pipe?

16-5. What are the advantages and disadvantages of heat pipes?

16-6. What are the advantages and disadvantages of water cooling?

16-7. What are the advantages and disadvantages of oil cooling?

16-8. Why is it necessary to determine the instantaneous junction temperature of a device?

16-9. What is a polarized snubber?

16-10. What is a nonpolarized snubber?

16-11. What is the cause of reverse recovery transient voltage?

16-12. What is the typical value of damping factor for an RC snubber?

16-13. What are the considerations for the design of optimum RC-snubber components?

16-14. What is the cause of load-side transient voltages?

16-15. What is the cause of supply-side transient voltages?

16-16. What are the characteristics of selenium diodes?

16-17. What are the advantages and disadvantages of selenium voltage suppressors?

16-18. What are the characteristics of varistors?

16-19. What are the advantages and disadvantages of varistors in voltage suppressions?

16-20. What is a melting time of a fuse?

16-21. What is an arcing time of a fuse?

16-22. What is a clearing time of a fuse?

16-23. What is a prospective fault current?

16-24. What are the considerations in selecting a fuse for a semiconductor device?

16-25. What is a crowbar?

16-26. What are the problems of protecting bipolar transistors by fuses?

16-27. What are the problems of fusing dc circuits?

PROBLEMS

16-1. The power loss in a device is shown in Fig. P16-1. Plot the instantaneous junction temperature rise above case. For $t_1 = t_3 = t_5 = t_7 = 0.5$ ms, $Z_1 = Z_3 = Z_5 = Z_7 = 0.025°C/W$.

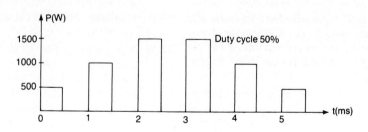

Figure P16-1

16-2. The power loss in a device is shown in Fig. P16-2. Plot the instantaneous junction temperature rise above the case temperature. For $t_1 = t_2 = \cdots = t_9 = t_{10} = 1$ ms, $Z_1 = Z_2 = \cdots = Z_9 = Z_{10} = 0.035°C/W$. (*Hint*: Approximate by five rectangular pulses of equal duration.)

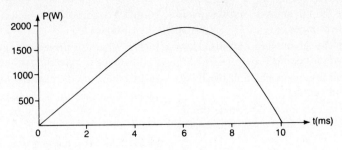

Figure P16-2

16-3. The current waveform through an IR thyristor, type S30EF, is shown in Fig. 4-19. Plot **(a)** the power loss against time, and **(b)** the instantaneous junction temperature rise above case. (*Hint*: Assume power loss during turn-on and turn-off as rectangles.)

16-4. The recovery current of a device as shown in Fig. 16-12 is $I_R = 30$ A and the circuit inductance is $L = 20\ \mu$H. The input voltage is $V_s = 200$ V. If it is necessary to limit the peak transient voltage to 1.8 times the input voltage, determine the **(a)** the optimum value of current ratio, d_o; **(b)** the optimum damping factor δ_o; **(c)** the snubber capacitance C; **(d)** the snubber resistance R; **(e)** the average dv/dt; and **(f)** the initial reverse voltage.

16-5. The recovery current of a device as shown in Fig. 16-12 is $I_R = 10$ A and the circuit inductance is $L = 80\ \mu$H. The input voltage is $V_s = 200$ V. The snubber resistance is $R = 2\ \Omega$ and snubber capacitance is $C = 50\ \mu$F. Determine the **(a)** the damping ratio δ, **(b)** the peak transient voltage V_p, **(c)** the current ratio d, **(d)** the average dv/dt, and **(e)** the initial reverse voltage.

16-6. An RC-snubber circuit as shown in Fig. 16-11c has: $C = 1.5\ \mu$F, $R = 4.5\ \Omega$, and the input voltage is $V_s = 220$ V. The circuit inductance is $L = 20\ \mu$H. Determine **(a)** the peak forward voltage V_p, **(b)** the initial dv/dt, and **(c)** the maximum dv/dt.

16-7. An RC-snubber circuit as shown in Fig. 16-11c has a circuit inductance of $L = 20\ \mu$H. The input voltage, $V_s = 200$ V. If it is necessary to limit the maximum dv/dt to $20\text{V}/\mu$s and damping factor is $\delta = 0.4$, determine **(a)** the snubber capacitance C, and **(b)** the snubber resistance R.

16-8. An RC-snubber circuit as shown in Fig. 16-11c has circuit inductance of $L = 50\ \mu$H. The input voltage $V_s = 220$ V. If it is necessary to limit the peak voltage to 1.5 times the input voltage, and the damping factor, $\alpha = 9500$, determine **(a)** the snubber capacitance C, and **(b)** the snubber resistance, R.

16-9. A capacitor is connected to the secondary of an input transformer as shown in Fig. 16-16a, with zero damping resistance $R = 0$. The secondary voltage, $V_s = 208$ V, 60 Hz, and the magnetizing inductance referred to the secondary is $L_m = 3.5$ mH. If the input supply to the transformer primary is disconnected at an angle of $\theta = 120°$ of the input ac voltage, determine **(a)** the initial capacitor value V_0, **(b)** the magnetizing current I_0, and **(c)** the capacitor value to limit the maximum transient capacitor voltage to $V_p = 350$ V.

16-10. The circuit in Fig. 16-18 has a load current of $I_L = 10$ A and the circuit inductance is $L = 50\ \mu$H. The input voltage is dc with $V_s = 200$ V. The snubber resistance is $R = 1.5\ \Omega$ and snubber capacitance is $C = 50\ \mu$F. If the load is disconnected, determine **(a)** the damping factor δ, and **(b)** the peak transient voltage V_p.

16-11. Selenium diodes are used to protect a three-phase circuit as shown in Fig. 16-20c. The three-phase voltage is 208 V, 60 Hz. If the voltage of each cell is 25 V, determine the number of diodes.

16-12. The load current at the beginning of a fault in Fig. 16-29 is $I_0 = 10$ A. The ac voltage is 208 V, 60 Hz. The resistance and inductance of the fault circuit are $L = 5$ mH and $R = 1.5$ Ω, respectively. If the fault occurs at an angle of $\theta = 45°$, determine the peak value of prospective current in the first half-cycle.

16-13. Repeat Prob. 16-12 if $R = 0$.

16-14. The current through a fuse is shown in Fig. P16-14. The total i^2t of the fuse is 5400 A²s. If the arcing time, $t_a = 0.1$ s and the melting time, $t_m = 0.05$ s, determine the peak let-through current, I_p.

16-15. The load current in Fig. 16-32 is $I_0 = 0$ A. The dc input voltage is $V_s = 220$ V. The fault circuit has an inductance of $L = 2$ mH and negligible resistance. The total i^2t of the fuse is 4500 A²s. The arcing time is 1.5 times the melting time. Deter-

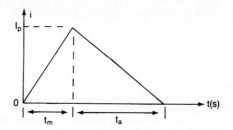

Figure P16-14

mine **(a)** the melting time t_m, **(b)** the clearing time t_c, and **(c)** the peak let-through current I_p.

16-16. Use PSpice to verify the design in Prob. 16-7.

16-17. Use PSpice to verify the results of Prob. 16-9.

16-18. Use PSpice to verify the results of Prob. 16-10.

A

Three-phase circuits

In a single-phase circuit as shown in Fig. A-1a, the current is expressed as

$$\bar{\mathbf{I}} = \frac{V\,\underline{/\alpha}}{R + jX} = \frac{V\,\underline{/\alpha - \theta}}{Z} \tag{A-1}$$

where $Z = (R^2 + X^2)^{1/2}$ and $\theta = \tan^{-1}(X/R)$. The power can be found from

$$P = VI \cos\theta \tag{A-2}$$

where $\cos\theta$ is called the *power factor,* and θ, which is the angle of the load impedance, is known as the *power factor angle.*

A three-phase circuit consists of three sinusoidal voltages of equal magnitudes and the phase angles between the individual voltages are 120°. A wye-connected load connected to a three-phase source is shown in Fig. A-2a. If the three phase voltages are

$$\bar{\mathbf{V}}_a = V_p\,\underline{/0}$$

$$\bar{\mathbf{V}}_b = V_p\,\underline{/-120°}$$

$$\bar{\mathbf{V}}_c = V_p\,\underline{/-240°}$$

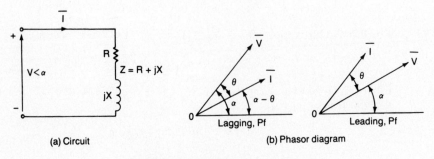

(a) Circuit   (b) Phasor diagram

Figure A-1 Single-phase circuit.

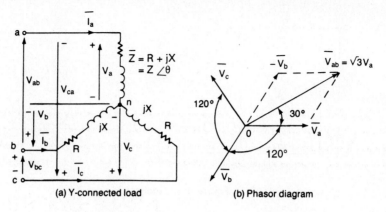

(a) Y-connected load   (b) Phasor diagram

Figure A-2 Wye-connected three-phase circuit.

the line-to-line voltages are

$$\overline{\mathbf{V}}_{ab} = \overline{\mathbf{V}}_a - \overline{\mathbf{V}}_b = \sqrt{3}\, V_p\ \underline{/30°} = V_L\ \underline{/30°}$$

$$\overline{\mathbf{V}}_{bc} = \overline{\mathbf{V}}_b - \overline{\mathbf{V}}_c = \sqrt{3}\, V_p\ \underline{/-90°} = V_L\ \underline{/-90°}$$

$$\overline{\mathbf{V}}_{ca} = \overline{\mathbf{V}}_c - \overline{\mathbf{V}}_a = \sqrt{3}\, V_p\ \underline{/-210°} = V_L\ \underline{/-210°}$$

Thus a line-to-line voltage, V_L, is $\sqrt{3}$ times of a phase voltage, V_p. The three line currents, which are the same as phase currents, are

$$\overline{\mathbf{I}}_a = \frac{\overline{\mathbf{V}}_a}{Z_a\ \underline{/\theta_a}} = \frac{V_p}{Z_a}\ \underline{/-\theta_a}$$

$$\overline{\mathbf{I}}_b = \frac{\overline{\mathbf{V}}_b}{Z_b\ \underline{/\theta_b}} = \frac{V_p}{Z_b}\ \underline{/-120° - \theta_b}$$

$$\overline{\mathbf{I}}_c = \frac{\overline{\mathbf{V}}_c}{Z_c\ \underline{/\theta_c}} = \frac{V_p}{Z_c}\ \underline{/-240° - \theta_c}$$

The input power to the load is

$$P = V_a I_a \cos\theta_a + V_b I_b \cos\theta_b + V_c I_c \cos\theta_c \qquad (A\text{-}3)$$

For a balanced supply, $V_a = V_b = V_c = V_p$. Eq. (A-3) becomes

$$P = V_p(I_a \cos\theta_a + I_b \cos\theta_b + I_c \cos\theta_c) \qquad (A\text{-}4)$$

For a balanced load, $Z_a = Z_b = Z_c = Z$, $\theta_a = \theta_b = \theta_c = \theta$, and $I_a = I_b = I_c = I_p = I_L$, Eq. (A-4) becomes

$$P = 3V_p I_p \cos\theta$$

$$= 3\,\frac{V_L}{\sqrt{3}}\, I_L \cos\theta = \sqrt{3} V_L I_L \cos\theta \qquad (A\text{-}5)$$

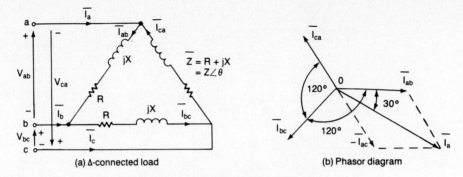

(a) Δ-connected load   (b) Phasor diagram

Figure A-3 Delta-connected load.

A delta-connected load is shown in Fig. A-3a, where the line voltages are the same as the phase voltages. If the three phase voltages are

$$\overline{\mathbf{V}}_a = \overline{\mathbf{V}}_{ab} = V_L \, \underline{/0} = V_p \, \underline{/0}$$

$$\overline{\mathbf{V}}_b = \overline{\mathbf{V}}_{bc} = V_L \, \underline{/-120°} = V_p \, \underline{/-120°}$$

$$\overline{\mathbf{V}}_c = \overline{\mathbf{V}}_{ca} = V_L \, \underline{/-240°} = V_p \, \underline{/-240°}$$

the three phase currents are

$$\overline{\mathbf{I}}_{ab} = \frac{\overline{\mathbf{V}}_a}{Z_a \, \underline{/\theta_a}} = \frac{V_L}{Z_a} \, \underline{/-\theta_a} = I_p \, \underline{/-\theta_a}$$

$$\overline{\mathbf{I}}_{bc} = \frac{\overline{\mathbf{V}}_b}{Z_b \, \underline{/\theta_b}} = \frac{V_L}{Z_b} \, \underline{/-120° - \theta_b} = I_p \, \underline{/-120° - \theta_b}$$

$$\overline{\mathbf{I}}_{ca} = \frac{\overline{\mathbf{V}}_c}{Z_c \, \underline{/\theta_c}} = \frac{V_L}{Z_c} \, \underline{/-240° - \theta_c} = I_p \, \underline{/-240° - \theta_c}$$

and the three line currents are

$$\overline{\mathbf{I}}_a = \overline{\mathbf{I}}_{ab} - \overline{\mathbf{I}}_{ca} = \sqrt{3} \, I_p \, \underline{/-30° - \theta_a} = I_L \, \underline{/-30° - \theta_a}$$

$$\overline{\mathbf{I}}_b = \overline{\mathbf{I}}_{bc} - \overline{\mathbf{I}}_{ab} = \sqrt{3} \, I_p \, \underline{/-150° - \theta_b} = I_L \, \underline{/-150° - \theta_b}$$

$$\overline{\mathbf{I}}_c = \overline{\mathbf{I}}_{ca} - \overline{\mathbf{I}}_{bc} = \sqrt{3} \, I_p \, \underline{/-270° - \theta_c} = I_L \, \underline{/-270° - \theta_c}$$

Therefore, in a delta-connected load, a line current is $\sqrt{3}$ times of a phase current.

The input power to the load is

$$P = V_{ab}I_{ab} \cos \theta_a + V_{bc}I_{bc} \cos \theta_b + V_{ca}I_{ca} \cos \theta_c \qquad \text{(A-6)}$$

For a balanced supply, $V_{ab} = V_{bc} = V_{ca} = V_L$, Eq. (A-6) becomes

$$P = V_L(I_{ab} \cos \theta_a + I_{bc} \cos \theta_b + I_{ca} \cos \theta_c) \qquad \text{(A-7)}$$

For a balanced load, $Z_a = Z_b = Z_c = Z$, $\theta_a = \theta_b = \theta_c = \theta$, and $I_{ab} = I_{bc} = I_{ca} = I_p$, Eq. (A-7) becomes

$$P = 3V_p I_p \cos \theta$$

$$= 3V_L \frac{I_L}{\sqrt{3}} \cos \theta = \sqrt{3} \, V_L I_L \cos \theta \qquad \text{(A-8)}$$

Note. Equations (A-5) and (A-8), which express power in a three-phase circuit, are the same. For the same phase voltages, the line currents in a delta-connected load are $\sqrt{3}$ times that of a wye-connected load.

Magnetic circuits

A magnetic ring is shown in Fig. B-1. If the magnetic field is uniform and normal to the area under consideration, a magnetic circuit is characterized by the following equations:

$$\phi = BA \tag{B-1}$$

$$B = \mu H \tag{B-2}$$

$$\mu = \mu_r \mu_0 \tag{B-3}$$

$$\mathscr{F} = NI = Hl \tag{B-4}$$

where ϕ = flux, webers
B = flux density, webers/m^2 (or teslas)
H = magnetizing force, ampere-turns/meter
μ = permeability of the magnetic material
μ_0 = permeability of the air (= $4\pi \times 10^{-7}$)
μ_r = relative permeability of the material
$\mathscr{F}$ = magnetomotive force, ampere-turns (At)
N = number of turns in the winding
I = current through the winding, amperes
l = length of the magnetic circuit, meters

If the magnetic circuit consists of different sections, Eq. (B-4) becomes

$$\mathscr{F} = NI = \Sigma \, H_i l_i \tag{B-5}$$

where H_i and l_i are the magnetizing force and length of ith section, respectively.

The reluctance of a magnetic circuit is related to the magnetomotive force and flux by

$$\mathscr{R} = \frac{\mathscr{F}}{\phi} = \frac{NI}{\phi} \tag{B-6}$$

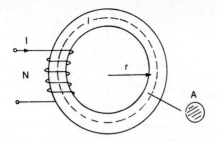

Figure B-1 Magnetic ring.

and $\mathcal{R}$ depends on the type and dimensions of the core,

$$\mathcal{R} = \frac{l}{\mu_r \theta_0 A} \qquad \text{(B-7)}$$

The permeability depends on the $B-H$ characteristic and is normally much larger than that of the air. A typical $B-H$ characteristic which is nonlinear is shown in Fig. B-2. For a large value of μ, $\mathcal{R}$ becomes very small, resulting in high value of flux. An air gap is normally introduced to limit the amount of flux flow.

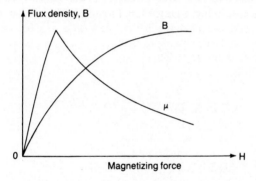

Figure B-2 Typical $B-H$ characteristic.

A magnetic circuit with an air gap is shown in Fig. B-3a and the analogous electric circuit is shown in Fig. B-3b. The reluctance of the air gap is

$$\mathcal{R}_g = \frac{l_g}{\mu_0 A_g} \qquad \text{(B-8)}$$

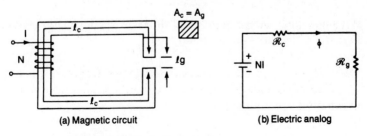

Figure B-3 Magnetic circuit with air gap.

and the reluctance of the core is

$$\mathscr{R}_c = \frac{l_c}{\mu_r \mu_0 A_c} \qquad \text{(B-9)}$$

where l_g = length of the air gap
  l_c = length of the core
  A_g = area of the cross section of the air gap
  A_c = area of the cross section of the core

The total reluctance of the magnetic circuit is

$$\mathscr{R} = \mathscr{R}_g + \mathscr{R}_c$$

Inductance is defined as the flux linkage (λ) per ampere,

$$L = \frac{\lambda}{I} = \frac{N\phi}{I} \qquad \text{(B-10)}$$

$$= \frac{N^2\phi}{NI} = \frac{N^2}{\mathscr{R}} \qquad \text{(B-11)}$$

Example B-1

The parameters of the core in Fig. B-3a are l_g = 1 mm, l_c = 30 cm, $A_g = A_c = 5 \times 10^{-3}$ m^2, N = 350, and I = 2 A. Calculate the inductance if (a) μ_r = 3500, and (b) the core is ideal, namely μ_r is very large tending to infinite.
Solution $\mu_0 = 4\pi \times 10^{-7}$ and N = 350.
  (a) From Eq. (B-8),

$$\mathscr{R}_g = \frac{1 \times 10^{-3}}{4\pi \times 10^{-7} \times 5 \times 10^{-3}} = 159{,}155$$

From Eq. (B-9),

$$\mathscr{R}_c = \frac{30 \times 10^{-2}}{3500 \times 4\pi \times 10^{-7} \times 5 \times 10^{-3}} = 13{,}641$$

$$\mathscr{R} = 159{,}155 + 13{,}641 = 172{,}796$$

From Eq. (B-11), $L = 350^2/172{,}796 = 0.71$ H.
  (b) If $\mu_r \approx \infty$, $\mathscr{R}_c = 0$ and $\mathscr{R} = \mathscr{R}_g = 159{,}155$, $L = 350^2/159{,}155 = 0.77$ H.

B-1 SINUSOIDAL EXCITATION

If a sinusoidal voltage of $v_s = V_m \sin \omega t = \sqrt{2} \, V_s \sin \omega t$, is applied to the core in Fig. B-3a, the flux can be found from

$$V_m \sin \omega t = -N \frac{d\phi}{dt} \qquad \text{(B-12)}$$

which after integrating gives

$$\phi = \phi_m \cos \omega t = \frac{V_m}{N\omega} \cos \omega t \qquad \text{(B-13)}$$

Thus

$$\phi_m = \frac{V_m}{2\pi fN} = \frac{\sqrt{2}\,V_s}{2\pi fN} = \frac{V_s}{4.44fN} \tag{B-14}$$

The peak flux, ϕ_m, depends on the voltage, frequency, and the number of turns. Equation (B-14) is valid if the core is not saturated. If peak flux is high, the core may saturate and the flux will not be sinusoidal. If the ratio of voltage to frequency is maintained constant, the flux will remain constant, provided that the number of turns is unchanged.

B-2 TRANSFORMER

If a second winding, called the *secondary winding,* is added to the core in Fig. B-3 and the core is excited from sinusoidal voltage, a voltage will be induced in the secondary winding. This is shown in Fig. B-4. If N_p and N_s are turns on the primary and secondary windings, respectively, the primary voltage V_p and secondary voltage V_s are related to each other as

$$\frac{V_p}{V_s} = \frac{I_s}{I_p} = \frac{N_p}{N_s} = a \tag{B-15}$$

where a is the turns ratio.

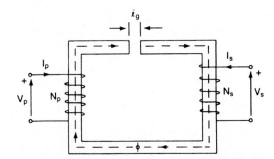

Figure B-4   Transformer core.

The equivalent circuit of a transformer is shown in Fig. B-5, where all the parameters are referred to the primary. To refer a secondary parameter to the primary side, the parameter is multiplied by a^2. The equivalent circuit can be referred to the secondary side by dividing all parameters of the circuit in Fig. B-5

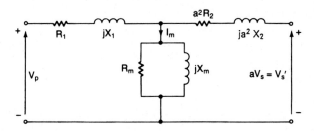

Figure B-5   Equivalent circuit of transformer.

by a^2. X_1 and X_2 are the leakage reactances of the primary and secondary windings, respectively. R_1 and R_2 are the resistances of the primary and secondary winding, X_m is the magnetizing reactance, and R_m represents the core loss.

The variations of the flux due to ac excitation cause two types of losses in the core: (a) hysteresis loss, and (2) eddy-current loss. The hysteresis loss is expressed empirically as

$$P_h = K_h f B_{max}^z \qquad \text{(B-16)}$$

where K_h is a hysteresis constant which depends on the material and B_{max} is the peak flux density. z is the Steinmetz constant, which has a value of 1.6 to 2. The eddy-current loss is expressed empirically as

$$P_e = K_e f^2 B_{max}^2 \qquad \text{(B-17)}$$

where K_e is the eddy-current constant and depends on the material. The total core loss is

$$P_c = K_h f B_{max}^2 + K_e f^2 B_{max}^2 \qquad \text{(B-18)}$$

Note. If a transformer is designed to operate at 60 Hz and it is operated at a higher frequency, the core loss will increase significantly.

Switching functions of converters

The output of a converter depends on the switching pattern of the converter switches and the input voltage (or current). Similar to a linear system, the output quantities of a converter can be expressed in terms of the input quantities, by spectrum multiplication. The arrangement of a single-phase converter is shown in Fig. C-1a. If $V_i(\theta)$ and $I_i(\theta)$ are the input voltage and current, respectively, the corresponding output voltage and current are $V_o(\theta)$ and $I_o(\theta)$, respectively. The input could be either a voltage source or a current source.

Voltage source. For a voltage source, the output voltage $V_o(\theta)$ can be related to input voltage $V_i(\theta)$ by

$$V_o(\theta) = S(\theta)V_i(\theta) \qquad \text{(C-1)}$$

where $S(\theta)$ is the switching function of the converter as shown in Fig. C-1b. $S(\theta)$ depends on the type of converter and the gating pattern of the switches. If g_1, g_2, g_3, and g_4 are the gating signals for switches Q_1, Q_2, Q_3, and Q_4, respectively, the switching function is

$$S(\theta) = g_1 - g_4 = g^2 - g^3$$

Neglecting the losses in the converter switches and using power balance give us

$$V_i(\theta)I_i(\theta) = V_o(\theta)I_o(\theta)$$

$$S(\theta) = \frac{V_o(\theta)}{V_i(\theta)} = \frac{I_i(\theta)}{I_o(\theta)} \qquad \text{(C-2)}$$

$$I_i(\theta) = S(\theta)I_o(\theta) \qquad \text{(C-3)}$$

Once $S(\theta)$ is known, $V_o(\theta)$ can be determined. $V_o(\theta)$ divided by the load impedance, gives $I_o(\theta)$; and then $I_i(\theta)$ can be found from Eq. (C-3).

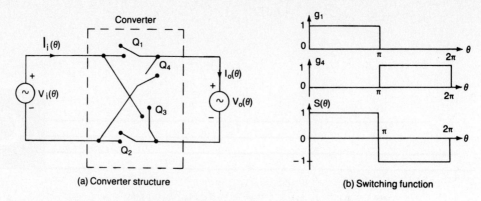

(a) Converter structure

(b) Switching function

Figure C-1 Single-phase converter structure.

Current source. In the case of current source, the input current remains constant, $I_i(\theta) = I_i$ and the output current $I_o(\theta)$ can be related to input current I_i,

$$I_o(\theta) = S(\theta)I_i$$

$$V_o(\theta)I_o(\theta) = V_i(\theta)I_i(\theta) \tag{C-4}$$

which gives

$$V_i(\theta) = S(\theta)V_o(\theta) \tag{C-5}$$

$$S(\theta) = \frac{V_i(\theta)}{V_o(\theta)} = \frac{I_o(\theta)}{I_i(\theta)} \tag{C-6}$$

C-1 SINGLE-PHASE FULL-BRIDGE INVERTERS

The switching function of a single-phase full-bridge inverter in Fig. 10-2a, is shown in Fig. C-2. If g_1 and g_4 are the gating signals for switches Q_1 and Q_4, respectively, the switching function is

$$S(\theta) = g_1 - g_4$$

$$= 1 \qquad \text{for } 0 \leq \theta \leq \pi$$

$$= -1 \qquad \text{for } \pi \leq \theta \leq 2\pi$$

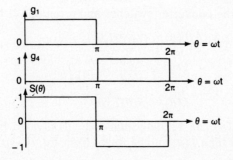

Figure C-2 Switching function of single-phase full-bridge inverter.

If f_0 is the fundamental frequency of the inverter,

$$\theta = \omega t = 2\pi f_0 t \tag{C-7}$$

$S(\theta)$ can be expressed in a Fourier series as

$$S(\theta) = \frac{A_0}{2} + \sum_{n=1,2,\ldots}^{\infty} (A_n \cos n\theta + B_n \sin n\theta)$$

$$B_n = \frac{2}{\pi} \int_0^{\pi} S(\theta) \sin n\theta \, d\theta = \frac{4}{n\pi} \qquad \text{for } n = 1, 3, \ldots \tag{C-8}$$

Due to half-wave symmetry, $A_0 = A_n = 0$.

Substituting A_0, A_n, and B_n in Eq. (C-8) yields

$$S(\theta) = \frac{4}{\pi} \sum_{n=1,3,5,\ldots}^{\infty} \frac{\sin n\theta}{n} \tag{C-9}$$

If the input voltage, which is dc, is $V_i(\theta) = V_s$, Eq. (C-1) gives the output voltage as

$$V_o(\theta) = S(\theta) V_i(\theta) = \frac{4 V_s}{\pi} \sum_{n=1,3,5,\ldots}^{\infty} \frac{\sin n\theta}{n} \tag{C-10}$$

which is the same as Eq. (10-11). For a three phase voltage-source inverter in Fig. 10-5, there are three switching functions: $S_1(\theta) = g_1 - g_4$, $S_2(\theta) = g_3 - g_6$, and $S_3(\theta) = g_5 - g_2$. There will be three line-to-line output voltages corresponding to three switching voltages, namely $V_{ab}(\theta) = S_1(\theta) V_i(\theta)$, $V_{bc}(\theta) = S_2(\theta) V_i(\theta)$, and $V_{ca}(\theta) = S_3(\theta) V_i(\theta)$.

C-2 SINGLE-PHASE BRIDGE RECTIFIERS

The switching function of a single-phase bridge rectifier is the same as that of the single-phase full-bridge inverter. If the input voltage is $V_i(\theta) = V_m \sin \theta$, Eqs. (C-1) and (C-9) give the output voltage as

$$V_o(\theta) = S(\theta) V_i(\theta) = \frac{4 V_m}{\pi} \sum_{n=1,3,5,\ldots}^{\infty} \frac{\sin \theta \sin n\theta}{n} \tag{C-11}$$

$$= \frac{4 V_m}{\pi} \sum_{n=1,3,5,\ldots}^{\infty} \frac{\cos(n-1)\theta - \cos(n+1)\theta}{2n} \tag{C-12}$$

$$= \frac{2 V_m}{\pi} \left[1 - \cos 2\theta + \frac{1}{3} \cos 2\theta - \frac{1}{3} \cos 4\theta \right.$$

$$\left. + \frac{1}{5} \cos 4\theta - \frac{1}{5} \cos 6\theta + \frac{1}{7} \cos 6\theta - \frac{1}{7} \cos 8\theta + \cdots \right]$$

$$= \frac{2V_m}{\pi} \left[1 - \frac{2}{3} \cos 2\theta - \frac{2}{15} \cos 4\theta - \frac{2}{35} \cos 6\theta - \cdots \right]$$

$$= \frac{2V_m}{\pi} - \frac{4V_m}{\pi} \sum_{m=1}^{\infty} \frac{\cos 2m\theta}{4m^2 - 1} \tag{C-13}$$

Equation (C-13) is the same as Eq. (3-63). The first part of Eq. (C-13) is the average output voltage and the second part is the ripple content on the output voltage.

For a three-phase rectifier in Figs. 3-25a and 5-10a, the switching functions are $S_1(\theta) = g_1 - g_4$, $S_2(\theta) = g_3 - g_6$, and $S_3(\theta) = g_5 - g_2$. If the three input phase voltages are $V_{an}(\theta)$, $V_{bn}(\theta)$, and $V_{cn}(\theta)$, the output voltage becomes

$$V_o(\theta) = S_1(\theta) V_{an}(\theta) + S_2(\theta) V_{bn}(\theta) + S_3(\theta) V_{cn}(\theta) \tag{C-14}$$

C-3 SINGLE-PHASE FULL-BRIDGE INVERTERS WITH SPWM

The switching function of a single-phase full-bridge inverter with sinusoidal pulse-width modulation (SPWM) is shown in Fig. C-3. The gating pulses are generated by comparing a cosine wave with triangular pulses. If g_1 and g_4 are the gating signals for switches Q_1 and Q_4, respectively, the switching function is

$$S(\theta) = g_1 - g_4$$

$S(\theta)$ can be expressed in a Fourier series as

$$S(\theta) = \frac{A_0}{2} + \sum_{n=1,2\ldots}^{\infty} (A_n \cos n\theta + B_n \sin n\theta) \tag{C-15}$$

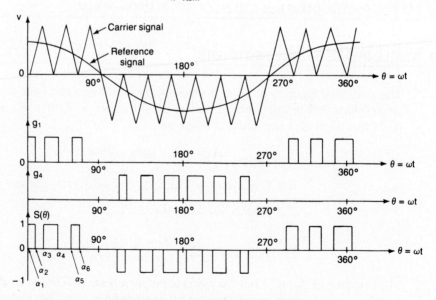

Figure C-3 Switching function with SPWM.

If there are p pulses per quarter cycle and p is an even number,

$$A_n = \frac{2}{\pi} \int_0^\pi S(\theta) \cos n\theta \, d\theta$$

$$= \frac{4}{\pi} \int_0^{\pi/2} S(\theta) \cos n\theta \, d\theta$$

$$= \frac{4}{\pi} \left[\int_{\alpha 1}^{\alpha 2} \cos n\theta \, d\theta + \int_{\alpha 3}^{\alpha 4} \cos n\theta \, d\theta + \int_{\alpha 5}^{\alpha 6} \cos n\theta \, d\theta + \cdots \right)$$

$$= \frac{4}{n\pi} \sum_{m=1,2,3,\ldots}^{p} [(-1)^m \sin n\alpha_m] \tag{C-16}$$

Due to quarter-wave symmetry, $B_n = A_0 = 0$. Substituting A_0, A_n, and B_n in Eq. (C-15) yields

$$S(\theta) = \sum_{n=1,3,5,\ldots}^{\infty} A_n \cos n\theta$$

$$= \frac{4}{n\pi} \sum_{n=1,3,5,\ldots}^{\infty} \left[\sum_{m=1,2,3,\ldots}^{p} (-1)^m \sin n\alpha_m \cos n\theta \right] \tag{C-17}$$

If the input voltage is $V_i(\theta) = V_s$, Eqs. (C-1) and (C-17) give the output voltage as

$$V_o(\theta) = V_s \sum_{n=1,3,5,\ldots}^{\infty} A_n \cos n\theta \tag{C-18}$$

C-4 SINGLE-PHASE CONTROLLED RECTIFIERS WITH SPWM

If the input voltage is $V_i(\theta) = V_m \cos \theta$, Eqs. (C-1) and (C-17) give the output voltage as

$$V_o(\theta) = V_m \sum_{n=1,3,5,\ldots}^{\infty} A_n \cos n\theta \cos \theta \tag{C-19}$$

$$= \frac{V_m}{2} \sum_{n=1,3,5,\ldots}^{\infty} A_n [\cos(n-1)\theta + \cos(n+1)\theta]$$

$$= 0.5 V_m [A_1(\cos 0 + \cos 2\theta) + A_3(\cos 2\theta + \cos 4\theta)$$

$$+ A_5(\cos 4\theta + \cos 6\theta) + \cdots]$$

$$= \frac{V_m A_1}{2} + V_m \sum_{n=2,4,6,\ldots}^{\infty} \frac{A_{n-1} + A_{n+1}}{2} \cos n\theta \tag{C-20}$$

The first part of Eq. (C-20) is the average output voltage and the second part is the ripple voltage. Equation (C-20) is valid, provided that the input voltage and the switching function are cosine waveforms.

In the case of sine waves, the input voltage is $V_i(\theta) = V_m \sin \theta$ and the switching function is

$$S(\theta) = \sum_{n=1,3,5,\ldots}^{\infty} A_n \sin n\theta \qquad \text{(C-21)}$$

Equations (C-1) and (C-21) give the output voltage as

$$V_o(\theta) = V_m \sum_{n=1,3,5,\ldots}^{\infty} A_n \sin \theta \sin n\theta \qquad \text{(C-22)}$$

$$= \frac{V_m}{2} \sum_{n=1,3,5,\ldots}^{\infty} A_n[\cos(n-1)\theta - \cos(n+1)\theta]$$

$$= 0.5 V_m[A_1(\cos 0 - \cos 2\theta) + A_3(\cos 2\theta - \cos 4\theta)$$

$$+ A_5(\cos 4\theta - \cos 6\theta) + \cdots]$$

$$= \frac{V_m A_1}{2} - V_m \sum_{n=2,4,6,\ldots}^{\infty} \frac{A_{n-1} - A_{n+1}}{2} \cos n\theta \qquad \text{(C-23)}$$

D

Dc transient analysis

D-1 *RC* CIRCUIT WITH STEP INPUT

When the switch S_1 in Fig. 3-1a is closed at $t = 0$, the charging current of the capacitor can be found from

$$V_s = v_R + v_c = Ri + \frac{1}{C} \int i \, dt + v_c(t = 0) \tag{D-1}$$

with initial condition: $v_c(t = 0) = 0$. Using Table D-1, Eq. (D-1) can be transformed into Laplace's domain of s:

$$\frac{V_s}{s} = RI(s) + \frac{1}{Cs} I(s)$$

which after solving for the current $I(s)$ gives

$$I(s) = \frac{V_s}{R(s + \alpha)} \tag{D-2}$$

where $\alpha = 1/RC$. Inverse transform of Eq. (D-2) in the time domain yields

$$i(t) = \frac{V_s}{R} e^{-\alpha t} \tag{D-3}$$

and the voltage across the capacitor is obtained as

$$v_c(t) = \frac{1}{C} \int_0^t i \, dt = V_s(1 - e^{-\alpha t}) \tag{D-4}$$

TABLE D-1 SOME LAPLACE TRANSFORMATIONS

$f(t)$	$F(s)$
1	$\dfrac{1}{s}$
t	$\dfrac{1}{s^2}$
$e^{-\alpha t}$	$\dfrac{1}{s + \alpha}$
$\sin \alpha t$	$\dfrac{\alpha}{s^2 + \alpha^2}$
$\cos \alpha t$	$\dfrac{s}{s^2 + \alpha^2}$
$f'(t)$	$sF(s) - F(0)$
$f''(t)$	$s^2F(s) - sF(s) - F'(0)$

In the steady state (at $t = \infty$),

$$I_s = i(t = \infty) = 0$$

$$V_c = v_c(t = \infty) = \frac{V_s}{R}$$

D-2 *RL* CIRCUIT WITH STEP INPUT

Two typical *RL* circuits are shown in Figs. 3-2a and 9-3a. The transient current through the inductor in Fig. 9-3a can be expressed as

$$V_s = v_L + v_R + E = L\frac{di}{dt} + Ri + E \tag{D-5}$$

with initial condition: $i(t = 0) = I_1$. In Laplace's domain of s, Eq. (D-5) becomes

$$\frac{V_s}{s} = L\,sI(s) - LI_1 + RI(s) + \frac{E}{s}$$

and solving for $I(s)$ gives

$$I(s) = \frac{V_s - E}{L\,s(s + \beta)} + \frac{I_1}{s + \beta}$$

$$= \frac{V_s - E}{R}\left(\frac{1}{s} - \frac{1}{s + \beta}\right) + \frac{I_1}{s + \beta} \tag{D-6}$$

where $\beta = R/L$. Taking inverse transform of Eq. (D-6) yields

$$i(t) = \frac{V_s}{R}(1 - e^{-\beta t}) + I_1 e^{-\beta t} \qquad \text{(D-7)}$$

If there is no initial current on the inductor (i.e., $I_1 = 0$), Eq. (D-7) becomes

$$i(t) = \frac{V_s}{R}(1 - e^{-\beta t}) \qquad \text{(D-8)}$$

In the steady state (at $t = \infty$), $I_s = i(t = \infty) = V_s/R$.

D-3 *LC* CIRCUIT WITH STEP INPUT

The transient current through the capacitor in Figs. 3-4a and 7-2a is expressed as

$$V_s = v_L + v_c = L\frac{di}{dt} + \frac{1}{C}\int i\,dt + v_c(t = 0) \qquad \text{(D-9)}$$

with initial conditions: $v_c(t = 0) = 0$ and $i(t = 0) = 0$. In Laplace's transform, Eq. (D-9) becomes

$$\frac{V_s}{s} = L\,sI(s) + \frac{1}{Cs}I(s)$$

and solving for $I(s)$ gives

$$I(s) = \frac{V_s}{L(s^2 + \omega_m^2)} \qquad \text{(D-10)}$$

where $\omega_m = 1/\sqrt{LC}$. The inverse transform of Eq. (D-10) yields the charging current as

$$i(t) = V_s\sqrt{\frac{C}{L}}\sin(\omega_m t) \qquad \text{(D-11)}$$

and the capacitor voltage is

$$v_c(t) = \frac{1}{C}\int_0^t i(t)\,dt = V_s[1 - \cos(\omega_m t)] \qquad \text{(D-12)}$$

An *LC* circuit with an initial inductor current of I_m and an initial capacitor voltage of V_o is shown in Fig. 7-18. The capacitor current is expressed as

$$V_s = L\frac{di}{dt} + \frac{1}{C}\int i\,dt + v_c(t = 0) \qquad \text{(D-13)}$$

with initial condition $i(t = 0) = Im$ and $vc(t = 0) = V_o$. Note: In Fig. 7-18, V_o is shown as equal to $-2V_s$. In Laplace's domain of s, Eq. (D-13) becomes

$$\frac{V_s}{s} = L\,sI(s) - LI_m + \frac{1}{Cs} + \frac{V_o}{s}$$

and solving for the current $I(s)$ gives

$$I(s) = \frac{V_s - V_o}{L(s^2 + \omega_m^2)} + \frac{sI_m}{s^2 + \omega_m^2} \qquad \text{(D-14)}$$

where $\omega_m = 1/\sqrt{LC}$. The inverse transform of Eq. (D-14) yields

$$i(t) = (V_s - V_o) \sqrt{\frac{C}{L}} \sin(\omega_m t) + I_m \cos(\omega_m t) \qquad \text{(D-15)}$$

and the capacitor voltage as

$$v_c(t) = \frac{1}{C} \int_0^t i(t) \, dt + V_o$$

$$= I_m \sqrt{\frac{L}{C}} \sin(\omega_m t) - (V_s - V_o) \cos(\omega_m t) + V_s$$

Fourier analysis

Under steady-state conditions, the output voltage of power converters is, generally, a periodic function of time defined by

$$v_o(t) = v_o(t + T) \tag{E-1}$$

where T is the periodic time. If f is the frequency of the output voltage in hertz, the angular frequency is

$$\omega = \frac{2\pi}{T} = 2\pi f \tag{E-2}$$

and Eq. (E-1) can be rewritten as

$$v_o(\omega t) = v_o(\omega t + 2\pi) \tag{E-3}$$

The Fourier theorem states that a periodic function $v_o(t)$ can be described by a constant term plus an infinite series of sine and cosine terms of frequency $n\omega$, where n is an integer. Therefore, $v_o(t)$ can be expressed as

$$v_o(t) = \frac{a_o}{2} + \sum_{n=1,2,\ldots}^{\infty} (a_n \cos n\omega t + b_n \sin n\omega t) \tag{E-4}$$

where $a_o/2$ is the average value of the output voltage, $v_o(t)$. The constant a_o, a_n and b_n can be determined from the following expressions:

$$a_o = \frac{2}{T} \int_0^T v_o(t)\, dt = \frac{1}{\pi} \int_0^{2\pi} v_o(\omega t)\, d(\omega t) \tag{E-5}$$

$$a_n = \frac{2}{T} \int_0^T v_o(t) \cos n\omega t\, dt = \frac{1}{\pi} \int_0^{2\pi} v_o(\omega t) \cos n\omega t\, d(\omega t) \tag{E-6}$$

$$b_n = \frac{2}{T} \int_0^T v_o(t) \sin n\omega t\, dt = \frac{1}{\pi} \int_0^{2\pi} v_o(\omega t) \sin n\omega t\, d(\omega t) \tag{E-7}$$

643

If $v_o(t)$ can be expressed as an analytical function, these constants can be determined by a single integration. If $v_o(t)$ is discontinuous, which is usually the case for the output of converters, several integrations (over the whole period of the output voltage) must be performed to determine the constants, a_o, a_n, and b_n.

$$a_n \cos n\omega t + b_n \sin n\omega t$$

$$= (a_n^2 + b_n^2)^{1/2} \left(\frac{a_n}{\sqrt{a_n^2 + b_n^2}} \cos n\omega t + \frac{b_n}{\sqrt{a_n^2 + b_n^2}} \sin n\omega t \right) \quad \text{(E-8)}$$

Let us define an angle ϕ_n, whose adjacent side is b_n, opposite side is a_n, and the hypotenuse is $(a_n^2 + b_n^2)^{1/2}$. As a result, Eq. (E-8) becomes

$$a_n \cos n\omega t + b_n \sin n\omega t = (a_n^2 + b_n^2)^{1/2}(\sin \phi_n \cos n\omega t + \cos \phi_n \sin n\omega t)$$

$$= (a_n^2 + b_n^2)^{1/2} \sin(n\omega t + \phi_n) \quad \text{(E-9)}$$

where

$$\phi_n = \tan^{-1} \frac{a_n}{b_n} \quad \text{(E-10)}$$

Substituting Eq. (E-9) into Eq. (E-4), the series may also be written as

$$v_o(t) = \frac{a_o}{2} + \sum_{n=1,2,\ldots}^{\infty} C_n \sin(n\omega t + \phi_n) \quad \text{(E-11)}$$

where

$$C_n = (a_n^2 + b_n^2)^{1/2} \quad \text{(E-12)}$$

C_n and ϕ_n are the peak magnitude and the delay angle of the nth harmonic component of the output voltage, $v_o(t)$, respectively.

If the output voltage has a *half-wave symmetry*, the number of integrations within the entire period can be reduced significantly. A waveform has the property of a half-wave symmetry if the waveform satisfies the following conditions:

$$v_o(t) = -v_o \left(t + \frac{T}{2} \right) \quad \text{(E-13)}$$

or

$$v_o(\omega t) = -v_o(\omega t + \pi) \quad \text{(E-14)}$$

In a waveform with a half-wave symmetry, the negative half-wave is the mirror image of the positive half-wave, but phase shifted by $T/2$ s(or π rad) from the positive half-wave. A waveform with a half-wave symmetry does not have the even harmonics (i.e., $n = 2, 4, 6, \ldots$) and possess only the odd harmonics (i.e., $n = 1, 3, 5, \ldots$). Due to the half-wave symmetry, the average value is zero (i.e., $a_o = 0$). Equations (E-6), (E-7), and (E-11) become

$$a_n = \frac{2}{T} \int_0^T v_o(t) \cos n\omega t \, dt = \frac{1}{\pi} \int_0^{2\pi} v_o(\omega t) \cos n\omega t \, d(\omega t), \quad n = 1, 3, 5, \ldots$$

$$b_n = \frac{2}{T} \int_0^T v_o(t) \sin n\omega t \, dt = \frac{1}{\pi} \int_0^{2\pi} v_o(\omega t) \sin n\omega t \, d(\omega t), \qquad n = 1, 3, 5, \ldots$$

$$v_o(t) = \sum_{n=1,3,5,\ldots}^{\infty} C_n \sin(n\omega t + \phi_n)$$

In general, with a half-wave symmetry, $a_o = a_n = 0$, and with a quarter-wave symmetry, $a_o = b_n = 0$.

F

Listing of computer programs in IBM-PC BASICA

```
********* PROG-1 **********

5   CLS
10  REM "PROG-1"
30  DIM ALFA(50), ALFAD(150), B(100)
40  REM UNIFORM PWM ANGLE CALCULATION
50  PRINT "No. of pulses per half cycle less than 50 ?"
60  INPUT NP
70  PRINT "Modulation index less than 1 ?"
80  INPUT AMF
82  PRINT "List of Fourier coefficients? For YES '1' and for NO '2' "
83  INPUT NC
84  PRINT "Highest desired harmonic component less than 100"
85  INPUT NM
87  CLS
90  PI=4!*ATN(1!)
97  PRINT "Copyright 1993, Power Electronics by Muhammad H Rashid"
98  PRINT "Fig. 5-16 AC-DC converter with uniform PWM"
100 DELTAM=(PI/NP)*AMF
110 CENTR=(PI/NP)/2
120 ALFA(1)=CENTR-DELTAM/2
130 ALFAD(1)=ALFA(1)*180/PI
140 DELTAD=DELTAM*180/PI
170 FOR M=2 TO NP
180 ALFA(M)=ALFA(1)+(PI/NP)*(M-1)
190 ALFAD(M)=ALFA(M)*180/PI
210 NEXT M
220 VMAX=1
230 V=0
240 FOR M=1 TO NP
250 V=(COS(ALFA(M))-COS(ALFA(M)+DELTAM))+V
260 NEXT M
270 VDC=(VMAX/PI)*V/SQR(2)
290 AN=0
300 IA=1
310 FOR N=1 TO NM
```

646

```
320 C=0
330 FOR M=1 TO NP
340 C=COS(N*ALFA(M))-COS(N*(ALFA(M)+DELTAM))+C
350 NEXT M
360 B(N)=(2*IA/(N*PI))*C
365 IF NC=1 THEN PRINT "B("N") = "B(N)
380 NEXT N
381 PRINT "No. of pulses per half-cycle = "NP
382 PRINT "Modulation index = "AMF
383 PRINT "Pulse width in degrees = "DELTAD
384 FOR M=1 TO NP
386 PRINT "'"M"in degrees = "ALFAD(M)
387 NEXT M
388 PRINT "Average output voltage in % of RMS input voltage = "VDC*100
390 I1=B(1)/SQR(2)
400 PRINT "RMS fundamental input current as % of dc load current = "I1*100
405 SUM=0
410 FOR N=1 TO NM
420 SUM=SUM+B(N)*B(N)/2
430 NEXT N
440 IS=SQR(SUM)
450 PRINT "RMS input current as % of dc load current = "IS*100
460 PF=I1/IS
463 DF=1
465 PRINT "Displacent factor = "DF
470 PRINT "Power factor ="  PF
500 END


             ********* PROG-2 **********

5   CLS
10  REM "PROG-2"
30  DIM SL(50), C(50), X(90), Y(90), ALFA(50), ALFAD(50), B(100)
40  REM AC-DC CONVERTER WITH SINUSOIDAL PWM
50  PRINT "No. of pulses per half cycle ?"
60  INPUT NP
70  PRINT "Modulation index less than 1 ?"
80  INPUT AMF
82  PRINT "List of Fourier coefficients? For YES '1' and for NO '2' "
83  INPUT NC
84  PRINT "Highest desired harmonic component less than 100"
85  INPUT NM
87  CLS
90  PI=4!*ATN(1!)
95  PRINT "Figure 5-17"
97  PRINT "Copyright 1993, Power Electronics by Muhammad H Rashid"
98  PRINT "Fig. 5-17 AC-DC converter with sinusoidal PWM"
100 NS=2*NP
102 TOL=.001
104 FOR M=1 TO NS
106 SL(M)=((-1)^M)*NS/PI
108 NEXT M
109 FOR M=1 TO NS STEP 2
110 C(M)=M
111 NEXT M
112 FOR M=2 TO NS STEP 2
113 C(M)=-(M-1)
114 NEXT M
116 Y(1)=AMF
118 X(1)=PI/2
119 FOR M=1 TO NS
```

```
120 FOR I-1 TO 50
122 K-I+1
124 X(K)-(Y(I)-C(M))/SL(M)
126 Y(K)-AMF*SIN(X(K))
128 XX-ABS(X(K)-X(K-1))
130 IF XX<TOL THEN 134
132 NEXT I
134 ALFA(M)-X(I)
136 NEXT M
138 FOR M-1 TO NS
140 ALFAD(M)-ALFA(M)*180/PI
142 NEXT M
220 VMAX-1
230 V-0
240 FOR M-1 TO NS STEP 2
250 V-(COS(ALFA(M))-COS(ALFA(M+1)))+V
260 NEXT M
270 VDC-(VMAX/PI)*V/SQR(2)
290 AN-0
300 IA-1
310 FOR N-1 TO NM
320 D-0
330 FOR M-1 TO NS STEP 2
340 D-(COS(N*ALFA(M))-COS(N*ALFA(M+1)))+D
350 NEXT M
360 B(N)-(2*IA/(N*PI))*D
365 IF NC-1 THEN PRINT "B("N") - "B(N)
380 NEXT N
381 PRINT "No. of pulses per half-cycle - "NP
382 PRINT "Modulation index - "AMF
384 FOR M-1 TO NS
386 PRINT "'"M"in degrees - "ALFAD(M)
387 NEXT M
388 PRINT "Average output voltage in % of RMS input voltage - "VDC*100
390 I1-B(1)/SQR(2)
400 PRINT "RMS fundamental input current as % of dc load current - "I1*100
405 SUM-0
410 FOR N-1 TO NM
420 SUM-SUM+B(N)*B(N)/2
430 NEXT N
440 IS-SQR(SUM)
450 PRINT "RMS input current as % of dc load current - "IS*100
460 PF-I1/IS
463 DF-1
465 PRINT "Displacent factor - "DF
470 PRINT "Power factor -" PF
500 END

                    ********* PROG-3 **********

2   CLS
3 PRINT "Example 6-4"
4 PRINT "Copyright 1993, Power Electronics by Muhammad H Rashid"
5 INPUT "DELAY ANGLE IN DEGREES ? ", ALP1
10 ALP-ALP1*3.1415927#/180
15 INPUT "INPUT VOLTAGE in rms ? ", VS
20 INPUT "INPUT FREQUENCY IN HZ ? ", F
30 INPUT "LOAD RESITANCE in j ? ", R
40 INPUT "LOAD INDUCTANCE IN mH ? ", L
42 PI-4*ATN(1!)
45 W-2*PI*F
50 XL-W*L*.001
55 PRINT "IMPEDANCE OF LOAD INDUCTANCE - ", XL
```

```
105 PH-ATN(XL/R)
110 PH1-180*PH/PI
120 PRINT "LOAD ANGLE -", PH1
130 DELB-.25
140 DB-DELB*PI/180
152 BET1-PH
170 N-1
180 Y1-(R/XL)*(ALP-BET1)
190 Y2-EXP(Y1)
195 YX-SIN(ALP-PH)
200 Y3-Y2*YX
210 Y4-SIN(BET1-PH)
215 N-N+1
217 IF N-1000 THEN 360
220 YY-ABS(ABS(Y3)-ABS(Y4))
222 IF YY<.001 THEN   360
230 BET1-BET1+DB
245 IF Y4<0 THEN   Y4-0
250 GOTO 180
360 BET-180*BET1/PI
365 PRINT "EXTINCTION ANGLE -", BET
380 J-0
384 DEL-(BET1-ALP)/180
390 SUM-0
400 SUMM-0
402 L-L*.001
410 WT-ALP+DEL*J
420 T-WT/W
430 X--T+ALP/W
440 X-X*R/L
450 X1-EXP(X)
460 X2-SIN(ALP-PH)
470 X3-X2*X1
480 X4-SIN(WT-PH)
490 X5-X4-X3
495 IF X5<0 THEN X5-0
500 XX-X5*X5
510 YY-X5
540 SUM-SUM+XX
550 SUMM-SUMM+YY
555 J-J+1
560 IF WT>BET1 THEN   800
570 IF WT-BET1 THEN 800
580 GOTO 410
800 SUM-SUM*DEL/PI
810 SUMM-SUMM*DEL/PI
811 SUMM-SUMM/SQR(2)
820 Y1-R*R+XL*XL
830 Y1-SQR(Y1)
840 IR-(VS/Y1)*SQR(SUM)
850 ID-(VS/Y1)*SUMM
860 PRINT "IR -  ", IR, "ID - ", ID
870 X-BET1-ALP+.5*SIN(2*ALP)-.5*SIN(2*BET1)
872 IF X>0 THEN 880
875 PRINT "TOO LOW INDUCTANCE"
880 VO-VS*SQR(X/PI)
890 PRINT "RMS OF OUTPUT VOLTAGE - ", VO
900 PO-2*IR*IR*R
905 PRINT "Output power  - ", PO
910 PF-PO/(VS*IR*SQR(2))
920 PRINT "Power factor - ", PF
930 END
```

********* PROG-4 **********

```
5 CLS
8 PRINT "Example 6-7"
9 PRINT "Copyright 1993, Power Electronics by Muhammad H Rashid"
10 INPUT "INPUT LINE VOLTAGE in rms ? ", VS
30 INPUT "LOAD RESISTANCE PER PHASE in j ? ", R
45 INPUT "DELAY ANGLE ' in DEGREES ? ", ALP1
70 X-4*ATN(1!)
80 ALP-ALP1*X/180
90 IM-SQR(2)*VS/R
91 AX-X-ALP+SIN(2*ALP)/2
92 VO-VS*SQR(AX/X)
93 PRINT "OUTPUT VOLTAGE FROM Eq. (6-35) ", VO
100 NS-360
110 DEL-X/NS
120 SUM-0
122 SUA-0
124 SUC-0
130 J-0
138 T1-ALP
139 IF T1>X/3 OR T1-X/3 THEN T1-X/3
140 Y-DEL*J
150 IA-0
155 IC-IM*SIN(Y-4*X/3)
156 IF ALP>2*X/3 AND Y<(X-ALP) THEN IC=0
158 YY-IA-IC
160 SUM-SUM+YY*YY
162 SUA-SUA+IA*IA
164 SUC-SUC+IC*IC
170 J-J+1
180 IF Y>T1 OR Y-T1 THEN 300
200 GOTO 140
300 T2-X/3
320 Y-DEL*J
322 IA-IM*SIN(Y)
324 IC-IM*SIN(Y-4*X/3)
330 YY-IA-IC
340 SUM-SUM+YY*YY
342 SUA-SUA+IA*IA
344 SUC-SUC+IC*IC
350 J-J+1
360 IF Y>T2 OR Y-T2 THEN  400
370 GOTO 320
400 T3-ALP+X/3
405 Y-DEL*J
410 IA-IM*SIN(Y)
411 IF ALP>X/3 AND Y<ALP THEN  IA-0
412 IC-0
416 YY-IA-IC
420 SUM-SUM+YY*YY
422 SUA-SUA+IA*IA
424 SUC-SUC+IC*IC
430 J-J+1
440 IF Y>T3 OR Y-T3 THEN  500
450 IF Y>X OR Y-X THEN  500
460 GOTO 405
500 T4-X
510 Y-DEL*J
512 IA-IM*SIN(Y)
514 IC-IM*SIN(Y-4*X/3)
520 YY-IA-IC
530 SUM-SUM+YY*YY
```

```
532 SUA-SUA+IA*IA
534 SUC-SUC+IC*IC
540 J-J+1
550 IF Y>T4 OR Y-T4 THEN  600
560 IF Y>X OR Y-X THEN  600
570 GOTO  510
600 PRINT "NO OF SAMPLES -  ", J
620 IA-SQR(SUA/J)
625 PRINT "RMS VALUE OF PHASE A - ", IA
630 IC-SQR(SUC/J)
635 PRINT "RMS CURRENT OF PHASE C - ", IC
640 IL-SQR(SUM/J)
645 PRINT "RMS VALUE OF LINE CURRENT A - ", IL
648 VO-IA*R
649 PRINT "RMS VALUEE OF OUTPUT PHASE VOLTAGE - ", VO
650 PO-IA*IA*R*3
660 PRINT "OUTPUT POWER" , PO
670 VA-3*VS*IA
680 PRINT "VOLT-AMP - ", VA
690 PF-PO/VA
700 PRINT "POWER FACTOR - ", PF
900 END



              ********* PROG-5 *********

5 CLS
10 REM "PROG-5"
30 DIM ALFAM(20), ALFAD(20), BN(100), AN(100), V(100)
40 REM UNIFORM PWM ANGLE CALCULATION
50 PRINT "No. of pulses per half cycle ?"
60 INPUT NP
65 PRINT "List of Fourier coefficients? For YES '1' and for NO '2' "
70 INPUT NC
75 PRINT "Highest desired harmonic component less than 100"
80 INPUT NM
82 PRINT "Modulation index less than 1 ?"
85 INPUT AMF
87 CLS
89 AMF-1.0*NM
90 PI-4!*ATN(1!)
97 PRINT "Copyright 1993, Power Electronics by Muhammad H Rashid"
98 PRINT "Fig. 10-13 inverter with uniform PWM"
100 DELTA-(PI/NP)*AMF
105 DELTAD-DELTA*180/PI
110 CENTR-(PI/NP)/2
120 ALFAM(1)-CENTR-DELTA/2
130 ALFAD(1)-ALFAM(1)*180/PI
170 FOR M-2 TO NP
180 ALFAM(M)-ALFAM(1)+(PI/NP)*(M-1)
190 ALFAD(M)-ALFAM(M)*180/PI
210 NEXT M
220 VS-1!
225 FOR N-1 TO NM STEP 2
230 A-0
234 B-0
240 FOR M-1 TO NP
250 B-SIN(N*(ALFAM(M)+DELTA/2))-SIN(N*(PI+ALFAM(M)+DELTA/2))+B
253 A-COS(N*(ALFAM(M)+DELTA/2))-COS(N*(PI+ALFAM(M)+DELTA/2))+A
260 NEXT M
265 AN(N)-0
270 BN(N)-((2*VS/(N*PI))*SIN(N*DELTA/2))*B
```

```
272 V(N)-SQR(AN(N)*AN(N)+BN(N)*BN(N))/SQR(2)
273 IF NC-1 THEN PRINT "AN("N") - "AN(N), "BN("N") - "BN(N), "V("N") - "V(N)
275 NEXT N
381 PRINT "No. of pulses per half-cycle - "NP
382 PRINT "Modulation index - "AMF
383 PRINT "Equal Pulse widths in degrees k - "DELTAD
384 FOR M-1 TO NP
386 PRINT "'"M"in degrees - "ALFAD(M)
387 NEXT M
388 PRINT "Fundamental RMS output voltage in % of dc input voltage - "V(1)*100
390 VX-VS*SQR((NP*DELTAD)/180)
400 PRINT "RMS output voltage as % of dc input voltage - " VX*100
410 HF-(SQR(VX*VX-V(1)*V(1)))/V(1)
420 PRINT "Harmonic factor in % - " HF*100
425 SUM-0
430 FOR N-2 TO NM
440 SUM-SUM+(V(N)/(N^2))^2
450 NEXT N
460 DF-SQR(SUM)/V(1)
465 PRINT "Distortion factor in % - " DF*100
500 END


                    ********* PROG-6 **********

5 CLS
10 REM "PROG-6"
30 DIM SL(25),C(25),V(100),Y(90),X(90),ALFA(25),ALFAD(25),AN(50),BN(50)
40 REM SPWM ANGLE CALCULATION
50 PRINT "No. of pulses per half cycle ?"
60 INPUT NP
70 PRINT "Modulation index less than 1 ?"
80 INPUT NM
82 PRINT "List of Fourier coefficients? For YES '1' and for NO '2' "
83 INPUT NC
84 PRINT "Highest desired harmonic component less than 100"
87 CLS
89 AMF-1.0*NM
90 PI-4!*ATN(1!)
97 PRINT "Copyright 1993, Power Electronics by Muhammad H Rashid"
98 PRINT "Fig. 10-15 inverter with SPWM"
100 NS-2*NP
105 TOL-.0001
110 FOR M-1 TO NS
115 SL(M)-((-1)^M)*((NS+2)/PI)
120 NEXT M
121 FOR M-1 TO NS STEP 2
122 C(M)-M+1
124 NEXT M
125 FOR M-2 TO NS STEP 2
130 C(M)--M
135 NEXT M
140 Y(1)-AMF
145 X(1)-PI/2
150 FOR M-1 TO NS
155 FOR I-1 TO 90
160 K-I+1
165 X(K)-(Y(I)-C(M))/SL(M)
170 Y(K)-AMF*SIN(X(K))
175 XX-ABS(X(K)-X(K-1))
180 IF XX<TOL THEN GOTO 190
185 NEXT I
190 ALFA(M)-X(I)
```

```
195 NEXT M
200 FOR M-1 TO NP
202 ALFAM(M)-ALFA(2*M-1)
204 DELTAM(M)-ALFA(2*M)-ALFA(2*M-1)
206 ALFAD(M)-ALFAM(M)*180/PI
208 DELTAD(M)-DELTAM(M)*180/PI
218 NEXT M
220 VS-1
225 FOR N-1 TO NM STEP 2
230 A-0
234 B-0
240 FOR M-1 TO NP
250 X-SIN(N*(ALFAM(M)+DELTAM(M)/2))-SIN(N*(PI+ALFAM(M)+DELTAM(M)/2))
251 B-SIN(N*DELTAM(M)/2)*X+B
255 Y-COS(N*(ALFAM(M)+DELTAM(M)/2))-COS(N*(PI+ALFAM(M)+DELTAM(M)/2))
256 A-0
260 NEXT M
265 AN(N)-((2*VS)/(N*PI))*A
270 BN(N)-((2*VS)/(N*PI))*B
272 V(N)-SQR(AN(N)*AN(N)+BN(N)*BN(N))/SQR(2)
273 IF NC-1 THEN PRINT "AN("N") - "AN(N), "BN("N") - "BN(N), "V("N") - "V(N)
275 NEXT N
381 PRINT "No. of pulses per half-cycle - "NP
382 PRINT "Modulation index - "AMF
383 PRINT "Unequal Pulse widths in degrees "
384 FOR M-1 TO NP
386 PRINT "'"M"in degrees - "ALFAD(M), "k"M"in degrees - "DELTAD(M)
387 NEXT M
388 PRINT "Fundamental RMS output voltage in % of dc input voltage - "V(1)*100
389 VX-0
390 VS-1
391 FOR M-1 TO NP
392 VX-DELTAM(M)+VX
394 NEXT M
396 VX-VS*SQR(VX/PI)
400 PRINT "RMS output voltage as % of dc input voltage - " VX*100
408 HF-(SQR(VX*VX-V(1)*V(1)))/V(1)
420 PRINT "Harmonic factor in % - " HF*100
425 SUM-0
430 FOR N-2 TO NM
440 SUM-SUM+(V(N)/(N^2))^2
450 NEXT N
460 DF-SQR(SUM)/V(1)
465 PRINT "Distortion factor in % - " DF*100
500 END


                    ********* PROG-7 **********

5 CLS
10 REM "PROG-7"
30 DIM SL(25),C(25),V(100),Y(90),X(90),ALFA(25),ALFAD(25),AN(50),BN(50)
40 REM SPWM ANGLE CALCULATION
50 PRINT "No. of pulses in the first 60 degrees ?"
60 INPUT NP
70 PRINT "Modulation index less than 1 ?"
80 INPUT NM
82 PRINT "List of Fourier coefficients? For YES '1' and for NO '2' "
83 INPUT NC
84 PRINT "Highest desired harmonic component less than 100"
87 CLS
89 AMF-1.0*NM
90 PI-4!*ATN(1!)
```

```
97 PRINT "Copyright 1993, Power Electronics by Muhammad H Rashid"
98 PRINT "Fig. 10-18 inverter with modified SPWM"
100 NS=2*NP
105 TOL=.0001
110 FOR M=1 TO NS
115 SL(M)=((-1)^M)*(3*(NS+1)/PI)
120 NEXT M
121 FOR M=1 TO NS STEP 2
122 C(M)=M+1
124 NEXT M
125 FOR M=2 TO NS STEP 2
130 C(M)=-M
135 NEXT M
136 NAT=2*NS+2
138 NPT=NAT/2
140 Y(1)=AMF
145 X(1)=PI/2
150 FOR M=1 TO NS
155 FOR I=1 TO 90
160 K=I+1
165 X(K)=(Y(I)-C(M))/SL(M)
170 Y(K)=AMF*SIN(X(K))
175 XX=ABS(X(K)-X(K-1))
180 IF XX<TOL THEN GOTO 184
182 NEXT I
184 ALFA(M)=X(I)
186 NEXT M
188 ALFA(NS+1)=PI/3
190 ALFA(NS+2)=2*PI/3
192 NAS=NS+1
194 FOR M=NAS TO NAT
195 J=NAT-M+1
196 ALFA(M)=PI-ALFA(J)
198 NEXT M
200 FOR M=1 TO NPT
202 ALFAM(M)=ALFA(2*M-1)
204 DELTAM(M)=ALFA(2*M)-ALFA(2*M-1)
206 ALFAD(M)=ALFAM(M)*180/PI
208 DELTAD(M)=DELTAM(M)*180/PI
218 NEXT M
220 VS=1
225 FOR N=1 TO NM STEP 2
230 A=0
234 B=0
240 FOR M=1 TO NPT
250 X=SIN(N*(ALFAM(M)+DELTAM(M)/2))-SIN(N*(PI+ALFAM(M)+DELTAM(M)/2))
251 B=SIN(N*DELTAM(M)/2)*X+B
255 Y=COS(N*(ALFAM(M)+DELTAM(M)/2))-COS(N*(PI+ALFAM(M)+DELTAM(M)/2))
256 A=0.
260 NEXT M
265 AN(N)=((2*VS)/(N*PI))*A
270 BN(N)=((2*VS)/(N*PI))*B
272 V(N)=SQR(AN(N)*AN(N)+BN(N)*BN(N))/SQR(2)
273 IF NC=1 THEN PRINT "AN("N") = "AN(N), "BN("N") = "BN(N), "V("N") = "V(N)
275 NEXT N
381 PRINT "No. of pulses per half-cycle = "NPT
382 PRINT "Modulation index = "AMF
384 PRINT "Angles are in degrees "
385 FOR M=1 TO NPT
386 PRINT "'"M"in degrees = "ALFAD(M), "k"M"in degrees = "DELTAD(M)
387 NEXT M
388 PRINT "Fundamental RMS output voltage in % of dc input voltage = "V(1)*100
389 VX=0
```

```
390 VS-1
391 FOR M-1 TO NPT
392 VX-DELTAM(M)+VX
394 NEXT M
396 VX-VS*SQR(VX/PI)
400 PRINT "RMS output voltage as % of dc input voltage = " VX*100
408 HF-(SQR(VX*VX-V(1)*V(1)))/V(1)
420 PRINT "Harmonic factor in % - " HF*100
425 SUM-0
430 FOR N-2 TO NM
440 SUM-SUM+(V(N)/(N^2))^2
450 NEXT N
460 DF-SQR(SUM)/V(1)
465 PRINT "Distortion factor in % - " DF*100
500 END
```

G

Data sheets

Data Sheet No. PD-2.110

R23AF SERIES
800–600 VOLTS RANGE
REVERSE RECOVERY TIME 0.9μs
300 AMP AVG HOCKEY PUK
SOFT FAST RECOVERY RECTIFIER DIODES

VOLTAGE RATINGS

VOLTAGE CODE (1)	V_{RRM}, V_R – (V) Max. rep. peak reverse and direct voltage	V_{RSM} – (V) Max. non-rep. peak reverse voltage
	T_J = –40° to 125°C	T_J = 25° to 125°C
8	800	900
6	600	700

MAXIMUM ALLOWABLE RATINGS

PARAMETER		VALUE	UNITS	NOTES	
T_J	Junction temperature	–40 to 125	°C		
T_{stg}	Storage temperature	–40 to 150	°C		
$I_{F(AV)}$	Max. av. current	300	A	180° half sine wave	
	@ Max. T_C	85	°C		
$I_{F(RMS)}$	Nom. RMS current	470	A		
I_{FSM}	Max. peak non-rep. surge current	4960	A	50Hz half cycle sine wave	Initial T_J = 125°C, rated V_{RRM} applied after surge.
		5200	A	60Hz half cycle sine wave	
		5900	A	50Hz half cycle sine wave	Initial T_J = 125°C, no voltage applied after surge.
		6180	A	60Hz half cycle sine wave	
I^2t	Max. I^2t capability	124	kA^2s	t = 10ms	Initial T_J = 125°C, rated V_{RRM} applied after surge.
		113	kA^2s	t = 8.3ms	
		175	kA^2s	t = 10ms	Initial T_J = 125°C, no voltage applied after surge.
		160	kA^2s	t = 8.3ms	
$I^2\sqrt{t}$	Max. $I^2\sqrt{t}$ capability	1750	$kA^2\sqrt{s}$	Initial T_J = 125°C, no voltage applied after surge. I^2t for time $t_x = I^2\sqrt{t} \cdot \sqrt{t_x}$. $0.1 \leq t_x \leq 10ms$.	
F	Mounting force	4450(1000) ± 10%	N(lbf)		

Figure G-1 Data sheet for IR diode, type R23AF. (Courtesy of International Rectifier.)

CHARACTERISTICS

	PARAMETER	MIN.	TYP.	MAX.	UNITS	TEST CONDITIONS
V_{FM}	Peak forward voltage	----	1.50	1.63	V	Initial T_J = 25°C, 50-60Hz half sine, I_{peak} = 940A.
$V_{F(TO)1}$	Low-level threshold	----	----	0.962	V	T_J = 125°C. Av. power = $V_{F(TO)} \cdot I_{F(AV)} + r_F \cdot (I_{F(RMS)})^2$
$V_{F(TO)2}$	High-level threshold	----	----	1.30		Use low level values for $I_{FM} \leq \pi$ rated $I_{F(AV)}$
r_{F1}	Low-level resistance	----	----	0.657	mΩ	
r_{F2}	High-level resistance	----	----	0.353		
t_{rr}	Reverse recovery time					
	"A" suffix	----	0.9	----	µs	T_J = 25°C, I_{FM} = 750A. di_R/dt = 25A/µs for sinusoidal pulse.
	"B" suffix	----	1.1	----		
t_{rr}	Reverse recovery time					
	"A" suffix	----	----	2.20	µs	T_J = 125°C, I_{FM} = 750A. di_R/dt = 25A/µs for sinusoidal pulse.
	"B" suffix	----	----	2.50		
S	"S" Factor (t_b/t_a)					
	"A" suffix	0.59	----	----		
	"B" suffix	0.56	----	----		
$I_{RM(REC)}$	Reverse current					
	"A" suffix	----	----	33	A	
	"B" suffix	----	----	36		
Q_{RR}	Recovered charge					
	"A" suffix	----	----	37	µC	
	"B" suffix	----	----	45		
I_{RM}	Peak reverse current	----	----	35	mA	T_J = 125°C. Max. rated V_{RRM}.
R_{thJC}	Thermal resistance, junction-to-case	----	----	0.08	°C/W	DC operation, double side cooled.
		----	----	0.09	°C/W	180° sine wave, double side cooled.
		----	----	0.09	°C/W	120° rectangular wave, double side cooled.
R_{thCS}	Thermal resistance, case-to-sink	----	----	0.06	°C/W	Mtg. surface smooth, flat and greased. Single side cooled. For double side, divide value by 2.
wt	Weight	----	57(2,0)	----	g(oz.)	
	Case Style		DO-200AA		JEDEC	

I_{FM}, t_a, t_b, i, t, $\dfrac{di_R}{dt}$, Q_{RR}, $I_{RM(REC)}$

$$t_{rr} = t_a + t_b$$

Figure G-1 *(continued)*

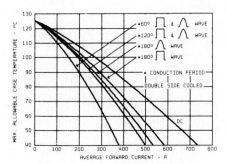

Fig. 1 — Case Temperature Ratings

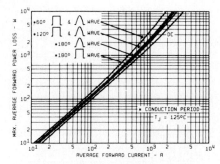

Fig. 2 — Power Loss Characteristics

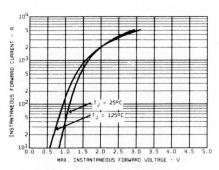

Fig. 3 — Forward Characteristics

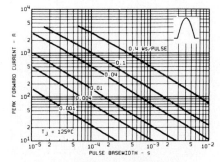

Fig. 4 — Max. Energy Loss Per Pulse —
Sinusoidal Waveforms

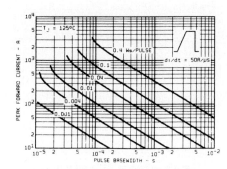

Fig. 5 — Max. Energy Loss Per Pulse —
Trapezoidal Waveforms

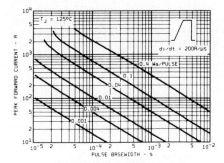

Fig. 6 — Max. Energy Loss Per Pulse —
Trapezoidal Waveforms

Figure G-1 *(continued)*

R23AF SERIES
800–600 VOLTS RANGE

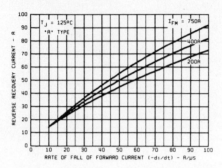

Fig. 7 — Typical Reverse Recovery Current

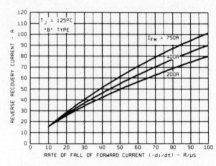

Fig. 7a — Typical Reverse Recovery Current

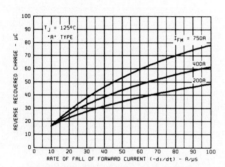

Fig. 8 — Typical Recovered Charge

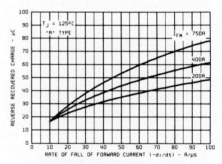

Fig. 8a — Typical Recovered Charge

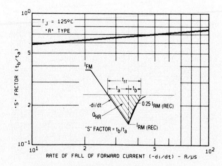

Fig. 9 — Typical "S" Factor (t_b/t_a)

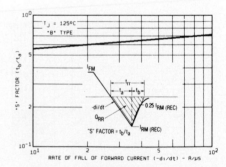

Fig. 9a — Typical "S" Factor (t_b/t_a)

Figure G-1 *(continued)*

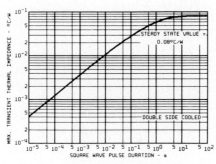

Fig. 10 — Transient Thermal Impedance, Junction-to-Case

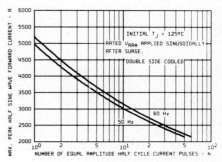

Fig. 11 — Non-Repetitive Surge Current Ratings

ORDERING INFORMATION

TYPE	VOLTAGE		RECOVERY TIME	
	CODE	V_{RRM}	CODE	t_{rr}
R23AF	8	800V	A	0.9µs
	6	600V	B	1.1µs

(1) t_{rr} is typical at 25°C.
Max. t_{rr} is guaranteed at 125°C. See table of Characteristics.

For example, for a device with t_{rr} = 0.9µs, V_{RRM} = 600V, order as R23AF6A.

Figure G-1 *(continued)*

R23AF SERIES
800–600 VOLTS RANGE

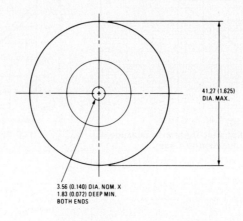

41.27 (1.625)
DIA. MAX.

3.56 (0.140) DIA. NOM. X
1.83 (0.072) DEEP MIN.
BOTH ENDS

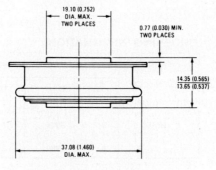

19.10 (0.752)
DIA. MAX.
TWO PLACES

0.77 (0.030) MIN.
TWO PLACES

14.35 (0.565)
13.65 (0.537)

37.08 (1.460)
DIA. MAX.

CREEPAGE DISTANCE: 12.45 (0.490) MIN.
STRIKE DISTANCE: 10.42 (0.410) MIN.

Conforms to JEDEC Outline DO-200AA
All Dimensions in Millimeters and (Inches)

Figure G-1 *(continued)*

S30EF & S30EFH SERIES
800-600 VOLTS RANGE
STANDARD TURN-OFF TIME 12 μs
850 AMP RMS, RING AMPLIFYING GATE
INVERTER TYPE HOCKEY PUK SCRs

VOLTAGE RATINGS

VOLTAGE CODE (1)	V_{RRM}, V_{DRM} - (V) Max. rep. peak reverse and off-state voltage	V_{RSM} - (V) Max. non-rep. peak reverse voltage $t_p \leq 5ms$	NOTES
	T_J = -40° to max. rated	T_J = 25° to max. rated	
8	800	900	Gate open
6	600	700	

MAXIMUM ALLOWABLE RATINGS

PARAMETER		SERIES	VALUE	UNITS	NOTES
T_J	Junction temperature	S30EF	-40 to 125	°C	
		S30EFH	-40 to 140		
T_{stg}	Storage temperature	ALL	-40 to 150	°C	
$I_{T(AV)}$	Max. av. current	ALL	540	A	180° half sine wave
	● Max. T_C	S30EF	70	°C	
		S30EFH	85		
$I_{T(RMS)}$	Nom. RMS current	ALL	850	A	
I_{TSM}	Max. peak non-repetitive surge current	ALL	8100	A	50Hz half cycle sine wave Initial T_J = 125°C, rated V_{RRM} applied after surge.
			8500		60Hz half cycle sine wave
			9650		50Hz half cycle sine wave Initial T_J = 125°C, no voltage applied after surge.
			10000		60Hz half cycle sine wave
I^2t	Max. I^2t capability	ALL	330	kA²s	t = 10ms Initial T_J = 125°C, rated V_{RRM} applied after surge.
			300		t = 8.3ms
			465		t = 10ms Initial T_J = 125°C, no voltage applied after surge.
			425		t = 8.3ms
$I^2\sqrt{t}$	Max. $I^2\sqrt{t}$ capability	ALL	4650	kA²√s	Initial T_J = 125°C, no voltage applied after surge. I^2t for time $t_x = I^2\sqrt{t} \cdot \sqrt{t_x}$. 0.1 $\leq t_x \leq$ 10ms.
di/dt	Max. non-repetitive rate-of-rise of current	ALL	800	A/μs	T_J = 125°C, $V_D = V_{DRM}$, I_{TM} = 1800A. Gate pulse: 20V, 20Ω, 10μs, 0.5μs rise time. Max. repetitive di/dt is approximately 40% of non-repetitive value.
P_{GM}	Max. peak gate power	ALL	10	W	$t_p \leq$ 5ms
$P_{G(AV)}$	Max. av. gate power	ALL	2	W	
$+I_{GM}$	Max. peak gate current	ALL	3	A	$t_p \leq$ 5ms
$-V_{GM}$	Max. peak negative gate voltage	ALL	15	V	
F	Mounting force	ALL	8900(2000) ± 10%	N(lbf)	

(1) To complete the part number, refer to the Ordering Information table.

Figure G-2   Data sheet for SCR, type IR-S30EF, S30EFH. (Courtesy of International Rectifier.)

CHARACTERISTICS

	PARAMETER	SERIES	MIN.	TYP.	MAX.	UNITS	TEST CONDITIONS
V_{TM}	Peak on-state voltage	ALL	—	2.15	2.23	V	Initial T_J = 25°C, 50-60Hz half sine, I_{peak} = 1700A.
$V_{T(TO)1}$	Low-level threshold	ALL	—	—	1.24	V	T_J = max. rated
$V_{T(TO)2}$	High-level threshold		—	—	1.56		Av. power = $V_{T(TO)} \cdot I_{T(AV)} + r_T \cdot [I_{T(RMS)}]^2$
r_{T1}	Low-level resistance	ALL	—	—	0.57	mΩ	Use low level values for $I_{TM} \leq \pi$ rated $I_{T(AV)}$
r_{T2}	High-level resistance		—	—	0.41		
I_L	Latching current	ALL	—	270	—	mA	T_C = 25°C, 12V anode. Gate pulse: 10V, 20Ω, 100µs.
I_H	Holding current	ALL	—	90	500	mA	T_C = 25°C, 12V anode. Initial I_T = 10A.
t_d	Delay time	ALL	—	0.5	1.5	µs	T_C = 25°C, V_D = rated V_{DRM}, 50A resistive load. Gate pulse: 10V, 20Ω, 10µs, 1µs rise time.
t_q	Turn-off time					µs	T_J = max. rated. I_{TM} = 500A, di_R/dt = 25A/µs, V_R = 50V,
	"A" suffix	S30EF	—	—	12		dv/dt = 200V/µs lin. to 80% rated V_{DRM}. Gate: 0V, 100Ω.
	"B" suffix	S30EF/H	—	—	15		
$t_{q(diode)}$	Turn-off time with feedback diode					µs	T_J = max. rated. I_{TM} = 500A, di_R/dt = 25A/µs, V_R = 1V,
	"A" suffix	S30EF	—	—	15		dv/dt = 600V/µs lin. to 40% V_{DRM}. Gate: 0V, 100Ω.
	"B" suffix	S30EF/H	—	—	20		
$I_{RM(REC)}$	Recovery current	ALL	—	47	—	A	T_J = 125°C, I_{TM} = 500A, di_R/dt = 50A/µs.
Q_{RR}	Recovered charge	ALL	—	46	—	µC	
dv/dt	Critical rate-of-rise of off-state voltage	ALL	500	700	—	V/µs	T_J = 125°C. Exp. to 100% or lin. Higher dv/dt values to 80% V_{DRM}, gate open. available.
			1000	—	—		T_J = 125°C. Exp. to 67% V_{DRM}, gate open.
I_{RM}, I_{DM}	Peak reverse and off-state current	S30EF	—	15	40	mA	T_J = max. rated. Rated V_{RRM} and V_{DRM}, gate open.
		S30EFH	—	20	80		
I_{GT}	DC gate current to trigger	ALL	—	—	300	mA	T_C = -40°C. +12V anode-to-cathode. For recommended gate drive see "Gate Characteristics" figure.
			50	70	150		T_C = 25°C
V_{GT}	DC gate voltage to trigger	ALL	—	—	3.3	V	T_C = -40°C
			—	1.2	2.5		T_C = 25°C
V_{GD}	DC gate voltage not to trigger	ALL	—	—	0.3	V	T_C = 125°C. Max. value which will not trigger with rated V_{DRM} anode-to-cathode.
R_{thJC}	Thermal resistance, junction-to-case	ALL	—	—	0.040	°C/W	DC operation, double side cooled.
			—	—	0.050	°C/W	180° sine wave, double side cooled.
			—	—	0.053	°C/W	120° rectangular wave, double side cooled.
R_{thCS}	Thermal resistance, case-to-sink	ALL	—	—	0.040	°C/W	Mtg. surface smooth, flat and greased. Single side cooled. For double side, divide value by 2.
wt	Weight	ALL	—	85(3.0)	—	g(oz.)	
	Case Style	ALL	IR A-29				

Figure G-2 *(continued)*

S30EF & S30EFH SERIES
800-600 VOLTS RANGE

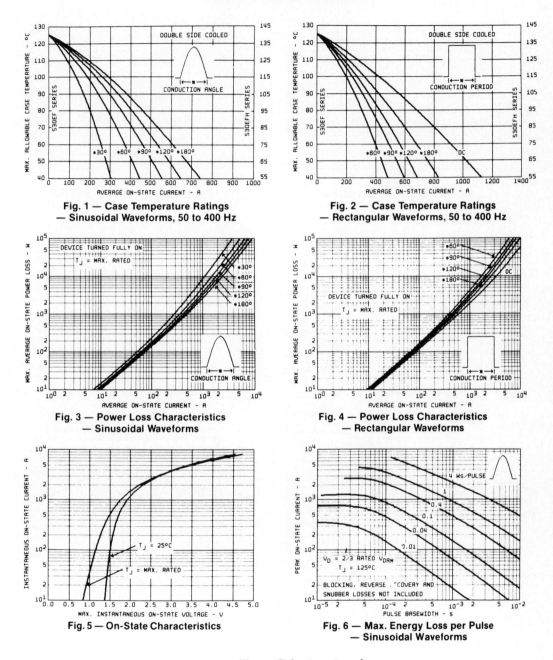

Fig. 1 — Case Temperature Ratings
— Sinusoidal Waveforms, 50 to 400 Hz

Fig. 2 — Case Temperature Ratings
— Rectangular Waveforms, 50 to 400 Hz

Fig. 3 — Power Loss Characteristics
— Sinusoidal Waveforms

Fig. 4 — Power Loss Characteristics
— Rectangular Waveforms

Fig. 5 — On-State Characteristics

Fig. 6 — Max. Energy Loss per Pulse
— Sinusoidal Waveforms

Figure G-2 *(continued)*

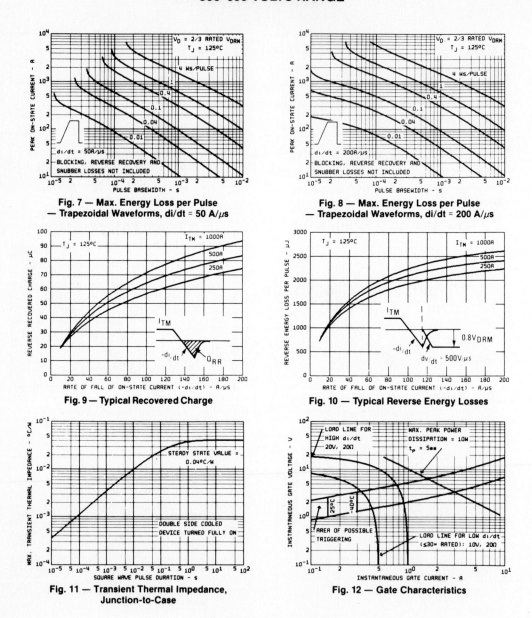

Fig. 7 — Max. Energy Loss per Pulse
— Trapezoidal Waveforms, di/dt = 50 A/μs

Fig. 8 — Max. Energy Loss per Pulse
— Trapezoidal Waveforms, di/dt = 200 A/μs

Fig. 9 — Typical Recovered Charge

Fig. 10 — Typical Reverse Energy Losses

Fig. 11 — Transient Thermal Impedance,
Junction-to-Case

Fig. 12 — Gate Characteristics

Figure G-2 *(continued)*

S30EF & S30EFH SERIES
800-600 VOLTS RANGE

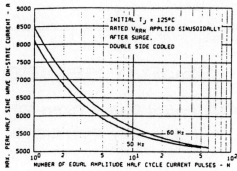

Fig. 13 — Non-Repetitive Surge Current Ratings

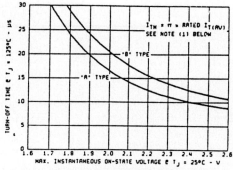

Fig. 14 — Trend for Turn-Off
Time vs. On-State Voltage

(1) These curves are intended as a guideline. To specify
non-standard t_q/V_{TM} contact factory.

ORDERING INFORMATION

TYPE	TEMPERATURE		VOLTAGE		TURN-OFF	
	CODE	MAX. T_J	CODE	V_{DRM}	CODE	MAX. t_q
S30EF	—	125°C	8	800V	A	12µs
	H	140°C	6	600V	B	15µs

For example, for a device with max. T_J = 125°C, V_{DRM} = 800V,
max. t_q = 12µs, order as: S30EF8A.

Figure G-2 *(continued)*

350PJT SERIES

1200A I_{TGQ} Gate Turn–Off Hockey Puk SCRs

Major Ratings

	350PJT	Units
I_{TGQ}	1200	A
$I_{T(RMS)}$	550	A
$I_{T(AV)}$	350	A
@ Max. T_C	80	°C
I_{TSM} @ 50 Hz	4500	A
@ 60 Hz	4700	
I^2t @ 50 Hz	101,000	A^2s
@ 60 Hz	92,000	
I_{GT}	2	A
dv/dt	1000	V/µs
di/dt	600	A/µs
t_{gq}	15	µs
T_J	–40 to 125	°C
V_{RRM}, V_{DRM}	1000 to 1600V	V

Description/Features

The 350PJT Series of GTO (gate turn-off) thyristors is designed for power control applications such as uninterruptible power supplies (UPS), variable speed ac motor drives, etc. Since they can be turned off by a negative current pulse to the gate, devices in the 350PJT Series allow reductions in overall size, weight, cost and acoustical noise when compared to conventional thyristors that require bulky commutating circuits.

- 350A average current.

- 1200A controllable on-state current.

- Maximum turn-off time of 15 µsec.

- Critical dv/dt of 1000 V/µsec.

- Available with maximum repetitive peak off-state voltage (V_{DRM}) to 1600V.

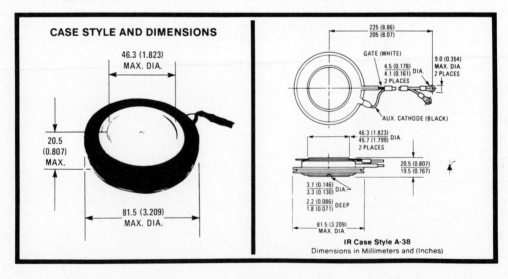

Figure G-3 Data sheet for GTO, type 350PJT. (Courtesy of International Rectifier.)

VOLTAGE RATINGS [1]

Part Number	V_{RRM}, V_{DRM} — Max. Repetitive Peak Reverse and Off-State Voltage (V) [3]	V_{RSM}, V_{DSM} — Max. Non-Repetitive Peak Reverse and Off-State Voltage $t_p \leqslant 5$ ms (V)
	$T_J = -40°C$ to $125°C$	$T_J = 25°C$ to $125°C$
350PJT100	1000	1200
350PJT120	1200	1400
350PJT140	1400	1600
350PJT160	1600	1750

ELECTRICAL SPECIFICATIONS

		350PJT	Units	Conditions
	ON-STATE			
$I_{T(RMS)}$	Nominal RMS on-state current	550	A	
$I_{T(AV)}$	Max. average on-state current	350	A	180° half sine wave conduction.
	@ Max. T_C	80	°C	
I_{TGQ}	Max. controllable peak on-state current	1200	A	$T_J = 125°C$, $V_{DM} = 1/2$ V_{DRM}. $G_{GQ} = 5$, $C_S = 3.0\,\mu F$. [2] Note: $V_S \leqslant 600V$ @ $T_J = 25°C$. $V_S \leqslant 500V$ @ $T_J = 125°C$. (V_S is the voltage spike which appears on the dynamic on-state voltage trace during fall time.)
I_{TSM}	Max. peak one cycle, non-repetitive surge current	4500	A	50 Hz half cycle sine wave or 6 ms rectangular pulse — Following any rated load condition, and with rated V_{RRM} applied following surge. SCR turned fully on.
		4700		60 Hz half cycle sine wave or 5 ms rectangular pulse
I^2t	Max. I^2t capability for fusing	101,000	A^2s	$t = 10$ ms — Rated V_{RRM} applied following surge, initial $T_J \leqslant 125°C$.
		92,000		$t = 8.3$ ms
V_{TM}	Max. peak on-state voltage	3.42	V	$T_J = 25°C$, $I_{T(AV)} = 350A$ (1100A peak), $I_G' = 4A$
I_L	Typical latching current	30	A	$T_J = 25°C$
I_H	Typical holding current	30	A	$T_J = 25°C$
	BLOCKING			
dv/dt	Min. critical rate-of-rise of off-state voltage	1000	V/μs	Gate voltage = –2V — $T_J = 125°C$
		400		Gate-to-cathode resistance = 2Ω — $V_D = 1/2$ V_{DRM}
I_{DM} & I_{RM}	Max. peak off-state and reverse current	80	mA	$T_J = 125°C$, $V_{DM} = $ rated V_{DRM}. Peak off-state current applies for –2V or more negative gate voltage or for gate-to-cathode resistance = 2Ω.
	SWITCHING			
di/dt	Max. repetitive rate-of-rise of turned-on current	600	A/μs	$di_G/dt \geqslant 5$ A/μs, $+I_{GM} \geqslant 10A$, $I_{TM} \leqslant 1200A$, $V_D \leqslant 1/2$ V_{DRM}.
t_{gt}	Max. turn-on time	8	μs	t_{gt} is measured from instant at which $i_G = 0.1I_{GM}$ to instant at which $v_D = 0.1V_D$ with resistive load. $T_J = 125°C$, $I_T = 1200A$, $+I_{GM} = 10A$, $di_G/dt = 5$ A/μs, $V_D = 1/2$ V_{DRM}.
t_{on}	Min. permissible on-time	16	μs	t_{on} is the time necessary to ensure that all cathode islands are in conduction. $T_J = 125°C$, $I_T = 1200A$, $V_D = 1/2$ V_{DRM}, $I_{GM} = 10A$, $di_G/dt = 60$ A/μs.
t_{gq}	Max. gate-controlled turn-off time	15	μs	t_{gq} is measured from instant at which $I_G = 24A$ to instant at which $I_T = 120A$ with resistive load. $T_J = 125°C$, $I_T = 1200A$, $di_G/dt = 60$ A/μs, $G_{GQ} = 5$. [2]

[1] Peak off-state voltages apply for –2V or more negative gate voltage or for gate-to-cathode resistance = 2Ω.

[2] $G_{GQ} = \dfrac{I_T}{\text{applied } I_{GQ}}$ forced turn-off gain. I_T = on-state current. Applied I_{GQ} = maximum negative gate current during turn-off interval.

[3] Peak reverse voltages apply for zero or negative gate voltage.

Figure G-3 *(continued)*

ELECTRICAL SPECIFICATIONS (Continued)

		350PJT	Units	Conditions
	SWITCHING (Continued)			
t_f	Max. fall time	1.2	μs	t_f is measured from instant at which I_T = 1080A to instant at which I_T = 120A with resistive load. T_J = 125°C, I_T = 1200A, V_D = 1/2 V_{DRM}, di_G/dt = 60 A/μs, G_{GQ} = 5. ①
t_{off}	Min. permissible off-time	80	μs	t_{off} is measured from the instant at which the turn-off pulse is ① applied to the gate to the earliest instant at which the GTO may be retriggered. T_J = 125°C, I_T = 1200A, di_G/dt = 60 A/μs, G_{GQ} = 5.
	TRIGGERING			
$P_{GF(AV)}$	Max. average forward gate power	30	W	Forward gate power is produced by positive gate current, reverse gate power is produced by negative gate current.
P_{GRM}	Max. peak reverse gate power	18,000	W	$t_p \leqslant 5~\mu s$.
$P_{GR(AV)}$	Max. average reverse gate power	80	W	
$+I_{GM}$	Max. peak positive gate current	100	A	$t_p \leqslant 100~\mu s$. Positive gate current may not be applied during reverse recovery interval.
$-I_{GM}$	Max. peak negative gate current	50	mA	T_J = 125°C, $-V_{GM}$ = rated $-V_{GRM}$. SCR blocking.
$-V_{GRM}$	Max. repetitive peak negative gate voltage	20	V	SCR blocking.
I_{GT}	Max. required DC gate current to trigger	4.6	A	T_C = -40°C — Max. required gate trigger current is the lowest value which will trigger all units with +12 volts anode-to-cathode and I_T = 50A after triggering
		2.0		T_C = 25°C
		0.5		T_C = 125°C
V_{GT}	Max. required DC gate voltage to trigger	1.25	V	T_C = -40°C — Max. required gate trigger voltage is the lowest value which will trigger all units with +12 volts anode-to-cathode and I_T = 50A after triggering
		1.0		T_C = 25°C

THERMAL-MECHANICAL SPECIFICATIONS

			Units	
T_J	Junction operating temperature range	-40 to 125	°C	
T_{stg}	Storage temperature range	-40 to 125	°C	
R_{thJC}	Max. internal thermal resistance, junction-to-case	0.035	deg. C/W	DC operation; double side cooled, mounting force = 11750N (2650lbf).
R_{thCS}	Thermal resistance, one pole piece to one heat dissipator	0.02	deg. C/W	Mounting surface smooth, flat and greased.
T	Mounting force Min.	10,600 (2400)	N	
	Max.	12,900 (2900)	(lbf)	
wt	Approximate weight	360 (12.7)	g (oz.)	
	Case Style	IR: A-38		

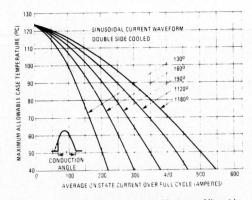

Fig. 1 — Average On-State Current Vs. Maximum Allowable Case Temperature (Sinusoidal Current Waveform)

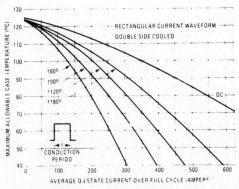

Fig. 2 — Average On-State Current Vs. Maximum Allowable Case Temperature (Rectangular Current Waveform)

Figure G-3 *(continued)*

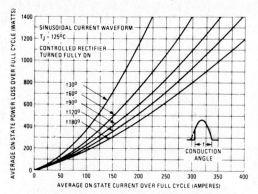

Fig. 3 — Maximum Low Level On-State Power Loss Vs. Average On-State Current (Sinusoidal Current Waveform)

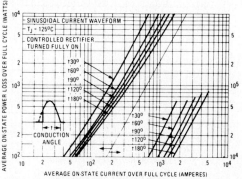

Fig. 4 — Maximum High Level On-State Power Loss Vs. Average On-State Current (Sinusoidal Current Waveform)

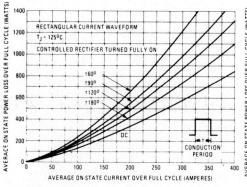

Fig. 5 — Maximum Low Level On-State Power Loss Vs. Average On-State Current (Rectangular Current Waveform)

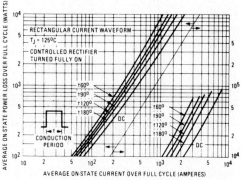

Fig. 6 — Maximum High Level On-State Power Loss Vs. Average On-State Current (Rectangular Current Waveform)

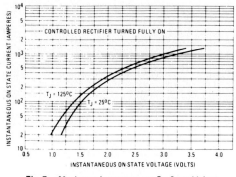

Fig. 7 — Maximum Instantaneous On-State Voltage Vs. Instantaneous On-State Current

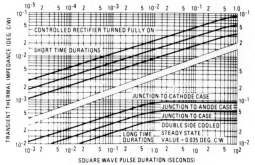

Fig. 8 — Maximum Transient Thermal Impedance Vs. Square Wave Pulse Duration

Figure G-3 (*continued*)

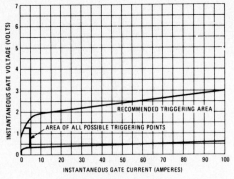

Fig. 9 — Gate Characteristics

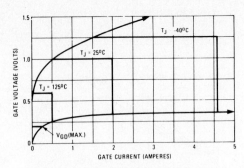

Fig. 9a — Areas of All Possible Triggering Points

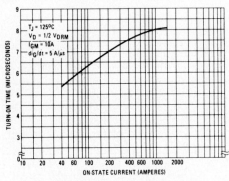

Fig. 10 — Turn-On Time Vs. On-State Current

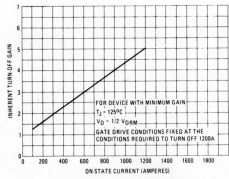

Fig. 11 — Inherent Turn-Off Gain Vs. Instantaneous On-State Current

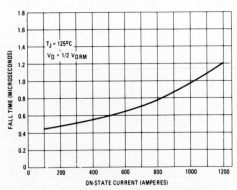

Fig. 12 — Maximum Fall Time Vs. On-State Current

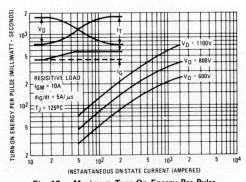

Fig. 13 — Maximum Turn-On Energy Per Pulse Vs. On-State Current

Figure G-3 (*continued*)

Data Sheets   App. G

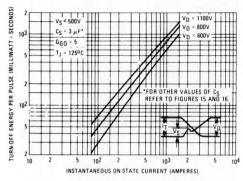

Fig. 14 — Maximum Turn-Off Energy Per Pulse
Vs. On-State Current, V_D = 600, 800 & 1100V;
C_S = 3 μF

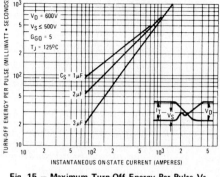

Fig. 15 — Maximum Turn-Off Energy Per Pulse Vs.
On-State Current, V_D = 600V; C_S =1, 2, & 3 μF

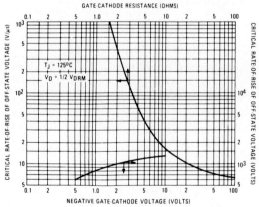

Fig. 16 — Maximum Turn-Off Energy Per Pulse Vs.
On-State Current, V_D = 800V; C_S = 1, 2, & 3 μF

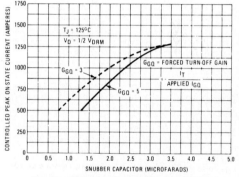

Fig. 17 — Maximum Controllable Peak On-State
Current Vs. Snubber Capacitor Value

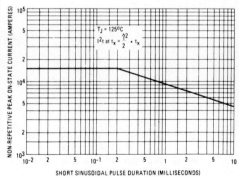

Fig. 18 — Minimum Critical Rate-of-Rise Off-State Voltage
Vs. Negative Gate-Cathode Voltage and Vs.
Gate-Cathode Resistance

Fig. 19 — Non-Repetitive Peak On-State Current
Vs. Sinusoidal Pulse Duration

Figure G-3 *(continued)*

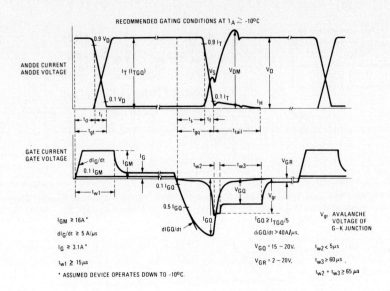

RECOMMENDED GATING CONDITIONS AT $T_A \geq -10^{o}C$

$I_{GM} \geq 16A$ *

$dI_G/dt \geq 5 A/\mu s$

$I_G \geq 3.1A$ *

$t_{w1} \geq 15\mu s$

* ASSUMED DEVICE OPERATES DOWN TO -10°C.

$I_{GQ} \geq I_{TGQ}/5$

$diGQ/dt > 40A/\mu s,$

$V_{GQ} = 15 \sim 20V.$

$V_{GR} = 2 \sim 20V,$

V_{gr}: AVALANCHE VOLTAGE OF G-K JUNCTION

$t_{w2} < 5\mu s$

$t_{w3} \geq 60 \mu s$

$t_{w2} + t_{w3} \geq 65 \mu s$

SNUBBER CAPACITOR Cs (μF)	SNUBBER RESISTOR Rs (Ω)	MINIMUM ON-TIME (μs)
3.0	10	75
	5	45
2.0	10	50
	5	30
1.5	10	38
	5	23

Fig. 20 — Recommended Gating Conditions at $T_A \geq -10^{o}C$

Figure G-3 *(continued)*

Power Transistors

D67DE

HI-LINE®
Power Darlington
150 A-PEAK   UP TO 500 V$_{CEO(SUS)}$ & 700V$_{CEV}$

The General Electric D67DE is a new High Current Power Darlington. It features collector isolation from the heat sink, an internal construction designed for stress-free operation at temperature extremes, hefty screw terminals for emitter and collector connection and quick electrical terminals for B1 and B2. The device is designed to meet UL creep, strike and isolation voltage. Major applications are for motor controls, switching power supplies and UPS systems.

UL RECOGNIZED
- Isolation- 2500 V$_{RMS}$
- Strike & creep for 460V systems
- File no. E60692

268

High Voltage: 400 – 500 V$_{CEO(SUS)}$; 500 – 700 V$_{CEV}$
High Current: 150 Amperes, I$_C$ (Peak)
High Gain: h$_{FE}$ 50 Minimum @ 100 Amperes I$_C$ (h$_{FE}$ 200 typical)

absolute maximum ratings: (T$_C$ = 25°C Unless Otherwise Specified)

Voltages

		D67DE5	D67DE6	D67DE7	Units
Collector Emitter	V$_{CEV}$	500	600	700	Volts
Collector Emitter	V$_{CEO (SUS)}$	400	450	500	Volts
Emitter Base	V$_{EBO}$	← 8 →			Volts

Currents

Collector Current (continuous)	I$_C$	← 100 →	Amps
Collector Current (peak)	I$_C$	← 150 →	Amps
Collector Current (Non-Repetitive)	I$_{CSM}$	← 250 →	Amps
Base Current (continuous)	I$_B$	← 10 →	Amps
Base Current (peak)	I$_B$	← 20 →	Amps

Dissipation

Power Dissipation (T$_C$ = 25°C)	P$_D$	← 312.5 →	Watts

Temperatures

Storage	T$_{stg}$	-40°C to +150°C	
Operating Junction	T$_J$	-40°C to +150°C	

Isolation Voltage	V$_{ISOL}$	← 2500 →	Volts (RMS)

Terminal & Mounting Torque Limits Units[1]

Thermal Resistance	R$_{\Theta JC}$	← .4 →	°C/W

(1) see back page for mounting considerations

electrical characteristics: (T$_C$ = 25°C Unless Otherwise Specified)

STATIC CHARACTERISTICS		SYMBOL	MIN.	TYP.	MAX.	UNITS
Collector-Emitter Sustaining Voltage						
(I$_C$ = 1A, I$_{B1}$=I$_{B2}$=0,) — D67DE5		V$_{CEO (SUS)}$	400	–	–	Volts
V$_{CLAMP}$ = V$_{CEO}$) — D67DE6		V$_{CEO (SUS)}$	450	–	–	Volts
— D67DE7		V$_{CEO (SUS)}$	500	–	–	Volts
Collector Cut-Off Current						
(V$_{CEV}$ = Rated Value, — T$_J$ = 25°C		I$_{CEV}$	–	–	2.5	mA
V$_{B1E}$ (off) = -1.5V) — T$_J$ = 150°C		I$_{CEV}$	–	–	10.0	mA
Emitter-Base Cut-Off Current						
(V$_{EB1}$ = 3.5 V, I$_C$ = O)		I$_{EBO}$	–	–	500	mA

Figure G-4 Data sheet for GE transistor, type D67DE. (Courtesy of General Electric Company.)

STATIC CHARACTERISTICS CONTINUED	SYMBOL	MIN.	TYP.	MAX.	UNITS
DC Current Gain					
(I_C = 150A, V_{CE} = 5V)	h_{FE}	25	90	–	
(I_C = 100A, V_{CE} = 5V)	h_{FE}	50	200	–	
(I_C = 40A, V_{CE} = 5V)	h_{FE}	100	275	–	
Collector-Emitter Saturation Voltage					
(I_C = 150A, I_B = 10A)	$V_{CE(SAT)}$	–	1.9	3.0	Volts
(I_C = 100A, I_B = 8A)	$V_{CE(SAT)}$	–	1.4	2.0	Volts
(I_C = 40A, I_B = 4A)	$V_{CE(SAT)}$	–	1.0	1.5	Volts
Base-Emitter Saturation Voltage					
(I_C = 150A, I_B = 10A)	$V_{BE(SAT)}$	–	2.75	3.5	Volts
(I_C = 100A, I_B = 8A)	$V_{BE(SAT)}$	–	2.3	3.0	Volts

SWITCHING CHARACTERISTICS (Reference Figure 21, Page 474)

Resistive (V_{CC} = 250V, I_C = 100A, I_{B1} = 5A, $-I_{B1}$ = 10A)					
Delay Time	t_d	–	.105	0.5	μs
Rise Time	t_r	–	.45	1.0	μs
Storage Time	t_s	–	3.2	5.0	μs
Fall Time	t_f	–	1.1	3.0	μs
Inductive (I_C = 100A, V_{CLAMP} = 250V, I_{B1} = 5A, $-I_{B1}$ = 10A, L=100μH)					
Storage Time	t_s	–	3.2	5.0	μs
Fall Time	t_f	–	.6	3.0	μs
Crossover Time	t_c	–	1.8	–	μs
Storage Time (T_J = 150°C)	t_s	—	5.8	—	μs
Fall Time (T_J = 150°C)	t_f	—	1.1	—	μs
Crossover Time (T_J = 150°C)	t_c	—	3.7	—	μs

DIODE CHARACTERISTICS

Diode Forward Voltage (I_F = 100A)					
— T_J = 25°C	V_F	—	1.9	3.25	Volts
— T_J = 150°C	V_F	—	1.75	3.00	Volts
Diode Reverse Recovery Time (T_J = 25°C)					
(I_F = 100A, di/dt = 25A/μsec, R_{B1E} = .25)	t_{rr}	—	4.5	10.0	μsec
Diode Forward Turn-on Time (T_J = 25°C)					
(I_F = 100A, di/dt = 100A/μsec)	t_{on}	—	1.7	2.5	μsec
Thermal Resistance	$R_{\theta JC}$	—	—	0.4	°C/W

DIMENSIONAL OUTLINE

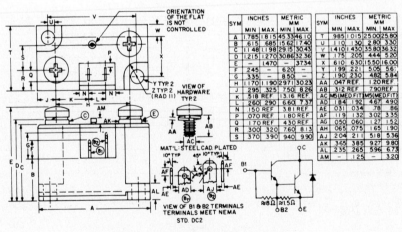

SYM	INCHES MIN	INCHES MAX	METRIC MM MIN	METRIC MM MAX
A	1.785	1.815	45.33	46.10
B	.615	.685	15.62	17.40
C	1.148	1.198	29.15	30.43
D	1.215	1.270	30.86	32.36
E	—	1.470	—	37.34
F	.245	—	6.20	—
G	.335	—	8.50	—
H	1.170	1.190	29.71	30.23
J	.295	.325	7.50	8.26
K	.518 REF		13.16 REF	
L	.260	.290	6.60	7.37
N	.150 REF		3.81 REF	
P	.070 REF		1.80 REF	
Q	.170 REF		4.30 REF	
R	.300	.320	7.60	8.13
S	.370	.390	9.40	9.90

SYM	INCHES MIN	INCHES MAX	METRIC MM MIN	METRIC MM MAX
T	.985	1.015	25.00	25.80
U	.110	.130	2.80	3.30
V	1.410	1.430	35.80	36.32
W	.175	.205	4.44	5.20
X	.610	.630	15.50	16.00
Y	.199	.221	5.05	5.61
Z	.190	.230	4.82	5.84
AA	.047 REF		1.20 REF	
AB	.312 REF		7.90 REF	
AC	M5(MED.FIT)		M5(MED.FIT)	
AD	.184	.192	4.67	4.90
AE	.031	.034	.78	.86
AF	.119	.132	3.02	3.35
AG	.050	.060	1.27	1.52
AH	.065	.075	1.65	1.90
AJ	.204	.211	5.18	5.36
AK	.365	.385	9.27	9.80
AL	.235	.265	5.96	6.73
AM	—	.125	—	3.20

Figure G-4 (*continued*)

TYPICAL CHARACTERISTICS

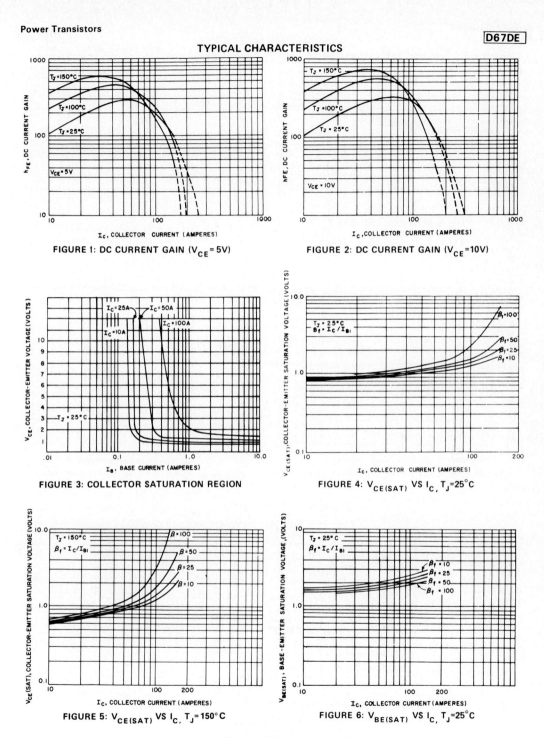

FIGURE 1: DC CURRENT GAIN (V_{CE} = 5V)

FIGURE 2: DC CURRENT GAIN (V_{CE} = 10V)

FIGURE 3: COLLECTOR SATURATION REGION

FIGURE 4: $V_{CE(SAT)}$ VS I_C, T_J = 25°C

FIGURE 5: $V_{CE(SAT)}$ VS I_C, T_J = 150°C

FIGURE 6: $V_{BE(SAT)}$ VS I_C, T_J = 25°C

Figure G-4 *(continued)*

TYPICAL CHARACTERISTICS

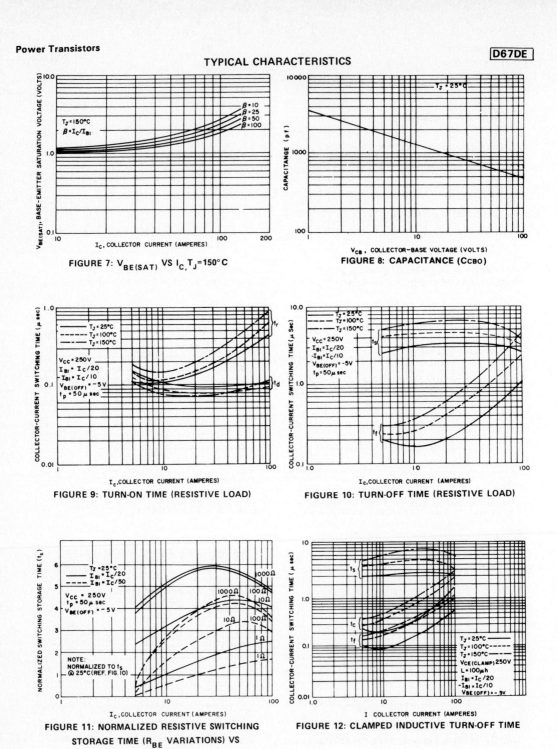

FIGURE 7: $V_{BE(SAT)}$ VS I_C, T_J=150°C

FIGURE 8: CAPACITANCE (C_{CBO})

FIGURE 9: TURN-ON TIME (RESISTIVE LOAD)

FIGURE 10: TURN-OFF TIME (RESISTIVE LOAD)

FIGURE 11: NORMALIZED RESISTIVE SWITCHING STORAGE TIME (R_{BE} VARIATIONS) VS COLLECTOR CURRENT

FIGURE 12: CLAMPED INDUCTIVE TURN-OFF TIME

Figure G-4 *(continued)*

TYPICAL CHARACTERISTICS

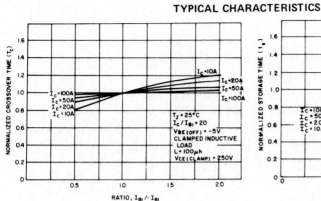

FIGURE 13: CROSSOVER TIME VARIATION WITH $-I_{B1}$

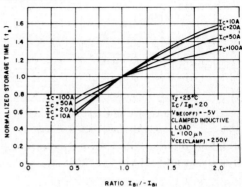

FIGURE 14: STORAGE TIME VARIATION WITH $-I_{B1}$

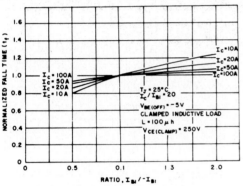

FIGURE 15: FALL TIME VARIATION WITH $-I_{B1}$

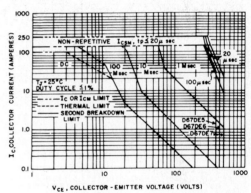

FIGURE 16: FORWARD BIAS SAFE OPERATING AREA

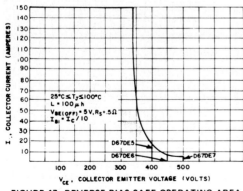

FIGURE 17: REVERSE BIAS SAFE OPERATING AREA
(CLAMPED)

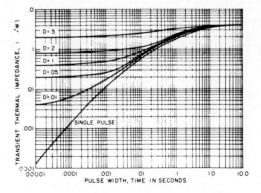

FIGURE 18: TRANSIENT THERMAL RESPONSE

Figure G-4 *(continued)*

TYPICAL CHARACTERISTICS

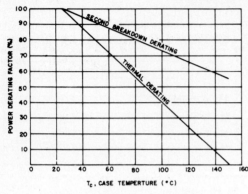

FIGURE 19: POWER DERATING

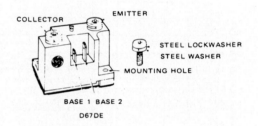

FIGURE 20: DIODE FORWARD CHARACTERISTICS

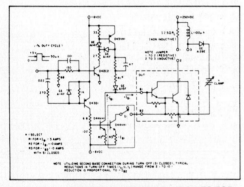

FIGURE 21: SWITCHING TIME TEST CIRCUITS FOR:
- **RESISTIVE & INDUCTIVE SWITCHING**
- **USING BASE 1 ONLY**
- **USING BASE 1 AND BASE 2**

MOUNTING AND ELECTRICAL TERMINATION PROCEDURES

HEAT SINK FLATNESS

Heat sink surfaces must be flat within ± 1.5 mils/inch (0.015mm/cm) over the mounting area and must have a surface finish of < 64 micro inches (1.62 microns).

THERMAL COMPOUND

To minimize the effects of flatness differential and/or voids between the base plate and the heat sink, apply a very thin layer of GE #6644 or Dow Corning #4 thermal compound to the back of the base plate and the heat sink. NOTE: excessive thermal compound *will not* squeeze out from underneath the device during mountdown. After applying thermal compound to the device and the heat sink, place the device on the heat sink and rotate slowly to distribute grease. Check both surfaces for uniform coverage before applying torque to mounting screws.

MOUNTING

HARDWARE: Standard #10 or M5
  $^{7}/_{16}$" – ½"OD
  (11 " 13mm)OD

TORQUE: 19-25 lb.-in. (2-3 NM)

ELECTRICAL TERMINATION

COLLECTOR & EMITTER:
Screw: M5 × 8mm
Lockwasher: 9.2 – 13mm OD
Torque: 25 – 28 lb. - in.
  (2.8 – 3.2 NM)

BASE:
Base 1: FASTON-AMP #640917-1
Base 2: FASTON-AMP #640903-1
  (or equivalents)

Figure G-4 *(continued)*

INTERNATIONAL RECTIFIER

HEXFET® TRANSISTORS IRFZ40 IRFZ42

N-Channel
50 VOLT
POWER MOSFETs

50 Volt, 0.028 Ohm HEXFET
TO-220AB Plastic Package

The HEXFET technology has expanded its product base to serve the low voltage, very low $R_{DS(on)}$ MOSFET transistor requirements. International Rectifier's highly efficient geometry and unique processing of the HEXFET have been combined to create the lowest on resistance per device performance. In addition to this feature all HEXFETs have documented reliability and parts per million quality!

The HEXFET transistors also offer all of the well established advantages of MOSFETs such as voltage control, freedom from second breakdown, very fast switching, ease of paralleling, and temperature stability of the electrical parameters.

They are well suited for applications such as switching power supplies, motor controls, inverters, choppers, audio amplifiers, high energy pulse circuits, and in systems that are operated from low voltage batteries, such as automotive, portable equipment, etc.

Features:
- Extremely Low $R_{DS(on)}$
- Compact Plastic Package
- Fast Switching
- Low Drive Current
- Ease of Paralleling
- No Second Breakdown
- Excellent Temperature Stability
- Parts Per Million Quality

Product Summary

PART NUMBER	V_{DS}	$R_{DS(ON)}$	I_D
IRFZ40	50V	0.028Ω	51A
IRFZ42	50V	0.035Ω	46A

CASE STYLE AND DIMENSIONS

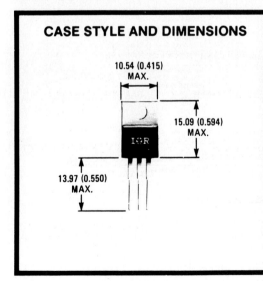

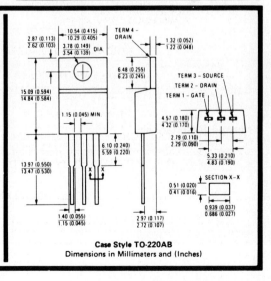

Case Style TO-220AB
Dimensions in Millimeters and (Inches)

Figure G-5　Data sheet for IR MOSFETs, typed IRFZ40 and IRFZ42. (Courtesy of International Rectifier.)

Absolute Maximum Ratings

	Parameter	IRFZ40	IRFZ42	Units
V_{DS}	Drain - Source Voltage ①	50	50	V
V_{DGR}	Drain - Gate Voltage (R_{GS} = 20 KΩ) ①	50	50	V
I_D @ T_C = 25°C	Continuous Drain Current	51	46	A
I_D @ T_C = 100°C	Continuous Drain Current	32	29	A
I_{DM}	Pulsed Drain Current ③	160	145	A
V_{GS}	Gate - Source Voltage	±20		V
P_D @ T_C = 25°C	Max. Power Dissipation	125 (See Fig. 14)		W
	Linear Derating Factor	1.0 (See Fig. 14)		W/K
I_{LM}	Inductive Current, Clamped	(See Fig. 15 and 16) L = 100μH 160 ǀ 145		A
T_J T_{stg}	Operating Junction and Storage Temperature Range	−55 to 150		°C
	Lead Temperature	300 (0.063 in. (1.6mm) from case for 10s)		°C

Electrical Characteristics @ T_C = 25°C (Unless Otherwise Specified)

	Parameter	Type	Min.	Typ.	Max.	Units	Test Conditions	
BV_{DSS}	Drain - Source Breakdown Voltage	IRFZ40	50	—	—	V	V_{GS} = 0V	
		IRFZ42	50	—	—	V	I_D = 250 μA	
$V_{GS(th)}$	Gate Threshold Voltage	ALL	2.0	—	4.0	V	V_{DS} = V_{GS}, I_D = 250 μA	
I_{GSS}	Gate-Source Leakage Forward	ALL	—	—	500	nA	V_{GS} = 20V	
I_{GSS}	Gate-Source Leakage Reverse	ALL	—	—	−500	nA	V_{GS} = −20V	
I_{DSS}	Zero Gate Voltage Drain Current	ALL	—	—	250	μA	V_{DS} = Max. Rating, V_{GS} = 0V	
			—	—	1000	μA	V_{DS} = Max. Rating × 0.8, V_{GS} = 0V, T_C = 125°C	
$I_{D(on)}$	On-State Drain Current ②	IRFZ40	51	—	—	A	V_{DS} > $I_{D(on)}$ × $R_{DS(on)max.}$, V_{GS} = 10V	
		IRFZ42	45	—	—	A		
$R_{DS(on)}$	Static Drain-Source On-State Resistance ②	IRFZ40	—	0.024	0.028	Ω	V_{GS} = 10V, I_D = 29A	
		IRFZ42	—	0.030	0.035	Ω		
g_{fs}	Forward Transconductance ②	ALL	17	22	—	S(Ω)	V_{DS} > $I_{D(on)}$ × $R_{DS(on)}$ max., I_D = 29A	
C_{iss}	Input Capacitance	ALL	—	2350	3000	pF	V_{GS} = 0V, V_{DS} = 25V, f = 1.0 MHz	
C_{oss}	Output Capacitance	ALL	—	920	1200	pF	See Fig. 10	
C_{rss}	Reverse Transfer Capacitance	ALL	—	250	400	pF		
$t_{d(on)}$	Turn-On Delay Time	ALL	—	18	25	ns	V_{DD} ≅ 25V, I_D = 29A, Z_0 = 4.7Ω	
t_r	Rise Time	ALL	—	25	60	ns	See Fig. 17	
$t_{d(off)}$	Turn-Off Delay Time	ALL	—	35	70	ns	(MOSFET switching times are essentially independent of	
t_f	Fall Time	ALL	—	12	25	ns	operating temperature.)	
Q_g	Total Gate Charge (Gate-Source Plus Gate-Drain)	ALL	—	40	60	nC	V_{GS} = 10V, I_D = 64A, V_{DS} = 0.8 Max. Rating. See Fig. 18 for test circuit. (Gate charge is essentially	
Q_{gs}	Gate-Source Charge	ALL	—	22	—	nC	independent of operating temperature.)	
Q_{gd}	Gate-Drain ("Miller") Charge	ALL	—	18	—	nC		
L_D	Internal Drain Inductance	ALL	—	3.5	—	nH	Measured from the contact screw on tab to center of die.	Modified MOSFET symbol showing the internal device inductances.
			—	4.5	—	nH	Measured from the drain lead, 6mm (0.25 in.) from package to center of die.	
L_S	Internal Source Inductance	ALL	—	7.5	—	nH	Measured from the source lead, 6mm (0.25 in.) from package to source bonding pad.	

Thermal Resistance

R_{thJC}	Junction-to-Case	ALL	—	—	1.0	K/W		
R_{thCS}	Case-to-Sink	ALL	—	1.0	—	K/W	Mounting surface flat, smooth, and greased.	
R_{thJA}	Junction-to-Ambient	ALL	—	—	80	K/W	Free Air Operation	

Figure G-5 *(continued)*

Source-Drain Diode Ratings and Characteristics

I_S	Continuous Source Current (Body Diode)	IRFZ40	—	—	51	A	Modified MOSFET symbol showing the integral reverse P-N junction rectifier.
		IRFZ42	—	—	46	A	
I_{SM}	Pulse Source Current (Body Diode) ③	IRFZ40	—	—	160	A	
		IRFZ42	—	—	145	A	
V_{SD}	Diode Forward Voltage ②	IRFZ40	—	—	2.5	V	T_C = 25°C, I_S = 51A, V_{GS} = 0V
		IRFZ42	—	—	2.2	V	T_C = 25°C, I_S = 46A, V_{GS} = 0V
t_{rr}	Reverse Recovery Time	ALL	—	350	—	ns	T_J = 150°C, I_F = 51A, dI_F/dt = 100A/μs
Q_{RR}	Reverse Recovered Charge	ALL	—	2.1	—	μC	T_J = 150°C, I_F = 51A, dI_F/dt = 100A/μs
t_{on}	Forward Turn-on Time	ALL	Intrinsic turn-on time is negligible. Turn-on speed is substantially controlled by L_S + L_D.				

① T_J = 25°C to 150°C.   ② Pulse Test: Pulse width ≤ 300μs, Duty Cycle ≤ 2%.   ③ Repetitive Rating: Pulse width limited by max. junction temperature. See Transient Thermal Impedance Curve (Fig. 5).

TO-220A

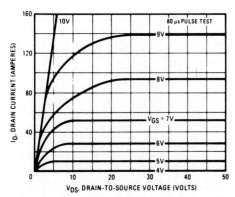

Fig. 1 — Typical Output Characteristics

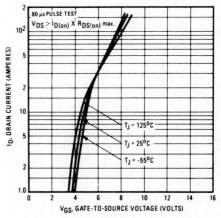

Fig. 2 — Typical Transfer Characteristics

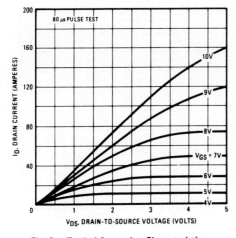

Fig. 3 — Typical Saturation Characteristics

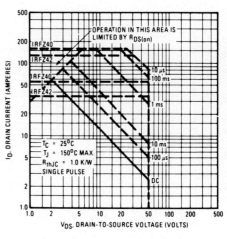

Fig. 4 — Maximum Safe Operating Area

Figure G-5 (*continued*)

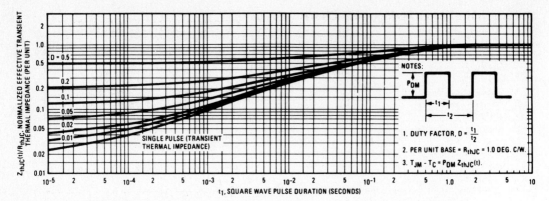

Fig. 5 — Maximum Effective Transient Thermal Impedance, Junction-toCase Vs. Pulse Duration

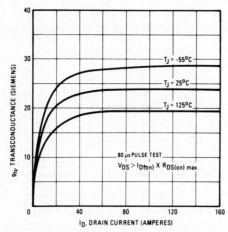

Fig. 6 — Typical Transconductance Vs. Drain Current

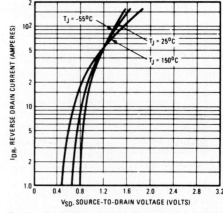

Fig. 7 — Typical Source-Drain Diode Forward Voltage

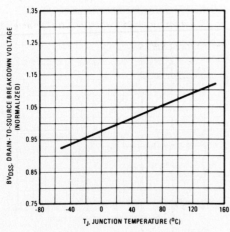

Fig. 8 — Breakdown Voltage Vs. Temperature

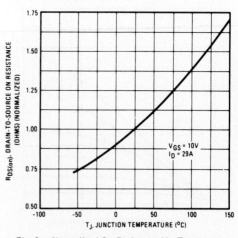

Fig. 9 — Normalized On-Resistance Vs. Temperature

Figure G-5 *(continued)*

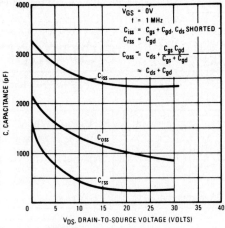

Fig. 10 — Typical Capacitance Vs. Drain-to-Source Voltage

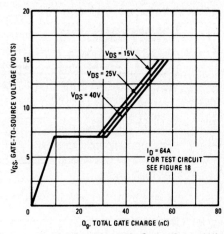

Fig. 11 — Typical Gate Charge Vs. Gate-to-Source Voltage

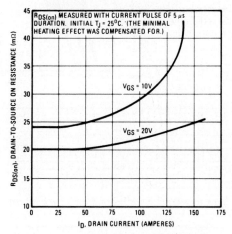

Fig. 12 — Typical On-Resistance Vs. Drain Current

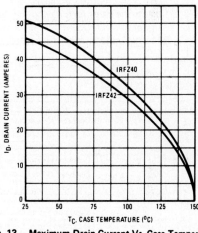

Fig. 13 — Maximum Drain Current Vs. Case Temperature

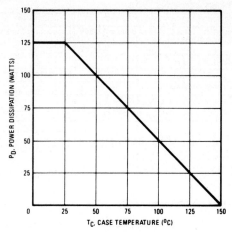

Fig. 14 — Power Vs. Temperature Derating Curve

Figure G-5 (*continued*)

International
IGR Rectifier

IRGPC50F

INSULATED GATE BIPOLAR TRANSISTOR

Fast-Speed IGBT

- Latch-proof
- Simple gate-drive
- Fast operation 3kHz~8kHz
- Switching-Loss Rating includes all "tail" losses

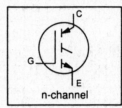

$V_{CEO} = 600$ V
$I_{C(DC)} = 70$ A
$V_{CE(sat)} \leq 1.7$V

$E_{TS} \leq 10$ mJ

n-channel

Description

Insulated Gate Bipolar Transistors (IGBTs) from International Rectifier have higher current densities than comparable bipolar transistors, while at the same time having simpler gate-drive requirements of the familiar power MOSFET. They provide substantial benefits to a host of higher-voltage, higher-current applications.

TO-247AC

Absolute Maximum Ratings

	Parameter	Max.	Units
I_C @ T_C = 25°C	Continuous Collector Current	70	A
I_C @ T_C = 100°C	Continuous Collector Current	39	
I_{CM}	Pulsed Collector Current ①	280	
V_{CE}	Collector-to-Emitter Breakdown Voltage	600	V
V_{GE}	Gate-to-Emitter Voltage	±20	
I_{LM}	Clamped Inductive Load Current ②	280	A
E_{ARV}	Reverse Voltage Avalanche Energy ③	20	mJ
P_D @ T_C = 25°C	Maximum Power Dissipation	200	W
P_D @ T_C = 100°C	Maximum Power Dissipation	78	
T_J T_{STG}	Operating Junction and Storage Temperature Range	-55 to +150	°C
	Soldering Temperature, for 10 sec.	300 (0.063 in. (1.6mm) from case)	
	Mounting Torque, 6-32 or 3mm MA screw	10 in·lbs (11.5 kg·cm)	

Thermal Resistance

	Parameter	Min.	Typ.	Max.	Units
$R_{\theta JC}$	Junction-to-Case	---	---	0.64	K/W ⑥
$R_{\theta CS}$	Case-to-Sink, flat, greased surface	---	0.24	---	
$R_{\theta JA}$	Junction-to-Ambient, typical socket mount	---	---	40	

Figure G-6 Data sheet for insulated gate bipolar transistor, type IRGPC50F. (Courtesy of International Rectifier.)

IRGPC50F

Electrical Characteristic @ T_J = 25°C (unless otherwise specified)

	Parameter	Min.	Typ.	Max.	Units	Test Conditions
BV_{CES}	Collector-to-Emitter Breakdown Voltage	600	---	---	V	$V_{GE}=0V$, $I_C=250\mu A$
BV_{ECS}	Emitter-to-Collector Breakdown Volt. ④	25	---	---		$V_{GE}=0V$, $I_C=1.0A$
$\Delta BV_{CES}/\Delta T_J$	Temp. Coeff. of Breakdown Voltage	---	0.62	---	V/°C	$V_{GE}=0V$, $I_C=1.0mA$
		---	---	1.7		$V_{GE}=15V$, $I_C=39A$ See fig 4.
$V_{CE(on)}$	Collector-to-Emitter Saturation Voltage	---	2.0	---	V	$V_{GE}=15V$, $I_C=70A$
		---	1.7	---		$V_{CE}=15V$, $I_C=39A$, $T_J=150°C$
$V_{GE(th)}$	Gate Threshold Voltage	3.0	---	5.5		$V_{CE}=V_{GE}$, $I_C=250\mu A$
$\Delta BV_{GE(th)}/\Delta T_J$	Temp. Coeff. of Threshold Voltage	---	-14	---	mV/°C	$V_{CE}=V_{GE}$, $I_C=250\mu A$
g_{fe}	Forward Transconductance ⑤	21	---	39	S	$V_{CE}=100V$, $I_C=39A$
I_{CES}	Zero Gate Voltage Collector Current	---	---	250	μA	$V_{GE}=0V$, $V_{CE}=600V$, $T_J=25°C$
		---	---	2000		$V_{GE}=0V$, $V_{CE}=600V$, $T_J=150°C$
I_{GES}	Gate-to-Emitter Leakage Current	---	---	±500	nA	$V_{GE}=\pm20V$

Switching Characteristics @ T_J = 25°C (unless otherwise specified)

	Parameter	Min.	Typ.	Max.	Units	Test Conditions
Q_G	Total Gate Charge (turn-on)	67	---	100	nC	$I_C=39A$, $V_{CC}=480V$
Q_{GE}	Gate - Emitter Charge (turn-on)	14	---	25		See Figure 6.
Q_{GC}	Gate - Collector Charge (turn-on)	35	---	67		
$t_{d(on)}$	Turn-On Delay Time	---	24	---		See test circuit, figure 13.
t_r	Rise Time	---	50	---	ns	$I_C=39A$, $V_{CC}=480V$
$t_{d(off)}$	Turn-off Delay Time	---	---	540		$T_J=25°C$
t_f	Fall Time	---	---	360		$V_{GE}=15V$, $R_G=2.0\Omega$
E_{on}	Turn-On Switching Loss	---	0.20	---		Energy losses include "tail".
E_{off}	Turn-Off Switching Loss	---	5.8	---	mJ	Also see figures 9, 10, & 11.
E_{ts}	Total Switching Loss	---	6.0	10		
$t_{d(on)}$	Turn-On Delay Time	---	25	---		$I_C=39A$, $V_{CC}=480V$
t_r	Rise Time	---	49	---	ns	$T_J=150°C$
$t_{d(off)}$	Turn-Off Delay Time	---	440	---		$V_{GE}=15V$
t_f	Fall Time	---	410	---		$R_G=2.0\Omega$
E_{ts}	Total Switching Loss	---	10	---	mJ	
L_E	Internal Emitter Inductance	---	13	---	nH	Measured 5mm from package.
C_{ies}	Input Capacitance	---	3000	---		$V_{GE}=0V$
C_{oes}	Output Capacitance	---	340	---	pF	$V_{CC}=30V$ See fig 5.
C_{res}	Reverse Transfer Capacitance	---	40	---		$f=1.0MHz$

Notes:

① Repetitive rating; $V_{GE}=20V$, pulse width limited by max. junction temperature (See figure 12b).

② $V_{CC}=80\%(BV_{CES})$, $V_{GE}=20V$, $L=10\mu H$, $R_G=10\Omega$, (See figure 12a).

③ Repetitive rating; pulse width limited by maximum junction temperature.

④ Pulse width ≤ 80μs; duty factor ≤0.1%.

⑤ Pulse width ≤ 5μs, single shot.

⑥ K/W equivalent to °C/W.

Figure G-6 (*continued*)

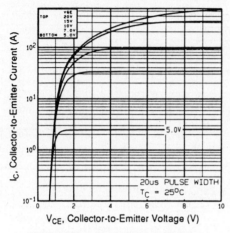

Fig 1. Typical Output Characteristics,
$T_J = 25°C$

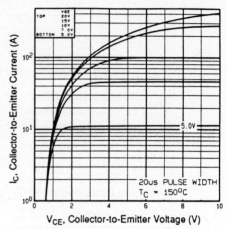

Fig 2. Typical Output Characteristics,
$T_J = 150°C$

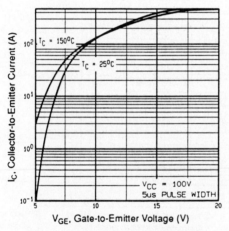

Fig 3. Typical Transfer Characteristics

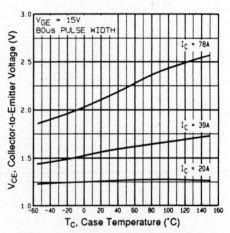

Fig 4. Collector-to-Emitter Saturation
Voltage vs. Case Temperature

Graphs indicate performance of typical devices

Figure G-6 *(continued)*

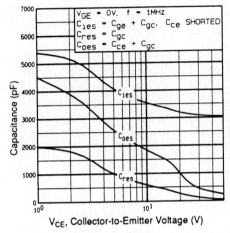

Fig 5. Typical Capacitance vs. Collector-to-Emitter Voltage

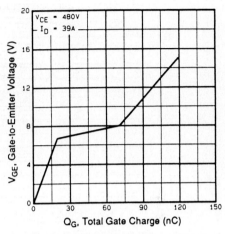

Fig 6. Typical Gate Charge vs. Gate-to-Emitter Voltage

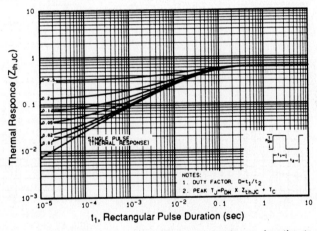

Fig 7. Maximum Effective Transient Thermal Impedance, Junction-to-Case

Graphs indicate performance of typical devices

Figure G-6 *(continued)*

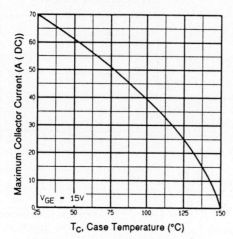

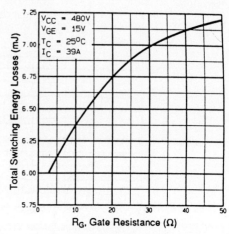

Fig 8. Maximum Collector Current vs. Case Temperature

Fig 9. Typical Switching Losses vs. Gate Resistance

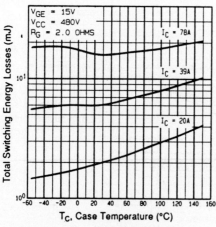

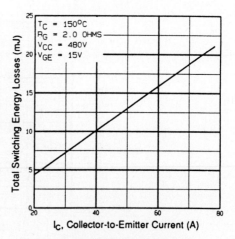

Fig 10. Typical Switching Losses vs. Case Temperature

Fig 11. Typical Switching Losses vs. Collector-to-Emitter Current

Graphs indicate performance of typical devices

Figure G-6   *(continued)*

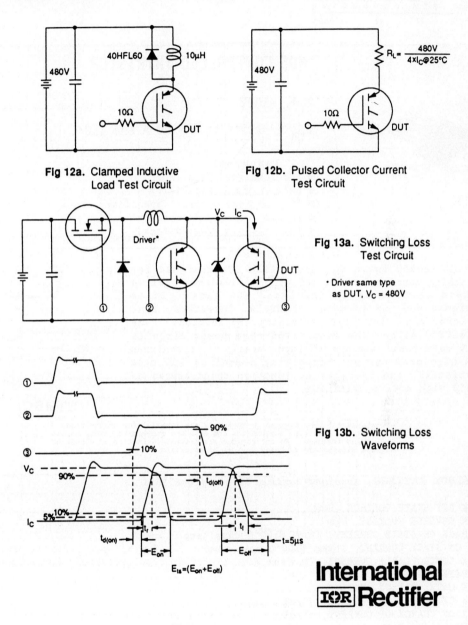

Fig 12a. Clamped Inductive Load Test Circuit

Fig 12b. Pulsed Collector Current Test Circuit

Fig 13a. Switching Loss Test Circuit

· Driver same type as DUT, V_C = 480V

Fig 13b. Switching Loss Waveforms

International IQR Rectifier

Figure G-6 *(continued)*

HARRIS
SEMICONDUCTOR

HARRIS · RCA · GE · INTERSIL

MODIFIED
TO-218

MOS CONTROLLED THYRISTOR

20A, -500V

V_{tm} = -1.5 V at I = 40 A and 150°C T_J

Features:

- *Mos Insulated Gate Input*
- *Gate Turn-Off Capability*
- *500 amp Peak Current Capability*
- *40 amp Turn-Off Capability*
- *500 volt Blocking Voltage*

The MCTA20P50 is an MOS Controlled SCR designed for switching currents on and off by positive and negative pulsed control of an insulated MOS gate. It is designed for use in motor controls, inverters, line switches and other power switching applications. The MCTA20P50 allows the control of high power circuits with very small amounts of input energy. It features the high peak current capability common to SCR type thyristors, and operates at junction temperatures of 150°C with active switching.

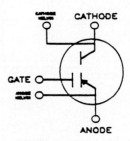

SYMBOL

MAXIMUM RATINGS, *Absolute-Maximum Values (Tc=25°C):*

PEAK OFF-STATE VOLTAGE, V_{DRM}	-500	V
PEAK REVERSE VOLTAGE, V_{RRM}	5	V
AVERAGE ON-STATE CURRENT, $I_{T(AV)}$ @100°C case temp.	13	A
RMS ON-STATE CURRENT, $I_{T(RMS)}$ @100°C case temp.	20	A
PEAK CONTROLLABLE CURRENT, I_{TC}.(see Fig. 1)	40	A
NONREPETITIVE PEAK CURRENT, I_{TSM}.....................................	500	A
PEAK GATE VOLTAGE, V_{GA}	±20	V
RATE OF CHANGE OF VOLTAGE, dV/dT	1000	v/us
RATE OF CHANGE OF CURRENT, dI/dT	500	a/us
POWER DISSIPATION, P_T:		
At Tc=90°C	100	W
Derated above 90°C	1.67	W/°C
OPERATING AND STORAGE TEMPERATURE, Tjc, Tstg	-55 to +150	°C

DEVELOPMENTAL

Figure G-7 Data sheet for MOS-controlled thyristor, type MCTA20P50. (Courtesy of Harris Semiconductor.)

ELECTRICAL CHARACTERISTICS, *Case Temperature (Tc)-25°C unless otherwise specified*

CHARACTERISTIC		TEST CONDITIONS	LIMITS		UNITS
			Min	Max	
Peak Off-State Blocking Current	Idrm	Vak= -500v, Vga= +10v	-	100	uA
Peak Off-State Blocking Current	Idrm	Vak= -500v, Vga= +10v Tj= 150°C	-	2	mA
Peak Reverse Blocking Current	Irrm	Vak= 5v, Vga= +10v	-	100	uA
Peak Reverse Blocking Current	Irrm	Vak= 5v, Vga= +10v Tj= 150°C	-	5	mA
Gate Leakage Current	Igas	Vga= ±20v	-	±200	nA
On-State Voltage	Vtm	Vga= -10v, It=40a	-	-1.5	V
On-State Voltage	Vtm	Vga= -10v, It=40a Tj= 150°C	-	-1.5	V
Latching Current	I_L	Vka= -30v, Vga1= -10v Vga2= 0v	-	1	A
Peak Controllable Current	I(off)	It=40a, Rg=5ohms vga= ±10v, Vd= -400v L= 50uH, R1=10ohms Tj= 150°C see Figure 1	40	-	A
Turn-on Delay	Td(on)i		0.6(typ)		us
Rise time Time	Tri		0.4(typ)		us
Turn-off Delay	Td(off)i		0.6(typ)		us
Fall Time	Tfi		3.7(typ)		us
Turn-off Switching Loss per cycle	Eoff		10(typ)		mj
Thermal Resistance Junction to Case	RθJC		-	0.6	°C/W
Thermal Resistance Junction to Ambient	RθJA		-	80	°C/W

Figure G-7 *(continued)*

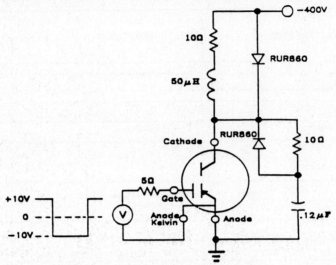

Fig. 1 - Switching Circuit

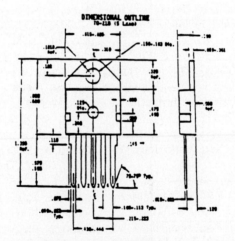

TERMINAL CONNECTIONS

Lead No. 1 - Gate
Lead No. 2 - Anode Kelvin
Lead No. 3 - Cathode Kelvin
Lead No. 4 - Anode Current
Lead No. 5 - Anode Current
Mounting Flange - Cathode Current

Figure G-7 (*continued*)

Bibliography

BEDFORD, F. E., and R. G. HOFT, *Principles of Inverter Circuits*. New York: John Wiley & Sons, Inc., 1964.

BIRD, B. M., and K. G. KING, *An Introduction to Power Electronics*. Chichester, West Sussex, England: John Wiley & Sons Ltd., 1983.

CSAKI, F., K. GANSZKY, I. IPSITS, and S. MARTI, *Power Electronics*. Budapest: Akadémiai Kiadó, 1980.

DATTA, S. M., *Power Electronics & Control*. Reston, Va.: Reston Publishing Co., Inc., 1985.

DAVIS, R. M., *Power Diode and Thyristor Circuits*. Stevenage, Herts, England: Institution of Electrical Engineers, 1979.

DEWAN, S. B., and A. STRAUGHEN, *Power Semiconductor Circuits*. New York: John Wiley & Sons, Inc., 1984.

DEWAN, S. B., G. R. SLEMON, and A. STRAUGHEN, *Power Semiconductor Drives*. New York: John Wiley & Sons, Inc. 1975.

DUBEY, G. K., *Power Semiconductor Controlled Drives*. Englewood Cliffs, N.J.: Prentice Hall, 1989.

General Electric, GRAFHAN, D. R., and F. B. GOLDEN, eds., *SCR Manual,* 6th ed., Englewood Cliffs, N.J.: Prentice Hall, 1982.

GOTTLIEB, I. M., *Power Control with Solid State Devices*. Reston, Va.: Reston Publishing Co., Inc., 1985.

HEUMANN, K., *Basic Principles of Power Electronics*. New York: Springer-Verlag, 1986.

HNATEK, E. R., *Design of Solid State Power Supplies*. New York: Van Nostrand Reinhold Company, Inc., 1981.

HOFT, R. G., *SCR Applications Handbook*. El Segundo, Calif.: International Rectifier Corporation, 1974.

HOFT, R. G., *Semiconductor Power Electronics*. New York: Van Nostrand Reinhold Company, Inc., 1986.

KASSAKIAN, J. G., M. SCHLECHT, and G. C. VERGHESE, *Principles of Power Electronics*. Reading, Mass.: Addison-Wesley Publishing Co., Inc., 1991.

KLOSS, A., *A Basic Guide to Power Electronics*. New York: John Wiley & Sons, Inc., 1984.

KUSKO, A., *Solid State DC Motor Drives*. Cambridge, Mass.: The MIT Press, 1969.

LANDER, C. W., *Power Electronics*. Maidenhead, Berkshire, England: McGraw-Hill Book Company (U.K.) Ltd., 1981.

LEONARD, W., *Control of Electrical Drives*. New York: Springer-Verlag, 1985.

LINDSAY, J. F., and M. H. RASHID, *Electromechanics and Electrical Machinery*. Englewood Cliffs, N.J.: Prentice Hall, 1986.

LYE, R. W., *Power Converter Handbook*. Peterborough, Ont.: Canadian General Electric Company Ltd., 1976.

MAZDA, F. F., *Thyristor Control*. Chichester, West Sussex, England: John Wiley & Sons Ltd., 1973.

McMurry, W., *The Theory and Design of Cycloconverters*. Cambridge, Mass.: The MIT Press, 1972.

Mohan, M., T. M. Undeland, and W. P. Robbins, *Power Electronics: Converters, Applications and Design*. New York: John Wiley & Sons, Inc., 1989.

Murphy, J. M. D., *Thyristor Control of AC Motors*. Oxford: Pergamon Press Ltd., 1973.

Pearman, R. A., *Power Electronics: Solid State Motor Control*. Reston, Va.: Reston Publishing Co., Inc., 1980.

Pelly, B. R., *Thyristor Phase Controlled Converters and Cycloconverters*. New York: John Wiley & Sons, Inc., 1971.

Ramamoorty, M., *An Introduction to Thyristors and Their Applications*. London: Macmillan Publishers Ltd., 1978.

Ramshaw, R. S., *Power Electronics: Thyristor Controlled Power for Electric Motors*. London: Chapman & Hall Ltd., 1982.

Rashid, M. H., *SPICE for Power Electronics and Electric Power*. Englewood Cliffs, N.J.: Prentice Hall, 1993.

Rice, L. R., *SCR Designers Handbook*. Pittsburgh, Pa.: Westinghouse Electric Corporation, 1970.

Rose, M. J., *Power Engineering Using Thyristors,* Vol. 1. London: Mullard Ltd., 1970.

Schaefer, J., *Rectifier Circuits: Theory and Design*. New York: John Wiley & Sons, Inc., 1965.

Sen, P. C., *Thyristor DC Drives*. New York: John Wiley & Sons, Inc., 1981.

Severns, R. P., and G. Bloom, *Modern DC-to-DC Switchmode Power Converter Circuits*. New York: Van Nostrand Reinhold Company, Inc., 1985.

Steven, R. E., *Electrical Machines and Power Electronics*. Wakingham, Berkshire, England: Van Nostrand Reinhold Ltd., 1983.

Sugandhi, R. K., and K. K. Sugandhi. *Thyristors: Theory and Applications*. New York: Halsted Press, 1984.

Tarter, R. E., *Principles of Solid-State Power Conversion*. Indianapolis, Ind.: Howard W. Sams & Company, Publishers, Inc., 1985.

Wells, R., *Static Power Converters*. New York: John Wiley & Sons, Inc., 1962.

Williams, B. W., *Power Electronics, Devices, Drivers and Applications*. New York: Halsted Press, 1987.

Wood, P., *Switching Power Converters*. New York: Van Nostrand Reinhold Company, Inc., 1981.

Index

Torque
  angle, 577
  developed, 544
  pull-out or breakdown, 546, 577
Transconductance, 268, 283
Transformer, 631
Transients
  reverse recovery, 597
  supply and load side, 603
Transient analysis, dc, 639
Transistors
  bipolar, 5, 8, 263
  IGBTs, 287
  isolation of gate and base drives of,
    276, 294
  MOSFETs, 5, 8, 280
  series and parallel operation of, 289
  SIT, 5, 8, 286
  unijunction, 120

Transistor saturation, 266
TRIAC, 110

U

Unijunction transistor, 120
  programmable, 123
UPS, 484

W

Winding, feedback, 48

Z

Zero current switching, 446
Zero voltage switching, 451